yala

WAVES AND FIELDS
IN
INHOMOGENEOUS MEDIA

IEEE PRESS Series on Electromagnetic Waves

The IEEE Press Series on Electromagnetic Waves consists of new titles as well as reprints and revisions of recognized classics that maintain long-term archival significance in electromagnetic waves and applications.

Series Editor
Donald G. Dudley
University of Arizona

Advisory Board
Robert E. Collin
Case Western University

Akira Ishimaru
University of Washington

Associate Editors
Electromagnetic Theory, Scattering, and Diffraction
Ehud Heyman
Tel-Aviv University

Differential Equation Methods
Andreas C. Cangellaris
University of Arizona

Integral Equation Methods
Donald R. Wilton
University of Houston

Antennas, Propagation, and Microwaves
David R. Jackson
University of Houston

Books in the Series
Chew, W. C., *Waves and Fields in Inhomogeneous Media*, 1995
Collin, R. E., *Field Theory of Guided Waves*, Second Edition, 1991
Dudley, D. G., *Mathematical Foundations for Electromagnetic Theory*, 1994
Elliott, R. S., *Electromagnetics: History, Theory, and Applications*, 1993
Felsen, L. B. and Marcuvitz, N., *Radiation and Scattering of Waves*, 1994
Harrington, R. F., *Field Computation by Moment Methods*, 1993
Tai, C. T., *Generalized Vector and Dyadic Analysis: Applied Mathematics in Field Theory*, 1991
Tai, C. T., *Dyadic Green Functions in Electromagnetic Theory*, Second Edition, 1994

WAVES AND FIELDS
IN
INHOMOGENEOUS MEDIA

WENG CHO CHEW

UNIVERSITY OF ILLINOIS
URBANA-CHAMPAIGN

**IEEE PRESS Series on
Electromagnetic Waves**

Donald G. Dudley, Series Editor

IEEE Antennas and Propagation Society, *Sponsor*

The Institute of Electrical and Electronics Engineers, Inc., New York

IEEE PRESS
445 Hoes Lane, PO Box 1331
Piscataway, NJ 08855-1331

This book may be purchased at a discount from the publisher when ordered in bulk quantities.
For more information contact:

IEEE PRESS Marketing
Attn: Special Sales
PO Box 1331
445 Hoes Lane
Piscataway, NJ 08855-1331
Fax: (908) 981-8062

This is the 1995 IEEE reprinting of a book originally published by
Van Nostrand Reinhold.
© 1990 by Van Nostrand Reinhold
© 1995 by the Institute of Electrical and Electronics Engineers, Inc.
345 East 47th Street, New York, NY 10017-2394

Printed in the United States of America
10 9 8 7 6 5 4 3 2 1

ISBN 0-7803-1116-7
IEEE Order Number: PC4846

Library of Congress Cataloging-in-Publication Data

Chew, Weng Cho.
　　Waves and fields in inhomogeneous media / Weng Cho Chew.
　　p.　　cm. — (Electromagnetic waves)
　　Originally published: New York : Van Nostrand Reinhold, c1990.
　　Includes bibliographical references and index.
　　ISBN 0-7803-1116-7
　　1. Electromagnetic fields.　2. Electromagnetic waves.　I. Title.
II. Series.
QC665.E4C48　1994　　　　　　　　　　　　　　　　94-32641
530. 1'41—dc20　　　　　　　　　　　　　　　　　　CIP

To

*Chew Chin, Huibin, Shinen,
and My Parents*

CONTENTS

PREFACE

This book is an outgrowth of a graduate-level course taught at the University of Illinois, Urbana-Champaign. The course was conceived to fill a need for educating students working with waves and fields in their research. As technology grows more sophisticated, the need for solutions to more complex problems becomes greater. Fortunately, in the last two decades, the development of computer technology has provided a new means for solving many of these problems. This book is intended to elucidate the methods and techniques pertinent to wave and field problems in inhomogeneous media, so that students may apply them in order to solve practical problems. Hence, the reader may use a computer to implement many of the techniques described in this book.

This is not a book on numerical techniques, however. For that, the reader is referred to many excellent works written on numerical methods. The emphasis here is on understanding the basic foundations of the many mathematical methods employed in solving problems of waves and fields in inhomogeneous media. Hence, the reader shall learn to formulate many complicated problems using such methods. A straightforward application of numerical methods to many wave and field problems involves using intensive computation and extensive computer time, and may saturate the resources of even present-day supercomputers. Good judgement that adapts a combination of numerical and analytical methods, however, can usually generate computer codes that are many times more efficient. The details of the numerical analyses of these techniques, namely, numerical integration, matrix analysis, or finite-element analysis, are beyond the scope of this book, but they can be gleaned from many other relevant sources. Furthermore, this book does not discuss random inhomogeneous media, a subject that could span another volume of similar size.

Analytic techniques were in vogue before computers became available. They were often the sole technique for analyses in the days of "pencil-and-paper" physics or engineering. For instance, many classical topics, like waves in planarly layered media, the Cagniard-de Hoop method for transient analysis, and the Gel'fand-Levitan-Marchenko methods for inverse scattering, are elegantly developed for solving one-dimensional, inhomogeneous-medium problems. Such techniques usually have limited scope for applications; however, to ignore the discussion of such topics in this book would render it incomplete. They are included to pay homage to those predecessors who developed them and to give the reader a historical perspective.

Despite the preponderance of electromagnetic examples contained herein, many of the techniques are easily adapted to other wave phenomena, such as acoustic waves, Schrödinger waves, or elastic waves. In order to readily convey the underlying concept of a technique to the reader, some are illustrated in the form of scalar wave problems. The extension to the vector electromagnetic wave case is often almost perfunctory.

Throughout the book, it is assumed that the reader knows the rudiments of electromagnetic theory. Chapter 1 reviews this required background cursorily. Planarly layered media remain the most studied of the inhomogeneous media because of its simplicity. Many results can be obtained without computationally intensive calculations. Furthermore, closed-form solutions in terms of spectral integrals (Sommerfeld integrals) allow for their asymptotic approximations, providing more physical insight into the problems. Hence, this topic, with its longer history, abounds in analytical techniques. Consequently, Chapter 2 will concentrate on waves in planarly layered media. The relatively greater length of this chapter reflects the abundance of interesting material on planarly layered media that has been developed over the years. Although this topic has been covered by many other books, sometimes as the sole subject, only a few of the more important topics are included here. Moreover, numerical schemes like the propagator-matrix approach and the numerical integration of Sommerfeld integrals are also discussed.

Chapter 3 covers waves in cylindrically and spherically layered media. This kind of wave problem finds applications in optics, geophysical probing, and propagation of radio waves on the earth's surface and the ionosphere. It also finds application in multilayer coatings on aircraft to reduce the radar cross-sections of aircraft. It is surprising to observe that this subject has not been developed as extensively as waves in planarly layered media.

Transient measurements are routine in radar, bioacoustic, or sonar measurements. Chapter 4 considers the propagation of transient waves in inhomogeneous media. First, the Cagniard-de Hoop method is presented, which finds closed-form solutions for the response of a line source on top of a layered, dispersionless medium. Then, for a general inhomogeneous medium, the celebrated time-domain finite-difference method is discussed in conjunction with the use of absorbing boundary conditions to obtain transient solutions.

Chapter 5 presents the variational methods for the scalar wave equation and the electromagnetic wave equation. Such methods are often the foundation of many modern numerical methods in waves and fields, such as the finite-element method and the method of moments. Moreover, many of these numerical methods can be applied in a recipe-like manner to obtain the solution to a problem. It is still important, however, that a student understands the foundations upon which these numerical methods are based. Therefore, the beginning of the chapter explains the basic concepts of a linear vector space. Subsequently, the formulations of variational expressions for self-adjoint as well as non-self-adjoint problems are presented followed by a step-by-step derivation of the matrix equation from a variational expression (using the Rayleigh-Ritz method). This derivation is the foundation of the finite-element method, and the reader, by following it, will understand the basis of the finite-element method. But the detailed implementation of the finite-element method, and the many bookkeeping techniques developed, are beyond the scope of this book.

Chapter 6 describes the mode matching technique. This technique is particularly useful for studying scattering at junction discontinuities encountered in dielectric slab waveguides or many geophysics problems. In order to understand this technique, the readers are first introduced to the concepts of the radiation modes and the guided modes of a dielectric slab. Then, the canonical scattering solution from a single junction discontinuity is discussed. This canonical solution is, in turn, used to find the scattering solution from many junction discontinuities in a composite problem using a recursive algorithm. Here, vector notations are used to write the solutions in a compact fashion.

The dyadic Green's function has been an interesting topic in electromagnetic theory. In this subject, many scalar wave concepts cannot simply be extended to the vector wave case. Even though most problems can be solved without the use of dyadic Green's functions, the symbolic simplicity with which they could be used to express relationships makes the formulations of many problems simpler and more compact. Moreover, it is easier to conceptualize many problems with the dyadic Green's functions. Consequently, the dyadic Green's functions in layered inhomogeneous media are discussed in Chapter 7, beginning with the spatial and spectral representations of the dyadic Green's function in a homogeneous medium. Next, the singularity of the dyadic Green's function is discussed; this has been a topic of heated debate in recent years. Then, vector wave functions are introduced to derive the dyadic Green's functions in different coordinate systems. Finally, the dyadic Green's function is generalized to layered media of planar, cylindrical, and spherical configurations.

Chapter 8 presents integral equation methods. These methods are especially useful for inhomogeneous media that are piecewise constant or when the inhomogeneous region has a finite support in space. Surface integral equations and the way they are solved (using the method of moments and the extended-boundary-condition method) are described. Along with the extended-boundary-condition method, the $\overline{\mathbf{T}}$ matrix and the $\overline{\mathbf{S}}$ matrix for the scattering due to one scatterer are derived. Then, it is shown how such a solution can be used to find the scattering from many scatterers, or multilayered scatterers. Moreover, the next topic covers the case of the unimoment method, a hybrid method, which combines the use of the finite-element method and the surface integral equation method. Finally, the volume integral equation and techniques for its solution are included here. The regimes of validity of the approximate solution technique, like the Born or the Rytov approximations, are also considered.

Many people who study waves in inhomogeneous media are undoubtedly interested in the inverse scattering problem, which finds many applications in biological sensing, geophysics, remote sensing, nondestructive evaluation, target identifications, and so on. Chapter 9 presents the solution techniques of several inverse scattering problems. The first discussion covers the area of linear inverse scattering, which has had a tremendous impact on medical tomography. In particular, back-projection tomography and diffraction tomography are reviewed. The scattered field of a scattering experiment is, however, more often nonlinearly related to the object that causes the scattering. Hence, different techniques for obtaining solutions to such nonlinear inverse scattering problems are given. Most advances in the past have been made in the area of the one-dimensional inverse scattering problem. Consequently, the method of characteristics, the Gel'fand-Levitan, and the Marchenko methods in solving such one-dimensional problems are discussed here. For higher dimensions, however, one has to resort to quasi-Newton type methods in solving the nonlinear inverse scattering problem. In addition, the use of the distorted Born iterative method, and the Born iterative method in seeking the solutions to such problems, are presented. It is imperative to develop new ideas and make advances in the multidimensional problem, for many technologies can benefit greatly from them.

Many topics within this book are from the published literature. Some were developed while writing the book. In some cases, the formulations are altered slightly—but without sacrificing accuracy—from the published literature to conform to the uniformity of presentation and style.

A book of this size cannot cover all the topics that have been written on waves and fields in inhomogeneous media, but it will make many more topics accessible to the reader. Also, the reader is encouraged to make use of the reference sections for further study in any of the topics presented.

ACKNOWLEDGEMENTS

I am indebted to all colleagues who have contributed in this area. I am also indebted to many from whom I have had the opportunity to learn about the theory of waves and fields. I am particularly grateful to Professor Jin Au Kong, and also to Leung Tsang, who taught me much about electromagnetic theory when I was at MIT. At Schlumberger-Doll Research, I had the opportunity to learn the Cagniard-de Hoop method first-hand from Adrianus de Hoop. Much of the information on the finite-difference method was brought to my attention by Curt Randall. I also owe a great deal to Mike Ekstrom, Bob Kleinberg, and Pabitra Sen, who influenced my view on science. Feedbacks on the first draft of the book from Allen Howard, Adrianus de Hoop, John Lovell, and Jim Wait are much appreciated. At the University of Illinois, I am grateful for productive interactions with all my colleagues.

The support for my research from the National Science Foundation through the Presidential Young Investigator Program, the Army Research Office through the Advanced Construction Technology Center at the University of Illinois, and the Office of Naval Research are gratefully acknowledged. I am also thankful to support from a number of industrial organizations, especially Schlumberger, General Electric, Northrop, and TRW.

Finally, I wish to thank Li Tong for working relentlessly and skillfully to typeset the manuscript of this book. I am also thankful to many of my students, who help proofread the book (they are Jim Friedrich, Levent Gurel, Mahta Moghaddam, Qinghuo Liu, Muhammad Nasir, Greg Otto, Rob Wagner, and Yiming Wang), and in particular to Jim Friedrich and Qinghuo Liu, who proofread every chapter. Fred Daab is gratefully acknowledged for drafting all the figures. I am also deeply grateful to my wife and two children, who displayed such patience and love while I wrote this book.

For the second printing of the book, I would like to thank Jiun-Hwa Lin, Qinghuo Liu, Caicheng Lu, Zaiping Nie, Greg Otto, Yiming Wang, and Bill Weedon for pointing out numerous typographic errors and suggestions for improvements in the first printing of the book.

WAVES AND FIELDS
IN
INHOMOGENEOUS MEDIA

CHAPTER 1

PRELIMINARY BACKGROUND

This chapter presents the fundamentals of electromagnetic theory necessary for reading this book. Many mathematical techniques discussed herein could be adapted for other kinds of waves. We will, however, illustrate most of the techniques with electromagnetic waves and fields. The material in this chapter is also discussed in many textbooks which are given in the reference list.

The electromagnetic field can sometimes be described by the scalar wave equation, but in most cases, it can only be described by the vector wave equation. In many instances, the mathematical techniques explained in this book can be illustrated more clearly using scalar wave equations. Since acoustic waves are always described by the scalar wave equation, the derivation of the acoustic wave equation for inhomogeneous medium is also given in Section 1.2 (on the topic of scalar wave equation).

§1.1 Maxwell's Equations

Maxwell's equations were established by James Clerk Maxwell in 1873. Prior to that time, the equations existed in incomplete forms as a result of the work of Faraday, Ampere, Gauss, and Poisson. Later, Maxwell added a displacement current term to the equations. Also, this was important to prove that an electromagnetic field could exist as waves. Finally, the wave nature of Maxwell's equations was verified experimentally by Heinrich Hertz in 1888. Even though the earth's surface is curved, with the aid of the ionosphere which reflects radio waves, Guglielmo Marconi was able to send a radio wave across the Atlantic Ocean in 1901. Since then, the importance of Maxwell's equations has been demonstrated in optics, microwaves, antennas, communications, radar, and many sensing applications.

§§1.1.1 Differential Representations

In vector notation and SI units, Maxwell's equations in differential representations are

$$\nabla \times \mathbf{E}(\mathbf{r}, t) = -\frac{\partial}{\partial t} \mathbf{B}(\mathbf{r}, t), \tag{1.1.1}$$

$$\nabla \times \mathbf{H}(\mathbf{r}, t) = \frac{\partial}{\partial t} \mathbf{D}(\mathbf{r}, t) + \mathbf{J}(\mathbf{r}, t), \tag{1.1.2}$$

$$\nabla \cdot \mathbf{B}(\mathbf{r}, t) = 0, \tag{1.1.3}$$

$$\nabla \cdot \mathbf{D}(\mathbf{r}, t) = \varrho(\mathbf{r}, t), \tag{1.1.4}$$

where $\mathbf{E}$ is the electric field in volts/m, $\mathbf{H}$ is the magnetic field in amperes/m, $\mathbf{D}$ is the electric flux in coulombs/m^2, $\mathbf{B}$ is the magnetic flux in webers/m^2,[1] $\mathbf{J}(\mathbf{r}, t)$ is the current density in amperes/m^2, and $\varrho(\mathbf{r}, t)$ is the charge density in coulombs/m^3.

For a time-varying electromagnetic field, Equations (3) and (4) of the above Maxwell's equations can be derived from Equations (1) and (2). For example, taking the divergence of (1) gives rise to (3). Taking the divergence of (2) and using the continuity equation,

$$\nabla \cdot \mathbf{J}(\mathbf{r}, t) + \frac{\partial \varrho(\mathbf{r}, t)}{\partial t} = 0, \tag{1.1.5}$$

we arrive at (4).

For static problems where $\partial/\partial t = 0$, the electric field and the magnetic field are decoupled. In this case, Equations (3) and (4) cannot be derived from Equations (1) and (2). Then, the electric field Equations (1) and (4) are to be solved independently from the magnetic field Equations (2) and (3). However, in practice, a current is carried by a conductor. Unless the conductor is a superconductor, the current would have to be driven by an electric field or a voltage. Therefore, the magnetic field may never be completely decoupled from the electric field in statics.

The curl operator $\nabla\times$ is a measure of field rotation. Hence, Equation (1) indicates that a time-varying magnetic flux generates an electric field with rotation. Moreover, Equation (2) indicates that a current or a time-varying electric flux (also known as displacement current) generates a magnetic field with rotation.

The divergence operator $\nabla\cdot$ is a measure of the total flux exuding from a point. If there is no source or sink at a point, the divergence of the flux at that point should be zero. Therefore, Equation (3) says that the divergence of the magnetic flux is always zero, since a source or a sink of magnetic flux (namely, magnetic charges) has not been found to date. Furthermore, Equation (4) states that the divergence of the electric flux at a point is proportional to the positive charge density present at the point.

Equation (1), which was discovered by Michael Faraday, is also known as Faraday's Law. Equation (2), without the $\partial\mathbf{D}/\partial t$ term, or the displacement current term, is also known as Ampere's Law. The displacement current term, discovered by Maxwell later, is very important because it couples the magnetic field to the time-varying electric flux. Moreover, it also allows for the possible existence of electromagnetic waves which were later shown to be the same as light waves. Equations (3) and (4) are the consequences of Gauss' Law, which is a statement of the conservation of flux. More specifically,

---

[1] 1 weber/m^2 = 1 Tesla = 10^4 Gauss. The earth's field is about 0.5 Gauss.

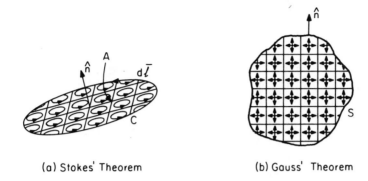

(a) Stokes' Theorem   (b) Gauss' Theorem

Figure 1.1.1 The parameters in the application of Stokes' theorem and Gauss' theorem.

Equation (4) implies that the electric flux $\mathbf{D}$ is produced by a charge density ϱ.

§§1.1.2 Integral Representations

Different insights sometimes result if we look at Maxwell's equations in their integral representations. To derive the integral forms of Equations (1) and (2), we integrate them over a cross-sectional area A and make use of Stokes' theorem,

$$\int_A d\mathbf{S} \cdot \nabla \times \mathbf{E}(\mathbf{r}, t) = \int_C d\mathbf{l} \cdot \mathbf{E}(\mathbf{r}, t). \qquad (1.1.6)$$

In (6), C is a contour that forms the perimeter of the area A (Figure 1.1.1a). Moreover, (6) is a statement that the sum of all the rotations due to the field $\mathbf{E}$ over the cross-sectional area A is equal to the "torque" produced by these rotations on the perimeter of A which is C: The left-hand side of (6) is the summation over all the rotations, while the right-hand side of (6) is the evaluation of the net "torque" on the perimeter C. The fact is that neighboring rotations within the area C cancel each other, leaving a net rotation on the perimeter.

Using Stokes' theorem, we can then convert (1) and (2) to

$$\int_C d\mathbf{l} \cdot \mathbf{E}(\mathbf{r}, t) = -\frac{\partial}{\partial t} \int_A d\mathbf{S} \cdot \mathbf{B}(\mathbf{r}, t), \qquad (1.1.7)$$

$$\int_C d\mathbf{l} \cdot \mathbf{H}(\mathbf{r}, t) = \frac{\partial}{\partial t} \int_A d\mathbf{S} \cdot \mathbf{D}(\mathbf{r}, t) + \int_A d\mathbf{S} \cdot \mathbf{J}(\mathbf{r}, t). \qquad (1.1.8)$$

But to convert Equations (3) and (4) into integral forms, we integrate them over a volume V and make use of Gauss' theorem, which states that

$$\int_V dV \, \nabla \cdot \mathbf{B}(\mathbf{r}, t) = \int_S d\mathbf{S} \cdot \mathbf{B}(\mathbf{r}, t). \qquad (1.1.9)$$

This is a mere statement that the sum of all divergences of a flux $\mathbf{B}$ in a volume V is equal to the net flux which is leaving the volume V through the surface S. In other words, neighboring divergences tend to cancel each other within a volume V (Figure 1.1.1b).

Consequently, (3) and (4) become

$$\int_S d\mathbf{S} \cdot \mathbf{B}(\mathbf{r}, t) = 0, \qquad (1.1.10)$$

$$\int_S d\mathbf{S} \cdot \mathbf{D}(\mathbf{r}, t) = \int_V \varrho(\mathbf{r}, t) \, dV = Q, \qquad (1.1.11)$$

where Q is the total charge in volume V.

The left-hand side of Equation (7) is also the definition of an electromotive force. Hence, Equation (7) implies that a time-varying magnetic flux through an area A generates an electromotive force around a loop C. For instance, if C is replaced with a metallic conductor, the electromotive force will drive a current through this metallic conductor.

By the same token, Equation (8) implies that a time-varying electric flux (displacement current) or a current will generate a magnetomotive force, or simply, a magnetic field that loops around the currents.

On the other hand, Equations (10) and (11) are mere statements of the conservation of fluxes. Equation (11) implies that the net flux through a surface S equals the total charge Q inside S.

§§1.1.3 Time Harmonic Forms

Maxwell's equations can be further simplified if we assume that the field is time harmonic. A time harmonic field can be expressed as

$$\mathbf{A}(\mathbf{r}, t) = \Re e[\mathbf{A}(\mathbf{r}) \, e^{-i\omega t}],$$

where $i = \sqrt{-1}$, ω is frequency in radians/second, and $\mathbf{A}(\mathbf{r})$ is a complex vector. This is also commonly referred to as the $e^{-i\omega t}$ time convention. (The $e^{j\omega t}$ time convention is sometimes used. Here, letting $-i \to j$ will make the two conventions equivalent.) In this case, $\mathbf{A}(\mathbf{r}, t)$ is a sinusoidal function of time—in other words, it is time harmonic. If this is in fact the case, it is easy to show that

$$\frac{\partial}{\partial t} \mathbf{A}(\mathbf{r}, t) = \Re e[-i\omega \mathbf{A}(\mathbf{r}) \, e^{-i\omega t}].$$

Subsequently, Equations (1) to (4) become

$$\nabla \times \mathbf{E}(\mathbf{r}) = i\omega \, \mathbf{B}(\mathbf{r}), \tag{1.1.12}$$

$$\nabla \times \mathbf{H}(\mathbf{r}) = -i\omega \, \mathbf{D}(\mathbf{r}) + \mathbf{J}(\mathbf{r}), \tag{1.1.13}$$

$$\nabla \cdot \mathbf{B}(\mathbf{r}) = 0, \tag{1.1.14}$$

$$\nabla \cdot \mathbf{D}(\mathbf{r}) = \varrho(\mathbf{r}). \tag{1.1.15}$$

In the above, $\mathbf{E}(\mathbf{r})$, $\mathbf{H}(\mathbf{r})$, $\mathbf{D}(\mathbf{r})$, $\mathbf{B}(\mathbf{r})$, $\mathbf{J}(\mathbf{r})$, and $\varrho(\mathbf{r})$ are complex vector or scalar functions known as *phasors*. Better still, a simple rule of thumb of obtaining (12) to (15) from (1) to (4) is to replace $\partial/\partial t$ with $-i\omega$, and vice versa if we were to obtain (1) to (4) from (12) to (15). Alternatively, Equations (12) to (15) can also be obtained by Fourier transforming Equations (1) to (4) with respect to time. In this case, the phasors are actually the Fourier transforms of the fields in the time domain, and they are functions of frequency as well (see Exercise 1.1). Hence, the phasors are also known as the *frequency domain* solutions of the field. Likewise, the solutions of (1) to (4) are the *time domain* solutions. Obviously, the advantage of working with (12) to (15) is the absence of the time dependence and time derivatives.

§§1.1.4 Constitutive Relations

Since only two of the four Maxwell's equations are independent in electrodynamics, we need only work with the first two: Equations (12) and (13). However, there are four vector unknowns, $\mathbf{E}$, $\mathbf{H}$, $\mathbf{B}$, and $\mathbf{D}$, with only two vector equations. Hence, in order to have a sufficient number of equations for the four unknowns, two more equations relating $\mathbf{E}$, $\mathbf{H}$, $\mathbf{B}$, and $\mathbf{D}$ are needed. This can be obtained from the constitutive relations.[2]

The electric and magnetic fluxes are related to the electric and magnetic fields via the constitutive relations. These general constitutive relations in the frequency domain have the form

$$\mathbf{D}(\mathbf{r},\omega) = \overline{\epsilon}(\mathbf{r},\omega) \cdot \mathbf{E}(\mathbf{r},\omega) + \overline{\xi}(\mathbf{r},\omega) \cdot \mathbf{H}(\mathbf{r},\omega), \tag{1.1.16a}$$

$$\mathbf{B}(\mathbf{r},\omega) = \overline{\mu}(\mathbf{r},\omega) \cdot \mathbf{H}(\mathbf{r},\omega) + \overline{\zeta}(\mathbf{r},\omega) \cdot \mathbf{E}(\mathbf{r},\omega), \tag{1.1.16b}$$

where $\overline{\epsilon}$, $\overline{\xi}$, $\overline{\mu}$, and $\overline{\zeta}$ are 3×3 tensors.[3] In addition, the constitutive relations also characterize the medium that we are describing. In fact, the above medium is also known as a *bianisotropic* medium because $\mathbf{D}$ and $\mathbf{B}$ are related to both $\mathbf{E}$ and $\mathbf{H}$. In contrast, a medium that is *anisotropic* has constitutive relations where $\mathbf{D}$ is only related to $\mathbf{E}$, and $\mathbf{B}$ is only related to $\mathbf{H}$, that is,

$$\mathbf{D} = \overline{\epsilon} \cdot \mathbf{E}, \tag{1.1.17a}$$

$$\mathbf{B} = \overline{\mu} \cdot \mathbf{H}. \tag{1.1.17b}$$

---

[2] Extended discussion of this topic is given in Kong (1986).

[3] For a review of tensors see Appendix B.

And the word "anisotropy" implies that relationships (16) and (17) are functions of the field directions.

When $\overline{\epsilon}$, $\overline{\xi}$, $\overline{\mu}$, or $\overline{\zeta}$ are functions of space, the medium is also known as an *inhomogeneous* medium. But when they are functions of frequency, the medium is *frequency dispersive*. In this case, the relations in the time domain correspond to convolutions. On the other hand, when the relations are convolutions over space, the medium is *spatially dispersive*. Moreover, when the tensors are functions of the fields, the medium is *nonlinear*. For *isotropic* media, however, relationship (17) is independent of field polarizations and the constitutive relations simply become

$$\mathbf{D} = \epsilon\, \mathbf{E}, \ \ \mathbf{B} = \mu\, \mathbf{H}. \tag{1.1.18}$$

In free-space, $\epsilon = \epsilon_0 = 8.854 \times 10^{-12}$ farad/m, while $\mu = \mu_0 = 4\pi \times 10^{-7}$ henry/m. The constant $c = 1/\sqrt{\mu_0 \epsilon_0}$ is the velocity of light, which has been measured very accurately. The value of μ_0 is assigned, while the value of ϵ_0 is calculated from c. In fact, the value of μ_0 is chosen so that the units of voltage and current in a laboratory experiment are not inordinately large or small (see Exercise 1.2). More recently, a study (Cohen and Taylor 1986) recommends that the unit of meter be redefined so that the velocity of light is exactly 299,792,458 m/s.

§§1.1.5 Poynting Theorem and Lossless Conditions

(a) Poynting Theorem

It can be easily shown that the vector $\mathbf{E}(\mathbf{r}, t) \times \mathbf{H}(\mathbf{r}, t)$ has a dimension of watts/m^2 which is that of power density. Therefore, it may be associated with the direction of power flow. If the fields are time harmonic, a time average of the vector can be defined as

$$\langle \mathbf{E}(\mathbf{r}, t) \times \mathbf{H}(\mathbf{r}, t) \rangle = \lim_{T \to \infty} \frac{1}{T} \int_0^T \mathbf{E}(\mathbf{r}, t) \times \mathbf{H}(\mathbf{r}, t)\, dt. \tag{1.1.19}$$

Given the phasors of time harmonic fields $\mathbf{E}(\mathbf{r}, t)$ and $\mathbf{H}(\mathbf{r}, t)$, namely, $\mathbf{E}(\mathbf{r})$ and $\mathbf{H}(\mathbf{r})$ respectively, we can show that (see Exercise 1.3)

$$\langle \mathbf{E}(\mathbf{r}, t) \times \mathbf{H}(\mathbf{r}, t) \rangle = \frac{1}{2} \Re e\{\mathbf{E}(\mathbf{r}) \times \mathbf{H}^*(\mathbf{r})\}. \tag{1.1.20}$$

Here, the vector $\mathbf{E}(\mathbf{r}) \times \mathbf{H}^*(\mathbf{r})$ is also known as the complex Poynting vector. Moreover, because of its aforementioned property, and its dimension of power density, we will study its conservative property. To do so, we take its divergence and use the appropriate vector identity to obtain

$$\nabla \cdot (\mathbf{E} \times \mathbf{H}^*) = \mathbf{H}^* \cdot \nabla \times \mathbf{E} - \mathbf{E} \cdot \nabla \times \mathbf{H}^*. \tag{1.1.21}$$

Next, using Maxwell's equations for $\nabla \times \mathbf{E}$ and $\nabla \times \mathbf{H}^*$ and the constitutive relations for anisotropic media, we have

$$\nabla \cdot (\mathbf{E} \times \mathbf{H}^*) = i\omega\, \mathbf{H}^* \cdot \mathbf{B} - i\omega\, \mathbf{E} \cdot \mathbf{D}^* - \mathbf{E} \cdot \mathbf{J}^*$$
$$= i\omega\, \mathbf{H}^* \cdot \overline{\mu} \cdot \mathbf{H} - i\omega\, \mathbf{E} \cdot \overline{\epsilon}^* \cdot \mathbf{E}^* - \mathbf{E} \cdot \mathbf{J}^*.$$

$$(1.1.22)$$

The above is also known as the complex Poynting theorem. It can also be written in an integral form using Gauss' theorem, namely,

$$\int d\mathbf{S} \cdot (\mathbf{E} \times \mathbf{H}^*) = i\omega \int_V dV (\mathbf{H}^* \cdot \overline{\mu} \cdot \mathbf{H} - \mathbf{E} \cdot \overline{\epsilon}^* \cdot \mathbf{E}^*) - \int_V dV\, \mathbf{E} \cdot \mathbf{J}^*.$$

$$(1.1.23)$$

(b) Lossless Conditions

For a region V that is lossless and source-free, $\mathbf{J} = 0$ and

$$\Re e \int_S d\mathbf{S} \cdot (\mathbf{E} \times \mathbf{H}^*) = 0,$$

because there is no net time-averaged power-flow out of or into this region V. Therefore, because of energy conservation, the real part of the right-hand side of (22), without the $\mathbf{E} \cdot \mathbf{J}^*$ term, must be zero. In other words,

$$\int_V dV (\mathbf{H}^* \cdot \overline{\mu} \cdot \mathbf{H} - \mathbf{E} \cdot \overline{\epsilon}^* \cdot \mathbf{E}^*)$$

$$(1.1.24)$$

must be a real quantity.

Other than the possibility that the above is zero, the general requirement for it to be real is that $\mathbf{H}^* \cdot \overline{\mu} \cdot \mathbf{H}$ and $\mathbf{E} \cdot \overline{\epsilon}^* \cdot \mathbf{E}^*$ are real quantities. But since the conjugate transpose of a real number is itself, we have $(\mathbf{H}^* \cdot \overline{\mu} \cdot \mathbf{H})^\dagger = \mathbf{H}^* \cdot \overline{\mu} \cdot \mathbf{H}$. Therefore,

$$(\mathbf{H}^* \cdot \overline{\mu} \cdot \mathbf{H})^\dagger = (\mathbf{H} \cdot \overline{\mu}^* \cdot \mathbf{H}^*)^t = \mathbf{H}^* \cdot \overline{\mu}^\dagger \cdot \mathbf{H} = \mathbf{H}^* \cdot \overline{\mu} \cdot \mathbf{H}. \qquad (1.1.25)$$

The last equality in the above is possible only if $\overline{\mu} = \overline{\mu}^\dagger$ (where the $\dagger$ implies conjugate transpose and t implies transpose), or that $\overline{\mu}$ is Hermitian. Therefore, the conditions for anisotropic media to be lossless are

$$\overline{\mu} = \overline{\mu}^\dagger, \qquad \overline{\epsilon} = \overline{\epsilon}^\dagger, \qquad (1.1.26)$$

requiring the permittivity and permeability tensors to be Hermitian. Then, for an isotropic medium, the lossless conditions are simply that $\Im m(\mu) = 0$ and $\Im m(\epsilon) = 0$.

If a medium is source-free, but lossy, then $\Re e \int d\mathbf{S} \cdot (\mathbf{E} \times \mathbf{H}^*) < 0$. Consequently, from (23), this implies

$$\Im m \int_V dV (\mathbf{H}^* \cdot \overline{\mu} \cdot \mathbf{H} - \mathbf{E} \cdot \overline{\epsilon}^* \cdot \mathbf{E}^*) > 0. \qquad (1.1.27)$$

But the above is the same as

$$i \int_V dV [\mathbf{H}^* \cdot (\overline{\mu}^\dagger - \overline{\mu}) \cdot \mathbf{H} + \mathbf{E}^* \cdot (\overline{\epsilon}^\dagger - \overline{\epsilon}) \cdot \mathbf{E}] > 0. \qquad (1.1.28)$$

Therefore, for a medium to be lossy, $i(\overline{\mu}^\dagger - \overline{\mu})$ and $i(\overline{\epsilon}^\dagger - \overline{\epsilon})$ must be Hermitian, **positive definite** matrices, to ensure the inequality in (28). Similarly, for an active medium, $i(\overline{\mu}^\dagger - \overline{\mu})$ and $i(\overline{\epsilon}^\dagger - \overline{\epsilon})$ must be Hermitian, **negative definite** matrices (see Exercise 1.6).

For a lossy medium which is conductive, we may define $\mathbf{J} = \overline{\sigma} \cdot \mathbf{E}$ where $\overline{\sigma}$ is a conductivity tensor. In this case, Equation (23), after combining the last two terms, may be written as

$$\int_S d\mathbf{S} \cdot (\mathbf{E} \times \mathbf{H}^*) = i\omega \int_V dV \left[\mathbf{H}^* \cdot \overline{\mu} \cdot \mathbf{H} - \mathbf{E} \cdot \left(\overline{\epsilon}^* - \frac{i\overline{\sigma}^*}{\omega} \right) \cdot \mathbf{E}^* \right]$$

$$= i\omega \int dV [\mathbf{H}^* \cdot \overline{\mu} \cdot \mathbf{H} - \mathbf{E} \cdot \widetilde{\overline{\epsilon}}^* \cdot \mathbf{E}^*], \qquad (1.1.29)$$

where $\widetilde{\overline{\epsilon}} = \overline{\epsilon} + \frac{i\overline{\sigma}}{\omega}$ is known as the **complex permittivity tensor**. In this manner, (29) has the same structure as the source-free Poynting theorem.

The quantity $\mathbf{H}^* \cdot \overline{\mu} \cdot \mathbf{H}$ for lossless media is associated with the time-averaged energy density stored in the magnetic field, while the quantity $\mathbf{E} \cdot \overline{\epsilon}^* \cdot \mathbf{E}^*$ for lossless media is associated with the time-averaged energy density stored in the electric field. Then, for lossless, source-free media, (23) implies that

$$\Im m \int_S d\mathbf{S} \cdot (\mathbf{E} \times \mathbf{H}^*) = \omega \int_V dV (\mathbf{H}^* \cdot \overline{\mu} \cdot \mathbf{H} - \mathbf{E} \cdot \overline{\epsilon}^* \cdot \mathbf{E}^*), \qquad (1.1.30)$$

or that

$$\Im m \int_S d\mathbf{S} \cdot (\mathbf{E} \times \mathbf{H}^*)$$

is proportional to the time rate of change of the difference of the time-averaged energy stored in the magnetic field and the electric field. Since this power is nondissipative, it is also known as the **reactive power** (see Exercise 1.5). Hence, the imaginary part of $\mathbf{E} \times \mathbf{H}^*$ may be associated with the reactive power density.

§§1.1.6 Duality Principle

Maxwell's equations exhibit a certain symmetry between $\mathbf{E}$ and $\mathbf{H}$, and $\mathbf{D}$ and $\mathbf{B}$. However, the absence of magnetic charges destroys the symmetry for the sources. Nevertheless, from a mathematical viewpoint, Maxwell's equations can be made symmetrical by introducing a magnetic current density $\mathbf{M}$ and a magnetic charge density ϱ_m. In this case, (1) to (4) become

$$\nabla \times \mathbf{E}(\mathbf{r}, t) = -\frac{\partial}{\partial t} \mathbf{B}(\mathbf{r}, t) - \mathbf{M}(\mathbf{r}, t), \qquad (1.1.31)$$

$$\nabla \times \mathbf{H}(\mathbf{r}, t) = \frac{\partial}{\partial t} \mathbf{D}(\mathbf{r}, t) + \mathbf{J}(\mathbf{r}, t), \qquad (1.1.32)$$

$$\nabla \cdot \mathbf{B}(\mathbf{r}, t) = \varrho_m(\mathbf{r}, t), \qquad (1.1.33)$$

$$\nabla \cdot \mathbf{D}(\mathbf{r}, t) = \varrho(\mathbf{r}, t). \qquad (1.1.34)$$

The symmetry exhibited by the above equations implies that given a solution to Maxwell's equations, with $\mathbf{E}$, $\mathbf{D}$, $\mathbf{H}$, $\mathbf{B}$, $\mathbf{M}$, $\mathbf{J}$, ϱ_m, and ϱ, another solution can be obtained by the following replacements:

$$\mathbf{E} \to \mathbf{H}, \quad \mathbf{H} \to \mathbf{E}, \quad \mathbf{B} \to -\mathbf{D}, \quad \mathbf{D} \to -\mathbf{B},$$

$$\mathbf{M} \to -\mathbf{J}, \quad \mathbf{J} \to -\mathbf{M}, \quad \varrho_m \to -\varrho, \quad \varrho \to -\varrho_m. \qquad (1.1.35)$$

Notice that the above replacements are nonunique and any other replacements that make Equations (31) to (34) invariant will also suffice (see Exercise 1.7).

If the constitutive relations appear explicitly in (31) to (34), the rule for replacements can be altered accordingly. For example, for anisotropic media, $\mathbf{B} = \overline{\mu} \cdot \mathbf{H}$ and $\mathbf{D} = \overline{\epsilon} \cdot \mathbf{E}$, a possible set of replacement rules is

$$\mathbf{E} \to \mathbf{H}, \quad \mathbf{H} \to \mathbf{E}, \quad \overline{\mu} \to -\overline{\epsilon}, \quad \overline{\epsilon} \to -\overline{\mu},$$

$$\mathbf{M} \to -\mathbf{J}, \quad \mathbf{J} \to -\mathbf{M}, \quad \varrho_m \to -\varrho, \quad \varrho \to -\varrho_m. \qquad (1.1.36)$$

For source-free Maxwell's equations, this becomes simply

$$\mathbf{E} \to \mathbf{H}, \quad \mathbf{H} \to \mathbf{E}, \quad \overline{\mu} \to -\overline{\epsilon}, \quad \overline{\epsilon} \to -\overline{\mu}. \qquad (1.1.37)$$

Even though there is no true magnetic current, one can still speak of equivalent magnetic current. For instance, a current loop carrying an electric current generates a field that resembles that of a magnetic dipole. Consequently, in the limit when the electric current loop is very small, it is equivalent to a magnetic dipole. Then, a series of current loops such as a solenoid or a toroid generates a field similar to that generated by a magnetic current.

§1.2 Scalar Wave Equations

Certain physical phenomena can be described using only the scalar wave equation, for example, acoustic waves and Schrödinger waves. In fact, in certain situations, electromagnetic waves can also be described by the scalar wave equation. Hence, we shall study the scalar wave equation and first illustrate the derivation of the acoustic wave equation.

§§1.2.1 Acoustic Wave Equation

The acoustic wave equation can be derived based on the conservation of mass and conservation of momentum. Similar to the conservation of charge expressed by Equation (1.1.5), the conservation of mass for a fluid can be written as

$$\nabla \cdot (\mathbf{v}\varrho) + \frac{\partial \varrho}{\partial t} = 0, \tag{1.2.1}$$

where ϱ is the mass density and $\mathbf{v}$ is the velocity of the fluid particles. If no external force is acting on a fluid mass, the conservation law for the i-th component of the momentum can be written in a manner similar to (1), giving

$$\nabla \cdot (\mathbf{v}\varrho v_i) + \frac{\partial \varrho v_i}{\partial t} = 0, \tag{1.2.2}$$

where v_i is either v_x, v_y, or v_z. Next, by writing the conservation law for three components of the momentum simultaneously, we have

$$\nabla \cdot \varrho \mathbf{v}\mathbf{v} + \frac{\partial \varrho \mathbf{v}}{\partial t} = 0. \tag{1.2.3}$$

Now, if an external force density $\mathbf{F}$ is applied to the fluid mass, then (3) becomes

$$\nabla \cdot \varrho \mathbf{v}\mathbf{v} + \frac{\partial \varrho \mathbf{v}}{\partial t} = \mathbf{F}. \tag{1.2.4}$$

Furthermore, by using the conservation of mass equation given by (1), (4) can be rewritten more simply as (see Exercise 1.8)

$$\varrho \left[\mathbf{v} \cdot \nabla \mathbf{v} + \frac{\partial \mathbf{v}}{\partial t} \right] = \mathbf{F}. \tag{1.2.5}$$

Force in a fluid sets up a disturbance, giving rise to particle velocity $\mathbf{v}$, and changes in mass density ϱ and pressure p. Before proceeding with a perturbation analysis, we denote the equilibrium quantities by subscript 0, and the perturbed quantities by subscript 1 as follows:

$$\mathbf{v}(\mathbf{r}, t) = \mathbf{v}_1(\mathbf{r}, t), \tag{1.2.6a}$$

$$\varrho(\mathbf{r}, t) = \varrho_0(\mathbf{r}) + \varrho_1(\mathbf{r}, t), \tag{1.2.6b}$$

$$p(\mathbf{r}, t) = p_0(\mathbf{r}) + p_1(\mathbf{r}, t), \tag{1.2.6c}$$

where we assume $\mathbf{v}_0 = 0$, i.e., the fluid particles are at rest before a wave is established. Next, assuming $\varrho_1 \ll \varrho_0$, $p_1 \ll p_0$, and $\mathbf{v}_1$ to be a small quantity, on substituting (6) into (1) yields

$$\nabla \cdot (\mathbf{v}_1 \varrho_0) + \nabla \cdot (\mathbf{v}_1 \varrho_1) + \frac{\partial \varrho_1}{\partial t} = 0. \tag{1.2.7}$$

Then, keeping only the first order terms (assuming $\mathbf{v}_1 \varrho_1$ to be much smaller than the other terms), we have

$$\nabla \cdot \mathbf{v}_1 \varrho_0 + \frac{\partial \varrho_1}{\partial t} = 0. \tag{1.2.8}$$

The restoring force in a fluid is provided by the pressure differential set up in it. Therefore, the force density in (5) can be shown to be (see Exercise 1.8)

$$\mathbf{F} = -\nabla p = -\nabla p_0 - \nabla p_1. \tag{1.2.9}$$

Then, using (6) and (9) in (5), we have

$$[\varrho_0 + \varrho_1] \left[\mathbf{v}_1 \cdot \nabla \mathbf{v}_1 + \frac{\partial \mathbf{v}_1}{\partial t} \right] = -\nabla p_0 - \nabla p_1. \tag{1.2.10}$$

Now, by equating the leading-order term in (10), we have

$$\nabla p_0 = 0, \tag{1.2.11}$$

while by keeping the first order term, we have

$$\varrho_0 \frac{\partial \mathbf{v}_1}{\partial t} = -\nabla p_1. \tag{1.2.12}$$

Equation (11) implies that the pressure has to be uniform in the equilibrium state. Note that this quiescent pressure need not be a constant in the presence of a gravitational force. But over a short length-scale, the pressure gradient induced by gravity can be ignored.

In a compressible fluid, if $\mathbf{v}_1$, p_1, and ϱ_1 are small, and constant entropy is assumed for adiabatic compression and expansion, they can be further linearly related as (see Exercise 1.9, see also Pierce 1981)

$$\frac{\partial p_1}{\partial t} + \mathbf{v}_1 \cdot \nabla p_0 = c^2 \left(\frac{\partial \varrho_1}{\partial t} + \mathbf{v}_1 \cdot \nabla \varrho_0 \right). \tag{1.2.13}$$

Next, using (13) in (8), and making use of (11) we have

$$\varrho_0 \nabla \cdot \mathbf{v}_1 + \frac{1}{c^2} \frac{\partial p_1}{\partial t} = 0. \tag{1.2.14}$$

Then, after differentiating the above once with respect to t, we obtain

$$\varrho_0 \nabla \cdot \left(\frac{\partial \mathbf{v}_1}{\partial t} \right) + \frac{1}{c^2} \frac{\partial^2 p_1}{\partial t^2} = 0. \tag{1.2.15}$$

Finally, using (12) in (15) yields

$$\varrho_0 \nabla \cdot \varrho_0^{-1} \nabla p_1(\mathbf{r}, t) - \frac{1}{c^2} \frac{\partial^2}{\partial t^2} p_1(\mathbf{r}, t) = 0. \tag{1.2.16}$$

Equation (16) is a scalar wave equation for acoustic waves in inhomogeneous media. In addition, $\varrho_0(\mathbf{r})$ and $c(\mathbf{r})$ are both functions of position. Furthermore, in the case of a homogeneous medium where ϱ_0 and c are constants, (16) becomes

$$\nabla^2 p_1(\mathbf{r}, t) - \frac{1}{c^2} \frac{\partial^2}{\partial t^2} p_1(\mathbf{r}, t) = 0, \tag{1.2.17}$$

where c is the velocity of the wave. For a time harmonic field, however, (16) becomes

$$\varrho_0 \nabla \cdot \varrho_0^{-1} \nabla p_1(\mathbf{r}, \omega) + k^2 p_1(\mathbf{r}, \omega) = 0, \tag{1.2.18}$$

where $k = \omega/c$. The above is the Helmholtz wave equation for inhomogeneous acoustic media.

§§1.2.2 Scalar Wave Equation from Electromagnetics

Certain electromagnetic problems can even be described by the scalar wave equation. For instance, in three dimensions, the vector wave equations reduce to the scalar wave equations in a homogeneous, isotropic medium.

In a homogeneous, isotropic, and source-free medium, $\mathbf{B} = \mu \mathbf{H}$ and $\mathbf{D} = \epsilon \mathbf{E}$. Next, after taking the curl of Equation (1.1.12) and substituting it for $\nabla \times \mathbf{H}$ from Equation (1.1.13) without the current source, we have

$$\nabla \times \nabla \times \mathbf{E}(\mathbf{r}) - \omega^2 \mu\epsilon \, \mathbf{E}(\mathbf{r}) = 0, \tag{1.2.19}$$

which is the vector wave equation for a source-free homogeneous medium. Moreover, by using the vector identity that $\nabla \times \nabla \times \mathbf{E} = \nabla(\nabla \cdot \mathbf{E}) - \nabla^2 \mathbf{E}$ and that $\nabla \cdot \mathbf{E} = 0$ for a homogeneous, source-free medium, Equation (19) becomes

$$\nabla^2 \mathbf{E}(\mathbf{r}) + k^2 \, \mathbf{E}(\mathbf{r}) = 0, \tag{1.2.20}$$

where $k^2 = \omega^2 \mu\epsilon$. In Cartesian coordinates, $\mathbf{E}(\mathbf{r}) = \hat{x} E_x + \hat{y} E_y + \hat{z} E_z$, where $\hat{x}$, $\hat{y}$, and $\hat{z}$ are unit vectors independent of position. Hence, (20) consists of three scalar wave equations,

$$\left(\nabla^2 + k^2\right) \psi(\mathbf{r}) = 0, \tag{1.2.21}$$

where $\psi(\mathbf{r})$ can be either E_x, E_y, or E_z. [Note that this statement is not true in cylindrical or spherical coordinates (see Exercise 1.10).] However, Equation (20) must be solved with $\nabla \cdot \mathbf{E} = 0$ condition before the solution is also admissible for (19). Hence, only two out of the three equations in (20) are independent.

The above establishes the wave nature of Maxwell's equations, which is a consequence of the displacement current term discovered by Maxwell. Therefore, it is worthwhile to study more extensively the solutions of the scalar wave equation.

§§1.2.3 Cartesian Coordinates

In Cartesian coordinates, the Laplacian operator in (21) becomes

$$\left(\frac{\partial^2}{\partial x^2} + \frac{\partial^2}{\partial y^2} + \frac{\partial^2}{\partial z^2} + k^2\right) \psi(\mathbf{r}) = 0. \tag{1.2.22}$$

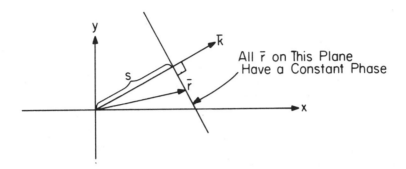

Figure 1.2.1 Constant phase front of a plane wave which is perpendicular to **k**.

Then, $\psi(\mathbf{r})$ has the general solution

$$\psi(\mathbf{r}) = A e^{i(k_x x + k_y y + k_z z)} = A e^{i\mathbf{k}\cdot\mathbf{r}}, \qquad (1.2.23)$$

where the vector $\mathbf{k} = \hat{x}k_x + \hat{y}k_y + \hat{z}k_z$ is known as the propagation vector, while the vector $\mathbf{r} = \hat{x}x + \hat{y}y + \hat{z}z$ is known as the position vector. Next, after substituting (23) into (22), we have

$$\left(-k_x^2 - k_y^2 - k_z^2 + k^2\right)\psi(\mathbf{r}) = 0. \qquad (1.2.24)$$

For nontrivial $\psi(\mathbf{r})$, we require that $k^2 = k_x^2 + k_y^2 + k_z^2$, which is also known as the **dispersion relation**. It implies that the propagation vector $\mathbf{k}$ is of a fixed length, i.e., $|\mathbf{k}| = k$, no matter what direction it is pointing.

Equation (23) denotes mathematically a plane wave propagating in the $\mathbf{k}$ direction. For example, for a plane wave propagating in the x direction, $\psi(\mathbf{r}) = A e^{ikx}$ and $\mathbf{k} = \hat{x}k$, a vector pointing in the x direction. More specifically, the function $e^{i\mathbf{k}\cdot\mathbf{r}}$ has a constant phase e^{iks} for all $\mathbf{r}$ such that $\mathbf{k}\cdot\mathbf{r} = ks$. The locus of the tips of all such $\mathbf{r}$'s is a plane perpendicular to $\mathbf{k}$ (see Figure 1.2.1).

In addition, assuming that $\mathbf{E}(\mathbf{r}) = \mathbf{E}_0 e^{i\mathbf{k}\cdot\mathbf{r}}$, we can show easily that $\nabla \rightarrow i\mathbf{k}$. Then, by using this fact in (19), we have

$$-\mathbf{k} \times \mathbf{k} \times \mathbf{E}(\mathbf{r}) + k^2\,\mathbf{E}(\mathbf{r}) = 0. \qquad (1.2.25)$$

Dotting the above with $\mathbf{k}$ implies that $\mathbf{k}\cdot\mathbf{E} = 0$ for all plane waves. Furthermore, for a homogeneous, isotropic and source-free medium, $\nabla \times \mathbf{E} = i\omega\mu\,\mathbf{H}$, implying that $\mathbf{k} \times \mathbf{E} = \omega\mu\,\mathbf{H}$ and $\mathbf{k}\cdot\mathbf{H} = 0$ for plane waves. Therefore, $\mathbf{E}$, $\mathbf{H}$, and $\mathbf{k}$ form a right-handed system: they are mutually orthogonal.

§§1.2.4 Cylindrical Coordinates

The scalar wave equation in cylindrical coordinates is

$$\left(\frac{1}{\rho}\frac{\partial}{\partial\rho}\rho\frac{\partial}{\partial\rho} + \frac{1}{\rho^2}\frac{\partial^2}{\partial\phi^2} + \frac{\partial^2}{\partial z^2} + k^2\right)\psi(\mathbf{r}) = 0. \qquad (1.2.26)$$

The above partial differential equation can be solved by the separation of variables. On the other hand, the simple $\partial^2/\partial\phi^2$ and $\partial^2/\partial z^2$ derivatives imply that the solutions are of the form[4]

$$\psi(\mathbf{r}) = F_n(\rho)e^{in\phi+ik_z z}, \qquad (1.2.27)$$

where n is an integer since the field has to be 2π periodic in ϕ. Then, substituting (27) into (26) gives rise to

$$\left(\frac{1}{\rho}\frac{d}{d\rho}\rho\frac{d}{d\rho} - \frac{n^2}{\rho^2} + k_\rho^2\right)F_n(\rho) = 0 \qquad (1.2.28)$$

where $k_\rho^2 = k^2 - k_z^2$. Notice that the above is just the Bessel equation with two linearly independent solutions. Its general solution is a linear superposition of any two of the following four special functions[5]:

(i)   the Bessel function $J_n(k_\rho\rho)$;

(ii)   the Neumann function $N_n(k_\rho\rho)$;

(iii) the Hankel function of the first kind $H_n^{(1)}(k_\rho\rho)$; and

(iv)   the Hankel function of the second kind $H_n^{(2)}(k_\rho\rho)$.

Since only two of these four special functions are independent, they are linearly related to each other, i.e.,

$$J_n(k_\rho\rho) = \frac{1}{2}\left[H_n^{(1)}(k_\rho\rho) + H_n^{(2)}(k_\rho\rho)\right], \qquad (1.2.29a)$$

$$N_n(k_\rho\rho) = \frac{1}{2i}\left[H_n^{(1)}(k_\rho\rho) - H_n^{(2)}(k_\rho\rho)\right], \qquad (1.2.29b)$$

or

$$H_n^{(1)}(k_\rho\rho) = J_n(k_\rho\rho) + iN_n(k_\rho\rho), \qquad (1.2.29c)$$

$$H_n^{(2)}(k_\rho\rho) = J_n(k_\rho\rho) - iN_n(k_\rho\rho). \qquad (1.2.29d)$$

These special functions behave differently around the origin when the argument $k_\rho\rho \to 0$. For instance, for $n = 0$, when $k_\rho\rho \to 0$,

$$J_0(k_\rho\rho) \sim 1, \qquad\qquad N_0(k_\rho\rho) \sim \frac{2}{\pi}\ln(k_\rho\rho), \qquad (1.2.30a)$$

$$H_0(k_\rho\rho) \sim \frac{2i}{\pi}\ln(k_\rho\rho), \qquad H_0^{(2)}(k_\rho\rho) \sim -\frac{2i}{\pi}\ln(k_\rho\rho). \qquad (1.2.30b)$$

---

[4] This form is also obtainable by separation of variables.

[5] See Abramowitz and Stegun (1965).

But for $n > 0$, when $k_\rho \rho \to 0$,

$$J_n(k_\rho \rho) \sim \frac{(k_\rho \rho/2)^n}{n!}, \qquad\qquad N_n(k_\rho \rho) \sim -\frac{(n-1)!}{\pi}\left(\frac{2}{k_\rho \rho}\right)^n,$$

(1.2.31a)

$$H_n^{(1)}(k_\rho \rho) \sim -\frac{i(n-1)!}{\pi}\left(\frac{2}{k_\rho \rho}\right)^n, \qquad H_n^{(2)}(k_\rho \rho) \sim \frac{i(n-1)!}{\pi}\left(\frac{2}{k_\rho \rho}\right)^n.$$

(1.2.31b)

Therefore, only the Bessel function is regular at the origin, while the other functions are singular.

On the other hand, when $k_\rho \rho \to \infty$,

$$J_n(k_\rho \rho) \sim \sqrt{\frac{2}{\pi k_\rho \rho}}\cos\left(k_\rho \rho - \frac{n\pi}{2} - \frac{\pi}{4}\right), \tag{1.2.32a}$$

$$N_n(k_\rho \rho) \sim \sqrt{\frac{2}{\pi k_\rho \rho}}\sin\left(k_\rho \rho - \frac{n\pi}{2} - \frac{\pi}{4}\right), \tag{1.2.32b}$$

$$H_n^{(1)}(k_\rho \rho) \sim \sqrt{\frac{2}{\pi k_\rho \rho}}e^{i\left(k_\rho \rho - \frac{n\pi}{2} - \frac{\pi}{4}\right)}, \tag{1.2.32c}$$

$$H_n^{(2)}(k_\rho \rho) \sim \sqrt{\frac{2}{\pi k_\rho \rho}}e^{-i\left(k_\rho \rho - \frac{n\pi}{2} - \frac{\pi}{4}\right)}. \tag{1.2.32d}$$

Therefore, the Bessel function and the Neumann function are standing waves. In contrast, $H_n^{(1)}(k_\rho \rho)$ is an outgoing wave, whereas $H_n^{(2)}(k_\rho \rho)$ is an incoming wave (assuming $e^{-i\omega t}$ dependence), when $k_\rho \rho \to \infty$. When $k_\rho \rho$ is real, $J_n(k_\rho \rho)$ and $N_n(k_\rho \rho)$ are real functions, whereas $H_n^{(1)}(k_\rho \rho)$ and $H_n^{(2)}(k_\rho \rho)$ are complex functions. Furthermore, $H_n^{(1)}(k_\rho \rho) = [H_n^{(2)}(k_\rho \rho)]^*$ in this case.

Equation (27) in general represents a cylindrical wave or a conical wave, since for large ρ, the wavefront has a cone shape (see Exercise 1.11). When $F_n(\rho)$ in (27) is a $H_n^{(1)}(k_\rho \rho)$, then

$$\psi(\mathbf{r}) \sim \sqrt{\frac{2}{\pi k_\rho \rho}}e^{-i\frac{n\pi}{2} - i\frac{\pi}{4}}e^{i\left(\frac{n}{\rho}\rho\phi + ik_z z + ik_\rho \rho\right)}. \tag{1.2.33}$$

In the above, $\rho\phi$ is the arc length in the ϕ direction, and n/ρ can be thought of as the ϕ component of the $\mathbf{k}$ vector if we compare (33) with (23). Consequently, (33) looks like a plane wave propagating mainly in the direction $\mathbf{k} = \hat{z}k_z + \hat{\rho}k_\rho$, when $\rho \to \infty$.

An important recurrence formula for solutions of the Bessel equation is

$$B_n'(k_\rho \rho) = B_{n-1}(k_\rho \rho) - \frac{n}{k_\rho \rho}B_n(k_\rho \rho)$$

$$= -B_{n+1}(k_\rho \rho) + \frac{n}{k_\rho \rho}B_n(k_\rho \rho), \tag{1.2.34}$$

where $B_n(k_\rho\rho)$ is either $J_n(k_\rho\rho)$, $N_n(k_\rho\rho)$, $H_n^{(1)}(k_\rho\rho)$, $H_n^{(2)}(k_\rho\rho)$, or a linear combination thereof.

§§1.2.5 Spherical Coordinates

In spherical coordinates, the scalar wave equation is

$$\left[\frac{1}{r^2}\frac{\partial}{\partial r}r^2\frac{\partial}{\partial r} + \frac{1}{r^2\sin\theta}\frac{\partial}{\partial\theta}\sin\theta\frac{\partial}{\partial\theta} + \frac{1}{r^2\sin^2\theta}\frac{\partial^2}{\partial\phi^2} + k^2\right]\psi(\mathbf{r}) = 0. \qquad (1.2.35)$$

Noting the $\partial^2/\partial\phi^2$ derivative, we assume that $\psi(\mathbf{r})$ is of the form

$$\psi(\mathbf{r}) = F(r,\theta)e^{im\phi}. \qquad (1.2.36)$$

Then, (35) becomes

$$\left[\frac{1}{r^2}\frac{\partial}{\partial r}r^2\frac{\partial}{\partial r} + \frac{1}{r^2\sin\theta}\frac{\partial}{\partial\theta}\sin\theta\frac{\partial}{\partial\theta} - \frac{m^2}{r^2\sin^2\theta} + k^2\right]F(r,\theta) = 0. \qquad (1.2.37)$$

The above can be further simplified by the separation of variables by letting

$$F(r,\theta) = b_n(kr)P_n^m(\cos\theta), \qquad (1.2.38)$$

where $P_n^m(\cos\theta)$ is the associate Legendre polynomial satisfying the equation

$$\left\{\frac{1}{\sin\theta}\frac{d}{d\theta}\sin\theta\frac{d}{d\theta} + \left[n(n+1) - \frac{m^2}{\sin^2\theta}\right]\right\}P_n^m(\cos\theta) = 0. \qquad (1.2.39)$$

Therefore, $b_n(kr)$ satisfies the equation

$$\left[\frac{1}{r^2}\frac{d}{dr}r^2\frac{d}{dr} + k^2 - \frac{n(n+1)}{r^2}\right]b_n(kr) = 0. \qquad (1.2.40)$$

The above is just the spherical Bessel equation, and $b_n(kr)$ is either the spherical Bessel function $j_n(kr)$, spherical Neumann function $n_n(kr)$, or the spherical Hankel functions $h_n^{(1)}(kr)$ and $h_n^{(2)}(kr)$.

The spherical functions are related to the cylindrical functions via (see Exercise 1.12)

$$b_n(kr) = \sqrt{\frac{\pi}{2kr}}B_{n+\frac{1}{2}}(kr), \qquad (1.2.40a)$$

where $b_n(kr)$ is either $j_n(kr)$, $n_n(kr)$, $h_n^{(1)}(kr)$, or $h_n^{(2)}(kr)$; while $B_{n+\frac{1}{2}}(kr)$ is either $J_{n+\frac{1}{2}}(kr)$, $N_{n+\frac{1}{2}}(kr)$, $H_{n+\frac{1}{2}}^{(1)}(kr)$, or $H_{n+\frac{1}{2}}^{(2)}(kr)$. More specifically,

$$h_0^{(1)}(kr) = \frac{e^{ikr}}{ikr}, \qquad h_1^{(1)}(kr) = -\left(1+\frac{i}{kr}\right)\frac{e^{ikr}}{kr}, \qquad (1.2.41a)$$

$$h_0^{(2)}(kr) = -\frac{e^{-ikr}}{ikr}, \qquad h_1^{(2)}(kr) = -\left(1-\frac{i}{kr}\right)\frac{e^{-ikr}}{kr}, \qquad (1.2.41b)$$

$$j_0(kr) = \frac{\sin kr}{kr}, \qquad j_1(kr) = -\frac{\cos kr}{kr} + \frac{\sin kr}{(kr)^2}, \qquad (1.2.41c)$$

$$n_0(kr) = -\frac{\cos kr}{kr}, \qquad n_1(kr) = -\frac{\sin kr}{kr} - \frac{\cos kr}{(kr)^2}. \qquad (1.2.41d)$$

Hence, the spherical functions represent spherical waves, which resemble plane waves when $r \to \infty$. Moreover, recurrence relations similar to (34) can be derived for spherical Bessel functions (see Exercise 1.13, also see Abramowitz and Stegun 1965)

§1.3 Vector Wave Equations

For an inhomogeneous, anisotropic medium, Maxwell's equations with a fictitious magnetic current density **M** could be written as

$$\nabla \times \mathbf{E}(\mathbf{r},t) = -\frac{\partial}{\partial t}\overline{\mu} \cdot \mathbf{H}(\mathbf{r},t) - \mathbf{M}(\mathbf{r},t), \qquad (1.3.1)$$

$$\nabla \times \mathbf{H}(\mathbf{r},t) = \frac{\partial}{\partial t}\overline{\epsilon} \cdot \mathbf{E}(\mathbf{r},t) + \mathbf{J}(\mathbf{r},t). \qquad (1.3.2)$$

Furthermore, if the fields are time harmonic, the above equations become

$$\nabla \times \mathbf{E}(\mathbf{r}) = i\omega \overline{\mu} \cdot \mathbf{H}(\mathbf{r}) - \mathbf{M}(\mathbf{r}), \qquad (1.3.3)$$

$$\nabla \times \mathbf{H}(\mathbf{r}) = -i\omega \overline{\epsilon} \cdot \mathbf{E}(\mathbf{r}) + \mathbf{J}(\mathbf{r}). \qquad (1.3.4)$$

Since electromagnetic fields are vector fields, the general wave equation is a vector wave equation. Hence, we will derive the general, time harmonic form of the vector wave equation first.

To do so, we take the curl of $\overline{\mu}^{-1} \cdot (3)$, and use (4), to obtain

$$\nabla \times \overline{\mu}^{-1} \cdot \nabla \times \mathbf{E}(\mathbf{r}) - \omega^2 \overline{\epsilon} \cdot \mathbf{E}(\mathbf{r}) = i\omega \mathbf{J}(\mathbf{r}) - \nabla \times \overline{\mu}^{-1} \cdot \mathbf{M}(\mathbf{r}). \qquad (1.3.5a)$$

Similarly,

$$\nabla \times \overline{\epsilon}^{-1} \cdot \nabla \times \mathbf{H}(\mathbf{r}) - \omega^2 \overline{\mu} \cdot \mathbf{H}(\mathbf{r}) = i\omega \mathbf{M}(\mathbf{r}) + \nabla \times \overline{\epsilon}^{-1} \cdot \mathbf{J}(\mathbf{r}). \qquad (1.3.5b)$$

The above also follows directly from the duality principle.

Equations (5a) and (5b) are two vector wave equations governing the solutions of an electromagnetic field in an inhomogeneous, anisotropic medium. Here, $\overline{\mu}$ and $\overline{\epsilon}$ are assumed to be functions of positions; hence, they do not commute with the ∇ operator. Moreover, for time-varying fields, **E** and **H** can be derived from each other; hence, only one of the two Equations (5a) and (5b) is needed to fully describe electromagnetic fields.

For an inhomogeneous isotropic medium, however, the above equations reduce to

$$\nabla \times \mu^{-1}\nabla \times \mathbf{E}(\mathbf{r}) - \omega^2 \epsilon \, \mathbf{E}(\mathbf{r}) = i\omega \, \mathbf{J}(\mathbf{r}) - \nabla \times \mu^{-1}\mathbf{M}(\mathbf{r}),$$
$$(1.3.6a)$$

$$\nabla \times \epsilon^{-1}\nabla \times \mathbf{H}(\mathbf{r}) - \omega^2 \mu \, \mathbf{H}(\mathbf{r}) = i\omega \, \mathbf{M}(\mathbf{r}) + \nabla \times \epsilon^{-1}\mathbf{J}(\mathbf{r}).$$
$$(1.3.6b)$$

As mentioned in the preceding paragraph, either one of the above equations is self-contained. Consequently, all phenomena of electrodynamic fields in

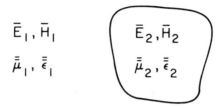

Figure 1.3.1 The solution to a piecewise-constant inhomogeneity can be obtained by first obtaining the solution in each region, and then patching the solutions together via boundary conditions.

inhomogeneous media are obtained by studying just one of them. In fact, the two equations are derivable from each other. Furthermore, since $\nabla \cdot \overline{\overline{\epsilon}} \cdot \mathbf{E} = \varrho$ and $\nabla \cdot \overline{\overline{\mu}} \cdot \mathbf{H} = \varrho_m$, the three components of $\mathbf{E}$ or $\mathbf{H}$ are not linearly independent of each other. Hence, many electromagnetic problems can be formulated in terms of only two of the six components in $\mathbf{E}$ and $\mathbf{H}$.

§§1.3.1 Boundary Conditions

Equation (5a) or (5b) describes all the phenomena of electrodynamic wave interaction with inhomogeneity. Therefore, either one of them can be considered as the basic equation rather than Maxwell's equations for electromagnetic phenomena. Moreover, what is derivable from Maxwell's equations is also derivable from the above equations. For instance, when solving problems involving piecewise-constant inhomogeneities, a common practice is to obtain first solutions to Maxwell's equations in each region, and later, match boundary conditions at the interfaces to obtain the solution valid everywhere (see Figure 1.3.1). We shall show that these boundary conditions can be derived from either one of the two vector wave equations.

To do so, we integrate (5a) about a small area between the interface of the two inhomogeneities (see Figure 1.3.2). Then, on invoking Stokes' theorem for the surface integral of a curl, (5a) becomes

$$\oint_C d\mathbf{l} \cdot (\overline{\overline{\mu}}^{-1} \cdot \nabla \times \mathbf{E}) - \omega^2 \int_A d\mathbf{S} \cdot \overline{\overline{\epsilon}} \cdot \mathbf{E} = i\omega \int_A d\mathbf{S} \cdot \mathbf{J} - \oint_C d\mathbf{l} \cdot \overline{\overline{\mu}}^{-1} \cdot \mathbf{M}. \tag{1.3.7}$$

But if $\mathbf{M}$ is a current sheet, the last term on the right-hand side of (7) vanishes because $\overline{\overline{\mu}}^{-1} \cdot \mathbf{M} = 0$ on C.

Consequently, when $\delta \to 0$ (see Figure 1.3.2), the area integral on the left-hand side of the above equation vanishes, because $\overline{\overline{\epsilon}} \cdot \mathbf{E}$ is nonsingular at

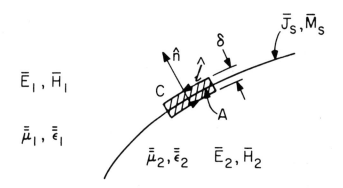

Figure 1.3.2 Derivation of the boundary conditions.

the interface. Moreover, if a current sheet $\mathbf{J}_s$ resides at the interface, then $\mathbf{J}$ is singular at the interface, and

$$\int_A d\mathbf{S} \cdot \mathbf{J} = \int_a^b dl \, (\hat{n} \times \hat{l} \cdot \mathbf{J}_s). \qquad (1.3.8)$$

Similarly,

$$\oint_C d\mathbf{l} \cdot (\overline{\boldsymbol{\mu}}^{-1} \cdot \nabla \times \mathbf{E}) = \int_a^b d\mathbf{l} \cdot (\overline{\boldsymbol{\mu}}_1^{-1} \cdot \nabla \times \mathbf{E}_1) - \int_a^b d\mathbf{l} \cdot (\overline{\boldsymbol{\mu}}_2^{-1} \cdot \nabla \times \mathbf{E}_2), \qquad (1.3.9)$$

where $\mathbf{E}_1$ and $\mathbf{E}_2$ are the fields and $\overline{\boldsymbol{\mu}}_1$ and $\overline{\boldsymbol{\mu}}_2$ are the permeability tensors in the two different regions. This further implies that

$$\hat{l} \cdot (\overline{\boldsymbol{\mu}}_1^{-1} \cdot \nabla \times \mathbf{E}_1) - \hat{l} \cdot (\overline{\boldsymbol{\mu}}_2^{-1} \cdot \nabla \times \mathbf{E}_2) = i\omega \, \hat{n} \times \hat{l} \cdot \mathbf{J}_s. \qquad (1.3.10)$$

On noticing that $\hat{l} = (\hat{n} \times \hat{l}) \times \hat{n}$ on the left-hand side, and using the appropriate vector identity, we have

$$\hat{n} \times (\overline{\boldsymbol{\mu}}_1^{-1} \cdot \nabla \times \mathbf{E}_1) - \hat{n} \times (\overline{\boldsymbol{\mu}}_2^{-1} \cdot \nabla \times \mathbf{E}_2) = i\omega \, \mathbf{J}_s. \qquad (1.3.11)$$

Since $\nabla \times \mathbf{E} = i\omega \, \overline{\boldsymbol{\mu}} \cdot \mathbf{H}$, the above is also the same as

$$\hat{n} \times \mathbf{H}_1 - \hat{n} \times \mathbf{H}_2 = \mathbf{J}_s, \qquad (1.3.12)$$

which states that the discontinuity in the tangential component of the magnetic field is proportional to the electric current sheet $\mathbf{J}_s$.

To derive another boundary condition, we rewrite Equation (5a) as

$$\nabla \times \overline{\boldsymbol{\mu}}^{-1} \cdot [\nabla \times \mathbf{E}(\mathbf{r}) + \mathbf{M}(\mathbf{r})] - \omega^2 \, \overline{\boldsymbol{\epsilon}} \cdot \mathbf{E}(\mathbf{r}) = i\omega \mathbf{J}(\mathbf{r}). \qquad (1.3.13)$$

Now, on the right-hand side, if $\mathbf{J}(\mathbf{r})$ is a current sheet $\mathbf{J}_s$, it will give rise to a discontinuity in $\overline{\mu}^{-1} \cdot [\nabla \times \mathbf{E}(\mathbf{r}) + \mathbf{M}]$. However, $\overline{\mu}^{-1} \cdot [\nabla \times \mathbf{E}(\mathbf{r}) + \mathbf{M}]$ must be regular or nonsingular, for if it is singular, its curl will make it doubly singular, which cannot be cancelled by any other terms in (13). But if $\overline{\mu}^{-1} \cdot [\nabla \times \mathbf{E}(\mathbf{r}) + \mathbf{M}(\mathbf{r})]$ is regular, so must $\nabla \times \mathbf{E}(\mathbf{r}) + \mathbf{M}$ since $\overline{\mu}^{-1}$ is nonsingular. Therefore, after integrating $\nabla \times \mathbf{E}(\mathbf{r}) + \mathbf{M}$ over A as in (7) and letting $\delta \to 0$, we conclude that

$$\hat{n} \times \mathbf{E}_1 - \hat{n} \times \mathbf{E}_2 = -\mathbf{M}_s, \qquad (1.3.14)$$

where $\mathbf{M}_s$ is a magnetic current sheet at the interface. Thus, the discontinuity in the tangential component of the electric field is proportional to the magnetic current sheet $\mathbf{M}_s$.

The boundary conditions (12) and (14) can also be derived more directly from Maxwell's equations. Though the derivation here is less direct, it illustrates that these boundary conditions are inherently buried in (5a). Similarly, they can also be extracted from (5b). In general, boundary conditions are buried in the partial differential equation that governs the field (see Exercise 1.14). This further reinforces the point that either (5a) or (5b) alone is sufficient to describe electrodynamic phenomena in an inhomogeneous, anisotropic medium.

§§1.3.2 Reciprocity Theorem

The reciprocity theorem relates in a simple manner the mutual interactions between two groups of sources, under certain conditions on the medium. Such a medium is then known as a reciprocal medium. We shall show how such a reciprocity relation can be derived from the vector wave equations.

If there are two groups of sources $\mathbf{J}_1, \mathbf{M}_1$; and $\mathbf{J}_2, \mathbf{M}_2$ radiating in an anisotropic, inhomogeneous medium, where $\mathbf{J}_1, \mathbf{M}_1$ produces the field $\mathbf{E}_1$, and $\mathbf{J}_2, \mathbf{M}_2$ produces the field $\mathbf{E}_2$, the vector wave equations that $\mathbf{E}_1$ and $\mathbf{E}_2$ satisfy are then

$$\nabla \times \overline{\mu}^{-1} \cdot \nabla \times \mathbf{E}_1 - \omega^2 \overline{\epsilon} \cdot \mathbf{E}_1 = i\omega \mathbf{J}_1 - \nabla \times \overline{\mu}^{-1} \cdot \mathbf{M}_1, \qquad (1.3.15a)$$

and

$$\nabla \times \overline{\mu}^{-1} \cdot \nabla \times \mathbf{E}_2 - \omega^2 \overline{\epsilon} \cdot \mathbf{E}_2 = i\omega \mathbf{J}_2 - \nabla \times \overline{\mu}^{-1} \cdot \mathbf{M}_2. \qquad (1.3.15b)$$

Next, by dotting (15a) by $\mathbf{E}_2$ and integrating over volume, we obtain

$$i\omega \langle \mathbf{E}_2, \mathbf{J}_1 \rangle - \langle \mathbf{E}_2, \nabla \times \overline{\mu}^{-1} \cdot \mathbf{M}_1 \rangle$$
$$= \langle \mathbf{E}_2, \nabla \times \overline{\mu}^{-1} \cdot \nabla \times \mathbf{E}_1 \rangle - \omega^2 \langle \mathbf{E}_2, \overline{\epsilon} \cdot \mathbf{E}_1 \rangle,$$
$$(1.3.16a)$$

where the reaction or the inner product $\langle \mathbf{A}, \mathbf{B} \rangle = \int d\mathbf{r} \mathbf{A} \cdot \mathbf{B}$ (Rumsey 1954). By the same token, from (15b),

$$i\omega \langle \mathbf{E}_1, \mathbf{J}_2 \rangle - \langle \mathbf{E}_1, \nabla \times \overline{\mu}^{-1} \cdot \mathbf{M}_2 \rangle$$
$$= \langle \mathbf{E}_1, \nabla \times \overline{\mu}^{-1} \cdot \nabla \times \mathbf{E}_2 \rangle - \omega^2 \langle \mathbf{E}_1, \overline{\epsilon} \cdot \mathbf{E}_2 \rangle.$$
$$(1.3.16b)$$

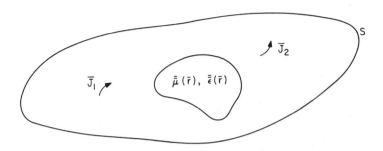

Figure 1.3.3 Derivation of the reciprocity relation.

The second terms on the right-hand side of (16a) and (16b) are equal if $\bar{\epsilon} = \bar{\epsilon}^t$. It is unclear if the first terms on the right-hand side of (16a) and (16b) are the same, but they are of the form

$$\langle \mathbf{E}_i, \nabla \times \overline{\boldsymbol{\mu}}^{-1} \cdot \nabla \times \mathbf{E}_j \rangle = \int_V d\mathbf{r}\, \mathbf{E}_i \cdot \nabla \times \overline{\boldsymbol{\mu}}^{-1} \cdot \nabla \times \mathbf{E}_j. \qquad (1.3.17)$$

The above can be rewritten using the vector identity $\nabla \cdot (\mathbf{A} \times \mathbf{B}) = \mathbf{B} \cdot \nabla \times \mathbf{A} - \mathbf{A} \cdot \nabla \times \mathbf{B}$ and Gauss' theorem[6]:

$$\langle \mathbf{E}_i, \nabla \times \overline{\boldsymbol{\mu}}^{-1} \cdot \nabla \times \mathbf{E}_j \rangle = \int_V d\mathbf{r}\, (\nabla \times \mathbf{E}_i) \cdot \overline{\boldsymbol{\mu}}^{-1} \cdot (\nabla \times \mathbf{E}_j)$$

$$+ \int_S dS\, \hat{n} \cdot (\overline{\boldsymbol{\mu}}^{-1} \cdot \nabla \times \mathbf{E}_j) \times \mathbf{E}_i, \qquad (1.3.18)$$

where V and S are volume and surface respectively, tending to infinity (see Figure 1.3.3). Note that now, the first term on the right-hand side of (18) is symmetric about $\mathbf{E}_i$ and $\mathbf{E}_j$ if $\overline{\boldsymbol{\mu}} = \overline{\boldsymbol{\mu}}^t$.

To show the symmetry of the second term, however, more manipulation is needed as follows: When $S \to \infty$, it is reasonable to assume that $\overline{\boldsymbol{\mu}}$ is isotropic and homogeneous. Furthermore, the fields, which are produced by sources of finite extent, become plane waves in the far field. Hence, $\nabla \to i\mathbf{k}$, which is the case for plane waves. Consequently,

$$(\overline{\boldsymbol{\mu}}^{-1} \cdot \nabla \times \mathbf{E}_j) \times \mathbf{E}_i = i\mu_0^{-1}(\mathbf{k} \times \mathbf{E}_j) \times \mathbf{E}_i = -i\mu_0^{-1}\mathbf{k}(\mathbf{E}_i \cdot \mathbf{E}_j). \qquad (1.3.19)$$

In arriving at the above, we have used $\mathbf{k} \cdot \mathbf{E}_i = 0$ because of the plane-wave assumption. In this manner, the surface integral in (18) is symmetric about $\mathbf{E}_i$ and $\mathbf{E}_j$.

---

[6] This manipulation is also referred to as integration by parts. It is the generalization of integration by parts in one dimension to higher dimensions and vector fields.

From the above, the right-hand sides of (16a) and (16b) are equal only if

$$\overline{\mu} = \overline{\mu}^t, \quad \overline{\epsilon} = \overline{\epsilon}^t, \tag{1.3.20}$$

i.e., when $\overline{\mu}$ and $\overline{\epsilon}$ are symmetric tensors. Consequently, Equation (20) implies that

$$i\omega\langle \mathbf{E}_2, \mathbf{J}_1\rangle - \langle \mathbf{E}_2, \nabla \times \overline{\mu}^{-1} \cdot \mathbf{M}_1\rangle = i\omega\langle \mathbf{E}_1, \mathbf{J}_2\rangle - \langle \mathbf{E}_1, \nabla \times \overline{\mu}^{-1} \cdot \mathbf{M}_2\rangle. \tag{1.3.21}$$

Moreover, using the vector identity used for Equation (18), we can show that

$$\langle \mathbf{E}_2, \nabla \times \overline{\mu}^{-1} \cdot \mathbf{M}_1\rangle = \int_V d\mathbf{r} \, (\nabla \times \mathbf{E}_2) \cdot \overline{\mu}^{-1} \cdot \mathbf{M}_1 = i\omega \int_V d\mathbf{r} \, \mathbf{H}_2 \cdot \mathbf{M}_1. \tag{1.3.22}$$

The last equality follows because $\overline{\mu} = \overline{\mu}^t$. Hence, (21) is identical to

$$\langle \mathbf{E}_2, \mathbf{J}_1\rangle - \langle \mathbf{H}_2, \mathbf{M}_1\rangle = \langle \mathbf{E}_1, \mathbf{J}_2\rangle - \langle \mathbf{H}_1, \mathbf{M}_2\rangle, \tag{1.3.23}$$

which is the reciprocal theorem. Note that the above describes a mutually reciprocal relationship.

One side of Equation (23) describes the mutual interaction between the field of one group of sources with another group of sources. This mutual interaction is only reciprocal if the medium satisfies the conditions of Equation (20). A medium for which conditions given by Equation (20) hold, implying the reciprocal relationship (23), is known as a *reciprocal medium*. We shall see later that the reciprocal nature of (23) is due to the symmetric nature of the vector wave Equation (15). Scalar wave equations with similar symmetry also have an analogous reciprocal relation (see Exercise 1.14).

The reaction $\langle \mathbf{E}_i, \mathbf{J}_j\rangle$ and $\langle \mathbf{H}_i, \mathbf{M}_j\rangle$ can be thought of as generalized measurements. Physically, Equation (23) states that the field resulting from $\mathbf{J}_1$, $\mathbf{M}_1$ measured by $\mathbf{J}_2$, $\mathbf{M}_2$ is the same as the field resulting from $\mathbf{J}_2$, $\mathbf{M}_2$ measured by $\mathbf{J}_1$, $\mathbf{M}_1$. Examples of reciprocal media are free-space and lossy media—a medium can be lossy and still be reciprocal! Examples of nonreciprocal media are plasma and ferrite media biased by a magnetic field.

§§1.3.3 Plane Wave in Homogeneous, Anisotropic Media

A plane wave is the simplest of the wave solutions. All wave types can be expanded in terms of plane waves as shall be shown later in Chapter 2. Therefore, we shall look for a plane-wave solution in a homogeneous, anisotropic and source-free medium. In such a medium, the vector wave equation from (5a) is

$$\nabla \times \overline{\mu}^{-1} \cdot \nabla \times \mathbf{E} - \omega^2 \overline{\epsilon} \cdot \mathbf{E} = 0. \tag{1.3.24}$$

To look for a plane-wave solution to (24), we assume $\mathbf{E}$ to be of the form

$$\mathbf{E} = \mathbf{E}_0 e^{i\mathbf{k}\cdot\mathbf{r}}, \tag{1.3.25}$$

where **k** is the **k** vector denoting the direction of propagation of the plane wave. Then, substituting (25) into (24) yields

$$\mathbf{k} \times \overline{\mu}^{-1} \cdot \mathbf{k} \times \mathbf{E}_0 + \omega^2 \overline{\epsilon} \cdot \mathbf{E}_0 = 0. \qquad (1.3.26)$$

Next, $\mathbf{k} \times \mathbf{E}_0$ can be written as $k\overline{\mathbf{K}} \cdot \mathbf{E}_0$ where $\overline{\mathbf{K}}$ is a tensor:

$$k\overline{\mathbf{K}} = \begin{bmatrix} 0 & -k_z & k_y \\ k_z & 0 & -k_x \\ -k_y & k_x & 0 \end{bmatrix}, \qquad (1.3.27)$$

which is an antisymmetric matrix in Cartesian coordinates (see Appendix B for a review of tensors). Moreover, we can write the Cartesian components of **k** in terms of direction cosines to obtain

$$k\overline{\mathbf{K}}(\theta, \phi) = k \begin{bmatrix} 0 & -\cos\theta & \sin\theta\sin\phi \\ \cos\theta & 0 & -\sin\theta\cos\phi \\ -\sin\theta\sin\phi & \sin\theta\cos\phi & 0 \end{bmatrix}, \qquad (1.3.28)$$

where k, which is yet to be found, is the length of the **k** vector. Alternatively, Equation (26) can be written as

$$\left[k^2 \overline{\mathbf{F}}(\theta, \phi) + \omega^2 \overline{\mathbf{I}} \right] \cdot \mathbf{D}_0 = 0, \qquad (1.3.29)$$

where

$$\overline{\mathbf{F}}(\theta, \phi) = \overline{\mathbf{K}}(\theta, \phi) \cdot \overline{\mu}^{-1} \cdot \overline{\mathbf{K}}(\theta, \phi) \cdot \overline{\epsilon}^{-1}, \qquad \mathbf{D}_0 = \overline{\epsilon} \cdot \mathbf{E}_0. \qquad (1.3.29a)$$

$\overline{\mathbf{F}}(\theta, \phi)$ is only a function of angles and tensors $\overline{\mu}$ and $\overline{\epsilon}$. Notice that for a plane-wave solution propagating in a fixed direction, $\overline{\mathbf{F}}(\theta, \phi)$ is a constant matrix.

Equation (29) corresponds to an eigenvalue problem where ω^2/k^2 is the eigenvalue and $\mathbf{D}_0$ is the eigenvector. Since $\overline{\mathbf{F}}$ is a 3×3 matrix, we expect the above equations to have three eigenvalues and three eigenvectors. However, from the $\nabla \cdot \mathbf{D} = 0$ condition, $\mathbf{k} \cdot \mathbf{D} = 0$. This can also be seen by dotting Equation (26) with **k** and noting that $\mathbf{k} \cdot \mathbf{k} \times \mathbf{A} = 0$. Therefore, only two out of three components of the electric flux **D** are independent, implying that only two equations in (29) are independent. Consequently, it will only yield two eigenvalues and two eigenvectors. This can be shown easily by expressing the field and the tensors in a coordinate system where the z axis corresponds to the **k** direction of the **k** vector (see Exercise 1.15). Hence, the general wave solution to (24) is of the form

$$\mathbf{E} = a_1 \mathbf{e}_1 e^{i\mathbf{k}_1 \cdot \mathbf{r}} + a_2 \mathbf{e}_2 e^{i\mathbf{k}_2 \cdot \mathbf{r}} = \sum_{j=1}^{2} a_j \mathbf{e}_j e^{i\mathbf{k}_j \cdot \mathbf{r}}, \qquad (1.3.30)$$

where $\mathbf{k}_j = k_j \hat{k}$, k_j is derived from the j-th eigenvalue, and $\mathbf{e}_j$ is derived from the j-th eigenvector of (29). Because k is different for the two waves, they

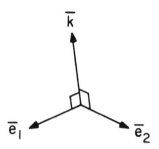

Figure 1.3.4 The electric field vectors and the **k** vector in an isotropic medium.

have **k** vectors of different lengths. Moreover, since the phase velocity for a plane wave is defined by $v_p = \omega/k$, the phase velocities for the two waves will be different. These two waves are generally known as type I and type II waves. Furthermore, their corresponding magnetic field can easily be derived to be

$$\mathbf{H} = \sum_{j=1}^{2} a_j (\omega \, \overline{\boldsymbol{\mu}})^{-1} \cdot \mathbf{k}_j \times \mathbf{e}_j e^{i\mathbf{k}_j \cdot \mathbf{r}} = \sum_{j=1}^{2} a_j \mathbf{h}_j e^{i\mathbf{k}_j \cdot \mathbf{r}}. \qquad (1.3.31)$$

Note that since $\overline{\mathbf{F}}$ is a function of angles, the eigenvalues k_j or the lengths of the $\mathbf{k}_j$ vectors change as a function of angle, i.e., $k_j(\theta, \phi)$. Therefore, the phase velocity in an anisotropic medium is also a function of angles for each of the two types of waves. With $\mathbf{k}_j$ changing as a function of angles, $\mathbf{e}_j$ also changes as a function of angles. Furthermore, since $\nabla \cdot \mathbf{E} \neq 0$ in general, $\mathbf{k} \cdot \mathbf{E} \neq 0$ for anisotropic media, unlike the case for isotropic media.

For a homogeneous, isotropic medium, the eigenvalues are degenerate where $k_j = \omega\sqrt{\mu\epsilon}$ and $\mathbf{k}_j \cdot \mathbf{e}_j = 0$ for $j = 1, 2$. The eigenvectors of (29) then are any two linearly independent vectors $\mathbf{e}_1$ and $\mathbf{e}_2$ that are orthogonal to **k** (Figure 1.3.4). Moreover, $\mathbf{e}_1$ and $\mathbf{e}_2$ can also be made orthogonal to each other without loss of generality. The general solution then becomes

$$\mathbf{E} = a_1 (\mathbf{k} \times \mathbf{c}) e^{i\mathbf{k}\cdot\mathbf{r}} + a_2 (\mathbf{k} \times \mathbf{k} \times \mathbf{c}) e^{i\mathbf{k}\cdot\mathbf{r}}, \qquad (1.3.32)$$

where **c** is an arbitrary constant vector. For instance, if $\mathbf{c} = \hat{z}$ which points upward, then the above two waves correspond to a horizontal polarization and a vertical polarization. The first corresponds to a transverse electric (TE) to z wave while the second corresponds to a transverse magnetic (TM) to z wave.

§§1.3.4 Green's Function

The Green's function of a wave equation is the solution of the wave equation for a point source. And when the solution to the wave equation due to a

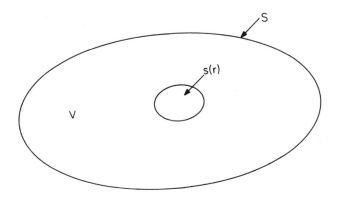

Figure 1.3.5 The radiation of a source $s(\mathbf{r})$ in a volume V.

point source is known, the solution due to a general source can be obtained by the principle of linear superposition (see Figure 1.3.5). This is merely a result of the linearity of the wave equation, and that a general source is just a linear superposition of point sources.

For example, to obtain the solution to the scalar wave equation in V in Figure 1.3.5,

$$(\nabla^2 + k^2)\,\psi(\mathbf{r}) = s(\mathbf{r}), \qquad (1.3.33)$$

we first seek the Green's function in the same V, which is the solution to the following equation:

$$(\nabla^2 + k^2)\,g(\mathbf{r}, \mathbf{r}') = -\delta(\mathbf{r} - \mathbf{r}'). \qquad (1.3.34)$$

Given $g(\mathbf{r}, \mathbf{r}')$, $\psi(\mathbf{r})$ can be found easily fron the principle of linear superposition, since $g(\mathbf{r}, \mathbf{r}')$ is the solution to (33) with a point source on the right-hand side. To see this more clearly, note that an arbitrary source $s(\mathbf{r})$ is just

$$s(\mathbf{r}) = \int d\mathbf{r}'s(\mathbf{r}')\,\delta(\mathbf{r} - \mathbf{r}'), \qquad (1.3.35)$$

which is actually a linear superposition of point sources in mathematical terms. Consequently, the solution to (33) is just

$$\psi(\mathbf{r}) = -\int_V d\mathbf{r}'g(\mathbf{r}, \mathbf{r}')\,s(\mathbf{r}'), \qquad (1.3.36)$$

which is an integral linear superposition of the solution of (34). Moreover, it can be seen that $g(\mathbf{r}, \mathbf{r}') = g(\mathbf{r}', \mathbf{r})$ from reciprocity irrespective of the shape of V (see Exercises 1.14, 1.17).

To find the solution of Equation (34) for an unbounded, homogeneous medium, one solves it in spherical coordinates with the origin at $\mathbf{r}'$. By so doing, (34) becomes

$$(\nabla^2 + k^2)\, g(\mathbf{r}) = -\delta(\mathbf{r}) = -\delta(x)\,\delta(y)\,\delta(z). \qquad (1.3.37)$$

But due to the spherical symmetry of a point source, $g(\mathbf{r})$ must also be spherically symmetric. Then, for $r \neq 0$, the homogeneous, spherically symmetric solution to (37) is given by

$$g(\mathbf{r}) = C\frac{e^{ikr}}{r} + D\frac{e^{-ikr}}{r}. \qquad (1.3.38)$$

Since sources are absent at infinity, physical grounds then imply that only an outgoing solution can exist; hence,

$$g(\mathbf{r}) = C\frac{e^{ikr}}{r}. \qquad (1.3.39)$$

The constant C is found by matching the singularities at the origin on both sides of (37). To do this, we substitute (39) into (37) and integrate Equation (37) over a small volume about the origin to yield

$$\int_{\Delta V} dV\, \nabla \cdot \nabla \frac{Ce^{ikr}}{r} + \int_{\Delta V} dV k^2 \frac{Ce^{ikr}}{r} = -1. \qquad (1.3.40)$$

Note that the second integral vanishes when $\Delta V \to 0$, because $dV = 4\pi r^2 dr$. Moreover, the first integral in (40) can be converted into a surface integral using Gauss' theorem to obtain

$$\lim_{r \to 0} 4\pi r^2 \frac{d}{dr} C\frac{e^{ikr}}{r} = -1, \qquad (1.3.41)$$

or $C = 1/4\pi$.

The solution to (34) must depend only on $|\mathbf{r} - \mathbf{r}'|$. Therefore, in general,

$$g(\mathbf{r}, \mathbf{r}') = g(\mathbf{r} - \mathbf{r}') = \frac{e^{ik|\mathbf{r}-\mathbf{r}'|}}{4\pi|\mathbf{r} - \mathbf{r}'|}, \qquad (1.3.42)$$

implying that $g(\mathbf{r}, \mathbf{r}')$ is translationally invariant for unbounded, homogeneous media. Consequently, the solution to (33), from Equation (36), is then

$$\psi(\mathbf{r}) = -\int_V d\mathbf{r}'\, \frac{e^{ik|\mathbf{r}-\mathbf{r}'|}}{4\pi|\mathbf{r} - \mathbf{r}'|}\, s(\mathbf{r}'). \qquad (1.3.43)$$

The Green's function for the scalar wave equation could be used to find the dyadic Green's function for the vector wave equation in a homogeneous, isotropic medium. First, notice that the vector wave equation in a homogeneous, isotropic medium is

$$\nabla \times \nabla \times \mathbf{E}(\mathbf{r}) - k^2 \, \mathbf{E}(\mathbf{r}) = i\omega\mu \, \mathbf{J}(\mathbf{r}). \qquad (1.3.44)$$

Then, by using the fact that $\nabla \times \nabla \times \mathbf{E} = -\nabla^2\mathbf{E} + \nabla\nabla \cdot \mathbf{E}$ and that $\nabla \cdot \mathbf{E} = \varrho/\epsilon = \nabla \cdot \mathbf{J}/i\omega\epsilon$, which follows from the continuity equation, we can rewrite (44) as

$$\nabla^2\mathbf{E}(\mathbf{r}) + k^2 \, \mathbf{E}(\mathbf{r}) = -i\omega\mu \left[\overline{\mathbf{I}} + \frac{\nabla\nabla}{k^2} \right] \cdot \mathbf{J}(\mathbf{r}), \qquad (1.3.45)$$

where $\overline{\mathbf{I}}$ is an identity operator. In Cartesian coordinates, there are actually three scalar wave equations embedded in the above vector equation, each of which can be solved easily in the manner of Equation (36). Consequently,

$$\mathbf{E}(\mathbf{r}) = i\omega\mu \int_V dr' g(\mathbf{r}' - \mathbf{r}) \left[\overline{\mathbf{I}} + \frac{\nabla'\nabla'}{k^2} \right] \cdot \mathbf{J}(\mathbf{r}') \qquad (1.3.46)$$

where $g(\mathbf{r}' - \mathbf{r})$ is the unbounded medium scalar Green's function. Moreover, by using the vector identities[7] $\nabla g f = f\nabla g + g\nabla f$ and $\nabla \cdot g\mathbf{F} = g\nabla \cdot \mathbf{F} + (\nabla g) \cdot \mathbf{F}$, it can be shown that

$$\int_V dr' g(\mathbf{r}' - \mathbf{r})\nabla' f(\mathbf{r}') = - \int_V dr' \left[\nabla' g(\mathbf{r}' - \mathbf{r}) \right] f(\mathbf{r}'), \qquad (1.3.47)$$

and

$$\int_V dr' \left[\nabla' g(\mathbf{r}' - \mathbf{r}) \right] \nabla' \cdot \mathbf{J}(\mathbf{r}') = - \int_V dr' \, \mathbf{J}(\mathbf{r}') \cdot \nabla'\nabla' g(\mathbf{r}' - \mathbf{r}). \qquad (1.3.48)$$

Hence, Equation (46) can be rewritten as

$$\mathbf{E}(\mathbf{r}) = i\omega\mu \int_V dr' \, \mathbf{J}(\mathbf{r}') \cdot \left[\overline{\mathbf{I}} + \frac{\nabla'\nabla'}{k^2} \right] g(\mathbf{r}' - \mathbf{r}). \qquad (1.3.49)$$

It can also be derived using scalar and vector potentials (see Exercise 1.16 and Chapter 7).

Alternatively, Equation (49) can be written as

$$\mathbf{E}(\mathbf{r}) = i\omega\mu \int_V dr' \, \mathbf{J}(\mathbf{r}') \cdot \overline{\mathbf{G}}_e(\mathbf{r}', \mathbf{r}), \qquad (1.3.50)$$

---

[7] The first identity is the same as the second identity if we think of $\mathbf{F}(\mathbf{r}) = \mathbf{a}f(\mathbf{r})$ where $\mathbf{a}$ is an arbitrary constant vector. Since $\mathbf{a}$ is an arbitrary constant vector, we can cancel it from both sides of the equation to obtain the first identity.

where

$$\overline{\mathbf{G}}_e(\mathbf{r}', \mathbf{r}) = \left[\overline{\mathbf{I}} + \frac{\nabla' \nabla'}{k^2} \right] g(\mathbf{r}' - \mathbf{r}) \qquad (1.3.51)$$

is a dyad known as the dyadic Green's function for the electric field in an unbounded, homogeneous medium. (A **dyad** is a 3×3 matrix that transforms a vector to a vector. It is also a second rank tensor. See Appendix B for details.) Even though (50) is established for an unbounded, homogeneous medium, such a general relationship also exists in a bounded, homogeneous medium. It could easily be shown from reciprocity that

$$\begin{aligned}
\left\langle \mathbf{J}_1(\mathbf{r}), \overline{\mathbf{G}}_e(\mathbf{r}, \mathbf{r}'), \mathbf{J}_2(\mathbf{r}') \right\rangle &= \left\langle \mathbf{J}_2(\mathbf{r}'), \overline{\mathbf{G}}_e(\mathbf{r}', \mathbf{r}), \mathbf{J}_1(\mathbf{r}) \right\rangle \\
&= \left\langle \mathbf{J}_1(\mathbf{r}), \overline{\mathbf{G}}_e^t(\mathbf{r}', \mathbf{r}), \mathbf{J}_2(\mathbf{r}') \right\rangle,
\end{aligned} \qquad (1.3.52)$$

where

$$\left\langle \mathbf{J}_i(\mathbf{r}'), \overline{\mathbf{G}}_e(\mathbf{r}', \mathbf{r}), \mathbf{J}_j(\mathbf{r}) \right\rangle = \int_V \int_V d\mathbf{r}' d\mathbf{r} \, \mathbf{J}_i(\mathbf{r}') \cdot \overline{\mathbf{G}}_e(\mathbf{r}', \mathbf{r}) \cdot \mathbf{J}_j(\mathbf{r}), \qquad (1.3.52a)$$

is the reaction between $\mathbf{J}_i$ and the electric field produced by $\mathbf{J}_j$. Notice that the above implies that[8]

$$\overline{\mathbf{G}}_e^t(\mathbf{r}', \mathbf{r}) = \overline{\mathbf{G}}_e(\mathbf{r}, \mathbf{r}'). \qquad (1.3.52b)$$

Then, by taking its transpose, (50) becomes

$$\mathbf{E}(\mathbf{r}) = i\omega\mu \int_V d\mathbf{r}' \, \overline{\mathbf{G}}_e(\mathbf{r}, \mathbf{r}') \cdot \mathbf{J}(\mathbf{r}'). \qquad (1.3.53)$$

Alternatively, the dyadic Green's function for an unbounded, homogeneous medium can also be written as

$$\overline{\mathbf{G}}_e(\mathbf{r}, \mathbf{r}') = \frac{1}{k^2} \left[\nabla \times \nabla \times \overline{\mathbf{I}} g(\mathbf{r} - \mathbf{r}') - \overline{\mathbf{I}} \delta(\mathbf{r} - \mathbf{r}') \right]. \qquad (1.3.54)$$

By substituting (53) back into (44) and writing

$$\mathbf{J}(\mathbf{r}) = \int d\mathbf{r}' \, \overline{\mathbf{I}} \delta(\mathbf{r} - \mathbf{r}') \cdot \mathbf{J}(\mathbf{r}'), \qquad (1.3.55)$$

we can show quite easily that

$$\nabla \times \nabla \times \overline{\mathbf{G}}_e(\mathbf{r}, \mathbf{r}') - k^2 \overline{\mathbf{G}}_e(\mathbf{r}, \mathbf{r}') = \overline{\mathbf{I}} \delta(\mathbf{r} - \mathbf{r}'). \qquad (1.3.56)$$

Equation (50) or (53), due to the $\nabla\nabla$ operator inside the integration operating on $g(\mathbf{r} - \mathbf{r}')$, has a singularity of $1/|\mathbf{r} - \mathbf{r}'|^3$ when $\mathbf{r} \to \mathbf{r}'$. Consequently, it has to be redefined in this case for it does not converge uniformly,

---

[8] Similar relations also hold for scalar wave equation (see Exercise 1.17).

specifically, when $\mathbf{r}$ is also in the source region occupied by $\mathbf{J}(\mathbf{r})$. Hence, at this point, the evaluation of Equation (53) in a source region is undefined. This singular nature of the dyadic Green's function will be addressed later in Chapter 7.

§1.4 Huygens' Principle

Huygens' principle shows how a wave field on a surface S determines the wave field off the surface S. This concept can be expressed mathematically for both scalar and vector waves. We shall first discuss the scalar wave case first, followed by the electromagnetic wave case.

§§1.4.1 Scalar Waves

For a $\psi(\mathbf{r})$ that satisfies the scalar wave equation

$$(\nabla^2 + k^2)\,\psi(\mathbf{r}) = 0, \qquad (1.4.1)$$

the corresponding scalar Green's function $g(\mathbf{r}, \mathbf{r}')$ satisfies

$$(\nabla^2 + k^2)\,g(\mathbf{r}, \mathbf{r}') = -\delta(\mathbf{r} - \mathbf{r}'). \qquad (1.4.2)$$

Next, on multiplying (1) by $g(\mathbf{r}, \mathbf{r}')$ and (2) by $\psi(\mathbf{r})$, subtracting the resultant equations and integrating over a volume containing $\mathbf{r}'$, we have

$$\int_V d\mathbf{r}\,[g(\mathbf{r}, \mathbf{r}')\nabla^2\psi(\mathbf{r}) - \psi(\mathbf{r})\nabla^2 g(\mathbf{r}, \mathbf{r}')] = \psi(\mathbf{r}'). \qquad (1.4.3)$$

Since $g\nabla^2\psi - \psi\nabla^2 g = \nabla \cdot (g\nabla\psi - \psi\nabla g)$, the left-hand side of (3) can be rewritten using Gauss' divergence theorem, giving[9]

$$\psi(\mathbf{r}') = \oint_S dS\,\hat{n} \cdot [g(\mathbf{r}, \mathbf{r}')\nabla\psi(\mathbf{r}) - \psi(\mathbf{r})\nabla g(\mathbf{r}, \mathbf{r}')], \qquad (1.4.4)$$

where S is the surface bounding V. The above is the mathematical expression that once $\psi(\mathbf{r})$ and $\hat{n} \cdot \nabla\psi(\mathbf{r})$ are known on S, then $\psi(\mathbf{r}')$ away from S could be found.

If the volume V is bounded by S and S_{inf} as shown in Figure 1.4.1, then the surface integral in (4) should include an integral over S_{inf}. But when $S_{inf} \to \infty$, all fields look like plane wave, and $\nabla \to \hat{r}ik$ on S_{inf}. Furthermore, $g(\mathbf{r} - \mathbf{r}') \sim O(1/r)$,[10] when $r \to \infty$, and $\psi(\mathbf{r}) \sim O(1/r)$, when $r \to \infty$, if $\psi(\mathbf{r})$ is due to a source of finite extent. Then, the integral over S_{inf} in (4) vanishes,

---

[9] The equivalence of the volume integral in (3) to the surface integral in (4) is also known as Green's theorem.

[10] The symbol "O" means "of the order."

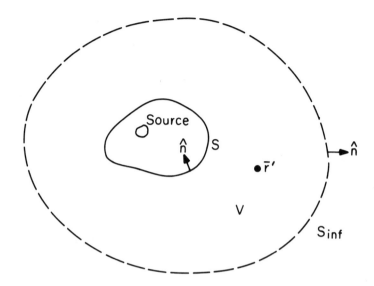

Figure 1.4.1 The geometry for the derivation of Huygens' principle.

and (4) is valid for the case shown in Figure 1.4.1 as well. Here, the field outside S at $\mathbf{r}'$ is expressible in terms of the field on S.

Notice that in deriving (4), $g(\mathbf{r}, \mathbf{r}')$ has only to satisfy (2) for both $\mathbf{r}$ and $\mathbf{r}'$ in V but no boundary condition has yet been imposed on $g(\mathbf{r}, \mathbf{r}')$. Therefore, if we further require that $g(\mathbf{r}, \mathbf{r}') = 0$ for $\mathbf{r} \in S$, then (4) becomes

$$\psi(\mathbf{r}') = - \oint_S dS\, \psi(\mathbf{r})\, \hat{n} \cdot \nabla g(\mathbf{r}, \mathbf{r}'). \qquad (1.4.5)$$

On the other hand, if require additionally that $g(\mathbf{r}, \mathbf{r}')$ satisfies (2) with the boundary condition $\hat{n} \cdot \nabla g(\mathbf{r}, \mathbf{r}') = 0$ for $\mathbf{r} \in S$, then (4) becomes

$$\psi(\mathbf{r}') = \oint_S dS\, g(\mathbf{r}, \mathbf{r}')\, \hat{n} \cdot \nabla \psi(\mathbf{r}). \qquad (1.4.6)$$

Equations (4), (5), and (6) are various forms of Huygens' principle depending on the definition of $g(\mathbf{r}, \mathbf{r}')$. Equations (5) and (6) stipulate that only $\psi(\mathbf{r})$ or $\hat{n} \cdot \nabla \psi(\mathbf{r})$ need be known on the surface S in order to determine $\psi(\mathbf{r}')$. (Note that in the above derivation, k^2 could be a function of position as well.)

§§1.4.2 Electromagnetic Waves

In a source-free region, an electromagnetic wave satisfies the vector wave equation

$$\nabla \times \nabla \times \mathbf{E}(\mathbf{r}) - k^2\, \mathbf{E}(\mathbf{r}) = 0. \tag{1.4.7}$$

Moreover, the dyadic Green's function satisfies the equation

$$\nabla \times \nabla \times \overline{\mathbf{G}}_e(\mathbf{r}, \mathbf{r}') - k^2\, \overline{\mathbf{G}}_e(\mathbf{r}, \mathbf{r}') = \overline{\mathbf{I}}\, \delta(\mathbf{r} - \mathbf{r}'). \tag{1.4.8}$$

Then, after post-multiplying (7) by $\overline{\mathbf{G}}_e(\mathbf{r}, \mathbf{r}')$, pre-multiplying (8) by $\mathbf{E}(\mathbf{r})$, subtracting the resultant equations and integrating the difference over volume V, we have

$$\mathbf{E}(\mathbf{r}') = \int_V dV\, \left[\mathbf{E}(\mathbf{r}) \cdot \nabla \times \nabla \times \overline{\mathbf{G}}_e(\mathbf{r}, \mathbf{r}') + \nabla \times \nabla \times \mathbf{E}(\mathbf{r}) \cdot \overline{\mathbf{G}}_e(\mathbf{r}, \mathbf{r}') \right]. \tag{1.4.9}$$

Next, using the vector identity that[11]

$$\begin{aligned} -\nabla \cdot &\left[\mathbf{E}(\mathbf{r}) \times \nabla \times \overline{\mathbf{G}}_e(\mathbf{r}, \mathbf{r}') + \nabla \times \mathbf{E}(\mathbf{r}) \times \overline{\mathbf{G}}_e(\mathbf{r}, \mathbf{r}') \right] \\ &= \mathbf{E}(\mathbf{r}) \cdot \nabla \times \nabla \times \overline{\mathbf{G}}_e(\mathbf{r}, \mathbf{r}') - \nabla \times \nabla \times \mathbf{E}(\mathbf{r}) \cdot \overline{\mathbf{G}}_e(\mathbf{r}, \mathbf{r}'), \end{aligned} \tag{1.4.10}$$

Equation (9), with the help of Gauss' divergence theorem, can be written as

$$\begin{aligned} \mathbf{E}(\mathbf{r}') &= - \oint_S dS\, \hat{n} \cdot \left[\mathbf{E}(\mathbf{r}) \times \nabla \times \overline{\mathbf{G}}_e(\mathbf{r}, \mathbf{r}') + \nabla \times \mathbf{E}(\mathbf{r}) \times \overline{\mathbf{G}}_e(\mathbf{r}, \mathbf{r}') \right] \\ &= - \oint_S dS\, \left[\hat{n} \times \mathbf{E}(\mathbf{r}) \cdot \nabla \times \overline{\mathbf{G}}_e(\mathbf{r}, \mathbf{r}') + i\omega\mu\, \hat{n} \times \mathbf{H}(\mathbf{r}) \cdot \overline{\mathbf{G}}_e(\mathbf{r}, \mathbf{r}') \right]. \end{aligned} \tag{1.4.11}$$

The above is just the vector analogue of (4). Again, notice that (11) is derived via the use of (8), but no boundary condition has yet been imposed on $\overline{\mathbf{G}}_e(\mathbf{r}, \mathbf{r}')$ on S. Now, if we require that $\hat{n} \times \overline{\mathbf{G}}_e(\mathbf{r}, \mathbf{r}') = 0$ for $\mathbf{r} \in S$, then (11) becomes

$$\mathbf{E}(\mathbf{r}') = - \oint_S dS\, \hat{n} \times \mathbf{E}(\mathbf{r}) \cdot \nabla \times \overline{\mathbf{G}}_e(\mathbf{r}, \mathbf{r}'), \tag{1.4.12}$$

for it could be shown that $\hat{n} \times \mathbf{H} \cdot \overline{\mathbf{G}}_e = \mathbf{H} \cdot \hat{n} \times \overline{\mathbf{G}}_e$ implying that the second term in (11) is zero. On the other hand, if we require that $\hat{n} \times \nabla \times \overline{\mathbf{G}}_e(\mathbf{r}, \mathbf{r}') = 0$ for $\mathbf{r} \in S$, then (11) becomes

$$\mathbf{E}(\mathbf{r}') = -i\omega\mu \oint_S dS\, \hat{n} \times \mathbf{H}(\mathbf{r}) \cdot \overline{\mathbf{G}}_e(\mathbf{r}, \mathbf{r}'). \tag{1.4.13}$$

---

[11] This identity can be established by using the identity $\nabla \cdot (\mathbf{A} \times \mathbf{B}) = \mathbf{B} \cdot \nabla \times \mathbf{A} - \mathbf{A} \cdot \nabla \times \mathbf{B}$. The equality of the volume integral in (9) to the surface integral in (11) is also known as vector Green's theorem.

Equations (12) and (13) state that $\mathbf{E}(\mathbf{r}')$ is determined if either $\hat{n} \times \mathbf{E}(\mathbf{r})$ or $\hat{n} \times \mathbf{H}(\mathbf{r})$ is specified on S.

It can be shown from reciprocity that (Exercise 1.18)

$$\left[\, \overline{\mathbf{G}}_e(\mathbf{r}, \mathbf{r}')\right]^t = \overline{\mathbf{G}}_e(\mathbf{r}', \mathbf{r}), \tag{1.4.14a}$$

$$\left[\nabla \times \overline{\mathbf{G}}_e(\mathbf{r}, \mathbf{r}')\right]^t = \nabla' \times \overline{\mathbf{G}}_m(\mathbf{r}', \mathbf{r}), \tag{1.4.14b}$$

where $\overline{\mathbf{G}}_m(\mathbf{r}, \mathbf{r}')$ is the dyadic Green's function for magnetic field, and $\overline{\mathbf{G}}_e(\mathbf{r}, \mathbf{r}')$ is the dyadic Green's function for electric field. Then, by taking its transpose, Equation (11) becomes

$$\mathbf{E}(\mathbf{r}') = -\nabla' \times \oint_S dS \, \overline{\mathbf{G}}_m(\mathbf{r}', \mathbf{r}) \cdot \hat{n} \times \mathbf{E}(\mathbf{r}) - i\omega\mu \oint_S dS \, \overline{\mathbf{G}}_e(\mathbf{r}', \mathbf{r}) \cdot \hat{n} \times \mathbf{H}(\mathbf{r}). \tag{1.4.15}$$

Moreover, Equation (12) then becomes

$$\mathbf{E}(\mathbf{r}') = -\nabla' \times \oint_S dS \, \overline{\mathbf{G}}_m(\mathbf{r}', \mathbf{r}) \cdot \hat{n} \times \mathbf{E}(\mathbf{r}), \tag{1.4.16}$$

while Equation (13) becomes

$$\mathbf{E}(\mathbf{r}') = -i\omega\mu \oint_S dS \, \overline{\mathbf{G}}_e(\mathbf{r}', \mathbf{r}) \cdot \hat{n} \times \mathbf{H}(\mathbf{r}). \tag{1.4.17}$$

The dyadic Green's functions in (12), (13), (16), and (17) are for a closed cavity since boundary conditions are imposed on S for them. But the dyadic Green's function for an unbounded, homogeneous medium can be written as

$$\overline{\mathbf{G}}(\mathbf{r}, \mathbf{r}') = \frac{1}{k^2}[\nabla \times \nabla \times \overline{\mathbf{I}}\, g(\mathbf{r} - \mathbf{r}') - \overline{\mathbf{I}}\, \delta(\mathbf{r} - \mathbf{r}')], \tag{1.4.18}$$

$$\nabla \times \overline{\mathbf{G}}(\mathbf{r}, \mathbf{r}') = \nabla \times \overline{\mathbf{I}}\, g(\mathbf{r} - \mathbf{r}'). \tag{1.4.19}$$

Also, for unbounded homogeneous medium, $\overline{\mathbf{G}}_e(\mathbf{r}, \mathbf{r}') = \overline{\mathbf{G}}_m(\mathbf{r}, \mathbf{r}')$. Then, (15) becomes

$$\mathbf{E}(\mathbf{r}') = -\nabla' \times \oint_S dS \, g(\mathbf{r} - \mathbf{r}')\, \hat{n} \times \mathbf{E}(\mathbf{r}) + \frac{1}{i\omega\epsilon}\nabla' \times \nabla' \times \oint_S dS \, g(\mathbf{r} - \mathbf{r}')\, \hat{n} \times \mathbf{H}(\mathbf{r}). \tag{1.4.20}$$

The above can be applied to the geometry in Figure 1.4.1 where $\mathbf{r}'$ is enclosed in S and S_{inf}. However, the integral over S_{inf} vanishes by virtue of the radiation condition as for (4). Then, (20) relates the field outside S at $\mathbf{r}'$ in terms of only the field on S.

§1.5 Uniqueness Theorem

The uniqueness theorem provides conditions under which the solution to the wave equation is unique. This is especially important because the solutions to a problem should not be indeterminate. These conditions under which a solution to a wave equation is unique are the boundary conditions and the radiation condition. Uniqueness also allows one to construct solutions by inspections; if a candidate solution satisfies the conditions of uniqueness, it is the unique solution. Because of its simplicity, the scalar wave equation shall be examined first for greater insight into this problem.

§§1.5.1 Scalar Wave Equation

Given a scalar wave equation with a source term on the right-hand side, we shall derive the conditions under which a solution is unique. First, assume that there are two different solutions to the scalar wave equation, namely,

$$[\nabla^2 + k^2(\mathbf{r})]\,\phi_1(\mathbf{r}) = s(\mathbf{r}), \tag{1.5.1a}$$

$$[\nabla^2 + k^2(\mathbf{r})]\,\phi_2(\mathbf{r}) = s(\mathbf{r}), \tag{1.5.1b}$$

where $k^2(\mathbf{r})$ includes inhomogeneities of finite extent. Then, on subtracting the two equations, we have

$$[\nabla^2 + k^2(\mathbf{r})]\,\delta\phi(\mathbf{r}) = 0, \tag{1.5.2}$$

where $\delta\phi(\mathbf{r}) = \phi_1(\mathbf{r}) - \phi_2(\mathbf{r})$. Note that the solution is unique if and only if $\delta\phi = 0$ for all $\mathbf{r}$.

Then, after multiplying (2) by $\delta\phi^*$, integrating over volume, and using the vector identity $\nabla \cdot \psi\mathbf{A} = \mathbf{A} \cdot \nabla\psi + \psi\nabla \cdot \mathbf{A}$, we have

$$\int_S \hat{n} \cdot (\delta\phi^*\nabla\delta\phi)\,dS - \int_V |\nabla\delta\phi|^2 dV + \int_V k^2|\delta\phi|^2 dV = 0, \tag{1.5.3}$$

where $\hat{n}$ is a unit normal to the surface S. Then, the imaginary part of the above equation is

$$\Im m \int_S \hat{n} \cdot (\delta\phi^*\nabla\delta\phi)\,dS + \int_V \Im m(k^2)|\delta\phi|^2 dV = 0. \tag{1.5.4}$$

Hence, if $\Im m[k^2(\mathbf{r})] \neq 0$ in V, and

(i) $\delta\phi = 0$ or $\hat{n} \cdot \nabla\delta\phi = 0$ on S, or

(ii) $\delta\phi = 0$ on part of S and $\hat{n} \cdot \nabla\delta\phi = 0$ on the rest of S,

then

$$\int_V \Im m[k^2(\mathbf{r})]|\delta\phi|^2 dV = 0. \tag{1.5.5}$$

Since $|\delta\phi|^2$ is positive definite for $\delta\phi \neq 0$, and $\Im m(k^2) \neq 0$ in V,[12] the above is only possible if $\delta\phi = 0$ everywhere inside V.

---

[12] More specifically, $\Im m[k^2(\mathbf{r})] > 0, \forall\, \mathbf{r} \in V$, or $\Im m[k^2(\mathbf{r})] < 0, \forall\, \mathbf{r} \in V$.

Therefore, in order to guarantee uniqueness, so that $\phi_1 = \phi_2$ in V, either

(i) $\phi_1 = \phi_2$ on S or $\hat{n} \cdot \nabla\phi_1 = \hat{n} \cdot \nabla\phi_2$ on S, or

(ii) $\phi_1 = \phi_2$ on one part of S, and $\hat{n} \cdot \nabla\phi_1 = \hat{n} \cdot \nabla\phi_2$ on the rest of S.

The specification of ϕ on S is also known as the **Dirichlet** boundary condition, while the specification of $\hat{n} \cdot \nabla\phi$, namely, the normal derivative, is also known as the **Neumann** boundary condition. In words, the uniqueness theorem says that if two solutions satisfy the same Dirichlet or Neumann boundary condition or a mixture thereof on S, the two solutions must be identical.

When $\Im m(k^2) = 0$, i.e., when k^2 is real, the condition $\delta\phi = 0$ or $\hat{n} \cdot \nabla\delta\phi = 0$ on S in (3) does not necessarily lead to $\delta\phi = 0$ in V, or uniqueness. The reason is that solutions for $\delta\phi = \phi_1 - \phi_2$ where

$$\int_V |\nabla\delta\phi|^2 \, dV = \int_V k^2 |\delta\phi|^2 \, dV \qquad (1.5.6)$$

can exist. These are the resonance solutions in the volume V (see Exercise 1.19). These resonance solutions are the homogeneous solutions[13] to the wave Equation (1) at the real resonance frequencies of the volume V. Because the medium is lossless, they are time harmonic solutions which satisfies the boundary conditions, and hence, can be added to the particular solution of (1). In fact, the particular solution usually becomes infinite at these resonance frequencies if $S(\mathbf{r}) \neq 0$.

Equation (6) implies the balance of two energies. In the case of acoustic waves, for example, it represents the balance of the kinetic energy and the potential energy in a volume V.

When $\Im m(k^2) \neq 0$, however, the resonance solutions of the volume V are exponentially decaying with time for a lossy medium [$\Im m(k^2) > 0$], and they are exponentially growing with time for an active medium [$\Im m(k^2) < 0$]. But if only time harmonic solutions ϕ_1 and ϕ_2 are permitted in (1), these resonance solutions are automatically eliminated from the class of permissible solutions. Hence, for a lossy medium [$\Im m(k^2) > 0$] or an active medium [$\Im m(k^2) < 0$], the uniqueness of the solution is guaranteed if we consider only time harmonic solutions, namely, two solutions will be identical if they have the same boundary conditions for ϕ and $\hat{n} \cdot \nabla\phi$ on S.[14]

When $S \to \infty$ or $V \to \infty$, the number of resonance frequencies of V becomes denser. In fact, when $S \to \infty$, the resonance frequencies of V become a

---

[13] "Homogeneous solutions" is a mathematical parlance for solutions to (1) without the source term.

[14] The nonuniqueness associated with the resonance solution for a lossless medium can be eliminated if we consider time domain solutions. In the time domain, we can set up an initial value problem in time, e.g., by requiring all fields be zero for $t < 0$; thus, the nonuniqueness problem can be removed via the causality requirement. The resonance solution, being time harmonic, is noncausal.

continuum implying that any real frequency could be the resonant frequency of V. Hence, if the medium is lossless, the uniqueness of the solution is not guaranteed at any frequency, even appropriate boundary conditions on S at infinity, as a result of the presence of the continuum of resonance frequencies. One remedy then is to introduce a small loss. With this small loss [$\Im m(k) > 0$], the solution is either exponentially small when $r \to \infty$ (if a solution corresponds to an outgoing wave, e^{ikr}), or exponentially large when $r \to \infty$ (if a solution corresponds to an incoming wave, e^{-ikr}). Now, if the solution is exponentially small, namely, keeping only the outgoing wave solutions, it is clear that the surface integral term in (4) vanishes when $S \to \infty$, and the uniqueness of the solution is guaranteed. This manner of imposing the outgoing wave condition at infinity is also known as the **Sommerfeld radiation condition** (Sommerfeld 1949, p. 188). This radiation condition can be used in the limit of a vanishing loss for an unbounded medium to guarantee uniqueness.

§§1.5.2 Vector Wave Equation

Similar to the uniqueness conditions for the scalar wave equation, analogous conditions for the vector wave equation can also be derived. First, assume that there are two different solutions to a vector wave Equation (1.3.5), i.e.,

$$\nabla \times \overline{\mu}^{-1} \cdot \nabla \times \mathbf{E}_1(\mathbf{r}) - \omega^2 \overline{\epsilon} \cdot \mathbf{E}_1(\mathbf{r}) = \mathbf{S}(\mathbf{r}), \qquad (1.5.7a)$$

$$\nabla \times \overline{\mu}^{-1} \cdot \nabla \times \mathbf{E}_2(\mathbf{r}) - \omega^2 \overline{\epsilon} \cdot \mathbf{E}_2(\mathbf{r}) = \mathbf{S}(\mathbf{r}), \qquad (1.5.7b)$$

where $\mathbf{S}(\mathbf{r}) = i\omega \mathbf{J}(\mathbf{r}) - \nabla \times \overline{\mu}^{-1} \cdot \mathbf{M}(\mathbf{r})$ corresponds to a source of finite extent. Similarly, $\overline{\mu}$ and $\overline{\epsilon}$ correspond to an inhomogeneity of finite extent. Subtracting (7a) from (7b) then yields

$$\nabla \times \overline{\mu}^{-1} \cdot \nabla \times \delta \mathbf{E} - \omega^2 \overline{\epsilon} \cdot \delta \mathbf{E} = 0, \qquad (1.5.8)$$

where $\delta \mathbf{E} = \mathbf{E}_1 - \mathbf{E}_2$. The solution is unique if and only if $\delta \mathbf{E} = 0$. Next, on multiplying the above by $\delta \mathbf{E}^*$, integrating over volume V, and using the vector identity $\mathbf{A} \cdot \nabla \times \mathbf{B} = -\nabla \cdot (\mathbf{A} \times \mathbf{B}) + \mathbf{B} \cdot \nabla \times \mathbf{A}$, we have

$$-\int_S \hat{n} \cdot (\delta \mathbf{E}^* \times \overline{\mu}^{-1} \cdot \nabla \times \delta \mathbf{E}) \, dS + \int_V \nabla \times \delta \mathbf{E}^* \cdot \overline{\mu}^{-1} \cdot \nabla \times \delta \mathbf{E} \, dV$$

$$-\omega^2 \int_V \delta \mathbf{E}^* \cdot \overline{\epsilon} \cdot \delta \mathbf{E} \, dV = 0. \qquad (1.5.9)$$

Since $\nabla \times \delta \mathbf{E} = i\omega \overline{\mu} \cdot \delta \mathbf{H}$, the above can be rewritten as

$$i\omega \int_S \hat{n} \cdot (\delta \mathbf{E}^* \times \delta \mathbf{H}) \, dS + \omega^2 \int_V (\delta \mathbf{H}^* \cdot \overline{\mu}^\dagger \cdot \delta \mathbf{H} - \delta \mathbf{E}^* \cdot \overline{\epsilon} \cdot \delta \mathbf{E}) \, dV = 0. \qquad (1.5.10)$$

Then, taking the imaginary part of (10) yields

$$\Im m \left\{ i\omega \int_S \hat{n} \cdot (\delta \mathbf{E}^* \times \delta \mathbf{H}) \, dS \right\}$$
$$- \frac{i\omega^2}{2} \int_V [\delta \mathbf{H}^* \cdot (\overline{\mu}^\dagger - \overline{\mu}) \cdot \delta \mathbf{H} + \delta \mathbf{E}^* \cdot (\overline{\epsilon}^\dagger - \overline{\epsilon}) \cdot \delta \mathbf{E}] \, dV = 0.$$

$$(1.5.11)$$

But if the medium is not lossless (either lossy or active), then $\overline{\mu}^\dagger \neq \overline{\mu}$ and $\overline{\epsilon}^\dagger \neq \overline{\epsilon}$ [see (1.1.2b)], and the second integral in (11) may not be zero. Moreover, if

(i) $\hat{n} \times \delta \mathbf{E} = 0$ or $\hat{n} \times \delta \mathbf{H} = 0$ on S, or

(ii) $\hat{n} \times \delta \mathbf{E} = 0$ on one part of S and $\hat{n} \times \delta \mathbf{H} = 0$ on the rest of S,

the first integral in (11) vanishes. In this case,

$$\frac{\omega^2}{2} \int_V [\delta \mathbf{H}^* \cdot i(\overline{\mu}^\dagger - \overline{\mu}) \cdot \delta \mathbf{H} + \delta \mathbf{E}^* \cdot i(\overline{\epsilon}^\dagger - \overline{\epsilon}) \cdot \delta \mathbf{E}] \, dV = 0. \qquad (1.5.12)$$

In the above, $i(\overline{\mu}^\dagger - \overline{\mu})$ and $i(\overline{\epsilon}^\dagger - \overline{\epsilon})$ are Hermitian matrices. Moreover, the integrand will be positive definite if both $\overline{\mu}$ and $\overline{\epsilon}$ are lossy, and the integrand will be negative definite if both $\overline{\mu}$ and $\overline{\epsilon}$ are active (see Subsection 1.1.5). Hence, the only way for (12) to be satisfied is for $\delta \mathbf{E} = 0$ and $\delta \mathbf{H} = 0$, or that $\mathbf{E}_1 = \mathbf{E}_2$ and $\mathbf{H}_1 = \mathbf{H}_2$, implying uniqueness.

Consequently, in order for uniqueness to be guaranteed, either

(i) $\hat{n} \times \mathbf{E}_1 = \hat{n} \times \mathbf{E}_2$ on S or $\hat{n} \times \mathbf{H}_1 = \hat{n} \times \mathbf{H}_2$ on S, or

(ii) $\hat{n} \times \mathbf{E}_1 = \hat{n} \times \mathbf{E}_2$ on a part of S while $\hat{n} \times \mathbf{H}_1 = \hat{n} \times \mathbf{H}_2$ on the rest of S.

In other words, if two solutions satisfy the same boundary conditions for tangential $\mathbf{E}$ or tangential $\mathbf{H}$, or a mixture thereof on S, the two solutions must be identical.

Again, the requirement for a nonlossless condition is to eliminate the real resonance solutions which could otherwise be time harmonic, homogeneous solutions to (7) satisfying the boundary conditions. For example, if the appropriate boundary conditions for $\delta \mathbf{E}$ and $\delta \mathbf{H}$ are imposed so that the first term of (10) is zero, then

$$\int_V (\delta \mathbf{H}^* \cdot \overline{\mu}^\dagger \cdot \delta \mathbf{H} - \delta \mathbf{E}^* \cdot \overline{\epsilon} \cdot \delta \mathbf{E}) \, dV = 0. \qquad (1.5.13)$$

The above does not imply that $\delta \mathbf{E}$ or $\delta \mathbf{H}$ equals zero, because at resonances, a perfect balance between the energy stored in the electric field and the energy stored in the magnetic field is maintained. As a result, the left-hand side of the above could vanish without having $\delta \mathbf{E}$ and $\delta \mathbf{H}$ be zero, which is necessary for uniqueness. But away from the resonances of the volume V, the energy stored in the electric field is not equal to that stored in the magnetic field. Hence, in order for (13) to be satisfied, $\delta \mathbf{E}$ and $\delta \mathbf{H}$ have to be zero since each term in (13) is positive definite for lossless media due to the Hermitian nature of $\overline{\mu}$ and $\overline{\epsilon}$.

When $V \to \infty$, as in the scalar wave equation case, some loss has to be imposed to guarantee uniqueness. This is the same as requiring the wave to be outgoing at infinity, namely, the radiation condition. Again, the radiation condition can be imposed for an unbounded medium with vanishing loss to guarantee uniqueness.

Exercises for Chapter 1

1.1 Show that Equations (1.1.12) to (1.1.15) can also be obtained from Equations (1.1.1) to (1.1.4) by Fourier transforms. In this case, we define a field $\mathbf{A}(\mathbf{r}, t) = \int\limits_{-\infty}^{\infty} d\omega \, e^{-i\omega t} \mathbf{A}(\mathbf{r}, \omega)$.

1.2 (a) The fundamental units in electromagnetics can be considered to be meter, kilogram, second, and coulomb. Show that 1 volt, which is 1 watt/amp, has the dimension of (kilogram meter^2)/(coulomb sec^2).

(b) From Maxwell's equations, show that μ_0 has the dimension of (second volt)/(meter amp), and hence, its dimension is (kilogram meter)/ (coulomb^2) in the more fundamental units.

(c) If we assign the value of μ_0 to be 4π instead of $4\pi \times 10^{-7}$, what would be the unit of coulomb in this new assignment compared to the old unit? What would be the present value of 1 volt and 1 amp in this new assignment?

1.3 Show that for two time harmonic functions,

$$\langle A(\mathbf{r}, t), B(\mathbf{r}, t) \rangle = \frac{1}{2} \, \Re e[A(\mathbf{r}) B^*(\mathbf{r})],$$

where $A(\mathbf{r})$ and $B(\mathbf{r})$ are the phasors of $A(\mathbf{r}, t)$ and $B(\mathbf{r}, t)$.

1.4 By putting an electrostatic field next to a magnetostatic field, show that $\mathbf{E} \times \mathbf{H}$ is not zero, but the quantity cannot possibly correspond to power flow.

1.5 Assume that a voltage is time harmonic, i.e., $V(t) = V_0 \cos \omega t$, and that a current $I(t) = I_I \cos \omega t + I_Q \sin \omega t$, i.e., it consists of an in-phase and a quadrature component.

(a) Find the instantaneous power due to this voltage and current, namely, $V(t) I(t)$.

(b) Find the phasor representations of the voltage and current and the complex power due to this voltage and current.

(c) Establish a relationship between the real part and reactive part of the complex power to the instantaneous power.

(d) Show that the reactive power is due to the quadrature component of the current, which is related to a time-varying part of the instantaneous power with zero-time average.

1.6 Show that the matrices $i(\overline{\overline{\mu}}^\dagger - \overline{\overline{\mu}})$ and $i(\overline{\overline{\epsilon}}^\dagger - \overline{\overline{\epsilon}})$ are either zero, positive, or negative definite. Explain the physical interpretation of each case.

1.7 Find another set of replacement rules different from Equation (1.1.36) that will leave Maxwell's equations invariant.

1.8 (a) Derive Equation (1.2.5).

(b) In one dimension, a pressure gradient $p(x)$ (force/unit area) is established. Show that the force on an elemental sheet between x and $x+\Delta x$ is $[p(x)-p(x+\Delta x)]A$ where A is the area of the elemental sheet. Hence, show that the force per unit volume is $[p(x) - p(x + \Delta x)]/\Delta x$, implying that the force density $F_x = -\partial p/\partial x$.

(c) Apply the same derivation to a cube and show that $\mathbf{F} = -\nabla p$.

1.9 (a) In hydrodynamic problems, it is easier to formulate a concept if one moves with the particles in a fluid. For example, if the density is described by $\varrho(\mathbf{r}, t)$, in the coordinate system which moves with a fluid particle, then $\mathbf{r}(t)$ is a function of time as well. Consequently, the total change in density in the neighborhood of the particle that one observes is affected by $\mathbf{r}$ being a function of t as well. This total change of ϱ with respect to t is usually denoted $\frac{D}{Dt}\varrho(\mathbf{r}, t)$. Show that

$$\frac{D}{Dt}\varrho(\mathbf{r}, t) = \frac{\partial \varrho}{\partial t} + \mathbf{v} \cdot \nabla \varrho.$$

(b) The pressure in a fluid is a function of both the density ϱ and entropy S, i.e., $p(\varrho, S)$. If one follows a fluid particle's motion, the entropy in the vicinity of the fluid particle is constant. This can be denoted by $\frac{DS}{Dt} = 0$. From this, deduce that

$$\frac{Dp}{Dt} = \frac{\partial p}{\partial \varrho}\frac{D\varrho}{Dt} + \frac{\partial p}{\partial S}\frac{DS}{Dt}$$

and then

$$\frac{\partial p}{\partial t} + \mathbf{v} \cdot \nabla p = \frac{\partial p}{\partial \varrho}\left[\frac{\partial \varrho}{\partial t} + \mathbf{v} \cdot \nabla \varrho\right].$$

Hence, derive Equation (1.2.13).

1.10 Explain why Equation (1.2.20) is not equivalent to three scalar wave equations if $\mathbf{E}$ is decomposed into three components not in the Cartesian coordinates.

1.11 Sketch the wavefront of Equation (1.2.33) in three dimensions and explain why it is called a conical wave.

1.12 Using Equations (1.2.28) and (1.2.40), establish the relationship in Equation (1.2.40a).

1.13 Show that a recurrence relationship similar to (1.2.34) for the solutions of (1.2.40) is

$$b'_n(kr) = b_{n-1}(kr) - \frac{n+1}{kr}b_n(kr) = -b_{n+1}(kr) + \frac{n}{kr}b_n(kr).$$

1.14 For a scalar wave equation, $\nabla \cdot p^{-1}(\mathbf{r})\nabla\phi(\mathbf{r}) + k^2\phi(\mathbf{r}) = s(\mathbf{r})$:

(a) What is the boundary condition at an interface where p is discontinuous?

(b) Show that a reciprocal relationship $\langle \phi_1(\mathbf{r}), s_2(\mathbf{r}) \rangle = \langle \phi_2(\mathbf{r}), s_1(\mathbf{r}) \rangle$ exists.

1.15 By considering the case where $\mathbf{k}$ is pointing in the z direction, prove that Equation (1.3.29) has only two nontrivial eigenvalues, and hence, only two nontrivial eigenvectors. Find these eigenvalues and eigenvectors.

1.16 By letting $\mathbf{B} = \nabla \times \mathbf{A}$ and $\mathbf{E} = -\nabla\phi + i\omega\mathbf{A}$, and starting from Maxwell's equations, derive an expression similar to (1.3.49).

1.17 (a) Define a Green's function to be a solution of $\nabla \cdot p^{-1}(\mathbf{r})\nabla g(\mathbf{r}, \mathbf{r}') + k^2 g(\mathbf{r}, \mathbf{r}') = -\delta(\mathbf{r} - \mathbf{r}')$ and show that the solution to the equation $\nabla \cdot p^{-1}(\mathbf{r})\nabla\psi(\mathbf{r}) + k^2\psi(\mathbf{r}) = s(\mathbf{r})$ can be written as

$$\psi(\mathbf{r}) = -\int d\mathbf{r}' \, g(\mathbf{r}, \mathbf{r}') \, s(\mathbf{r}').$$

(b) Using the result of Exercise 1.14, show that $g(\mathbf{r}, \mathbf{r}') = g(\mathbf{r}', \mathbf{r})$.

1.18 (a) In the manner of Equation (1.3.52), show that $[\overline{\mathbf{G}}(\mathbf{r}, \mathbf{r}')]^t = \overline{\mathbf{G}}(\mathbf{r}', \mathbf{r})$ for a dyadic Green's function defined over a bounded region.

(b) Define a magnetic field dyadic Green's function such that

$$\mathbf{H}(\mathbf{r}) = i\omega\epsilon \int_V d\mathbf{r}' \, \overline{\mathbf{G}}_m(\mathbf{r}, \mathbf{r}') \cdot \mathbf{M}(\mathbf{r}').$$

From the reciprocity requirement that $\langle \mathbf{M}_2, \mathbf{H}_1 \rangle = -\langle \mathbf{J}_1, \mathbf{E}_2 \rangle$, show that $[\nabla \times \overline{\mathbf{G}}_e(\mathbf{r}, \mathbf{r}')]^t = \nabla' \times \overline{\mathbf{G}}_m(\mathbf{r}', \mathbf{r})$.

1.19 For the lossless scalar wave equation in a homogeneous medium like (1.5.1):

(a) Find the resonance solutions to a box of dimension $a \times b \times d$, with homogeneous Neumann boundary condition ($\hat{n} \cdot \nabla\phi = 0$) on the sides of the box.

(b) Show that the resonance solutions satisfy (1.5.6).

(c) At resonance, show that the nontrivial difference between two solutions is still a solution satisfying the boundary condition.

(d) Describe what happens to the resonance solutions when a, b, and $d \rightarrow \infty$.

References for Chapter 1

Abramowitz, M., and I. A. Stegun. 1965. *Handbook of Mathematical Functions.* New York: Dover Publications.

Cohen, E. R., and B. N. Taylor. 1986. *The 1986 Adjustment of the Fundamental Physical Constants.* CODATA Bulletin 63. Elmsford, New York: Pergamon Press.

Kong, J. A. 1986. *Electromagnetic Wave Theory.* New York: John Wiley & Sons.

Pierce, A. D. 1981. *Acoustics.* New York: McGraw-Hill.

Rumsey, V. H. 1954. "Reaction concept in electromagnetic theory." *Phys. Rev.* 94: 1483–91; 95: 1706.

Sommerfeld, A. 1949. *Partial Differential Equation.* New York: Academic Press.

Further Readings for Chapter 1

Abraham, A., and R. Becker. 1932. *The Classical Theory of Electricity.* Glasgow: Blackie & Son, Ltd.

Achenbach, J. D. 1973. *Wave Propagation in Elastic Solids.* Amsterdam: North-Holland.

Adams, A. T. 1971. *Electromagnetics for Engineers.* New York: Renold Press.

Barut, A. O. 1964. *Electrodynamics and Classical Theory of Fields and Particles.* New York: Macmillan.

Booker, H. G. 1982. *Energy in Electromagnetism.* New York: Peter Peregrinus.

Born, M., and E. Wolf. 1970. *Principles of Optics.* New York: Pergamon Press.

Corson, D., and P. Lorrain. 1962. *Electromagnetic Waves and Fields.* San Francisco: Freeman.

DeGroot, S. R., and L. G. Suttorp. 1972. *Foundations of Electrodynamics.* Amsterdam: North-Holland.

Fano, F. M., L. J. Chu, and R. B. Adler. 1960. *Electromagnetic Fields, Energy, and Forces.* New York: Wiley, and Cambridge, Mass.: M.I.T. Press.

Felsen, L. B., and N. Marcuvitz. 1973. *Radiation and Scattering of Electromagnetic Waves.* New Jersey: Prentice-Hall.

Grant, I. S., and W. R. Phillips. 1975. *Electromagnetism.* New York: John Wiley & Sons.

Harrington, R. F. 1961. *Time-Harmonic Electromagnetic Fields.* New York: McGraw-Hill.

Heaviside, O. *Electromagnetic Theory.* 1950. Reprint. New York: Dover Publications.

Jackson, J. D. 1962. *Classical Electrodynamics.* New York: John Wiley & Sons.

Jeans, J. 1933. *Electric and Magnetic Fields.* London: Cambridge University Press.

Jones, D. S. 1964. *The Theory of Electromagnetism.* New York: Macmillan.

King, R. W. P. 1953. *Electromagnetic Engineering.* New York: McGraw-Hill.

King, R. W. P., and C. W. H. Harrison, Jr. 1969. *Antennas and Waves.* Cambridge, Mass.: M.I.T. Press.

Kong, J. A. 1975. *Theory of Electromagnetic Waves.* New York: John Wiley & Sons.

Kraus, J. D. 1984. *Electromagnetics*. New York: McGraw-Hill.

Landau, L. D., and E. M. Lifshitz. 1960. *Electrodynamics of Continuous Media*. Reading, Mass.: Addison-Wesley.

Magnus, W., and F. Oberhettinger. 1954. *Formulas and Theorems for the Special Functions of Mathematical Physics*. New York: Chelsea Publishing Co.

Marion, J. B., and M. A. Heald. 1980. *Classical Electromagnetic Radiation*, 2nd ed. New York: Academic Press.

Maxwell, J. C. 1954. *A Treatise on Electricity and Magnetism*. New York: Dover Publications.

Morse, P. 1948. *Vibration and Sound*. New York: McGraw-Hill.

Morse, P. M., and H. Feshbach. 1953. *Methods of Theoretical Physics*. New York: McGraw-Hill.

Panofsky, W. K. H., and M. Phillips. 1962. *Classical Electricity and Magnetism*, 2nd ed. Reading, Mass.: Addison-Wesley.

Papas, C. H. 1965. *Theory of Electromagnetic Wave Propagation*. New York: McGraw-Hill.

Paris, D. T., and G. K. Hurd. 1969. *Basic Electromagnetic Theory*. New York: McGraw-Hill.

Plonsey, R., and R. E. Collin. 1961. *Principles and Applications of Electromagnetic Fields*. New York: McGraw-Hill.

Plonus, M. A. 1978. *Applied Electromagnetics*. New York: McGraw-Hill.

Popovic, B. D. 1971. *Introductory Engineering Electromagnetic*. Reading, Mass.: Addison-Wesley.

Purcell, E. M. 1963. *Electricity and Magnetism*. New York: McGraw-Hill.

Ramo, S., J. R. Whinnery, and T. Van Duzer. 1970. *Fields and Waves in Communication Electronics*. New York: Pergamon Press.

Rao, N. N. 1972. *Basic Electromagnetism with Applications*. New Jersey: Prentice-Hall.

Read, F. H. 1980. *Electromagnetic Radiation*. New York: John Wiley & Sons.

Schelkunoff, S. A. 1943. *Electromagnetic Waves*. New York: D. Van Nostrand.

Schelkunoff, S. A. 1963. *Electromagnetic Fields*. Waltham, Mass.: Blaisdell Publishing Co.

Shen, L. C., and J. A. Kong. 1983. *Electromagnetism*. California: Grooks/Cole.

Sommerfeld, A. 1952. *Electrodynamics*. New York: Academic Press.

Temkin, S. 1981. *Elements of Acoustics.* New York: John Wiley & Sons.

Titchmarsh, E. C. 1950. *The Theory of Functions*, 2nd ed. New York: Oxford University Press.

Towne, D. H. 1976. *Wave Phenomena.* Reading, Mass.: Addison-Wesley.

Van Bladel, J. 1964. *Electromagnetic Fields.* New York: McGraw-Hill.

Wait, J. R. 1986. *Introduction to Antennas and Propagation.* London: Peter Peregrinus Ltd.

Watson, G. N. 1944. *A Treatise on the Theory of Bessel Functions*, 2nd ed. New York: Pergamon Press.

CHAPTER 2

Compared with other inhomogeneous media, waves in planarly layered media have been studied most intensively because they are the simplest of the inhomogeneous media. Electromagnetic waves propagating in a planarly layered, isotropic medium can be reduced to the study of two uncoupled scalar wave equations. These scalar wave equations can be reduced to one-dimensional wave equations. Furthermore, their simple nature also allows for the use of many mathematically elegant techniques for their solutions. For example, the field resulting from a source excitation in a planarly layered medium can be expressed in terms of Fourier-type integrals. Hence, asymptotic expansion methods can be used to simplify such integrals, providing further insight into the problem.

Examined in this chapter are the solutions to waves in planarly layered media and the associated analytic techniques often used in conjunction with them. Many books have been written on this and related subjects (Sommerfeld 1949; Brekhovskikh 1960; Baños 1966; Clemmow 1966; Tyras 1969; Wait 1970, 1971, 1982; Yeh and Liu 1972; Felsen and Marcuvitz 1973; Kong 1975, 1986; Yeh 1988). Many other works are listed in the further readings for this chapter. The reader is assumed here to have the basic knowledge of complex variables (e.g., Hildebrand 1976).

§2.1 One-Dimensional Planar Inhomogeneity

When the electromagnetic properties of an isotropic medium, μ and ϵ, are varying only in one direction, e.g., the z direction, the vector wave equations need not be solved in their full forms. In fact, for the source-free case, the vector wave equations can be reduced to two scalar equations that are decoupled from each other. Moreover, they characterize two types of waves, namely, the *transverse electric* (TE) waves and the *transverse magnetic* (TM) waves. More specifically, for the TE waves, the electric field is transverse to the z direction, while for the TM waves, the magnetic field is transverse to the z direction.

§§2.1.1 Derivation of the Scalar Wave Equations

The electric field in the TE waves is directed in the xy plane. Assuming that the electric field is linearly polarized so that it points only in a fixed direction, then the coordinates can always be rotated about the z axis to make the electric field point only, e.g., in the y direction. In this case, $\mathbf{E} = \mathbf{E}_y$

and the vector wave equation for the electric field in a source-free medium becomes

$$\mu \nabla \times \mu^{-1} \nabla \times \mathbf{E}_y - \omega^2 \mu \epsilon \, \mathbf{E}_y = 0. \qquad (2.1.1)$$

Since

$$\nabla \cdot \epsilon \mathbf{E}_y = \frac{\partial}{\partial y} \epsilon(z) E_y = 0, \qquad (2.1.2)$$

this implies that $\partial/\partial y E_y = 0$. After extracting the y component of (1), we consequently obtain

$$\left[\frac{\partial^2}{\partial x^2} + \mu(z) \frac{\partial}{\partial z} \mu^{-1}(z) \frac{\partial}{\partial z} + \omega^2 \mu \epsilon \right] E_y = 0, \qquad (2.1.3)$$

which is the scalar wave equation describing linearly polarized TE waves in planarly layered media.

Yet another way of characterizing a TE wave is to use the H_z component of the magnetic field, since for TE waves, $H_z \neq 0$ while $E_z = 0$. To derive the corresponding equation for H_z, we first note that the vector wave equation for the magnetic field in an inhomogeneous medium is

$$\epsilon \nabla \times \epsilon^{-1} \nabla \times \mathbf{H} - \omega^2 \mu \epsilon \, \mathbf{H} = 0. \qquad (2.1.4)$$

Then, assuming that μ and ϵ are functions of z only, we can extract the z component of (4) to show that (see Exercise 2.1)

$$\left[\frac{\partial^2}{\partial x^2} + \frac{\partial^2}{\partial y^2} + \mu(z) \frac{\partial}{\partial z} \mu^{-1}(z) \frac{\partial}{\partial z} + \omega^2 \mu \epsilon \right] \mu H_z = 0. \qquad (2.1.5)$$

If $\partial^2/\partial y^2 = 0$, however, which is the case for the TE wave considered in (3), then Equations (3) and (5) are the same partial differential equations. The reason is that E_y and μH_z describe the same physical phenomenon, namely, the TE waves.

Similarly, from the duality principle, for TM waves, the governing equation for H_y is

$$\left[\frac{\partial^2}{\partial x^2} + \epsilon(z) \frac{\partial}{\partial z} \epsilon^{-1}(z) \frac{\partial}{\partial z} + \omega^2 \mu \epsilon \right] H_y = 0. \qquad (2.1.6)$$

Moreover, the TM waves can also be characterized by E_z, which is governed by

$$\left[\frac{\partial^2}{\partial x^2} + \frac{\partial^2}{\partial y^2} + \epsilon(z) \frac{\partial}{\partial z} \epsilon^{-1}(z) \frac{\partial}{\partial z} + \omega^2 \mu \epsilon \right] \epsilon E_z = 0. \qquad (2.1.7)$$

Note that the TE and TM wave equations are decoupled from each other. Hence, in a planar, one-dimensional inhomogeneity, the vector wave equations reduce to two simpler scalar wave equations, describing TE and TM waves separately.

Because of the $\partial^2/\partial x^2$ terms in (3) and (6), they must admit solutions of the form

$$\begin{bmatrix} E_y \\ H_y \end{bmatrix} = \begin{bmatrix} e_y(z) \\ h_y(z) \end{bmatrix} e^{\pm ik_x x}, \tag{2.1.8}$$

for all z. As the medium is translationally invariant in the x direction, the phase matching condition has to be satisfied in the x direction; namely, the solution for all z's must have the same phase variation, $e^{\pm ik_x x}$, in the x direction. Alternatively, the form of (8) can be derived from (3) and (6) using the separation of variables.

Under this assumption, (3) and (6) become ordinary differential equations

$$\left[\mu \frac{d}{dz} \mu^{-1} \frac{d}{dz} + \omega^2 \mu\epsilon - k_x^2 \right] e_y = 0, \tag{2.1.8a}$$

$$\left[\epsilon \frac{d}{dz} \epsilon^{-1} \frac{d}{dz} + \omega^2 \mu\epsilon - k_x^2 \right] h_y = 0, \tag{2.1.8b}$$

which correspond to one-dimensional problems. Furthermore, when μ and ϵ are arbitrary functions of z, many different methods may be used to solve the above equations. For instance, they may be solved with a finite-difference method.

Consider first the case in which μ and ϵ are piecewise constant functions of z. In this case, (8) can be solved first in each piecewise constant region and a unique solution is found later by matching boundary conditions across the discontinuities at the interfaces. For example, the solution of (8), when μ and ϵ are constants, is a linear superposition of $\exp(\pm ik_z z)$ where $k_z = (\omega^2 \mu\epsilon - k_x^2)^{1/2}$.

The boundary conditions for e_y across a discontinuity could be derived from (8a). To do so, first note that the term involving the derivatives has to be finite since all the other terms are finite in the equation. This is only possible if $(d/dz)\mu^{-1}(d/dz)e_y$ is a finite quantity at the interface of the two regions, or that e_y and $\mu^{-1}(d/dz)e_y$ are continuous quantities at the interface. Hence, at an interface between region 1 and region 2, the following boundary conditions should hold, namely,

$$e_{1y} = e_{2y}, \quad \mu_1^{-1} \frac{d}{dz} e_{1y} = \mu_2^{-1} \frac{d}{dz} e_{2y}. \tag{2.1.9a}$$

Similarly, for h_y the boundary conditions are

$$h_{1y} = h_{2y}, \quad \epsilon_1^{-1} \frac{d}{dz} h_{1y} = \epsilon_2^{-1} \frac{d}{dz} h_{2y}. \tag{2.1.9b}$$

These boundary conditions can also be derived from the fundamental boundary conditions of Maxwell's equations, which require that $\hat{n} \times \mathbf{E}$ and

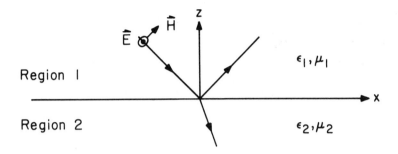

Figure 2.1.1 Reflection and transmission of a plane wave at an interface.

$\hat{n} \times \mathbf{H}$ be continuous in a source free region. However, the above illustrates that these boundary conditions are, in fact, buried in (8a) and (8b).

§§2.1.2 Reflection from a Half-Space

A half-space with two piecewise constant regions is the simplest example of a planar, one-dimensional inhomogeneity. Consider first a TE wave incident on a half-space as shown in Figure 2.1.1; the problem can be reduced to a one-dimensional problem by letting $E_y = e_y(z)e^{ik_x x}$ in both regions as in (8). But in the upper half-space, both incident and reflected waves are present; hence, the general expression in the upper half-space is a linear superposition of $\exp(\pm ik_{1z}z)$, which can be written as

$$e_{1y}(z) = e_0 e^{-ik_{1z}z} + R^{TE}e_0 e^{ik_{1z}z}, \qquad (2.1.10)$$

where R^{TE} is the ratio of the amplitude of the reflected wave to the amplitude of the incident wave. In the lower half-space, however, only a transmitted wave is present; hence, the general expression is

$$e_{2y}(z) = T^{TE}e_0 e^{-ik_{2z}z}, \qquad (2.1.11)$$

where T^{TE} is the ratio of the amplitude of the transmitted wave to the amplitude of the incident wave at $z = 0$. Note that in (10) and (11), $k_{iz} = (k_i^2 - k_x^2)^{1/2}$, where $k_i^2 = \omega^2 \mu_i \epsilon_i$.

On invoking boundary conditions (9a), at $z = 0$, we arrive at

$$1 + R^{TE} = T^{TE}, \qquad (2.1.12a)$$

$$\frac{k_{1z}}{\mu_1}(1 - R^{TE}) = \frac{k_{2z}}{\mu_2}T^{TE}. \qquad (2.1.12b)$$

Solving the preceding two equations gives us the **Fresnel reflection** and **transmission coefficients**,

$$R^{TE} = \frac{\mu_2 k_{1z} - \mu_1 k_{2z}}{\mu_2 k_{1z} + \mu_1 k_{2z}}, \tag{2.1.13a}$$

$$T^{TE} = \frac{2\mu_2 k_{1z}}{\mu_2 k_{1z} + \mu_1 k_{2z}}. \tag{2.1.13b}$$

These coefficients were first derived by Fresnel in 1823 in a slightly less general form (see Born and Wolf 1980). Similarly, for the TM fields, the reflection and transmission coefficients are

$$R^{TM} = \frac{\epsilon_2 k_{1z} - \epsilon_1 k_{2z}}{\epsilon_2 k_{1z} + \epsilon_1 k_{2z}}, \tag{2.1.14a}$$

$$T^{TM} = \frac{2\epsilon_2 k_{1z}}{\epsilon_2 k_{1z} + \epsilon_1 k_{2z}}. \tag{2.1.14b}$$

Alternatively, the above can be derived from the duality principle. It is important to note that R^{TE} and T^{TE} are reflection and transmission coefficients for the electric field, while R^{TM} and T^{TM} are the reflection and transmission coefficients for the magnetic field.

If $k_1 > k_2$, there exist values of k_x such that $k_1 > k_x > k_2$, implying that $k_{2z} = (k_2^2 - k_x^2)^{1/2}$ is purely imaginary, while $k_{1z} = (k_1^2 - k_x^2)^{1/2}$ is purely real. In this case, the numerator of (13a) or (14a) is a complex number that is the complex conjugate of the denominator. Hence, the magnitude of R^{TE} or R^{TM} equals 1. In other words, all the energy of the incident wave is reflected. This phenomenon is known as **total internal reflection**.

For $\mu_2/\mu_1 \neq 1$ or $\epsilon_2/\epsilon_1 \neq 1$, there exist values of k_x such that R^{TE} or R^{TM} equals zero. Then, the corresponding angle for which the reflection coefficient equals zero is known as the **Brewster angle** (see Exercise 2.2), since it was first noted by Brewster in 1815. Since most materials are nonmagnetic, i.e., $\mu_2/\mu_1 = 1$, the Brewster angle effect is more prevalent for TM waves than for TE waves.

§§2.1.3 Reflection and Transmission in a Multilayered Medium

A planar, inhomogeneous half-space with μ and ϵ varying as a function of z only can be modeled by a finely layered medium, where the electromagnetic property is piecewise constant in each region (see Figure 2.1.2). Before solving the general problem, consider first the more specific problem of the reflection of a TE wave from a three-layer medium (see Figure 2.1.3); the wave in region 1 can, as before, be written as

$$e_{1y} = A_1 \left[e^{-ik_{1z}z} + \tilde{R}_{12} e^{2ik_{1z}d_1 + ik_{1z}z} \right], \tag{2.1.15}$$

where $\tilde{R}_{12}$ is a reflection coefficient that is the ratio of the upgoing wave amplitude and the downgoing wave amplitude at the first interface $z = -d_1$.

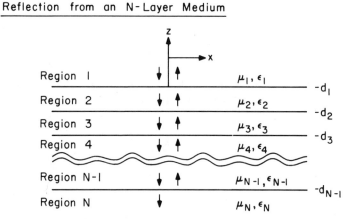

Figure 2.1.2 Reflection and transmission in a multilayered medium.

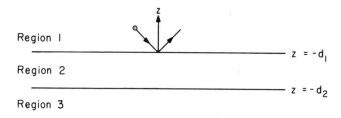

Figure 2.1.3 Reflection from a three-layer medium

Note that the extra phase factor in the second term in (15) ensures that this ratio is properly defined.

The wave in region 2 has a similar form,

$$e_{2y} = A_2\left[e^{-ik_{2z}z} + R_{23}e^{2ik_{2z}d_2+ik_{2z}z}\right], \qquad (2.1.16)$$

where R_{23} is the Fresnel reflection coefficient for a downgoing wave in region 2 reflected by region 3, because region 3 extends to infinity in the $-z$ direction. The superscript TE for R_{23} is ignored here, since it is understood from the context that TE waves are being discussed.

The wave in region 3 can be written as

$$e_{3y} = A_3 e^{-ik_{3z}z}. \qquad (2.1.17)$$

Since region 3 extends to infinity, there can only be a downgoing wave in this region.

The unknowns A_1, A_2, A_3, and $\tilde{R}_{12}$ are found by imposing **constraint conditions** at the interfaces. First, by tracing the propagation of waves, note that the downgoing wave in region 2 is a consequence of the transmission of the downgoing wave in region 1 plus a reflection of the upgoing wave in region 2; that is, at the top interface, $z = -d_1$, the constraint condition is

$$A_2 e^{ik_{2z}d_1} = A_1 e^{ik_{1z}d_1} T_{12} + R_{21} A_2 R_{23} e^{2ik_{2z}d_2 - ik_{2z}d_1}. \qquad (2.1.18)$$

Observe that the first term is the transmission of the downgoing wave amplitude in region 1 at $z = -d_1$, i.e., $A_1 e^{ik_{1z}d_1}$ via the Fresnel transmission coefficient T_{12}. Furthermore, the second term is the reflection of the upgoing wave amplitude in region 2 at $z = -d_1$, i.e., $A_2 R_{23} e^{2ik_{2z}d_2 - ik_{2z}d_1}$ via the Fresnel coefficient R_{21}.

Next, notice that the upgoing wave in region 1 is caused by the reflection of the downgoing wave in region 1 plus a transmission of the upgoing wave in region 2. Consequently, at the interface $z = -d_1$, we have the constraint condition

$$A_1 \tilde{R}_{12} e^{ik_{1z}d_1} = R_{12} A_1 e^{ik_{1z}d_1} + T_{21} A_2 R_{23} e^{2ik_{2z}d_2 - ik_{2z}d_1}. \qquad (2.1.19)$$

From (18), A_2 can be solved for in terms of A_1, yielding

$$A_2 = \frac{T_{12} A_1 e^{i(k_{1z} - k_{2z})d_1}}{1 - R_{21} R_{23} e^{2ik_{2z}(d_2 - d_1)}}. \qquad (2.1.20)$$

Then, on substituting (20) into (19), we derive that

$$\tilde{R}_{12} = R_{12} + \frac{T_{12} R_{23} T_{21} e^{2ik_{2z}(d_2 - d_1)}}{1 - R_{21} R_{23} e^{2ik_{2z}(d_2 - d_1)}}. \qquad (2.1.21)$$

Here, $\tilde{R}_{12}$ is the **generalized reflection coefficient** for the three-layer medium that relates the amplitude of the upgoing wave to the amplitude of the downgoing wave in region 1. It includes the effect of subsurface reflections as well as the reflection from the first interface.

To elucidate the physics better, Equation (21) can be expanded in terms of a series,

$$\tilde{R}_{12} = R_{12} + T_{12} R_{23} T_{21} e^{2ik_{2z}(d_2 - d_1)} + T_{12} R_{23}^2 R_{21} T_{21} e^{4ik_{2z}(d_2 - d_1)} + \cdots. \qquad (2.1.22)$$

The first term in the above is just the result of a single reflection off the first interface. The n-th term above is the consequence of the n-th reflection from the three-layer medium (see Figure 2.1.4). Hence, the expansion of (21) into (22) renders a lucid interpretation for the generalized reflection coefficient. Consequently, the series in (22) can be thought of as a **ray** series

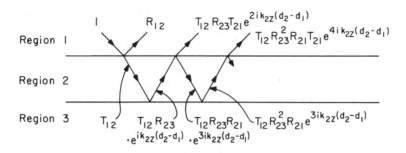

Figure 2.1.4 Geometric series of multiple reflections in a three-layer medium.

or a *geometrical optics* series. It is the consequence of multiple reflections and transmissions in region 2 of the three-layer medium. It is also the consequence of expanding the denominator of the second term in (21). Hence, the denominator of the second term in (21) can be physically interpreted as a consequence of multiple reflections within region 2.

Now, if an additional layer is added below region 3, we need only replace R_{23} in Equation (21) by $\tilde{R}_{23}$ (a generalized reflection coefficient that incorporates subsurface reflection), for in the previous derivation, R_{23} is the ratio between upgoing and downgoing waves in region 2. Therefore, if a subsurface layer is added in region 3, this ratio will just become $\tilde{R}_{23}$. In general, for an N-layer medium, the generalized reflection coefficient at the interface between region i and region $i + 1$, $\tilde{R}_{i,i+1}$, can be written as

$$\tilde{R}_{i,i+1} = R_{i,i+1} + \frac{T_{i,i+1}\tilde{R}_{i+1,i+2}\,T_{i+1,i}e^{2ik_{i+1,z}(d_{i+1}-d_i)}}{1 - R_{i+1,i}\tilde{R}_{i+1,i+2}e^{2ik_{i+1,z}(d_{i+1}-d_i)}}. \tag{2.1.23}$$

Furthermore, by using the facts that $T_{ij} = 1 + R_{ij}$ and $R_{ij} = -R_{ji}$, the above can be simplified as (see Exercise 2.3)

$$\tilde{R}_{i,i+1} = \frac{R_{i,i+1} + \tilde{R}_{i+1,i+2}e^{2ik_{i+1,z}(d_{i+1}-d_i)}}{1 + R_{i,i+1}\tilde{R}_{i+1,i+2}e^{2ik_{i+1,z}(d_{i+1}-d_i)}}. \tag{2.1.24}$$

Notice that the above are recursive relations that express $\tilde{R}_{i,i+1}$ in terms of $\tilde{R}_{i+1,i+2}$. Such recursive relations are attributed to Stokes (see Bellman and Wing 1975).

The wave in the i-th region assumes the form

$$e_{iy} = A_i \left[e^{-ik_{iz}z} + \tilde{R}_{i,i+1}e^{2ik_{iz}d_i+ik_{iz}z} \right], \tag{2.1.25}$$

which is the generalization of (15) and (16). Since $\tilde{R}_{N,N+1} = 0$, Equation (23) or (24) can be solved recursively for $\tilde{R}_{i,i+1}$ in all the regions.

In the same manner as Equation (20), we can also write A_i in terms of A_{i-1} of the adjacent layers as

$$A_i e^{ik_{iz}d_{i-1}} = \frac{T_{i-1,i}A_{i-1}e^{ik_{i-1,z}d_{i-1}}}{1 - R_{i,i-1}\tilde{R}_{i,i+1}e^{2ik_{iz}(d_i-d_{i-1})}} = A_{i-1}e^{ik_{i-1,z}d_{i-1}}S_{i-1,i},$$

$$(2.1.26)$$

where

$$S_{i-1,i} = \frac{T_{i-1,i}}{1 - R_{i,i-1}\tilde{R}_{i,i+1}e^{2ik_{iz}(d_i-d_{i-1})}}. \qquad (2.1.26a)$$

Here, $S_{i-1,i}$ is known for all the regions since $\tilde{R}_{i,i+1}$ is known for all the regions. Since A_1 is known in region 1, we can derive A_i for all the regions as

$$A_i e^{ik_{iz}d_{i-1}} = A_1 e^{ik_{1z}d_1} S_{12} e^{ik_{2z}(d_2-d_1)} S_{23} \cdots e^{ik_{i-1,z}(d_{i-1}-d_{i-2})}S_{i-1,i}$$

$$= A_1 e^{ik_{1z}d_1} \prod_{j=1}^{i-1} e^{ik_{jz}(d_j-d_{j-1})}S_{j,j+1}, \qquad (2.1.27)$$

where we assume $d_0 = d_1$. Once A_i and $\tilde{R}_{i,i+1}$ are known in all the regions, the field everywhere can be derived easily via Equation (25).

A generalized transmission coefficient for a layered slab can be defined as

$$\tilde{T}_{1N} = \prod_{j=1}^{N-1} e^{ik_{jz}(d_j-d_{j-1})}S_{j,j+1}, \qquad (2.1.28)$$

where we have assumed $d_0 = d_1$. In this manner,

$$A_N e^{ik_{Nz}d_{N-1}} = \tilde{T}_{1N} A_1 e^{ik_{1z}d_1}, \qquad (2.1.29)$$

that is, the downgoing wave amplitude in region N at $z = -d_{N-1}$ is just $\tilde{T}_{1N}$ times the downgoing wave amplitude in region 1 at $z = -d_1$. Moreover, reciprocal property can be derived for $\tilde{T}_{1N}$ (see Exercise 2.4).

A general inhomogeneous profile can be replaced by many fine layers of piecewise constant regions and the field obtained everywhere via the above method. The rule of thumb is then to approximate the layers such that they are much thinner than the wavelength of the wave in the medium.

Notice that in the above example, TE waves have been assumed. Therefore, all the reflection and transmission coefficients are of the TE type. But the same analysis holds true for TM waves, where the reflection and transmission coefficients would just be of the TM type.

§§2.1.4 Ricatti Equation for Reflection Coefficients

Equation (23) is the formula for the generalized reflection coefficient that relates the upgoing wave to the downgoing wave at $z = -d_i$. In the limit when the inhomogeneous medium is very finely layered, Equation (23) is reducible to an ordinary differential equation. To see this, consider the case of a finely layered medium where the thickness of each layer is Δ. Then, by expressing the reflection coefficient and the generalized reflection coefficient at the boundary $z = -d_i$ as $R(z)$ and $\tilde{R}(z)$ respectively, and $k_{i+1,z}$ as $k_z(z - \Delta)$, Equation (24) can be written as

$$\tilde{R}(z) = \frac{R(z) + \tilde{R}(z - \Delta)e^{2ik_z(z-\Delta)\Delta}}{1 + R(z)\tilde{R}(z - \Delta)e^{2ik_z(z-\Delta)\Delta}}, \tag{2.1.30}$$

where from (13) and (14),

$$R(z) = \frac{k_z(z)/p(z) - k_z(z - \Delta)/p(z - \Delta)}{k_z(z)/p(z) + k_z(z - \Delta)/p(z - \Delta)}, \tag{2.1.30a}$$

and $p(z) = \mu(z)$ for TE waves while $p(z) = \epsilon(z)$ for TM waves. In the limit when $\Delta \to 0$ to recover a continuous profile, we have

$$R(z) \simeq \frac{(k_z/p)'\Delta}{2(k_z/p)}, \qquad e^{2ik_z(z-\Delta)\Delta} \simeq 1 + \Delta 2ik_z(z),$$

and

$$\tilde{R}(z - \Delta) \simeq \tilde{R}(z) - \Delta\tilde{R}'(z), \tag{2.1.31}$$

where the prime denotes the derivative with respect to the z variable. Then, in this limit, (30) becomes

$$\tilde{R}(z) \simeq \tilde{R}(z) - \Delta\tilde{R}'(z) + \Delta 2ik_z(z)\tilde{R}(z) + \Delta\frac{(k_z/p)'}{2(k_z/p)} - \Delta\frac{(k_z/p)'}{2(k_z/p)}\tilde{R}^2(z). \tag{2.1.32}$$

Next, by equating the first order terms, we conclude that

$$\tilde{R}'(z) = 2ik_z(z)\tilde{R}(z) + \frac{(k_z/p)'}{2(k_z/p)}[1 - \tilde{R}^2(z)]. \tag{2.1.33}$$

The above is now an ordinary differential equation for $\tilde{R}(z)$. Because of the $\tilde{R}^2(z)$ term, it is a nonlinear differential equation known as the **Ricatti equation**. It can be solved, for example, by the Runge-Kutta method (Hildebrand 1976) as an initial value problem with $\tilde{R}(z = z_m) = 0$ where z_m is the lower boundary of the inhomogeneous slab (see Figure 2.1.5).

In the above, $\tilde{R}(z')$ is the generalized reflection coefficient at z' assuming that $\mu(z)$ and $\epsilon(z)$ are constants for $z > z'$ and that the constants are the values $\mu(z')$ and $\epsilon(z')$ (see Figure 2.1.5). However, it is sometimes desirable to derive an ordinary differential equation for $\tilde{R}_0(z')$, a generalized reflection coefficient which assumes that $\mu(z) = \mu_0$ and $\epsilon(z) = \epsilon_0$ for $z > z'$. In other words, free-space is present for $z > z'$ (see Figure 2.1.5). Moreover, given $\tilde{R}(z)$, one can find $\tilde{R}_0(z)$ easily via the relationship

$$\tilde{R}_0(z) = \frac{R_0(z) + \tilde{R}(z)}{1 + R_0(z)\tilde{R}(z)}, \tag{2.1.34}$$

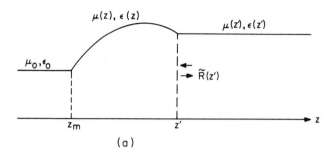

(a)

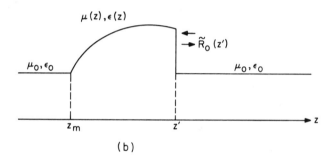

(b)

Figure 2.1.5 (a) Construction of $\tilde{R}(z)$ in Equation (33) assumes that the μ and ϵ are constants $\mu(z')$, $\epsilon(z')$ to the right of z'. (b) Construction of $\tilde{R}_0(z)$ in Equation (36) assumes that μ and ϵ are μ_0 and ϵ_0 to the right of z'.

where

$$R_0(z) = \frac{k_{0z}/p_0 - k_z(z)/p(z)}{k_{0z}/p_0 + k_z(z)/p(z)}, \tag{2.1.34a}$$

and k_{0z} and p_0 are the values of k_z and p in free-space. In view of (30), (34) is arrived at by adding a slab of zero thickness that has properties of $\mu(z)$, $\epsilon(z)$. Obviously, Equation (34) can be inverted, yielding

$$\tilde{R}(z) = \frac{-R_0(z) + \tilde{R}_0(z)}{1 - R_0(z)\tilde{R}_0(z)}. \tag{2.1.35}$$

Next, by substituting (35) into (33), we arrive at (see Exercise 2.5)

$$\tilde{R}_0'(z) = \frac{2ik_z(z)}{1 - R_0^2(z)} \left[\tilde{R}_0(z) - R_0(z) \right] \left[1 - R_0(z)\tilde{R}_0(z) \right]. \tag{2.1.36}$$

The above is another Ricatti equation for $\tilde{R}_0(z)$. It can also be solved as an initial value problem using the Runge-Kutta method (see Exercise 2.6). It has the advantage of not involving derivatives of (k_z/p). An equation similar to (36) was derived by Barrar and Redheffer (1955) using transmission line theory.

§§2.1.5 Specific Inhomogeneous Profiles

When the inhomogeneous, planar profile is analytic and continuous, the ordinary differential equations in (8a) and (8b) may have closed-form solutions in terms of special functions. This is especially true for the TE case where the medium may be assumed to be nonmagnetic, that is, $\mu = \mu_0$ is a constant. Various forms of $\epsilon(z)$ in (8a) that admit closed-form solutions have been tabulated in the literature (Tyras 1969).

Consider the case of a linear profile as an example. In this case, assume that

$$\epsilon(z) = a + bz. \tag{2.1.37}$$

Equation (8a) for a nonmagnetic material then becomes

$$\left[\frac{d^2}{dz^2} + A + Bz\right] e_y = 0, \tag{2.1.38}$$

where $A = \omega^2 \mu a - k_x^2$ and $B = \omega^2 \mu b$. The above can be simplified with a change of variable by letting

$$\eta = B^{\frac{1}{3}}\left(z + \frac{A}{B}\right).$$

Then, (38) will become

$$\left[\frac{d^2}{d\eta^2} + \eta\right] e_y = 0. \tag{2.1.39}$$

The above differential equation is known as the Airy equation, and the corresponding solutions are Airy functions $Ai(-\eta)$ and $Bi(-\eta)$ (see Abramowitz and Stegun 1965; Olver 1974). Hence,

$$e_y = C_1 Ai(-\eta) + C_2 Bi(-\eta), \tag{2.1.40}$$

where $Ai(-\eta)$ and $Bi(-\eta)$ are Airy functions of the first and second type.[1] Moreover, when the arguments of the Airy functions are large, they have asymptotic expansions of the form

$$Ai(x) \sim \tfrac{1}{2}\pi^{-\frac{1}{2}}x^{-\frac{1}{4}}e^{-\frac{2}{3}x^{\frac{3}{2}}}, \qquad\qquad x \to \infty, \tag{2.1.41a}$$

$$Ai(-x) \sim \pi^{-\frac{1}{2}}x^{-\frac{1}{4}}\sin\left(\tfrac{2}{3}x^{\frac{3}{2}} + \tfrac{\pi}{4}\right), \quad x \to \infty, \tag{2.1.41b}$$

$$Bi(x) \sim \pi^{-\frac{1}{2}}x^{-\frac{1}{4}}e^{\frac{2}{3}x^{\frac{3}{2}}}, \qquad\qquad x \to \infty, \tag{2.1.41c}$$

$$Bi(-x) \sim \pi^{-\frac{1}{2}}x^{-\frac{1}{4}}\cos\left(\tfrac{2}{3}x^{\frac{3}{2}} + \tfrac{\pi}{4}\right), \quad x \to \infty. \tag{2.1.41d}$$

---

[1] Airy functions can also be related to Bessel functions of fractional order.

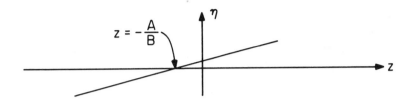

Figure 2.1.6 A linear profile.

Here, "$\sim$" means "asymptotic to". Notice that η as a function of z is as shown in Figure 2.1.6; hence, $\eta \to -\infty$ when $z \to -\infty$, and $\eta \to +\infty$ when $z \to +\infty$. Therefore, when $z \to -\infty$, we have

$$Ai(-\eta) \sim \tfrac{1}{2}\pi^{-\frac{1}{2}}(-\eta)^{-\frac{1}{4}}e^{-\frac{2}{3}(-\eta)^{\frac{3}{2}}}, \quad \eta \to -\infty, \qquad (2.1.42a)$$

$$Bi(-\eta) \sim \pi^{-\frac{1}{2}}(-\eta)^{-\frac{1}{4}}e^{\frac{2}{3}(-\eta)^{\frac{3}{2}}}, \qquad \eta \to -\infty. \qquad (2.1.42b)$$

In other words, $Ai(-\eta)$ corresponds to an exponentially decaying wave while $Bi(-\eta)$ corresponds to an exponentially growing wave. But if the wave source is at $z = +\infty$, an exponentially growing solution is not expected when $z \to -\infty$. As a result, $C_2 = 0$ in Equation (40). Consequently, the solution is

$$e_y = C_1 Ai(-\eta). \qquad (2.1.43)$$

Similarly, when $z \to +\infty$, $\eta \to +\infty$, and we have

$$Ai(-\eta) \sim \pi^{-\frac{1}{2}}\eta^{-\frac{1}{4}}\sin\left(\tfrac{2}{3}\eta^{\frac{3}{2}} + \tfrac{\pi}{4}\right), \quad \eta \to +\infty, \qquad (2.1.44)$$

which corresponds to a standing wave resulting from a superposition of incident and reflected waves on the far right.

The physical interpretation of the above observation is that when a wave is incident from the right into the profile shown in Figure 2.1.6, the wave sees an optically less and less dense medium. Consequently, the wave is slowly refracted, and finally at z where $\eta = 0$, total internal reflection occurs, and the wave is totally reflected. Therefore, this gives rise to a standing wave to the right and an evanescent wave to the left.

§2.2 Spectral Representations of Sources

The previous section shows how the reflection of a plane wave from a layered medium can be calculated. However, a plane wave is an idealization that does not exist in the real world. In practice, waves are nonplanar in nature as they are generated by finite sources, such as antennas and scatterers.

Fortunately, these waves can be expanded in terms of plane waves. Once this is done, then the study of non-plane-wave reflections from a layered medium becomes routine. In the following, we show how waves resulting from a line source or a point source can be expanded in terms of plane waves.

§§2.2.1 A Line Source

Consider a scalar wave equation with a line source,

$$\left[\frac{\partial^2}{\partial x^2} + \frac{\partial^2}{\partial y^2} + k_\rho^2\right]\phi(x,y) = -\delta(x)\,\delta(y). \tag{2.2.1}$$

Because of the cylindrical symmetry of the problem, the above equation is solved most conveniently in the cylindrical coordinates, i.e.,

$$\left[\frac{\partial^2}{\partial\rho^2} + \frac{1}{\rho}\frac{\partial}{\partial\rho} + k_\rho^2\right]\phi(\rho) = -\delta(\boldsymbol{\rho}), \tag{2.2.2}$$

where $\delta(\boldsymbol{\rho}) = \delta(x)\delta(y)$.

Outside the source region, the right-hand side of (2) is zero, and we have a Bessel equation of zeroth order. Therefore, in order to have an outgoing-wave solution that satisfies the radiation condition, the Hankel function of the first kind is chosen for $\phi(\rho)$ with $e^{-i\omega t}$ time dependence. In other words,

$$\phi(\rho) = C H_0^{(1)}(k_\rho\rho) \sim C\sqrt{\frac{2}{i\pi k_\rho\rho}}\,e^{ik_\rho\rho}, \; k_\rho\rho \to \infty. \tag{2.2.3}$$

The constant C may be found by matching the singularity of the Hankel function at $\rho = 0$ to the line source.[2] Hence, it can be shown that

$$\phi(\rho) = \frac{i}{4}H_0^{(1)}(k_\rho\rho). \tag{2.2.4}$$

In addition, Equation (1) can be solved by the Fourier transform technique. Hence, for a fixed y, assuming that the Fourier transform of $\phi(x,y)$ exists, then $\phi(x,y)$ is expressible as a Fourier inverse transform integral such that

$$\phi(x,y) = \frac{1}{2\pi}\int_{-\infty}^{\infty} dk_x\, e^{ik_x x}\tilde{\phi}(k_x,y). \tag{2.2.5}$$

Consequently, on substituting (5) into (1), and using the fact that

$$\delta(x) = \frac{1}{2\pi}\int_{-\infty}^{\infty} dk_x\, e^{ik_x x}, \tag{2.2.6}$$

---

[2] See Equation (1.3.41) on determining C [see also Exercise 2.8(a)].

one obtains

$$\frac{1}{2\pi} \int\limits_{-\infty}^{\infty} dk_x\, e^{ik_x x} \left[\frac{\partial^2}{\partial y^2} + k_\rho^2 - k_x^2 \right] \tilde{\phi}(k_x, y) = -\frac{1}{2\pi} \int\limits_{-\infty}^{\infty} dk_x\, e^{ik_x x} \delta(y).$$
(2.2.7)

Since (7) is satisfied for all x, we must have

$$\left[\frac{d^2}{dy^2} + k_y^2 \right] \tilde{\phi}(k_x, y) = -\delta(y),$$
(2.2.8)

where $k_y^2 = k_\rho^2 - k_x^2$.

Outside the source for $|y| > 0$, the solution of (8) is the homogeneous solution $\exp(\pm i k_y y)$. The solution, however, must have discontinuous second derivatives at $y = 0$ in order to yield the singularity on the right in (8). Therefore, by matching the singularity at $y = 0$, the above yields the solution

$$\tilde{\phi}(k_x, y) = \frac{i e^{ik_y |y|}}{2 k_y},$$
(2.2.9)

if only the outgoing-wave solution is considered. This is necessary to satisfy the radiation condition at infinity. Hence, (5) becomes

$$\phi(x, y) = \frac{i}{4\pi} \int\limits_{-\infty}^{\infty} dk_x\, \frac{e^{ik_x x + ik_y |y|}}{k_y}.$$
(2.2.10)

The above integral is undefined for the following reason: Since $k_y = (k_\rho^2 - k_x^2)^{1/2}$, there are branch-point singularities at $k_x = \pm k_\rho$. But the path of integration of the Fourier inverse transform is on the real axis on the complex k_x-plane. So if k_ρ is real, then the branch-point singularities are on the real axis, rendering the integral in (10) undefined. To overcome this ambiguity, loss is assumed in the medium so that $k_\rho = k_\rho' + i k_\rho''$ has a small positive imaginary part. In this case, (4) will correspond to an outgoing, decaying wave. Furthermore, the branch-point singularities are now located off the real axis (see Figure 2.2.1), and the Fourier inverse transform in (10) becomes unambiguous.

By the uniqueness of the solution to the partial differential equation (1),[3] (10) must also be equal to (4) since both of them satisfy (1). Moreover, with the introduction of a small loss, both are exponentially small at infinity, satisfying the radiation condition. Hence, we arrive at the identity (see a related formula in Watson 1944, p. 178)

$$H_0^{(1)}(k_\rho \rho) = \frac{1}{\pi} \int\limits_{-\infty}^{\infty} dk_x\, \frac{e^{ik_x x + ik_y |y|}}{k_y}.$$
(2.2.11)

---

[3] See Chapter 1, Subsection 1.5.1.

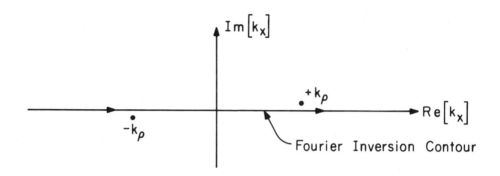

Figure 2.2.1 Fourier inversion contour for Equation (11).

The right-hand side of (11) could be interpreted as an integral summation of plane waves propagating in different directions including evanescent waves. Furthermore, these plane waves satisfy the dispersion relation $k_\rho^2 = k_x^2 + k_y^2$. Hence, Equation (11) is also the plane-wave expansion of a cylindrical wave. Also, a similar expression can be derived for the Hankel function of the second kind (Exercise 2.8).

A small loss is assumed in the derivation because the wave field (4) is actually not absolutely integrable over all x if the medium is lossless, i.e.,

$$\int_{-\infty}^{\infty} dx \, |\phi(x,y)| \to \infty.$$

This can be seen from (3) where the field decays only as $1/\rho^{1/2}$ when $\rho \to \infty$. Hence, strictly speaking, the Fourier transform $\tilde{\phi}(k_x, y)$ may not exist, which is further born out by the fact that $\tilde{\phi}(k_x, y) \to \infty$ when $k_x \to k_\rho$ (see Papoulis 1962, p. 9).[4] If we assume a small loss, however, the wave field becomes absolutely integrable, and (5) is then well-defined. Despite this fact, the identity (11) can still be used for a lossless medium provided that the path of integration is stipulated as shown in Figure 2.2.2. With this path of integration, the lossless solution can be thought of as the limiting case of the lossy medium solution, and the integral in (11) is well-defined even for a lossless medium.

Furthermore, since $k_y = (k_\rho^2 - k_x^2)^{1/2}$, k_y could be a complex number. Hence, in order to satisfy radiation conditions of having only outgoing waves

---

[4] Note that the real and imaginary parts of $H_0^{(1)}(k_\rho\rho)$ satisfy condition 2 on page 9 of Papoulis (1962). Hence, a Cauchy principal value integral may be used to yield a unique value for the singular integral.

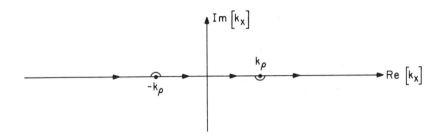

Figure 2.2.2 Path of integration for a lossless medium.

in the integrand, we have to ensure that $\Im m[k_y] > 0$ and $\Re e[k_y] > 0$ over this path of integration. Observe that the introduction of a small loss ensures that the radiation condition is satisfied. Because if a medium is lossy, any incoming wave must be exponentially large at infinity. Such a wave cannot be included in the Fourier representation (5) as it is not absolutely integrable. Hence, the representation (5) excludes incoming waves, further ensuring the radiation condition which is necessary for the uniqueness of the solution (see also Chapter 1, Section 1.5).

In addition, other integral identities for Bessel functions are derivable from (11). To do so, first assume that $|y| = 0$. Then, $x = \rho$ in the cylindrical coordinates, and Equation (11) becomes

$$H_0^{(1)}(k_\rho \rho) = \frac{1}{\pi} \int_{-\infty}^{\infty} dk_x \frac{e^{ik_x \rho}}{k_y} = \frac{2}{\pi} \int_{0}^{\infty} dk_x \frac{\cos k_x \rho}{k_y}. \qquad (2.2.11a)$$

The second equality just follows from folding the integral from $-\infty$ to ∞ to an integral from 0 to ∞. Observe that the real part of the left-hand side is $J_0(k_\rho \rho)$ when k_ρ is pure-real, since $H_0^{(1)}(x) = J_0(x) + iN_0(x)$, where $J_0(x)$ (Bessel function) and $N_0(x)$ (Neumann function) are real-value functions when x is real. It can be shown, however, that when $|k_x| > k_\rho$, the right-hand side is pure-imaginary; hence, taking only the real part of the right-hand side, we have

$$J_0(k_\rho \rho) = \frac{1}{\pi} \int_{-k_\rho}^{k_\rho} dk_x \frac{e^{ik_x \rho}}{k_y} = \frac{2}{\pi} \int_{0}^{k_\rho} dk_x \frac{\cos k_x \rho}{k_y}. \qquad (2.2.12)$$

Next, by letting $k_x = k_\rho \sin \alpha$, then $k_y = \sqrt{k_\rho^2 - k_\rho^2 \sin^2 \alpha} = k_\rho \cos \alpha$, and

$dk_x = k_\rho \cos \alpha \, d\alpha$. Consequently, we have

$$J_0(k_\rho \rho) = \frac{1}{\pi} \int_{-\frac{\pi}{2}}^{\frac{\pi}{2}} d\alpha \, e^{ik_\rho \rho \sin \alpha} = \frac{2}{\pi} \int_0^{\frac{\pi}{2}} d\alpha \, \cos[k_\rho \rho \sin \alpha]. \qquad (2.2.13)$$

Equation (13) is now an integral identity that is often used for Bessel functions. Furthermore, it can be written as

$$J_0(k_\rho \rho) = \frac{1}{2\pi} \int_{-\pi}^{\pi} d\alpha \, e^{ik_\rho \rho \sin \alpha}. \qquad (2.2.14)$$

Since the integrand is a 2π periodic function and the integration is over 2π, the starting point of the integral is unimportant as long as the interval of integration is 2π. Hence,

$$J_0(k_\rho \rho) = \frac{1}{2\pi} \int_0^{2\pi} d\alpha \, e^{ik_\rho \rho \sin \alpha} = \frac{1}{2\pi} \int_0^{2\pi} d\alpha \, e^{ik_\rho \rho \cos(\alpha - \phi)}. \qquad (2.2.15)$$

Because $e^{ik_\rho \rho \cos(\alpha - \phi)} = e^{i\mathbf{k}_\rho \cdot \boldsymbol{\rho}}$ (where $\mathbf{k}_\rho = \hat{x} k_\rho \cos \alpha + \hat{y} k_\rho \sin \alpha$, $\boldsymbol{\rho} = \hat{x} \rho \cos \phi + \hat{y} \rho \sin \phi$, and $0 < \alpha < 2\pi$), it is a plane wave propagating in the direction $\mathbf{k}_\rho$. Consequently, the last integral has the pleasing physical interpretation that a Bessel function, which represents a standing cylindrical wave, is an integral linear superposition of plane waves propagating in all directions.

Furthermore, using the property of the raising operator that (see Exercise 2.9)

$$-\frac{1}{k_\rho} \left[\frac{\partial}{\partial x} + i \frac{\partial}{\partial y} \right] B_n(k_\rho \rho) e^{in\phi} = B_{n+1}(k_\rho \rho) e^{i(n+1)\phi}, \qquad (2.2.16)$$

where $B_n(x)$ is a solution of the n-th order Bessel equation, we can show that

$$J_n(k_\rho \rho) e^{in\phi} = \frac{1}{2\pi} \int_0^{2\pi} d\alpha \, e^{ik_\rho \rho \cos(\alpha - \phi) + in\alpha - in\frac{\pi}{2}}, \qquad (2.2.17)$$

which is also an integral summation of plane waves. The above is an important integral identity for $J_n(x)$. Alternatively, via a change of variable, it can be written as

$$J_n(k_\rho \rho) = \frac{1}{2\pi} \int_0^{2\pi} d\alpha \, e^{ik_\rho \rho \cos \alpha + in\alpha - in\frac{\pi}{2}}. \qquad (2.2.18)$$

The above formula is also given by Whitaker and Watson (1927). In addition, it is useful for deriving Hankel transforms from Fourier transforms (Exercise 2.10).

§§2.2.2 A Point Source

The spectral decomposition or the plane-wave expansion of the field due to a point source could be derived in a manner similar to that of a line source. A different approach will be taken here, however, in order to illustrate a new technique.

First, notice that the scalar wave equation with a point source is

$$\left[\frac{\partial^2}{\partial x^2} + \frac{\partial^2}{\partial y^2} + \frac{\partial^2}{\partial z^2} + k_0^2\right] \phi(x,y,z) = -\delta(x)\,\delta(y)\,\delta(z). \tag{2.2.19}$$

The above equation could then be solved in the spherical coordinates, yielding the solution (see Chapter 1, Subsection 1.3.4)

$$\phi(r) = \frac{e^{ik_0 r}}{4\pi r}. \tag{2.2.20}$$

Next, assuming that the Fourier transform of $\phi(x,y,z)$ exists, we can write

$$\phi(x,y,z) = \frac{1}{(2\pi)^3} \iiint\limits_{-\infty}^{\infty} dk_x dk_y dk_z\, \tilde{\phi}(k_x,k_y,k_z) e^{ik_x x + ik_y y + ik_z z}. \tag{2.2.21}$$

Then, using the above, together with the Fourier representation of the delta function, we convert (19) into

$$\iiint\limits_{-\infty}^{\infty} dk_x dk_y dk_z\, [k_0^2 - k_x^2 - k_y^2 - k_z^2]\tilde{\phi}(k_x,k_y,k_z)e^{ik_x x + ik_y y + ik_z z}$$

$$= -\iiint\limits_{-\infty}^{\infty} dk_x dk_y dk_z\, e^{ik_x x + ik_y y + ik_z z}. \tag{2.2.22}$$

Since the above is equal for all x, y, and z, we must have

$$\tilde{\phi}(k_x,k_y,k_z) = \frac{-1}{k_0^2 - k_x^2 - k_y^2 - k_z^2}. \tag{2.2.23}$$

Consequently, we have

$$\phi(x,y,z) = \frac{-1}{(2\pi)^3} \iiint\limits_{-\infty}^{\infty} d\mathbf{k}\, \frac{e^{ik_x x + ik_y y + ik_z z}}{k_0^2 - k_x^2 - k_y^2 - k_z^2}. \tag{2.2.24}$$

In the above, if we examine the k_z integral first, then the integrand has poles at $k_z = \pm(k_0^2 - k_x^2 - k_y^2)^{1/2}$. Moreover, for real k_0, and real values of k_x and k_y, these two poles lie on the real axis, rendering the integral in (24) undefined. However, if a small loss is assumed in k_0 such that $k_0 = k_0' + ik_0''$,

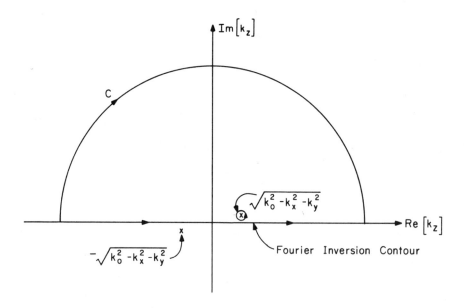

Figure 2.2.3 The integration along the real axis is equal to the integration along C plus the residue of the pole at $(k_0^2 - k_x^2 - k_y^2)^{1/2}$.

then the poles are off the real axis (see Figure 2.2.3), and the integrals in (24) are well-defined. The reason is that $\phi(x, y, z)$ is not strictly absolutely integrable for a lossless medium, and hence, its Fourier transform may not exist. But the introduction of a small loss also guarantees the radiation condition and the uniqueness of the solution to (19) (see Section 1.5), and therefore, the equality of (20) and (24), just as in the case of Equation (11).

Observe that in (24), when $z > 0$, the integrand is exponentially small when $\Im m[k_z] \to \infty$. Therefore, by Jordan's lemma, the integration for k_z over the contour C as shown in Figure 2.2.3 vanishes. Then, by Cauchy's theorem, the integration over the Fourier inversion contour on the real axis is the same as integrating over the pole singularity located at $(k_0^2 - k_x^2 - k_y^2)^{1/2}$, yielding the residue of the pole (see Figure 2.2.3). Consequently, we have

$$\phi(x, y, z) = \frac{i}{2(2\pi)^2} \int\!\!\!\int\limits_{-\infty}^{\infty} dk_x dk_y \frac{e^{ik_x x + ik_y y + ik_z' z}}{k_z'}, \quad z > 0, \qquad (2.2.25)$$

where $k_z' = (k_0^2 - k_x^2 - k_y^2)^{1/2}$. Similarly, for $z < 0$, the integral is equal to the pole contribution at $-(k_0^2 - k_x^2 - k_y^2)^{1/2}$. As such, the result for all z can be

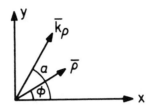

Figure 2.2.4 The $\mathbf{k}_\rho$ and the $\boldsymbol{\rho}$ vector on the xy plane.

written as

$$\phi(x, y, z) = \frac{i}{2(2\pi)^2} \iint\limits_{-\infty}^{\infty} dk_x dk_y \, \frac{e^{ik_x x + ik_y y + ik_z'|z|}}{k_z'}, \qquad \text{all } z. \qquad (2.2.26)$$

By the uniqueness of the solution to the partial differential equation (19) satisfying radiation condition at infinity, we can equate (20) and (26), yielding the identity

$$\frac{e^{ik_0 r}}{r} = \frac{i}{2\pi} \iint\limits_{-\infty}^{\infty} dk_x dk_y \, \frac{e^{ik_x x + ik_y y + ik_z|z|}}{k_z}, \qquad (2.2.27)$$

where $k_x^2 + k_y^2 + k_z^2 = k_0^2$, or $k_z = (k_0^2 - k_x^2 - k_y^2)^{1/2}$. The above is known as the **Weyl identity** (Weyl 1919). To ensure the radiation condition, we require that $\Im m[k_z] > 0$ and $\Re e[k_z] > 0$ over all values of k_x and k_y in the integration. Furthermore, Equation (27) could be interpreted as an integral summation of plane waves propagating in all directions, including evanescent waves. It is the plane-wave expansion of a spherical wave.

In (27), we can write $\mathbf{k}_\rho = \hat{x} k_\rho \cos\alpha + \hat{y} k_\rho \sin\alpha$, $\boldsymbol{\rho} = \hat{x}\rho \cos\phi + \hat{y}\rho \sin\phi$ (see Figure 2.2.4), and $dk_x dk_y = k_\rho dk_\rho \, d\alpha$. Then, $k_x x + k_y y = \mathbf{k}_\rho \cdot \boldsymbol{\rho} = k_\rho \cos(\alpha - \phi)$, and we have

$$\frac{e^{ik_0 r}}{r} = \frac{i}{2\pi} \int\limits_0^{\infty} k_\rho dk_\rho \int_0^{2\pi} d\alpha \frac{e^{ik_\rho \rho \cos(\alpha - \phi) + ik_z|z|}}{k_z}, \qquad (2.2.28)$$

where $k_z = (k_0^2 - k_x^2 - k_y^2)^{1/2} = (k_0^2 - k_\rho^2)^{1/2}$. Then, using the integral identity for Bessel functions given by (15), namely,

$$J_0(k_\rho \rho) = \frac{1}{2\pi} \int\limits_0^{2\pi} d\alpha \, e^{ik_\rho \rho \cos(\alpha - \phi)}, \qquad (2.2.29)$$

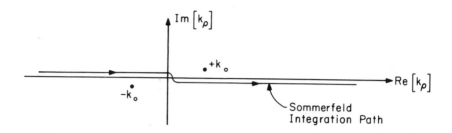

Figure 2.2.5 Sommerfeld integration path.

(28) becomes

$$\frac{e^{ik_0 r}}{r} = i \int_0^\infty dk_\rho \frac{k_\rho}{k_z} J_0(k_\rho \rho) e^{ik_z |z|}. \tag{2.2.30}$$

The above is also known as the **Sommerfeld identity** (Sommerfeld 1909; 1949, p. 242). Its physical interpretation is that a spherical wave can be expanded as an integral summation of conical waves or cylindrical waves in the ρ direction, times a plane wave in the z direction over all wave numbers k_ρ. This wave is evanescent in the $\pm z$ direction when $k_\rho > k_0$.

By using the fact that $J_0(k_\rho \rho) = 1/2[H_0^{(1)}(k_\rho \rho) + H_0^{(2)}(k_\rho \rho)]$, and the reflection formula that $H_0^{(1)}(e^{i\pi} x) = -H_0^{(2)}(x)$, a variation of the above identity can be derived as

$$\frac{e^{ik_0 r}}{r} = \frac{i}{2} \int_{-\infty}^\infty dk_\rho \frac{k_\rho}{k_z} H_0^{(1)}(k_\rho \rho) e^{ik_z |z|}. \tag{2.2.31}$$

Since $H_0^{(1)}(x)$ has a logarithmic branch-point singularity at $x = 0$, and $k_z = (k_0^2 - k_\rho^2)^{1/2}$ has algebraic branch-point singularities at $k_\rho = \pm k_0$, the integral in Equation (31) is undefined unless we stipulate also the path of integration. Hence, a path of integration adopted by Sommerfeld, which is even good for a lossless medium, is shown in Figure 2.2.5. Because of the manner in which we have selected the reflection formula for Hankel functions, i.e., $H_0^{(1)}(e^{i\pi} x) = -H_0^{(2)}(x)$, the path of integration should be above the logarithmic branch-point singularity at the origin.

§§2.2.3 Riemann Sheets and Branch Cuts

The functions $k_y = (k_\rho^2 - k_x^2)^{1/2}$ in (11) and $k_z = (k_0^2 - k_\rho^2)^{1/2}$ in (31) are double-value functions because, in taking the square root of a number, two values are possible. In particular, k_y is a double-value function of k_x, while k_z

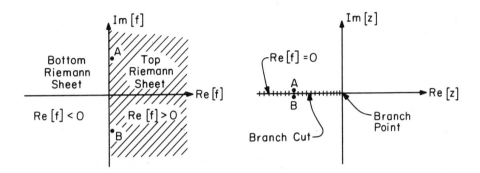

Figure 2.2.6 The complex f and z planes where $f = z^{1/2}$. The right-half plane of the f plane is mapped onto the top Riemann sheet while the left-half plane is mapped onto the bottom Riemann sheet. The points A and B are mapped as shown.

is a double-value function of k_ρ. Consequently, for every point on a complex k_ρ plane in Figure 2.2.5, there are two possible values of k_z. Therefore, the integral (31) is undefined unless we stipulate which of the two values of k_z is adopted in performing the integration.

A multivalue function is denoted on a complex plane with the help of **Riemann sheets**. For instance, a double-value function such as k_z is assigned two Riemann sheets to a single complex plane. On one of these Riemann sheets, k_z assumes a value just opposite in sign to the value on the other Riemann sheet.

As an illustration, consider the function $f(z) = z^{\frac{1}{2}}$, which is a double-value function. By letting $z = ce^{i\theta}$, where c and θ are real, then $f(z) = \sqrt{c}\,e^{i\frac{\theta}{2}}$, where c and $\sqrt{c}$ are positive real constants. Hence, $f(z)$ is a periodic function of θ with period 4π. Note that for $\theta = \frac{3\pi}{2}$ and $\theta = -\frac{\pi}{2}$, which denote the same point on the complex z plane, $f(z)$ has two possible values of $\sqrt{c}\,e^{i\frac{3\pi}{4}}$ and $\sqrt{c}\,e^{-i\frac{\pi}{4}}$ respectively. For the sake of clarity, Riemann sheets are used on the complex z plane to distinguish between these two values: the values of $f(z)$ associated with $-\pi < \theta < \pi$ are assigned to the top Riemann sheet, while the values of $f(z)$ associated with $\pi < \theta < 3\pi$ are assigned to the bottom Riemann sheet on the complex z plane.

For $-\pi < \theta < \pi$, the argument of $f(z)$ varies between $-\frac{\pi}{2}$ and $\frac{\pi}{2}$. Consequently, the right-half plane of the complex f plane maps to the top Riemann sheet. Similarly, the left half of the complex f plane maps to the bottom Riemann sheet (see Figure 2.2.6). Since the points $z = ce^{-i\pi}$ and $z = ce^{i\pi}$, which denote the same point on the complex z plane, are assigned two different values of $f(z)$, as shown by the mappings of the points A and B, there is a discontinuity in $f(z)$ at $z = ce^{\pm i\pi}$. Moreover, this discontinuity is denoted by a branch cut. A branch cut always emanates from a branch point and ends at infinity or another branch point. Since $\Re e[f] = 0$ on this branch cut, this branch cut is also denoted as the $\Re e[f] = 0$ branch cut. Topologically, one may think of the top Riemann sheet on the complex z plane a consequence of stretching the right-half plane of the complex f plane, and sewing together the $\Re e[f] = 0$ line, forming a seam as shown in Figure 2.2.6.

Despite the aforementioned discontinuity, $f(z)$ can still be a continuous function of θ when θ varies continuously from $-\infty$ to ∞. For example, imagine that one walks on the top Riemann sheet. When θ varies from 0 to π, the branch cut is then encountered at $\theta = \pi$. But to maintain the continuity of $f(z)$, one needs to switch from the top Riemann sheet when $\theta = \pi^-$ to the lower Riemann sheet when $\theta = \pi^+$. Then, the values of $f(z)$ between these two points are now continuous. To maintain the continuity of $f(z)$ when θ varies from $3\pi^-$ to $3\pi^+$ (when $\pi < \theta < 3\pi$, one is in the lower Riemann sheet), again, one switches to the top Riemann sheet from the bottom Riemann sheet. Hence, for $3\pi < \theta < 5\pi$, the values of $f(z)$ are again the same as for $-\pi < \theta < \pi$ since they are on the same Riemann sheet. That is to say, the values of $f(z)$ for every value of z do not assume more than two values. On the other hand, for a multivalue function such as $\ln z$, which can assume infinitely many values for one point on the complex z plane, an infinite number of Riemann sheets is needed to define the function uniquely.

In the integral (31), the values of k_z have to be such that $\Im m[k_z] > 0$ over the path of integration in order to satisfy the radiation condition. Hence, it is expedient to define a top Riemann sheet on the complex k_ρ plane on which $\Im m[k_z] > 0$. The path of integration and the integrand are then unambiguously defined over this Riemann sheet. The corresponding bottom Riemann sheet is then defined by $\Im m[k_z] < 0$, and the branch cut that separates these sheets is defined by $\Im m[k_z] = 0$. To find such a branch cut, first assume that $k_0 = k_0' + ik_0''$, $k_\rho = k_\rho' + ik_\rho''$ so that k_0 and k_ρ are both complex numbers. Then,

$$k_z = (k_0^2 - k_\rho^2)^{\frac{1}{2}} = (k_0'^2 - k_0''^2 + 2ik_0'k_0'' - k_\rho'^2 + k_\rho''^2 - 2ik_\rho'k_\rho'')^{\frac{1}{2}}.$$

$$(2.2.32)$$

In order for k_z to be real, or $\Im m[k_z] = 0$, the conditions are that

$$k_\rho'k_\rho'' = k_0'k_0'', \qquad k_\rho'^2 - k_\rho''^2 < k_0'^2 - k_0''^2. \qquad (2.2.33)$$

Consequently, the above equations define rectangular hyperbolas that pass through the branch points $\pm k_0$ as shown in Figure 2.2.7. They define the branch cuts as shown.

The upper-half plane where $\Im m[k_z] > 0$ on the complex k_z plane maps to the upper Riemann sheet on the complex k_ρ plane. Furthermore, the real axis of the complex k_z plane defined by $\Im m[k_z] = 0$ maps to the branch cuts

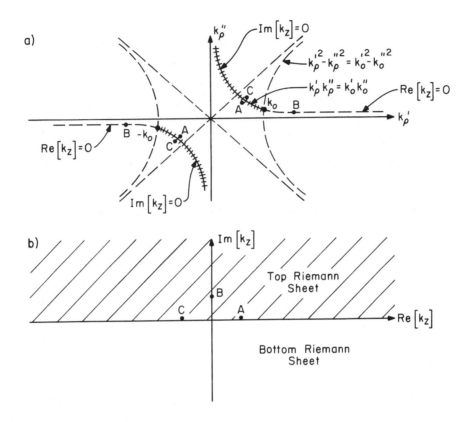

Figure 2.2.7 The definition of the $\Im m[k_z] = 0$ branch cuts.

on the complex k_ρ plane, and the imaginary axis of the complex k_z plane maps to the loci defined by $\Re e[k_z] = 0$ on the complex k_ρ plane. Because $k_z = (k_0^2 - k_\rho^2)^{\frac{1}{2}}$, there are two values of k_ρ, namely, $\pm k_\rho$, that correspond to each value of k_z. Therefore, the real k_z axis maps to two branch cuts while the imaginary k_z axis maps to two loci on the complex k_ρ plane, as shown in Figure 2.2.7.[5] Moreover, the points A, B, and C on the complex k_z plane map to the top Riemann sheet on the complex k_ρ plane as shown (see Exercise 2.11).

With this mapping, the transformation from the upper-half complex k_z-plane to the top Riemann sheet of the complex k_ρ-plane is clear—the first

---

[5] Another way of saying this is that since $k_\rho = (k_0^2 - k_z^2)^{\frac{1}{2}}$, i.e., k_ρ is a double-value function of k_z, there will be two values of k_ρ for every value of k_z. It is important to note that we have the symmetric property that k_ρ is a double-value function of k_z while k_z is also a double-value function of k_ρ.

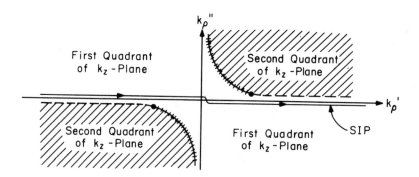

Figure 2.2.8 The mapping of the first and second quadrant in the upper-half complex k_z-plane onto the complex k_ρ-plane.

quadrant of the complex k_z-plane maps to the white area as shown in Figure 2.2.8. The second quadrant of the complex k_z-plane maps to the shaded area as shown. Therefore, with this choice of the branch cut, the Sommerfeld integration path is always in the first quadrant of the complex k_z-plane. Better still, in this quadrant, $\Re e[k_z] > 0$, $\Im m[k_z] > 0$, and the radiation condition is always ensured. This choice of the branch cut is sometimes referred to as the Sommerfeld branch cut. Alternatively, a $\Re e[k_z] = 0$ branch cut can also be chosen for this type of problem if one so wishes (Exercise 2.12). Finally, note that for (31), there exists a logarithmic singularity at the origin due to the Hankel function. The Sommerfeld integration path is always on the principal branch of this logarithmic singularity (see Subsection 2.2.2).

Another useful integral identity similar to (31) is (see Exercise 2.13)

$$\frac{e^{ik_0 r}}{r} = \frac{i}{2} \int\limits_{-\infty}^{\infty} dk_z \, e^{ik_z z} H_0^{(1)}(k_\rho \rho). \qquad (2.2.34)$$

The above is obtained by deforming the contour of integration from the Sommerfeld integration path to wrap around the branch cut. Alternatively, it can also be derived by Fourier transforming (19) in the z direction first, and obtaining the identity in a manner similar to that in Subsection 2.2.1.

In addition to using Fourier transforms to obtain integral identities, Hankel transforms sometimes can be used directly (Exercise 2.14).

§2.3 A Source on Top of a Layered Medium

Section 2.1 indicates that plane waves reflecting from a layered medium can be decomposed into TE-type plane waves, where $E_z = 0$, $H_z \neq 0$, and

TM-type plane waves, where $H_z = 0$, $E_z \neq 0$. One also sees how the field due to a point source can be expanded into plane waves in Section 2.2. In view of the above observations, when a point source is on top of a layered medium, it is then best to decompose its field in terms of waves of TE-type and TM-type. Then, the nonzero component of E_z characterizes TM waves, while the nonzero component of H_z characterizes TE waves. Hence, given a field, its TM and TE components can be extracted readily. Furthermore, if these TM and TE components are expanded in terms of plane waves, their propagations in a layered medium can be studied easily.

The problem of a vertical electric dipole on top of a half space was first solved by Sommerfeld (1909) using Hertzian potentials, which are related to the z components of the electromagnetic field. The work is later generalized to layered media, as discussed in the books listed at the beginning of the chapter. Later, Kong (1972) suggested the use of the z components of the electromagnetic field instead of the Hertzian potentials.

§§2.3.1 Electric Dipole Fields

The $\mathbf{E}$ field in a homogeneous medium due to a point current source directed in the $\hat{\alpha}$ direction, $\mathbf{J} = \hat{\alpha} I\ell\, \delta(\mathbf{r})$, is derivable via the vector potential method or the dyadic Green's function approach. Such a source is also known as a Hertzian dipole. Then, using the dyadic Green's function approach, the field due to a Hertzian dipole is given by [see Equation (1.3.50)]

$$\mathbf{E}(\mathbf{r}) = i\omega\mu \left(\overline{\mathbf{I}} + \frac{\nabla\nabla}{k^2} \right) \cdot \hat{\alpha} I\ell \, \frac{e^{ikr}}{4\pi r}, \tag{2.3.1}$$

where $I\ell$ is the current moment and $k = \omega\sqrt{\mu\epsilon}$, the wave number of the homogeneous medium. Furthermore, from $\nabla \times \mathbf{E} = i\omega\mu\mathbf{H}$, the magnetic field due to a Hertzian dipole is given by

$$\mathbf{H}(\mathbf{r}) = \nabla \times \hat{\alpha} I\ell \, \frac{e^{ikr}}{4\pi r}. \tag{2.3.2}$$

With the above fields, their TM and TE components can be derived easily.

(a) Vertical Electric Dipole (VED)

A vertical electric dipole shown in Figure 2.3.1 has $\hat{\alpha} = \hat{z}$; hence, the TM component of the field is characterized by

$$E_z = \frac{i\omega\mu I\ell}{4\pi k^2} \left(k^2 + \frac{\partial^2}{\partial z^2} \right) \frac{e^{ikr}}{r}, \tag{2.3.3a}$$

and the TE component of the field is characterized by

$$H_z = 0, \tag{2.3.3b}$$

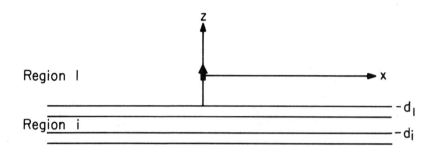

Figure 2.3.1 A vertical electric dipole over a layered medium.

implying the absence of the TE field. Next, using the Sommerfeld identity
(2.2.31) in the above, and after exchanging the order of integration and differentiation, we have[6]

$$E_z = \frac{-I\ell}{8\pi\omega\epsilon} \int\limits_{-\infty}^{\infty} dk_\rho \frac{k_\rho^3}{k_z} H_0^{(1)}(k_\rho\rho) e^{ik_z|z|}, \qquad (2.3.4)$$

after noting that $k_\rho^2 + k_z^2 = k^2$. Notice that now Equation (4) expands the z
component of the electric field in terms of cylindrical waves in the ρ direction
and a plane wave in the z direction. Since cylindrical waves actually are
linear superpositions of plane waves [see Equation (2.2.11)], the integrand in
(4) in fact consists of a linear superposition of TM-type plane waves. The
above is also the ***primary field*** generated by the source.

Consequently, for a VED on top of a stratified medium as shown, the
downgoing plane wave from the point source will be reflected like TM waves
with the generalized reflection coefficient $\tilde{R}_{12}^{TM}$. Hence, over a stratified
medium, the field in region 1 can be written as

$$E_{1z} = \frac{-I\ell}{8\pi\omega\epsilon_1} \int\limits_{-\infty}^{\infty} dk_\rho \frac{k_\rho^3}{k_{1z}} H_0^{(1)}(k_\rho\rho) \left[e^{ik_{1z}|z|} + \tilde{R}_{12}^{TM} e^{ik_{1z}z + 2ik_{1z}d_1} \right],$$
$$\qquad (2.3.5)$$

where $k_{1z} = (k_1^2 - k_\rho^2)^{\frac{1}{2}}$, and $k_1^2 = \omega^2\mu_1\epsilon_1$, the wave number in region 1.

The phase-matching condition dictates that the transverse variation of
the field in all the regions must be the same. Consequently, in the i-th

---

[6] By using (2.2.31) in (3a), the $\partial^2/\partial z^2$ operating on $e^{ik_z|z|}$ produces a Dirac delta function
singularity. We will postpone the discussion of this singularity until the chapter on
dyadic Green's function.

region, the solution becomes

$$\epsilon_i E_{iz} = \frac{-I\ell}{8\pi\omega} \int\limits_{-\infty}^{\infty} dk_\rho \frac{k_\rho^3}{k_{1z}} H_0^{(1)}(k_\rho\rho) A_i \left[e^{-ik_{iz}z} + \tilde{R}_{i,i+1}^{TM} e^{ik_{iz}z + 2ik_{iz}d_i} \right].$$
$$(2.3.6)$$

Notice that Equation (6) is now expressed in terms of $\epsilon_i E_{iz}$ because $\epsilon_i E_{iz}$ reflects and transmits like H_{iy}, the transverse component of the magnetic field or TM waves (see Section 2.1). Therefore, $\tilde{R}_{i,i+1}^{TM}$ can be calculated using (2.1.24), and A_i could be obtained using (2.1.26) (see Exercise 2.15).

This completes the derivation of the integral representation of the electric field everywhere in the stratified medium. These integrals are sometimes known as Sommerfeld integrals. The case when the source is embedded in a layered medium can be derived similarly (Exercise 2.16, Section 2.4).

(b) Horizontal Electric Dipole (HED)

For a horizontal electric dipole pointing in the x direction, $\hat{\alpha} = \hat{x}$; hence, (1) and (2) give the TM and the TE components as

$$E_z = \frac{iI\ell}{4\pi\omega\epsilon} \frac{\partial^2}{\partial z \partial x} \frac{e^{ikr}}{r}, \qquad (2.3.7a)$$

$$H_z = -\frac{I\ell}{4\pi} \frac{\partial}{\partial y} \frac{e^{ikr}}{r}. \qquad (2.3.7b)$$

Then, with the Sommerfeld identity (2.2.31), we can expand the above as

$$E_z = \pm \frac{iI\ell}{8\pi\omega\epsilon} \cos\phi \int\limits_{-\infty}^{\infty} dk_\rho\, k_\rho^2 H_1^{(1)}(k_\rho\rho) e^{ik_z|z|}, \qquad (2.3.8a)$$

$$H_z = i\frac{I\ell}{8\pi} \sin\phi \int\limits_{-\infty}^{\infty} dk_\rho \frac{k_\rho^2}{k_z} H_1^{(1)}(k_\rho\rho) e^{ik_z|z|}. \qquad (2.3.8b)$$

Now, Equation (8a) represents the wave expansion of the TM field, while (8b) represents the wave expansion of the TE field. Observe that because E_z is odd about $z = 0$ in (8a), the downgoing wave has an opposite sign from the upgoing wave. At this point, the above are just the primary field generated by the source.

On top of a stratified medium, the downgoing wave is reflected accordingly, depending on its wave type. Consequently, we have

$$E_{1z} = \frac{iI\ell}{8\pi\omega\epsilon_1} \cos\phi \int_{-\infty}^{\infty} dk_\rho \, k_\rho^2 H_1^{(1)}(k_\rho\rho) \left[\pm e^{ik_{1z}|z|} - \tilde{R}_{12}^{TM} e^{ik_{1z}(z+2d_1)} \right],$$

(2.3.9a)

$$H_{1z} = \frac{iI\ell}{8\pi} \sin\phi \int_{-\infty}^{\infty} dk_\rho \, \frac{k_\rho^2}{k_{1z}} H_1^{(1)}(k_\rho\rho) \left[e^{ik_{1z}|z|} + \tilde{R}_{12}^{TE} e^{ik_{1z}(z+2d_1)} \right].$$

(2.3.9b)

Notice that the negative sign in front of $\tilde{R}_{12}^{TM}$ in (9a) follows because the downgoing wave in the primary field has a negative sign.

§§2.3.2 Magnetic Dipole Fields

For the sake of completeness, we shall also give the expressions due to point magnetic dipoles. A point magnetic dipole can be simulated by a small electric current loop antenna. Furthermore, the field it generates is dual to that of a Hertzian dipole. The dual of an electric dipole with moment P is $\mu I A$ where IA is the magnetic moment of a loop of area A carrying a current I. But for a Hertzian dipole, $I\ell = \ell \, dQ/dt = -i\omega Q\ell = -i\omega P$. Hence, the replacement of $I\ell \rightarrow -i\omega\mu I A$, and the consequent use of duality will yield the magnetic dipole results.

(a) Vertical Magnetic Dipole (VMD)

A vertical magnetic dipole on top of a layered medium produces only a TE field. Consequently, we have

$$H_{1z} = -\frac{iIA}{8\pi} \int_{-\infty}^{\infty} dk_\rho \, \frac{k_\rho^3}{k_{1z}} H_0^{(1)}(k_\rho\rho) \left[e^{ik_{1z}|z|} + \tilde{R}_{12}^{TE} e^{ik_{1z}z+2ik_{1z}d_1} \right],$$

(2.3.10a)

$$E_{1z} = 0,$$

(2.3.10b)

where A is the area of the small electric current loop simulating a magnetic dipole.

(b) Horizontal Magnetic Dipole (HMD)

For a horizontal magnetic dipole pointing in the x direction, the expressions for the fields are

$$H_{1z} = \frac{-IA}{8\pi} \cos\phi \int_{-\infty}^{\infty} dk_\rho \, k_\rho^2 H_1^{(1)}(k_\rho\rho) \left[\pm e^{ik_{1z}|z|} - \tilde{R}_{12}^{TE} e^{ik_{1z}z+2ik_{1z}d_1} \right],$$

(2.3.11a)

$$E_{1z} = \frac{\omega\mu IA}{8\pi} \sin\phi \int_{-\infty}^{\infty} dk_\rho \frac{k_\rho^2}{k_{1z}} H_1^{(1)}(k_\rho\rho) \left[e^{ik_{1z}|z|} + \tilde{R}_{12}^{TM} e^{ik_{1z}z+2ik_{1z}d_1} \right].$$

(2.3.11b)

The field in each region for each of the above cases can be derived by a method similar to (6). This follows from the fact that the propagation of the

TE and TM waves through a layered medium are completely decoupled from each other—they are only coupled at the source.

The above are field solutions written in cylindrical coordinates. When the solutions in Cartesian coordinates are required, we need only use the Weyl identity instead of the Sommerfeld identity in (1) and (2) (see Exercise 2.17). Finally, given the field due to a point source, the field due to a sheet source can be easily obtained by convolution (Exercises 2.17, 2.27).

§§2.3.3 The Transverse Field Components

Given the integral representations of the z components of the electromagnetic field, the integral representation of the other components of the electromagnetic fields can also be derived in each of the homogeneous layers (Kong 1972). In general, the field in a homogeneous layer is of the form

$$\mathbf{E}(\mathbf{r}) = \int_{-\infty}^{\infty} dk_\rho\, \tilde{\mathbf{E}}(k_\rho, \mathbf{r}), \qquad (2.3.12a)$$

$$\mathbf{H}(\mathbf{r}) = \int_{-\infty}^{\infty} dk_\rho\, \tilde{\mathbf{H}}(k_\rho, \mathbf{r}). \qquad (2.3.12b)$$

In the above, each spectral component, $\tilde{\mathbf{E}}(k_\rho, \mathbf{r})$ or $\tilde{\mathbf{H}}(k_\rho, \mathbf{r})$, is also a solution of Maxwell's equations. Consequently, by writing

$$\tilde{\mathbf{E}}(k_\rho, \mathbf{r}) = \tilde{\mathbf{E}}_s(k_\rho, \mathbf{r}) + \hat{z}\tilde{E}_z(k_\rho, \mathbf{r}), \qquad (2.3.13a)$$

$$\tilde{\mathbf{H}}(k_\rho, \mathbf{r}) = \tilde{\mathbf{H}}_s(k_\rho, \mathbf{r}) + \hat{z}\tilde{H}_z(k_\rho, \mathbf{r}), \qquad (2.3.13b)$$

and letting $\nabla = \nabla_s + \hat{z}\frac{\partial}{\partial z}$, where the subscript s indicates transverse to the z components of the vector, Maxwell's equations become

$$\left(\nabla_s + \hat{z}\frac{\partial}{\partial z}\right) \times (\tilde{\mathbf{E}}_s + \hat{z}\tilde{E}_z) = i\omega\mu\,(\tilde{\mathbf{H}}_s + \hat{z}\tilde{H}_z), \qquad (2.3.14a)$$

$$\left(\nabla_s + \hat{z}\frac{\partial}{\partial z}\right) \times (\tilde{\mathbf{H}}_s + \hat{z}\tilde{H}_z) = -i\omega\epsilon\,(\tilde{\mathbf{E}}_s + \hat{z}\tilde{E}_z), \qquad (2.3.14b)$$

where μ and ϵ are constants. Moreover, after equating the transverse components in (14a) and (14b), we have

$$\nabla_s \times \hat{z}\tilde{E}_z + \frac{\partial}{\partial z}\hat{z} \times \tilde{\mathbf{E}}_s = i\omega\mu\,\tilde{\mathbf{H}}_s, \qquad (2.3.15a)$$

$$\nabla_s \times \hat{z}\tilde{H}_z + \frac{\partial}{\partial z}\hat{z} \times \tilde{\mathbf{H}}_s = -i\omega\epsilon\,\tilde{\mathbf{E}}_s. \qquad (2.3.15b)$$

Then, on taking $\frac{\partial}{\partial z}\hat{z}\times$(15a) and eliminating $\frac{\partial}{\partial z}\hat{z}\times\tilde{\mathbf{H}}_s$ from the resultant equation with (15b), we obtain

$$\frac{\partial}{\partial z}\hat{z}\times\nabla_s\times\hat{z}\tilde{E}_z + \frac{\partial^2}{\partial z^2}\hat{z}\times\hat{z}\times\tilde{\mathbf{E}}_s = \omega^2\mu\epsilon\,\tilde{\mathbf{E}}_s - i\omega\mu\nabla_s\times\hat{z}\tilde{H}_z.$$
$$(2.3.16)$$

Since $\tilde{\mathbf{E}}_s$ in general has $e^{\pm ik_z z}$ dependence, $\partial^2/\partial z^2 \to -k_z^2$ in the above. Consequently, after simplifying the above with the appropriate vector identity, we have

$$\tilde{\mathbf{E}}_s(k_\rho,\mathbf{r}) = \frac{1}{\omega^2\mu\epsilon - k_z^2}\left[\nabla_s\frac{\partial\tilde{E}_z}{\partial z} - i\omega\mu\,\hat{z}\times\nabla_s\tilde{H}_z\right].$$
$$(2.3.17a)$$

Furthermore, from duality, it follows that

$$\tilde{\mathbf{H}}_s(k_\rho,\mathbf{r}) = \frac{1}{\omega^2\mu\epsilon - k_z^2}\left[\nabla_s\frac{\partial\tilde{H}_z}{\partial z} + i\omega\epsilon\,\hat{z}\times\nabla_s\tilde{E}_z\right].$$
$$(2.3.17b)$$

It is important to emphasize that the quantities in (17a) and (17b) are the integrands of the integral representations of the field. Hence, the equations are to be applied to each spectral component individually, but not to the integral as a whole (see Exercise 2.15).

§2.4 A Source Embedded in a Layered Medium

From the above analysis, it is apparent that the problem of a source on top of a layered medium is equivalent to a one-dimensional problem. This is even the case when a source is embedded in a layered medium. Consider first a point source radiating in an unbounded medium, the z variation of the solution, as exemplified by (2.3.4) and (2.3.8), is of the form

$$F(z,z') = e^{ik_z|z-z'|}.$$
$$(2.4.1)$$

If, instead, the source is now in region m as shown in Figure 2.4.1, then the z variation of the solution in region m is augmented by an upgoing wave plus a downgoing wave, namely,

$$F(z,z') = e^{ik_{mz}|z-z'|} + B_m e^{-ik_{mz}z} + D_m e^{ik_{mz}z}.$$
$$(2.4.2)$$

The last two terms are due to reflected waves at the boundaries $z = -d_{m-1}$ and $z = -d_m$. To find B_m and D_m, constraint conditions can be derived for the waves at $z = -d_{m-1}$ and $z = -d_m$.

The downgoing wave for $z > z'$ is a consequence of the reflection of the upgoing wave for $z > z'$, at $z = -d_{m-1}$. Mathematically, this relationship can be expressed as

$$B_m e^{ik_{mz}d_{m-1}} = \tilde{R}_{m,m-1}\left[e^{ik_{mz}|d_{m-1}+z'|} + D_m e^{-ik_{mz}d_{m-1}}\right].$$
$$(2.4.3)$$

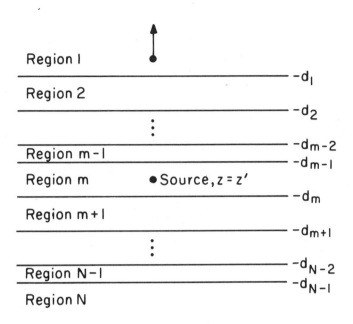

Figure 2.4.1 A source embedded in a multilayered medium.

Similarly, for $z < z'$, a relationship can be written between the upgoing wave and the downgoing wave at $z = -d_m$, namely,

$$D_m e^{-ik_{mz}d_m} = \tilde{R}_{m,m+1} \left[e^{ik_{mz}|d_m+z'|} + B_m e^{ik_{mz}d_m} \right]. \qquad (2.4.4)$$

Then, solving for B_m and D_m from (3) and (4) yields

$$B_m e^{ik_{mz}d_{m-1}} = \tilde{R}_{m,m-1} \left[e^{ik_{mz}|d_{m-1}+z'|} + e^{ik_{mz}(d_m-d_{m-1})} \tilde{R}_{m,m+1} e^{ik_{mz}|d_m+z'|} \right] \tilde{M}_m,$$
$$(2.4.5a)$$

$$D_m e^{-ik_{mz}d_m} = \tilde{R}_{m,m+1} \left[e^{ik_{mz}|d_m+z'|} + e^{ik_{mz}(d_m-d_{m-1})} \tilde{R}_{m,m-1} e^{ik_{mz}|d_{m-1}+z'|} \right] \tilde{M}_m,$$
$$(2.4.5b)$$

where $\tilde{M}_m = \left[1 - \tilde{R}_{m,m+1}\tilde{R}_{m,m-1}e^{2ik_{mz}(d_m-d_{m-1})} \right]^{-1}$.

After substituting (5) back into (2), and bringing all the terms under a common denominator, we can show that (2) becomes (Exercise 2.18)

$$F_+(z,z') = \left[e^{-ik_{mz}z'} + e^{ik_{mz}(z'+2d_m)} \tilde{R}_{m,m+1} \right]$$
$$\cdot \left[e^{ik_{mz}z} + e^{-ik_{mz}(z+2d_{m-1})} \tilde{R}_{m,m-1} \right] \tilde{M}_m, \quad z > z', \qquad (2.4.6a)$$

$$F_-(z,z') = \left[e^{ik_{mz}z'} + e^{-ik_{mz}(z'+2d_{m-1})} \tilde{R}_{m,m-1} \right]$$
$$\cdot \left[e^{-ik_{mz}z} + e^{ik_{mz}(z+2d_m)} \tilde{R}_{m,m+1} \right] \tilde{M}_m, \quad z < z'. \qquad (2.4.6b)$$

Note that $F_+(z, z') = F_-(z', z)$, implying the satisfaction of reciprocity.

Furthermore, if the field in region $n < m$ is desired, the recursive method of Subsection 2.1.3 can be used to find the field in region n. Hence, for region $n < m$, the field could be written as

$$A_n^+ \left[e^{ik_{nz}z} + \tilde{R}_{n,n-1} e^{-2ik_{nz}d_{n-1}-ik_{nz}z} \right]. \qquad (2.4.7)$$

Moreover, using the fact that

$$A_i^+ e^{-ik_{iz}d_i} = A_{i+1}^+ e^{-ik_{i+1,z}d_i} \frac{T_{i+1,i}}{1 - R_{i,i+1}\tilde{R}_{i,i-1} e^{2ik_{iz}(d_i - d_{i-1})}}$$
$$= A_{i+1}^+ e^{-ik_{i+1,z}d_i} S_{i+1,i}^+, \qquad (2.4.8)$$

an equation derived similar to (2.1.26), we can recursively relate the upgoing wave amplitude in region n to the upgoing wave amplitude in region m at $z = -d_{m-1}$, which is obtained from (6a) as

$$A_m^+ = \left[e^{-ik_{mz}z'} + e^{ik_{mz}(z'+2d_m)} \tilde{R}_{m,m+1} \right] \tilde{M}_m. \qquad (2.4.9)$$

Similarly, for region $n > m$, the field is

$$A_n^- \left[e^{-ik_{nz}z} + \tilde{R}_{n,n+1} e^{2ik_{nz}d_n+ik_{nz}z} \right]. \qquad (2.4.10)$$

A recursive relation similar to (8) is then

$$A_i^- e^{ik_{iz}d_i} = A_{i-1}^- e^{ik_{i-1,z}d_{i-1}} S_{i-1,i}^-, \qquad (2.4.11)$$

where $S_{i-1,i}^-$ is the same as $S_{i-1,i}$ defined in (2.1.26a), and A_i^- is the amplitude of the downgoing wave in region $i > m$. Consequently, using (11), this downgoing wave amplitude can be recursively related to that in region m at $z = -d_m$, which, from (6b), is

$$A_m^- = \left[e^{ik_{mz}z'} + e^{-ik_{mz}(z'+2d_{m-1})} \tilde{R}_{m,m-1} \right] \tilde{M}_m. \qquad (2.4.12)$$

Yet another method exists for finding A_n^+ or A_n^- in region n using the generalized transmission coefficient $\tilde{T}_{mn}$ defined in (2.1.28). For example, it can be used to determine A_n^+ by imposing a constraint condition at $z = -d_{m-1}$. To do so, note that at $z = -d_n$, A_n^+ is a consequence of the transmission of the upgoing wave at region m at $z = -d_{m-1}$ given by (9) plus a reflection of the downgoing wave in region n, namely,

$$A_n^+ e^{-ik_{nz}d_n} = \tilde{T}_{mn} A_m^+ e^{-ik_{mz}d_{m-1}} + A_n^+ \tilde{R}_{n,n+1}\tilde{R}_{n,n-1} e^{-2ik_{nz}d_{n-1}+ik_{nz}d_n}. \qquad (2.4.13)$$

Then, solving for A_n^+ in terms of A_m^+ given by (9) yields

$$A_n^+ = e^{ik_{nz}d_n} \tilde{T}_{mn} e^{-ik_{mz}d_{m-1}} \left[e^{-ik_{mz}z'} + e^{ik_{mz}(z'+2d_m)} \tilde{R}_{m,m+1} \right] \tilde{M}_m \tilde{M}_n. \qquad (2.4.14)$$

Consequently, the field in region n at z due to a source in region m at z' is

$$F_+(z, z') = \left[e^{ik_{nz}z} + e^{-ik_{nz}(z+2d_{n-1})}\tilde{R}_{n,n-1}\right] e^{ik_{nz}d_n}\tilde{T}_{mn}e^{-ik_{mz}d_{m-1}}$$
$$\cdot \left[e^{-ik_{mz}z'} + e^{ik_{mz}(z'+2d_m)}\tilde{R}_{m,m+1}\right] \tilde{M}_m\tilde{M}_n,$$
$$z \in \text{region } n, \quad z' \in \text{region } m, \quad n < m. \tag{2.4.15}$$

This form of the solution has the advantage of being more symmetrical than those given in the literature (Kong 1986, p. 309).

In a similar manner, A_n^- for $n > m$ can be determined, yielding

$$F_-(z, z') = \left[e^{-ik_{nz}z} + e^{ik_{nz}(z+2d_n)}\tilde{R}_{n,n+1}\right] e^{-ik_{nz}d_{n-1}}\tilde{T}_{mn}e^{ik_{mz}d_m}$$
$$\cdot \left[e^{ik_{mz}z'} + e^{-ik_{mz}(z'+2d_{m-1})}\tilde{R}_{m,m-1}\right] \tilde{M}_m\tilde{M}_n,$$
$$z \in \text{region } n, \quad z' \in \text{region } m, \quad n > m. \tag{2.4.16}$$

From the above, we can further show that $\frac{\mu_m}{k_{mz}}F_+(z, z') = \frac{\mu_n}{k_{nz}}F_-(z', z)$ where $z' \in R_m$, $z \in R_n$, and R_i stands for region i (see Exercise 2.18).

Some sources generate a field that is odd-symmetric about z' in a homogeneous medium. So, instead of (1), the field is

$$F(z, z') = \pm e^{ik_z|z-z'|}, \tag{2.4.17}$$

The solution for this case can be obtained from the above solution by differentiating with respect to z'.

§2.5 Asymptotic Expansions of Integrals

Even though the solutions of a source over or embedded in a layered medium have been expressed in terms of Fourier-type integrals, not much physical insight into these integrals is gained other than that they are integral summations of transmitted and reflected plane or cylindrical waves. These integrals can, however, be subject to approximations, giving rise to simpler expressions. These simpler expressions offer a better understanding of these waves. They are the asymptotic approximations that can be derived via mathematical techniques which we shall describe next.

A number of these mathematical techniques are also discussed by many authors (Brekhovskikh 1960; Tyras 1969; Wait 1970, 1971; Yeh and Liu 1972; Kong 1975, 1986). Furthermore, many scientists have also used such techniques to obtain better physical interpretation of such integrals. Some of these works are listed in the references.

§§2.5.1 Method of Stationary Phase

The method of stationary phase is useful in deriving the leading-order approximation to an integral whose integrand is rapidly oscillating. Such

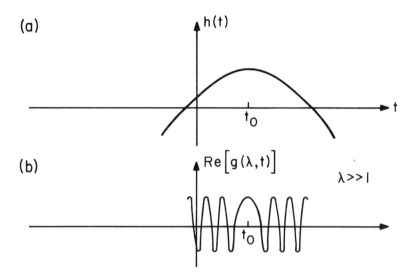

Figure 2.5.1 A rapid oscillation of the integrand as a function of t because λ is large.

rapid oscillations render the numerical integration of the integral difficult. In this case, an asymptotic approximation may be the best way to obtain numerical values for the integral.

As an example, consider an integral

$$I = \int_{-\infty}^{\infty} dt\, f(t)\, g(\lambda, t), \qquad (2.5.1)$$

where $g(\lambda, t)$ can be integrated in closed form, but not $f(t)g(\lambda, t)$. Furthermore, if

$$g(\lambda, t) \sim e^{i\lambda h(t)}, \quad \lambda \to \infty, \qquad (2.5.2)$$

("$\sim$" means "asymptotic to") then when λ is large, $g(\lambda, t)$ is a rapidly oscillating function of h, and hence, of t, where $h(t)$ may be a function of t as shown in Figure 2.5.1a. As a result, the real part of the function $g(\lambda, t)$ will look like that shown in Figure 2.5.1b. Also, the imaginary part of $g(\lambda, t)$ has the same feature. If $h(t)$ has a stationary point at $t = t_0$, i.e.,

$$h'(t_0) = 0, \qquad (2.5.3)$$

it varies least slowly as a function of t at $t = t_0$; hence, $g(\lambda, t)$ is least rapidly oscillating at $t = t_0$. Furthermore, we assume that when $\lambda \to \infty$, $f(t)$ is a more slowly varying function compared to $g(\lambda, t)$.

If the integrand is rapidly oscillating, the contribution to the integration is small because of the cancellation of the positive and negative parts of the integrand. However, this cancellation is least at t_0 where the integrand is least rapidly oscillating. Therefore, most of the contribution to the integration will come from the neighborhood of t_0. Consequently, the integral can be approximated as (see Chew 1988)

$$I \sim f(t_0) \int_{-\infty}^{\infty} dt\, g(\lambda, t), \quad \lambda \to \infty, \qquad (2.5.4)$$

because $f(t)$ is approximately a constant in the neighborhood of t_0. Moreover, since $g(\lambda, t)$ has a closed-form solution, a simple algebraic approximation is obtained for I. This approximation is asymptotic in the sense that it becomes better when λ becomes larger.

As an example, let us consider the asymptotic approximation of the integral (2.3.4)

$$I = \int_{-\infty}^{\infty} dk_\rho \frac{k_\rho^3}{k_z} H_0^{(1)}(k_\rho \rho) e^{ik_z |z|}. \qquad (2.5.5)$$

Note that the integrand resembles the integrand of the Sommerfeld identity. Furthermore, when ρ and $z \to \infty$, the integrand becomes a rapidly oscillating function of k_ρ, because

$$H_0^{(1)}(k_\rho \rho) \sim \sqrt{\frac{2}{\pi k_\rho \rho}} e^{ik_\rho \rho - i\frac{\pi}{4}}, \quad \rho \to \infty, \qquad (2.5.6)$$

and $e^{ik_z |z|}$ are both rapidly oscillating functions of k_ρ. Hence, the rapidly oscillating part of the integrand becomes

$$H_0^{(1)}(k_\rho \rho) e^{ik_z |z|} \sim \sqrt{\frac{2}{\pi k_\rho \rho}} e^{-i\frac{\pi}{4} + ik_\rho \rho + ik_z |z|}, \qquad \rho, z \to \infty. \qquad (2.5.7)$$

The stationary-phase point, i.e., the point of least rapid oscillation, is found from

$$\frac{d}{dk_\rho} [k_\rho \rho + k_z |z|] = 0. \qquad (2.5.8)$$

Then, using the fact that $k_z = (k^2 - k_\rho^2)^{\frac{1}{2}}$, the stationary-phase point is found at

$$k_{\rho s} = k \frac{\rho}{(\rho^2 + z^2)^{\frac{1}{2}}} = k \sin \theta, \qquad (2.5.9)$$

and $\theta = \sin^{-1}(\rho/r)$, $r = (\rho^2 + z^2)^{\frac{1}{2}}$. Note that θ is the observation angle in the spherical coordinates. Moreover, in (5), $k_\rho^3 \simeq k_{\rho s}^2 k_\rho$ around the stationary-phase point, and it can be rewritten

$$I \sim k_{\rho s}^2 \int_{-\infty}^{\infty} dk_\rho \frac{k_\rho}{k_z} H_0^{(1)}(k_\rho \rho) e^{ik_z z} = \frac{2k^2 \sin^2 \theta}{ir} e^{ikr}, \quad r \to \infty. \qquad (2.5.10)$$

The latter equality is a consequence of the Sommerfeld identity.

In fact, Equation (2.3.4) has a closed-form expression given by (2.3.3a). The above is only the leading-order approximation of (2.3.3a) when $r \to \infty$. The method of stationary phase offers a quick way to obtain a leading-order approximation of an integral. For example, it can be used to derive the approximations for (2.3.9a) and (2.3.9b) (see Exercise 2.19). But to obtain higher-order approximations, we use the method of steepest descent. The method of stationary phase physically implies that only the spectra in the vicinity of the stationary phase point interfere constructively to contribute to the field at an observation point.

§§2.5.2 *Method of Steepest Descent*

(a) Infinite Integrals

The method of steepest descent, or the saddle-point method, deals with the asymptotic expansion of an integral of the kind (see also Erdelyi 1956; Copson 1965)

$$I = \int_C e^{\lambda h(t)} f(t) \, dt, \qquad (2.5.11)$$

where C is an infinite contour on a complex plane. The stationary-phase point in this case may be complex; hence, the intuitive argument of the stationary phase method may not apply here.

If $h(t)$ has a stationary point at $t = t_0$, implying $h'(t_0) = 0$, then,

$$h(t) - h(t_0) \approx (t - t_0)^2 \frac{h''(t_0)}{2} + \cdots, \quad t \to t_0, \qquad (2.5.12)$$

i.e., $h(t) - h(t_0)$ is quadratic around the stationary point. This then suggests a change of variable to s such that

$$-s^2 = h(t) - h(t_0). \qquad (2.5.13)$$

Note that Equation (13) performs a mapping from the complex t plane to the complex s plane (see Figure 2.5.2). Moreover, the point t_0 is mapped to $s = 0$ on the complex s plane, while the contour C on the t plane maps to the contour C' on the complex s plane.

On the real axis of s, the function $h(t) - h(t_0)$ is purely real and negative; hence,

$$e^{\lambda h(t)} = e^{-\lambda s^2 + \lambda h(t_0)}. \qquad (2.5.14)$$

Notice that on this path, which corresponds to the path P on the complex t plane, $e^{\lambda h(t)}$ has a constant phase. Hence, P is also called the constant-phase path. Furthermore, $e^{\lambda h(t)}$ has a maximum at $s = 0$, or $t = t_0$, and is exponentially small on the real axis away from $s = 0$, or away from $t = t_0$ on the constant phase path P. But on the imaginary axis on the complex s plane, $e^{-\lambda s^2}$ becomes exponentially large, and so does $e^{\lambda h(t)}$. Therefore, the function $e^{\lambda h(t)}$, when $\lambda \to \infty$, looks like a saddle at the point $t = t_0$ on the complex t plane and $s = 0$ on the complex s plane respectively. Hence, this point is also known as the **saddle point**. The constant-phase path on which

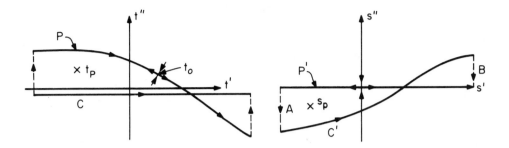

Figure 2.5.2 The mapping from the complex t plane to the complex s plane.

the function $e^{\lambda h(t)}$ descends steeply away from the saddle point is also known as the **steepest-descent path**.

With the change of variable in (13), (11) becomes

$$I = e^{\lambda h(t_0)} \int_{C'} e^{-\lambda s^2} F(s)\, ds, \qquad (2.5.15)$$

where

$$F(s) = f(t) \frac{dt}{ds}. \qquad (2.5.16)$$

The path of integration can be further deformed from C' to the real axis provided that the contributions to the integral over the paths A and B in Figure 2.5.2 are vanishingly small at infinity by invoking Jordan's lemma. However, in deforming the contour, the contribution due to any singularity enclosed between the paths C' and P' needs to be included. Hence, (15) becomes

$$I = e^{\lambda h(t_0)} \int_{-\infty}^{\infty} e^{-\lambda s^2} F(s)\, ds + I_p, \qquad (2.5.17)$$

where I_p is the contribution from the singularities enclosed between the contours C' and P'.

More specifically, the saddle-point contribution to I is

$$I_s = e^{\lambda h(t_0)} \int_{-\infty}^{\infty} e^{-\lambda s^2} F(s)\, ds. \qquad (2.5.18)$$

Observe that when $\lambda \to \infty$, $e^{-\lambda s^2}$ is exponentially small away from $s = 0$ on the path of integration. This implies that most of the contribution to the integral is from around $s = 0$. Hence, $F(s)$ can be expanded into a Taylor series about the origin to obtain

$$I_s \sim e^{\lambda h(t_0)} \int_{-\infty}^{\infty} e^{-\lambda s^2} \sum_{n=0}^{\infty} s^n F^{(n)}(0)/n!\, ds. \qquad (2.5.19)$$

The above is only an approximation because a Taylor series usually has a finite radius of convergence, while the integration above is over an infinite range.

Consequently, by integrating the above term by term, and applying the formulas

$$\int_{-\infty}^{\infty} s^{2m+1} e^{-\lambda s^2}\, ds = 0, \ m = 0, 1, 2, \cdots \qquad (2.5.20a)$$

$$\int_{-\infty}^{\infty} s^{2m} e^{-\lambda s^2}\, ds = \frac{(2m)!}{m!\, 2^{2m} \lambda^m} \sqrt{\frac{\pi}{\lambda}}, \ m = 0, 1, 2, \cdots \qquad (2.5.20b)$$

[which is obtainable by converting the above into a gamma function integral (Abramowitz and Stegun 1965; also see Exercise 2.20)], we obtain the asymptotic expansion of I_s as

$$I_s \sim e^{\lambda h(t_0)} \sqrt{\frac{\pi}{\lambda}} \left[F(0) + \frac{F''(0)}{4\lambda} + \frac{F^{(iv)}(0)}{32\lambda^2} + \cdots \right], \ \lambda \to \infty. \qquad (2.5.21)$$

Furthermore, from (12), (13), and (16), one can show that $\frac{dt}{ds}\big|_{s=0} = \lim_{s \to 0} \frac{t-t_0}{s} = \sqrt{\frac{-2}{h''(t_0)}}$, and

$$F(0) = f(t_0) \sqrt{\frac{-2}{h''(t_0)}}. \qquad (2.5.22)$$

Hence, the leading-order approximation to (21) is

$$I_s \sim e^{\lambda h(t_0)} f(t_0) \sqrt{\frac{-2\pi}{\lambda h''(t_0)}}, \ \lambda \to \infty, \qquad (2.5.23)$$

which usually agrees with the approximation given by the stationary phase method. The above method is also known as the saddle-point method.

(b) Semi-Infinite Integrals

For a semi-infinite integral,

$$I = \int_C e^{\lambda h(t)} f(t)\, dt, \qquad (2.5.24)$$

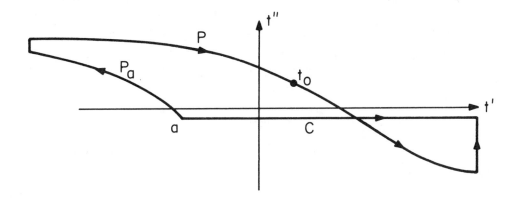

Figure 2.5.3 The case for a semi-infinite integral. P_a is the steepest-descent path passing through a while P is the steepest-descent path passing through the saddle point at t_0.

where C is as shown in Figure 2.5.3, the complete path of integration may not be deformed to pass through the saddle point. The path, however, may be deformed through a constant-phase path that passes through a, i.e., P_a plus a path P that passes through the saddle point at t_0. Assuming that no singularity is enclosed in the course of deforming the integration path, we then have

$$I = \int_{P_a} e^{\lambda h(t)} f(t)\, dt + \int_P e^{\lambda h(t)} f(t)\, dt. \qquad (2.5.25)$$

The second integral above, which is an infinite integral, can be evaluated by the saddle-point method as described previously. To evaluate the first integral, which is a semi-infinite integral, we perform the following transformation:

$$-s = h(t) - h(a). \qquad (2.5.26)$$

Unlike (13), a linear transformation is chosen because $h(t)-h(a) \approx (t-a)h'(a)$ when $t \to a$; namely, $h(t) - h(a)$ is linear around a. Then, the first integral becomes

$$I_a = e^{\lambda h(a)} \int_0^\infty e^{-\lambda s} s^\alpha F(s)\, ds, \qquad (2.5.27)$$

where $-1 < \alpha < 1$ and

$$s^\alpha F(s) = f(t)\frac{dt}{ds}. \qquad (2.5.28)$$

In (27), the possibility of a factorizable algebraic branch point at $t = a$ is allowed, i.e., when $t \to a$, we assume that $f(t) \sim (t - a)^\alpha B$. In such a case, transformation (26) leads to an integrand of the form as shown. Moreover, the positive real axis of the s plane maps onto the path P_a on the complex t plane since P_a is the constant-phase path. On the real s axis, $h(t) - h(a)$ is purely real from (2b), and hence, $e^{\lambda h(t)}$ has only a constant phase.

When $\lambda \to \infty$, most of the contributions to the integral would come from around $s = 0$. Hence, one can Taylor series expand $F(s)$ about $s = 0$, and obtain a series expansion for I_a as in (19), yielding

$$I_a \sim e^{\lambda h(a)} \int_0^\infty e^{-\lambda s} \sum_{n=0}^\infty s^{\alpha+n} F^{(n)}(0)/n! \, ds. \qquad (2.5.29)$$

Then, using the fact that

$$\int_0^\infty e^{-\lambda s} s^{n+\alpha} ds = \frac{\Gamma(n+1+\alpha)}{\lambda^{n+1+\alpha}}, \qquad (2.5.30)$$

where $\Gamma(x)$ is a gamma function, we have

$$I_a \sim e^{\lambda h(a)} \left[F(0) \frac{\Gamma(1+\alpha)}{\lambda^{1+\alpha}} + F'(0) \frac{\Gamma(2+\alpha)}{\lambda^{2+\alpha}} + F''(0) \frac{\Gamma(3+\alpha)}{2\lambda^{3+\alpha}} + \cdots \right], \quad \lambda \to \infty, \qquad (2.5.31)$$

which is an asymptotic expansion of I_a when $\lambda \to \infty$ (see Exercise 2.21).

Since $f(t) \sim (t-a)^\alpha B$, when $t \to a$, it can be shown that

$$\frac{dt}{ds}\bigg|_{s=0} = \lim_{t \to a} \frac{t-a}{s} = -\frac{1}{h'(a)}$$

which can be used to show that

$$F(0) = \left[\frac{-1}{h'(a)} \right]^{\alpha+1} B, \qquad (2.5.32)$$

where B and α are derived from the fact that $f(t) \sim (t-a)^\alpha B$, when $t \to a$. The above can then be used to find the leading-order term in (31).

A point is in order on the convergence of the series (21) and (31). These series are usually not convergent for the following reason. As mentioned before, when $F(s)$ is replaced by its Taylor series approximation, the Taylor series of a function usually has a finite radius of convergence. Hence, the Taylor series representation is only good over a finite domain in s. But the paths of integration in both (19) and (29) are over infinite domains. Therefore, the right-hand sides of (19) and (29) are not rigorously valid, and these series, in general, diverge. Such a series, which is characteristic of an asymptotic expansion, is also known as a *semiconvergent series*.

How would one know how many terms to include in the approximations? To gain familiarity with this, one examines the integrand of the n-th term in (19). Note that in the n-th term of the integrand, $s^n e^{-\lambda s^2}$ peaks around $s = \sqrt{\frac{n}{2\lambda}}$. Therefore, as the higher order terms in the series are computed, the contribution comes from points farther and farther away from the origin. If the radius of convergence of the Taylor series is R, then in order for the n-th term to be effective in the approximation, we require that

$$\sqrt{\frac{n}{2\lambda}} \ll R. \tag{2.5.33}$$

The radius of convergence of the Taylor series is given by the distance of the nearest singularity to the origin. Hence, the criterion (33) could also be written as

$$\sqrt{\frac{n}{2\lambda}} \ll |s_p|, \tag{2.5.34}$$

where $|s_p|$ is the distance of the nearest singularity from the origin on the complex s plane.

Similarly, for the n-th term in the series (31) to be effective, we require that

$$\frac{(n+\alpha)}{\lambda} \ll |s_p|. \tag{2.5.35}$$

It is clear now that if a singularity is close to the origin either in (18) or (27), it may render the asymptotic expansion useless unless λ is very large. In such a case, it will be better to use a uniform asymptotic expansion method to derive the asymptotic approximation of the integral. Uniform asymptotic expansions are usually expressed in terms of special functions.

§§2.5.3 Uniform Asymptotic Expansions

As observed previously, when a singularity moves close to a saddle point, the asymptotic expansion obtained by the method of steepest descent is not uniformly valid because the radius of convergence of the Taylor series diminishes in (19) or (29). Hence, to obtain an asymptotic expansion that is uniformly valid for all distances between the singularity and the saddle point, we resort to the uniform asymptotic expansion.

First, consider an integral of the form

$$I = \int_C (t - t_b)^k f(t) e^{\lambda h(t)} dt, \tag{2.5.36}$$

where $k < 1$ and C is an infinite contour. Notice that in the integrand, there could be an algebraic branch-point singularity or a pole singularity at $t = t_b$. For example, if $k = 1/2$, then a square-root branch-point singularity exists at t_b. On the other hand, if $k = -1$, then a simple pole singularity is at t_b. Furthermore, assume that a stationary-phase point exists at $t = t_0$ such that

$h'(t_0) = 0$. Then, using the transformation $-s^2 = h(t) - h(t_0)$, we obtain, via a change of variable,

$$I = e^{\lambda h(t_0)} \int_{C'} (s - s_b)^k F(s) e^{-\lambda s^2} ds, \qquad (2.5.37)$$

where

$$F(s) = \left(\frac{t - t_b}{s - s_b}\right)^k f(t) \frac{dt}{ds}. \qquad (2.5.38)$$

Notice that the point s_b is the image of t_b on the complex s plane. The contour C' is the image of C on the complex s plane. Because of the rational form $(t - t_b)/(s - s_b)$, $F(s)$ does not have a singularity at $s = s_b$. Furthermore, the stationary-phase point at t_0 on the complex t plane is now mapped to the origin on the complex s plane.

There are now two **critical points** on the complex s plane, the saddle point at $s = 0$, which is also the stationary-phase point, and the singularity at $s = s_b$. Both of these critical points are essential in affecting the value of the integral. Then, using a method developed by Bleistein (1966),[7] we approximate $F(s)$ with a polynomial that interpolates these critical points. Consequently,

$$F(s) = \gamma_0 + \gamma_1(s - s_b) + s(s - s_b)F_1(s). \qquad (2.5.39)$$

Notice that the last term vanishes at $s = 0$, the saddle point, and $s = s_b$, the singularity, implying that the polynomial in the first two terms exactly reproduces the value of $F(s)$ at the critical points. Then, solving for γ_0 and γ_1 gives

$$\gamma_0 = F(s_b), \qquad \gamma_1 = \frac{F(s_b) - F(0)}{s_b}. \qquad (2.5.40)$$

Consequently, (37) becomes

$$I = e^{\lambda h(t_0)} \int_{C'} (s - s_b)^k [\gamma_0 + \gamma_1(s - s_b)] e^{-\lambda s^2} ds$$

$$+ e^{\lambda h(t_0)} \int_{C'} (s - s_b)^{k+1} s F_1(s) e^{-\lambda s^2} ds. \qquad (2.5.41)$$

Furthermore, using integration by parts, we can show that the second integral is of the order $1/\lambda$ smaller than the first integral. In particular, the

---

[7] See also Bleistein and Handelsman (1975). Stickler (1976) also discussed a method to yield a uniform approximation.

second integral is

$$I_2 = e^{\lambda h(t_0)} \int\limits_{C'} (s - s_b)^{k+1} s F_1(s) e^{-\lambda s^2} ds$$

$$= \frac{e^{\lambda h(t_0)}}{2\lambda} \int\limits_{C'} (s - s_b)^k [(s - s_b) F_1'(s) + (k+1) F_1(s)] e^{-\lambda s^2} ds. \tag{2.5.42}$$

Hence, I_2 is of the same form as the first integral in (41) except that it is scaled down by $1/\lambda$. As a result, the leading-order approximation to I is

$$I \sim e^{\lambda h(t_0)} \int\limits_{C'} (s - s_b)^k [\gamma_0 + \gamma_1 (s - s_b)] e^{-\lambda s^2} ds, \qquad \lambda \to \infty. \tag{2.5.43}$$

Next, with the change of variable to $u = \sqrt{2\lambda}\,(s - s_b)$, we have

$$I \sim \frac{e^{\lambda [h(t_0) - s_b^2]}}{\sqrt{2\lambda}} \int\limits_{-\infty}^{\infty} \left[\frac{\gamma_0 u^k}{(2\lambda)^{k/2}} + \frac{\gamma_1 u^{k+1}}{(2\lambda)^{(k+1)/2}} \right] e^{-\left(\frac{u^2}{2} + \sqrt{2\lambda}\,s_b u\right)} du, \qquad \lambda \to \infty. \tag{2.5.44}$$

This change of variable shifts the location of the singularity to the origin. Moreover, the saddle point is now located at $u = -\sqrt{2\lambda}\,s_b$ (see Figure 2.5.4).

In (44), the path of integration is above the origin if the original path C' in (43) is above the singularity; and it is below the origin if the original path C' in (43) is below the singularity.[8] Finally, the above integrals can be evaluated in terms of W functions, namely,[9]

$$W_k(\sqrt{2\lambda}\,s_b) = \int\limits_{-\infty}^{\infty} u^k e^{-\left(\frac{u^2}{2} + \sqrt{2\lambda}\,s_b u\right)} du. \tag{2.5.45}$$

The preceding function will have differing values depending on whether the path of integration is above the origin or below the origin. In particular, the W function is related to the parabolic cylinder function via

$$W_k\left(\sqrt{2\lambda}\,s_b\right) = \sqrt{2\pi}\, e^{i\frac{\pi}{2}k + \lambda \frac{s_b^2}{2}} D_k\left(i\sqrt{2\lambda}\,s_b\right), \tag{2.5.46a}$$

if the integration path is above the origin, and

$$W_k\left(\sqrt{2\lambda}\,s_b\right) = \sqrt{2\pi}\, e^{-i\frac{\pi}{2}k + \lambda \frac{s_b^2}{2}} D_k\left(-i\sqrt{2\lambda}\,s_b\right), \tag{2.5.46b}$$

---

[8] This assumes that λ is positive real.

[9] Bleistein refers to this function as a Weber function. Since "Weber function" is the name reserved for another function (see Abramowitz and Stegun 1965), we shall call this the W function.

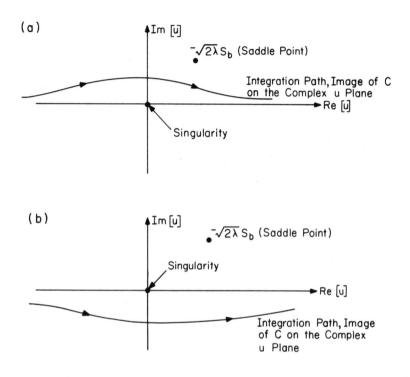

Figure 2.5.4 Two possible cases of integration paths, which are the images of C on the complex u plane. (a) Integration path is above the singularity, and (b) integration path is below the singularity.

if the integration path is below the origin (Gradshteyn and Ryzhik 1965, p. 1064). Hence, the leading-order uniform asymptotic expansion to I is

$$I \sim \frac{e^{\lambda\left[h(t_0)-s_b^2\right]}}{\sqrt{2\lambda}}\left[\frac{\gamma_0}{(2\lambda)^{k/2}}W_k\left(\sqrt{2\lambda}\,s_b\right) + \frac{\gamma_1}{(2\lambda)^{(k+1)/2}}W_{k+1}\left(\sqrt{2\lambda}\,s_b\right)\right], \quad \lambda \to \infty. \tag{2.5.47}$$

Since the effect of the singularity is taken into account in the above approximation, it is uniformly valid for all s_b, the distance between the saddle point and the branch point.

When $\sqrt{2\lambda}\,s_b \gg 1$, or the distance between the saddle point and the singularity is large compared to $1/\sqrt{2\lambda}$, Equation (47) can be simplified using the large argument expansion of the parabolic cylinder function. However, a parabolic cylinder function has different large argument expansions depending on the phase of its argument. Then, assuming that $|z| \gg 1$ and $|z| \gg k$

(Gradshteyn and Ryzhik 1965, p. 1065),

$$D_k(z) \sim e^{-\frac{z^2}{4}} z^k, \qquad\qquad |\arg z| < \frac{3\pi}{4}, \qquad (2.5.48a)$$

$$D_k(z) \sim e^{-\frac{z^2}{4}} z^k - \frac{\sqrt{2\pi}}{\Gamma(-k)} e^{i\pi k} e^{\frac{z^2}{4}} z^{-k-1}, \qquad \frac{\pi}{4} < \arg z < \frac{5\pi}{4}, \qquad (2.5.48b)$$

$$D_k(z) \sim e^{-\frac{z^2}{4}} z^k - \frac{\sqrt{2\pi}}{\Gamma(-k)} e^{-i\pi k} e^{\frac{z^2}{4}} z^{-k-1}, \qquad -\frac{\pi}{4} > \arg z > -\frac{5\pi}{4}. \qquad (2.5.48c)$$

The phenomenon of different large argument behavior depending on the phase of the argument is known as ***Stokes' phenomenon*** (Olver 1974, p. 240). Note that the second terms in (48b) and (48c) are similar except for a phase of $e^{2i\pi k}$. Hence, if k is an integer, corresponding to a pole type singularity in (45), then (48b) and (48c) are identical. If k is fractional, however, implying an algebraic branch-point singularity in (45), then (48b) and (48c) would be different. This difference is the result of a branch-cut integral because cases (a) and (b) in Figure 2.5.4 differ by a branch-cut integral. Moreover, the second term in (48b) and (48c) will only dominate if $|\arg z| > \frac{3\pi}{4}$; otherwise, it would be exponentially small compared to the first term.

Now, assume that the path of integration is above the origin on the complex u plane. Then, the W function is represented by (46a), and the location of the saddle point, $u = -\sqrt{2\lambda}\, s_b$, is in the first quadrant on the complex u plane (see Figure 2.5.4a), i.e., $\frac{\pi}{2} > \arg\left(-\sqrt{2\lambda}\, s_b\right) > 0$. Moreover, the argument of the parabolic cylinder function in (46a) is in the fourth quadrant. Consequently, a large argument approximation for the parabolic cylinder function that is uniformly valid in the fourth quadrant is given by (48a). As a result,

$$W_k\left(\sqrt{2\lambda}\, s_b\right) \sim \sqrt{2\pi}\, e^{\lambda s_b^2} \left(-\sqrt{2\lambda}\, s_b\right)^k. \qquad (2.5.49)$$

Next, by substituting the above into (47), we have

$$I \sim \frac{e^{\lambda h(t_0)}}{\sqrt{2\lambda}} \left[\gamma_0 \sqrt{2\pi}\, (-s_b)^k + \gamma_1 \sqrt{2\pi}\, (-s_b)^{k+1} \right]. \qquad (2.5.50)$$

Using the definitions of γ_0 and γ_1 given in (40), the above becomes

$$I \sim \sqrt{\frac{\pi}{\lambda}}\, e^{\lambda h(t_0)} F(0) (-s_b)^k. \qquad (2.5.51)$$

On comparing (51) with the leading-order term in (21), we see that the above is just the leading-order saddle-point approximation of the integral in (37) (see Exercise 2.23). Hence, the leading-order uniform asymptotic expansion reduces to the leading-order term in the asymptotic expansion derived by the steepest-descent method when s_b is large. Notice that there is only a saddle-point contribution because, in the method of steepest descent, no singularity is enclosed in the course of deforming the original integration path to the steepest-descent path, since the original path and the saddle point are both above the origin.

On the other hand, if the original integration path is below the singularity as show in Figure 2.5.4b, then (46b) would be used for the W function. Furthermore, if the saddle point is in the first quadrant, then the argument of the parabolic cylinder function in (46b) is in the second quadrant. Consequently, a large argument expansion for the parabolic cylinder function which is uniformly valid in the second quadrant is given by (48b). In this case,

$$W_k\left(\sqrt{2\lambda}\,s_b\right) \sim \sqrt{2\pi}\,e^{\lambda s_b^2}\left(-\sqrt{2\lambda}\,s_b\right)^k + \frac{2\pi i}{\Gamma(-k)}\left(-\sqrt{2\lambda}\,s_b\right)^{-k-1}, \quad \sqrt{2\lambda}\,s_b \to \infty.$$
(2.5.52)

Finally, using the above in (47), we have

$$I \sim \frac{e^{\lambda h(t_0)}}{\sqrt{2\lambda}}\left[\gamma_0\sqrt{2\pi}\,(-s_b)^k + \gamma_1\sqrt{2\pi}\,(-s_b)^{k+1} + \frac{2\pi i \gamma_0 e^{-\lambda s_b^2}}{\Gamma(-k)(2\lambda)^{k+\frac{1}{2}}(-s_b)^{k+1}}\right.$$
$$\left. + \frac{2\pi i \gamma_1 e^{-\lambda s_b^2}}{\Gamma(-k-1)(2\lambda)^{k+\frac{3}{2}}(-s_b)^{k+2}}\right], \quad \sqrt{2\lambda}\,s_b \to \infty.$$
(2.5.53)

But when $\lambda \to \infty$, the last term in the above can be neglected. Hence, using the definition of γ_0 and γ_1, and the fact that $h(t_0) - s_b^2 = h(t_b)$, we have eventually

$$I \sim \sqrt{\frac{\pi}{\lambda}}\,e^{\lambda h(t_0)}F(0)(-s_b)^k + \frac{e^{\lambda h(t_b)}}{\sqrt{2\lambda}}\frac{2\pi i}{\Gamma(-k)(2\lambda)^{k+\frac{1}{2}}(-s_b)^{k+1}}F(s_b), \quad \sqrt{2\lambda}\,s_b \to \infty.$$
(2.5.54)

The first term is again the saddle-point contribution. Now, the second term arises because when the original path of integration is deformed to the steepest-descent path that passes through the saddle point, the contribution from the singularity has to be included, since the original path of integration is below the origin (see Exercise 2.23).

In the above, note that when s_b is small, the uniform asymptotic expansion does not treat the saddle-point and the singularity contributions separately. In fact, they are indistinguishably buried in (47). However, when the saddle point and the singularity are far apart, the large argument expansion of the W function can be used to arrive at (51) or (54), where a saddle-point contribution and a singularity contribution are identifiable in the context of the steepest-descent method. Furthermore, whether a singularity will contribute is accounted for by the definition of the W function for two different cases in (46a) and (46b). Note further that the singularity contribution in (54) is the leading-order contribution from the singularity; it can be an algebraic branch-point singularity or a pole singularity depending on the value of k.

§2.6 Dipole Over Layered Media—Asymptotic Expansions

Having shown the theory behind asymptotic expansions of integrals, we shall next show the use of the technique for the asymptotic expansions of Sommerfeld integrals. Sommerfeld integrals in general contain branch-point and pole singularities in addition to saddle points. Because of the presence of branch-points on the complex plane, the analysis of Sommerfeld integrals is often complicated by the presence of Riemann sheets. Hence, a good grasp of the Riemann sheet concept in addition to the asymptotic expansion theory is essential here.

We shall first illustrate the asymptotic expansions with an integral that has only branch-point singularities. Next, integrals with pole singularities in addition to branch-point singularities will be considered. Later, the case where a singularity could be close to the saddle point, hence necessitating the use of uniform asymptotic expansions, will be illustrated.

§§2.6.1 Dipole Over Half-Space (VMD)

For a nonmagnetic medium where $\mu = \mu_0$ is a constant, the asymptotic expansion of the TE wave due to a point source is more benign than that of the TM wave. Hence, let us consider a source that will generate only TE waves, namely, a vertical magnetic dipole. Because of the TE nature of its field, a VMD over a half-space has a field that can be characterized by H_z only. Consequently, we have [see (2.3.10a)],

$$H_{1z} = -\frac{iIA}{8\pi} \int_{-\infty}^{\infty} dk_\rho \frac{k_\rho^3}{k_{1z}} H_0^{(1)}(k_\rho\rho) \left[e^{ik_{1z}|z|} + R_{12}^{TE} e^{ik_{1z}z + 2ik_{1z}d_1} \right],$$
$$(2.6.1)$$

$$H_{2z} = -\frac{iIA}{8\pi} \int_{-\infty}^{\infty} dk_\rho \frac{k_\rho^3}{k_{1z}} H_0^{(1)}(k_\rho\rho) T_{12}^{TE} e^{-ik_{2z}(z+d_1) + ik_{1z}d_1}, \qquad (2.6.2)$$

where

$$R_{12}^{TE} = \frac{k_{1z} - k_{2z}}{k_{1z} + k_{2z}}, \quad T_{12}^{TE} = \frac{2k_{1z}}{k_{1z} + k_{2z}}, \qquad (2.6.3)$$

and $k_{iz} = (k_i^2 - k_\rho^2)^{1/2}$.

Notice that the asymptotic expansion of the primary field in (1) gives rise to a spherical wave as in (2.5.10). The reflected wave field in (1) is, however, given by

$$H_{1z}^R = -\frac{iIA}{8\pi} \int_{-\infty}^{\infty} dk_\rho \frac{k_\rho^3}{k_{1z}} H_0^{(1)}(k_\rho\rho) R_{12}^{TE} e^{ik_{1z}z + 2ik_{1z}d_1}, \qquad (2.6.4)$$

which is more complex than that for the primary field. There are two branch points in this integrand, and the branch cuts associated with them are as

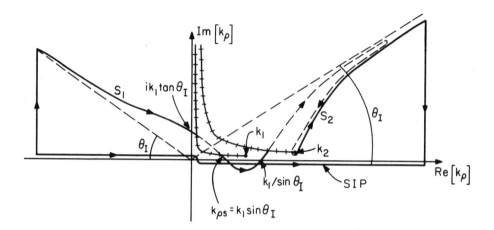

Figure 2.6.1 $\Im m[k_{iz}] = 0$ branch cuts, the steepest-descent paths through the saddle point, the branch point, and the Sommerfeld integration path.

shown in Figure 2.6.1. Here, only the branch cuts in the first quadrant are considered, as it will become evident that only they are the important ones. Then, when ρ, z, and d_1 are large, a stationary-phase point, or a saddle point is located at [analogous to (2.5.9)]

$$k_{\rho s} = k_1 \frac{\rho}{\sqrt{\rho^2 + (z + 2d_1)^2}} = k_1 \sin \theta_I, \qquad (2.6.5)$$

where $\theta_I = \sin^{-1}(\frac{\rho}{r_I})$, and $r_I = [\rho^2 + (z + 2d_1)^2]^{1/2}$. Therefore, the stationary-phase point is located at $k_{\rho s} < k_1$. The constant-phase path, or the steepest-descent path that passes through the saddle point, is shown in Figure 2.6.1 as S_1 (see Exercise 2.24). Furthermore, the constant-phase path that passes through the branch point k_2 is S_2.

By virtue of Jordan's lemma and Cauchy's theorem, we can deform the contour from the Sommerfeld integration path (SIP) to the paths S_1 and S_2. Notice that the region enclosed by S_1, S_2, and SIP is analytic except for the possible occurrence of pole singularities. But it can be proved that for the integrand of (4), no poles are in this enclosed region. Hence, the integration over the SIP is the same as the integration over S_1 and S_2. Moreover, the contribution from the two vertical paths vanishes by virtue of Jordan's lemma when they tend to infinity. Consequently, the contribution to H_{1z}^R has two parts, one from the saddle point, and the other from the branch point at k_2, given by

$$H_{1z}^R = H_{1z}^{RS} + H_{1z}^{RB}. \qquad (2.6.5a)$$

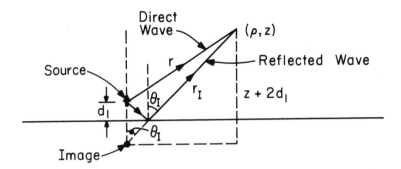

Figure 2.6.2 Direct wave, reflected wave, and the image point for a dipole over a half-space.

The leading-order saddle-point contribution is the same as that one would have obtained using the method of stationary phase. Hence, by inspection of the integral in (4), and comparison with the integral in (2.5.5), we have

$$H_{1z}^{RS} \sim \frac{-IA}{4\pi} k_1^2 \sin^2 \theta_I R_{12}^{TE}(k_{\rho s}) \frac{e^{ik_1 r_I}}{r_I}, \qquad (2.6.6)$$

where $k_{\rho s}$ is given in (5) and $r_I = [\rho^2 + (z + 2d_1)^2]^{1/2}$.

In the above, R_{12}^{TE} is evaluated with $k_\rho = k_{\rho s} = k_1 \sin \theta_I$. Therefore, expression (6) represents a spherical wave that emanates from a point at $(\rho, z) = (0, -2d_1)$, which is the image point of the source at $(\rho, z) = (0,0)$. Moreover, the wave is attenuated by the Fresnel reflection coefficient for a plane wave incident at angle θ_I on the half-space, as shown in Figure 2.6.2.

If $k_1/\sin \theta_I < k_2$ as shown in Figure 2.6.1, the branch-point contribution, which manifests itself as a branch-cut integral, has to be included. This branch-cut integral is deformed to integrate along S_2 as shown in the figure. To calculate the branch-cut integral, note that the function k_{2z} in R_{12}^{TE} assumes opposite signs on the parts of S_2 that are on different Riemann sheets. Therefore, the integration on different Riemann sheets can be combined into a single integral, yielding

$$H_{1z}^{RB} = -\frac{iIA}{8\pi} \int_{S_2} dk_\rho \frac{k_\rho^3}{k_{1z}} H_0^{(1)}(k_\rho \rho) e^{ik_{1z}(z+2d_1)} \left[R_{12}^{TE+} - R_{12}^{TE-} \right], \qquad (2.6.7)$$

where S_2 starts at k_2 and extends to infinity, and

$$R_{12}^{TE+} - R_{12}^{TE-} = \frac{k_{1z} - k_{2z}}{k_{1z} + k_{2z}} - \frac{k_{1z} + k_{2z}}{k_{1z} - k_{2z}} = \frac{-4k_{1z}k_{2z}}{k_{1z}^2 - k_{2z}^2} = \frac{-4k_{1z}k_{2z}}{k_1^2 - k_2^2}. \qquad (2.6.8)$$

The superscripts $+$ and $-$ above indicate the values of R_{12}^{TE} on the top and the bottom Riemann sheets respectively.

But when ρ is large, the integrand in (7) can be replaced by

$$H_{1z}^{RB} \sim \frac{iIA}{2\pi} \sqrt{\frac{2}{i\pi\rho}} \int_{S_2} dk_\rho \, k_\rho^{\frac{5}{2}} e^{ik_\rho\rho + ik_{1z}(z+2d_1)} \frac{k_{2z}}{k_1^2 - k_2^2}, \quad \rho \to \infty. \tag{2.6.9}$$

Moreover, the integral in (9) is a semi-infinite integral. Then, on comparing (9) with the first integral in (2.5.25), we identify λ with r_I, and $h(t)$ with $i\left[k_\rho\rho/r_I + k_{1z}(z+2d_1)/r_I \right]$.

As a result, by letting $-s = h(t) - h(a)$, or equivalently,

$$-\lambda s = ik_\rho\rho + ik_{1z}(z+2d_1) - ik_2\rho - i\sqrt{k_1^2 - k_2^2}\,(z+2d_1), \tag{2.6.10}$$

Equation (9) becomes

$$H_{1z}^{RB} \sim \frac{IAe^{i\frac{\pi}{4}}}{\pi\sqrt{2\pi\rho}} e^{ik_2\rho - \sqrt{k_2^2 - k_1^2}\,(z+2d_1)} \int_0^\infty ds\, e^{-\lambda s} s^{\frac{1}{2}} F(s), \tag{2.6.11}$$

where

$$s^{\frac{1}{2}} F(s) = f(k_\rho) \frac{dk_\rho}{ds}, \tag{2.6.12a}$$

and

$$f(k_\rho) = k_\rho^{\frac{5}{2}} \frac{k_{2z}}{k_1^2 - k_2^2}. \tag{2.6.12b}$$

Since $k_{2z} = \sqrt{k_2^2 - k_\rho^2}$, $f(k_\rho)$ has a branch-point singularity at $k_\rho = k_2$. Moreover, from (12b), we note that

$$f(k_\rho) \sim \frac{ik_2^3\sqrt{2}\,(k_\rho - k_2)^{\frac{1}{2}}}{k_1^2 - k_2^2}, \quad k_\rho \to k_2. \tag{2.6.13}$$

Consequently, B in (2.5.32) is identified as

$$\frac{ik_2^3\sqrt{2}}{(k_1^2 - k_2^2)},$$

and $\alpha = 1/2$ in (2.5.32). Then, the leading-order expansion of (11) using the first term in (2.5.31) is (Exercise 2.25)

$$H_{1z}^{RB} \sim \frac{IAe^{i\frac{\pi}{4}}}{2\pi\sqrt{\rho}} e^{ik_2\rho - \sqrt{k_2^2 - k_1^2}\,(z+2d_1)} \left[\frac{i}{\rho + \frac{ik_2}{\sqrt{k_2^2 - k_1^2}}(z+2d_1)} \right]^{\frac{3}{2}} \frac{ik_2^3}{k_1^2 - k_2^2}. \tag{2.6.14}$$

This branch point contributes to a wave that has a special physical interpretation. Notice that this wave is exponentially decaying away from the surface at $z = -d_1$. Furthermore, in the ρ direction, it has the phase velocity

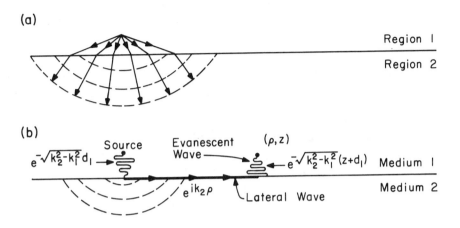

Figure 2.6.3 Physical interpretation of the lateral wave.

of medium 2. Since it is only observed close to the surface, it is a surface wave. If z and d_1 are small but ρ is large so that both the source and the observation points are close to the surface, then along the surface, the wave has a form of

$$H_{1z}^{RB} \sim C \frac{e^{ik_2\rho - \sqrt{k_2^2 - k_1^2}\,(z+2d_1)}}{\rho^2}, \qquad (2.6.15)$$

indicating an algebraic decay of $1/\rho^2$ in the ρ direction. This surface wave is also known as the **lateral wave** or the **head wave** (see Brekhovskikh 1960; Tyras 1969; Tamir 1972, 1973; Aki and Richards 1980).

Figure 2.6.3 illustrates the physical interpretation of such a wave. To interprete the result, recall that a point source emanates a spectrum of cylindrical waves or plane waves, some of which are propagating, while some are evanescent in the z direction. But the propagating part of the spectrum is always transmitted at an angle smaller than the angle of incidence because $k_2 > k_1$ (see Figure 2.6.3a). Hence, the largest transmitted angle due to a propagating wave in medium 1 is at θ_c, the critical angle due to a wave incident at grazing in region 1. If the wave is to be transmitted at an angle greater than θ_c, it has to come from an evanescent wave in region 1. Therefore, the evanescent part of the spectrum is needed to excite a wave propagating in the lateral direction parallel to the boundary, which is the lateral wave (see Figure 2.6.3b). But this wave is propagating in medium 2; hence, it has the phase velocity of medium 2. Moreover, it is evanescent in medium 1; therefore, it is observed in medium 1, and excited by a source in medium 1 if only both the source and the observation points are close to the interface.

Alternatively, the lateral wave could be thought of as a part of the spherical wave in medium 2 that is propagating along the interface, and is being refracted back as evanescent waves. It has an algebraic decay of ρ^{-2}, which is faster than the decay of a spherical wave, because of an interference effect near the interface. As a result of the ρ^{-2} decay, the branch-point contribution is of higher order compared to the saddle-point contribution.

The transmitted field given by (2) can be approximated in a comparable manner (see Exercise 2.26; also see Subsection 2.6.4). Similarly, integral representations of the Weyl type can also be approximated (Exercise 2.27).

§§2.6.2 Dipole Over Half-Space (VED)

A VED produces TM waves rather than TE waves. In this case, the field is characterized by E_z, and we have

$$E_{1z} = \frac{-I\ell}{8\pi\omega\epsilon_1} \int\limits_{-\infty}^{\infty} dk_\rho \frac{k_\rho^3}{k_{1z}} H_0^{(1)}(k_\rho\rho) \left[e^{ik_{1z}|z|} + R_{12}^{TM} e^{ik_{1z}z + 2ik_{1z}d_1} \right],$$

$$(2.6.16)$$

where

$$R_{12}^{TM} = \frac{\epsilon_2 k_{1z} - \epsilon_1 k_{2z}}{\epsilon_2 k_{1z} + \epsilon_1 k_{2z}}. \qquad (2.6.17)$$

The asymptotic expansion of E_{1z} above will be very similar to the asymptotic expansion of H_{1z} except that now, R_{12}^{TM} may have a pole. For instance, R_{12}^{TM} will be infinite if

$$\epsilon_2 k_{1z} = -\epsilon_1 k_{2z}. \qquad (2.6.18)$$

On squaring (18) and making use of the facts that $k_{1z}^2 = k_1^2 - k_\rho^2$ and that $k_{2z}^2 = k_2^2 - k_\rho^2$, we can solve for k_ρ to obtain

$$k_\rho = \pm\sqrt{\frac{\epsilon_2^2 k_1^2 - \epsilon_1^2 k_2^2}{\epsilon_2^2 - \epsilon_1^2}} = \pm k_P. \qquad (2.6.19)$$

Since the Brewster angle condition for which $R_{12}^{TM} = 0$ is given by $\epsilon_2 k_{1z} = \epsilon_1 k_{2z}$, (19) is also the condition for the zero of R_{12}^{TM}. Therefore, the pole and the zero of the function R_{12}^{TM} are at the same location on the complex k_ρ plane. Because the integrand of (16) is a multivalued function with four possible values for each value of k_ρ, the pole and zero must be on different Riemann sheets. Moreover, for $\mu_1 = \mu_2$, the Brewster angle is given by

$$k_P = k_1 \sin\theta_B = k_1\sqrt{\frac{\epsilon_2}{\epsilon_2 + \epsilon_1}} < k_1. \qquad (2.6.20)$$

For such a case, the pole or zero is located at $k_P < k_1$. However, for the pole to exist, the real parts and the imaginary parts of k_{1z} and k_{2z} have to be opposite in signs if ϵ_1 and ϵ_2 are predominantly real. As a result, the pole

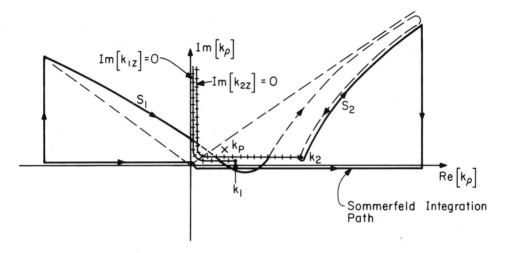

Figure 2.6.4 Location of the Sommerfeld poles at $k_\rho = k_P$. This is also the location of the zeroes but on different Riemann sheets.

cannot be concurrently on the top sheets of the Riemann sheets of k_{1z} and k_{2z}, or concurrently on the bottom sheets of the Riemann sheets of k_{1z} and k_{2z}.

The locations of the pole and the zero are shown in Figure 2.6.4. The pole is located on

(i) the top Riemann sheet of k_{1z} and the bottom Riemann sheet of k_{2z}; and

(ii) the bottom Riemann sheet of k_{1z} and the top Riemann sheet of k_{2z}.

The zero is located on

(i) both the bottom Riemann sheets of k_{1z} and k_{2z}; and

(ii) both the top Riemann sheets of k_{1z} and k_{2z}.

This pole is also sometimes known as the Sommerfeld pole, or the Zenneck surface wave pole, because it could be associated with the Zenneck surface wave.[10] When the Sommerfeld integration path is deformed to the path S_1, the dashed part of S_1 to the left of k_1 is on the bottom sheets of both the Riemann sheets of k_{1z} and k_{2z}. Hence, this pole does not contribute to the asymptotic expansion of E_{1z}.

However, when $\epsilon_2 < 0$, which is possible in a plasma medium, this pole

---

[10] An extensive discussion of this pole was given by Baños (1966). Discussion of the Zenneck wave was given by Tyras (1969), and Hill and Wait (1978).

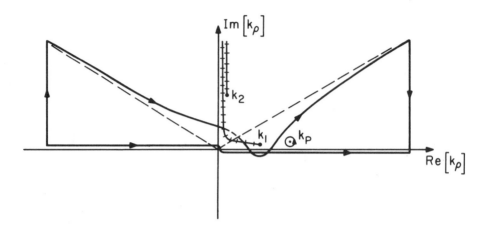

Figure 2.6.5 The surface plasmon case where $\epsilon_2 < 0$. k_P is now on the top Riemann sheet.

corresponds to the surface plasmon mode that is observed in optical scattering and absorption on metallic surfaces as discussed in Boardman (1982) and P. Yeh (1988, p. 332). In this case, if $|\epsilon_2| > \epsilon_1$, then $k_P > k_1$, and a careful study shows that the pole can in fact be on the topmost Riemann sheet. Hence, it may be enclosed in the course of deforming the contour of integration from the real axis to the steepest-descent path as shown in Figure 2.6.5. In this case, the asymptotic expansion of E_{1z} consists of a saddle-point contribution such as the spherical wave in (6) plus a residue of the pole contribution. Hence, if the residue of the reflected wave term in (16) is taken, we have the following term

$$E_{1z}^{RP} = \frac{-I\ell}{8\pi\omega\epsilon_1}\frac{k_P^3}{k_{1zP}}H_0^{(1)}(k_P\rho)e^{ik_{1zP}(z+2d_1)}A^{TM}, \qquad (2.6.21)$$

where $k_{1zP} = (k_1^2 - k_P^2)^{1/2} = i\alpha_{1P}$ is imaginary since $k_P > k_1$, and $A^{TM} = 2\pi i \mathrm{Res}\left[R_{12}^{TM}(k_P)\right]$, where "Res" stands for "residue of."

In the limit when $k_P\rho$ is large, this wave becomes

$$E_{1z}^{RP} \sim \frac{-I\ell}{8\pi\omega\epsilon_1}\frac{k_P^3}{k_{1zP}}\sqrt{\frac{2}{i\pi k_P\rho}}\,e^{ik_P\rho - \alpha_{1P}(z+2d_1)}A^{TM}. \qquad (2.6.22)$$

Since the field decays exponentially away from the interface, it is a surface wave. It is also evanescent in region 2 since $\epsilon_2 < 0$. Furthermore, if the media are lossless, this wave decays only algebraically in the ρ direction just like a cylindrical wave, because it is actually guided by the interface and does not radiate. Hence, the only decay in the ρ direction is from the algebraic spreading of the cylindrical wave which gives rise to a $1/\sqrt{\rho}$ decay. As mentioned

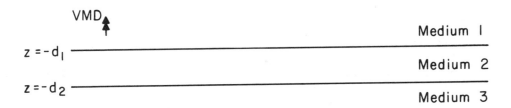

Figure 2.6.6 A vertical magnetic dipole (VMD) over a three-layer medium.

previously, this wave is also known as the surface plasmon mode. It is again excited by the evanescent part of the source spectrum. Hence, it will not be excited if d_1 is large or if the source is far away from the interface. It is a good example of an electromagnetic wave guidance by one interface alone.

From the above analysis, we observe that the Zenneck surface wave is difficult to excite since it is not on the appropriate Riemann sheet on the complex k_ρ plane. However, the proximity of the Zenneck surface wave pole (Sommerfeld pole) to the saddle point can affect the nature of the direct wave and reflected wave in Figure 2.6.2.

§§2.6.3 Dipole Over a Slab

Consider the case of a vertical magnetic dipole over a slab as shown in Figure 2.6.6. In this case, the field in the upper half-space is given by

$$H_{1z} = -\frac{iIA}{8\pi} \int\limits_{-\infty}^{\infty} dk_\rho \frac{k_\rho^3}{k_{1z}} H_0^{(1)}(k_\rho \rho) \left[e^{ik_{1z}|z|} + \tilde{R}_{12}^{TE} e^{ik_{1z}(z+2d_1)} \right], \qquad (2.6.23)$$

where

$$\tilde{R}_{12}^{TE} = \frac{R_{12}^{TE} + R_{23}^{TE} e^{2ik_{2z}(d_2-d_1)}}{1 - R_{21}^{TE} R_{23}^{TE} e^{2ik_{2z}(d_2-d_1)}}. \qquad (2.6.24)$$

Now, $\tilde{R}_{12}^{TE}$ is a function of k_{3z} as well as k_{1z} and k_{2z}. However, $\tilde{R}_{12}^{TE}$ can be shown to be an even function of k_{2z} despite the complicated appearance of (24).[11] Hence, (24) can only be a function of k_{2z}^2. Consequently, $\tilde{R}_{12}^{TE}$ is not a double-value function of k_{2z} but is only a double-value function of k_{1z} and a double-value function of k_{3z}. As a result, the multivalue function $\tilde{R}_{12}^{TE}$ can be properly defined with only branch cuts for k_{1z} and k_{3z} (considering only the first quadrant) and four Riemann sheets.

(a) Normal Mode Expansion

Next, the pole singularities of the integrand as a consequence of (24) have

---

[11] This point is discussed further in Subsection 2.7.1. See also Exercise 2.28.

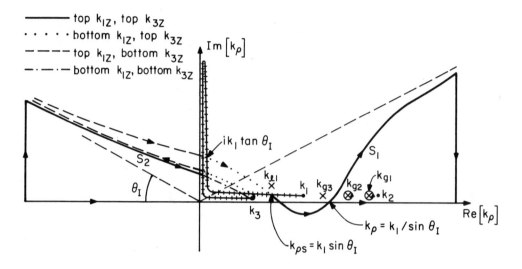

Figure 2.6.7 The locations of poles of a slab for guided modes, k_{gi}, and leaky modes, k_{li}, and the saddle point, $k_{\rho s}$, on the complex k_ρ plane.

to be identified in order to find the asymptotic expansion of H_{1z}. The poles of (24) are given by the equation

$$1 - R_{21}^{TE} R_{23}^{TE} e^{2ik_{2z}(d_2-d_1)} = 0. \qquad (2.6.25)$$

Physically, the above implies that a wave, after reflecting from the top and the bottom interfaces, together with a phase shift through the slab, should become in phase with itself again. This is precisely the guidance condition (sometimes referred to as the transverse resonance condition) for guided modes in a dielectric slab. Therefore, these poles on the complex k_ρ plane are actually related to the guided modes of a dielectric slab. Moreover, since $k_{iz} = (k_i^2 - k_\rho^2)^{1/2}$, Equation (25) is just a function of k_ρ. Hence, the roots of (25) can be solved for numerically on the complex k_ρ plane.

If $k_2 > k_1 > k_3$, however, the roots of (25) corresponding to guided modes at $k_\rho = k_{gi}$ have to be such that $k_1 < k_{gi} < k_2$, because a guided mode in medium 2 is evanescent both in media 1 and 3. In other words, k_{1z} and k_{3z} are pure imaginary and total internal reflections occur at both the top and bottom interfaces. The locations of these poles are shown in Figure 2.6.7 with $k_3 < k_1 < k_2$ [see also Exercise 2.27(f)].

The number of guided-mode poles of a slab depends on the frequency and the thickness of the slab. For instance, there are a large number of guided-mode poles at high frequencies and a fewer number of poles at low

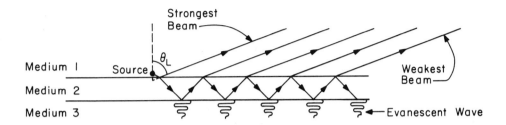

Figure 2.6.8 The physical picture of a leaky mode. The radiation damping on the mode gives it an apparent exponential growth in the z direction. The mode gets weaker as it propagates in the ρ direction due to radiation damping.

frequencies. Moreover, the thicker the slab, the more guided-mode poles there are. Hence, if a number of guided-mode poles exist between k_1 and k_2, this number must be smaller when the frequency is lowered. Therefore, as the frequency gradually lowers, some of the poles must disappear from the $k_1 < k_\rho < k_2$ region in Figure 2.6.7. This is impossible unless the poles migrate to the left. As a pole migrates gradually to the left, it reaches a point when the pole is at $k_{li} < k_1$. In such a circumstance, k_{1z} is no longer pure-imaginary, and total internal reflection disappears at the top interface. Furthermore, the mode can no longer be guided, as it must radiate into medium 1. Consequently, the corresponding mode decays in the ρ direction due to radiation damping. Therefore, k_{li} must have a positive imaginary part indicating a decaying mode. Such a mode is then known as a **leaky mode**, since it leaks energy into medium 1 (see Hessel 1969; Tamir 1969). Because it attenuates when it propagates, the "rays" that it radiates gradually become weaker, as shown in Figure 2.6.8. As a result, if we fix ρ to observe the field behavior in the z direction, an exponential growth is seen in the z direction. This is only possible if $\Im m[k_{1z}] < 0$ for this mode. Therefore, it has to be at the bottom Riemann sheet of k_{1z} while it could still be on the top Riemann sheet of k_{3z}. The proper location of the leaky mode k_{l1} is shown in Figure 2.6.7.

When the integration path is deformed from the original path of integration to the steepest-descent path passing through the saddle point at $k_\rho = k_1 \sin\theta_I$ [the saddle point is the same as given by Equation (5)], a k_3 branch-point contribution will be included if $k_1\sin\theta_I > k_3$. Furthermore, by noting the point where the steepest-descent path crosses the real axis, the guided-mode contribution will be included if $k_1/\sin\theta_I < k_{gi}$, the location of the i-th guided mode. On the other hand, the leaky mode will be included if

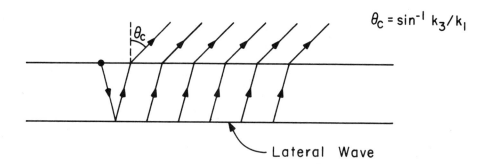

Figure 2.6.9 The lateral wave and its association with θ_c, the critical angle.

$k_1 \sin \theta_I > k_{li}$ where k_{li} is the location of the leaky mode.[12] Hence, the leaky mode is only observed for θ_I greater than or approximately equal to θ_L where

$$\theta_L = \sin^{-1} \left(\frac{k_{li}}{k_1} \right).$$

The angle θ_L is shown approximately in Figure 2.6.8 since to be strictly correct, it is a complex number. Consequently, even though the leaky mode is exponentially growing in the z direction for a fixed ρ, it is not observed for all z. (It should be noted here that if the saddle point is close to a singularity, the uniform asymptotic expansion of the integral needs to be used in deriving the asymptotic approximations.)

The k_3 branch point in Figure 2.6.7 gives rise to a lateral wave that propagates in medium 3 parallel to the interface and refracts back to medium 1 as shown in Figure 2.6.9. But it is only observed when

$$\theta > \sin^{-1} \left(\frac{k_3}{k_1} \right) = \theta_c.$$

The expansion of the field in terms of the guided mode of the slab is known as the ***normal-mode*** approach (see, e.g., Pierce 1965; Wait 1970; Tsang and Kong 1973).

(b) Geometrical Optics Series

When $d_2 - d_1$ is very large, or the frequency is high, there are many guided modes inside the slab. Consequently, the expansion of the field in

---

[12] The imaginary part of k_{li} is assumed to be small so that the inequality is still meaningful.

terms of the guided modes becomes laborious as it is necessary to search for the locations of these modes. In this case, it is more expedient to expand (24) in terms of a geometrical optics series in the manner of (2.1.22) and evaluate the integral term by term. By doing so, the reflected field becomes

$$
H_{1z}^R = -\frac{iIA}{8\pi} \int\limits_{-\infty}^{\infty} dk_\rho \frac{k_\rho^3}{k_{1z}} H_0^{(1)}(k_\rho\rho) e^{ik_{1z}(z+2d_1)} R_{12}
$$

$$
-\frac{iIA}{8\pi} \sum_{m=1}^{\infty} \int\limits_{-\infty}^{\infty} dk_\rho \frac{k_\rho^3}{k_{1z}} H_0^{(1)}(k_\rho\rho) e^{ik_{1z}(z+2d_1)}
$$

$$
\cdot T_{12} R_{23}^m R_{21}^{m-1} T_{21} e^{2imk_{2z}(d_2-d_1)}.
$$

$$(2.6.26)$$

The first term in the above is just the same as the reflected wave from a half-space whose asymptotic expansion has been found previously. The other terms are the consequence of multiple reflections.

To find the asymptotic expansions of the multiply reflected field in (26), the stationary-phase point or the saddle point of the integrand is required. Observe that if both the source point and the observation point are close to the top interface, so that z and d_1 are small, then the rapidly oscillating part of the integrand comes from $H_0^{(1)}(k_\rho\rho)$ and $e^{2imk_{2z}(d_2-d_1)}$. Then, using the large argument expansion of the Hankel function, the saddle point for the integrand of the m-th term of the series is at [analogous to (2.5.9)]

$$
k_{\rho sm} = k_2 \frac{\rho}{\sqrt{\rho^2 + 4m^2(d_2-d_1)^2}} = k_2 \frac{\rho}{r_m} = k_2 \sin\theta_m, \qquad (2.6.27)
$$

where

$$
\theta_m = \sin^{-1}\left(\frac{\rho}{r_m}\right),
$$

$$
r_m = \sqrt{\rho^2 + 4m^2(d_2-d_1)^2}.
$$

Consequently, the leading-order saddle-point contribution for the m-th term is

$$
H_{1z}^{RSm} \sim -\frac{IA}{4\pi} \left[k_\rho^2 \frac{k_{2z}}{k_{1z}} T_{12} R_{23}^m R_{21}^{m-1} T_{21} \right]\bigg|_{k_\rho = k_{\rho sm}} \frac{e^{ik_2 r_m}}{r_m}. \qquad (2.6.28)
$$

Physically, the saddle-point contribution of the m-th term in the series can be thought of as the contribution from a spherical wave with wave number k_2 emanating from the m-th image point of the source as shown in Figure 2.6.10. Here, r_m is the distance from the image point to the observation point. Moreover, the m-th image is due to m-th subsurface reflection of the spherical wave generated by the source. Note that all the reflection and transmission coefficients are evaluated at $k_\rho = k_{\rho sm} = k_2 \sin\theta_m$ for the m-th image.

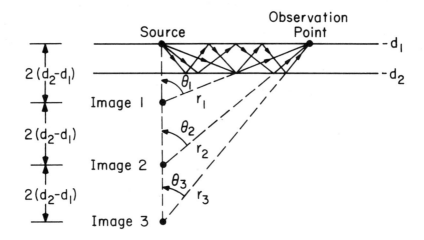

Figure 2.6.10 Physical interpretation of the geometric ray series.

In addition to the above contribution, there are branch-point contributions from k_1 and k_3 for each of the multiple reflection terms in (26), corresponding to lateral waves. [For an N-layer medium, there are branch-point contributions from only k_1 and k_N (see Subsection 2.7.1 for further discussions).]

Expanding the field in terms of images or rays is known as the *geometrical optics* or *ray optics* approach (Wait 1970). In contrast, the normal-mode approach is good when the wavelength is long compared to the slab thickness. But the geometrical optics approach is good for short wavelengths. When the wavelength is neither long nor short, a combination of the normal mode and geometrical optics approaches could be used and this is known as the *hybrid ray-mode* approach (Felsen 1981).

For a vertical electric dipole that excites TM waves, the corresponding R_{12} and R_{23} in (26) contain the Zenneck surface wave pole. Moreover, the order of the pole increases with increasing m. Even though the pole is not enclosed when the path of integration is deformed to the steepest-descent path, it influences the saddle-point contribution. In this case, uniform asymptotic expansions have to be used in deriving the approximation of the field [Chew and Kong (1981, 1982)].

§§2.6.4 Example of Uniform Asymptotic Expansion —Transmitted Wave in a Half-Space

In the previous examples, the steepest-descent or the stationary-phase method has been used to find the leading-order asymptotic expansion of the

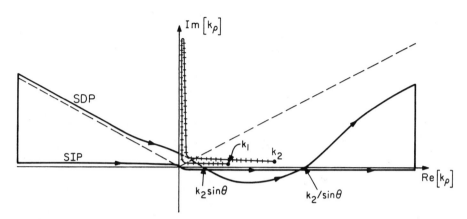

Figure 2.6.11 The steepest-descent path (SDP) through the saddle point at $k_\rho = k_2 \sin \theta$. When $k_2 \sin \theta \simeq k_1$, we would have to use the uniform asymptotic expansion for the integral.

field. In many cases, however, these leading-order approximations may not be valid if a singularity is close to the saddle point [see (2.5.34) and (2.5.35)]. For example, when the dipole is over a slab, as can be seen from Figure 2.6.7, a leaky pole or the branch point k_3 may be close to the saddle point. Another example is the expression given by (28), which may not be uniformly valid due to the presence of singularities in the vicinity of the saddle point. In these cases, uniform asymptotic expansions are needed to derive uniform approximations for the integrals.

Consider first the integral given by Equation (2). Assuming that $d_1 \approx 0$, or small, so that the source is close to the interface, then the dominant rapid oscillations in the integrand are coming from $H_0^{(1)}(k_\rho \rho)e^{-ik_{2z}z}$. Next, using the large argument approximation of the Hankel function, one can show that the saddle point is given by $k_\rho = k_{\rho s} = k_2 \sin \theta$, where $\theta = \sin^{-1}(\rho/r)$ and $r = (\rho^2 + z^2)^{1/2}$. Moreover, there are branch-point singularities at k_1 and k_2. The steepest-descent path through the saddle point together with the branch points are shown in Figure 2.6.11, where it is assumed that $k_1 < k_2$. Note that if $k_2 \sin \theta > k_1$, then the k_1 branch-point contribution has to be included similarly as in the case of the k_3 branch-point contribution in Figure 2.6.7. The fact is when $k_2 \sin \theta \simeq k_1$, i.e., the saddle point is in the vicinity of the branch-point singularity, the steepest-descent method breaks down. Hence, when $\theta \simeq \theta_c = \sin^{-1}(k_1/k_2)$, a uniform asymptotic approximation is needed.

Now, assuming that $\rho \to \infty$, one can use the large argument approximation for the Hankel function in (2) to arrive at

$$H_{2z} \sim -\frac{IA}{4\pi}\sqrt{\frac{2i}{\pi\rho}}\int_{-\infty}^{\infty}dk_\rho\,\frac{k_\rho^{5/2}}{k_{1z}+k_{2z}}e^{ik_\rho\rho-ik_{2z}z}, \qquad (2.6.29)$$

where it is assumed that $d_1 = 0$, or that the source is right at the interface. Next, we multiply the numerator and the denominator of the integrand by $k_{1z} - k_{2z}$ to arrive at

$$H_{2z} \sim -\frac{IA}{4\pi}\sqrt{\frac{2i}{\pi\rho}}\,\frac{1}{k_1^2-k_2^2}\int_{-\infty}^{\infty}dk_\rho\,k_\rho^{5/2}(k_{1z}-k_{2z})e^{ik_\rho\rho-ik_{2z}z}. \qquad (2.6.30)$$

In the above, the term involving k_{1z} has a branch-point singularity at k_1 and a saddle point at $k_2\sin\theta$ which can be close together. Hence, to arrive at an approximation valid for all θ, a uniform asymptotic approximation is required. Consequently, the integral involving only k_{1z} is expressible as

$$I = \int_{-\infty}^{\infty}dk_\rho\,k_\rho^{5/2}(k_1-k_\rho)^{1/2}(k_1+k_\rho)^{1/2}e^{ik_\rho\rho-ik_{2z}z}. \qquad (2.6.31)$$

Next, by comparing (31) with (2.5.36), k_ρ is identified with t, λ with $r = \sqrt{\rho^2+z^2}$, $h(t)$ with $ik_\rho(\rho/r) - ik_{2z}(z/r)$, t_b with k_1, k with $1/2$, and $f(t)$ with $ik_\rho^{5/2}(k_1+k_\rho)^{1/2}$. Furthermore, in (2.5.37), we identify $-\lambda s_b^2 = \lambda h(t_b) - \lambda h(t_0) = ik_1\rho - i\sqrt{k_2^2-k_1^2}\,z - ik_2 r$. Then, the uniform asymptotic approximation to I is given by (2.5.19):

$$I \sim \frac{e^{ik_1\rho-i\sqrt{k_2^2-k_1^2}\,z}}{\sqrt{2r}}\left[\frac{\gamma_0}{(2r)^{1/4}}W_{\frac{1}{2}}\left(\sqrt{2r}\,s_b\right) + \frac{\gamma_1}{(2r)^{3/4}}W_{\frac{3}{2}}\left(\sqrt{2r}\,s_b\right)\right]. \qquad (2.6.32)$$

In the above, because the original path of integration, the SIP in Figure 2.6.11, is below the singularity, (2.5.46b) is used for the definition of the W functions above.

When $\sqrt{2r}\,s_b$ is large, (2.5.48) will be used to obtain the approximation to Equation (32). Depending on whether the saddle point $k_2\sin\theta$ is larger or smaller than k_1, s_b will have different phases. The different phases imply that different large argument approximations are needed from Equation (2.5.48). Consequently, a branch-cut contribution will surface in the large argument approximation when $k_2\sin\theta > k_1$. From the second term of (2.5.54), however, it is seen that the branch-point contribution is of the form

$$I_b \sim \frac{e^{ik_1\rho-i\sqrt{k_2^2-k_1^2}\,z}}{(2r)^{3/2}}\,\frac{2\pi i F(s_b)}{\Gamma(-\frac{1}{2})(-s_b)^{3/2}}. \qquad (2.6.33)$$

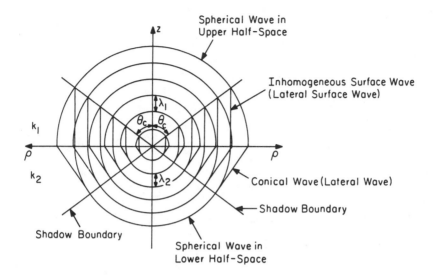

Figure 2.6.12 The behavior of waves generated by a point source placed on top of a half-space, where $k_1 < k_2$.

Notice that the above is a conical wave. It is observed only for $\theta > \theta_c = \sin^{-1}(k_1/k_2)$.

The other term in Equation (30) does not have a branch point close to the saddle point. A subsequent steepest-descent analysis of this term then yields a spherical wave $\exp(ik_2 r)/r$ analogous to (2.5.10) or (6) [see also Exercise 2.26(c)].

Hence, with the information from (2.5.10), (6), and the above, we see that when a dipole is at the interface of a half-space, it will generate waves, which are sketched approximately in Figure 2.6.12. The source will generate spherical waves, which are due to saddle-point contributions, both in medium 1 and medium 2. Since $k_2 > k_1$, the wavelength of the wave will be smaller in medium 2. Hence, at the interface, the phases of the spherical waves do not match with each other. To compensate for this, lateral waves, which are due to the branch-point contributions, are generated in each region. Consequently, the lateral wave observed in medium 1 travels in the ρ direction with the phase velocity of medium 2, and hence is evanescent in region 1, since $k_2 > k_1$. It is also known as the inhomogeneous surface wave. In medium 2, the lateral wave travels with the phase velocity of medium 1 in the ρ direction. Since $k_2 > k_1$, however, it is also propagating in the z direction. Hence, the lateral wave forms a conical wave in medium 2. The fact is that a lateral wave in a half-space is induced by the spherical wave in the other half-space.

In medium 2, the lateral wave is observed only for $\theta > \theta_c$, because only then the branch-point contribution at k_1 in Figure 2.6.11 needs to be included. Consequently, there is a boundary which delineates the existence and absence of the lateral wave. This is known as the **shadow boundary**. It is when the observation angle is at the shadow boundary, i.e., $\theta = \theta_c$, that the saddle point coincides with the branch point k_1. Therefore, in this vicinity, a uniform asymptotic approximation to the field is required. Moreover, it is well known that geometrical optics or ray optics theory breaks down at shadow boundaries. Hence, the simple stationary-phase method or steepest-descent method only reproduces geometrical optics theory. Note further that at the shadow boundaries, i.e., $\theta = \theta_c$, the lateral waves are "phase-matched" to the spherical waves.

Another place where uniform asymptotic approximation is useful is in the approximations of Equations (23) and (26) (see Exercise 2.29 and 2.30; also see Chew and Kong 1981, 1982; Chew and Gianzero 1981).

§§2.6.5 Angular Spectrum Representation

This section shall not be complete without mentioning the angular spectrum representation (Born and Wolf 1980, p. 562). Notice that if a large argument approximation to the Hankel function is used, Equation (4) for a VMD becomes

$$H_{1z}^R \sim -\frac{iIA}{8\pi}\sqrt{\frac{2}{i\pi\rho}}\int\limits_{-\infty}^{\infty} dk_\rho\, \frac{k_\rho^{\frac{5}{2}}}{k_{1z}} R_{12}^{TE} e^{ik_\rho\rho + ik_{1z}(z+2d_1)}. \tag{2.6.34}$$

By letting $k_\rho = k_1 \sin\alpha$, then $k_{1z} = (k_1^2 - k_\rho^2)^{1/2} = k_1 \cos\alpha$. Furthermore, by letting $\rho = r_I \sin\theta_I$, $(z + 2d_1) = r_I \cos\theta_I$, we convert (34) into

$$H_{1z}^R \sim -\frac{iIA}{8\pi}\sqrt{\frac{2}{i\pi\rho}}\int_\Gamma d\alpha\, (k_1 \sin\alpha)^{\frac{5}{2}} R_{12}^{TE} e^{ik_1 r_I \cos(\theta_I - \alpha)}. \tag{2.6.35}$$

The above is an integration over all possible angles of α that correspond to $-\infty < k_\rho < \infty$. Here, α is the angle the vector $\mathbf{k}_1 = \hat{\rho}k_1 \sin\alpha + \hat{z}k_1 \cos\alpha$ makes with the z axis. Hence, when $-k_1 < k_\rho < k_1$, α ranges between $-\frac{\pi}{2}$ to $\frac{\pi}{2}$. But when $|k_\rho| > k_1$, α becomes complex. Consequently, the path Γ is as shown in Figure 2.6.13. Moreover, it is easy to show that the saddle point of the integrand is at $\alpha = \theta_I$; the steepest-descent path that passes through the saddle point is shown in Figure 2.6.13.

Notice that in the angular spectrum representation, $k_{1z} = k_1 \cos\alpha$ is no longer a double-value function, implying that the branch point at $k_\rho = k_1$ is now removed. But the branch point at $k_\rho = k_2$ is transformed to

$$\alpha = \sin^{-1}\frac{k_2}{k_1} = \frac{\pi}{2} - i\theta_B,$$

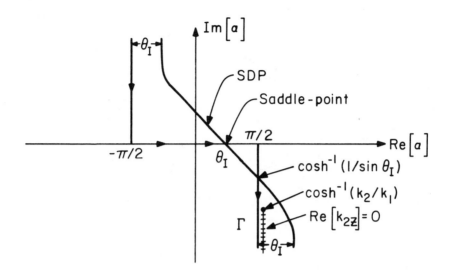

Figure 2.6.13 The steepest-descent path (SDP) on the complex α plane in angular spectrum representation.

where

$$\theta_B = \cosh^{-1}\left(\frac{k_2}{k_1}\right).$$

Figure 2.6.13 shows the $\Re e[k_{2z}] = 0$ branch cut. Hence, when the original path of integration is deformed to the steepest-descent path, the k_2 branch-point contribution has to be included if $k_2 \sin \theta_I > k_1$.

The result derived using the angular spectrum representation is the same as that from k_ρ representation (see Exercise 2.31). The advantage of the angular spectrum representation is that the exponential term in the integrand of (35) is simpler.

§2.7 Singularities of the Sommerfeld Integrals

The integrands of the Sommerfeld integrals have singularities on the complex k_ρ plane. As has been shown, these singularities affect the results of asymptotic expansions and uniform asymptotic expansions. They also affect the definition of the integration paths if the Sommerfeld integrals are to be evaluated numerically. Therefore, it is worthwhile to discuss the general properties of these singularities.

There are two basic types of singularities—the pole singularities and the branch-point singularities. The pole singularities correspond to guided modes in the layered medium. The branch points, as will be shown in Chapter 6,

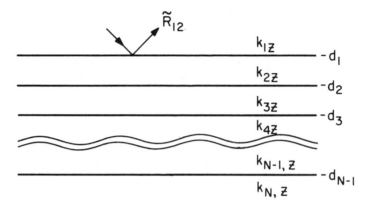

Figure 2.7.1 There could only be branch points associated with k_{1z} and k_{Nz} in an N-layer medium.

correspond to radiation modes. These radiation modes form a continuum of modes.

In addition, for the layered medium, the branch points are only associated with the outermost regions, as will be shown in the next subsection. Subsequently, we shall show bounds on the locations of the pole singularities of the integrand. This knowledge is useful when one wishes to numerically integrate the Sommerfeld integrals, or even when the asymptotic expansion of the integral is desired.

§§2.7.1 Absence of Branch Points

In a three-layer problem, the reflection coefficient is an even function of k_{2z}, and hence, is only a multivalue function due to k_{1z} and k_{3z}, which are double-value functions of k_ρ.[13] Therefore, branch points only exist at $k_\rho = k_1$ and $k_\rho = k_3$ (see Exercise 2.28). It turns out that even for an N-layer medium as shown in Figure 2.7.1, the generalized reflection coefficient $\tilde{R}_{12}$ is still only a multivalue function due to k_{1z} and k_{Nz}, although from Equation (2.1.23), it appears to be a double-value function due to all k_{iz}, as $k_{iz} = (k_i^2 - k_\rho^2)^{1/2}$ is a double-value function.

To prove that $\tilde{R}_{12}$ is not a double-value function due to k_{iz}, it has to be shown that the value of $\tilde{R}_{12}$ is unaffected when the sign of k_{iz} changes. This can be seen from the following argument. In solving for $\tilde{R}_{12}$, for instance, we

---

[13] Results in this subsection can also be proved using integral equations. Since integral equation will be introduced later, we provide an alternative proof here. Proof by direct algebraic manipulation is also given by Felsen and Marcuvitz (1973).

write down the general field expression in the i-th region as

$$e_i = A_i e^{-ik_{iz}z} + B_i e^{ik_{iz}z}. \tag{2.7.1}$$

Then, A_i and B_i are determined by matching boundary conditions at each interface, which in turn determine $\tilde{R}_{12}$. Now, if the problem were solved over again by switching the sign of k_{iz} in region i, where $i \neq 1$ and $i \neq N$, then the field in the i-th region is still represented by a superposition of upgoing and downgoing waves as in (1). Hence, if the solution of the problem is again solved for using the same set of boundary conditions. By the uniqueness principle, there could only be one unique solution, since the field in the i-th region is still represented by a complete set of waves, namely, upgoing and downgoing waves. Therefore, the value of $\tilde{R}_{12}$ found will be unchanged, and it will not depend on the sign of k_{iz} where $i \neq 1$ and $i \neq N$. In other words, $\tilde{R}_{12}$ is indeed not a double-value function due to k_{iz}, where $i \neq 1$ and $i \neq N$, even though k_{iz} is a double-value function of k_ρ.

In region N, however, the field expression is

$$e_N = A_N e^{-ik_{Nz}z}. \tag{2.7.2}$$

Hence, if the sign of k_{Nz} were switched, an upgoing wave in region N would ensue. Therefore, a completely different boundary value problem is being solved in this case. Consequently, the value of $\tilde{R}_{12}$ obtained would be different. Hence, $\tilde{R}_{12}$ is a double-value function of k_{Nz} since its value changes when k_{Nz} switches sign.

Similarly, if the sign of k_{1z} in region 1 is switched, the roles of the upgoing and the downgoing waves would be reversed. Hence, $\tilde{R}_{12}(-k_{1z}) \neq \tilde{R}_{12}(k_{1z})$. In particular, since the roles of upgoing and downgoing waves are swapped,

$$\tilde{R}_{12}(-k_{1z}) = \frac{1}{\tilde{R}_{12}(k_{1z})}. \tag{2.7.3}$$

By the same token, if the sign of k_{iz} is switched, the roles of A_i and B_i in (1) are reversed; therefore,

$$A_i(-k_{iz}) = B_i(k_{iz}). \tag{2.7.4}$$

Even for the case when the source is embedded in a layered medium (Exercises 2.15, 2.28), it follows from the uniqueness principle that the branch points are always associated with the outermost layers for the integrands of the Sommerfeld integrals.

The fact that branch points are only associated with the outermost layers also has a physical significance. It was seen previously that branch points are physically associated with lateral waves; hence, it implies that a lateral wave propagates only at the outermost boundaries because it is actually the part of a spherical wave close to the interface of a half-space. Therefore, a lateral wave cannot exist if a spherical wave cannot exist. Since a spherical wave exists only in an unbounded region, a lateral wave is always associated with an unbounded region 1 or N. Here is an example for which the results of mathematics and physics are in perfect harmony with each other!

As a final note, the above proof is also adaptable to cylindrically layered media (Chew 1983).

§§2.7.2 *Bounds on the Locations of Singularities*

In performing the asymptotic expansion and numerical integration of a Sommerfeld integral, it is imperative to know the locations of the singularities of the integrand on the complex k_ρ plane. Subsection 2.7.1 shows how the branch-point singularities are easily identified since they are related to the outermost, unbounded regions in a layered medium. The pole locations are more elusive. Their domain, however, can be ascertained via an energy conservation argument in a manner similar to that used in quantum scattering (Newton 1966, p. 367).

The poles of the integrand of a Sommerfeld integral correspond to guided modes of a layered medium. Furthermore, the integrand of a Sommerfeld integral satisfies (2.1.8a) or (2.1.8b) (depending on whether it is a TE or TM wave) with k_ρ replacing k_x. For example,

$$\left[\frac{d}{dz} \epsilon^{-1}(z) \frac{d}{dz} + k^2(z)\epsilon^{-1}(z) - k_\rho^2 \epsilon^{-1}(z) \right] \phi(z) = 0, \qquad (2.7.5)$$

where $\phi(z)$ represents the integrand of a TM wave as in (2.3.6). Then, on multiplying (5) by $\phi^*(z)$ and integrating from $-\infty$ to $+\infty$, and assuming $\phi(z)$ to be the field of a guided mode and k_ρ its corresponding wavenumber, we have

$$-\int_{-\infty}^{\infty} dz\, \epsilon^{-1} \left| \frac{d\phi}{dz} \right|^2 + \int_{-\infty}^{\infty} dz\, k^2 \epsilon^{-1} |\phi|^2 - k_\rho^2 \int_{-\infty}^{\infty} dz\, \epsilon^{-1}|\phi|^2 = 0, \qquad (2.7.6)$$

where integration by parts has been used to simplify the first term in (6). These integrals converge because $\phi(z)$ represents a guided mode whose field vanishes as $|z| \to \infty$.

Next, after taking the real part of (6), we have

$$\int_{-\infty}^{\infty} dz\, \Re e(k_\rho^2 \epsilon^{-1}) |\phi^2| = \int_{-\infty}^{\infty} dz\, \Re e(k^2 \epsilon^{-1}) |\phi|^2 - \int_{-\infty}^{\infty} dz\, \Re e(\epsilon^{-1}) \left| \frac{d\phi}{dz} \right|^2. \qquad (2.7.7)$$

Since $\Re e(\epsilon^{-1}) > 0$, then

$$\int_{-\infty}^{\infty} dz\, \Re e(k_\rho^2 \epsilon^{-1}) |\phi|^2 < \int_{-\infty}^{\infty} dz\, \Re e(k^2 \epsilon^{-1}) |\phi|^2. \qquad (2.7.8)$$

But

$$\int_{-\infty}^{\infty} dz\, \Re e(k_\rho^2 \epsilon^{-1}) |\phi|^2 \geq \min \Re e(k_\rho^2 \epsilon^{-1}) \int_{-\infty}^{\infty} dz\, |\phi|^2, \qquad (2.7.9a)$$

and

$$\int_{-\infty}^{\infty} dz\, \Re e(k^2 \epsilon^{-1}) |\phi|^2 \leq \max \Re e(k^2 \epsilon^{-1}) \int_{-\infty}^{\infty} dz\, |\phi|^2, \qquad (2.7.9b)$$

where min or max $f(z)$ is the minimum or maximum of $f(z)$ respectively for $-\infty < z < \infty$. Therefore, (8) implies that

$$\min \Re e(k_\rho^2 \epsilon^{-1}) < \max \Re e(k^2 \epsilon^{-1}). \qquad (2.7.10)$$

Because $\Re e(k_\rho^2 \epsilon^{-1}) = \Re e(k_\rho^2)\Re e(\epsilon^{-1}) - \Im m(k_\rho^2)\Im m(\epsilon^{-1})$, we have

$$\min \Re e(k_\rho^2 \epsilon^{-1}) \geq \Re e(k_\rho^2) \min \Re e(\epsilon^{-1}) - \Im m(k_\rho^2) \max \Im m(\epsilon^{-1}), \qquad (2.7.11)$$

if $\Re e(k_\rho^2)$ and $\Im m(k_\rho^2)$ are positive. In arriving at the above, we have used

$$\min [a(z) + b(z)] \geq \min a(z) + \min b(z),$$

and that

$$\min [Ab(z)] = \begin{cases} A \min b(z), & \text{if } A > 0, \\ A \max b(z), & \text{if } A < 0. \end{cases} \qquad (2.7.11a)$$

As $\Im m(\epsilon^{-1}) \leq 0$ for a passive medium, we have

$$\min \Re e(k_\rho^2 \epsilon^{-1}) \geq \Re e(k_\rho^2) \min \Re e(\epsilon^{-1}), \quad \text{if } \Im m(k_\rho^2) > 0 \qquad (2.7.12)$$

and (10) becomes

$$\Re e(k_\rho^2) \min \Re e(\epsilon^{-1}) < \max \Re e(k^2 \epsilon^{-1}), \qquad (2.7.13)$$

defining an upper bound for $\Re e(k_\rho^2)$ as shown in Figure 2.7.2(a) if k_ρ^2 is a solution of (5). A tighter bound than the above is arrived at if $\Re e(k_\rho^2) < 0$ but it is encompassed by the above bound.

Similarly, by taking the imaginary part of (6), we have

$$\int_{-\infty}^{\infty} dz\, \Im m(k_\rho^2 \epsilon^{-1}) |\phi|^2 = \int_{-\infty}^{\infty} dz\, \Im m(k^2 \epsilon^{-1}) |\phi|^2 - \int_{-\infty}^{\infty} dz\, \Im m(\epsilon^{-1}) \left| \frac{d\phi}{dz} \right|^2. \qquad (2.7.14)$$

Equation (14) implies that

$$\max \Im m(k_\rho^2 \epsilon^{-1}) \int_{-\infty}^{\infty} dz\, |\phi|^2 \geq \min \Im m(k^2 \epsilon^{-1}) \int_{-\infty}^{\infty} dz\, |\phi|^2$$

$$- \max \Im m(\epsilon^{-1}) \int_{-\infty}^{\infty} dz\, \left| \frac{d\phi}{dz} \right|^2. \qquad (2.7.15)$$

The preceding inequality also implies that

$$\max \Im m(k_\rho^2 \epsilon^{-1}) \geq \min \Im m(k^2 \epsilon^{-1}) - \max \Im m(\epsilon^{-1})\eta, \qquad (2.7.16)$$

where

$$\eta = \int_{-\infty}^{\infty} dz \left| \frac{d\phi}{dz} \right|^2 \bigg/ \int_{-\infty}^{\infty} dz \, |\phi|^2. \qquad (2.7.16a)$$

Since $\Im m(k_\rho^2 \epsilon^{-1}) = \Im m(k_\rho^2)\Re e(\epsilon^{-1}) + \Re e(k_\rho^2)\Im m(\epsilon^{-1})$, therefore,

$$\max \Im m(k_\rho^2 \epsilon^{-1}) \leq \Im m(k_\rho^2) \max \Re e(\epsilon^{-1}) + \Re e(k_\rho^2) \max \Im m(\epsilon^{-1}),$$
$$(2.7.17)$$

if $\Im m(k_\rho^2)$ and $\Re e(k_\rho^2)$ are positive. In arriving at the above, we have used $\max [a(z)+b(z)] \leq \max a(z) + \max b(z)$. Consequently, from (16) and (17),

$$\Im m(k_\rho^2) \max \Re e(\epsilon^{-1}) \geq \min \Im m(k^2 \epsilon^{-1}) - \max \Im m(\epsilon^{-1})[\eta + \Re e(k_\rho^2)].$$
$$(2.7.18)$$

If $\Re e(k_\rho^2) > 0$, the right-hand side of the above is positive because $\Im m(\epsilon^{-1}) \leq 0$ for a passive medium. Hence, $\Im m(k_\rho^2)$ is bounded from below by a positive number.

From (5) and (16a), note that $k_\rho^2 \to -|k_\rho|^2 \sim -\eta$ when $|k_\rho| \to \infty$. Hence, the above bound is invalid when $|k_\rho| \to \infty$ because $\Re e(k_\rho^2) < 0$. But when $\Re e(k_\rho^2)$ and $\Im m(k_\rho^2)$ are negative, using the equivalence of (11a) for "max", the equivalence of (18) is

$$\Im m(k_\rho^2) \min \Re e(\epsilon^{-1}) \geq \min \Im m(k^2 \epsilon^{-1}) - \max \Im m(\epsilon^{-1})\eta$$
$$- \min \Im m(\epsilon^{-1})\Re e(k_\rho^2). \qquad (2.7.19)$$

Since $k_\rho^2 \to -\eta$ when $|k_\rho| \to \infty$, the last two terms in the above become

$$\eta [\min \Im m(\epsilon^{-1}) - \max \Im m(\epsilon^{-1})], \quad |k_\rho| \to \infty.$$

Moreover, since η is a large number, the inequality (19) is approximately

$$\Im m(k_\rho^2) \min \Re e(\epsilon^{-1}) \geq \eta [\min \Im m(\epsilon^{-1}) - \max \Im m(\epsilon^{-1})], \quad |k_\rho| \to \infty.$$
$$(2.7.20)$$

Equations (18) and (20) together give us a lower bound locus as shown in Figure 2.7.2(a), since $\eta \sim -k_\rho^2$, when $|k_\rho| \to \infty$.

From (14), in a manner similar to deriving (16), we can further deduce that

$$\min \Im m(k_\rho^2 \epsilon^{-1}) \leq \max \Im m(k^2 \epsilon^{-1}) - \min \Im m(\epsilon^{-1})\eta \qquad (2.7.21)$$

Similar to (17), we have

$$\min \Im m(k_\rho^2 \epsilon^{-1}) \geq \Im m(k_\rho^2) \min \Re e(\epsilon^{-1}) + \Re e(k_\rho^2) \min \Im m(\epsilon^{-1}),$$
$$(2.7.22)$$

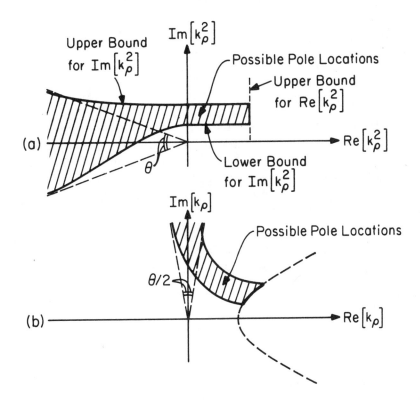

Figure 2.7.2 Possible pole locations on the complex k_ρ^2 and k_ρ planes.

if $\Im m(k_\rho^2)$ and $\Re e(k_\rho^2)$ are positive. Equation (21) together with (22) yields

$$\Im m(k_\rho^2)\, \min \Re e(\epsilon^{-1}) \leq \max \Im m(k^2\epsilon^{-1}) - \min \Im m(\epsilon^{-1})[\eta + \Re e(k_\rho^2)]. \tag{2.7.23}$$

When $\Re e(k_\rho^2) < 0$, similar to (19) and (20), we have approximately

$$\Im m(k_\rho^2)\, \min \Re e(\epsilon^{-1}) \leq \eta\,[\max \Im m(\epsilon^{-1}) - \min \Im m(\epsilon^{-1})]. \tag{2.7.24}$$

Equations (24) and (23) together define a upper bound locus for $\Im m(k_\rho^2)$ as shown in Figure 2.7.2(a).

The possible region of pole locations is then defined as shown in Figure 2.7.2(a). Since a straight line on k_ρ^2 plane maps into a hyperbola on a k_ρ plane, the possible pole locations on the complex k_ρ plane are as shown in Figure 2.7.2(b) (see Exercise 2.32). (These bounds simplify for TE waves in the case of nonmagnetic material as shown in Exercise 2.32.) If the $\Im m(\epsilon^{-1}) = 0$ but $\Im m(k^2) \neq 0$, then the bounds given by (16) and (21) simplify since the second terms on the right-hand sides equal zero. Consequently, the upper and lower bound loci for $\Im m(k_\rho^2)$ reduce to straight lines. As a result, the possible pole locations are as those shown in Figure 2.7.4.

Notice that the bounds for $\Im m(k_\rho^2)$ become looser when $\Re e(k_\rho^2) \to -\infty$, as shown in Figure 2.7.2(a). However, since $k_\rho^2 \to -\eta$ when $|k_\rho| \to \infty$, tighter bounds should exist for $\Im m(k_\rho^2)$ when $|k_\rho| \to \infty$.

§§2.7.3 Numerical Integration of Sommerfeld Integrals

Even though asymptotic methods greatly accelerate the process of field calculation, they have the weakness of being problem specific. For instance, asymptotic expansions often have to be rederived for different geometries. One way of computing the field in a robust manner so that the computer program works under a wide variety of conditions is by the numerical integration of Sommerfeld integrals.

To save computation time, Sommerfeld integrals are often folded to range from 0 to ∞. For example, Equation (2.3.5) can be folded by the reverse process of going from (2.2.30) to (2.2.31). Then it becomes

$$
E_{1z} = -\frac{I\ell}{4\pi\omega\epsilon_1} \int\limits_0^\infty dk_\rho \, \frac{k_\rho^3}{k_{1z}} J_0(k_\rho\rho) \left[e^{ik_{1z}|z|} + \tilde{R}_{12}^{TM} e^{ik_{1z}z + 2ik_{1z}d_1} \right].
$$

$$(2.7.25)$$

The integrand has branch points and poles which could be right on the real axis if the medium is lossless. But the error of an integration routine, e.g., the Simpson's rule (Hildebrand 1974), is proportional to the derivatives of the integrand. Therefore, at the singularities, the error of integration is intolerable. One way of defining a robust numerical integration path is shown in Figure 2.7.3. By Cauchy's theorem, the integral value is unchanged along this new integration path. Adaptive Simpson's rule or Romberg's rule (Hildebrand 1974) whereby the discretization of the integration step is chosen to adapt to the oscillations in the integral can be used to integrate such integrals.

Note that when $z = 0$, the first term of the integrand in (25) is nonconvergent; and hence, numerical integration is impossible. This term is the direct field term which has a closed-form expression because it is the response of a point source in a homogeneous medium. The reason for its nonconvergence is similar to that given after Equation (2.2.11). In other words, the $z = 0$ plane contains the source-point singularity rendering the Fourier transform of the field nonabsolutely integrable.

When the observation point is far from the source point, however, the integrand becomes a rapidly oscillating function of k_ρ, which is the motivation behind the stationary phase method. Numerically, such a rapidly oscillating integrand can be integrated with Filon quadrature (Hildebrand 1974). One way of avoiding this oscillation is to integrate the Sommerfeld integral along

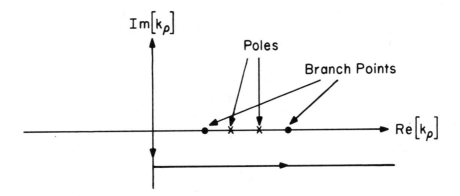

Figure 2.7.3 A robust way of defining a numerical integration path for the Sommerfeld integrals.

the steepest-descent path. For example, we can rewrite (2.6.4) as

$$H_{1z}^R = -\frac{iIA}{8\pi} \int_{-\infty}^{\infty} dk_\rho \frac{k_\rho^3}{k_{1z}} \hat{H}_0^{(1)}(k_\rho\rho) R_{12}^{TE} e^{ik_\rho\rho + ik_{1z}z + 2ik_{1z}d_1}, \qquad (2.7.26)$$

where $\hat{H}_0^{(1)}(k_\rho\rho) = H_0^{(1)}(k_\rho\rho)e^{-ik_\rho\rho}$. In the manner of (2.5.13), by letting

$$ik_\rho\rho + ik_{1z}(z + 2d_1) = -\lambda s^2 + ik_1 r_I, \qquad (2.7.27)$$

where $r_I = [\rho^2 + (z + 2d_1)^2]^{1/2}$, (27) becomes, via a change of variable,

$$H_{1z}^R = -\frac{iIA}{8\pi} e^{ik_1 r_I} \int_C ds \frac{dk_\rho}{ds} \frac{k_\rho^3}{k_{1z}} \hat{H}_0^{(1)}(k_\rho\rho) R_{12}^{TE} e^{-\lambda s^2}. \qquad (2.7.28)$$

Moreover, k_ρ can be found in closed form in terms of s via (27) and the integral in (28) is well defined. C is the mapping of the real axis on the complex k_ρ plane to the complex s plane. If (28) is now deformed to integrate along the real axis on the s plane, it would be equivalent to integrating along the steepest-descent path as shown in Figure 2.6.11 on the k_ρ plane. We must, however, include contributions from singularities encountered when deforming the integration path.

By integrating along the steepest-descent path, a certain amount of bookkeeping is still necessary because the contributing-pole locations need to be found. An alternative method is to deform the integration path to a path that asymptotically approaches the steepest-descent path when $|k_\rho| \rightarrow \infty$,

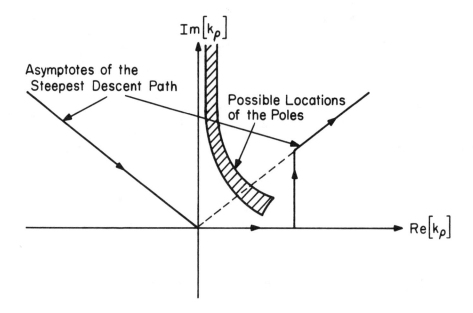

Figure 2.7.4 A fast convergent integration path where pole tracking is unnecessary.

but no singularities need to be enclosed in the course of deformation.[14] As shown in the previous subsection and Exercise 2.32, the poles can only be in the shaded region as shown in Figure 2.7.2(b). Hence, by deforming the integration path to one shown in Figure 2.7.4, where the end points of the integration are asymptotic to the steepest-descent path, we can avoid excessive bookkeeping of having to keep track of the poles on the complex k_ρ plane and, at the same time, obtain good convergence in the integral.

Another way of numerically integrating the Sommerfeld integral is to integrate along the branch cuts (which requires pole tracking). This avoids the slow convergence of the integral when z and z' are equal or are close to an interface. Yet another method is to subtract out the slowly convergent part of the integral and integrate it in closed form. Sometimes, a pole close to an integration path gives rise to poor numerical convergence, but it can be subtracted from the integrand to be evaluated in closed form to accelerate convergence (see Exercise 2.33).

The integral (26) resembles a Fourier transform with the kernel $e^{ik_\rho\rho}$. Hence, if the field values at numerous ρ are required, the fast Fourier trans-

---

[14] This path was first proposed by G. Minerbo.

form (FFT, see Oppenheim and Schafer 1975) can be used to evaluate the integral (Lee et al. 1986; Franke and Swenson 1989). Moreover, when ρ is large, $e^{ik_\rho \rho}$ is rapidly oscillating, requiring a high sampling rate to reduce the aliasing effect. Since the factor multiplying $e^{ik_\rho \rho}$ is costly to evaluate in the event of a high sampling rate, interpolations of the factor for high sampling rate can reduce the cost.

§2.8 WKB Method

The WKB method[15] is named after Wentzel, Kramers, and Brillouin even though others before them have used similar ideas. It is a high-frequency method of obtaining approximate solutions to an equation of the form

$$\frac{d^2\phi}{dz^2} + k_z^2(z)\,\phi = 0, \qquad (2.8.1)$$

where $k_z^2(z) = k^2(z) - k_x^2 = k^2(z)\cos^2\theta(z)$ if we let $k_x = k(z)\sin\theta(z)$. Notice that for a TE wave propagating in a nonmagnetic, one-dimensional inhomogeneity, the equation is already in the form of (1). For TM waves, however, the equation is

$$\left[\epsilon\frac{d}{dz}\epsilon^{-1}\frac{d}{dz} + k_z^2(z)\right]\psi = 0. \qquad (2.8.2)$$

The above could be cast into a form similar to (1) via the transformation $\phi = \epsilon^{-\frac{1}{2}}\psi$. Then, one can show that

$$\frac{d^2}{dz^2}\phi = \psi\frac{d^2}{dz^2}\epsilon^{-\frac{1}{2}} + 2\frac{d\psi}{dz}\frac{d}{dz}\epsilon^{-\frac{1}{2}} + \epsilon^{-\frac{1}{2}}\frac{d^2}{dz^2}\psi. \qquad (2.8.3)$$

Furthermore, we deduce from (2) that

$$\frac{d^2\psi}{dz^2} = -k_z^2\psi + \epsilon^{-1}\frac{d\epsilon}{dz}\frac{d\psi}{dz}. \qquad (2.8.4)$$

On substituting into (3), we have

$$\frac{d^2}{dz^2}\phi + \left[k_z^2 - \epsilon^{\frac{1}{2}}\frac{d^2}{dz^2}\epsilon^{-\frac{1}{2}}\right]\phi = 0. \qquad (2.8.5)$$

Equation (5) is similar to (1) except for the

$$\epsilon^{\frac{1}{2}}\frac{d^2}{dz^2}\epsilon^{-\frac{1}{2}}$$

term. This term appears only in TM waves because, unlike TE waves, TM waves have a z component of the electric field which gives rise to polarization charges due to the variation in $\epsilon(z)$. When the frequency is very high,

---

[15] This topic is discussed in books on layered media and quantum mechanics.

however, $k_z = \frac{\omega}{c}\cos\theta \to \infty$, and the third term in (5) could be neglected if $\epsilon(z)$ is sufficiently smooth, or

$$k_z^2 \gg \epsilon^{\frac{1}{2}}\frac{d^2}{dz^2}\epsilon^{-\frac{1}{2}}. \tag{2.8.6}$$

In this case, both TE and TM waves may be described by Equation (1).

§§2.8.1 Derivation of the WKB Solution

Equation (1) is a one-dimensional wave equation. The solutions are waves with exponential dependence. Hence, consider first a solution of the form

$$\phi(z) = Ae^{i\omega\tau(z)}. \tag{2.8.7}$$

Then, it follows that

$$\phi''(z) = \{i\omega\tau''(z) - [\omega\tau'(z)]^2\}Ae^{i\omega\tau(z)}, \tag{2.8.8}$$

and (1) becomes

$$i\omega\tau''(z) - [\omega\tau'(z)]^2 + k_z^2(z) = 0. \tag{2.8.9}$$

The above problem can be solved via the use of a perturbation series by letting

$$\tau(z) = \tau_0(z) + \frac{1}{\omega}\tau_1(z) + \ldots, \quad \omega \to \infty. \tag{2.8.10}$$

Note that

$$i\omega\tau_0''(z) \ll k_z^2(z), \quad \omega \to \infty, \tag{2.8.11}$$

because k_z^2 is proportional to ω^2. Therefore, after substituting (10) into (9), collecting leading-order terms when $\omega \to \infty$, we have

$$[\omega\tau_0'(z)]^2 = k_z^2(z) = \omega^2 s_z^2(z), \tag{2.8.12}$$

where $s_z = k_z/\omega$, which is independent of frequency. Also, s_z is sometimes known as the slowness of a wave. Consequently, on solving (12), one arrives at

$$\tau_0(z) = \pm \int_{z_0}^{z} dz'\, s_z(z') + C_0. \tag{2.8.13}$$

By collecting terms of first order in (9), we have

$$i\omega\tau_0''(z) - 2\omega\tau_0'(z)\tau_1'(z) = 0. \tag{2.8.14}$$

This gives

$$\tau_1(z) = \frac{i}{2}\ln\tau_0'(z) + C_1 = \frac{i}{2}\ln s_z(z) + C_{1\pm}. \tag{2.8.15}$$

Finally, using (13) and (15) in (10) results in

$$\tau(z) = \pm \int_{z_0}^{z} dz' \, s_z(z') + \frac{i}{2\omega} \ln s_z(z) + C_\pm, \qquad (2.8.16)$$

or that the general leading-order or the first-order approximation to (7) is of the form

$$\phi(z) \sim \frac{A_+}{\sqrt{s_z}} \exp\left[i\omega \int_{z_0}^{z} s_z(z') \, dz'\right] + \frac{A_-}{\sqrt{s_z}} \exp\left[-i\omega \int_{z_0}^{z} s_z(z') \, dz'\right]. \qquad (2.8.17)$$

The integration constants $C_\pm$ in (16) can be absorbed by $A_\pm$.

The first term in (17) corresponds to a right-going wave because its phase is increasing with distance. Similarly, the second term in (17) corresponds to a left-going wave. Moreover, the integral in the exponents implies that the phase gained by a wave going from z_0 to z is proportional to

$$i\omega \int_{z_0}^{z} s_z(z') \, dz', \qquad (2.8.18)$$

which is an integral summation of all the phases gained locally at z' over the range from z_0 to z. This physical picture is true only if the multiple reflections of the wave can be neglected as it is propagating. Furthermore, Equation (11) shows that this physical picture, which corresponds to the leading-order solution, is only valid if

$$\omega s_z'(z) \ll \omega^2 s_z^2(z), \qquad (2.8.19)$$

or

$$\frac{d}{dz} \ln k_z(z) \ll k_z = \frac{2\pi}{\lambda_z}. \qquad (2.8.20)$$

Hence, this physical picture breaks down if the frequency is not high or if $s_z \simeq 0$.

The factor $1/\sqrt{s_z}$ is essential for energy conservation. For instance, in the normal incidence case, where $s_z = s = \sqrt{\mu\epsilon(z)}$, the forward-going wave solution becomes

$$\phi_+(z) \sim \frac{A_+}{(\mu\epsilon)^{\frac{1}{4}}} \exp\left[i\omega \int_{z_0}^{z} s(z') \, dz'\right]. \qquad (2.8.21)$$

Moreover, if this is a TE wave, and $\phi_+(z)$ represents E_y, for example, then the power flow in this wave is

$$S = \frac{1}{2} \sqrt{\frac{\epsilon}{\mu}} \, |\phi_+|^2 = \frac{1}{2\mu} |A_+|^2, \qquad (2.8.22)$$

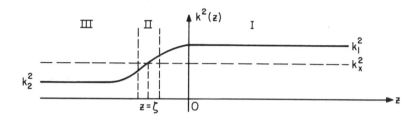

Figure 2.8.1 WKB solutions exist in regions I and III, but not in region II. However, a solution can be found by asymptotic matching.

which is a constant, since μ is a constant as nonmagnetic material is assumed. The above implies energy conservation, which is a consequence of including the $1/\sqrt{s_z}$ term in the solution.

Note that when $s_z = 0$, the solution becomes infinite because the inequality (19), the criterion for the validity of the WKB solution, is violated. This case corresponds to the $\theta = \pi/2$ case implying the onset of critical reflections. Hence, the WKB solution cannot be used at a place where critical reflection occurs.

Because of their physical interpretations, Equation (12) is also known as the *eikonal equation* as in ray optics, and Equation (14) is also known as the *transport equation*, which is essential for energy conservation. These equations can be generalized to higher dimensions albeit with more complexity (Exercise 2.34; also see Kline and Kay 1965; Bremmer and Lee 1984).

§§2.8.2 Asymptotic Matching

Consider a profile with $k^2(z)$ as shown in Figure 2.8.1, with an obliquely incident wave ($k_x > 0$) from the right. In this case, $k_z^2(z) = k^2(z) - k_x^2$. Here, k_x has to be a constant for all z due to the phase matching condition. Moreover, assuming that the wave satisfies Equation (1), then the WKB solution can be derived for the above problem. But for $k_x^2 > k_2^2$, $k_z = 0$ at $z = \zeta$, and the condition (11) or (19) is violated. In such a case, the WKB solution can only be written down in region I to the right of ζ, and in region III to the left, but not in region II; in region II, the WKB solution breaks down. Nevertheless, region II can be approximated very well by a linear profile region whose solution is expressible in terms of Airy functions. Then, the concept of asymptotic matching[16] can be used to patch the solutions in all the three regions together.

---

[16] The idea of asymptotic matching is embodied in the idea of matched asymptotic expansions (Van Dyke 1975; Bender and Orzag 1978).

Consequently, the WKB solution in region I is

$$\phi_I(z) \sim \frac{A_+}{\sqrt{s_z}} \exp\left[i\omega \int_\zeta^z s_z(z')\,dz'\right] + \frac{A_-}{\sqrt{s_z}} \exp\left[-i\omega \int_\zeta^z s_z(z')\,dz'\right]. \tag{2.8.23}$$

Since z_0 in (17) is arbitrary, it is chosen so that $z_0 = \zeta$ in (23). In region III, since $k_x^2 > k^2$, $s_z = i\alpha_z$ is pure-imaginary; however, the inequalities (19) and (20) can still be met. As such, we can write the WKB solution in region III as

$$\phi_{III}(z) \sim \frac{B_-}{\sqrt{\alpha_z}} \exp\left[\omega \int_\zeta^z \alpha_z(z')\,dz'\right]. \tag{2.8.24}$$

In the above, only the solution that is exponentially decaying in the $-z$ direction is included. Here, the exponentially growing solution in the $-z$ direction is eliminated since there is no source at $z \to -\infty$.

In the vicinity of ζ, k_z^2 is approaching zero linearly, and hence,

$$k_z^2 \sim \Omega\,(z - \zeta), \quad z \to \zeta. \tag{2.8.25}$$

Because k_z^2 is proportional to ω^2, Ω is also proportional to ω^2. Subsequently, around ζ, Equation (1) becomes

$$\left[\frac{d^2}{dz^2} + \Omega\,(z - \zeta)\right]\phi = 0. \tag{2.8.26}$$

Next, by letting $\eta = \Omega^{\frac{1}{3}}\,(z - \zeta)$, (26) becomes

$$\left[\frac{d^2}{d\eta^2} + \eta\right]\phi = 0, \tag{2.8.27}$$

which is the Airy equation we have already encountered in Subsection 2.1.5. Therefore, the solution in the vicinity of ζ, or region II, is

$$\phi_{II}(z) \sim C\,Ai(-\eta). \tag{2.8.28}$$

The $Bi(-\eta)$ solution is eliminated because it is exponentially growing when $\eta \to -\infty$, and hence, is unphysical due to the absence of a source at $z \to -\infty$. The unknowns $A_\pm$, B_-, and C in (23), (24), and (28) can be related to each other through asymptotic matching.

Since the profile tends to k_1 when $z \to +\infty$, the field to the far right is of the form

$$\phi_I(z) \sim \phi_0 e^{-ik_{1z}z} + R\phi_0 e^{ik_{1z}z}, \quad z \to +\infty, \tag{2.8.29}$$

where $k_{1z} = \sqrt{k_1^2 - k_x^2}$. Here, R, a reflection coefficient that relates the amplitude of the reflected wave to the incident wave at $z = 0$, is the unknown to be sought. Moreover, R and ϕ_0 can be related to $A_\pm$ in (23) via the following observation.

The integrals in the exponents in ϕ_I can be rewritten as

$$\omega \int_\zeta^z s_z(z')\,dz' = k_{1z}\int_\zeta^z dz' + \int_\zeta^z [\omega s_z(z') - k_{1z}]\,dz'. \qquad (2.8.30)$$

Note that the second integral tends to be a constant when $z \to \infty$, because $\omega s_z - k_{1z} \to 0$, when $z \to \infty$. Hence, (30) becomes

$$\omega \int_\zeta^z s_z(z')\,dz' \sim k_{1z}z + a, \quad z \to \infty, \qquad (2.8.31)$$

where

$$a = \int_\zeta^\infty [\omega s_z(z') - k_{1z}]\,dz' - k_{1z}\zeta \qquad (2.8.32)$$

is a constant independent of z. Therefore, for large z, (23) becomes

$$\phi_I(z) \sim \sqrt{\frac{\omega}{k_{1z}}}\, A_+ e^{ik_{1z}z + ia} + \sqrt{\frac{\omega}{k_{1z}}}\, A_- e^{-ik_{1z}z - ia}, \quad z \to \infty. \qquad (2.8.33)$$

On comparing (29) and (33), which are both valid in the overlapping regime of $z \to \infty$, we conclude that

$$\phi_0 = \sqrt{\frac{\omega}{k_{1z}}}\, A_- e^{-ia}, \quad R\phi_0 = \sqrt{\frac{\omega}{k_{1z}}}\, A_+ e^{ia}, \quad R = \frac{A_+}{A_-}e^{2ia}. \qquad (2.8.34)$$

Notice that A_- is known since it can be related to ϕ_0, the incident wave amplitude. However, A_+ is still an unknown at this point. To find R or A_+, the solution in region I needs to be matched to the solution in region II. To do this, a region where both the solutions ϕ_I and ϕ_{II} are valid is sought. Such a region exists when $\omega \to \infty$.

From (19), it is seen that the WKB method is valid when $\omega s_z'(z) \ll \omega^2 s_z^2(z)$. But when $\omega \to \infty$, the region over which this condition is violated is very narrow. Hence, region II shrinks when $\omega \to \infty$. Moreover, the WKB solutions for regions I and III are valid even very close to $z = \zeta$. Furthermore, the approximation of a linear profile in (25) is valid outside region II. In this case, the solution ϕ_{II} is valid outside region II as well. As a result, there is an overlapping region where both ϕ_I and ϕ_{II} are valid. Similarly, an overlapping region exists for ϕ_{II} and ϕ_{III}.

As a consequence of this, one can examine the form of the solution ϕ_{II} to the right of $z = \zeta$. Note that when $\omega \to \infty$, since Ω is proportional to

ω^2, $\eta = \Omega^{\frac{1}{3}}(z - \zeta) \gg 1$ to the right of ζ even though $z - \zeta$ need not be large. Therefore, the large argument approximation of the Airy function can be used to get [see (2.1.41b)]

$$\phi_{II} \sim C\pi^{-\frac{1}{2}}\eta^{-\frac{1}{4}}\sin(\tfrac{2}{3}\eta^{\frac{3}{2}} + \tfrac{\pi}{4})$$

$$= \frac{C}{2i\pi^{\frac{1}{2}}\Omega^{\frac{1}{12}}(z - \zeta)^{\frac{1}{4}}}\left[e^{i\frac{2}{3}\Omega^{\frac{1}{2}}(z-\zeta)^{\frac{3}{2}}+i\frac{\pi}{4}} - e^{-i\frac{2}{3}\Omega^{\frac{1}{2}}(z-\zeta)^{\frac{3}{2}}-i\frac{\pi}{4}}\right].$$

$$(2.8.35)$$

Equation (35) is next compared with ϕ_I in the neighborhood of region II. For high frequencies, region II becomes very small, and in its neighborhood, the approximation $\omega s_z \simeq \Omega^{\frac{1}{2}}(z - \zeta)^{\frac{1}{2}}$ yields

$$\omega \int_{\zeta}^{z} s_z(z')\,dz' \simeq \frac{2}{3}\Omega^{\frac{1}{2}}(z - \zeta)^{\frac{3}{2}}, \quad z \simeq \zeta. \qquad (2.8.36)$$

As a consequence, ϕ_I becomes

$$\phi_I \sim \frac{A_+\omega^{\frac{1}{2}}}{\Omega^{\frac{1}{4}}(z - \zeta)^{\frac{1}{4}}}e^{i\frac{2}{3}\Omega^{\frac{1}{2}}(z-\zeta)^{\frac{3}{2}}} + \frac{A_-\omega^{\frac{1}{2}}}{\Omega^{\frac{1}{4}}(z - \zeta)^{\frac{1}{4}}}e^{-i\frac{2}{3}\Omega^{\frac{1}{2}}(z-\zeta)^{\frac{3}{2}}}. \qquad (2.8.37)$$

Then, by comparing (35) and (37), we deduce that

$$\frac{A_+}{A_-} = -i, \quad R = -ie^{2ia}. \qquad (2.8.38)$$

Therefore, both A_+ and A_- are now known in region I. Furthermore, on comparing (35) and (37), C is found in terms of A_+, i.e.,

$$C = 2A_+\omega^{\frac{1}{2}}\pi^{\frac{1}{2}}e^{i\frac{\pi}{4}}\Omega^{-\frac{1}{6}}. \qquad (2.8.39)$$

Similarly, B_- can be found by matching ϕ_{II} and ϕ_{III} to the left of $z = \zeta$. Hence, when $-\eta \gg 1$, using the asymptotic expansion of the Airy function for negative η gives a ϕ_{II} which is [see (2.1.42a)]

$$\phi_{II}(z) \sim \frac{C}{2\pi^{\frac{1}{2}}\Omega^{\frac{1}{12}}(\zeta - z)^{\frac{1}{4}}}e^{-\frac{2}{3}\Omega^{\frac{1}{2}}(\zeta-z)^{\frac{3}{2}}}, \quad \Omega^{\frac{1}{3}}(\zeta - z) \gg 1. \qquad (2.8.40)$$

Next, on using (36) for $z < \zeta$, ϕ_{III} becomes

$$\phi_{III}(z) \sim \frac{B_-\omega^{\frac{1}{2}}}{\Omega^{\frac{1}{4}}(\zeta - z)^{\frac{1}{4}}}e^{-\frac{2}{3}\Omega^{\frac{1}{2}}(\zeta-z)^{\frac{3}{2}}}, \quad z \approx \zeta. \qquad (2.8.41)$$

Consequently, on comparing (40) and (41), we conclude that

$$C = 2B_-\pi^{\frac{1}{2}}\omega^{\frac{1}{2}}\Omega^{-\frac{1}{6}}, \qquad (2.8.42)$$

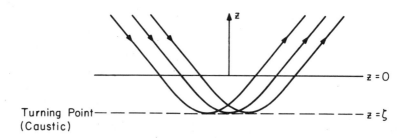

Figure 2.8.2 Propagation and refraction of waves through an inhomogeneous profile in Figure 2.8.1.

giving B_- in terms of C.

Therefore, given the amplitude of the incident wave ϕ_0, we can find the amplitude of the field everywhere via Equations (34), (38), (39), and (42). This solution is asymptotic in the sense that it becomes increasingly good when $\omega \to \infty$, because region II shrinks in size, and the matching regions or overlapping regions exist for the solutions ϕ_I, ϕ_{II}, and ϕ_{III}.

The propagation of the wave through the profile in Figure 2.8.1 is depicted in Figure 2.8.2, which shows an obliquely incident wave. Since the medium becomes optically less and less dense when the wave propagates downward, the ray is refracted to be more horizontal as it propagates. Finally, it is critically reflected at $z = \zeta$, where $s_z = 0$. Furthermore, the wave becomes evanescent below ζ, and it turns around (totally reflected) at $z = \zeta$. The point $z = \zeta$ is also known as the **turning point**. It is also a **caustic** since the rays that one would draw will bunch together at ζ. A caustic is often the place where geometrical optics rays bunch together. At a simple caustic, a ray undergoes a 90^0 phase shift as indicated by Equation (38). It is at this turning point or caustic that the WKB approximation breaks down.

The WKB method, together with asymptotic matching, has been used to analyze propagation of modes in optical waveguides (see Exercise 2.35; also see Yamada et al. 1978).

§2.9 Propagator Matrix

A matrix method is sometimes used in solving the one-dimensional inhomogeneous profile problem. Such a matrix method is also known varyingly as a propagation matrix, transition matrix, transfer matrix, chain matrix, or scattering matrix methods. The multiplicity of names exists because this method is used for electromagnetic waves, elastic waves, and Schrödinger waves. Moreover, the matrix method also allows for an easy extension to the numerical method. This method is attributed to Abelès (1950), Thomson

(1950), and Haskell (1953). Furthermore, variation of this method is also given by Born and Wolf (1980), Yeh and Liu (1972), Kong (1975, 1986), and Aki and Richards (1980). We shall illustrate it in the following subsections.

§§2.9.1 Derivation of the State Equation

Without loss of generality, consider first the differential equation that governs the propagation of TM waves, namely,

$$\left[\epsilon \frac{d}{dz}\epsilon^{-1}\frac{d}{dz} + k_z^2(z)\right]\phi = 0. \tag{2.9.1}$$

(The TE wave equation is of the same form as the above and need not be reconsidered.) This equation can be converted into a first-order differential equation so that a simple stepping procedure can be used to solve the equation. As shall be shown, this can be done via the *state-variable* approach.

To this end, we define a new function

$$\psi = \frac{1}{i\omega\epsilon}\frac{d}{dz}\phi, \tag{2.9.2}$$

Then, Equation (1) becomes

$$\frac{d}{dz}\psi = -\frac{k_z^2}{i\omega\epsilon}\phi. \tag{2.9.3}$$

Consequently, Equations (2) and (3) could be written in a matrix form as

$$\frac{d}{dz}\mathbf{V} = \overline{\mathbf{H}} \cdot \mathbf{V}, \tag{2.9.4}$$

where $\mathbf{V}^t = [\phi \quad \psi]$ is the state vector describing the state of the system, and the matrix

$$\overline{\mathbf{H}} = \begin{bmatrix} 0 & i\omega\epsilon \\ \frac{ik_z^2}{\omega\epsilon} & 0 \end{bmatrix}. \tag{2.9.5}$$

§§2.9.2 Solution of the State Equation

Consider the case where (4) has a closed-form solution, e.g., when ϵ and k_z are constants. In this case, let

$$\mathbf{V} = e^{\lambda z}\mathbf{V}_0. \tag{2.9.6}$$

Equation (4) then becomes

$$(\overline{\mathbf{H}} - \lambda\overline{\mathbf{I}}) \cdot \mathbf{V}_0 = 0. \tag{2.9.7}$$

For nontrivial $\mathbf{V}_0$, the determinant of $\overline{\mathbf{H}} - \lambda\overline{\mathbf{I}}$ has to be zero, giving $\lambda = \pm ik_z$. Hence,

$$\mathbf{V}(z) = A_+e^{ik_z z}\mathbf{a}_+ + A_-e^{-ik_z z}\mathbf{a}_-, \tag{2.9.8}$$

where $\mathbf{a}_\pm$ are eigenvectors corresponding to the eigenvalues $\pm i k_z$.

Since $\overline{\mathbf{H}}$ is not symmetric or Hermitian, these eigenvectors need not be orthogonal. But the eigenvectors $\mathbf{a}_\pm$ could be normalized if necessary. Also, (8) is expressible in a matrix form as

$$\mathbf{V}(z) = \overline{\mathbf{a}} \cdot e^{i\overline{\mathbf{K}}z} \cdot \mathbf{A}, \qquad (2.9.9)$$

where

$$\mathbf{A}^t = [\, A_+ \quad A_- \,], \qquad (2.9.10a)$$

$$e^{i\overline{\mathbf{K}}z} = \begin{bmatrix} e^{ik_z z} & 0 \\ 0 & e^{-ik_z z} \end{bmatrix}, \qquad (2.9.10b)$$

$$\overline{\mathbf{a}} = [\, \mathbf{a}_+ \quad \mathbf{a}_- \,]. \qquad (2.9.10c)$$

Next, using the fact $\overline{\mathbf{a}}^{-1} \cdot \overline{\mathbf{a}} = \overline{\mathbf{I}}$, we rewrite (9) as

$$\mathbf{V}(z) = \overline{\mathbf{a}} \cdot e^{i\overline{\mathbf{K}}(z-z')} \cdot \overline{\mathbf{a}}^{-1} \cdot \overline{\mathbf{a}} \cdot e^{i\overline{\mathbf{K}}z'} \cdot \mathbf{A} = \overline{\mathbf{a}} \cdot e^{i\overline{\mathbf{K}}(z-z')} \cdot \overline{\mathbf{a}}^{-1} \cdot \mathbf{V}(z').$$
$$(2.9.11)$$

Furthermore, if we define

$$\overline{\mathbf{P}}(z, z') = \overline{\mathbf{a}} \cdot e^{i\overline{\mathbf{K}}(z-z')} \overline{\mathbf{a}}^{-1}, \qquad (2.9.12)$$

then (11) becomes

$$\mathbf{V}(z) = \overline{\mathbf{P}}(z, z') \cdot \mathbf{V}(z'). \qquad (2.9.13)$$

The matrix $\overline{\mathbf{P}}$ is also known as the propagator matrix or the transition matrix. It relates the state vectors that describe the fields at two different locations z and z'.

§§2.9.3 Reflection from a Three-Layer Medium

We shall now show that the propagator matrix can be used to solve a three-layer problem as shown in Figure 2.9.1. First, the state vector in region 1 has been derived as

$$\mathbf{V}_1(z) = A_{1-}e^{-ik_{1z}z}\mathbf{a}_{1-} + RA_{1-}e^{ik_{1z}z}\mathbf{a}_{1+} = \overline{\mathbf{a}}_1 \cdot e^{i\overline{\mathbf{K}}_1 z} \cdot \begin{bmatrix} R \\ 1 \end{bmatrix} A_{1-}.$$
$$(2.9.14)$$

In order to render the definition of R unique, it is appropriate to normalize $\mathbf{a}_{1\pm}$ here. Because of the definitions of ϕ and ψ, they are continuous quantities across a discontinuity. Therefore, $\mathbf{V}_1(0) = \mathbf{V}_2(0)$, and $\mathbf{V}_3(-d) = \mathbf{V}_2(-d)$. Moreover, the $\overline{\mathbf{P}}$ matrix can be used to find $\mathbf{V}_2(-d)$ in terms of $\mathbf{V}_2(0)$. Consequently,

$$\mathbf{V}_3(-d) = \overline{\mathbf{P}}_2(-d, 0) \cdot \mathbf{V}_1(0) = \overline{\mathbf{P}}_2(-d, 0) \cdot \overline{\mathbf{a}}_1 \cdot \begin{bmatrix} R \\ 1 \end{bmatrix} A_{1-}. \qquad (2.9.15)$$

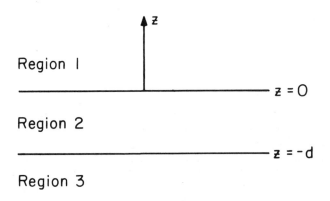

Figure 2.9.1 A three-layer problem.

Since region 3 should only have a transmitted wave, $\mathbf{V}_3(z)$ has to be of the form

$$\mathbf{V}_3(z) = TA_{1-}e^{-ik_{3z}(z+d)}\mathbf{a}_{3-} = \overline{\mathbf{a}}_3 \cdot e^{i\overline{\mathbf{K}}_3(z+d)} \cdot \begin{bmatrix} 0 \\ T \end{bmatrix} A_{1-}. \qquad (2.9.16)$$

Consequently, from (15) and (16), we conclude that

$$\overline{\mathbf{a}}_3^{-1} \cdot \overline{\mathbf{P}}_2(-d,0) \cdot \overline{\mathbf{a}}_1 \cdot \begin{bmatrix} R \\ 1 \end{bmatrix} = \begin{bmatrix} 0 \\ T \end{bmatrix}. \qquad (2.9.17)$$

There are two scalar equations with two scalar unknowns R and T in (17). By rearranging the equations, R and T can be solved for readily. A similar approach can be used to find reflection and transmission at a half-space (Exercise 2.36).

§§2.9.4 Reflection from an Inhomogeneous Slab

If region 2 is now an inhomogeneous region, the above method of finding R and T via Equation (17) will still apply if the $\overline{\mathbf{P}}$ matrix can be found for the inhomogeneous layer. One way would just be to find it numerically. To do this, the inhomogeneous layer is divided into many fine, homogeneous layers as shown in Figure 2.9.2, and the $\overline{\mathbf{P}}$ matrix for the total layer would just be the product of the $\overline{\mathbf{P}}$ matrix of each of the layers, because

$$
\begin{aligned}
\mathbf{V}_2(z = -d) &= \overline{\mathbf{P}}(-z_N, -z_{N-1}) \cdot \mathbf{V}_2(-z_{N-1}) \\
&= \overline{\mathbf{P}}(-z_N, -z_{N-1}) \cdot \overline{\mathbf{P}}(-z_{N-1}, -z_{N-2}) \cdot \mathbf{V}_2(-z_{N-2}) \\
&= \left[\prod_{i=N}^{1} \overline{\mathbf{P}}(-z_i, -z_{i-1}) \right] \cdot \mathbf{V}_2(0) \\
&= \overline{\mathbf{P}}(-d,0) \cdot \overline{\mathbf{V}}_2(0), \qquad (2.9.18)
\end{aligned}
$$

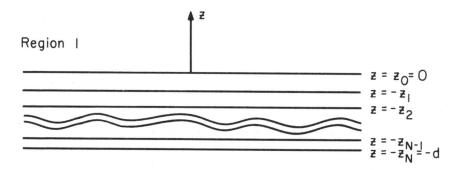

Figure 2.9.2 Inhomogeneous layer case.

where $\overline{\mathbf{P}}(-d, 0) = \prod_{i=N}^{1} \overline{\mathbf{P}}(-z_i, -z_{i-1})$. Hence, $\overline{\mathbf{P}}(-d, 0)$ can be found via this numerical approximation.

Alternatively, the $\overline{\mathbf{P}}$ matrix could be solved numerically and directly from Equation (4). A finite difference approximation of d/dz yields

$$\frac{d}{dz} \mathbf{V} \simeq \frac{[\mathbf{V}(z_j) - \mathbf{V}(z_{j-1})]}{\triangle}, \qquad (2.9.19)$$

where $\triangle = z_j - z_{j-1}$. Consequently, Equation (4) becomes

$$\mathbf{V}(z_j) - \mathbf{V}(z_{j-1}) = \triangle \overline{\mathbf{H}}(z_{j-1}) \cdot \mathbf{V}(z_{j-1}), \qquad (2.9.20)$$

which can be rearranged to give

$$\mathbf{V}(z_j) = [\overline{\mathbf{I}} + \triangle \overline{\mathbf{H}}(z_{j-1})] \cdot \mathbf{V}(z_{j-1}). \qquad (2.9.21)$$

Therefore,

$$\mathbf{V}(z_N) = \left\{ \prod_{j=N}^{1} [\overline{\mathbf{I}} + \triangle \overline{\mathbf{H}}(z_{j-1})] \right\} \cdot \mathbf{V}(z_0) = \overline{\mathbf{P}}(z_N, z_0) \cdot \mathbf{V}(z_0). \qquad (2.9.22)$$

Hence, the numerical approximation of the propagator matrix between two points z_N and z_0 can be found. This $\overline{\mathbf{P}}$ matrix could then be used in (17) to find R and T for an inhomogeneous layer. Other more sophisticated differencing schemes could also be used to solve (4). The rule of thumb is that the step size must be much smaller than the smallest wavelength in the inhomogeneous layer.

If the medium is lossy, or if the wave is evanescent, Equation (10b) implies that the $\overline{\mathbf{P}}$ matrix may have exponentially large terms as well as exponentially small terms. The existence of these exponentially large terms could make the method unstable. In this case, it is preferable to use the method outlined in Section 2.1. Alternatively, Equation (1) can be solved as a two-point boundary value problem.

§2.10 Waves in Anisotropic, Layered Media

In Chapter 1, we explain that in a homogeneous, anisotropic medium, there exist two types of waves. These two types of waves are, in general, decoupled in a homogeneous medium. In the presence of a planar interface, however, they will be coupled to each other at the interface—in other words, a type I wave may generate transmitted and reflected waves of type I and type II. Therefore, this is unlike a homogeneous and isotropic medium where the two types of waves, TE and TM, are still decoupled at the interface. Hence, the problem of waves in anisotropic, layered media cannot be treated as a scalar problem—it has to be regarded as a vector field problem.

Previously, it was shown in the state-variable approach for the homogeneous, isotropic medium that the problem could be reduced to two coupled first-order differential equations with a state vector that had two components. Similarly, the state variable approach can be used in solving for the waves in anisotropic, layered media. Because of the vector nature of the problem, a state vector with at least four components is required.

Works on anisotropically layered media were performed by Teitler and Henvis (1970); Berremen (1972); Barkovskii and Borzdov (1976, 1978); Damaskos et al. (1982); Graglia and Uslenghi (1984); and Morgan et al. (1987). Moreover, this topic is also discussed in books by Budden (1961); Wait (1968, 1970); and Yeh and Liu (1972).

§§2.10.1 Derivation of the State Equation

In order to facilitate the matching of boundary conditions at the interface of a layered medium, it is more expedient to work with the transverse to z components of the electric and magnetic fields, namely, $\mathbf{E}_s$ and $\mathbf{H}_s$.

Consequently, starting with Maxwell's equations for source-free, anisotropic media, we have

$$\nabla \times \mathbf{E} = i\omega\overline{\mu} \cdot \mathbf{H}, \tag{2.10.1}$$

$$\nabla \times \mathbf{H} = -i\omega\overline{\epsilon} \cdot \mathbf{E}. \tag{2.10.2}$$

Next, we can decompose $\nabla = \nabla_s + \hat{z}\frac{\partial}{\partial z}$, $\mathbf{E} = \mathbf{E}_s + \mathbf{E}_z$, $\mathbf{H} = \mathbf{H}_s + \mathbf{H}_z$, and the tensors $\overline{\mu}$ and $\overline{\epsilon}$ can be partitioned as

$$\overline{\mu} = \begin{bmatrix} \overline{\mu}_s & \overline{\mu}_{sz} \\ \overline{\mu}_{zs} & \overline{\mu}_{zz} \end{bmatrix}, \qquad \overline{\epsilon} = \begin{bmatrix} \overline{\epsilon}_s & \overline{\epsilon}_{sz} \\ \overline{\epsilon}_{zs} & \overline{\epsilon}_{zz} \end{bmatrix}. \tag{2.10.3}$$

In the above, the subscript s denotes quantities transverse to z. Moreover, $\overline{\mu}_s$ is a 2×2 tensor, $\overline{\mu}_{sz}$ is a 2×1 matrix, $\overline{\mu}_{zs}$ is a 1×2 matrix, and $\overline{\mu}_{zz}$ is a 1×1 matrix. Similar decomposition holds for $\overline{\epsilon}$.

After substituting the decomposition into (1), and equating the transverse and longitudinal components, we have

$$\frac{\partial}{\partial z}\hat{z} \times \mathbf{E}_s = i\omega\,\overline{\mu}_s \cdot \mathbf{H}_s + i\omega\,\overline{\mu}_{sz} \cdot \mathbf{H}_z - \nabla_s \times \mathbf{E}_z,$$

(2.10.4a)

$$\nabla_s \times \mathbf{E}_s = i\omega\,\overline{\mu}_{zs} \cdot \mathbf{H}_s + i\omega\mu_{zz}\mathbf{H}_z.$$

(2.10.4b)

In addition, by duality, or from (2), we have

$$\frac{\partial}{\partial z}\hat{z} \times \mathbf{H}_s = -i\omega\,\overline{\epsilon}_s \cdot \mathbf{E}_s - i\omega\,\overline{\epsilon}_{sz} \cdot \mathbf{E}_z - \nabla_s \times \mathbf{H}_z,$$

(2.10.5a)

$$\nabla_s \times \mathbf{H}_s = -i\omega\,\overline{\epsilon}_{zs} \cdot \mathbf{E}_s - i\omega\epsilon_{zz}\mathbf{E}_z.$$

(2.10.5b)

Then, using (4b) and (5b), we can remove $\mathbf{E}_z$ and $\mathbf{H}_z$ from (4a) and (5a) by expressing them in terms of $\mathbf{E}_s$ and $\mathbf{H}_s$, that is,

$$\mathbf{E}_z = -\frac{1}{i\omega}\kappa_{zz}\nabla_s \times \mathbf{H}_s - \kappa_{zz}\overline{\epsilon}_{zs} \cdot \mathbf{E}_s,$$

(2.10.6a)

$$\mathbf{H}_z = \frac{1}{i\omega}\nu_{zz}\nabla_s \times \mathbf{E}_s - \nu_{zz}\overline{\mu}_{zs} \cdot \mathbf{H}_s,$$

(2.10.6b)

where $\kappa_{zz} = \epsilon_{zz}^{-1}$, $\nu_{zz} = \mu_{zz}^{-1}$. Equations (6a) and (6b) can be used in (4a) to yield

$$\frac{\partial}{\partial z}\hat{z} \times \mathbf{E}_s = i\omega\,\overline{\mu}_s \cdot \mathbf{H}_s + \overline{\mu}_{sz} \cdot \nu_{zz}\nabla_s \times \mathbf{E}_s - i\omega\,\overline{\mu}_{sz} \cdot \overline{\mu}_{zs} \cdot \nu_{zz}\mathbf{H}_s$$

$$+\frac{1}{i\omega}\nabla_s \times \kappa_{zz}\nabla_s \times \mathbf{H}_s + \nabla_s \times \kappa_{zz}\overline{\epsilon}_{zs} \cdot \mathbf{E}_s.$$

(2.10.7a)

By duality, we then have

$$\frac{\partial}{\partial z}\hat{z} \times \mathbf{H}_s = -i\omega\,\overline{\epsilon}_s \cdot \mathbf{E}_s + \overline{\epsilon}_{sz} \cdot \kappa_{zz}\nabla_s \times \mathbf{H}_s + i\omega\,\overline{\epsilon}_{sz} \cdot \overline{\epsilon}_{zs} \cdot \kappa_{zz}\mathbf{E}_s$$

$$-\frac{1}{i\omega}\nabla_s \times \nu_{zz}\nabla_s \times \mathbf{E}_s + \nabla_s \times \nu_{zz}\overline{\mu}_{zs} \cdot \mathbf{H}_s.$$

(2.10.7b)

Moreover, by assuming that the fields $\mathbf{E}_s$ and $\mathbf{H}_s$ have $e^{i\mathbf{k}_s \cdot \mathbf{r}_s}$ dependence in the transverse direction for all z's (phase matching), (7a) and (7b), after crossing with $-\hat{z}$, can be written as

$$\frac{d}{dz}\mathbf{E}_s = \left[(-i\omega\,\hat{z} \times \overline{\mu}_s \cdot) + i\omega\,\hat{z} \times \overline{\mu}_{sz} \cdot \overline{\mu}_{zs} \cdot \nu_{zz} - \left(\frac{i\hat{z}}{\omega} \times \mathbf{k}_s \times \kappa_{zz}\mathbf{k}_s \times\right)\right]\mathbf{H}_s$$

$$+[(-i\hat{z} \times \overline{\mu}_{sz} \cdot \nu_{zz}\mathbf{k}_s\times) - i\hat{z} \times \mathbf{k}_s \times \kappa_{zz}\overline{\epsilon}_{zs} \cdot]\mathbf{E}_s$$

(2.10.8a)

$$\frac{d}{dz}\mathbf{H}_s = \left[(i\omega\,\hat{z} \times \overline{\epsilon}_s \cdot) - i\omega\,\hat{z} \times \overline{\epsilon}_{sz} \cdot \overline{\epsilon}_{zs} \cdot \kappa_{zz} + \left(\frac{i\hat{z}}{\omega} \times \mathbf{k}_s \times \nu_{zz}\mathbf{k}_s \times\right)\right]\mathbf{E}_s$$

$$+[(-i\hat{z} \times \overline{\epsilon}_{sz} \cdot \kappa_{zz}\mathbf{k}_s\times) - i\hat{z} \times \mathbf{k}_s \times \nu_{zz}\overline{\mu}_{zs} \cdot]\mathbf{H}_s,$$

(2.10.8b)

where ∇_s has been replaced by $i\mathbf{k}_s$, and we assume that $\overline{\mu}$ and $\overline{\epsilon}$ are functions of z only. Consequently, the preceding equations can be written in a matrix form as a state equation

$$\frac{d}{dz}\begin{bmatrix} \mathbf{E}_s \\ \mathbf{H}_s \end{bmatrix} = \begin{bmatrix} \overline{\mathbf{H}}_{11} & \overline{\mathbf{H}}_{12} \\ \overline{\mathbf{H}}_{21} & \overline{\mathbf{H}}_{22} \end{bmatrix} \cdot \begin{bmatrix} \mathbf{E}_s \\ \mathbf{H}_s \end{bmatrix}, \qquad (2.10.9)$$

where $\overline{\mathbf{H}}_{ij}$ are 2×2 matrices derivable from (8a) and (8b) (Exercise 2.37). The above state equation then becomes (see Morgan et al. 1987)

$$\frac{d}{dz}\mathbf{V} = \overline{\mathbf{H}} \cdot \mathbf{V}, \qquad (2.10.10)$$

where $\mathbf{V}^t = (\mathbf{E}_s, \mathbf{H}_s)$ is a four-component row vector and $\overline{\mathbf{H}}$ is a 4×4 matrix.

§§2.10.2 Solution of the State Equation

For a homogeneous medium, the eigensolutions of (10) can be found by letting

$$\mathbf{V} = \mathbf{V}_0 e^{\lambda z}. \qquad (2.10.11)$$

The subsequent use of (11) in (10) then yields

$$\left(\overline{\mathbf{H}} - \lambda \overline{\mathbf{I}}\right) \cdot \mathbf{V}_0 = 0, \qquad (2.10.12)$$

which is an eigenequation for the eigenvalue λ. Since $\overline{\mathbf{H}}$ is a 4×4 matrix, there will be four eigenvalues and four eigenvectors from the above equation because the equation $\det\left(\overline{\mathbf{H}} - \lambda \overline{\mathbf{I}}\right) = 0$ yields a quartic equation for λ. Therefore, the general solution to (11) is of the form

$$\mathbf{V}(z) = A_1\mathbf{a}_1 e^{i\beta_1 z} + A_2\mathbf{a}_2 e^{i\beta_2 z} + A_3\mathbf{a}_3 e^{-i\beta_3 z} + A_4\mathbf{a}_4 e^{-i\beta_4 z}, \qquad (2.10.13)$$

where $\mathbf{a}_i$ is the eigenvector corresponding to the i-th eigenvalue. Note that the $\mathbf{a}_i$'s are not orthogonal because, in general, $\overline{\mathbf{H}}$ is not symmetric or Hermitian. Furthermore, these four solutions correspond to type I and type II waves going in the positive and negative z directions.

Equation (13) is also more compactly written as

$$\mathbf{V}(z) = \overline{\mathbf{a}} \cdot e^{i\overline{\beta}z} \cdot \mathbf{A}, \qquad (2.10.14)$$

where $\overline{\mathbf{a}}$ is a 4×4 matrix containing the eigenvectors $\mathbf{a}_i$'s, i.e.,

$$\overline{\mathbf{a}} = [\mathbf{a}_1, \mathbf{a}_2, \mathbf{a}_3, \mathbf{a}_4],$$

and $\mathbf{A}$ is a column vector containing A_i's. Furthermore, $\overline{\beta}$ is a diagonal matrix where the i-th diagonal element corresponds to the i-th eigenvalue. Moreover, from the definition of the exponentiation of a matrix,

$$e^{i\overline{\beta}z} = \begin{bmatrix} e^{i\beta_1 z} & & & \\ & e^{i\beta_2 z} & & \\ & & e^{-i\beta_3 z} & \\ & & & e^{-i\beta_4 z} \end{bmatrix} \qquad (2.10.15)$$

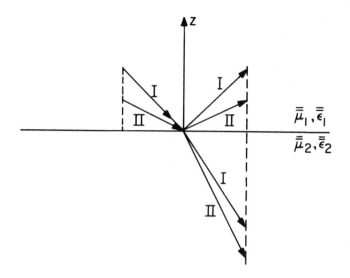

Figure 2.10.1 The **k** vectors of incident, reflected, and transmitted waves at an interface of two anisotropic media.

In addition these eigenvalues and the eigenvectors are ordered such that the first two elements of (15) correspond to an upgoing wave while the last two elements correspond to downgoing waves.

As a result, Equation (14) can be rewritten as

$$\mathbf{V}(z) = \overline{\mathbf{a}} \cdot e^{i\overline{\beta}(z-z')} \cdot \overline{\mathbf{a}}^{-1} \cdot \overline{\mathbf{a}} \cdot e^{i\overline{\beta}z'} \cdot \mathbf{A}$$
$$= \overline{\mathbf{P}}(z, z') \cdot \mathbf{V}(z'), \qquad (2.10.16)$$

where

$$\overline{\mathbf{P}}(z, z') = \overline{\mathbf{a}} \cdot e^{i\overline{\beta}(z-z')} \cdot \overline{\mathbf{a}}^{-1} \qquad (2.10.16a)$$

is a propagator matrix of the field from z' to z, analogous to (2.9.12).

§§2.10.3 Reflection from an Interface of Anisotropic Half-Spaces

We shall demonstrate how the above conclusions can be used to derive the reflection from an interface between two anisotropic media. In an anisotropic medium, the lengths of the **k** vector for type I and type II waves are different. In order for the phase-matching condition to be met, the components of these **k** vectors transverse to the z direction must be the same. In other words, $\mathbf{k}_s$ in (8) is a constant independent of z. Hence, a picture of the incident, reflected, and transmitted **k** vectors is as shown in Figure 2.10.1.

In region 1, the state vector can be expressed as

$$\mathbf{V}_1(z) = \overline{\mathbf{a}}_1 \cdot e^{i\overline{\beta}_1 z} \cdot \mathbf{A}_1. \qquad (2.10.17)$$

Equation (17) is similar to Equation (14) except for the subscripts 1 used to denote that (17) is for the solution in region 1. In the above, $\mathbf{A}_1^t = [A_{11}, A_{21}, A_{31}, A_{41}]$ where A_{31} and A_{41} are the amplitudes of the downgoing type I and type II waves, while A_{11} and A_{21} are the amplitudes of the upgoing type I and type II waves. A reflection matrix can be defined that relates upgoing waves to downgoing waves in region 1 as

$$\begin{bmatrix} A_{11} \\ A_{21} \end{bmatrix} = \overline{\mathbf{R}}_{12} \cdot \begin{bmatrix} A_{31} \\ A_{41} \end{bmatrix}. \tag{2.10.18}$$

Here, A_{31} and A_{41}, the downgoing or incident wave amplitudes, are assumed to be known. Hence, A_{11} and A_{21}, or $\overline{\mathbf{R}}_{12}$, are the unknowns to be sought in this problem.

As a result of this, (17) can be rewritten as

$$\mathbf{V}_1(z) = \overline{\mathbf{a}}_1 \cdot e^{i\overline{\beta}_1 z} \cdot \begin{bmatrix} \overline{\mathbf{R}}_{12} \\ \overline{\mathbf{I}} \end{bmatrix} \cdot \begin{bmatrix} A_{31} \\ A_{41} \end{bmatrix}. \tag{2.10.19}$$

To render the definition of $\overline{\mathbf{R}}_{12}$ unique, it is appropriate to normalize the eigenvectors $\mathbf{a}_i$'s.

In region 2, only downgoing waves exist. Similarly, a transmission matrix is defined to express the state vector in region 2 as

$$\mathbf{V}_2(z) = \overline{\mathbf{a}}_2 \cdot e^{i\overline{\beta}_2 z} \cdot \begin{bmatrix} 0 \\ \overline{\mathbf{T}}_{12} \end{bmatrix} \cdot \begin{bmatrix} A_{31} \\ A_{41} \end{bmatrix}. \tag{2.10.20}$$

Since $\mathbf{V}_1(z)$ and $\mathbf{V}_2(z)$ are the state vectors related to the tangential components of the electric and magnetic fields at the interface, the boundary conditions require that $\mathbf{V}_1(0) = \mathbf{V}_2(0)$, leading to

$$\overline{\mathbf{a}}_1 \cdot \begin{bmatrix} \overline{\mathbf{R}}_{12} \\ \overline{\mathbf{I}} \end{bmatrix} = \overline{\mathbf{a}}_2 \cdot \begin{bmatrix} 0 \\ \overline{\mathbf{T}}_{12} \end{bmatrix}. \tag{2.10.21}$$

The above constitutes two matrix equations with two matrix unknowns, $\overline{\mathbf{R}}_{12}$ and $\overline{\mathbf{T}}_{12}$. Hence, $\overline{\mathbf{R}}_{12}$ and $\overline{\mathbf{T}}_{12}$ can be found easily.

If a medium is isotropic instead, the type I and type II waves reduce to TE and TM waves. In this case, the lengths of their $\mathbf{k}$ vectors are the same (Exercise 2.38).

§§2.10.4 Reflection from a Slab

The reflection of a wave from an anisotropic slab as shown in Figure 2.10.2 can be found in a manner parallel to Subsections 2.9.3 and 2.9.4. In this case, the equation in place of (21) is similar to (2.9.17), that is,

$$\overline{\mathbf{a}}_3^{-1} \cdot \overline{\mathbf{P}}_2(-d, 0) \cdot \overline{\mathbf{a}}_1 \cdot \begin{bmatrix} \overline{\mathbf{R}} \\ \overline{\mathbf{I}} \end{bmatrix} = \begin{bmatrix} 0 \\ \overline{\mathbf{T}} \end{bmatrix}. \tag{2.10.22}$$

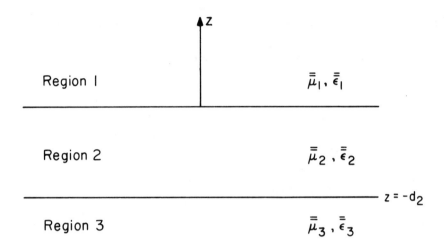

Figure 2.10.2 A three-layer medium where all the layers are anisotropic.

In the case of a homogeneous slab, the propagator matrix, $\overline{\mathbf{P}}_2(z, z')$, is given by (16a). But when the slab is inhomogeneous, it could be modeled as consisting of many fine layers, each of which is homogeneous. Then, the propagator matrix for an inhomogeneous, anisotropic slab is given by an equation similar to (2.9.18). Similarly, Equation (10) can be solved numerically to obtain an approximation of the propagator matrix in an inhomogeneous slab as in Equation (2.9.22).

§§2.10.5 Geometrical Optics Series

In the previous method, the propagator matrix may contain exponentially large terms when a lossy medium is involved or when we have an evanescent wave. In this case, the propagator matrix in (22) may become exponentially large. Such a problem can be overcome by proper bookkeeping; alternatively, a geometric optics series can be used to find the reflection matrix for a slab similar to Equation (2.1.21).

Suppose that the reflection and transmission matrices for a single interface are derived as in Subsection 2.10.3. Then, for a three-layer medium, the state vector in region 1 is

$$\mathbf{V}_1(z) = \mathbf{a}_{11}e^{i\beta_{11}z}A_{11} + \mathbf{a}_{21}e^{i\beta_{21}z}A_{21} + \mathbf{a}_{31}e^{-i\beta_{31}z}A_{31} + \mathbf{a}_{41}e^{-i\beta_{41}z}A_{41}.$$

$$(2.10.23)$$

The above can be written as

$$\mathbf{V}_1(z) = \overline{\mathbf{a}}_{1+} \cdot e^{i\overline{\beta}_{1+}z} \cdot \mathbf{A}_{1+} + \overline{\mathbf{a}}_{1-} \cdot e^{-i\overline{\beta}_{1-}z} \cdot \mathbf{A}_{1-}, \qquad (2.10.24)$$

where $\overline{\mathbf{a}}_{1+} = [\mathbf{a}_{11}, \ \mathbf{a}_{21}]$, $\overline{\mathbf{a}}_{1-} = [\mathbf{a}_{31}, \ \mathbf{a}_{41}]$, $\mathbf{A}_{1+}^t = [A_{11}, \ A_{21}]$, $\mathbf{A}_{1-}^t = [A_{31}, \ A_{41}]$, and $\overline{\beta}_{1+}$ is a diagonal matrix containing β_{11} and β_{21} on the diagonal, while $\overline{\beta}_{1-}$ is a diagonal matrix containing β_{31} and β_{41} on the diagonal. Next, by defining $\mathbf{A}_{1+} = \widetilde{\overline{\mathbf{R}}}_{12} \cdot \mathbf{A}_{1-}$, we can write the previous equation as

$$\mathbf{V}_1(z) = \overline{\mathbf{a}}_{1+} \cdot e^{i\overline{\beta}_{1+}z} \cdot \widetilde{\overline{\mathbf{R}}}_{12} \cdot \mathbf{A}_{1-} + \overline{\mathbf{a}}_{1-} \cdot e^{-i\overline{\beta}_{1-}z} \cdot \mathbf{A}_{1-}. \qquad (2.10.25)$$

Similarly, in region 2, the field becomes

$$\mathbf{V}_2(z) = \overline{\mathbf{a}}_{2+} \cdot e^{i\overline{\beta}_{2+}(z+d_2)} \cdot \overline{\mathbf{R}}_{23} \cdot \mathbf{A}_{2-} + \overline{\mathbf{a}}_{2-} \cdot e^{-i\overline{\beta}_{2-}(z+d_2)} \cdot \mathbf{A}_{2-}. \qquad (2.10.26)$$

But the upgoing wave in region 1 is a consequence of a reflection of the downgoing wave in region 1 plus a transmission of an upgoing wave in region 2. Hence,

$$\widetilde{\overline{\mathbf{R}}}_{12} \cdot \mathbf{A}_{1-} = \overline{\mathbf{R}}_{12} \cdot \mathbf{A}_{1-} + \overline{\mathbf{T}}_{21} \cdot e^{i\overline{\beta}_{2+}d_2} \cdot \overline{\mathbf{R}}_{23} \cdot \mathbf{A}_{2-}. \qquad (2.10.27)$$

Similarly, the downgoing wave in region 2 is a consequence of the transmission of the downgoing wave in region 1 plus a reflection of the upgoing wave in region 2, or

$$e^{-i\overline{\beta}_{2-}d_2} \cdot \mathbf{A}_{2-} = \overline{\mathbf{R}}_{21} \cdot e^{i\overline{\beta}_{2+}d_2} \cdot \overline{\mathbf{R}}_{23} \cdot \mathbf{A}_{2-} + \overline{\mathbf{T}}_{12} \cdot \mathbf{A}_{1-}. \qquad (2.10.28)$$

Equation (28) can be solved for $\mathbf{A}_{2-}$ to give

$$\mathbf{A}_{2-} = \left[\overline{\mathbf{I}} - e^{i\overline{\beta}_{2-}d_2} \cdot \overline{\mathbf{R}}_{21} \cdot e^{i\overline{\beta}_{2+}d_2} \cdot \overline{\mathbf{R}}_{23} \right]^{-1} \cdot e^{i\overline{\beta}_{2-}d_2} \cdot \overline{\mathbf{T}}_{12} \cdot \mathbf{A}_{1-}. \qquad (2.10.29)$$

Equation (29) used in (27) will then yield

$$\widetilde{\overline{\mathbf{R}}}_{12} = \overline{\mathbf{R}}_{12} + \overline{\mathbf{T}}_{21} \cdot e^{i\overline{\beta}_{2+}d_2} \cdot \overline{\mathbf{R}}_{23} \cdot \left[\overline{\mathbf{I}} - e^{i\overline{\beta}_{2-}d_2} \cdot \overline{\mathbf{R}}_{21} \cdot e^{i\overline{\beta}_{2+}d_2} \cdot \overline{\mathbf{R}}_{23} \right]^{-1} \cdot e^{i\overline{\beta}_{2-}d_2} \cdot \overline{\mathbf{T}}_{12}. \qquad (2.10.30)$$

The above is an equation for finding the generalized reflection matrix, $\widetilde{\overline{\mathbf{R}}}_{12}$, that incorporates subsurface reflections for a three-layer medium (Exercise 2.39). It is very similar to (2.1.21) in its physical interpretation.

Now, if another layer is added below region 3, only $\overline{\mathbf{R}}_{23}$ needs to be changed to $\widetilde{\overline{\mathbf{R}}}_{23}$ to get the correct $\widetilde{\overline{\mathbf{R}}}_{12}$. Hence, a general expression analogous to (2.1.23) is

$$\widetilde{\overline{\mathbf{R}}}_{i,i+1} = \overline{\mathbf{R}}_{i,i+1} + \overline{\mathbf{T}}_{i+1,i} \cdot e^{i\overline{\beta}_{i+1,+}t_{i+1}} \cdot \widetilde{\overline{\mathbf{R}}}_{i+1,i+2}$$
$$\cdot \left[\overline{\mathbf{I}} - e^{i\overline{\beta}_{i+1,-}t_{i+1}} \cdot \overline{\mathbf{R}}_{i+1,i} \cdot e^{i\overline{\beta}_{i+1,+}t_{i+1}} \cdot \widetilde{\overline{\mathbf{R}}}_{i+1,i+2} \right]^{-1} \cdot e^{i\overline{\beta}_{i+1,-}t_{i+1}} \cdot \overline{\mathbf{T}}_{i,i+1}, \qquad (2.10.31)$$

where $t_{i+1} = d_{i+1} - d_i$, the thickness of the $i+1$ region. The above allows one to find the generalized reflection matrix for all regions in an N-layer, anisotropic medium. Similarly, (29) can be generalized to

$$\mathbf{A}_{i+1,-} = \left[\overline{\mathbf{I}} - e^{i\overline{\beta}_{i+1,-}t_{i+1}} \cdot \overline{\mathbf{R}}_{i+1,i} \cdot e^{i\overline{\beta}_{i+1,+}t_{i+1}} \cdot \widetilde{\overline{\mathbf{R}}}_{i+1,i+2} \right]^{-1}$$
$$\cdot e^{i\overline{\beta}_{i+1,-}t_{i+1}} \cdot \overline{\mathbf{T}}_{i,i+1} \cdot \mathbf{A}_{i,-}$$
$$= \overline{\mathbf{S}}_{i+1,i} \cdot \mathbf{A}_{i-}. \qquad (2.10.32)$$

Equation (32) can be used recursively to give the wave amplitudes in all regions in an N-layer medium.

Now, in Equations (31) and (32), exponentially large terms are not encountered. Hence, the method is more stable.

Exercises for Chapter 2

2.1 For a one-dimensional profile where ϵ and μ are isotropic and are functions of z only, extract the z components of the vector wave equations for the **E** field and **H** field and show that

$$\nabla_s^2 E_z + \frac{\partial}{\partial z}\epsilon^{-1}\frac{\partial}{\partial z}\epsilon E_z + \omega^2\mu\epsilon E_z = 0,$$

$$\nabla_s^2 H_z + \frac{\partial}{\partial z}\mu^{-1}\frac{\partial}{\partial z}\mu H_z + \omega^2\mu\epsilon H_z = 0.$$

From these two equations, derive the boundary conditions for E_z and H_z for a step discontinuity in μ and ϵ.

2.2 Derive the conditions under which the Fresnel reflection coefficient for TM waves, i.e., R^{TM}, would vanish, and when it would have unit amplitude.

2.3 Derive Equation (2.1.24) from (2.1.23) by using the fact that $T_{ij} = 1 + R_{ij}$, and $R_{ij} = -R_{ji}$.

2.4 (a) Show that

$$\left[1 - R_{21}\tilde{R}_{23}e^{2ik_{2z}(d_2-d_1)}\right]\left[1 - R_{32}\tilde{R}_{34}e^{2ik_{3z}(d_3-d_2)}\right]$$

equals

$$\left[1 - R_{21}R_{23}e^{2ik_{2z}(d_2-d_1)}\right]\left[1 - \tilde{R}_{32}\tilde{R}_{34}e^{2ik_{3z}(d_3-d_2)}\right].$$

(b) Using the above idea, i.e., the tilde can be moved from R_{23} to R_{32} and the expression remains unchanged, show that for TE waves

$$\frac{\mu_1}{k_{1z}}\tilde{T}_{1N} = \frac{\mu_N}{k_{Nz}}\tilde{T}_{N1},$$

where $\tilde{T}_{ij}$ is defined in (2.1.28). The above is important for establishing the reciprocal nature of the generalized transmission coefficients.

2.5 Making use of (2.1.35), derive (2.1.36) from (2.1.33).

2.6 Using the Runge-Kutta method, write a first order approximate solution to Equation (2.1.36) which models a piecewise linear profile. If one chooses instead to model an inhomogeneous slab with a finely layered medium, Equation (2.1.24) would be used recursively to find the reflection coefficient. Count the number of multiplications and additions required in both cases, and estimate which algorithm will be more efficient.

Note: Now, computers take about the same time to perform a multiplication or a division compared to an addition; hence, the number of floating point additions and subtractions cannot be ignored.

2.7 An inhomogeneous profile is described by $\epsilon(z) = \epsilon_0$ for $z < 0$, and $\epsilon(z) = \epsilon_0 + az$ for $z > 0$ and $\mu = \mu_0$ (see figure). For a normally incident ($k_x = 0$) wave from the left-hand side ($z < 0$), find the reflection coefficient for the electric field at $z = 0$.

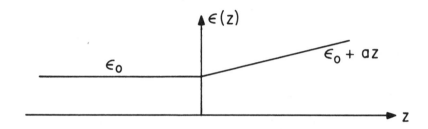

Figure for Exercise 2.7

Hint: With the introduction of some loss, make sure that the wave is exponentially small when $z \to \infty$.

2.8 The Hankel function of the first kind is obtained for (2.2.4) because we have made use of $\exp(-i\omega t)$ dependence.

(a) Show that if $\exp(j\omega t)$ dependence is used, the solution to (2.2.2) will be a Hankel function of the second kind. Obtain the constant C in this case by matching the singularity at the origin.

(b) Going through similar derivations as in (2.2.5) to (2.2.11), and introducing a small loss in the medium, derive a plane-wave expansion of the Hankel function of the second kind. Stipulate appropriately the path of integration of the Fourier inverse transform in the complex plane.

2.9 (a) By making use of the recurrence relations of Bessel functions [see Chapter 1, Equation (1.2.34)], prove the property of the raising operator in (2.2.16).

(b) Hence, prove the identity (2.2.17).

(c) Can you design a lowering operator?

2.10 The two-dimensional Fourier transform pair is given as

$$F(k_x, k_y) = \iint\limits_{-\infty}^{\infty} dx\, dy\, e^{-ik_x x - ik_y y} f(x, y),$$

$$f(x,y) = \frac{1}{(2\pi)^2} \int\!\!\int\limits_{-\infty}^{\infty} dk_x dk_y \, e^{ik_x x + ik_y y} F(k_x, k_y).$$

(a) By writing the above in cylindrical coordinates, i.e., let $x = \rho \cos \phi$, $y = \rho \sin \phi$, $k_x = k_\rho \cos \alpha$, $k_y = k_\rho \sin \alpha$, show that the above becomes

$$F(k_\rho, \alpha) = \int_0^\infty d\rho \, \rho \int_0^{2\pi} d\phi \, e^{-ik_\rho \rho \cos(\alpha - \phi)} f(\rho, \phi),$$

$$f(\rho, \phi) = \frac{1}{(2\pi)^2} \int_0^\infty dk_\rho \, k_\rho \int_0^{2\pi} d\alpha \, e^{ik_\rho \rho \cos(\alpha - \phi)} F(k_\rho, \alpha).$$

(b) By letting

$$F(k_\rho, \alpha) = \sum_{n=-\infty}^{\infty} F_n(k_\rho) e^{in\alpha},$$

$$f(\rho, \phi) = \frac{1}{2\pi} \sum_{n=-\infty}^{\infty} f_n(\rho) e^{in\phi + i\frac{n\pi}{2}},$$

and using the integral representation of Bessel function, show that

$$F_n(k_\rho) = \int_0^\infty d\rho \, \rho f_n(\rho) J_n(k_\rho \rho),$$

$$f_n(\rho) = \int_0^\infty dk_\rho \, k_\rho F_n(k_\rho) J_n(k_\rho \rho).$$

The above is known as the Hankel transform pair.

(c) From the Hankel transform pair, deduce that

$$\frac{\delta(k_\rho - k_\rho')}{k_\rho} = \int_0^\infty d\rho \, \rho J_n(k_\rho \rho) J_n(k_\rho' \rho),$$

$$\frac{\delta(\rho - \rho')}{\rho} = \int_0^\infty dk_\rho \, k_\rho J_n(k_\rho \rho) J_n(k_\rho \rho').$$

2.11 The mapping $k_z = (k_0^2 - k_\rho^2)^{1/2}$ maps from the complex k_ρ plane to the complex k_z plane. By letting $k_\rho = k_0 + \delta e^{i\theta}$, where δ is a small real number, and $k_0 = |k_0| e^{i\alpha}$, show that for the Sommerfeld branch cut

(Figure 2.2.7), the right seam of the branch cut maps to the negative real k_z axis, and the left shore of the branch cut maps to the positive real k_z axis of the complex k_z plane.

2.12 Define a $\Re e(k_z) = 0$ branch cut with the upper Riemann sheet as

$$\Re e(k_z) > 0.$$

Show where the real and imaginary axes on the complex k_z plane are mapped to on the complex k_ρ plane.

2.13 (a) First, write the solution to (2.2.19) as

$$\phi(x, y, z) = \frac{1}{2\pi} \int\limits_{-\infty}^{\infty} dk_z \, \tilde{\phi}(x, y, k_z) e^{ik_z z}.$$

By substituting the above into (2.2.19), derive an expression for

$$\tilde{\phi}(x, y, k_z),$$

and hence, the identity (2.2.34).

(b) Alternatively, derive (2.2.34) from (2.2.31) by contour deformation and a change of variable of integration.

2.14 When we have a current loop as a source, i.e., $\mathbf{J} = -\hat{\phi} I \, \delta(\rho - \rho') \, \delta(z - z')$:

(a) Show that the electric field in cylindrical coordinates has only a ϕ component, and that it satisfies

$$\left[\rho \frac{\partial}{\partial \rho} \frac{1}{\rho} \frac{\partial}{\partial \rho} + \frac{\partial^2}{\partial z^2} + k^2 \right] \rho E_\phi = i\omega \mu I \rho \, \delta(\rho - \rho') \, \delta(z - z');$$

(b) By using Hankel transform (Exercise 2.10), show that the solution is

$$E_\phi = \frac{\omega \mu I \rho'}{2} \int\limits_{0}^{\infty} dk_\rho \, \frac{k_\rho}{k_z} J_1(k_\rho \rho') J_1(k_\rho \rho) e^{ik_z |z - z'|};$$

(c) What would the solution be if we have a current disk instead, i.e., $\mathbf{J} = \hat{\phi} J(\rho) \, \delta(z - z')$?

2.15 Assume that a VED is over a three-layer medium:

(a) Write down the solution (2.3.6) for region 2, i.e., the slab region. Give explicitly the expression for A_2.

(b) Derive the electric and magnetic field components transverse to z in region 2.

2.16 (a) Assume that a VED is embedded in region 2 of a three-layer medium (see Figure 2.1.3). Then, the field in region 2 can be written as

$$E_{2z} = \frac{-I\ell}{8\pi\omega\epsilon_2} \int\limits_{-\infty}^{\infty} dk_\rho \frac{k_\rho^3}{k_{2z}} H_0^{(1)}(k_\rho\rho) \left[e^{ik_{2z}|z-z'|} + A_2 e^{-ik_{2z}z} + B_2 e^{ik_{2z}z} \right],$$

where the extra terms to the source term are due to multiple reflections of the waves from the top and the bottom boundaries. Find the coefficients A_2 and B_2.

(b) What is the expression when region 1 and region 3 are perfect conductors?

2.17 For certain problems, Cartesian coordinates may be more pertinent.

(a) Using the Weyl identity [Equation (2.2.27)], show that for an HED, where $\mathbf{J} = \hat{x} I\ell\, \delta(x - x')\delta(y - y')\delta(z)$, E_z is given as

$$E_z = \pm \frac{I\ell}{8\pi^2\omega\epsilon} \iint\limits_{-\infty}^{\infty} dk_x dk_y\, k_x e^{ik_x(x-x')+ik_y(y-y')+ik_z|z|}.$$

(b) A current sheet source $\mathbf{J} = \hat{x}\delta(z) J_s(x,y)$. Using the idea of convolution, show that E_z produced by the current sheet is

$$E_z = \pm \frac{1}{8\pi^2\omega\epsilon} \iint\limits_{-\infty}^{\infty} dk_x dk_y\, k_x \tilde{J}_s(k_x, k_y) e^{ik_x x + ik_y y + ik_z|z|},$$

where $\tilde{J}_s(k_x, k_y)$ is the two-dimensional Fourier transform of $J_s(x,y)$. Remember that convolutions in real space are the same as multiplications in Fourier space.

2.18 (a) Derive (2.4.6a) and (2.4.6b).

(b) Derive (2.4.15) and (2.4.16).

(c) Show that for TE waves

$$\frac{\mu_m}{k_{mz}} F_+(z, z') = \frac{\mu_n}{k_{nz}} F_-(z', z),$$

where $z' \in R_m$, $z \in R_n$, and R_i stands for region i.

2.19 Assuming that $d_1 = 0$, i.e., the dipole is right at the interface, derive the leading-order asymptotic expansion to the integrals (2.3.9a) and (2.3.9b) using the method of stationary phase.

Hint: Write $H_1^{(1)}(k_\rho\rho) = \left[H_1^{(1)}(k_\rho\rho)/H_0^{(1)}(k_\rho\rho) \right] H_0^{(1)}(k_\rho\rho)$. The ratio of Hankel functions can be treated as slowly varying, and hence, factored outside the integral.

2.20 (a) Derive (2.5.20) using the integral identity for gamma function or the identity that

$$\int\limits_{-\infty}^{\infty} e^{-\lambda s^2}\, ds = \sqrt{\pi/\lambda}.$$

(b) Hence, derive the expressions (2.5.21) and (2.5.23).

2.21 Derive the integral identity (2.5.30), and the expressions (2.5.31) and (2.5.32).

2.22 Show that $s^n e^{-\lambda s^2}$ in (2.5.19) peaks at $s = \sqrt{n/2\lambda}$, and $s^{n+\alpha} e^{-\lambda s}$ in (2.5.29) peaks at $(n+\alpha)/\lambda$, and hence, derive the conditions (2.5.34) and (2.5.35).

2.23 (a) Show that (2.5.51) is the leading-order saddle-point contribution of (2.5.37).

(b) Show that (2.5.54) is a consequence of the leading-order saddle-point contribution of (2.5.37), plus a leading-order contribution of the singularity. Show it for the case when $k = 1/2$ so that the singularity is an algebraic branch point, and when $k = -1$ so that the singularity is a pole.

2.24 The governing equation for the steepest-descent path that passes through a saddle point is given by (2.5.13) when s is real.

(a) Show that for (2.6.1), when $d_1 = 0$, and $z > 0$, the steepest-descent path is given by the equation

$$ik_\rho \rho + ik_{1z} z - ik_1 r = -\lambda s^2,$$

where $r = (\rho^2 + z^2)^{1/2}$.

(b) Show that the steepest-descent path intercepts the real axis at $k_1 \sin\theta$ and $k_1/\sin\theta$ where $\theta = \sin^{-1}(\rho/r)$, and that it intercepts the imaginary axis at $k_1 \tan\theta$.

(c) Show that the asymptotes of the steepest-descent path are

$$k_\rho \sim \lambda s^2 \frac{i\rho \pm z}{r^2}, \qquad s \to \pm\infty,$$

and hence, sketch the steepest-descent path on the complex k_ρ plane.

2.25 (a) Derive the branch-cut contribution given by (2.6.14).

(b) Give a physical interpretation of this branch-point contribution.

2.26 Find the asymptotic expansion of the transmitted field in (2.6.2) when $d_1 = 0$.

(a) Using the large argument expansion of the Hankel function, show that the stationary point or the saddle point of the integrand is given by

$$k_{\rho s} = k_2 \frac{\rho}{(\rho^2 + z^2)^{1/2}} = k_2 \sin\theta = k_2 \frac{\rho}{r}, \qquad \text{where } r = \sqrt{\rho^2 + z^2}.$$

(b) The constant-phase path that passes through the saddle point is governed by Equation (2.5.13) when s is real. Show that for (2.6.2), the constant-phase path is governed by the equation

$$ik_\rho\rho - ik_{2z}z - ik_2 r = -\lambda s^2, \qquad z < 0.$$

Solve this equation when $k_\rho \to \pm\infty$, or $s \to \pm\infty$, and show that

$$k_\rho \sim \lambda s^2 \frac{i\rho \mp z}{r^2}, \qquad s \to \pm\infty.$$

Therefore, the steepest-descent path has similar but not identical asymptotes as those shown in Figure 2.6.1. Sketch the steepest-descent path and the branch cuts for this problem.

(c) Find the leading-order expansion of (2.6.2) using the method of stationary phase (which is equal to the leading-order saddle-point contribution). Show that this is a spherical wave with $\exp(ik_2 r)/r$ dependence.

(d) (Optional) By deforming the contour from SIP to the steepest-descent paths or constant-phase paths through the saddle point and the branch point, show that there is a branch-point contribution when $k_2 \sin\theta > k_1$, and find this branch-point contribution. Give a physical interpretation to this branch-point contribution. Can you identify another name with the above inequality?

2.27 It was shown in Exercise 2.17 that for a current sheet $\mathbf{J} = \hat{x}\delta(z)J_x(x,y)$, its E_{1z} field could be derived.

(a) Derive H_{1z} for the same current sheet.

(b) Show that when we place the current sheet over a dielectric slab of thickness t, backed by a perfect electric conductor (ground plane at the bottom), as in microwave integrated circuits (see figure), the field in region 1 is

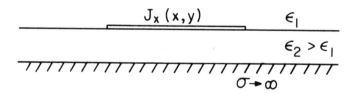

Figure for Exercise 2.27

$$E_{1z} = \frac{1}{8\pi^2\omega\epsilon_1} \int\!\!\int\limits_{-\infty}^{\infty} d\mathbf{k}_s\, k_x \tilde{J}_x(k_x, k_y) e^{i\mathbf{k}_1\cdot\mathbf{r}} \left(1 - R^{TM}\right),$$

$$H_{1z} = \frac{1}{8\pi^2} \int\!\!\int\limits_{-\infty}^{\infty} d\mathbf{k}_s\, \frac{k_y}{k_{1z}} \tilde{J}_x(k_x, k_y) e^{i\mathbf{k}_1\cdot\mathbf{r}} \left(1 + R^{TE}\right),$$

where $\mathbf{k}_1 \cdot \mathbf{r} = k_x x + k_y y + k_z z$.

(c) What should R^{TE} and R^{TM} be in this case?

(d) Find the far-field approximation of E_{1z} and H_{1z} using the method of stationary phase and the identity (2.2.27), assuming that t is small.

Hint: To find the far-field approximation, first find the stationary phase point where $\frac{\partial}{\partial k_x}\mathbf{k}_1 \cdot \mathbf{r} = 0$ and $\frac{\partial}{\partial k_y}\mathbf{k}_1 \cdot \mathbf{r} = 0$. Then, pull out the slowly varying part of the integrand and evaluate it at the stationary-phase point, leaving behind an integral such as (2.2.27) that can be evaluated in closed form.

(e) From E_{1z} and H_{1z}, find $E_{1\theta}$ and $H_{1\theta}$, and hence, $E_{1\phi}$ and $H_{1\phi}$. (These are useful for plotting the radiation patterns of an antenna.) Find the Poynting vector in the far field.

Hint: Use the plane-wave approximation.

(f) (Optional) Show that there is a TM_0 mode with no cut-off frequency for the ground plane backed substrate above. Hence, R^{TM} always has a pole. Assuming that the pole is contributing (when $z \approx 0$), find this pole contribution.

Hint: The answer will be in terms of an integral.

2.28 Show that when a VED, located at $z = 0$, is placed between two parallel metallic plates, located at $z = d_1$ and $z = -d_2$, the solution is

$$E_{2z} = \frac{-I\ell}{8\pi\omega\epsilon_2} \int_{-\infty}^{\infty} dk_\rho\, \frac{k_\rho^3}{k_{2z}} H_0^{(1)}(k_\rho\rho) \left[e^{ik_{2z}|z|} + A_2 e^{-ik_{2z}z} + B_2 e^{ik_{2z}z}\right],$$

where

$$A_2 = \frac{1 + e^{2ik_{2z}d_2}}{1 - e^{2ik_{2z}(d_1+d_2)}} e^{2ik_{2z}d_1}, \quad B_2 = \frac{1 + e^{2ik_{2z}d_1}}{1 - e^{2ik_{2z}(d_1+d_2)}} e^{2ik_{2z}d_2}.$$

(a) Show that the integrand has no branch point at $k_\rho = k_2$. Hence, the only singularities of the integrand are poles.

(b) Identify the locations of the poles (which are the guided modes of the parallel-plate waveguide) on the complex k_ρ plane. Discuss what happens to the poles when $d_1 + d_2 \to \infty$.

(c) Show that by deforming the SIP contour to infinity, the field E_{2z} could alternatively be expanded in terms of the modes of the waveguide.

2.29 The reflected field when a VMD is placed over a slab is given by the second term in (2.6.23), i.e.,

$$H_{1z}^R = -\frac{iIA}{8\pi} \int_{-\infty}^{\infty} dk_\rho \frac{k_\rho^3}{k_{1z}} H_0^{(1)}(k_\rho\rho) e^{ik_{1z}(z+2d_1)} \tilde{R}_{12}^{TE}.$$

The above integrand could have a leaky pole near the saddle point (see Figure 2.6.7). Derive a leading-order uniform asymptotic expansion for the integral by including the effect of a leaky pole.

2.30 The m-th image term in (2.6.26) is

$$H_{1z}^{Rm} = -\frac{iIA}{8\pi} \int_{-\infty}^{\infty} dk_\rho \frac{k_\rho^3}{k_{1z}} H_0^{(1)}(k_\rho\rho) e^{2imk_{2z}d_2} T_{12} R_{23}^m R_{21}^{m-1} T_{21},$$

where we have assumed that $z = d_1 = 0$.

(a) Show that the location of the saddle point can be close to the branch point at $k_\rho = k_1$.

(b) Derive a uniform asymptotic approximation for the above integral by including the effect of the branch point on the saddle-point contribution.

2.31 Using the steepest-descent method, derive the leading-order saddle-point approximation to (2.6.35).

2.32 A guided TE mode of an inhomogeneous slab satisfies the equation

$$\left[\mu \frac{d}{dz} \mu^{-1} \frac{d}{dz} + k^2 - k_\rho^2 \right] \phi(z) = 0.$$

(a) Assuming that μ is lossless but ϵ is lossy, deduce the region on the complex k_ρ^2 plane where the poles of the integrand of the Sommerfeld integral for a TE wave can be located.

(b) Show that $\Im m[k_\rho^2] = $ constant and $\Re e[k_\rho^2] = $ constant map into hyperbolas on the complex k_ρ plane. Hence, ascertain the region of pole locations on the complex k_ρ plane.

2.33 Given an integral of the form

$$\phi = \int_0^{\infty} dk_\rho \, k_\rho J_0(k_\rho\rho) G(k_\rho)$$

and that $G(k_\rho)$ has a pole at $k_\rho = k_P$, i.e., $G(k_\rho) \sim \frac{A}{k_\rho - k_P}$, $k_\rho \to k_P$:

(a) Show that the expression

$$F(k_\rho) = G(k_\rho) - \frac{2Ak_P}{k_\rho^2 - k_P^2} \frac{J_0(k_\rho a)}{J_0(k_P a)}$$

has no pole at $k_\rho = k_P$. Hence, ϕ can be written as

$$\phi = \int_0^\infty dk_\rho\, k_\rho J_0(k_\rho \rho) F(k_\rho) + \frac{2Ak_P}{J_0(k_P a)} \int_0^\infty dk_\rho\, \frac{k_\rho J_0(k_\rho \rho) J_0(k_\rho a)}{k_\rho^2 - k_P^2}.$$

(b) The first integral now does not have a pole at $k_\rho = k_P$. The second integral can be evaluated in closed form. By letting $J_0(k_\rho \rho) = \frac{1}{2}[H_0^{(1)}(k_\rho \rho) + H_0^{(2)}(k_\rho \rho)]$ and unfolding the integral to range from $-\infty$ to $+\infty$, use the contour integration technique to evaluate the second integral.

2.34 For a scalar wave equation in three dimensions, i.e.,

$$\nabla^2 \phi(\mathbf{r}) + k^2(\mathbf{r})\phi(\mathbf{r}) = 0,$$

we can make high-frequency approximations similar to the WKB approximations.

(a) Derive the corresponding eikonal equation and the transport equation for this wave equation in three dimensions.

(b) (Optional) Describe how you would solve these derived equations.

2.35 A profile is described by $\epsilon(z) = \frac{\epsilon_1 - \epsilon_2}{2} \tanh(\kappa z) + \frac{\epsilon_1 + \epsilon_2}{2}$, $\epsilon_1 > \epsilon_2$, for $z < d$. For $z \geq d$, $\epsilon = \epsilon_0$ (see figure). Assume further that $\mu = \mu_0$ is a constant:

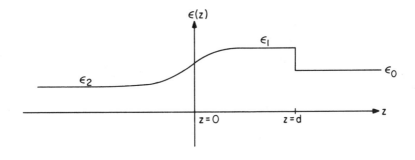

Figure for Exercise 2.35

(a) Using asymptotic matching and WKB method, find the reflection coefficient of a wave which is obliquely incident on this profile from $z = +\infty$, assuming that $\epsilon_2 < \epsilon_0$. Assume $\kappa d \gg 1$.

(b) Such a profile could approximate a profile obtained by doping a semi-conductor material and could be used for integrated optics applications. Assuming that $d \gg \lambda$, derive a guidance condition for an optical wave propagating in this waveguide.

2.36 Using a state variable approach and derive the reflected wave for a half-space similar to what has been done in Subsection 2.9.3 for a three-layer medium. Normalize your eigenvectors such that $\mathbf{a}^\dagger \cdot \mathbf{a} = 1$. Give the expression for the reflection coefficient explicitly.

2.37 Derive explicit expressions for the matrices $\overline{\mathbf{H}}_{ij}$ in (2.10.9).

2.38 Assume that two half-spaces are homogeneous and isotropic. In this case, the type I and type II waves are the TE and TM waves. Assume further that the $\mathbf{k}$ vector for the incident field is in the xz plane.

(a) Use the formulation of Subsection 2.10.3 to derive the reflection and transmission matrices

(b) Show that the reflection and transmission matrices are diagonal.

(c) How are they related to the Fresnel reflection coefficient?

2.39 Expand Equation (2.10.30) as a series and give physical interpretations to each term of the series.

References for Chapter 2

Abelès, F. 1950. *Ann. de Physique* 5: 596, 706.

Abramowitz, M., and I. A. Stegun. 1965. *Handbook of Mathematical Functions.* New York: Dover Publications.

Aki, K., and P. G. Richards. 1980. *Quantitative Seismology, Theory and Methods,* vols. I and II. New York: Freeman.

Baños, A., Jr. 1966. *Dipole Radiation in the Presence of a Conducting Half-Space.* New York: Pergamon Press.

Barkovskii, L. M., and G. N. Borzdov. 1976. "Electromagnetic waves in absorbing, plane-layered anisotropic and gyrotropic media." *J. Appl. Spectros.* 23(1): 985–91.

Barkovskii, L. M., and G. N. Borzdov. 1978. "Reflection of electromagnetic waves from layered continuously inhomogeneous anisotropic media: multiple-reflection method." *Opt. Spectrosc.* (USSR) 45(4): 701–05.

Barrar, R. B., and R. M. Redheffer. July 1955. "On nonuniform dielectric media." *IRE Trans. Antennas Propagat.*, 101–07.

Bellman, R., and G. M. Wing. 1975. *An Introduction to Invariant Imbedding.* New York: John Wiley & Sons.

Bender, C., and S. Orzag. 1978. *Advanced Mathematical Methods for Scientists and Engineers.* New York: McGraw-Hill.

Berreman, D. W. 1972. "Optics in stratified and anisotropic media: 4×4-matrix formulation." *J. Opt. Soc. Am.* 62: 502–10.

Bleistein, N. 1966. "Uniform asymptotic expansion of integrals with stationary point near algebraic singularity." *Comm. Pure and Appl. Math.* 19: 353–70.

Bleistein, N., and R. A. Handelsman. 1975. *Asymptotic Expansion of Integrals.* New York: Holt, Rinehart and Winston.

Boardman, A. D. 1982. *Electromagnetic Surface Modes.* New York: John Wiley & Sons.

Born, M., and E. Wolf. 1980. *Principles of Optics.* New York: Pergamon Press.

Brekhovskikh, L. M. 1960. *Waves in Layered Media.* New York: Academic Press.

Bremmer, H., and S. W. Lee. 1984. "Propagation of a geometrical field in an isotropic inhomogeneous medium." *Radio Science* 19(1): 243–57.

Budden, K. G. 1961. *Radio Waves in the Ionosphere.* Cambridge, England: Cambridge University Press.

Chew, W. C. 1983. "The singularities of a Fourier-type integral in a multi-cylindrically layered medium." *IEEE Trans. Antennas Propagat.* AP-31:653.

Chew, W. C. 1988. "A quick way to approximate a Sommerfeld-Weyl-type integral." *IEEE Trans. Antennas Propagat.* AP-36: 1654–57.

Chew, W. C., and S. C. Gianzero. 1981. "Theoretical investigation of the electromagnetic wave propagation tool." *IEEE Trans. Geosci. Remote Sensing* GE-19: 1–7.

Chew, W. C., and J. A. Kong. 1981. "Electromagnetic field of a dipole on a two-layer earth." *Geophysics* 46: 309–15.

Chew, W. C., and J. A. Kong. 1982. "Asymptotic approximation of waves due to a dipole on a two-layer medium." *Radio Sci.* 17: 509–13.

Clemmow, P. C. 1966. *The Plane Wave Spectrum Representation of Electromagnetic Fields.* New York: Pergamon Press.

Copson, E. T. 1965. *Asymptotic Expansions.* Cambridge, England: Cambridge University Press.

Damaskos, N. J., A. L. Mafett, and P. L. E. Uslenghi. 1982. "Dispersion relation for general anisotropic media." *IEEE Trans. Antennas Propagat.* AP-30(5): 991–93.

Erdelyi, A. 1956. *Asymptotic Expansions.* New York: Dover Publishing Co. 1956.

Felsen, L. B. 1981. "Hybrid ray-mode fields in inhomogeneous waveguides and ducts." *J. Acoust. Soc. Am.* 69: 352–61.

Felsen, L. B., and N. Marcuvitz. 1973. *Radiation and Scattering of Electromagnetic Waves.* New Jersey: Prentice-Hall.

Franke, S. J., and G. W. Swenson, Jr. 1989. "A brief tutorial on the fast field program (FFP) as applied to sound propagation in air." *Appl. Acoustics* 27: 203–15.

Gradshteyn, I. S., and I. M. Ryzhik. 1965. *Table of Integrals, Series and Products.* New York: Academic Press.

Graglia, R. D., and P. L. E. Uslenghi. 1984. "Electromagnetic scattering from anisotropic material, Part I: General theory." *IEEE Trans. Antennas Propagat.* AP-32(8): 867–69.

Haskell, N. A. 1953. "The dispersion of surface waves in multilayered media." *Bull. Seism. Soc. Am.* 43: 17.

Hessel, A. 1969. "General characteristics of travelling-wave antennas." *Antenna Theory*, part 2, eds. R. Collin and F. Zucker. New York: McGraw-Hill.

Hildebrand, F. B. 1974. *Introduction to Numerical Analysis.* New York: McGraw-Hill.

Hildebrand, F. B. 1976. *Advanced Calculus for Applications.* Englewood

Cliffs, New Jersey: Prentice-Hall.

Hill, D. A., and J. R. Wait. 1978. "Excitation of the Zenneck surface wave by a vertical aperture." *Radio Sci.* 13: 969–77.

Kline, M., and I. Kay. 1965. *Electromagnetic Theory and Geometrical Optics.* New York: Academic Press.

Kong, J. A. 1972. "Electromagnetic field due to dipole antennas over stratified anisotropic media." *Geophysics* 38: 985–96.

Kong, J. A. 1975. *Theory of Electromagnetic Waves.* New York: John Wiley & Sons.

Kong, J. A. 1986. *Electromagnetic Wave Theory.* New York: John Wiley & Sons.

Lee, S. W., N. Bong, W. F. Richards, and R. Raspet. 1986. "Impedance formulation of the fast field program for acoustic wave propagation in the atmosphere." *J. Acoust. Soc. Am.* 79: 628–34.

Morgan, M. A., D. L. Fisher, and E. A. Milne. 1987. "Electromagnetic scattering by stratified inhomogeneous anisotropic media." *IEEE Trans. Antennas Propagat.* AP-35: 191–97.

Newton, R. G. 1966. *Scattering Theory of Waves and Particles.* New York: McGraw-Hill.

Olver, F. W. J. 1974. *Introduction to Asymptotics and Special Functions.* New York: Academic Press.

Oppenheim, A. V., and R. W. Schafer. 1975. *Digital Signal Processing.* Englewood Cliff, N.J.: Prentice-Hall.

Papoulis, A. 1962. *The Fourier Integral and its Applications.* New York: McGraw-Hill.

Pierce, A. D. 1965. "Extension of the method of normal modes to sound propagation in an almost stratified medium." *J. Acoust. Soc. Am* 37: 19–27.

Sommerfeld, A. 1909. "Uber die Ausbreitung der Wellen in der drahtlosen Telegraphie." *Ann. Physik* 28: 665–737.

Sommerfeld, A. 1949. *Partial Differential Equations in Physics.* New York: Academic Press.

Stickler, D. C. 1976. "Reflected and lateral waves for the Sommerfeld model." *J. Acoust. Soc. Am.* 60: 1061.

Tamir, T. 1969. "Leaky wave antennas." *Antenna Theory*, part 2, eds. R. Collin and F. Zucker. New York: McGraw-Hill.

Tamir, T. 1972. "Inhomogeneous wave types at planar structures: I. The lateral wave." *Optik* 36: 209–32.

Tamir, T. 1973. "Inhomogeneous wave types at planar structures: II. Surface waves." *Optik* 37: 204–28.

Teitler, T., and B. W. Henvis. 1970. "Refraction in stratified, anisotropic media." *J. Opt. Soc. Am.* 60: 830–34.

Thomson, W. T. 1950. "Transmission of elastic waves through a stratified solid." *J. Appl. Phys.* 21: 89–93.

Tsang, L., and J. A. Kong. 1973. "Interference patterns of a horizontal electric dipole over layered dielectric media." *J. Geophys. Res.* 78: 3287–3300.

Tyras, G. 1969. *Radiation and Propagation of Electromagnetic Waves*. New York: Academic Press.

Van Dyke, M. 1975. *Perturbation Methods in Fluid Mechanics*. Stanford: Parabolic Press.

Wait, J. R. 1968. *Electromagnetics and Plasmas*. New York: Holt, Rinehart and Winston.

Wait, J. R. 1970. *Electromagnetic Waves in Stratified Media*, 2nd ed. New York: Pergamon Press.

Wait, J. R. 1971. *Electromagnetic Probing in Geophysics*. Boulder, Co: Golem Press.

Wait, J. R. 1982. *Geoelectromagnetism*. New York: Academic Press.

Watson, G. N. 1944. *A Treatise on the Theory of Bessel Functions*. Cambridge: Cambridge University Press.

Weyl, H. 1919. *Ann. d. Physik* 60: 481.

Whitaker, E. T., and G. N. Watson. 1927. *Modern Analysis*, 4th ed. Cambridge: Cambridge University Press.

Yamada, J. I., M. Saruwatari, K. Asatani, H. Tsuchiya, A. Kawana, K. Sugiyama, and T. Kimura, "High speed optical pulse transmission at 1.29 μm wavelength using low loss single mode fibers," *IEEE J. Quantum Electronics*, QE. 14, p. 791, 1978.

Yeh, K. C., and C. H. Liu. 1972. *Theory of Ionospheric Waves*. Orlando, Fla.: Academic.

Yeh, P. 1988. *Optical Waves in Layered Media*. New York: John Wiley & Sons.

Further Readings for Chapter 2

Achenbach, J. D. 1973. *Wave Propagation in Elastic Solids.* Amsterdam: North-Holland.

Agrawal, B. S., and E. Bahar. 1980. "Propagation of EM waves in imhomogeneous anisotropic media." *IEEE Trans. Antennas Propagat.* AP-28: 422–24.

Allis, W. P., S. J. Buchsbaum, and A. Bers. 1962. *Waves in Anisotropic Plasmas,* chapter 3. New York: John Wiley & Sons.

Altman, C., and K. Suchy. 1979. "Generalization of a scattering theorem for plane-stratified gyrotropic media." *J. Appl. Phys.* 19: 337–43.

Annan, A. P. 1973. "Radio interferometry depth sounding, part I–theoretical discussion." *Geophysics* 38: 557–80.

Annan, A. P. 1975. "The electromagnetic response of a low-loss, two-layer dielectric earth for horizontal electric dipole excitation." *Geophysics* 40: 285–98.

Arbel, E., and L. B. Felsen. 1963. "Theory of radiation from sources in anisotropic media. Part I—General sources in stratified media. Part II—Point source in an infinite homogeneous medium." *Electromagnetic Theory and Antennas,* ed. E. C. Jordan, 391–459. New York: Pergamon Press.

Bahar, E. 1986. "Full wave solutions for electromagnetic scattering and depolarization in irregular stratified media." *Radio Sci.* 21: 543.

Bannister, P. R. 1979. "Summary of image theory expressions for the quasi-static fields of antennas at or above the earth's surface." *Proc. IEEE* 67(7): 1001–08.

Bannister, P. R. 1982. "The image theory for electromagnetic fields of a horizontal electric dipole in the presence of a conducting half space." *Radio Sci.* 17: 1095–1102.

Bannister, P. R. 1986. "Applications of complex image theory." *Radio Sci.* 21: 605.

Bertoni, H., and A. Hessel. 1967. "Ray optics for radiation problems in anisotropic regions with boundaries. 2. Point source excitation." *Radio Sci.* 2: 793–812.

Bleistein, N. 1967. "Uniform asymptotic expansions of integrals with many nearby stationary points and algebraic singularities." *J. Math. Mech.* 17: 533–59.

Bremmer, H. 1951. "The WKB approximation as the first term of a geometric-optical series." *Comm. Pure and Appl. Math.* 4: 105.

Chang, H. C., S. K. Jeng, R. B. Wu, and C. H. Chen. 1986. "Propagation of waves through magnetoplasma slab within a parallel plate guide." *IEEE Trans. Microwave Theory Tech.,* in press.

Chen, C. H., C. D. Lien, and Y. W. Kiang. 1982. "Discrete energy conservation law and reciprocity relationship for one-dimensional wave propagation problems." *IEEE Trans. Antennas Propagat.* AP-30: 483–86.

Chester, C., B. Friedman, and F. Ursell. 1957. "An extension of the method of steepest descents." *Proc. of Cambridge Phil. Soc.* 53: 599–611.

Chew, W. C. 1983. "Mixed boundary value problem of two non-identical circular conducting disks: an electrode configuration suitable for probing stratified media." *J. of Electrostatics* 14: 59–72.

Chew, W. C. 1986. "A current emitting in the presence of an insulating and a conducting disk over stratified media." *J. of Electrostatics* 18: 273–87.

Chew, W. C., J. A. Kong, and L. C. Shen. 1980. "Radiation characteristics of a circular microstrip antenna." *J. Appl. Phys.* 51(7): 3907–15.

Chu, R. S., and J. A. Kong. 1977. "Diffraction of Gaussian beams by a periodically-modulated layer." *J. Opt. Soc. Am.* 67: 1555–61.

Chu, R. S., and J. A. Kong. 1980. "Diffraction of optical beams with arbitrary profiles by a perodically-modulated layer." *J. Opt. Soc. Am.* 70: 1–6.

Copson, E. T. 1935. *Theory of Functions of a Complex Variable.* London: Oxford University Press.

deBruijn, N. G. 1958. *Asymptotic Methods in Analysis.* chapters 4–6. New York: Interscience Publishing Co.

Dey, A., and S. H. Ward. 1970. "Inductive soundings of a layered earth with a horizontal magnetic dipole." *Geophysics* 35: 660.

El-Said, M. A. H. 1956. "Geophysical prospection of underground water in the desert by means of electromagnetic interference fringes." *Proc. IRE* 44: 24–30.

Epstein, P. S. 1930. "Reflection of waves in an inhomogeneous medium." *Proc. Natl. Acad. Sci. (USA)* 16: 627.

Erdelyi, A., A. W. Magnus, F. Oberhettinger, and F. Tricomi. 1953. *Higher Transcendental Functions*, vol. I, 142–44. New York: McGraw-Hill.

Esmersoy, C., and B. C. Levy. 1986. "Multidimensional Born inversion with a wide-band plane-wave source." *Proc. IEEE* 74: 466–75.

Ewing, W. M., W. S. Jardetzky, and F. Press. 1957. *Elastic Waves in Layered Media.* New York: McGraw-Hill.

Felsen, L. B. 1964. "Radiation from a uniaxially anisotropic plasma half space." *IEEE Trans. Antennas Propagat.* AP-11: 469–84.

Felsen, L. B. 1964. "Propagation and diffraction in uniaxially anisotropic regions. I—Theory. II—Applications." *Proc. IEE (London)* 111: 445–64.

Felsen, L. B. 1965. "On the use of refractive index diagrams for source excited anisotropic regions." *Radio Sci.* 69D: 155-69.

Felsen, L. B., and L. Levey. 1966. "A relation between a class of boundary value problems in a homogeneous and an inhomogeneous region." *IEEE Trans. Antennas Propagat.* AP-14: 308–17.

Felsen, L. B., and S. Rosenbaum. 1967. "Ray optics for radiation problems in anisotropic regions with boundaries. 1. Line source excitation." *Radio Sci.* 2 (new series): 767–91.

Fock, V. A. 1965. *Electromagnetic Diffraction and Propagation Problems.* New York: Pergamon Press.

Freedman, R., and J. P. Vogiatzis. 1979. "Theory of microwave dielectric constant logging using the electromagnetic wave propagation method." *Geophysics* 44: 969–86.

Fried, B. D., and S. D. Conte. 1961. *The Plasma Dispersion Function.* New York: Academic Press.

Frischknecht, F. C. 1967. "Fields about an oscillating magnetic dipole over a two-layer earth, and application to ground and airborne electromagnetic surveys." *Colorado School of Mines Quart* 62.

Ginzburg, V. L. 1964. *The Propagation of Electromagnetic Waves in Plasmas.* New York: Pergamon Press.

Habashy, T. M., J. A. Kong, and L. Tsang. 1985. "Quasi-static electromagnetic fields due to dipole antennas in bounded conducting media." *IEEE Trans. Geosci. Remote Sensing* GE-23: 325–33.

Handelsman, R. A., and N. Bleistein. 1969. "Uniform asymptotic expansion of integrals that arise in the analysis of precursors." *Arch. for Rat. Mech. and Analysis* 35: 267–83.

Hansen, P. M., and C. T. Tai. 1970. "Radiation from sources in the presence of a flat Earth." *IEEE Trans. Antennas Propagat.* AP-18: 423–24.

Hill, D. A., and J. R. Wait. 1986. "Anomalous vertical magnetic field for electromagnetic induction in a laterally varying thin conductive sheet." *Radio Sci.* 21: 617.

Ishimaru, A. 1963. "Unidirectional surface waves in anisotropic media." *Electromagnetic Theory and Antennas*, ed. E. C. Jordan. 591–601. New York: Pergamon Press.

Jeffreys, H. 1962. *Asymptotic Approximations*, chapter 2. London: Oxford University Press

Jeng, S. K., and C. H. Chen. 1984. "Variational finite element solution of electromagnetic wave propagation in a one-dimensional inhomogeneous anisotropic medium." *J. Appl. Phys.* 55: 630–36.

Jeng, S.-K., R.-B. Wu, and C. H. Chen. 1986. "Waves obliquely incident upon a stratified anisotropic slab: A variational reaction approach." *Radio Sci.* 21: 681.

Jones, D. S., and M. Kline. 1958. "Asymptotic expansion of multiple integrals and the method of stationary phase." *J. Math. Phys.* 37: 1–28.

Kazarinoff, N. D. 1959. "Asymptotic theory of second order differential equations with two simple turning points." *Arch. for Rat. Mech. and Anal.* 2: 129–50.

Kong, J. A. 1981. *Research Topics in Electromagnetic Wave Theory.* New York: Wiley-Interscience.

Kong, J. A., L. C. Shen, and L. Tsang. 1977. "Field of an antenna submerged in dissipative dielectric medium." *IEEE Trans. Antennas Propagat.* AP-25: 887–89.

Kong, J. A., L. Tsang, and G. Simmons. 1974. "Geophysical subsurface probing with radio-frequency interferometry." *IEEE Trans. Antennas Propagat.* AP-22: 616–20.

Lamb, H. 1904. "On the propagation of tremors over the surface of an elastic solid." *Phil. Trans. Roy. Soc. (London), Ser. A* 203: 1–42.

Langer, R. E. 1959. "The asymptotic solution of a linear differential equation of the second order with two turning points." *Trans. Amer. Math. Soc.* 90: 113–42.

Levey, L., and L. B. Felsen. 1969. "On incomplete Airy functions and their application to diffraction problems." *Radio Sci.* 4: 959–69.

Lewis, R. M., and B. Granoff. May 1969. "Asymptotic theory of electromagnetic wave propagation in an inhomogeneous anisotropic plasma." *Alta Frequenza* (special issue) 38: 51–59.

Lindell, I. V., and E. Alanen. 1984. "Exact image theory for the Sommerfeld half-space problem, I. Vertical magnetic dipole." *IEEE Trans. Antennas Propagat.* AP-32(2): 126–33.

Lindell, I. V., and E. Alanen. 1984. "Exact image theory for the Sommerfeld half-space problem, II. Vertical electric dipole." *IEEE Trans. Antennas Propagat.* AP-32(8): 841–47.

Lindell, I. V., and E. Alanen. 1984. "Exact image theory for the Sommerfeld half-space problem, III. General formulation." *IEEE Trans. Antennas Propagat.* AP-32(10): 1027–32.

Lindell, I. V., E. Alanen, and K. Mannersalo. 1985. "Exact image method for impedance computation of antennas above the ground." *IEEE Trans. Antennas Propagat.* AP-33: 937–45.

Ludwig, D. 1966. "Uniform asymptotic expansions at a caustic." *Comm. Pure and Appl. Math.* 19: 215–60.

Magnus, W., and F. Oberhettinger. 1954. *Formulas and Theorems for the Special Functions of Mathematical Physics.* New York: Chelsea Publishing Co.

Mahmoud, S. F., and A. D. Metwally. 1981a. "New image representation for dipoles near a dissipative earth, 1. Discrete images." *Radio Sci.* 16(6): 1271–75.

Mahmoud, S. F., and A. D. Metwally. 1981b. "New image representation for dipoles near a dissipative earth, 2. Discrete plus continuous images." *Radio Sci.* 16(6): 1277–83.

Mahmoud, S. F., and A. Mohsen. 1985. "Assessment of image theory for field evaluation over a multi-layer earth." *IEEE Trans. Antennas Propagat.* AP-33(10): 1054–58.

Miller, J. C. P. 1946. *The Airy Integral* (British Association Mathematical Tables). Cambridge, England: Cambridge University Press.

Mittra, R., P. Parhami, and Y. Rahmat-Samii. 1979. "Solving the current element problem over lossy half-space without Sommerfeld integrals." *IEEE Trans. Antennas Propagat.* AP-27(6): 778–82.

Olsen, R. G., and T. A. Pankaskie. 1983. "On the exact, Carson, and image theories for wires at or above the earth's surface." *IEEE Trans. Power Appar. Syst.* PAS-102(4): 769–78.

Olver, F. W. J. 1958. "Uniform asymptotic expansions of solutions of linear second-order differential equations for large values of a parameter." *Phil. Trans. Roy. Soc. London* 250A: 479.

Olver, F. W. J. 1959. "Uniform asymptotic expansions for Weber parabolic cylinder functions of large orders." *J. Res. NBS* 63B: 131.

Pearcey, T. 1946. "The structure of an electromagnetic field in the neighborhood of a cusp or caustic." *Phil. Mag.* 37: 311.

Rice, S. O. 1968. "Uniform asymptotic expansions for saddle point integrals." *Bell System Tech. Jour.* 47: 1971–2013.

Rydbeck, O. E. H. 1948. "On the propagation of waves in an inhomogeneous medium." *Trans. of Chalmers Univ. of Technology*, no. 74. Gothenburg, Sweden.

Seckler, B. D., and J. B. Keller. 1959a. "Geometrical theory of diffraction in inhomogeneous media." *J. Acoust. Soc. Am.* 31: 192–205.

Seckler, B. D., and J. B. Keller. 1959b. "Asymptotic theory of diffraction in inhomogeneous media." *J. Acoust. Soc.* 31: 206–16.

Seshadri, S. R. 1962. "Excitation of surface waves on a perfectly conducting screen covered with anisotropic plasma." *IRE Trans. on Microwave Theory and Techniques* MTT-10: 573–78.

Seshadri, S. R., and T. T. Wu. 1970. "Radiation condition for a magneto-plasma medium." *Quart. J. of Mech. and Appl. Math.* 23(2): 285–313.

Seshadri, S. R. 1971. *Fundamentals of Transmission Line and Electromagnetic Fields*. Reading, Mass.: Addison-Wesley.

Sinha, A. P. 1968. "Electromagnetic fields of an oscillating magnetic dipole over an anisotropic earth." *Geophysics* 33: 346–53.

Smith, G. S. 1984. "Directive properties of antennas for transmission into a material half-space." *IEEE Trans. Antennas Propagat.* AP-32: 232–46.

Titchmarsh, E. C. 1937. *Introduction to the Theory of Fourier Integrals*. London: Oxford University Press.

Titchmarsh, E. C. 1950. *The Theory of Functions*, 2nd ed. New York: Oxford University Press.

Tsang, L., J. A. Kong, and R. T. Shin. 1985. *Theory of Microwave Remote Sensing*. New York: Wiley-Interscience.

Wait, J. R. 1961. "The electromagnetic fields of a horizontal dipole in the presence of a conducting half-space." *Can. J. Phys.* 39(7): 1017–28.

Weaver, J. T. 1971. "Image theory for an arbitrary quasi-static field in the presence of a conducting half-space." *Radio Sci.* 6(6): 647–53.

Whitham, G. B. 1961. "Group velocity and energy propagation for three-dimensional waves." *Comm. on Pure and Applied Math.* 16: 657–91.

Yamazawa, M., N. Inagaki, and T. Sekiguchi. 1971. "Excitation of surface wave on circular-loop array." *IEEE Trans. Antennas Propagat.* AP-19: 433–35.

Yeh, C., and K. F. Casey. 1966. "Reflection and transmission of electromagnetic waves by a moving dielectric slab." *Phys. Rev.* 144: 665–69.

CHAPTER 3

CYLINDRICALLY AND SPHERICALLY LAYERED MEDIA

Information obtained from the study of waves in circular cylindrically, and spherically layered media can be applied to fiber optics, geophysical probing, radar cross-section studies, microstrip antennas, and other technologies. For example, in optical fibers, the doping of the fiber may have a gradual transition rather than a step transition. Such a gradual transition may be modeled with many thin layers of piecewise homogeneous layers.

In the geophysical exploration of the subsurface earth, a borehole drilled deep into the earth is often employed. Such boreholes are usually filled with fluid. The subsequent invasion of borehole fluid into the rock formation gives rise to an altered zone whose electromagnetic property varies radially away from the borehole. Such a zone may be modeled by a cylindrically layered medium.

Curved layers are also useful in modeling the scattering by curved surfaces, which are coated with dielectric materials in order to reduce their radar scattering cross sections. This has applications in reducing the radar cross sections of aircrafts, for example. Also, for an understanding of microstrip antennas mounted on curved surfaces, and dielectric coated circular waveguides, a knowledge of waves in cylindrically layered media is often necessary.

Works in this area are usually related to wave propagation on curved surfaces, scattering by cylindrical structures (for a review, see Pearson 1987), borehole geophysics (for a review, see Lovell and Chew 1987a), and dielectric optical waveguides (for a review, see Gronthoud and Blok 1978; and Yariv 1985).

§3.1 Cylindrically Layered Media—Single Interface Case

In a cylindrical geometry of circular symmetry, a wave can be broken down into cylindrical harmonics with different $e^{in\phi}$ dependence. In principle, only two of the six electromagnetic field components are needed to characterize the wave. Moreover, due to the preferred z direction of the medium, the electromagnetic field can be decomposed into TM to z and TE to z type waves. As a result, E_z and H_z components of the field can be used to characterize the TM and TE waves respectively. Unlike planar boundaries, however, both TM and TE waves are needed (in general) to match the boundary conditions at an interface between two cylindrical layers. The only exceptions are axially symmetric ($n = 0$) and z-invariant ($\partial/\partial z = 0$) TM and TE waves, whose boundary conditions can be satisfied independently of each other. Consequently, for nonaxially symmetric waves, both TM and TE waves must coexist in a circular, cylindrically layered medium. Such a wave is also known as a *hybrid wave*. In this presentation, we shall use the formulation of Chew (1984) and Lovell and Chew (1987a, 1987b).

§§3.1.1 Vector Wave Equation in Cylindrical Coordinates

The vector wave equations in a homogeneous, isotropic, source-free medium are

$$\nabla \times \nabla \times \mathbf{E} - k^2 \mathbf{E} = 0, \tag{3.1.1a}$$

$$\nabla \times \nabla \times \mathbf{H} - k^2 \mathbf{H} = 0. \tag{3.1.1b}$$

In cylindrical coordinates, it is most expedient to extract the z components of the above equations to yield (see Exercise 3.1)

$$(\nabla^2 + k^2)E_z = 0, \tag{3.1.2a}$$

$$(\nabla^2 + k^2)H_z = 0. \tag{3.1.2b}$$

Then, the general solutions to the above equations are

$$\begin{bmatrix} E_z \\ H_z \end{bmatrix} = [\mathbf{A}_n J_n(k_\rho \rho) + \mathbf{B}_n H_n^{(1)}(k_\rho \rho)]e^{ik_z z + in\phi}, \tag{3.1.3}$$

where $k_\rho^2 + k_z^2 = k^2$. Notice that the above can also be replaced by any other linear superposition of two linearly independent solutions of Bessel's equation (Exercise 3.2).

Furthermore, given the z components of the fields, the transverse components of the fields in cylindrical coordinates can be determined from the following equations[1]

$$\mathbf{E}_s = \frac{1}{k_\rho^2}[ik_z \nabla_s E_z - i\omega\mu\, \hat{z} \times \nabla_s H_z], \tag{3.1.4a}$$

$$\mathbf{H}_s = \frac{1}{k_\rho^2}[ik_z \nabla_s H_z + i\omega\epsilon\, \hat{z} \times \nabla_s E_z], \tag{3.1.4b}$$

where $\nabla_s = \hat{\rho}\frac{\partial}{\partial\rho} + \hat{\phi}\frac{1}{\rho}\frac{\partial}{\partial\phi} = \hat{\rho}\frac{\partial}{\partial\rho} + \hat{\phi}\frac{in}{\rho}$. Note that the above equations are only valid in a homogeneous region. Hence, with the above method of using E_z and H_z only, all six components of the electromagnetic field can be derived in cylindrical coordinates. In particular, E_z characterizes a wave TM to z, while H_z characterizes a wave TE to z.

---

[1] The derivation of these equations is similar to (2.3.17a) and (2.3.17b).

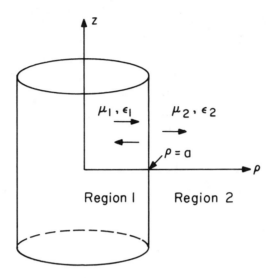

Figure 3.1.1 Reflection and transmission of an outgoing wave.

In general, at a cylindrical interface, TE waves and TM waves are coupled together. In other words, E_z and H_z have to be considered simultaneously. Hence, at least two components of the field are required, implying that the problem is vector in nature. Consequently, when studying reflections from a cylindrical boundary, we need to define 2×2 reflection and transmission matrices rather than reflection and transmission coefficients.

The reflection and transmission of waves at a cylindrical interface can be categorized into two cases:

(i) an outgoing wave incident on a cylindrical boundary, and

(ii) a standing wave incident on a cylindrical boundary.

We shall first consider the outgoing-wave case, followed by the standing-wave case.

§§3.1.2 Reflection and Transmission of an Outgoing Wave

First, assume that an outgoing ($+\rho$ direction) cylindrical wave (e.g., generated by a line source) is in region 1 as shown in Figure 3.1.1. Then, the outgoing wave for the n-th harmonic with $e^{in\phi}$ dependence is

$$\begin{bmatrix} E_{1z} \\ H_{1z} \end{bmatrix} = H_n^{(1)}(k_{1\rho}\rho) \begin{bmatrix} e_{1z} \\ h_{1z} \end{bmatrix} = H_n^{(1)}(k_{1\rho}\rho)\, \mathbf{a}_1, \qquad (3.1.5)$$

where $k_{1\rho} = \sqrt{k_1^2 - k_z^2}$. Here, the field has been assumed to have $e^{ik_z z}$ dependence in all regions due to the translational invariance of the problem in the

z direction (phase matching; see Exercise 3.3). Moreover, both the ϕ and z dependences are suppressed in (5).

In general, the wave in region 1 will have an incoming wave ($-\rho$ direction) as well. But because of the closed nature of a cylindrical geometry, the incoming wave, which is represented by a Hankel function of the second kind, is always perfectly reflected at the origin, resulting in a standing wave. Hence, the reflected wave in region 1 always becomes a standing wave. Consequently, the total field in region 1 can be expressed as

$$\begin{bmatrix} E_{1z} \\ H_{1z} \end{bmatrix} = H_n^{(1)}(k_{1\rho}\rho)\mathbf{a}_1 + J_n(k_{1\rho}\rho)\overline{\mathbf{R}}_{12} \cdot \mathbf{a}_1. \tag{3.1.6}$$

The field in region 2, however, consists of a transmitted outgoing wave. This wave has to be an outgoing wave in order to satisfy the radiation condition. Therefore, it is

$$\begin{bmatrix} E_{2z} \\ H_{2z} \end{bmatrix} = H_n^{(1)}(k_{2\rho}\rho)\overline{\mathbf{T}}_{12} \cdot \mathbf{a}_1, \tag{3.1.7}$$

where $k_{2\rho} = \sqrt{k_2^2 - k_z^2}$. At this point, $\overline{\mathbf{R}}_{12}$ and $\overline{\mathbf{T}}_{12}$ are 2×2 matrices yet to be determined. To find them, boundary conditions are imposed for the field at $\rho = a$. The requisite boundary conditions here are that the z component and the ϕ component of the field must be continuous at $\rho = a$.

Next, Equations (4a) and (4b) can be used in each of the homogeneous regions 1 and 2 to yield the ϕ components of the fields as (see Exercise 3.4)

$$\begin{bmatrix} H_{1\phi} \\ E_{1\phi} \end{bmatrix} = \overline{\mathbf{H}}_n^{(1)}(k_{1\rho}\rho) \cdot \mathbf{a}_1 + \overline{\mathbf{J}}_n(k_{1\rho}\rho) \cdot \overline{\mathbf{R}}_{12} \cdot \mathbf{a}_1, \tag{3.1.8a}$$

$$\begin{bmatrix} H_{2\phi} \\ E_{2\phi} \end{bmatrix} = \overline{\mathbf{H}}_n^{(1)}(k_{2\rho}\rho) \cdot \overline{\mathbf{T}}_{12} \cdot \mathbf{a}_1, \tag{3.1.8b}$$

where

$$\overline{\mathbf{B}}_n(k_{i\rho}\rho) = \frac{1}{k_{i\rho}^2\rho} \begin{bmatrix} i\omega\epsilon_i k_{i\rho}\rho B_n'(k_{i\rho}\rho) & -nk_z B_n(k_{i\rho}\rho) \\ -nk_z B_n(k_{i\rho}\rho) & -i\omega\mu_i k_{i\rho}\rho B_n'(k_{i\rho}\rho) \end{bmatrix}. \tag{3.1.9}$$

Here, B_n is either $H_n^{(1)}$ or J_n depending on whether we are defining the $\overline{\mathbf{H}}_n^{(1)}$ matrix or the $\overline{\mathbf{J}}_n$ matrix. Moreover, $\overline{\mathbf{B}}_n$ is diagonal when $n = 0$.

Then, the boundary conditions at $\rho = a$, which require the continuity of the tangential components of the electromagnetic field, yield

$$\left[H_n^{(1)}(k_{1\rho}a)\overline{\mathbf{I}} + J_n(k_{1\rho}a)\overline{\mathbf{R}}_{12} \right] \cdot \mathbf{a}_1 = H_n^{(1)}(k_{2\rho}a)\overline{\mathbf{T}}_{12} \cdot \mathbf{a}_1, \tag{3.1.10a}$$

$$\left[\overline{\mathbf{H}}_n^{(1)}(k_{1\rho}a) + \overline{\mathbf{J}}_n(k_{1\rho}a) \cdot \overline{\mathbf{R}}_{12} \right] \cdot \mathbf{a}_1 = \overline{\mathbf{H}}_n^{(1)}(k_{2\rho}a) \cdot \overline{\mathbf{T}}_{12} \cdot \mathbf{a}_1. \tag{3.1.10b}$$

Since $\mathbf{a}_1$ is nonzero and arbitrary at this point, it can be cancelled from both sides of the preceding equations and the resulting matrix equations solved for $\overline{\mathbf{R}}_{12}$ and $\overline{\mathbf{T}}_{12}$. As a result, they are

$$\overline{\mathbf{R}}_{12} = \overline{\mathbf{D}}^{-1} \cdot \left[H_n^{(1)}(k_{1\rho}a) \,\overline{\mathbf{H}}_n^{(1)}(k_{2\rho}a) - H_n^{(1)}(k_{2\rho}a) \,\overline{\mathbf{H}}_n^{(1)}(k_{1\rho}a) \right],$$

(3.1.11a)

$$\overline{\mathbf{T}}_{12} = \overline{\mathbf{D}}^{-1} \cdot \left[H_n^{(1)}(k_{1\rho}a) \,\overline{\mathbf{J}}_n(k_{1\rho}a) - J_n(k_{1\rho}a) \,\overline{\mathbf{H}}_n^{(1)}(k_{1\rho}a) \right],$$

(3.1.11b)

where

$$\overline{\mathbf{D}} = \left[\overline{\mathbf{J}}_n(k_{1\rho}a) H_n^{(1)}(k_{2\rho}a) - \overline{\mathbf{H}}_n^{(1)}(k_{2\rho}a) J_n(k_{1\rho}a) \right].$$

(3.1.11c)

Furthermore, the Wronskian for Hankel functions (see Exercise 3.5; also see Abramowitz and Stegun 1965), which is

$$H_n^{(1)}(x) J_n'(x) - J_n(x) H_n^{(1)'}(x) = -\frac{2i}{\pi x},$$

can be used to simplify $\overline{\mathbf{T}}_{12}$ to

$$\overline{\mathbf{T}}_{12} = \frac{2\omega}{\pi k_{1\rho}^2 a} \overline{\mathbf{D}}^{-1} \cdot \begin{bmatrix} \epsilon_1 & 0 \\ 0 & -\mu_1 \end{bmatrix}.$$

(3.1.12)

Note that $\overline{\mathbf{R}}_{12}$ and $\overline{\mathbf{T}}_{12}$ are both functions of n.

§§3.1.3 Reflection and Transmission of a Standing Wave

First, consider the geometry shown in Figure 3.1.2 but with region 1 the same as region 2. Then, an incoming wave, denoted by $H_n^{(2)}(k_{2\rho}\rho)$, is always totally reflected at the origin resulting in a standing wave, which can be denoted by $J_n(k_{2\rho}\rho)$. Hence, in the absence of region 1, only a standing wave exists in region 2. Now, if region 1 is inserted around the origin, it will disturb the standing wave, resulting in an outgoing wave in region 2 denoted by $H_n^{(1)}(k_{2\rho}\rho)$. Furthermore, in region 1, a different standing wave, denoted by $J_n(k_{1\rho}\rho)$, ensues.

At this point the amplitudes of the standing wave in region 1, $J_n(k_{1\rho}\rho)$, and the outgoing wave in region 2, $H_n^{(1)}(k_{2\rho}\rho)$, are unknowns yet to be found. Furthermore, they can be regarded as the disturbance of the original standing wave in region 2, $J_n(k_{2\rho}\rho)$. Mathematically, this is the same as the reflection of a standing wave into an outgoing wave in region 2 and the transmission of a standing wave from region 2 to region 1. As a consequence, one can think of the original standing wave as an "incoming" wave even though it is actually a standing wave. Now, if an "incoming" standing wave is in region 2, then the field in region 2 for the n-th harmonic becomes

$$\begin{bmatrix} E_{2z} \\ H_{2z} \end{bmatrix} = H_n^{(1)}(k_{2\rho}\rho) \,\overline{\mathbf{R}}_{21} \cdot \mathbf{a}_2 + J_n(k_{2\rho}\rho) \cdot \mathbf{a}_2.$$

(3.1.13)

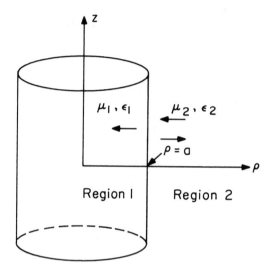

Figure 3.1.2 Reflection and transmission of a standing wave.

Moreover, the field in region 1 will consist of a standing wave, which is

$$
\begin{bmatrix} E_{1z} \\ H_{1z} \end{bmatrix} = J_n(k_{1\rho}\rho)\,\overline{\mathbf{T}}_{21}\cdot\mathbf{a}_2. \tag{3.1.14}
$$

Similar to the previous case, the ϕ components of the fields are derived from (13) and (14) via the use of (4a) and (4b), giving

$$
\begin{bmatrix} H_{2\phi} \\ E_{2\phi} \end{bmatrix} = \overline{\mathbf{H}}_n^{(1)}(k_{2\rho}\rho)\cdot\overline{\mathbf{R}}_{21}\cdot\mathbf{a}_2 + \overline{\mathbf{J}}_n(k_{2\rho}\rho)\cdot\mathbf{a}_2, \tag{3.1.15a}
$$

$$
\begin{bmatrix} H_{1\phi} \\ E_{1\phi} \end{bmatrix} = \overline{\mathbf{J}}_n(k_{1\rho}\rho)\cdot\overline{\mathbf{T}}_{21}\cdot\mathbf{a}_2. \tag{3.1.15b}
$$

Then, the boundary conditions at $\rho = a$ can be used to give

$$
\left[H_n^{(1)}(k_{2\rho}a)\,\overline{\mathbf{R}}_{21} + J_n(k_{2\rho}a)\overline{\mathbf{I}} \right]\cdot\mathbf{a}_2 = J_n(k_{1\rho}a)\,\overline{\mathbf{T}}_{21}\cdot\mathbf{a}_2, \tag{3.1.16a}
$$

$$
\left[\overline{\mathbf{H}}_n^{(1)}(k_{2\rho}a)\cdot\overline{\mathbf{R}}_{21} + \overline{\mathbf{J}}_n(k_{2\rho}a) \right]\cdot\mathbf{a}_2 = \overline{\mathbf{J}}_n(k_{1\rho}a)\cdot\overline{\mathbf{T}}_{21}\cdot\mathbf{a}_2. \tag{3.1.16b}
$$

Consequently, the solution of the above yields

$$
\overline{\mathbf{R}}_{21} = \overline{\mathbf{D}}^{-1}\cdot\left[J_n(k_{1\rho}a)\,\overline{\mathbf{J}}_n(k_{2\rho}a) - J_n(k_{2\rho}a)\,\overline{\mathbf{J}}_n(k_{1\rho}a) \right], \tag{3.1.17a}
$$

$$
\overline{\mathbf{T}}_{21} = \frac{2\omega}{\pi k_{2\rho}^2 a}\overline{\mathbf{D}}^{-1}\cdot\begin{bmatrix} \epsilon_2 & 0 \\ 0 & -\mu_2 \end{bmatrix}. \tag{3.1.17b}
$$

In the above, $\overline{\mathbf{D}}$ is as defined in (11c). Moreover, $\overline{\mathbf{R}}_{21}$ and $\overline{\mathbf{T}}_{21}$ are both functions of n.

The reflection and transmission matrices at an interface for outgoing and standing waves are now derived. The diagonal terms of $\overline{\mathbf{R}}_{ij}$ and $\overline{\mathbf{T}}_{ij}$ denote the self-coupling of the wave, that is, TM to TM, or TE to TE coupling. But the off-diagonal terms of $\overline{\mathbf{R}}_{ij}$ and $\overline{\mathbf{T}}_{ij}$ denote the coupling from the TE to TM and the TM to TE waves. Note that when $n = 0$ (axially symmetric case), or when $k_z = 0$ ($\partial/\partial z = 0$), $\overline{\mathbf{H}}_n(x)$ and $\overline{\mathbf{J}}_n(x)$ are diagonal; hence, $\overline{\mathbf{D}}$ is diagonal. Consequently, $\overline{\mathbf{D}}^{-1}$ is also diagonal. Therefore, when $n = 0$ or $k_z = 0$, $\overline{\mathbf{R}}_{ij}$ and $\overline{\mathbf{T}}_{ij}$ given by (11) and (17) are diagonal, implying that the TM and TE waves are decoupled. In this case, they can be treated independently of each other, implying a scalar problem. Furthermore, since $B_{-n}(x) = (-1)^n B_n(x)$, one can show that when $n \to -n$, only the diagonal elements of $\overline{\mathbf{R}}_{12}$, $\overline{\mathbf{R}}_{21}$, $\overline{\mathbf{T}}_{12}$, and $\overline{\mathbf{T}}_{21}$ change sign. This fact could be used to fold the summation over n, which ranges from $-\infty$ to ∞, to range from 0 to ∞ when summing over all n harmonics (see Lovell and Chew 1987b).

The above are the canonical problems which illustrate the ongoing physical processes at an interface of cylindrical layers. Consequently, the solution for more complex cylindrical layers can be built upon these canonical solutions. This shall be shown in the next section.

§3.2 Cylindrically Layered Media—Multi-interface Case

Given the reflection and transmission matrices for a single interface, it is possible to derive reflection and transmission matrices for multicylindrically layered media. The reason being that the reflection and transmission of waves through a multilayered medium are a result of multiple single-interface reflections and transmissions.

§§3.2.1 The Outgoing-Wave Case

Consider first a three-layer medium as shown in Figure 3.2.1, where an outgoing wave exists in region 1. Upon impinging on the multilayer medium, this wave is reflected in region 1. Then, in region 2, there are both standing and outgoing waves. And in region 3, there is only an outgoing wave. Hence, the field in region 1 is

$$\begin{bmatrix} E_{1z} \\ H_{1z} \end{bmatrix} = \left[H_n^{(1)}(k_{1\rho}\rho)\,\overline{\mathbf{I}} + J_n(k_{1\rho}\rho)\,\widetilde{\overline{\mathbf{R}}}_{12} \right] \cdot \mathbf{a}_1, \qquad (3.2.1)$$

where $\widetilde{\overline{\mathbf{R}}}_{12}$ is a generalized reflection matrix yet to be derived that relates standing wave to outgoing wave. Moreover, the field in region 2 can be written as

$$\begin{bmatrix} E_{2z} \\ H_{2z} \end{bmatrix} = \left[H_n^{(1)}(k_{2\rho}\rho)\,\overline{\mathbf{I}} + J_n(k_{2\rho}\rho)\,\overline{\mathbf{R}}_{23} \right] \cdot \mathbf{a}_2, \qquad (3.2.2)$$

where $\overline{\mathbf{R}}_{23}$ is the local reflection matrix between regions 2 and 3 derived in the previous section. The absence of source or standing wave in region 3

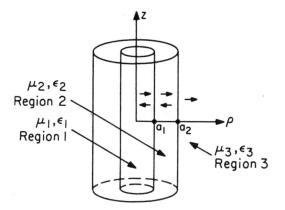

Figure 3.2.1 Reflection and transmission through a three-layer cylindrical medium—the outgoing-wave case.

dictates that the standing and outgoing waves in region 2 be related by such a "ratio." Finally, in region 3, we have

$$\begin{bmatrix} E_{3z} \\ H_{3z} \end{bmatrix} = H_n^{(1)}(k_{3\rho}\rho)\, \mathbf{a}_3. \tag{3.2.3}$$

In the above, $\widetilde{\overline{\mathbf{R}}}_{12}$, $\mathbf{a}_2$, and $\mathbf{a}_3$ are unknowns yet to be determined. Since an outgoing wave in region 2 is a consequence of the transmission of an outgoing wave in region 1 plus the reflection of a standing wave in region 2, we have

$$\mathbf{a}_2 = \overline{\mathbf{T}}_{12} \cdot \mathbf{a}_1 + \overline{\mathbf{R}}_{21} \cdot \overline{\mathbf{R}}_{23} \cdot \mathbf{a}_2. \tag{3.2.4}$$

Furthermore, the standing wave in region 1 is a result of the reflection of an outgoing wave in region 1 plus the transmission of a standing wave in region 2. Therefore, we have

$$\widetilde{\overline{\mathbf{R}}}_{12} \cdot \mathbf{a}_1 = \overline{\mathbf{R}}_{12} \cdot \mathbf{a}_1 + \overline{\mathbf{T}}_{21} \cdot \overline{\mathbf{R}}_{23} \cdot \mathbf{a}_2. \tag{3.2.5}$$

Solving (4) yields

$$\mathbf{a}_2 = \left(\overline{\mathbf{I}} - \overline{\mathbf{R}}_{21} \cdot \overline{\mathbf{R}}_{23}\right)^{-1} \cdot \overline{\mathbf{T}}_{12} \cdot \mathbf{a}_1. \tag{3.2.6}$$

Then, (6) can be used in (5) to give

$$\widetilde{\overline{\mathbf{R}}}_{12} = \overline{\mathbf{R}}_{12} + \overline{\mathbf{T}}_{21} \cdot \overline{\mathbf{R}}_{23} \cdot \left(\overline{\mathbf{I}} - \overline{\mathbf{R}}_{21} \cdot \overline{\mathbf{R}}_{23}\right)^{-1} \cdot \overline{\mathbf{T}}_{12}. \tag{3.2.7}$$

Equation (7) expresses $\widetilde{\overline{\mathbf{R}}}_{12}$ in terms of $\overline{\mathbf{R}}_{23}$ and the reflection and transmission matrices between regions 1 and 2. The fact is that it includes the

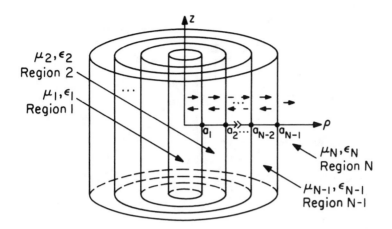

Figure 3.2.2 Reflection and transmission through an N-layer cylindrical medium.

physics of the multiple reflections in region 2. Now, if another layer beyond region 3 is added, only $\overline{\mathbf{R}}_{23}$, a local reflection matrix, needs to be altered to $\widetilde{\mathbf{R}}_{23}$, a generalized reflection matrix. Consequently, Equation (7) is expressible as a general recursive relation,

$$\widetilde{\overline{\mathbf{R}}}_{i,i+1} = \overline{\mathbf{R}}_{i,i+1} + \overline{\mathbf{T}}_{i+1,i} \cdot \widetilde{\overline{\mathbf{R}}}_{i+1,i+2} \cdot \left(\overline{\mathbf{I}} - \overline{\mathbf{R}}_{i+1,i} \cdot \widetilde{\overline{\mathbf{R}}}_{i+1,i+2}\right)^{-1} \cdot \overline{\mathbf{T}}_{i,i+1}. \qquad (3.2.8)$$

Equation (8) can be used to find the generalized reflection matrices for an N-layer medium as shown in Figure 3.2.2. In general, the field in region i can be written as

$$\begin{bmatrix} E_{iz} \\ H_{iz} \end{bmatrix} = \left[H_n^{(1)}(k_{i\rho}\rho)\overline{\mathbf{I}} + J_n(k_{i\rho}\rho)\widetilde{\overline{\mathbf{R}}}_{i,i+1} \right] \cdot \mathbf{a}_i, \qquad (3.2.9)$$

where $\widetilde{\overline{\mathbf{R}}}_{N,N+1} = 0$, if $i = N$, since no reflection exists beyond region N. Hence, one can use (8) recursively to find $\widetilde{\overline{\mathbf{R}}}_{i,i+1}$ for all the regions. But in order to know the solutions for all the regions, $\mathbf{a}_i$ is needed for all i as well. Then, from Equation (6), one can deduce that in general,

$$\mathbf{a}_{i+1} = \left(\overline{\mathbf{I}} - \overline{\mathbf{R}}_{i+1,i} \cdot \widetilde{\overline{\mathbf{R}}}_{i+1,i+2}\right)^{-1} \cdot \overline{\mathbf{T}}_{i,i+1} \cdot \mathbf{a}_i = \overline{\mathbf{S}}_{i,i+1} \cdot \mathbf{a}_i, \qquad (3.2.10)$$

where

$$\overline{\mathbf{S}}_{i,i+1} = \left(\overline{\mathbf{I}} - \overline{\mathbf{R}}_{i+1,i} \cdot \widetilde{\overline{\mathbf{R}}}_{i+1,i+2}\right)^{-1} \cdot \overline{\mathbf{T}}_{i,i+1}. \qquad (3.2.10a)$$

Since $\mathbf{a}_1$ is known, (10) can be used recursively to find $\mathbf{a}_i$ for all i. In general,

$$\mathbf{a}_N = \overline{\mathbf{T}}_{N-1,N} \cdot \overline{\mathbf{S}}_{N-2,N-1} \cdots \overline{\mathbf{S}}_{12} \cdot \mathbf{a}_1 = \widetilde{\overline{\mathbf{T}}}_{1N} \cdot \mathbf{a}_1, \qquad (3.2.11)$$

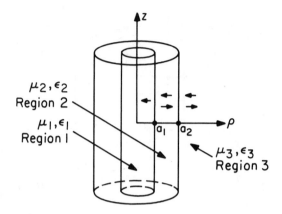

Figure 3.2.3 Reflection and transmission through a three-layer cylindrical medium—the standing-wave case.

where

$$\widetilde{\overline{\mathbf{T}}}_{1N} = \overline{\mathbf{T}}_{N-1,N} \cdot \overline{\mathbf{S}}_{N-2,N-1} \cdots \overline{\mathbf{S}}_{12}. \tag{3.2.11a}$$

It can be regarded as a generalized transmission matrix for a layered slab.

§§3.2.2 The Standing-Wave Case

If a standing wave is in region 3 as shown in Figure 3.2.3, it will set up both standing and outgoing waves in all the regions except for region 1, where only a standing wave exists because the field has to be regular at the origin. The field in region 3 then becomes

$$\begin{bmatrix} E_{3z} \\ H_{3z} \end{bmatrix} = \left[H_n^{(1)}(k_{3\rho}\rho)\,\widetilde{\overline{\mathbf{R}}}_{32} + J_n(k_{3\rho}\rho)\,\overline{\mathbf{I}} \right] \cdot \mathbf{a}_3. \tag{3.2.12}$$

In the above, $\mathbf{a}_3$ is the amplitude of the standing wave, which is known, while $\widetilde{\overline{\mathbf{R}}}_{32}$ is the generalized reflection matrix that relates an outgoing wave to a standing wave. The field in region 2 is just

$$\begin{bmatrix} E_{2z} \\ H_{2z} \end{bmatrix} = \left[H_n^{(1)}(k_{2\rho}\rho)\,\overline{\mathbf{R}}_{21} + J_n(k_{2\rho}\rho)\,\overline{\mathbf{I}} \right] \cdot \mathbf{a}_2. \tag{3.2.13}$$

Finally, the field in region 1 is

$$\begin{bmatrix} E_{1z} \\ H_{1z} \end{bmatrix} = J_n(k_{1\rho}\rho)\,\mathbf{a}_1. \tag{3.2.14}$$

Note that the Hankel function solution has to be excluded here because it is singular at $\rho = 0$.

Since a standing wave in region 2 is a consequence of the transmission of the standing wave in region 3 plus the reflection of the outgoing wave in region 2, we have

$$\mathbf{a}_2 = \overline{\mathbf{T}}_{32} \cdot \mathbf{a}_3 + \overline{\mathbf{R}}_{23} \cdot \overline{\mathbf{R}}_{21} \cdot \mathbf{a}_2. \tag{3.2.15}$$

Furthermore, the outgoing wave in region 3 is just a consequence of the reflection of the standing wave in region 3 plus the transmission of the outgoing wave in region 2. Therefore,

$$\widetilde{\overline{\mathbf{R}}}_{32} \cdot \mathbf{a}_3 = \overline{\mathbf{R}}_{32} \cdot \mathbf{a}_3 + \overline{\mathbf{T}}_{23} \cdot \overline{\mathbf{R}}_{21} \cdot \mathbf{a}_2. \tag{3.2.16}$$

Equation (15) can be solved to yield

$$\mathbf{a}_2 = \left(\overline{\mathbf{I}} - \overline{\mathbf{R}}_{23} \cdot \overline{\mathbf{R}}_{21} \right)^{-1} \cdot \overline{\mathbf{T}}_{32} \cdot \mathbf{a}_3. \tag{3.2.17}$$

Moreover, using the above in (16), we have

$$\widetilde{\overline{\mathbf{R}}}_{32} = \overline{\mathbf{R}}_{32} + \overline{\mathbf{T}}_{23} \cdot \overline{\mathbf{R}}_{21} \cdot \left(\overline{\mathbf{I}} - \overline{\mathbf{R}}_{23} \cdot \overline{\mathbf{R}}_{21} \right)^{-1} \cdot \overline{\mathbf{T}}_{32}. \tag{3.2.18}$$

Now, if another region is added in region 1, we need only change $\overline{\mathbf{R}}_{21}$ to $\widetilde{\overline{\mathbf{R}}}_{21}$. Consequently, for an N-layer medium, the above could be generalized to a recursive relation:

$$\widetilde{\overline{\mathbf{R}}}_{i,i-1} = \overline{\mathbf{R}}_{i,i-1} + \overline{\mathbf{T}}_{i-1,i} \cdot \widetilde{\overline{\mathbf{R}}}_{i-1,i-2} \cdot \left(\overline{\mathbf{I}} - \overline{\mathbf{R}}_{i-1,i} \cdot \widetilde{\overline{\mathbf{R}}}_{i-1,i-2} \right)^{-1} \cdot \overline{\mathbf{T}}_{i,i-1}. \tag{3.2.19}$$

In general, the field in region i can be written as

$$\begin{bmatrix} E_{iz} \\ H_{iz} \end{bmatrix} = \left[H_n^{(1)}(k_{i\rho}\rho)\, \widetilde{\overline{\mathbf{R}}}_{i,i-1} + J_n(k_{i\rho}\rho)\, \overline{\mathbf{I}} \right] \cdot \mathbf{a}_i. \tag{3.2.20}$$

Since $\widetilde{\overline{\mathbf{R}}}_{10} = 0$ in the above recursive relation, one can use (19) recursively to find $\widetilde{\overline{\mathbf{R}}}_{i,i-1}$ for all i, starting from the innermost layer. Furthermore, by generalizing Equation (17), we have a recursive relation,

$$\mathbf{a}_{i-1} = \left(\overline{\mathbf{I}} - \overline{\mathbf{R}}_{i-1,i} \cdot \widetilde{\overline{\mathbf{R}}}_{i-1,i-2} \right)^{-1} \cdot \overline{\mathbf{T}}_{i,i-1} \cdot \mathbf{a}_i = \overline{\mathbf{S}}_{i,i-1} \cdot \mathbf{a}_i, \tag{3.2.21}$$

where

$$\overline{\mathbf{S}}_{i,i-1} = \left(\overline{\mathbf{I}} - \overline{\mathbf{R}}_{i-1,i} \cdot \widetilde{\overline{\mathbf{R}}}_{i-1,i-2} \right)^{-1} \cdot \widetilde{\overline{\mathbf{T}}}_{i,i-1}. \tag{3.2.21a}$$

The above can be used to find $\mathbf{a}_i$ for all i. Consequently, the fields in all the regions are known in this problem. In general,

$$\mathbf{a}_1 = \overline{\mathbf{T}}_{21} \cdot \overline{\mathbf{S}}_{32} \cdots \overline{\mathbf{S}}_{N,N-1} \cdot \mathbf{a}_N = \widetilde{\overline{\mathbf{T}}}_{N1} \cdot \mathbf{a}_N, \tag{3.2.22}$$

where

$$\widetilde{\overline{\mathbf{T}}}_{N1} = \overline{\mathbf{T}}_{21} \cdot \overline{\mathbf{S}}_{32} \cdots \overline{\mathbf{S}}_{N,N-1} \tag{3.2.22a}$$

is a generalized transmission matrix for a cylindrically layered slab similar to that in (11a).

The recursive relations in (8), (10), (18), and (21) do not have the exponential phase factors compared to the recursive relations derived in Section 2.1. The reason being that the phase factors are buried in the definitions of the transmission and reflection matrices. But in a lossy medium or when a wave is evanescent, the Bessel functions in these reflection and transmission matrices may become exponentially large. In this case, it is more appropriate to renormalize the Bessel functions in the definitions of these matrices to prevent numerical overflows. Furthermore, when $n \to -n$, using the property of $\overline{\mathbf{R}}_{12}$, $\overline{\mathbf{R}}_{21}$, $\overline{\mathbf{T}}_{12}$, and $\overline{\mathbf{T}}_{21}$ under this transformation, one can show that only the off-diagonal elements of $\widetilde{\overline{\mathbf{R}}}_{12}$, $\widetilde{\overline{\mathbf{R}}}_{21}$, $\widetilde{\overline{\mathbf{T}}}_{12}$, and $\widetilde{\overline{\mathbf{T}}}_{21}$ would change sign (see Exercise 3.8). This fact can be used to fold the summation over n, which is from $-\infty$ to $+\infty$, to range from 0 to ∞ when summing over all cylindrical harmonics.

Once the reflection matrices are known, the solution of guided modes along a multilayered structure can be easily found (Exercise 3.6). For the same reason, the solution of wave scattering by a multilayer cylinder is also easily sought (Exercise 3.7).

§3.3 Source in a Cylindrically Layered Medium

To derive the field due to a point source in a cylindrically layered medium, we first need to expand the field due to a point source in terms of cylindrical wave functions. This can be done as follows: If a point source is in a homogeneous medium, using the identity given in (2.2.34), then

$$\frac{e^{ik|\mathbf{r}-\mathbf{r}'|}}{|\mathbf{r}-\mathbf{r}'|} = \frac{i}{2} \int_{-\infty}^{\infty} dk_z\, e^{ik_z(z-z')} H_0^{(1)}(k_\rho|\boldsymbol{\rho}-\boldsymbol{\rho}'|), \qquad (3.3.1)$$

where $k_\rho = \sqrt{k^2 - k_z^2}$. Furthermore, since the addition theorem implies that (Exercise 3.9)

$$H_0^{(1)}(k_\rho|\boldsymbol{\rho}-\boldsymbol{\rho}'|) = \sum_{n=-\infty}^{\infty} J_n(k_\rho\rho_<)H_n^{(1)}(k_\rho\rho_>)e^{in(\phi-\phi')}, \qquad (3.3.2)$$

where $\rho_<$ is the smaller of ρ and ρ', and $\rho_>$ is the larger of ρ and ρ', we deduce that

$$\frac{e^{ik|\mathbf{r}-\mathbf{r}'|}}{|\mathbf{r}-\mathbf{r}'|} = \sum_{n=-\infty}^{\infty} \frac{ie^{in(\phi-\phi')}}{2} \int_{-\infty}^{\infty} dk_z\, e^{ik_z(z-z')} J_n(k_\rho\rho_<)H_n^{(1)}(k_\rho\rho_>). \qquad (3.3.3)$$

Next, we shall put this point source in a layered medium to derive the resultant field. We shall present the discrete angular-wave-number representation first, followed by the continuum angular-wave number representation.

§§*3.3.1 Discrete, Angular-Wave-Number Representation*

Given the representation (3) in cylindrical coordinates, the E_z and H_z components of the fields due to any point sources can be derived as in Chapter 2. In general, for a point electric dipole pointing in the $\hat{\alpha}$ direction denoted by a current

$$\mathbf{J}(\mathbf{r}) = I\ell\hat{\alpha}\,\delta(\mathbf{r}-\mathbf{r}'), \tag{3.3.4}$$

the z component of the electric field can be obtained from (1.3.50) of Chapter 1, yielding

$$E_z = \frac{iI\ell}{\omega\epsilon}\left[\hat{z}\cdot\hat{\alpha}k^2 + \frac{\partial}{\partial z'}\nabla'\cdot\hat{\alpha}\right]\frac{e^{ik|\mathbf{r}-\mathbf{r}'|}}{4\pi|\mathbf{r}-\mathbf{r}'|}. \tag{3.3.5a}$$

Since $\mathbf{H} = (\nabla\times\mathbf{E})/(i\omega\mu)$, the z component of H_z, using (1.3.50), is

$$H_z = -\hat{z}\cdot\nabla'\times\hat{\alpha}I\ell\frac{e^{ik|\mathbf{r}-\mathbf{r}'|}}{4\pi|\mathbf{r}-\mathbf{r}'|}. \tag{3.3.5b}$$

Consequently, using (3), the primary field generated by the source (4) is

$$\begin{bmatrix}E_z\\H_z\end{bmatrix} = \frac{iI\ell}{4\pi\omega\epsilon}\mathbf{D}'\sum_{n=-\infty}^{\infty}e^{in(\phi-\phi')}\int_{-\infty}^{\infty}dk_z\,e^{ik_z(z-z')}J_n(k_\rho\rho_<)H_n^{(1)}(k_\rho\rho_>), \tag{3.3.6}$$

where

$$\mathbf{D}' = \frac{i}{2}\begin{bmatrix}(\hat{z}k^2+\frac{\partial}{\partial z'}\nabla')\cdot\hat{\alpha}\\i\omega\epsilon\hat{\alpha}\cdot\hat{z}\times\nabla'\end{bmatrix}. \tag{3.3.6a}$$

A dk_z integration is used here rather than a dk_ρ integration so that the integral is equivalent to a Fourier transform in the z direction. Moreover, because of the translational invariance of the inhomogeneity in the z direction, the fields in all regions have to be phase-matched: they all have $e^{ik_z z}$ dependence.

If the point source is now put in region j as in Figure 3.3.1, the boundaries at $\rho = a_j$ and $\rho = a_{j-1}$ reflect waves. As a result, the field in region j looks like

$$\begin{bmatrix}E_{jz}\\H_{jz}\end{bmatrix} = \frac{iI\ell}{4\pi\omega\epsilon_j}\sum_{n=-\infty}^{\infty}e^{in(\phi-\phi')}\int_{-\infty}^{\infty}dk_z\,e^{ik_z(z-z')}\left\{J_n(k_{j\rho}\rho_<)H_n^{(1)}(k_{j\rho}\rho_>)\mathbf{\bar{I}}\right.$$

$$\left.+H_n^{(1)}(k_{j\rho}\rho)\,\mathbf{\bar{a}}_{jn}(\rho') + J_n(k_{j\rho}\rho)\,\mathbf{\bar{b}}_{jn}(\rho')\right\}\cdot\overleftarrow{\mathbf{D}}'_j, \tag{3.3.7}$$

where

$$\overleftarrow{\mathbf{D}}'_j = \begin{bmatrix}(\hat{z}k_j^2 - ik_z\nabla')\cdot\hat{\alpha}\\i\omega\epsilon_j\,\hat{\alpha}\cdot\hat{z}\times\nabla'\end{bmatrix} \tag{3.3.8}$$

is an operator that acts on functions to its left. As a consequence of the $e^{-in\phi'-ik_z z'}$ dependence, $\nabla' = \hat{\rho}'\frac{\partial}{\partial\rho'} - \hat{\phi}'\frac{in}{\rho'} - \hat{z}ik_z$. Note that additional standing and outgoing waves are generated in region j as a result of reflections of the primary field.

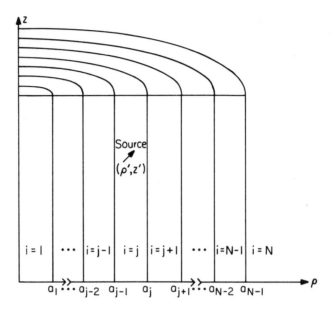

Figure 3.3.1 A source in a cylindrically layered medium.

The quantity to the right of "{" in (7) behaves like the $[E_z, H_z]^t$ with $e^{in\phi + ik_z z}$ dependence. Hence, the result of the previous section can be used to find the unknowns $\overline{\mathbf{a}}_{jn}$ and $\overline{\mathbf{b}}_{jn}$. For instance, to find $\overline{\mathbf{a}}_{jn}$ and $\overline{\mathbf{b}}_{jn}$, constraint conditions are imposed at $\rho = a_{j-1}$ and $\rho = a_j$. Consequently, at $\rho = a_j$, the standing wave is related to the outgoing wave via $\widetilde{\overline{\mathbf{R}}}_{j,j+1}$, i.e.,

$$\overline{\mathbf{b}}_{jn} \cdot \overleftarrow{\mathbf{D}}'_j = \widetilde{\overline{\mathbf{R}}}_{j,j+1} \cdot \left[J_n(k_{j\rho}\rho')\,\overline{\mathbf{I}} + \overline{\mathbf{a}}_{jn} \right] \cdot \overleftarrow{\mathbf{D}}'_j. \tag{3.3.9}$$

Similarly, at $\rho = a_{j-1}$, the outgoing wave is related to the standing wave via $\widetilde{\overline{\mathbf{R}}}_{j,j-1}$, i.e.,

$$\overline{\mathbf{a}}_{jn} \cdot \overleftarrow{\mathbf{D}}'_j = \widetilde{\overline{\mathbf{R}}}_{j,j-1} \cdot \left[H_n^{(1)}(k_{j\rho}\rho')\,\overline{\mathbf{I}} + \overline{\mathbf{b}}_{jn} \right] \cdot \overleftarrow{\mathbf{D}}'_j. \tag{3.3.10}$$

Here, $\widetilde{\overline{\mathbf{R}}}_{j,j+1}$ and $\widetilde{\overline{\mathbf{R}}}_{j,j-1}$ were derived in the previous sections. Then, solving (9) and (10) yields

$$\overline{\mathbf{a}}_{jn} = \left[\overline{\mathbf{I}} - \widetilde{\overline{\mathbf{R}}}_{j,j-1} \cdot \widetilde{\overline{\mathbf{R}}}_{j,j+1} \right]^{-1} \cdot \widetilde{\overline{\mathbf{R}}}_{j,j-1} \cdot \left[H_n^{(1)}(k_{j\rho}\rho')\,\overline{\mathbf{I}} + J_n(k_{j\rho}\rho')\widetilde{\overline{\mathbf{R}}}_{j,j+1} \right],$$
$$\tag{3.3.11}$$

$$\overline{\mathbf{b}}_{jn} = \left[\overline{\mathbf{I}} - \widetilde{\overline{\mathbf{R}}}_{j,j+1} \cdot \widetilde{\overline{\mathbf{R}}}_{j,j-1} \right]^{-1} \cdot \widetilde{\overline{\mathbf{R}}}_{j,j+1} \cdot \left[J_n(k_{j\rho}\rho')\,\overline{\mathbf{I}} + H_n^{(1)}(k_{j\rho}\rho')\widetilde{\overline{\mathbf{R}}}_{j,j-1} \right].$$
$$\tag{3.3.12}$$

Note that in the above, the reflection matrices for the appropriate n have to be chosen since they are dependent on n. Equations (11) and (12) can be substituted back into (7). After some simplifications, it can be shown that (Exercise 3.11)

$$
\left[J_n(k_{j\rho}\rho_<) H_n^{(1)}(k_{j\rho}\rho_>) \,\overline{\mathbf{I}} + H_n^{(1)}(k_{j\rho}\rho)\,\overline{\mathbf{a}}_{jn} + J_n(k_{j\rho}\rho)\,\overline{\mathbf{b}}_{jn} \right] \cdot \overleftarrow{\mathbf{D}}'_j
$$

$$
= \begin{cases}
\left[H_n^{(1)}(k_{j\rho}\rho)\,\overline{\mathbf{I}} + J_n(k_{j\rho}\rho)\,\widetilde{\overline{\mathbf{R}}}_{j,j+1} \right] \cdot \widetilde{\overline{\mathbf{M}}}_{j+} \\
\qquad \cdot \left[J_n(k_{j\rho}\rho')\,\overline{\mathbf{I}} + H_n^{(1)}(k_{j\rho}\rho')\,\widetilde{\overline{\mathbf{R}}}_{j,j-1} \right] \cdot \overleftarrow{\mathbf{D}}'_j, \quad \rho > \rho', \\[2mm]
\left[J_n(k_{j\rho}\rho)\,\overline{\mathbf{I}} + H_n^{(1)}(k_{j\rho}\rho)\,\widetilde{\overline{\mathbf{R}}}_{j,j-1} \right] \cdot \widetilde{\overline{\mathbf{M}}}_{j-} \\
\qquad \cdot \left[H_n^{(1)}(k_{j\rho}\rho')\,\overline{\mathbf{I}} + J_n(k_{j\rho}\rho')\,\widetilde{\overline{\mathbf{R}}}_{j,j+1} \right] \cdot \overleftarrow{\mathbf{D}}'_j, \quad \rho < \rho'.
\end{cases}
\tag{3.3.13}
$$

In the above,

$$
\widetilde{\overline{\mathbf{M}}}_{j\mp} = \left(\overline{\mathbf{I}} - \widetilde{\overline{\mathbf{R}}}_{j,j\pm1} \cdot \widetilde{\overline{\mathbf{R}}}_{j,j\mp1} \right)^{-1}
\tag{3.3.14}
$$

is a factor accounting for multiple reflections in region j. Equation (13) has the advantage of being symmetrical.

For $i > j$, the field in region i can be written as

$$
\begin{bmatrix} E_{iz} \\ H_{iz} \end{bmatrix} = \frac{iI\ell}{4\pi\omega\epsilon_j} \sum_{n=-\infty}^{\infty} e^{in(\phi-\phi')} \int_{-\infty}^{\infty} dk_z\, e^{ik_z(z-z')} \left[H_n^{(1)}(k_{i\rho}\rho)\,\overline{\mathbf{I}} \right.
$$
$$
\left. + J_n(k_{i\rho}\rho)\,\widetilde{\overline{\mathbf{R}}}_{i,i+1} \right] \cdot \mathbf{a}_{in}.
\tag{3.3.15}
$$

Moreover, $\widetilde{\overline{\mathbf{R}}}_{i,i+1}$ can be found by using the recursion relation in Equation (3.2.8). Here, $\mathbf{a}_{in}$ is the amplitude of the outgoing wave in region i. Therefore, the relation in (3.2.11) can be used to find it. In other words (Exercise 3.12),

$$
\mathbf{a}_{in} = \widetilde{\overline{\mathbf{M}}}_{i+} \cdot \widetilde{\overline{\mathbf{T}}}_{ji} \cdot \mathbf{a}_{jn},
\tag{3.3.16}
$$

where $\mathbf{a}_{jn}$ is the amplitude of the outgoing wave in region j. Here, $\widetilde{\overline{\mathbf{T}}}_{ji}$ is the generalized transmission matrix[2] through a cylindrically layered slab between region i and region j. From (13), we identify

$$
\mathbf{a}_{jn} = \widetilde{\overline{\mathbf{M}}}_{j+} \cdot \left[J_n(k_{j\rho}\rho')\overline{\mathbf{I}} + H_n^{(1)}(k_{j\rho}\rho')\,\widetilde{\overline{\mathbf{R}}}_{j,j-1} \right] \cdot \overleftarrow{\mathbf{D}}'_j.
\tag{3.3.17}
$$

Note that the derivation of (16) and (17) is similar to the derivation of (2.4.14) of Chapter 2. Furthermore, Equations (16) and (17), when used in (15), give (Exercise 3.12)

$$
\begin{bmatrix} E_{iz} \\ H_{iz} \end{bmatrix} = \frac{iI\ell}{4\pi\omega\epsilon_j} \sum_{n=-\infty}^{\infty} e^{in(\phi-\phi')} \int_{-\infty}^{\infty} dk_z\, e^{ik_z(z-z')} \left[H_n^{(1)}(k_{i\rho}\rho)\,\overline{\mathbf{I}} + J_n(k_{i\rho}\rho)\widetilde{\overline{\mathbf{R}}}_{i,i+1} \right]
$$
$$
\cdot \widetilde{\overline{\mathbf{M}}}_{i+} \cdot \widetilde{\overline{\mathbf{T}}}_{ji} \cdot \widetilde{\overline{\mathbf{M}}}_{j+} \cdot \left[J_n(k_{j\rho}\rho')\,\overline{\mathbf{I}} + H_n^{(1)}(k_{j\rho}\rho')\widetilde{\overline{\mathbf{R}}}_{j,j-1} \right] \cdot \overleftarrow{\mathbf{D}}'_j.
\tag{3.3.18}
$$

---

[2] Since $\overline{\mathbf{T}}_{ji}$ in (3.2.11) is defined without multiple reflections in the i-th region, the $\widetilde{\overline{\mathbf{M}}}_i$ factor is needed in (16) to include these multiple reflections.

Notice that Equation (18) has the advantage of being symmetrical, showing that the solution satisfies reciprocity.

For $i < j$, the field in region i is of the form

$$\begin{bmatrix} E_{iz} \\ H_{iz} \end{bmatrix} = \frac{iI\ell}{4\pi\omega\epsilon_j} \sum_{n=-\infty}^{\infty} e^{in(\phi-\phi')} \int_{-\infty}^{\infty} dk_z\, e^{ik_z(z-z')} \left[J_n(k_{i\rho}\rho)\, \overline{\mathbf{I}} \right.$$

$$\left. + H_n^{(1)}(k_{i\rho}\rho)\, \widetilde{\overline{\mathbf{R}}}_{i,i-1} \right] \cdot \mathbf{a}_{in}. \qquad (3.3.19)$$

A similar analysis yields a result similar to (18). In this case, using $\widetilde{\overline{\mathbf{T}}}_{ji}$ for a cylindrically layered slab given in (3.2.22a), we can show that (Exercise 3.12)

$$\begin{bmatrix} E_{iz} \\ H_{iz} \end{bmatrix} = \frac{iI\ell}{4\pi\omega\epsilon_j} \sum_{n=-\infty}^{\infty} e^{in(\phi-\phi')} \int_{-\infty}^{\infty} dk_z\, e^{ik_z(z-z')} \left[J_n(k_{i\rho}\rho)\, \overline{\mathbf{I}} + H_n^{(1)}(k_{i\rho}\rho)\, \widetilde{\overline{\mathbf{R}}}_{i,i-1} \right]$$

$$\cdot\, \widetilde{\overline{\mathbf{M}}}_{i-} \cdot \widetilde{\overline{\mathbf{T}}}_{ji} \cdot \widetilde{\overline{\mathbf{M}}}_{j-} \cdot \left[H_n^{(1)}(k_{j\rho}\rho')\, \overline{\mathbf{I}} + J_n(k_{j\rho}\rho')\, \widetilde{\overline{\mathbf{R}}}_{j,j+1} \right] \cdot \overleftarrow{\mathbf{D}}'_j. \qquad (3.3.20)$$

In summary,

$$\begin{bmatrix} E_z \\ H_z \end{bmatrix} = \frac{iI\ell}{4\pi\omega\epsilon_j} \sum_{n=-\infty}^{\infty} e^{in(\phi-\phi')} \int_{-\infty}^{\infty} dk_z\, e^{ik_z(z-z')} \overline{\mathbf{F}}_n(\rho,\rho') \cdot \overleftarrow{\mathbf{D}}'_j, \qquad (3.3.21)$$

where

$$\overline{\mathbf{F}}_n(\rho,\rho') = \begin{cases} \left[H_n^{(1)}(k_{j\rho}\rho)\, \overline{\mathbf{I}} + J_n(k_{j\rho}\rho)\, \widetilde{\overline{\mathbf{R}}}_{j,j+1} \right] \cdot \widetilde{\overline{\mathbf{M}}}_{j+} \\ \qquad\qquad \cdot \left[J_n(k_{j\rho}\rho')\, \overline{\mathbf{I}} + H_n^{(1)}(k_{j\rho}\rho')\, \widetilde{\overline{\mathbf{R}}}_{j,j-1} \right], \quad \rho > \rho', \\ \left[J_n(k_{j\rho}\rho)\, \overline{\mathbf{I}} + H_n^{(1)}(k_{j\rho}\rho)\, \widetilde{\overline{\mathbf{R}}}_{j,j-1} \right] \cdot \widetilde{\overline{\mathbf{M}}}_{j-} \\ \qquad\qquad \cdot \left[H_n^{(1)}(k_{j\rho}\rho')\, \overline{\mathbf{I}} + J_n(k_{j\rho}\rho')\, \widetilde{\overline{\mathbf{R}}}_{j,j+1} \right], \quad \rho < \rho', \end{cases}$$

$$(3.3.22)$$

when both ρ and ρ' are of the same region j. But when $i > j$, $\rho \in$ region i, and $\rho' \in$ region j,

$$\overline{\mathbf{F}}_n(\rho,\rho') = \left[H_n^{(1)}(k_{i\rho}\rho)\, \overline{\mathbf{I}} + J_n(k_{i\rho}\rho)\, \widetilde{\overline{\mathbf{R}}}_{i,i+1} \right] \cdot \widetilde{\overline{\mathbf{M}}}_{i+} \cdot \widetilde{\overline{\mathbf{T}}}_{ji} \cdot \widetilde{\overline{\mathbf{M}}}_{j+}$$

$$\cdot \left[J_n(k_{j\rho}\rho')\, \overline{\mathbf{I}} + H_n(k_{j\rho}\rho')\, \widetilde{\overline{\mathbf{R}}}_{j,j-1} \right]. \qquad (3.3.23)$$

Furthermore, when $i < j$, $\rho \in$ region i, and $\rho' \in$ region j,

$$\overline{\mathbf{F}}_n(\rho,\rho') = \left[J_n(k_{i\rho}\rho)\, \overline{\mathbf{I}} + H_n^{(1)}(k_{i\rho}\rho)\, \widetilde{\overline{\mathbf{R}}}_{i,i-1} \right] \cdot \widetilde{\overline{\mathbf{M}}}_{i-} \cdot \widetilde{\overline{\mathbf{T}}}_{ji} \cdot \widetilde{\overline{\mathbf{M}}}_{j-}$$

$$\cdot \left[H_n^{(1)}(k_{j\rho}\rho')\, \overline{\mathbf{I}} + J_n(k_{j\rho}\rho')\, \widetilde{\overline{\mathbf{R}}}_{j,j+1} \right]. \qquad (3.3.24)$$

In these integral representations, since $k_{i\rho} = \sqrt{k_i^2 - k_z^2}$, it seems that there are branch points at k_i's. But using the uniqueness principle as in Chapter 2, one can show that the branch points exist only at $k_z = \pm k_N$, where N is the outermost layer (see Exercise 3.13; also see Chew 1983). The integral in (21) can be numerically evaluated in the same manner as discussed in Chapter 2, Subsection 2.7.3. Moreover, when a large parameter exists, asymptotic expansion of the integral is derivable.

§§*3.3.2 Continuum, Angular-Wave-Number Representation*

In the previous subsection, notice that when an eccentered point source is in a cylindrically layered medium, it generates $e^{in\phi}$ harmonics of all orders. Hence, the solution involves the discrete summation over all integer n of the $e^{in\phi}$ harmonics. Sometimes, however, it is advantageous to express the ϕ harmonics of the field not as a discrete sum, but as an integral over the angular wave number. An integral allows the use of contour integration technique, whereas a discrete summation does not directly lend itself to such a technique.

Equation (3) consists of an expansion in terms of discrete n's. But it can be written as an integral summation of continuum angular wave numbers by the use of Dirac delta functions; hence, we have

$$\frac{e^{ik|\mathbf{r}-\mathbf{r}'|}}{|\mathbf{r}-\mathbf{r}'|} = \frac{i}{2} \int_{-\infty}^{\infty} dk_z\, e^{ik_z z} \int_{-\infty}^{\infty} d\nu\, e^{i\nu(\phi-\phi')} J_\nu(k_\rho \rho_<) H_\nu^{(1)}(k_\rho \rho_>) \sum_{n=-\infty}^{\infty} \delta(\nu - n).$$

(3.3.25)

Moreover, by using the Fourier series expansion (see Appendix C), one can show that

$$\sum_{n=-\infty}^{\infty} \delta(\nu - n) = \sum_{n=-\infty}^{\infty} e^{2i\pi n\nu}.$$

(3.3.26)

Note that both sides of the equation are periodic functions. Consequently, using (26) in (25), and after exchanging the order of integration and summation, we have

$$\frac{e^{ik|\mathbf{r}-\mathbf{r}'|}}{|\mathbf{r}-\mathbf{r}'|} = \frac{i}{2} \int_{-\infty}^{\infty} dk_z\, e^{ik_z z} \sum_{n=-\infty}^{\infty} \int_{-\infty}^{\infty} d\nu\, e^{i\nu(\phi-\phi')+2in\pi\nu} H_\nu^{(1)}(k_\rho \rho_>) J_\nu(k_\rho \rho_<).$$

(3.3.27)

By the same token, we conclude that

$$H_0^{(1)}(k_\rho|\boldsymbol{\rho}-\boldsymbol{\rho}'|) = \sum_{n=-\infty}^{\infty} \int_{-\infty}^{\infty} d\nu\, e^{i\nu(\phi-\phi')+2in\nu\pi} H_\nu^{(1)}(k_\rho \rho_>) J_\nu(k_\rho \rho_<).$$

(3.3.28)

Notice that by changing ν from discrete numbers to continuum numbers, $e^{i\nu\phi}$ is no more 2π periodic. Moreover, the above is now physically equivalent to putting the source at $\phi' - 2n\pi$, $-\infty < n < \infty$, in a ϕ space ranging from $-\infty$ to $+\infty$. In other words, the summation over the integer n in the above now is to include contributions from the image terms.

These continuum angular-wave-number representations of a source can be similarly used to derive the equivalence of (7). Hence, we arrive at

$$
\begin{bmatrix} E_{jz} \\ H_{jz} \end{bmatrix} = \frac{iI\ell}{4\pi\omega\epsilon_j} \int\limits_{-\infty}^{\infty} dk_z\, e^{ik_z(z-z')} \sum_{n=-\infty}^{\infty} \int\limits_{-\infty}^{\infty} d\nu\, e^{i\nu(\phi-\phi')+i2n\nu\pi}
$$

$$
\cdot \left\{ J_\nu(k_{j\rho}\rho_<)H_\nu^{(1)}(k_{j\rho}\rho_>) + H_\nu^{(1)}(k_{j\rho}\rho)\,\overline{\mathbf{a}}_{j\nu}(\rho') + J_\nu(k_{j\rho}\rho)\,\overline{\mathbf{b}}_{j\nu}(\rho') \right\} \cdot \overleftarrow{\mathbf{D}}'_j. \tag{3.3.29}
$$

Note that the operator $\overleftarrow{\mathbf{D}}'_j$ is dependent on the kind of source considered. Furthermore, $\overline{\mathbf{a}}_{j\nu}$ and $\overline{\mathbf{b}}_{j\nu}$ are determined by Equations (11) and (12) (Exercise 3.14). They include the important physics of the layered medium, since they are due to reflected waves from the layered media. Such an integral representation has also been given by Felsen and Marcuvitz (1973), but with a different derivation.

The reflected wave terms in (29) comprise

$$
\begin{bmatrix} E_{jz}^R \\ H_{jz}^R \end{bmatrix} = \int\limits_{-\infty}^{\infty} dk_z\, e^{ik_z(z-z')} \sum_{n=-\infty}^{\infty} \int\limits_{-\infty}^{\infty} d\nu\, e^{i\nu(\phi-\phi')+i2n\nu\pi}
$$

$$
\cdot \left[H_\nu^{(1)}(k_{j\rho}\rho)\,\overline{\mathbf{a}}_{j\nu} + J_\nu(k_{j\rho}\rho)\,\overline{\mathbf{b}}_{j\nu} \right] \cdot \overleftarrow{\mathbf{D}}_j. \tag{3.3.30}
$$

Here, $\overline{\mathbf{a}}_{j\nu}$ and $\overline{\mathbf{b}}_{j\nu}$ often have pole singularities on the complex ν plane. Consequently, the path of integration on the real axis of the complex ν plane can often be deformed to enclose such singularities. For $\phi - \phi' + 2n\pi > 0$, we can deform the path to the upper half complex ν plane, while for $\phi - \phi' + 2n\pi < 0$, we can deform it to the lower half complex ν plane. Moreover, it can be shown easily that the bracketed term in (30) is an even function of ν, and hence, the pole singularities are symmetrically located. Consequently, through residue calculus, the reflected wave term is of the form (Exercise 3.14)

$$
\begin{bmatrix} E_{jz}^R \\ H_{jz}^R \end{bmatrix} = \int\limits_{-\infty}^{\infty} dk_z\, e^{ik_z(z-z')} \sum_{n=-\infty}^{\infty} \sum_{p=1}^{\infty} e^{i\nu_p|\phi-\phi'+2n\pi|}\,\boldsymbol{\beta}_p
$$

$$
\approx \int\limits_{-\infty}^{\infty} dk_z\, e^{ik_z(z-z')} \left[e^{i\nu_1(\phi-\phi')} + e^{i\nu_1(2\pi-\phi+\phi')} \right] \boldsymbol{\beta}_1, \quad \phi > \phi', \tag{3.3.30a}
$$

where $\boldsymbol{\beta}_p$ is the residue of the p-th pole. The last approximation follows if we keep only the $n = 0$ and $n = -1$ terms and $p = 1$ term,[3] assuming that ν_p has a positive imaginary part; hence $e^{2in\nu_p\pi}$ for $|n| > 1$ is often small. The

---

[3] Note that for $\phi < \phi'$, $n = 0$ and $n = +1$ terms are the dominant terms. In other words, we keep only the image term at $\phi' \pm 2\pi$ depending on the relative location of ϕ and ϕ'.

higher p terms may also have larger imaginary parts, and hence, are more attenuative (also see Exercise 3.14).

For a fixed k_z, Equation (30a) describes a wave propagating azimuthally. Moreover, $e^{i\nu_p|\phi-\phi'+2n\pi|}$ corresponds to a wave emanating from the source and its images at $\phi' - 2n\pi$ for $-\infty < n < \infty$. Since ν_p is assumed attenuative, we have kept only terms in (30a) where the distance travelled by the wave is less than 2π. One of these reaches the observation point in the clockwise direction, while the other one reaches the observation point in the counter-clockwise direction. Such a physical picture is important when the radius of curvature of the surface is large or the frequency is high.

These azimuthally propagating waves are known as **creeping waves** when they are travelling on a convex surface. But they are known as **whispering gallery waves** when they are travelling on a concave surface. On a convex surface, the wave radiates as it propagates along. As a result, creeping waves are usually attenuative. On a concave surface, however, the wave can propagate long distances with little attenuations since it cannot radiate. Such a phenomenon is often observed on concave surfaces of galleries, and hence, its name. As a final note, the above method of analysis is related to Watson's transformation; Watson studied the propagation of waves on the earth's surface (Watson 1918). Moreover, Elliott (1955), Helstrom (1963), Einziger and Felsen (1983), and Pearson (1986, 1987) have also studied azimuthal propagation of waves.

§3.4 Propagator Matrix—Cylindrical Layers

The cylindrically layered-medium solution can also be arrived at via the propagator matrix approach. In this approach, we first need to derive a state equation. Because the problem is inherently vector, the state vector in the state equation requires at least four components. In order to facilitate matching the boundary conditions easily, E_z, E_ϕ, H_z, and H_ϕ components are used in a state vector. Gronthoud and Blok (1978) and Pearson (1986) have also studied the propagation of waves in cylindrical layers using similar matrix methods.

§§3.4.1 Isotropic, Layered Media

The state equation for a cylindrically layered medium can be derived from Maxwell's equations, which are

$$\nabla \times \mathbf{E} = i\omega\mu\,\mathbf{H}, \tag{3.4.1}$$

$$\nabla \times \mathbf{H} = -i\omega\epsilon\,\mathbf{E}. \tag{3.4.2}$$

First, we decompose $\nabla = \nabla_s + \nabla_\rho$, $\mathbf{E} = \mathbf{E}_s + \mathbf{E}_\rho$, $\mathbf{H} = \mathbf{H}_s + \mathbf{H}_\rho$ where the subscript s denotes a vector transverse to ρ. Then, equating the transverse and longitudinal components of (1), we obtain

$$\nabla_\rho \times \mathbf{E}_s = i\omega\mu\,\mathbf{H}_s - \nabla_s \times \mathbf{E}_\rho, \tag{3.4.3a}$$

$$\nabla_s \times \mathbf{E}_s = i\omega\mu\,\mathbf{H}_\rho. \tag{3.4.3b}$$

Similarly, from (2), we have

$$\nabla_\rho \times \mathbf{H}_s = -i\omega\epsilon\,\mathbf{E}_s - \nabla_s \times \mathbf{H}_\rho, \tag{3.4.4a}$$

$$\nabla_s \times \mathbf{H}_s = -i\omega\epsilon\,\mathbf{E}_\rho. \tag{3.4.4b}$$

Then, using (4b) to eliminate $\mathbf{E}_\rho$ in (3a), one obtains

$$\nabla_\rho \times \mathbf{E}_s = i\omega\mu\,\mathbf{H}_s + \nabla_s \times \frac{1}{i\omega\epsilon}\nabla_s \times \mathbf{H}_s. \tag{3.4.5a}$$

Moreover, by duality, we also have

$$\nabla_\rho \times \mathbf{H}_s = -i\omega\epsilon\,\mathbf{E}_s - \nabla_s \times \frac{1}{i\omega\mu}\nabla_s \times \mathbf{E}_s. \tag{3.4.5b}$$

From (5a), it follows that

$$\frac{\partial}{\partial\rho}E_z = -i\omega\mu H_\phi - \frac{\partial}{\partial z}\frac{1}{i\omega\epsilon}\left(\frac{1}{\rho}\frac{\partial}{\partial\phi}H_z - \frac{\partial}{\partial z}H_\phi\right), \tag{3.4.6a}$$

$$\frac{1}{\rho}\frac{\partial}{\partial\rho}\rho E_\phi = i\omega\mu H_z - \frac{1}{\rho}\frac{\partial}{\partial\phi}\frac{1}{i\omega\epsilon}\left(\frac{1}{\rho}\frac{\partial}{\partial\phi}H_z - \frac{\partial}{\partial z}H_\phi\right). \tag{3.4.6b}$$

Now, with the assumption that the field has $e^{in\phi}$ and $e^{ik_z z}$ dependences, and that ϵ and μ are functions of ρ only, then

$$\frac{d}{d\rho}E_z = -i\omega\mu H_\phi - \frac{ik_z}{\omega\epsilon}\left(\frac{n}{\rho}H_z - k_z H_\phi\right), \tag{3.4.7a}$$

$$\frac{1}{\rho}\frac{d}{d\rho}\rho E_\phi = i\omega\mu H_z - \frac{in}{\omega\epsilon\rho}\left(\frac{n}{\rho}H_z - k_z H_\phi\right). \tag{3.4.7b}$$

Similarly, from (5b),

$$\frac{d}{d\rho}H_z = i\omega\epsilon E_\phi + \frac{ik_z}{\omega\mu}\left(\frac{n}{\rho}E_z - k_z E_\phi\right), \tag{3.4.8a}$$

$$\frac{1}{\rho}\frac{d}{d\rho}\rho H_\phi = -i\omega\epsilon E_z + \frac{in}{\omega\mu\rho}\left(\frac{n}{\rho}E_z - k_z E_\phi\right). \tag{3.4.8b}$$

Next, by noting that $\frac{1}{\rho}\frac{d}{d\rho}\rho = \frac{1}{\rho} + \frac{d}{d\rho}$, the above could be written as

$$\frac{d}{d\rho}\begin{bmatrix} E_z \\ E_\phi \\ H_z \\ H_\phi \end{bmatrix} = \overline{\mathbf{H}}(\rho)\cdot\begin{bmatrix} E_z \\ E_\phi \\ H_z \\ H_\phi \end{bmatrix}, \quad \text{or} \quad \frac{d\mathbf{V}}{d\rho} = \overline{\mathbf{H}}(\rho)\cdot\mathbf{V}, \tag{3.4.9}$$

where $\overline{\mathbf{H}}(\rho)$ is a 4×4 matrix derivable from (7) and (8) (Exercise 3.15; also see Gronthoud and Blok 1978). $\overline{\mathbf{H}}$ is a function of ρ, however, and hence,

(9) does not admit exponential functions as solutions. But when ϵ and μ are constants, it is known from Section 3.1 that Maxwell's equations in cylindrical coordinates have closed-form solutions. Hence, this knowledge can be used to derive closed-form solutions to (9) when μ and ϵ are constants. They are the TM to z and TE to z standing and outgoing waves. Hence, there are four independent eigensolutions to (9) in this case. In fact, these four eigensolutions are

$$
\mathbf{a}_1 = \begin{bmatrix} J_n(k_\rho\rho) \\ -\frac{nk_z}{k_\rho^2\rho} J_n(k_\rho\rho) \\ 0 \\ \frac{i\omega\epsilon}{k_\rho} J_n'(k_\rho\rho) \end{bmatrix}, \qquad
\mathbf{a}_2 = \begin{bmatrix} 0 \\ -\frac{i\omega\mu}{k_\rho} J_n'(k_\rho\rho) \\ J_n(k_\rho\rho) \\ -\frac{nk_z}{k_\rho^2\rho} J_n(k_\rho\rho) \end{bmatrix},
$$

$$
\mathbf{a}_3 = \begin{bmatrix} H_n^{(1)}(k_\rho\rho) \\ -\frac{nk_z}{k_\rho^2\rho} H_n^{(1)}(k_\rho\rho) \\ 0 \\ \frac{i\omega\epsilon}{k_\rho} H_n^{(1)'}(k_\rho\rho) \end{bmatrix}, \qquad
\mathbf{a}_4 = \begin{bmatrix} 0 \\ -\frac{i\omega\mu}{k_\rho} H_n^{(1)'}(k_\rho\rho) \\ H_n^{(1)}(k_\rho\rho) \\ -\frac{nk_z}{k_\rho^2\rho} H_n^{(1)}(k_\rho\rho) \end{bmatrix}. \qquad (3.4.10)
$$

Note that $E_z \neq 0$, $H_z = 0$, for $\mathbf{a}_1$ and $\mathbf{a}_3$; hence, they correspond to waves TM to z. Similarly, $\mathbf{a}_2$ and $\mathbf{a}_4$ correspond to waves TE to z. Moreover, $\mathbf{a}_1$ and $\mathbf{a}_2$ are standing waves, while $\mathbf{a}_3$ and $\mathbf{a}_4$ are outgoing waves.

Therefore, the general solution to (9) is of the form

$$
\mathbf{V}(\rho) = \sum_{i=1}^{4} A_i \mathbf{a}_i(\rho) = \overline{\mathbf{a}}(\rho) \cdot \mathbf{A}, \qquad (3.4.11)
$$

where $\overline{\mathbf{a}}(\rho)$ is a 4×4 matrix containing the eigenvectors $\mathbf{a}_i$'s, i.e., $\overline{\mathbf{a}} = [\mathbf{a}_1, \mathbf{a}_2, \mathbf{a}_3, \mathbf{a}_4]$, and $\mathbf{A}$ is a column vector containing A_i's. Alternatively, Equation (11) can be written as

$$
\begin{aligned}
\mathbf{V}(\rho) &= \overline{\mathbf{a}}(\rho) \cdot \overline{\mathbf{a}}^{-1}(\rho') \cdot \overline{\mathbf{a}}(\rho') \cdot \mathbf{A} \\
&= \overline{\mathbf{P}}(\rho, \rho') \cdot \mathbf{V}(\rho'),
\end{aligned} \qquad (3.4.12)
$$

where the propagator matrix is[4]

$$
\overline{\mathbf{P}}(\rho, \rho') = \overline{\mathbf{a}}(\rho) \cdot \overline{\mathbf{a}}^{-1}(\rho'). \qquad (3.4.12\text{a})
$$

Now, equipped with the propagator, we can solve for the transmission and reflection of cylindrical waves through a layered medium as in Section 2.7 in Chapter 2 (Exercise 3.17). Note that Equation (9) can also be solved numerically to obtain the propagator matrix in an inhomogeneous layer as in (2.9.22) in Chapter 2 (Exercise 3.18).

---

[4] The matrix $\overline{\mathbf{a}}^{-1}(\rho)$ has a closed form (see Exercise 3.16; also see Pearson 1986).

§§3.4.2 Anisotropic, Layered Media

If a cylindrically layered medium is anisotropic, the tensors $\overline{\mu}$ and $\overline{\epsilon}$ can be decomposed as

$$\overline{\mu} = \begin{bmatrix} \mu_{\rho\rho} & \overline{\mu}_{\rho s} \\ \overline{\mu}_{s\rho} & \overline{\mu}_s \end{bmatrix}, \qquad \overline{\epsilon} = \begin{bmatrix} \epsilon_{\rho\rho} & \overline{\epsilon}_{\rho s} \\ \overline{\epsilon}_{s\rho} & \overline{\epsilon}_s \end{bmatrix}, \qquad (3.4.13)$$

where $\overline{\mu}_s$ and $\overline{\epsilon}_s$ are 2×2 tensors, $\overline{\mu}_{s\rho}$ and $\overline{\epsilon}_{s\rho}$ are 2×1 matrices, $\overline{\mu}_{\rho s}$ and $\overline{\epsilon}_{\rho s}$ are 1×2 matrices, and $\mu_{\rho\rho}$ and $\epsilon_{\rho\rho}$ are scalars. Here, the subscript s denotes components transverse to ρ. In addition, we can also decompose the field as $\mathbf{E} = \mathbf{E}_\rho + \mathbf{E}_s$, $\mathbf{H} = \mathbf{H}_\rho + \mathbf{H}_s$, and the operator $\nabla = \nabla_\rho + \nabla_s$. Consequently, from Maxwell's equations, we can derive equations similar to (2.10.4) and (2.10.5). Hence,

$$\nabla_\rho \times \mathbf{E}_s = i\omega\, \overline{\mu}_s \cdot \mathbf{H}_s + i\omega\, \overline{\mu}_{s\rho} \cdot \mathbf{H}_\rho - \nabla_s \times \mathbf{E}_\rho, \qquad (3.4.14\text{a})$$

$$\nabla_s \times \mathbf{E}_s = i\omega\, \overline{\mu}_{\rho s} \cdot \mathbf{H}_s + i\omega\mu_{\rho\rho}\mathbf{H}_\rho \qquad (3.4.14\text{b})$$

and

$$\nabla_\rho \times \mathbf{H}_s = -i\omega\, \overline{\epsilon}_s \cdot \mathbf{E}_s - i\omega\, \overline{\epsilon}_{s\rho} \cdot \mathbf{E}_\rho - \nabla_s \times \mathbf{H}_\rho, \qquad (3.4.15\text{a})$$

$$\nabla_s \times \mathbf{H}_s = -i\omega\, \overline{\epsilon}_{\rho s} \cdot \mathbf{E}_s - i\omega\epsilon_{\rho\rho}\mathbf{E}_\rho. \qquad (3.4.15\text{b})$$

Moreover, $\mathbf{E}_\rho$ and $\mathbf{H}_\rho$ can be eliminated in (14a) and (15a) using (14b) and (15b) to obtain

$$\nabla_\rho \times \mathbf{E}_s = i\omega\, \overline{\mu}_s \cdot \mathbf{H}_s + \overline{\mu}_{s\rho} \cdot \nu_{\rho\rho}\nabla_s \times \mathbf{E}_s - i\omega\, \overline{\mu}_{s\rho} \cdot \overline{\mu}_{\rho s} \cdot \nu_{\rho\rho}\mathbf{H}_s$$
$$+ \frac{1}{i\omega}\nabla_s \times \kappa_{\rho\rho}\nabla_s \times \mathbf{H}_s + \nabla_s \times \kappa_{\rho\rho}\overline{\epsilon}_{\rho s} \cdot \mathbf{E}_s, \qquad (3.4.16\text{a})$$

$$\nabla_\rho \times \mathbf{H}_s = -i\omega\, \overline{\epsilon}_s \cdot \mathbf{E}_s + \overline{\epsilon}_{s\rho} \cdot \kappa_{\rho\rho}\nabla_s \times \mathbf{H}_s + i\omega\, \overline{\epsilon}_{s\rho} \cdot \overline{\epsilon}_{\rho s} \cdot \kappa_{\rho\rho}\mathbf{E}_s$$
$$- \frac{1}{i\omega}\nabla_s \times \nu_{\rho\rho}\nabla_s \times \mathbf{E}_s + \nabla_s \times \nu_{\rho\rho}\overline{\mu}_{\rho s} \cdot \mathbf{H}_s, \qquad (3.4.16\text{b})$$

where $\nu_{\rho\rho} = \mu_{\rho\rho}^{-1}$, $\kappa_{\rho\rho} = \epsilon_{\rho\rho}^{-1}$. If, however, the field is assumed to have $e^{in\phi}$ and $e^{ik_z z}$ dependences, and that $\overline{\epsilon}$ and $\overline{\mu}$ are independent of ϕ and z, then $\nabla_s \to i\mathbf{k}_s = \hat{\phi}\frac{in}{\rho} + \hat{z}ik_z$. In this case, the above becomes

$$\frac{d}{d\rho}\begin{bmatrix} \mathbf{E}_s \\ \mathbf{H}_s \end{bmatrix} = \overline{\mathbf{H}}(\rho) \cdot \begin{bmatrix} \mathbf{E}_s \\ \mathbf{H}_s \end{bmatrix}, \qquad (3.4.17)$$

as in (9), where $\overline{\mathbf{H}}(\rho)$ is derivable from (16a) and (16b) (Exercise 3.19). Except for the case when $n = 0$, the above does not have a closed-form solution. Hence, in order to solve for the propagator matrix, one resorts to a numerical method as in (2.9.22).

In the above, it is assumed that $\overline{\mu}_s$, $\overline{\mu}_{s\rho}$, $\overline{\mu}_{\rho s}$, $\mu_{\rho\rho}$, and $\overline{\epsilon}_s$, $\overline{\epsilon}_{s\rho}$, $\overline{\epsilon}_{\rho s}$, $\epsilon_{\rho\rho}$ are independent of ϕ. This is not true in general. For example, in Cartesian coordinates, the most general anisotropic medium has $\overline{\mu}$ and $\overline{\epsilon}$, which are

$$\overline{\mu} = \begin{bmatrix} \mu_{xx} & \mu_{xy} & \mu_{xz} \\ \mu_{yx} & \mu_{yy} & \mu_{yz} \\ \mu_{zx} & \mu_{zy} & \mu_{zz} \end{bmatrix}, \quad \overline{\epsilon} = \begin{bmatrix} \epsilon_{xx} & \epsilon_{xy} & \epsilon_{xz} \\ \epsilon_{yx} & \epsilon_{yy} & \epsilon_{yz} \\ \epsilon_{zx} & \epsilon_{zy} & \epsilon_{zz} \end{bmatrix}, \quad (3.4.18)$$

where μ_{ij} and ϵ_{ij} are all constants of positions. But when $\overline{\mu}$ and $\overline{\epsilon}$ are expressed in cylindrical coordinates as in (13), they will be functions of ϕ. To see this, first, note that

$$\begin{bmatrix} D_x \\ D_y \\ D_z \end{bmatrix} = \overline{\epsilon} \cdot \begin{bmatrix} E_x \\ E_y \\ E_z \end{bmatrix}. \quad (3.4.19)$$

Moreover, it can easily be shown that

$$\begin{bmatrix} E_x \\ E_y \\ E_z \end{bmatrix} = \begin{bmatrix} \cos\phi & -\sin\phi & 0 \\ \sin\phi & \cos\phi & 0 \\ 0 & 0 & 1 \end{bmatrix} \begin{bmatrix} E_\rho \\ E_\phi \\ E_z \end{bmatrix} = \overline{\mathbf{T}} \cdot \begin{bmatrix} E_\rho \\ E_\phi \\ E_z \end{bmatrix}. \quad (3.4.20)$$

Furthermore, $\overline{\mathbf{T}}^{-1} = \overline{\mathbf{T}}^t$. Hence, (19) can be written as

$$\begin{bmatrix} D_\rho \\ D_\phi \\ D_z \end{bmatrix} = \overline{\mathbf{T}}^t \cdot \overline{\epsilon} \cdot \overline{\mathbf{T}} \cdot \begin{bmatrix} E_\rho \\ E_\phi \\ E_z \end{bmatrix} = \widetilde{\overline{\epsilon}} \cdot \begin{bmatrix} E_\rho \\ E_\phi \\ E_z \end{bmatrix}, \quad (3.4.21)$$

where $\widetilde{\overline{\epsilon}} = \overline{\mathbf{T}}^t \cdot \overline{\epsilon} \cdot \overline{\mathbf{T}}$ is just the permittivity tensor expressed in cylindrical coordinates.

Since $\overline{\mathbf{T}}$ is a function of ϕ, $\widetilde{\overline{\epsilon}}$ in cylindrical coordinates is now a function of ϕ. But if the partitioned matrices in (13) are functions of ϕ, Equations (16a) and (16b) are more difficult to solve. In this case, all the azimuthal harmonics, $e^{in\phi}$, are coupled together. Therefore, unlike (17), where each $e^{in\phi}$ harmonic is considered independently, all azimuthal harmonics need to be considered simultaneously. To this end, we let

$$\mathbf{E}_s = \sum_{n=-\infty}^{\infty} e^{in\phi} \mathbf{E}_{ns}, \quad \mathbf{H}_s = \sum_{n=-\infty}^{\infty} e^{in\phi} \mathbf{H}_{ns}, \quad (3.4.22)$$

where $\mathbf{E}_{ns}$ and $\mathbf{H}_{ns}$ are independent of ϕ. Then, after substituting the above into (16), assuming that the field has $e^{ik_z z}$ dependence, we arrive at

$$\sum_{n=-\infty}^{\infty} e^{in\phi} \nabla_\rho \times \mathbf{E}_{ns} = \sum_{n=-\infty}^{\infty} \left[i\omega \overline{\mu}_s \cdot \mathbf{H}_{ns} + i\overline{\mu}_{s\rho} \cdot \nu_{\rho\rho} \mathbf{k}_s \times \mathbf{E}_{ns} \right.$$

$$- i\omega \overline{\mu}_{s\rho} \cdot \overline{\mu}_{\rho s} \cdot \nu_{\rho\rho} \mathbf{H}_{ns} + \frac{1}{i\omega} \left(\hat{\phi} \frac{1}{\rho} \frac{\partial}{\partial\phi} + \hat{z} ik_z \right) \times \kappa_{\rho\rho} \mathbf{k}_s \times \mathbf{H}_{ns}$$

$$\left. + \left(\hat{\phi} \frac{1}{\rho} \frac{\partial}{\partial\phi} + \hat{z} ik_z \right) \times \kappa_{\rho\rho} \overline{\epsilon}_{\rho s} \cdot \mathbf{E}_{ns} \right] e^{in\phi}, \quad (3.4.23a)$$

$$\sum_{n=-\infty}^{\infty} e^{in\phi} \nabla_\rho \times \mathbf{H}_{ns} = \sum_{n=-\infty}^{\infty} \left[-i\omega \,\overline{\boldsymbol{\epsilon}}_s \cdot \mathbf{E}_{ns} + i\overline{\boldsymbol{\epsilon}}_{s\rho} \cdot \kappa_{\rho\rho} \mathbf{k}_s \times \mathbf{H}_{ns} \right.$$

$$+ i\omega \,\overline{\boldsymbol{\epsilon}}_{s\rho} \cdot \overline{\boldsymbol{\epsilon}}_{\rho s} \cdot \kappa_{\rho\rho} \mathbf{E}_{ns} - \frac{1}{i\omega} \left(\hat{\phi} \frac{1}{\rho} \frac{\partial}{\partial \phi} + \hat{z} i k_z \right) \times \nu_{\rho\rho} \mathbf{k}_s \times \mathbf{E}_{ns}$$

$$\left. + \left(\hat{\phi} \frac{1}{\rho} \frac{\partial}{\partial \phi} + \hat{z} i k_z \right) \times \nu_{\rho\rho} \overline{\boldsymbol{\mu}}_{\rho s} \cdot \mathbf{H}_{ns} \right] e^{in\phi}, \qquad (3.4.23b)$$

where $\mathbf{k}_s = \hat{\phi}\frac{n}{\rho} + \hat{z}k_z$. Next, on multiplying the above by $e^{-im\phi}$, and integrating over ϕ from 0 to 2π, we arrive at

$$\nabla_\rho \times \mathbf{E}_{ms} = \sum_{n=-\infty}^{\infty} \left[\overline{\mathbf{A}}_{mn}^{11} \cdot \mathbf{E}_{ns} + \overline{\mathbf{A}}_{mn}^{12} \cdot \mathbf{H}_{ns} \right], \qquad (3.4.24a)$$

$$\nabla_\rho \times \mathbf{H}_{ms} = \sum_{n=-\infty}^{\infty} \left[\overline{\mathbf{A}}_{mn}^{21} \cdot \mathbf{E}_{ns} + \overline{\mathbf{A}}_{mn}^{22} \cdot \mathbf{H}_{ns} \right], \qquad (3.4.24b)$$

where $\overline{\mathbf{A}}_{mn}^{ij}$ can be found from (23). In general, they are nondiagonal because $\overline{\mu}$ and $\overline{\epsilon}$ are functions of ϕ in cylindrical coordinates.

The above can be written more compactly as

$$\frac{d}{d\rho} \begin{bmatrix} \mathbf{E}_{ms} \\ \mathbf{H}_{ms} \end{bmatrix} = \sum_{n=-\infty}^{\infty} \overline{\mathbf{H}}_{mn} \cdot \begin{bmatrix} \mathbf{E}_{ns} \\ \mathbf{H}_{ns} \end{bmatrix}, \qquad (3.4.25)$$

or

$$\frac{d}{d\rho} \mathbf{V}_m = \sum_{n=-\infty}^{\infty} \overline{\mathbf{H}}_{mn} \cdot \mathbf{V}_n, \qquad (3.4.25a)$$

as in Equations (9) and (17). In the above elements of $\overline{\mathbf{H}}_{mn}$ are to be gleaned from (24). Moreover, by defining an infinite-dimensional state vector

$$\mathbf{V}^t = [\cdots, \mathbf{V}_{-2}, \mathbf{V}_{-1}, \mathbf{V}_0, \mathbf{V}_1, \mathbf{V}_2, \cdots],$$

and a corresponding infinite-dimensional matrix $\overline{\mathbf{H}}$, (25a) becomes

$$\frac{d}{d\rho} \mathbf{V} = \overline{\mathbf{H}} \cdot \mathbf{V}. \qquad (3.4.26)$$

Equation (26) has to be solved numerically to obtain the propagator matrix for an anisotropic, inhomogeneous layer. Once this propagator is known, the reflection and transmission of waves through an inhomogeneous, anisotropic layer can be readily found (see Exercise 3.20). Since all harmonics are involved in (26), the reflected and transmitted waves will consist of all harmonics. This phenomenon is also known as *mode conversion*.

§3.5 Spherically Layered Media—Single Interface Case

In spherical coordinates, it is most convenient to decompose the fields of a homogeneous medium into fields TM to r and TE to r due to the preferred

r direction. But unlike the cylindrical interface case, the TM and TE waves are decoupled at a spherical interface. Consequently, we shall discuss first solutions to the vector wave equations in spherical coordinates for a homogeneous medium and show how these waves are decoupled even in a layered medium.

Scattering of plane waves by a sphere, also known as Mie scattering, was studied by Mie in 1908 (see Logan 1965 for a discussion). Subsequently, propagation of radio waves on the curved earth surface was considered by Watson in 1918. In 1951, Marcuvitz studied propagation of electromagnetic waves in spherically layered media. Moreover, Wait (1956, 1963) also performed studies in this area. Many later works are listed in the readings for this chapter.

§§3.5.1 Vector Wave Equation in Spherical Coordinates

To obtain the solutions to the vector wave equation in a homogeneous, isotropic medium, it is best to introduce the Debye potential π_e and π_m (Debye 1909) to characterize the TM and TE waves respectively. Consequently, for TM waves,

$$\mathbf{H}^{TM} = \nabla \times \mathbf{r}\pi_e \qquad (3.5.1)$$

and for TE waves,

$$\mathbf{E}^{TE} = \nabla \times \mathbf{r}\pi_m. \qquad (3.5.2)$$

Next, by substituting the above into the source-free Maxwell's equations, or the vector wave equations for $\mathbf{H}$ and $\mathbf{E}$ fields, it follows that (Exercise 3.21)

$$\left(\nabla^2 + k^2\right)\pi_e = 0, \qquad (3.5.3a)$$

$$\left(\nabla^2 + k^2\right)\pi_m = 0. \qquad (3.5.3b)$$

In general, in a source-free region,

$$\mathbf{H} = \nabla \times \mathbf{r}\pi_e + \frac{1}{i\omega\mu}\nabla \times \nabla \times \mathbf{r}\pi_m, \qquad (3.5.4a)$$

$$\mathbf{E} = \nabla \times \mathbf{r}\pi_m - \frac{1}{i\omega\epsilon}\nabla \times \nabla \times \mathbf{r}\pi_e. \qquad (3.5.4b)$$

Subsequently, we can extract the r components of $\mathbf{H}$ and $\mathbf{E}$ in the above (Exercise 3.22) to yield

$$H_r = \frac{1}{i\omega\mu}\left[\frac{\partial^2}{\partial r^2}r\pi_m + k^2 r\pi_m\right], \qquad (3.5.5a)$$

$$E_r = -\frac{1}{i\omega\epsilon}\left[\frac{\partial^2}{\partial r^2}r\pi_e + k^2 r\pi_e\right]. \qquad (3.5.5b)$$

Since π_m and π_e are solutions of the scalar wave equation, their general solutions are of the form

$$\left\{\begin{array}{c} j_n(kr) \\ h_n^{(1)}(kr) \end{array}\right\} P_n^m(\cos\theta) \left\{\begin{array}{c} \cos m\phi \\ \sin m\phi \end{array}\right\}, \qquad (3.5.6)$$

where the braces imply "linear superpositions of." If, however, π_m and π_e are of the form given in (6), then from the scalar wave equation expressed in spherical coordinates (see Chapter 1), we have

$$\frac{\partial^2}{\partial r^2}\begin{bmatrix} r\pi_m \\ r\pi_e \end{bmatrix} = -\left[k^2 - \frac{n(n+1)}{r^2} \right]\begin{bmatrix} r\pi_m \\ r\pi_e \end{bmatrix}. \qquad (3.5.7)$$

Therefore, for the n-th harmonic,[5]

$$H_r = \frac{1}{i\omega\mu}\frac{n(n+1)}{r}\pi_m, \qquad (3.5.8a)$$

$$E_r = -\frac{1}{i\omega\epsilon}\frac{n(n+1)}{r}\pi_e. \qquad (3.5.8b)$$

Next, if the transverse to r components of (4a) and (4b) are extracted (Exercise 3.22), then

$$\mathbf{H}_s = -\mathbf{r} \times \nabla_s\pi_e + \frac{1}{i\omega\mu}\frac{1}{r}\frac{\partial}{\partial r}r^2\nabla_s\pi_m, \qquad (3.5.9a)$$

$$\mathbf{E}_s = -\mathbf{r} \times \nabla_s\pi_m - \frac{1}{i\omega\epsilon}\frac{1}{r}\frac{\partial}{\partial r}r^2\nabla_s\pi_e. \qquad (3.5.9b)$$

Equations (8) and (9) imply that once the radial components of the fields are known, the transverse components are readily found for the n-th harmonic.

The general solution to (3a) and (3b) are of the form

$$\begin{bmatrix} \pi_e \\ \pi_m \end{bmatrix} = \left[\mathbf{a}j_n(kr) + \mathbf{b}h_n^{(1)}(kr) \right] P_n^m(\cos\theta)e^{im\phi}, \qquad (3.5.10)$$

where $\mathbf{a}$ and $\mathbf{b}$ are two-component column vectors. Then, from (9a) and (9b), it can be shown that

$$\mathbf{E}_s = \begin{bmatrix} E_\theta \\ E_\phi \end{bmatrix} = \begin{bmatrix} \frac{i}{\omega\epsilon}\frac{1}{r}\frac{\partial^2}{\partial r\partial\theta}r & \frac{1}{\sin\theta}\frac{\partial}{\partial\phi} \\ \frac{i}{\omega\epsilon}\frac{1}{r\sin\theta}\frac{\partial^2}{\partial r\partial\phi}r & -\frac{\partial}{\partial\theta} \end{bmatrix}\begin{bmatrix} \pi_e \\ \pi_m \end{bmatrix},$$

$$(3.5.11a)$$

$$\mathbf{H}_s = \begin{bmatrix} H_\theta \\ H_\phi \end{bmatrix} = \begin{bmatrix} \frac{1}{\sin\theta}\frac{\partial}{\partial\phi} & -\frac{i}{\omega\mu}\frac{1}{r}\frac{\partial^2}{\partial r\partial\theta}r \\ -\frac{\partial}{\partial\theta} & -\frac{i}{\omega\mu}\frac{1}{r\sin\theta}\frac{\partial^2}{\partial r\partial\phi}r \end{bmatrix}\begin{bmatrix} \pi_e \\ \pi_m \end{bmatrix}.$$

$$(3.5.11b)$$

Moreover, Equations (11a) and (11b) can be written as

$$\mathbf{E}_s = \begin{bmatrix} \frac{\partial}{\partial\theta} & -\frac{1}{\sin\theta}\frac{\partial}{\partial\phi} \\ \frac{1}{\sin\theta}\frac{\partial}{\partial\phi} & \frac{\partial}{\partial\theta} \end{bmatrix}\begin{bmatrix} \frac{i}{\omega\epsilon}\frac{1}{r}\frac{\partial}{\partial r}r & 0 \\ 0 & -1 \end{bmatrix}\begin{bmatrix} \pi_e \\ \pi_m \end{bmatrix},$$

$$(3.5.12a)$$

$$\mathbf{H}_s = \begin{bmatrix} \frac{\partial}{\partial\theta} & -\frac{1}{\sin\theta}\frac{\partial}{\partial\phi} \\ \frac{1}{\sin\theta}\frac{\partial}{\partial\phi} & \frac{\partial}{\partial\theta} \end{bmatrix}\begin{bmatrix} 0 & -\frac{i}{\omega\mu}\frac{1}{r}\frac{\partial}{\partial r}r \\ -1 & 0 \end{bmatrix}\begin{bmatrix} \pi_e \\ \pi_m \end{bmatrix}.$$

$$(3.5.12b)$$

---

[5] Note that B_r and D_r could have been used to characterize the TE and TM waves respectively just as B_z and D_z were used in the planarly layered media.

On substituting (10) into (12a) and (12b), they become

$$\mathbf{E}_s = \overline{\mathbf{P}}_n^m(\cos\theta) \cdot \left[\overline{\mathbf{j}}_{ne}(kr) \cdot \mathbf{a} + \overline{\mathbf{h}}_{ne}^{(1)}(kr) \cdot \mathbf{b}\right] e^{im\phi}, \tag{3.5.13a}$$

$$\mathbf{H}_s = \overline{\mathbf{P}}_n^m(\cos\theta) \cdot \left[\overline{\mathbf{j}}_{nh}(kr) \cdot \mathbf{a} + \overline{\mathbf{h}}_{nh}^{(1)}(kr) \cdot \mathbf{b}\right] e^{im\phi}, \tag{3.5.13b}$$

where

$$\overline{\mathbf{P}}_n^m(\cos\theta) = \begin{bmatrix} \frac{\partial}{\partial\theta}P_n^m(\cos\theta) & -\frac{im}{\sin\theta}P_n^m(\cos\theta) \\ \frac{im}{\sin\theta}P_n^m(\cos\theta) & \frac{\partial}{\partial\theta}P_n^m(\cos\theta) \end{bmatrix}, \tag{3.5.14a}$$

$$\overline{\mathbf{b}}_{ne}(kr) = \begin{bmatrix} \frac{i}{\omega\epsilon}\frac{1}{r}\frac{\partial}{\partial r}rb_n(kr) & 0 \\ 0 & -b_n(kr) \end{bmatrix}, \tag{3.5.14b}$$

and

$$\overline{\mathbf{b}}_{nh}(kr) = \begin{bmatrix} 0 & -\frac{i}{\omega\mu}\frac{1}{r}\frac{\partial}{\partial r}rb_n(kr) \\ -b_n(kr) & 0 \end{bmatrix}. \tag{3.5.14c}$$

In the above, b_n is either j_n or $h_n^{(1)}$ depending on the matrix being defined. But it can be shown that (Exercise 3.23; also see Stratton 1941, p. 417)

$$\int_0^\pi d\theta \sin\theta\, \overline{\mathbf{P}}_n^m(\cos\theta)\, \overline{\mathbf{P}}_{n'}^m(\cos\theta) = \delta_{nn'}\frac{2n(n+1)(n+m)!}{(2n+1)(n-m)!}\overline{\mathbf{I}}. \tag{3.5.15}$$

As a consequence of this, and that all the $e^{im\phi}$ harmonics are orthogonal, all the n and m harmonics are decoupled in a spherically symmetric region.

§§3.5.2 Reflection and Transmission of an Outgoing Wave

Unlike cylindrically layered media, the TM and TE to r waves in a spherically layered medium are decoupled. This is actually the result of the identity (15). At first, we can consider the TM and TE waves together and describe their reflections and transmissions using reflection and transmission matrices. As shall be shown, however, such matrices are diagonal, implying the decoupling of TM and TE waves.

Consider first an outgoing wave in region 1 (see Figure 3.5.1) which impinges on a spherical interface. Then, the reflected waves will set up a standing wave in region 1. Hence, the field in region 1 is of the form

$$\mathbf{E}_{1s} = \overline{\mathbf{P}}_n^m(\cos\theta) \cdot \left[\overline{\mathbf{h}}_{ne}^{(1)}(k_1r) + \overline{\mathbf{j}}_{ne}(k_1r) \cdot \overline{\mathbf{R}}_{12}\right] \cdot \mathbf{a}, \tag{3.5.16a}$$

$$\mathbf{H}_{1s} = \overline{\mathbf{P}}_n^m(\cos\theta) \cdot \left[\overline{\mathbf{h}}_{nh}^{(1)}(k_1r) + \overline{\mathbf{j}}_{nh}(k_1r) \cdot \overline{\mathbf{R}}_{12}\right] \cdot \mathbf{a}. \tag{3.5.16b}$$

In region 2, however, the transmitted wave becomes an outgoing wave. Hence, the fields in region 2 are

$$\mathbf{E}_{2s} = \overline{\mathbf{P}}_n^m(\cos\theta) \cdot \overline{\mathbf{h}}_{ne}^{(1)}(k_2r) \cdot \overline{\mathbf{T}}_{12} \cdot \mathbf{a}, \tag{3.5.17a}$$

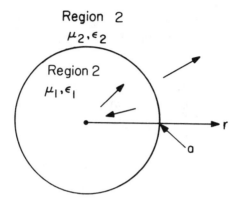

Figure 3.5.1 Reflection of an outgoing wave at a spherical boundary.

$$\mathbf{H}_{2s} = \overline{\mathbf{P}}_n^m(\cos\theta) \cdot \overline{\mathbf{h}}_{nh}^{(1)}(k_2 r) \cdot \overline{\mathbf{T}}_{12} \cdot \mathbf{a}. \tag{3.5.17b}$$

Notice that in the above, we have suppressed the $e^{im\phi}$ dependence. Moreover, the orthogonality of the $\overline{\mathbf{P}}_n^m(\cos\theta)$ harmonics warranted by (15) implies that all the harmonics are decoupled. Hence, only one harmonic needs to be considered at a time in (16) and (17).

Furthermore, the boundary conditions at $r = a$, give rise to

$$\overline{\mathbf{h}}_{ne}^{(1)}(k_1 a) + \overline{\mathbf{j}}_{ne}(k_1 a) \cdot \overline{\mathbf{R}}_{12} = \overline{\mathbf{h}}_{ne}^{(1)}(k_2 a) \cdot \overline{\mathbf{T}}_{12}, \tag{3.5.18a}$$

$$\overline{\mathbf{h}}_{nh}^{(1)}(k_1 a) + \overline{\mathbf{j}}_{nh}(k_1 a) \cdot \overline{\mathbf{R}}_{12} = \overline{\mathbf{h}}_{nh}^{(1)}(k_2 a) \cdot \overline{\mathbf{T}}_{12}. \tag{3.5.18b}$$

As a result, the above can be solved to yield

$$\overline{\mathbf{R}}_{12} = -\left[\overline{\mathbf{h}}_{ne}^{(1)^{-1}}(k_2 a) \cdot \overline{\mathbf{j}}_{ne}(k_1 a) - \overline{\mathbf{h}}_{nh}^{(1)^{-1}}(k_2 a) \cdot \overline{\mathbf{j}}_{nh}(k_1 a)\right]^{-1} \cdot \\ \cdot \left[\overline{\mathbf{h}}_{ne}^{(1)^{-1}}(k_2 a) \cdot \overline{\mathbf{h}}_{ne}^{(1)}(k_1 a) - \overline{\mathbf{h}}_{nh}^{(1)^{-1}}(k_2 a) \cdot \overline{\mathbf{h}}_{nh}^{(1)}(k_1 a)\right], \tag{3.5.19a}$$

$$\overline{\mathbf{T}}_{12} = \left[\overline{\mathbf{j}}_{ne}^{-1}(k_1 a) \cdot \overline{\mathbf{h}}_{ne}^{(1)}(k_2 a) - \overline{\mathbf{j}}_{nh}^{-1}(k_1 a) \cdot \overline{\mathbf{h}}_{nh}^{(1)}(k_2 a)\right]^{-1} \cdot \\ \cdot \left[\overline{\mathbf{j}}_{ne}^{-1}(k_1 a) \cdot \overline{\mathbf{h}}_{ne}^{(1)}(k_1 a) - \overline{\mathbf{j}}_{nh}^{-1}(k_1 a) \cdot \overline{\mathbf{h}}_{nh}^{(1)}(k_1 a)\right]. \tag{3.5.19b}$$

Since $\overline{\mathbf{h}}_{ne}^{(1)}$ and $\overline{\mathbf{j}}_{ne}$ are diagonal while $\overline{\mathbf{h}}_{nh}^{(1)}$ and $\overline{\mathbf{j}}_{nh}$ are off-diagonal, it can be shown easily that $\overline{\mathbf{R}}_{12}$ and $\overline{\mathbf{T}}_{12}$ are also diagonal, implying the decoupling of TM and TE waves (see Exercise 3.24).

Consequently, one can define

$$\overline{\mathbf{R}}_{12} = \begin{bmatrix} R_{12}^{TM} & 0 \\ 0 & R_{12}^{TE} \end{bmatrix}, \qquad \overline{\mathbf{T}}_{12} = \begin{bmatrix} T_{12}^{TM} & 0 \\ 0 & T_{12}^{TE} \end{bmatrix}, \qquad (3.5.20)$$

and show that (Exercise 3.24)

$$R_{12}^{TM} = \frac{\sqrt{\epsilon_2 \mu_1}\, \hat{H}_n^{(1)}(k_2 a) \hat{H}_n^{(1)'}(k_1 a) - \sqrt{\epsilon_1 \mu_2}\, \hat{H}_n^{(1)'}(k_2 a) \hat{H}_n^{(1)}(k_1 a)}{\sqrt{\epsilon_1 \mu_2}\, \hat{J}_n(k_1 a) \hat{H}_n^{(1)'}(k_2 a) - \sqrt{\epsilon_2 \mu_1}\, \hat{H}_n^{(1)}(k_2 a) \hat{J}_n'(k_1 a)}, \qquad (3.5.21a)$$

$$T_{12}^{TM} = \frac{i\epsilon_2 \sqrt{\frac{\mu_2}{\epsilon_1}}}{\sqrt{\epsilon_1 \mu_2}\, \hat{J}_n(k_1 a) \hat{H}_n^{(1)'}(k_2 a) - \sqrt{\epsilon_2 \mu_1}\, \hat{H}_n^{(1)}(k_2 a) \hat{J}_n'(k_1 a)}, \qquad (3.5.21b)$$

where $\hat{J}_n(x) = x j_n(x)$, $\hat{H}_n^{(1)}(x) = x h_n^{(1)}(x)$. In the above, the Wronskian of spherical Bessel functions,

$$\hat{J}_n(k_1 a) \hat{H}_n^{(1)'}(k_1 a) - \hat{J}_n'(k_1 a) \hat{H}_n^{(1)}(k_1 a) = i,$$

has been used to further simplify the numerator of T_{12}^{TM}. By duality, then

$$R_{12}^{TE} = \frac{\sqrt{\epsilon_1 \mu_2}\, \hat{H}_n^{(1)}(k_2 a) \hat{H}_n^{(1)'}(k_1 a) - \sqrt{\epsilon_2 \mu_1}\, \hat{H}_n^{(1)'}(k_2 a) \hat{H}_n^{(1)}(k_1 a)}{\sqrt{\epsilon_2 \mu_1}\, \hat{J}_n(k_1 a) \hat{H}_n^{(1)'}(k_2 a) - \sqrt{\epsilon_1 \mu_2}\, \hat{H}_n^{(1)}(k_2 a) \hat{J}_n'(k_1 a)}, \qquad (3.5.22a)$$

$$T_{12}^{TE} = \frac{i\mu_2 \sqrt{\frac{\epsilon_2}{\mu_1}}}{\sqrt{\epsilon_2 \mu_1}\, \hat{J}_n(k_1 a) \hat{H}_n^{(1)'}(k_2 a) - \sqrt{\epsilon_1 \mu_2}\, \hat{H}_n^{(1)}(k_2 a) \hat{J}_n'(k_1 a)}. \qquad (3.5.22b)$$

Note that R_{12} and T_{12} are functions of n but not m.

§§3.5.3 Reflection and Transmission of a Standing Wave

It has been shown previously that TM and TE waves are decoupled at a spherical boundary. Hence, we need only consider first the TM wave reflection and transmission and subsequently derive the TE wave reflection and transmission by duality.

When an "incoming" standing wave impinges on a spherical interface (see Figure 3.5.2), it generates a reflected outgoing wave in region 2 plus a transmitted standing wave in region 1. Hence, in region 2, we can write

$$\begin{bmatrix} H_{2\theta} \\ H_{2\phi} \end{bmatrix} = \begin{bmatrix} \frac{im}{\sin\theta} P_n^m(\cos\theta) \\ -\frac{\partial}{\partial\theta} P_n^m(\cos\theta) \end{bmatrix} \left[R_{21}^{TM} h_n^{(1)}(k_2 r) + j_n(k_2 r) \right], \qquad (3.5.23a)$$

$$\begin{bmatrix} E_{2\theta} \\ E_{2\phi} \end{bmatrix} = \frac{-1}{i\omega\epsilon_2} \begin{bmatrix} \frac{\partial}{\partial\theta} P_n^m(\cos\theta) \\ \frac{im}{\sin\theta} P_n^m(\cos\theta) \end{bmatrix} \left[R_{21}^{TM} \frac{1}{r} \frac{\partial}{\partial r} r h_n^{(1)}(k_2 r) + \frac{1}{r} \frac{\partial}{\partial r} r j_n(k_2 r) \right]. \qquad (3.5.23b)$$

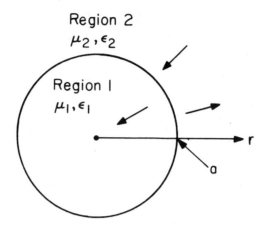

Figure 3.5.2 Reflection of an "incoming" standing wave at a spherical boundary.

Moreover, in region 1, there is a transmitted wave, or

$$\begin{bmatrix} H_{1\theta} \\ H_{1\phi} \end{bmatrix} = \begin{bmatrix} \frac{im}{\sin\theta} P_n^m(\cos\theta) \\ -\frac{\partial}{\partial\theta} P_n^m(\cos\theta) \end{bmatrix} T_{21}^{TM} j_n(k_1 r), \tag{3.5.24a}$$

$$\begin{bmatrix} E_{1\theta} \\ E_{1\phi} \end{bmatrix} = \frac{-1}{i\omega\epsilon_1} \begin{bmatrix} \frac{\partial}{\partial\theta} P_n^m(\cos\theta) \\ \frac{im}{\sin\theta} P_n^m(\cos\theta) \end{bmatrix} T_{21}^{TM} \frac{1}{r}\frac{\partial}{\partial r} r j_n(k_1 r). \tag{3.5.24b}$$

Again, the $e^{im\phi}$ dependence is suppressed in the above. Next, by matching boundary conditions, we obtain

$$R_{21}^{TM} h_n^{(1)}(k_2 a) + j_n(k_2 a) = T_{21}^{TM} j_n(k_1 a), \tag{3.5.25a}$$

$$\epsilon_1 \left[R_{21}^{TM} \frac{1}{a}\frac{\partial}{\partial a} a h_n^{(1)}(k_2 a) + \frac{1}{a}\frac{\partial}{\partial a} a j_n(k_2 a) \right] = \epsilon_2 T_{21}^{TM} \frac{1}{a}\frac{\partial}{\partial a} a j_n(k_1 a). \tag{3.5.25b}$$

Then, the above can be solved to yield

$$R_{21}^{TM} = \frac{j_n(k_2 a)\frac{\epsilon_2}{a}\frac{\partial}{\partial a} a j_n(k_1 a) - j_n(k_1 a)\frac{\epsilon_1}{a}\frac{\partial}{\partial a} a j_n(k_2 a)}{j_n(k_1 a)\frac{\epsilon_1}{a}\frac{\partial}{\partial a} a h_n^{(1)}(k_2 a) - h_n^{(1)}(k_2 a)\frac{\epsilon_2}{a}\frac{\partial}{\partial a} a j_n(k_1 a)}, \tag{3.5.26a}$$

$$T_{21}^{TM} = \frac{j_n(k_2 a)\frac{\epsilon_1}{a}\frac{\partial}{\partial a} a h_n^{(1)}(k_2 a) - h_n^{(1)}(k_2 a)\frac{\epsilon_1}{a}\frac{\partial}{\partial a} a j_n(k_2 a)}{j_n(k_1 a)\frac{\epsilon_1}{a}\frac{\partial}{\partial a} a h_n^{(1)}(k_2 a) - h_n^{(1)}(k_2 a)\frac{\epsilon_2}{a}\frac{\partial}{\partial a} a j_n(k_1 a)}. \tag{3.5.26b}$$

Moreover, the above can be written as (Exercise 3.25; also see Harrington 1961, p. 297)

$$R_{21}^{TM} = \frac{\sqrt{\epsilon_2\mu_1}\,\hat{J}_n(k_2 a)\hat{J}_n'(k_1 a) - \sqrt{\epsilon_1\mu_2}\,\hat{J}_n(k_1 a)\hat{J}_n'(k_2 a)}{\sqrt{\epsilon_1\mu_2}\,\hat{J}_n(k_1 a)\hat{H}_n^{(1)'}(k_2 a) - \sqrt{\epsilon_2\mu_1}\,\hat{H}_n^{(1)}(k_2 a)\hat{J}_n'(k_1 a)}, \tag{3.5.27a}$$

$$T_{21}^{TM} = \frac{i\epsilon_1 \sqrt{\frac{\mu_1}{\epsilon_2}}}{\sqrt{\epsilon_1\mu_2}\,\hat{J}_n(k_1 a)\hat{H}_n^{(1)'}(k_2 a) - \sqrt{\epsilon_2\mu_1}\,\hat{H}_n^{(1)}(k_2 a)\hat{J}_n'(k_1 a)}, \qquad (3.5.27b)$$

where $\hat{J}_n(x) = x j_n(x)$, $\hat{H}_n^{(1)}(x) = x h_n^{(1)}(x)$, and the Wronskian for spherical Bessel functions has been used to simplify T_{21}^{TM}. By duality, then

$$R_{21}^{TE} = \frac{\sqrt{\epsilon_1\mu_2}\,\hat{J}_n(k_2 a)\hat{J}_n'(k_1 a) - \sqrt{\epsilon_2\mu_1}\,\hat{J}_n(k_1 a)\hat{J}_n'(k_2 a)}{\sqrt{\epsilon_2\mu_1}\,\hat{J}_n(k_1 a)\hat{H}_n^{(1)'}(k_2 a) - \sqrt{\epsilon_1\mu_2}\,\hat{H}_n^{(1)}(k_2 a)\hat{J}_n'(k_1 a)}, \qquad (3.5.28a)$$

$$T_{21}^{TE} = \frac{i\mu_1 \sqrt{\frac{\epsilon_1}{\mu_2}}}{\sqrt{\epsilon_2\mu_1}\,\hat{J}_n(k_1 a)\hat{H}_n^{(1)'}(k_2 a) - \sqrt{\epsilon_1\mu_2}\,\hat{H}_n^{(1)}(k_2 a)\hat{J}_n'(k_1 a)}. \qquad (3.5.28b)$$

Notice that the reflection and transmission coefficients are functions of n but not m.

§3.6 Spherically Layered Media—Multi-interface Case

With the canonical reflection and transmission coefficients for a single interface derived, we can derive the reflection and transmission of waves through a multispherically layered medium. This follows from that transmission and reflection through a layered medium are consequences of multiple reflections and transmissions through many single interfaces.

§§3.6.1 The Outgoing-Wave Case

The derivation for an N-layer medium can be arrived at by first considering the case of a three-layer medium as shown in Figure 3.6.1. Indeed, the derivation is very similar to that presented in Subsection 3.2.1, except that now the problem is scalar rather than vector. As a result, Debye potentials are used to characterize the scalar waves.

For instance, in region 1, if there is an outgoing wave, the reflected wave will give rise to a standing wave. Hence, the Debye potential is of the form

$$\pi_1 = a_1 \left[h_n^{(1)}(k_1 r) + \tilde{R}_{12} j_n(k_1 r) \right], \qquad (3.6.1)$$

where we have suppressed the $P_n^m(\cos\theta)$ and $e^{im\phi}$ dependences. In the above, π_1 and R_{12} are for either the TM or the TE case, since the two derivations are identical to each other. By the same token, in region 2, we have

$$\pi_2 = a_2 \left[h_n^{(1)}(k_2 r) + R_{23} j_n(k_2 r) \right]. \qquad (3.6.2)$$

Finally, in region 3, there is only an outgoing wave; therefore,

$$\pi_3 = a_3 h_n^{(1)}(k_3 r). \qquad (3.6.3)$$

In the above, $\tilde{R}_{12}$ is a generalized reflection coefficient including multiple subsurface reflections, while R_{23} is a local, single-interface reflection coefficient. Consequently, in a manner similar to deriving (3.2.7), it follows that

$$\tilde{R}_{12} = R_{12} + \frac{T_{21} R_{23} T_{12}}{1 - R_{21} R_{23}}. \qquad (3.6.4)$$

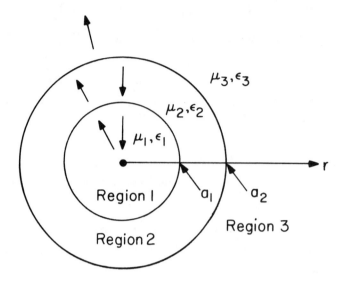

Figure 3.6.1 Reflection and transmission through a three-layer medium.

Similar to (3.2.6), it also follows that

$$a_2 = \frac{T_{12}}{1 - R_{21}R_{23}} a_1. \tag{3.6.5}$$

Now, if another region is added beyond region 3, we need only change R_{23} in the above to $\tilde{R}_{23}$ to incorporate multiple reflections. Moreover, Equations (4) and (5) can be written as a general recursive relation, yielding

$$\tilde{R}_{i,i+1} = R_{i,i+1} + \frac{T_{i+1,i}\tilde{R}_{i+1,i+2}T_{i,i+1}}{1 - R_{i+1,i}\tilde{R}_{i+1,i+2}}, \tag{3.6.6}$$

$$a_{i+1} = \frac{T_{i,i+1}}{1 - R_{i+1,i}\tilde{R}_{i+1,i+2}} a_i = S_{i,i+1}a_i. \tag{3.6.7}$$

For an N-layer medium, the above can be used recursively to find the field in all regions. In addition, a generalized transmission coefficient can be defined such that

$$\tilde{T}_{1N} = S_{12}S_{23}\cdots T_{N-1,N}, \tag{3.6.7a}$$

which relates the outgoing-wave amplitude in region N to that in region 1.

§§3.6.2 The Standing-Wave Case

The generalized reflection coefficient for a standing wave impinging on a layered medium can be derived in a manner similar to that shown in Subsection 3.2.2. Consequently, the results, similar to (3.2.19) and (3.2.21), are

$$\tilde{R}_{i,i-1} = R_{i,i-1} + \frac{T_{i-1,i}\tilde{R}_{i-1,i-2}T_{i,i-1}}{1 - R_{i-1,i}\tilde{R}_{i-1,i-2}}, \qquad (3.6.8)$$

$$a_{i-1} = \frac{T_{i,i-1}}{1 - R_{i-1,i}\tilde{R}_{i-1,i-2}}a_{i,n} = S_{i,i-1}a_i. \qquad (3.6.9)$$

Moreover, a generalized transmission coefficient can be defined here such that

$$\tilde{T}_{N1} = S_{N,N-1}S_{N-1,N-2}\cdots T_{21} \qquad (3.6.9a)$$

similar to (7a). Notice that the recursive relations (6), (7), (8), and (9) are very similar to those given in Section 2.1 in Chapter 2, except for the conspicuous absence of the exponential phase factors. The reason being that the phase factors are buried in the definition of the reflection and transmission coefficients; hence, it is not necessary to express them explicitly. But when a medium is lossy or the waves highly evanescent, these buried phase factors may make the spherical Bessel functions in the reflection and transmission coefficients inordinately large. In this case, it is necessary to renormalize the spherical Bessel functions to prevent numerical overflows.

§3.7 Source in a Spherically Layered Medium

In order to obtain the solution due to a point source in a spherically layered medium, it is important first to expand the field of a point source in a homogeneous, unbounded medium in spherical harmonics. Consequently, using the fact that $h_0^{(1)}(kr) = \frac{e^{ikr}}{ikr}$, and the addition theorem for spherical harmonics and spherical Bessel functions (see Exercise 3.27), we have

$$\frac{e^{ik|\mathbf{r}-\mathbf{r}'|}}{ik|\mathbf{r}-\mathbf{r}'|} = h_0^{(1)}(k|\mathbf{r}-\mathbf{r}'|) = \sum_{n=0}^{\infty}\sum_{m=-n}^{n}\frac{(2n+1)(n-|m|)!}{(n+|m|)!}$$
$$j_n(kr_<)h_n^{(1)}(kr_>)P_n^{|m|}(\cos\theta')P_n^{|m|}(\cos\theta)e^{im(\phi-\phi')}, \qquad (3.7.1)$$

where $r_>$ is the larger of r and r' and $r_<$ is the smaller of r and r'. Moreover, the above can be readily established by solving the following equation in spherical coordinates:

$$(\nabla^2 + k^2)\,\phi(r,\theta,\phi) = -\delta(\mathbf{r}-\mathbf{r}') = -\frac{\delta(r-r')}{r^2\sin\theta}\,\delta(\theta-\theta')\,\delta(\phi-\phi'). \qquad (3.7.2)$$

Then, using the definition that[6]

$$Y_{nm}(\theta,\phi) = \sqrt{\frac{(n-m)!}{(n+m)!}\frac{2n+1}{4\pi}}\,P_n^m(\cos\theta)e^{im\phi}, \qquad (3.7.3)$$

---

[6] Here, we define $P_n^m(x) = (-1)^m(1-x^2)^{m/2}d^m P_n^m(x)/dx^m$, and $P_n^{-m}(x) = (-1)^m\frac{(n-m)!}{(n+m)!}$ $P_n^m(x)$ for $m > 0$ (see Jackson 1975, p. 98-99).

(1) can be rewritten as

$$\frac{e^{ik|\mathbf{r}-\mathbf{r}'|}}{ik|\mathbf{r}-\mathbf{r}'|} = 4\pi \sum_{n=0}^{\infty} \sum_{m=-n}^{n} j_n(kr_<)\, h_n^{(1)}(kr_>)\, Y_{nm}(\theta,\phi)\, Y_{nm}^*(\theta',\phi'). \qquad (3.7.4)$$

With the identity (4), the field due to any point source in a homogeneous, unbounded medium can be expressed in terms of spherical harmonics. For example, given a point source pointing in an arbitrary direction $\hat{\alpha}$, the following equations,

$$\mathbf{E} = i\omega\mu \left(\overline{\mathbf{I}} + \frac{\nabla\nabla}{k^2}\right) \cdot \hat{\alpha} I\ell\, g(\mathbf{r}-\mathbf{r}'), \qquad (3.7.5a)$$

$$\mathbf{H} = \nabla \times \hat{\alpha} I\ell\, g(\mathbf{r}-\mathbf{r}'), \qquad (3.7.5b)$$

and the identity (4) can be used to derive the $\mathbf{E}$ and $\mathbf{H}$ fields of a point source. Note that in the above, $g(\mathbf{r}-\mathbf{r}') = e^{ik|\mathbf{r}-\mathbf{r}'|}/4\pi|\mathbf{r}-\mathbf{r}'|$. Consequently, the TM and TE waves can be obtained by extracting the H_r and the E_r components of the fields. Once the r components of the fields are established, the Debye potential can be derived via the use of Equations (3.5.8a) and (3.5.8b). Then, with these Debye potentials for the homogeneous medium case, those in the presence of a layered medium are easily sought.

Outside the source point, however, Equation (5a) can be alternatively written as (Exercise 3.28)

$$\mathbf{E} = \frac{iI\ell}{\omega\epsilon} \nabla \times \nabla \times \hat{\alpha}\, g(\mathbf{r}-\mathbf{r}'). \qquad (3.7.6)$$

Moreover, the $\hat{r}$ component of the above equation can be extracted to yield

$$rE_r = \frac{iI\ell}{\omega\epsilon} \mathbf{r} \cdot \nabla \times \nabla \times \hat{\alpha}\, g(\mathbf{r}-\mathbf{r}'). \qquad (3.7.7)$$

But by reciprocity, the above is also equal to (Exercise 3.28)

$$rE_r = \frac{iI\ell}{\omega\epsilon} \hat{\alpha} \cdot \nabla' \times \nabla' \times \mathbf{r} g(\mathbf{r}-\mathbf{r}'). \qquad (3.7.8)$$

Furthermore, it can be shown that (Exercise 3.28)

$$\nabla' \times (\mathbf{r}-\mathbf{r}')\, g(\mathbf{r}-\mathbf{r}') = 0. \qquad (3.7.9)$$

Hence, using (9) in (8), we have

$$rE_r = \frac{iI\ell}{\omega\epsilon} \hat{\alpha} \cdot \nabla' \times \nabla' \times \mathbf{r}' g(\mathbf{r}-\mathbf{r}'). \qquad (3.7.10)$$

Next, the representation (4) can be used to substitute for the scalar Green's function in (10) to give

$$rE_r = \frac{-kI\ell}{\omega\epsilon} \hat{\alpha} \cdot \nabla' \times \nabla' \times \mathbf{r}' \sum_{n=0}^{\infty} j_n(kr_<) h_n^{(1)}(kr_>) A_n(\theta,\phi;\theta',\phi'), \qquad (3.7.11)$$

where $A_n(\theta, \phi; \theta', \phi') = \sum\limits_{m=-n}^{n} Y_{nm}(\theta, \phi) Y_{nm}^*(\theta', \phi')$. Then, by way of Equation (3.5.8b),

$$\pi_e = ikI\ell\,\hat{\alpha}\cdot\nabla'\times\nabla'\times\mathbf{r}'\sum_{n=0}^{\infty} j_n(kr_<)\,h_n^{(1)}(kr_>)A_n(\theta,\phi;\theta',\phi')/n(n+1).$$

(3.7.12)

By the same token, extracting the $\hat{r}$ component of (5b) yields

$$rH_r = I\ell\,\hat{\alpha}\cdot\nabla'\times\mathbf{r}'g(\mathbf{r}-\mathbf{r}')$$

$$= ikI\ell\,\hat{\alpha}\cdot\nabla'\times\mathbf{r}'\sum_{n=0}^{\infty} j_n(kr_<)\,h_n^{(1)}(kr_>)A_n(\theta,\phi;\theta',\phi').$$

(3.7.13)

Consequently, by way of Equation (3.5.8a),

$$\pi_m = -\omega\mu kI\ell\,\hat{\alpha}\cdot\nabla'\times\mathbf{r}'\sum_{n=0}^{\infty} j_n(kr_<)\,h_n^{(1)}(kr_>)A_n(\theta,\phi;\theta',\phi')/n(n+1).$$

(3.7.14)

Therefore, the Debye potentials due to an arbitrary point source can be derived (Exercise 3.29). In general, the Debye potential due to a point source has the form

$$\pi = D'\sum_{n=0}^{\infty}\frac{j_n(kr_<)\,h_n^{(1)}(kr_>)}{n(n+1)}A_n(\theta,\phi;\theta',\phi'),$$

(3.7.15)

where D' is a linear differential operator acting only on the prime coordinates.

When the point source is put in the j-th region of a spherically layered medium, additional reflected waves are generated. Then,

$$\pi_j = D_j'\sum_{n=0}^{\infty}\left[j_n(k_jr_<)h_n^{(1)}(k_jr_>) + a_{jn}h_n^{(1)}(k_jr) + b_{jn}j_n(k_jr)\right]\frac{A_n(\theta,\phi;\theta',\phi')}{n(n+1)}.$$

(3.7.16)

Here, a_{jn} and b_{jn} are determined in a manner similar to (3.3.11) and (3.3.12), yielding (Exercise 3.29)

$$a_{jn} = \tilde{R}_{j,j-1}\left[h_n^{(1)}(k_jr') + \tilde{R}_{j,j+1}j_n(k_jr')\right]\tilde{M}_j,$$

(3.7.17a)

$$b_{jn} = \tilde{R}_{j,j+1}\left[j_n(k_jr') + \tilde{R}_{j,j-1}h_n^{(1)}(k_jr')\right]\tilde{M}_j,$$

(3.7.17b)

where $\tilde{M}_j = \left(1 - \tilde{R}_{j,j-1}\tilde{R}_{j,j+1}\right)^{-1}$ is a factor accounting for multiple reflections in region j. Moreover, the generalized reflection coefficients $\tilde{R}$ can be determined via the method in Section 3.6. Then, using (17a) and (17b) in

(16), we have, after some simplifications (see Exercise 3.29),

$$\pi_j = D'_j \sum_{n=0}^{\infty} \left[h_n^{(1)}(k_j r_>) + \tilde{R}_{j,j+1} j_n(k_j r_>) \right] \left[j_n(k_j r_<) + \tilde{R}_{j,j-1} h_n^{(1)}(k_j r_<) \right] \tilde{M}_j$$
$$\cdot \frac{A_n(\theta, \phi; \theta', \phi')}{n(n+1)}. \qquad (3.7.18)$$

When $r \in$ region i where $i > j$, the Debye potential in region i can be expressed as

$$\pi_i = D'_j \sum_{n=0}^{\infty} a_{in}(\mathbf{r}') \left[h_n^{(1)}(k_i r) + \tilde{R}_{i,i+1} j_n(k_i r) \right] \frac{A_n(\theta, \phi; \theta', \phi')}{n(n+1)}, \qquad (3.7.19)$$

where

$$D'_j a_{in}(r') = \tilde{T}_{ji} \tilde{M}_i D'_j a_{jn}(r'). \qquad (3.7.20)$$

Here, $\tilde{T}_{ji}$ is the generalized transmission coefficient through the layered slab between region j and i equivalent to (3.6.7a). Since $\tilde{T}_{ji}$ does not include multiple reflections in region i, the factor $\tilde{M}_i$ is required (Exercise 3.29). In addition, $D'_j a_{jn}(r')$ is the amplitude of the outgoing wave in region j. As a result, from (18), we identify

$$a_{jn}(r') = \left[j_n(k_j r') + \tilde{R}_{j,j-1} h_n^{(1)}(k_j r') \right] \tilde{M}_j. \qquad (3.7.21)$$

Consequently, (19) becomes

$$\pi_i = D'_j \sum_{n=0}^{\infty} \left[h_n^{(1)}(k_i r) + \tilde{R}_{i,i+1} j_n(k_i r) \right] \left[j_n(k_j r') + \tilde{R}_{j,j-1} h_n^{(1)}(k_j r') \right]$$
$$\cdot \tilde{T}_{ji} \tilde{M}_i \tilde{M}_j \frac{A_n(\theta, \phi; \theta', \phi')}{n(n+1)}. \qquad (3.7.22)$$

When $r \in$ region i where $i < j$, the Debye potential in region i is of the form

$$\pi_i = D'_j \sum_{n=0}^{\infty} a_{in}(r') \left[j_n(k_i r) + \tilde{R}_{i,i-1} h_n^{(1)}(k_i r) \right] \frac{A_n(\theta, \phi; \theta', \phi')}{n(n+1)}. \qquad (3.7.23)$$

Here, $a_{in}(r')$ can be determined via the use of the generalized transmission coefficient for a standing wave given by (3.6.9a). Consequently,

$$\pi_i = D'_j \sum_{n=0}^{\infty} \left[j_n(k_i r) + \tilde{R}_{i,i-1} h_n^{(1)}(k_i r) \right] \left[h_n^{(1)}(k_j r') + \tilde{R}_{j,j+1} j_n(k_j r') \right]$$
$$\cdot \tilde{T}_{ji} \tilde{M}_i \tilde{M}_j \frac{A_n(\theta, \phi; \theta', \phi')}{n(n+1)}. \qquad (3.7.24)$$

In summary, for a point source in region j, i.e., $r' \in$ region j,

$$\pi = D'_j \sum_{n=0}^{\infty} F_n(r, r') \frac{A_n(\theta, \phi; \theta', \phi')}{n(n+1)}, \qquad (3.7.25)$$

where

$$F_n(r, r') = \left[h_n^{(1)}(k_j r_>) + \tilde{R}_{j,j+1} j_n(k_j r_>) \right] \left[j_n(k_j r_<) + \tilde{R}_{j,j-1} h_n^{(1)}(k_j r_<) \right] \tilde{M}_j \qquad (3.7.26)$$

when $r \in$ region j. But when $r \in$ region $i > j$,

$$F_n(r, r') = \left[h_n^{(1)}(k_i r) + \tilde{R}_{i,i+1} j_n(k_i r) \right] \left[j_n(k_j r') + \tilde{R}_{j,j-1} h_n^{(1)}(k_j r') \right] \tilde{T}_{ji} \tilde{M}_i \tilde{M}_j. \qquad (3.7.27)$$

Moreover, when $r \in$ region $i < j$,

$$F_n(r, r') = \left[j_n(k_i r) + \tilde{R}_{i,i-1} h_n^{(1)}(k_i r) \right] \left[h_n^{(1)}(k_j r') + \tilde{R}_{j,j+1} j_n(k_j r') \right] \tilde{T}_{ji} \tilde{M}_i \tilde{M}_j. \qquad (3.7.28)$$

In the above, $\tilde{R}_{i,i+1}$ and $\tilde{R}_{i,i-1}$ are determined recursively using (3.6.6) and (3.6.8). Furthermore, they have to be chosen for the appropriate n's, since they are functions of n's. In this manner, the fields in all regions in the layered medium can be determined.

Equation (1) can also be cast into a form where the angular wave numbers are expressed as an integral over a continuum. Such a representation then allows the use of contour integration techniques to extract the creeping waves and whispering gallery waves on a spherical surface, and hence, to further elucidate the physics of the problem.

§3.8 Propagator Matrix—Spherical Layers

The transmission and reflection of waves through a spherically layered medium is scalar. Therefore, if the problem is solved with the state variable approach, the state vector will consist of at least two elements. But in order to derive the state equation, we need to first derive the equations satisfied by the Debye potentials. Better still, when μ and ϵ are functions of r only, partial differential equations satisfied by the Debye potential can be derived.

From (3.5.11a), continuity of $\mathbf{E}_s$ and $\mathbf{H}_s$ at an interface implies that π_e and $\frac{1}{\epsilon} \frac{\partial}{\partial r} r \pi_e$ are continuous at an interface, and that π_m and $\frac{1}{\mu} \frac{\partial}{\partial r} r \pi_m$ are continuous at an interface. Since for a homogeneous medium, π_e and π_m satisfy Equations (3.5.3a) and (3.5.3b), by inspection, the equations to be satisfied by π_e and π_m in a spherically layered medium where ϵ and μ are functions of r only are

$$\left[\nabla_s^2 + \frac{\epsilon}{r} \frac{\partial}{\partial r} \frac{1}{\epsilon} \frac{\partial}{\partial r} r + k^2 \right] \pi_e = 0, \qquad (3.8.1a)$$

$$\left[\nabla_s^2 + \frac{\mu}{r} \frac{\partial}{\partial r} \frac{1}{\mu} \frac{\partial}{\partial r} r + k^2 \right] \pi_m = 0, \qquad (3.8.1b)$$

where ∇_s^2 is the portion of the Laplacian operator that is transverse to r. By inspection, solutions to the preceding equations have to satisfy the aforementioned boundary conditions, and yet they reduce to (3.5.3) for a homogeneous medium. More rigorously, these equations can be established from Maxwell's equations directly (Exercise 3.30). Furthermore, if the Debye potentials are assumed to have

$$R_n(r)P_n^m(\cos\theta)e^{im\phi}$$

dependences, then $R_n(\mathbf{r})$ satisfies the equation

$$\frac{p}{r}\frac{d}{dr}\frac{1}{p}\frac{d}{dr}rR_n(r) + \left[k^2 - \frac{n(n+1)}{r^2}\right]R_n(r) = 0, \qquad (3.8.2)$$

where $p = \epsilon$ for TM waves, while $p = \mu$ for TE waves. Next, on defining $S_n(r) = \frac{1}{p}\frac{d}{dr}rR_n(r)$, (2) becomes

$$\frac{d}{dr}S_n(r) = \left[k^2 - \frac{n(n+1)}{r^2}\right]\frac{r}{p}R_n(r), \qquad (3.8.3a)$$

where

$$\frac{d}{dr}rR_n(r) = pS_n(r). \qquad (3.8.3b)$$

Consequently, the above can be written as a state equation

$$\frac{d}{dr}\mathbf{V}(r) = \overline{\mathbf{H}}(r)\cdot\mathbf{V}(r), \qquad (3.8.4)$$

where $\mathbf{V}^t(r) = [S_n(r), rR_n(r)]$ is the state vector and

$$\overline{\mathbf{H}}(r) = \begin{bmatrix} 0 & k^2 - \frac{n(n+1)}{r^2} \\ p & 0 \end{bmatrix}. \qquad (3.8.5)$$

When p is a constant, from (1a) and (1b), the general solution for $R_n(r)$ is a linear superposition of $j_n(kr)$ and $h_n^{(1)}(kr)$. Hence, the general solution to $\mathbf{V}(r)$ is

$$\mathbf{V}(r) = A_1\mathbf{a}_1(r) + A_2\mathbf{a}_2(r), \qquad (3.8.6)$$

where

$$\mathbf{a}_1(r) = \begin{bmatrix} \frac{1}{p}\frac{d}{dr}rj_n(kr) \\ rj_n(kr) \end{bmatrix}, \quad \mathbf{a}_2(r) = \begin{bmatrix} \frac{1}{p}\frac{d}{dr}rh_n^{(1)}(kr) \\ rh_n^{(1)}(kr) \end{bmatrix}. \qquad (3.8.7)$$

Then, Equation (6) can be expressed as

$$\mathbf{V}(r) = \overline{\mathbf{a}}(r)\cdot\mathbf{A}, \qquad (3.8.8)$$

where $\overline{\mathbf{a}}(r) = [\mathbf{a}_1(r), \mathbf{a}_2(r)]$ and $\mathbf{A}^t = [A_1, A_2]$. Alternatively, the above can be written as

$$\begin{aligned}\mathbf{V}(r) &= \overline{\mathbf{a}}(r)\cdot\overline{\mathbf{a}}^{-1}(r')\cdot\overline{\mathbf{a}}(r')\cdot\mathbf{A} \\ &= \overline{\mathbf{P}}(r,r')\cdot\mathbf{V}(r'), \qquad (3.8.9)\end{aligned}$$

where the propagator matrix between two points r and r' is[7]

$$\overline{\mathbf{P}}(r, r') = \overline{\mathbf{a}}(r) \cdot \overline{\mathbf{a}}^{-1}(r'). \tag{3.8.10}$$

Knowing the propagator, we can use it to solve the transmission and reflection of spherical waves through a layered medium as in Section 2.8. Alternatively, Equation (4) can also be solved numerically to obtain the propagator matrix for a general, radially inhomogeneous profile.

Exercises for Chapter 3

3.1 Show that the z component of Equation (3.1.1) satisfies Equation (3.1.2), but the ϕ and ρ components do not satisfy the same equation.

3.2 Explain why for a given n, any linear superposition of two independent solutions to Bessel's equation is complete for representing the general solution to (3.1.2).

3.3 In a cylindrically layered medium, if one region has $e^{ik_z z}$ dependence, convince yourself that all other regions must have $e^{ik_z z}$ dependence when the problem is translationally invariant in the z direction.

3.4 Using (3.1.4), derive Equations (3.1.8a), (3.1.8b), and (3.1.9).

3.5 (a) Using Bessel's equation, show that the Wronskian of two independent solutions $J_n(x)$ and $H_n^{(1)}(x)$ is

$$H_n^{(1)}(x) J_n'(x) - J_n(x) H_n^{(1)'}(x) = \frac{\text{const}}{x}.$$

     (b) Using the small argument approximation of $J_n(x)$ and $H_n^{(1)}(x)$ (see Chapter 1, Subsection 1.2.4), show that the value of the constant is $-2i/\pi$.

3.6 A guided mode in a cylinder is defined to be a nontrivial solution that exists in a cylinder without the need for an external excitation.

     (a) Show that this statement is the same as requiring that the reflection matrices (3.1.11a) and (3.1.17a) have infinite determinantal values.

     (b) Show that this is the same as requiring that the determinant of $\overline{\mathbf{D}}$ be zero in Equation (3.1.17a).

     (c) Why do we obtain the same guidance condition with (3.1.11a) and (3.1.17a)?

     (d) Looking at Equation (3.2.18), what do you think is the guidance condition if there is a three-layer medium?

---

[7] $\overline{\mathbf{a}}^{-1}(r)$ has a closed form (see Exercise 3.31).

3.7 Sometimes, the scattering solution of a plane wave by a dielectric cylinder is needed. In this case, it is best to expand the plane wave into a linear superposition of standing cylindrical waves, i.e.,

$$e^{ikx} = e^{ik\rho\cos\phi} = \sum_{n=-\infty}^{\infty} a_n e^{in\phi} J_n(k\rho).$$

(a) Show that the coefficient a_n is $e^{in\pi/2}$.

Hint: Use the plane wave expansion formula for Bessel functions in Chapter 2, Equation (2.2.17).

(b) Show that an arbitrarily polarized plane wave can be decomposed into TM to z and TE to z plane waves. Hence, it can be represented as

$$\begin{bmatrix} E_z \\ H_z \end{bmatrix} = \begin{bmatrix} E_0 \\ H_0 \end{bmatrix} e^{ik_x x + ik_y y + ik_z z}.$$

(c) Find the reflected wave when such a plane wave impinges on a multi-layer cylinder.

3.8 (a) Show that when $n \to -n$ or $k_z \to -k_z$, only the off-diagonal elements of $\overline{\mathbf{R}}_{12}$, $\overline{\mathbf{R}}_{21}$, $\overline{\mathbf{T}}_{12}$, and $\overline{\mathbf{T}}_{21}$, and $\widetilde{\overline{\mathbf{R}}}_{12}$, $\widetilde{\overline{\mathbf{R}}}_{21}$, $\widetilde{\overline{\mathbf{T}}}_{12}$, and $\widetilde{\overline{\mathbf{T}}}_{21}$ change sign, after using the fact that $B_{-n}(x) = (-1)^n B_n(x)$.

(b) Show that this is equivalent to the fact that an odd-symmetric (even-symmetric) TM wave (TE wave) about ϕ or z will couple only to an even-symmetric (odd-symmetric) TE wave (TM wave) about ϕ or z at the circular, cylindrical interface.

3.9 (a) By solving for the solution of the wave equation

$$[\nabla_s^2 + k_\rho^2]\Phi(\rho, \phi) = -\delta(\boldsymbol{\rho} - \boldsymbol{\rho}') = -\frac{\delta(\rho - \rho')}{\rho'}\delta(\phi - \phi')$$

in cylindrical coordinates using cylindrical harmonics, prove the addition theorem in (3.3.2).

Hint: You need to break the solution into two regions, $\rho > \rho'$ and $\rho < \rho'$, and match the solution across the discontinuity at $\rho = \rho'$. The Wronskian property of Bessel functions is useful here.

(b) By using the raising operator derived in Chapter 2, Equation (2.2.16), derive the addition theorem for the n-th cylindrical wave function.

3.10 Derive the expression corresponding to (3.3.6) if the source at $\mathbf{r}'$ is:

(a) A vertical electric dipole pointing in the z direction.

(b) A horizontal electric dipole pointing in the x direction and located at $\phi' = 0$.

3.11 Derive Equation (3.3.13) using Equations (3.3.7), (3.3.11), and (3.3.12).

3.12 (a) Show that the factor $\widetilde{\overline{\mathbf{M}}}_{i+}$ in (3.3.16) is needed to correctly account for multiple reflections.

   (b) Derive Equation (3.3.18).

   (c) Derive Equation (3.3.20).

3.13 Using the uniqueness of the solution argument (see Subsection 2.7.1 of Chapter 2), show that the solution of a source in a cylindrically layered medium has branch points only at $k_\rho = \pm k_N$ on the complex k_z plane, where k_N is the wave number of the outermost region.

3.14 If a vertical electric dipole is radiating near a perfectly conducting cylinder:

   (a) Show that in (3.3.30), $\overline{\mathbf{b}}_{j\nu} = 0$. What would $\overline{\mathbf{a}}_{j\nu}$ be in this case?

   (b) Show that the poles on the complex ν plane are the zeroes of $H_\nu^{(1)}$ $(k_\rho a)$, where a is the radius of the cylinder. Show that if ν_p is a pole, so is $-\nu_p$.

   (c) Show that the zeroes of $H_\nu^{(1)}(k_\rho a)$ are approximately at $\nu \simeq k_\rho a$ (see Abramowitz and Stegun 1965).

   (d) Verify the expression (3.3.30a) by contour integration technique.

   (e) Derive the corresponding approximation in (3.3.30a) for part (a) and give a physical interpretation to the result.

3.15 Find explicitly the expressions for the matrix elements of $\overline{\mathbf{H}}(\rho)$ in (3.4.9).

3.16 It can be shown that $\overline{\mathbf{a}}(\rho)$ defined in (3.4.10) and (3.4.11) has a closed-form inverse by first finding a vector $\mathbf{b}_1(\rho)$ that is orthogonal to $\mathbf{a}_2$, $\mathbf{a}_3$, and $\mathbf{a}_4$.

   (a) Show that such a $\mathbf{b}_1(\rho)$ is

$$
\mathbf{b}_1(\rho) = \begin{bmatrix} -\frac{i\omega\epsilon}{k_\rho} H_n^{(1)'}(k_\rho\rho) \\ 0 \\ \frac{nk_z}{k_\rho^2\rho} H_n^{(1)}(k_\rho\rho) \\ H_n^{(1)}(k_\rho\rho) \end{bmatrix}.
$$

   (b) Derive $\mathbf{b}_2(\rho)$ that is orthogonal to $\mathbf{a}_1$, $\mathbf{a}_3$, and $\mathbf{a}_4$, and, therefore, derive also $\mathbf{b}_3(\rho)$ and $\mathbf{b}_4(\rho)$.

   (c) Construct a matrix that is the left inverse of $\overline{\mathbf{a}}(\rho)$. Simplify the result by using the Wronskian properties of Bessel functions (see Exercise 3.5; also see Abramowitz and Stegun). Hence, find $\overline{\mathbf{P}}(\rho, \rho')$.

3.17 Show that once $\overline{\mathbf{P}}(\rho, \rho')$ in an inhomogeneous layer is known, it can be used to derive the transmission and reflection of waves through layered media for:

(a) The outgoing-wave case.

(b) The standing-wave case.

3.18 Describe a numerical scheme to find $\overline{\mathbf{P}}(\rho, \rho')$, when a layer is inhomogeneous.

3.19 Derive the expression for the matrix elements of $\overline{\mathbf{H}}(\rho)$ in (3.4.17).

3.20 (a) Describe a numerical scheme to solve (3.4.26) and obtain a propagator matrix for an anisotropic layer.

(b) With the propagator matrix, show how one can study the transmission and reflection of waves through the anisotropic layer.

3.21 Show that if the Debye potentials satisfy (3.5.3a) and (3.5.3b), then the $\mathbf{H}$ and $\mathbf{E}$ fields defined by (3.5.1) and (3.5.2) satisfy the vector wave equation.

3.22 (a) Extract the r components of (3.5.4a) and (3.5.4b), and verify (3.5.5a), (3.5.5b), (3.5.8a), and (3.5.8b).

(b) Extract the transverse to r components of (3.5.4a) and (3.5.4b), and verify (3.5.9a) and (3.5.9b).

3.23 Equation (3.5.15) is important to prove the decoupling of TE and TM waves in a spherically layered medium. Verify this equation.

3.24 Show that $\overline{\mathbf{R}}_{12}$ and $\overline{\mathbf{T}}_{12}$ defined in (3.5.19) are diagonal, and hence, verify Equations (3.5.21) and (3.5.22).

3.25 Derive Equations (3.5.27) and (3.5.28).

3.26 Derive Equations (3.6.6), (3.6.7), (3.6.8), and (3.6.9).

3.27 By solving the equation

$$(\nabla^2 + k^2)\psi(r, \theta, \phi) = -\delta(\mathbf{r} - \mathbf{r}') = -\frac{\delta(r - r')}{r'^2 \sin\theta}\,\delta(\theta - \theta')\,\delta(\phi - \phi')$$

in spherical coordinates, verify the addition theorem for spherical Bessel functions given by (3.7.1) (also see Exercise 3.9).

3.28 (a) Verify Equation (3.7.6) from (3.7.5a).

(b) Using reciprocity, show that (3.7.8) follows from (3.7.7).

(c) Show that (3.7.9) is true by showing that $(\mathbf{r} - \mathbf{r}')g(\mathbf{r} - \mathbf{r}')$ is the gradient of a scalar function, and hence, verify (3.7.10).

3.29 (a) For an electric dipole with $\hat{\alpha} = \hat{r}$ in (3.7.5a) and (3.7.5b), derive the Debye potential associated with this dipole in the manner of (3.7.12) and (3.7.14). Are both Debye potentials necessary here?

(b) Derive the expressions (3.7.17a) and (3.7.17b), and hence (3.7.18).

(c) Derive the expression (3.7.20), and explain why the factor $\tilde{M}_i$ is needed.

3.30 Establish (3.8.1) more directly from Maxwell's equations and the definition of Debye potentials.

3.31 Derive the closed-form inverse of $\bar{\mathbf{a}}^{-1}(r)$ in (3.8.9) (see Exercise 3.16).

References for Chapter 3

Abramowitz, M., and I. A. Stegun. 1965. *Handbook of Mathematical Functions*. New York: Dover Publications.

Chew, W. C. 1983. "The singularities of a Fourier type integral in a multicylindrical layer problem." *IEEE Trans. Antennas Propagat.* AP-31: 653.

Chew, W. C. 1984. "Response of a current loop antenna in an invaded borehole." *Geophysics* 49(1): 81–91.

Debye, P. 1909. *Ann. d. Physik* 4(30): 57.

Einziger, P. D., and L. B. Felsen. 1983. "Rigorous asymptotic analysis of transmission through a curved dielectric slab." *IEEE Trans. Antennas Propagat.* AP-31(6): 863–69.

Elliott, R. S. 1955. "Azimuthal surface waves on circular cylinders." *J. Appl. Phys.* 26(4): 368–76.

Felsen, L. B., and N. Marcuvitz. 1973. *Radiation and Scattering of Electromagnetic Waves*. Englewood Cliffs, New Jersey: Prentice Hall.

Gronthoud, A. G., and H. Blok. 1978. "The influence of bulk losses and bulk dispersion on the propagation properties of surface waves in a radially inhomogeneous optical waveguide." *Optical and Quantum Electronics* 10: 95–106.

Harrington, R. F. 1961. *Time-Harmonic Electromagnetic Fields*. New York: McGraw-Hill.

Helstrom, C. W. 1963. "Scattering from a cylinder coated with a dielectric material." In *Electromagnetic Theory and Antennas*, ed. E. C. Jordan. New York: Macmillan.

Jackson, J. D. 1975. *Classical Electrodynamics*. New York: John Wiley & Sons.

Logan, N. A. 1965. "Survey of some early studies of the scattering of plane waves by a sphere." *Proceedings IEEE.* 53(8): 773-85.

Lovell, J. R., and W. C. Chew. 1987a. "Response of a point source in a multicylindrically layered medium." *IEEE Trans. Geosci. Remote Sensing* GE-25: 850–58.

Lovell, J. R., and W. C. Chew. 1987b. "Effect of tool eccentricity on some electrical well logging tools." Presented at the IEEE Geosci. Remote Sensing Symp., Ann Arbor.

Marcuvitz, N. 1951. "Field representations in spherically stratified regions." *Comm. Pure and Appl. Math.* 4: 263–315.

Mie, G. 1908. "Beiträge zur optik trüber medien speziel kolloidaler metallösungen." *Ann. Phys. (Leipzig)* 25: 377.

Pearson, L. W. 1986. "A construction of the fields radiated by z directed point sources of current in the presence of a cylindrically layered obstacle." *Radio Sci.* 21: 559.

Pearson, L. W. 1987. "A ray representation of surface diffraction by a multilayer cylinder." *IEEE Trans. Antennas Propagat.* AP-35: 698.

Stratton, J. A. 1941. *Electromagnetic Theory.* New York: McGraw-Hill.

Wait, J. R. 1956. "Radiation from a vertical antenna over a curved stratified ground." *J. Res. NBS* 56: 237.

Wait, J. R. 1963. "Electromagnetic scattering from a radially inhomogeneous sphere." *Appl. Sci. Res. B* 10: 441.

Watson, G. N. 1918. "The diffraction of electric waves by the earth." *Proc. Roy. Soc. (London)* A95: 83–99.

Yariv, A. 1985. *Optical Electronics.* New York: Holt, Rinehart & Winston.

Further Readings for Chapter 3

Alden, A. L., and M. Kerker. 1951. "Scattering of electromagnetic waves from two concentric spheres." *J. Appl. Phys.* 22: 1242.

Bahar, E. 1975. "Propagation in irregular multilayered cylindrical structures of finite conductivity—Full wave solutions." *Can. J. Phys.* 53(11): 1088–96.

Born, M., and E. Wolf. 1970. *Principle of Optics.* New York: Pergamon Press.

Bowman, J. J., and V. H. Weston. 1966. "The effect of curvature on the reflection coefficient of layered absorbers." *IEEE Trans. Antennas Propagat.* AP-14(6): 760–67.

Bowman, J. J., T. B. A. Senior, and P. L. E. Uslenghi. 1969. *Electromagnetic and Acoustic Scattering by Simple Shapes.* Amsterdam: North-Holland.

Bremmer, H. 1949. *Terrestrial Radio Waves.* New York: Elsevier Publishing Co.

Budden, K. G. 1961. *Radio Waves in the Ionosphere,* Chapter 18. Cambridge, England: Cambridge University Press.

Bussey, H. E., and J. H. Richmond. 1975. "Scattering by a lossy dielectric circular cylindrical multilayer, numerical values." *IEEE Trans. Antennas Propagat.* AP-23(5): 723–25.

Chew, W. C., S. Gianzero, and K. J. Kaplin. 1981. "Transient response of an induction logging tool in a borehole." *Geophysics* 46: 1291–1300.

Chew, W. C., and J. R. Lovell. 1985. "Radiation of a point source in the vicinity of a multicylindrically layered geometry." North American Radio Science Meeting, Univ. of B. C., Vancouver, B. C.

Clarricoats, P. J. B., and K. B. Chan. 1973. "Propagation behaviour of cylindrical dielectric-rod waveguides." *Proc. IEEE* 120(11): 1371–78.

Duesterhoeft, W. C. 1963. "Propagation effects on radial response in induction logging." *Geophysics* 27: 463–69.

Friedlander, F. G. 1954. "Diffraction of pulses by a circular cylinder." *Comm. Pure Appl. Math.* 7: 705–32.

Gianzero, S. C. 1978. "Effect of sonde eccentricity on responses of conventional induction logging tools." *IEEE Trans. Geosci. Remote Sensing* GE-16: 332.

Gianzero, S., and B. Anderson. 1984. "Mathematical theory for the fields due to a finite A.C. coil in an infinitely thick bed with an arbitrary number of coaxial layers." *The Log Analyst* 25(2): 25–32.

Gloge, D. 1971. "Weakly guiding fibers." *Appl. Opt.* 10: 2252.

Hjelt, S. E. 1971. "The transient electromagnetic field of a two layer sphere."

Geoexploration 9: 213–30.

Hobson, E. W. 1931. *The Theory of Spherical and Ellipsoidal Harmonics.* Cambridge, England: Cambridge University Press.

Howard, A. Q. 1972. "The electromagnetic fields of a subterranean cylindrical inhomogeneity excited by a line source." *Geophysics* 37: 975–84.

Kao, C. K., and T. W. Davies. 1968. "Spectroscopic studies of ultra low loss optical glasses." *J. Sci. Instrum.* 1(2): 1063.

Kapron, F. P., D. B. Keck, and R. D. Maurer. 1970. "Radiation losses in glass optical waveguides." *Appl. Phys. Lett.* 17: 423.

Keller, J. B. 1956. "Diffraction by a convex cylinder." *IRE Trans. on Antennas and Propagation* AP-4: 312–21.

Keller, J. B., S. I. Rubinow, and M. Goldstein. June 1963. "Zeros of Hankel functions and poles of scattering amplitudes." *J. of Math. Phys.* 4.

Kikuchi, H., and E. Yamachita. 1959. "Theory of dielectric waveguides and some experiments at 50 KMC/sec." *Proc. Symp. Millimeter Waves*, p. 619. Brooklyn, NY: Polytechnic Institute of Brooklyn.

Kodis, R. D. 1962. "On the theory of diffraction by a composite cylinder." *J. Res. Natl. Bur. Stand., Sect D* 65(1): 19–33.

Kodis, R. D. 1963. "The scattering cross-section of a composite cylinder, geometrical optics." *IEEE Trans. Antennas Propagat.* AP-11: 86–93.

Lee, T. 1975. "Transient electromagnetic response of a sphere in a layered medium." *Geophysical Prospecting* 23: 492–512.

Lewin, L., D. C. Chang, and E. F. Kuester. 1977. *Electromagnetic Waves and Curved Structures.* London: Peter Perigrinus.

Magnus, W., and L. Kotin. 1960. "The zeros of the Hankel function as a function of its order." *Numerische Mathematik* 2: 228–44.

Marcatilli, E. A. J. 1969. "Bends in optical dielectric guides." *Bell. Syst. Tech. J.* 48(7): 2103–32.

Mieras, H. Nov. 1982. "Radiation pattern computation of a spherical lens using Mie series." *IEEE Trans. Antennas Propagat.* AP-30: 2–9.

Moran, J. H., and K. S. Kunz. 1962. "Basic theory of induction logging and application to study of two coil sondes." *Geophysics* 27: 829–58.

Nabighian, M. N. 1971. "Quasistatic response of a conducting permeable two layered sphere in a dipolar field." *Geophysics* 36: 25–37.

Nitikina, W. N. 1960. "The general solution of an axially symmetric problem in induction logging theory." *Izv. Akad. SSSR Sev. Fiz.*, pp. 607–16.

Paknys, R., and N. Wang. 1986. "Creeping wave propagation constants and modal impedance for a dielectric coated cylinder." *IEEE Trans. Antennas Propagat.* AP-34: 674.

Pathak, P. H. 1979. "An asymptotic analysis of the scattering of plane waves by a smooth convex cylinder." *Radio Sci.* 14: 419.

Rheinstein, J. 1964. "Scattering of electromagnetic waves from dielectric coated conducting cylinders." *IEEE Trans. Antennas Propagat.* AP-12(3): 334–40.

Ruck, G. T., D. E. Barrick, W. D. Stuart, and C. K. Krichbaum. 1970. *Radar Cross Section Handbook*, Chapters 3 and 6. New York: Plenum Press.

Safaazi-Jazi, A., and G. L. Yip. 1980. "Scattering from an arbitrarily located off-axis inhomogeneity in a step-index optical fiber." *IEEE Trans. Microwave Theory Tech.* MTT-28(1): 24–32.

Senior, T. B. A., and R. F. Goodrich. 1964. "Scattering by a sphere." *Proc. IEEE* 111: 907–16.

Shakir, S. A., and A. F. Turner. 1982. "Method of poles for multilayer thin-film waveguides." *Appl. Phys.* A-29: 151–55.

Singh, R. N. 1972. "Transient electromagnetic screening in a two concentric spherical shell model." *Pure and Applied Geophysics* 94: 226–32.

Singh, R. N. 1973. "Electromagnetic transient response of a conducting sphere in a conductive medium." *Geophysics* 38: 864–93.

Snitzer, E. 1961. "Cylindrical dielectric waveguide modes." *J. Opt. Soc. Am.* 51: 491.

Tang, C. C. 1957. "Backscattering from dielectric coated infinite cylindrical obstacles." *J. Appl. Phys.* 28(5): 628–33.

Verma, S. K. 1972. "Transient electromagnetic response of a conducting sphere excited by different types of input pulses." *Geophysical Prospecting* 20: 752–71.

Verma, S. K., and R. N. Singh. 1970. "Transient electromagnetic response of an inhomogeneous conducting sphere" *Geophysics* 35: 331–36.

Wait, J. R. 1955. "Scattering of plane wave from a circular dielectric cylinder at oblique incidence" *Can. J. Phys.* 33(5): 189–95.

Wait, J. R. 1958. "Pattern of an antenna on a curved lossy surface." *IRE Trans. on Antennas and Propagation* AP-6: 348–59.

Wait, J. R. 1959. *Electromagnetic Radiation from Cylindrical Structures.* New York: Pergamon Press.

Wait, J. R. 1984. "General formulation of the induction logging problem for concentric layers about the borehole." *IEEE Trans. Geosci. Remote Sensing* GE-22: 34–42.

Wait, J. R. 1986. "Impedance conditions for a coated cylindrical conductor." *Radio Sci.* 21: 623.

Wait, J. R., and D. A. Hill. 1979. "Theory of transmission of electromagnetic waves along a drill rod in a conducting rock." *IEEE Trans. Geosci. Remote Sensing* GE-17: 21–24.

Wang, D.-S. 1988. "Asymptotic behavior of the scattering solutions for a multilayered sphere." *IEEE Trans. Antennas Propagat.* AP-36: 1594–1601.

Wasylkiwskyj, W. 1975. "Diffraction by a concave perfectly conducting circular cylinder." *IEEE Trans. Antennas Propagat.* AP-23(4): 480–92.

Weston, V. H., and R. Hemenger. 1962. "High-frequency scattering from a coated sphere." *J. Res. NBS* 66D: 613.

Wimp, J. 1984. *Computation with Recurrence Relations.* New York: Pitman.

Yamada, J. I., M. Saruwatari, K. Asatani, H. Tsuchiya, A. Kawana, K. Sugiyama, and T. Kimura. 1978. "High speed optical pulse transmission at $1.29\mu m$ wavelength using low loss single mode fibers." *IEEE J. Quantum Electronics* QE. 14: 791.

CHAPTER 4

TRANSIENTS

The study of the transient response of fields in an inhomogeneous medium is important in a number of applications. An obvious advantage of transient measurement is its rapid acquisition of data over a large bandwidth. Moreover, it also allows the separation of different events by time gating in the time domain. Hence, many measurements are inherently transient in nature, some of which are radar, numerous sounding measurements, and tomograms. As a matter of fact, these measurements are usually performed with pulses of signals. Unfortunately, many transient electromagnetic measurements generally require larger bandwidths compared to acoustic measurements. Since electromagnetic waves travel 30 cm in 1 ns, in order to resolve structures of the order of 30 cm, a 1 ns pulse is required, implying a bandwidth of over 1 GHz. A consequence of the large bandwidths in transient electromagnetic measurements is their poor signal-to-noise ratio compared to narrow-band, time harmonic measurements. Nevertheless, this can be remedied by a strong signal source, or by time averaging a measurement repeated many times, which is also known as multishot averaging.

In acoustic measurements, where the velocity of the wave is slower and therefore, the wavelength shorter, transient measurements can be achieved with less bandwidth. Consequently, transient type measurements are quite popular in acoustics.

In this chapter, we will discuss some techniques for calculating the transient response of electromagnetic fields in inhomogeneous media. Such techniques, after some minor modifications, can evidently be used for other waves.

§4.1 Causality of Transient Response

One way to obtain the transient solution of a problem is by first solving the problem in the frequency domain, and then obtaining the time-domain solution by Fourier inverse transforming the solution. Physical grounds require that all such solutions satisfy the law of causality. In other words, a response does not exist until after a signal is applied. Furthermore, if a receiver is at a certain distance from the transmitter, the receiver shall not sense a signal until after the time it takes for the signal to reach the receiver from the transmitter. Thus, energy carrying signals cannot travel faster than the speed of light, which is the consequence of the law of special relativity. Moreover, the law of causality imposes analytic properties on the constitutive parameters that describe a medium. We shall first study these analytic properties which relate the real and imaginary part of a constitutive parameter. The relation is also known as the Kramers-Kronig relation.

§§4.1.1 The Kramers-Kronig Relation

The constitutive relations we have encountered in Chapter 1 are expressed in the frequency domain. For example, the relationship between an electric flux and an electric field is more appropriately written as

$$\mathbf{D}(\omega) = \epsilon(\omega)\mathbf{E}(\omega). \tag{4.1.1}$$

In the time domain, however, it becomes

$$\mathbf{D}(t) = \int_{-\infty}^{t} \epsilon(t - \tau)\mathbf{E}(\tau)\, d\tau. \tag{4.1.2}$$

In order for $\mathbf{D}(t)$ to be a causal function, $\epsilon(t)$ and $\mathbf{E}(t)$ must both be causal functions. The fact that $\epsilon(t)$ is causal puts a constraint on $\epsilon(\omega)$, the Fourier transform of $\epsilon(t)$. More specifically, the relationship between $\epsilon(t)$ and $\epsilon(\omega)$ is

$$\epsilon(t) = \frac{1}{2\pi} \int_{-\infty}^{\infty} d\omega\, e^{-i\omega t}\epsilon(\omega). \tag{4.1.3}$$

But if $\epsilon(\omega) \to \epsilon(\infty)$, when $|\omega| \to \infty$, then $\epsilon(\omega) = [\epsilon(\omega) - \epsilon(\infty)] + \epsilon(\infty)$, and the above becomes

$$\epsilon(t) = \frac{1}{2\pi} \int_{-\infty}^{\infty} d\omega\, e^{-i\omega t}[\epsilon(\omega) - \epsilon(\infty)] + \delta(t)\epsilon(\infty), \tag{4.1.4}$$

where the last Fourier integral has been performed to yield the delta function for the last term. Since $\epsilon(t)$ is a real-valued function, $\epsilon(\infty)$ must also be a real-valued constant. But from causality, $\epsilon(t) = 0$ for $t < 0$. This imposes an analytic property on $\epsilon(\omega)$ as follows: Since $\epsilon(\omega) \to \epsilon(\infty)$ when $\omega \to +i\infty$, then for $t < 0$, the integral on the complex ω plane can be deformed to $\omega \to i\infty$. The fact that $\epsilon(t) = 0$ for $t < 0$ implies that the integrand must be analytic in the upper half of the complex ω plane. This implies that $\epsilon(\omega)$ can have singularities only on the lower half of the complex ω plane.

Since $\epsilon(\omega) - \epsilon(\infty)$ is analytic on the upper half of the complex ω plane, using Cauchy's theorem, we can write

$$\epsilon(\omega) - \epsilon(\infty) = \frac{1}{2\pi i} \int_{C-C_\infty} d\omega'\, \frac{\epsilon(\omega') - \epsilon(\infty)}{\omega' - \omega}, \tag{4.1.5}$$

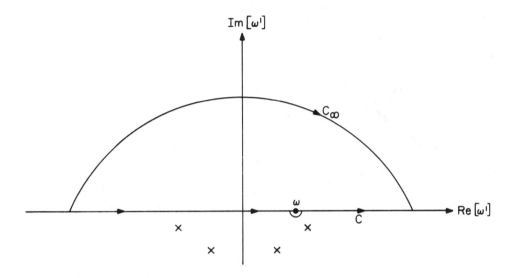

Figure 4.1.1 The C and C_∞ contours on the complex ω' plane.

where C and C_∞ are as shown in Figure 4.1.1. But since $\epsilon(\omega) - \epsilon(\infty) \to 0$ when $\omega \to \infty$, the integral over C_∞ vanishes. Consequently, (5) becomes

$$\epsilon(\omega) - \epsilon(\infty) = \frac{1}{2\pi i} \int_C d\omega' \frac{\epsilon(\omega') - \epsilon(\infty)}{\omega' - \omega}. \tag{4.1.6}$$

Now, if both ω and ω' are real, then the above can alternatively be written as

$$\epsilon(\omega) - \epsilon(\infty) = \frac{1}{\pi i} P.V. \int_{-\infty}^{\infty} d\omega' \frac{\epsilon(\omega') - \epsilon(\infty)}{\omega' - \omega}, \tag{4.1.7}$$

where $P.V.$ stands for principal value integral. Furthermore, by letting $\epsilon(\omega) = \epsilon'(\omega) + i\epsilon''(\omega)$, where $\epsilon'(\omega)$ and $\epsilon''(\omega)$ are the real and the imaginary parts of $\epsilon(\omega)$ respectively, and then equating the real and imaginary parts of (7), we have

$$\epsilon'(\omega) - \epsilon(\infty) = \frac{1}{\pi} P.V. \int_{-\infty}^{\infty} \frac{\epsilon''(\omega')}{\omega' - \omega} d\omega', \tag{4.1.8}$$

$$\epsilon''(\omega) = -\frac{1}{\pi} P.V. \int_{-\infty}^{\infty} \frac{\epsilon'(\omega') - \epsilon(\infty)}{\omega' - \omega} d\omega'. \tag{4.1.9}$$

Equations (8) or (9) are the **Kramers-Kronig relations** (Kramers 1927) that relate the real and the imaginary parts of $\epsilon(\omega)$. In other words, once the real or the imaginary part of $\epsilon(\omega)$ is determined, the other part is readily deducible from (8) or (9). Moreover, Equations (8) and (9) are also the consequences of the causality requirement on $\epsilon(t)$. They also constitute a transform pair known as the **Hilbert transform** (Titchmarsh 1948, Chapter 5; see also Appendix A). Consequently, this relationship applies to the Fourier transform of any causal function (Exercises 4.1, 4.2, and 4.3). That is to say, given any causal transfer function $H(\omega)$ which relates an input $X(\omega)$ to an output $Y(\omega)$ via $Y(\omega) = H(\omega)X(\omega)$, the real and imaginary parts of $H(\omega)$ are related via the Hilbert transforms as in (8) and (9).

§§4.1.2 Causality and Contour of Integration

As mentioned before, one way of obtaining the time domain solution is to Fourier inverse transform the frequency domain solution, i.e.,

$$\phi(t) = \frac{1}{2\pi} \int_{-\infty}^{\infty} d\omega \, e^{-i\omega t} \, \tilde{\phi}(\omega). \qquad (4.1.10)$$

More importantly, the Fourier inverse transform should not be performed at the expense of violating causality. To ensure this, the Fourier inversion contour in (10) is defined to be above all the singularities of $\tilde{\phi}(\omega)$. So, if $\tilde{\phi}(\omega) \to 0$ when $|\omega| \to \infty$, then for $t < 0$, the path of integration in (10) can be deformed to $\omega \to i\infty$. Consequently, $\phi(t) = 0$, for $t < 0$, ensuring that causality is not violated.

For $t > 0$, the path of integration can be deformed to $\omega \to -i\infty$, thereby picking up the singularity contributions of the integral (see Figure 4.1.2). By Jordan's lemma, the contribution of the contour at infinity vanishes. In this manner, (10) becomes

$$\phi(t) = \frac{1}{2\pi} \sum_{i=1}^{N} A_i e^{-i\omega_{pi} t} + \sum_{i=1}^{M} \frac{1}{2\pi} \int_{C_i} d\omega \, e^{-i\omega t} \tilde{\phi}(\omega). \qquad (4.1.11)$$

The first term in (11) is the residue contributions from the pole singularities at $\omega = \omega_{pi}$. But the second term is from the branch-point contributions. Furthermore, by letting $\omega = \omega_{bi} - i\eta$, a branch-cut integral can be altered to

$$\int_{C_i} d\omega \, e^{-i\omega t} \tilde{\phi}(\omega) = e^{-i\omega_{bi} t} \int_{0}^{\infty} i d\eta \, e^{-\eta t} \left[\tilde{\phi}_+(\omega_{bi} - i\eta) - \tilde{\phi}_-(\omega_{bi} - i\eta) \right]$$

$$= e^{-i\omega_{bi} t} f_i(t). \qquad (4.1.12)$$

where $\tilde{\phi}_+$ and $\tilde{\phi}_-$ are the values of $\tilde{\phi}$ on the two different shores of the branch cut. Therefore, Equation (11) is of the form

$$\phi(t) = \frac{1}{2\pi} \sum_{i=1}^{N} A_i e^{-i\omega_{pi} t} + \frac{1}{2\pi} \sum_{i=1}^{N} f_i(t) e^{-i\omega_{bi} t}. \qquad (4.1.13)$$

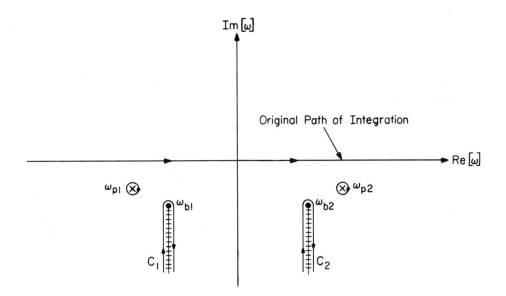

Figure 4.1.2 The original path of integration and the singularity contributions when we deform from the real axis to $\omega \to -i\infty$.

Now, if ω_{pi} and ω_{bi} are below the real axis as in the case of lossy media, then the above are decaying sinusoids. For large t, the solution will be dominated by the most slowly decaying sinusoids.

When a medium is active, however, some of the singularities may be above the real axis. In this case, the Fourier inversion contour will have to be above the real axis in order to preserve causality (Exercise 4.4).

§4.2 The Cagniard-de Hoop Method

One way to calculate the transient response of a source over a planarly layered medium is to Fourier inverse transform the frequency domain solutions obtained in Chapter 2 to obtain the time domain solution. Since this involves a double integral, it could be quite computationally intensive. When the medium is dispersionless, however, the Cagniard-de Hoop method provides the closed-form time domain solution of a line source over a layered medium, while for a point source, the solution is in terms of just one integral. As a result, this greatly expedites the computation of transient fields over a dispersionless layered medium. The Cagniard-de Hoop method was first suggested by Cagniard (1939), and later refined by de Hoop (1960). Moreover, a similar method was discussed by Doak (1952). In addition, the Cagniard-de Hoop method is also discussed in textbooks (Achenbach 1973; Aki and Richards 1980; Tygel and Hubral 1987).

Consider the frequency domain solution of a line-source response, or the two-dimensional scalar Green's function

$$\tilde{g}(\omega, \rho) = \frac{i}{4} H_0^{(1)}(k_0 \rho) = \frac{i}{4\pi} \int\limits_{-\infty}^{\infty} dk_x \frac{e^{ik_x x + ik_y |y|}}{k_y}, \qquad (4.2.1)$$

where $k_y = \sqrt{k_0^2 - k_x^2}$, $k_0 = \omega/c_0$, and c_0 is the velocity of the wave in the medium. Notice that the above is a solution to the equation

$$\left(\frac{\partial^2}{\partial x^2} + \frac{\partial^2}{\partial y^2} + \frac{\omega^2}{c_0^2} \right) \tilde{g}(\omega, \rho) = -\delta(x)\,\delta(y). \qquad (4.2.2)$$

Then, by Fourier transforming the above with respect to ω, we obtain the following equation,

$$\left(\frac{\partial^2}{\partial x^2} + \frac{\partial^2}{\partial y^2} - \frac{1}{c_0^2} \frac{\partial^2}{\partial t^2} \right) g(t, \rho) = -\delta(x)\,\delta(y)\,\delta(t), \qquad (4.2.3)$$

where

$$g(t, \rho) = \frac{1}{2\pi} \int\limits_{-\infty}^{\infty} d\omega\, e^{-i\omega t} \tilde{g}(\omega, \rho) = \frac{1}{\pi} \Re e \int\limits_{0}^{\infty} d\omega\, e^{-i\omega t} \tilde{g}(\omega, \rho). \qquad (4.2.4)$$

The last equality follows from the solution that $\tilde{g}(\omega, \rho) = \tilde{g}^*(-\omega, \rho)$ since $g(t, \rho)$ is real. We will next show how to obtain a closed-form solution for $g(t, \rho)$ via the Cagniard-de Hoop method.

§§4.2.1 Line Source in Free-Space —Two-Dimensional Green's Function

The Cagniard-de Hoop method derives closed-form solutions for the responses of a line source on top of a dispersionless, layered medium. First, consider the integral

$$\tilde{g}(\omega, \rho) = \frac{i}{4\pi} \int\limits_{-\infty}^{\infty} dk_x \frac{e^{ik_x x + ik_y |y|}}{k_y}. \qquad (4.2.5)$$

After letting $k_x = \omega s_x$, $k_0 = \omega s_0$, where $s_0 = 1/c_0$ is the slowness of the wave, Equation (5) becomes

$$\tilde{g}(\omega, \rho) = \frac{i}{4\pi} \int\limits_{-\infty}^{\infty} ds_x \frac{e^{i\omega\left(s_x x + \sqrt{s_0^2 - s_x^2}\,|y|\right)}}{\sqrt{s_0^2 - s_x^2}}. \qquad (4.2.6)$$

Then, by a change of variable such that

$$t = s_x x + (s_0^2 - s_x^2)^{\frac{1}{2}} |y|, \qquad (4.2.7)$$

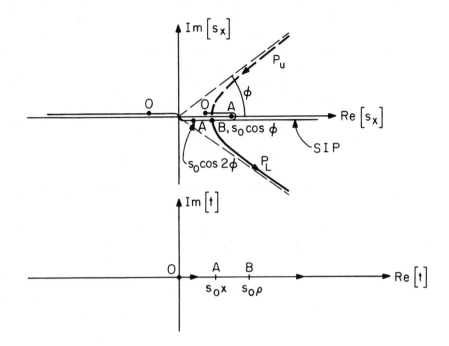

Figure 4.2.1 The contours on the complex s_x plane and complex t plane in the Cagniard-de Hoop method.

Equation (6) becomes

$$\tilde{g}(\omega, \rho) = \frac{i}{4\pi} \int\limits_{C'} dt \left(\frac{ds_x}{dt} \frac{1}{\sqrt{s_0^2 - s_x^2}} \right) e^{i\omega t}, \qquad (4.2.8)$$

which is very much like a Fourier transform integral except that the contour C' is not on the real axis, but it is the image of the real s_x axis on the complex t plane via the transformation (7).

One would like to convert (8) into an integration on the real axis so that the term in the parentheses in (8) can be related to $g(t, \rho)$. Hence, one should know the image of the real axis of the complex t plane on the complex s_x plane. After solving (7) for s_x in terms of t, we have

$$s_x = \frac{t}{\rho} \cos \phi \pm \sin \phi \left(s_0^2 - \frac{t^2}{\rho^2} \right)^{\frac{1}{2}}, \qquad (4.2.9)$$

where $\rho = \sqrt{x^2 + y^2}$, $\cos \phi = x/\rho$, and $\sin \phi = y/\rho$. Notice that there are two values of s_x for every value of t. Moreover, the mappings of the points O, A, and B from the real t axis to their two values on the complex s_x plane are as shown in Figure 4.2.1. Furthermore, when $t = s_0\rho$, the two values in (9) reduce to one value. This happens at B, where the two points meet. However, they split again for larger t and map into P_U and P_L in Figure 4.2.1. Nevertheless, the path of integration can be deformed from the Sommerfeld integration path to the hyperbola P defined by Equation (9). This hyperbola is the mapping of the real t axis, where $s_0\rho < t < \infty$, to the complex s_x plane (Exercise 4.5). Then, Equation (8) becomes

$$\tilde{g}(\omega, \rho) = \frac{i}{4\pi} \int_{P_U} ds_x \frac{e^{i\omega\left(s_x x + \sqrt{s_0^2 - s_x^2}\, |y|\right)}}{\sqrt{s_0^2 - s_x^2}} + \frac{i}{4\pi} \int_{P_L} ds_x \frac{e^{i\omega\left(s_x x + \sqrt{s_0^2 - s_x^2}\, |y|\right)}}{\sqrt{s_0^2 - s_x^2}}. \tag{4.2.10a}$$

Subsequently, via the transformation (9), the corresponding integral on the t plane is now

$$\tilde{g}(\omega, \rho) = \frac{i}{4\pi} \int_{s_0\rho}^{\infty} dt \left[\left(\frac{ds_x}{dt} \frac{1}{\sqrt{s_0^2 - s_x^2}} \right)_L - \left(\frac{ds_x}{dt} \frac{1}{\sqrt{s_0^2 - s_x^2}} \right)_U \right] e^{i\omega t}, \tag{4.2.10b}$$

where the subscripts U and L indicate mapping of the integration along P_U and P_L respectively; hence, the values of the associated terms have to be evaluated accordingly. Since it can be shown that

$$\left(\frac{ds_x}{dt} \right)_U = \left(\frac{ds_x}{dt} \right)^*, \qquad (s_0^2 - s_x^2)^{\frac{1}{2}}_U = (s_0^2 - s_x^2)^{\frac{1}{2}*}_L, \tag{4.2.11}$$

Equation (10) can be written as

$$\tilde{g}(\omega, \rho) = -\frac{1}{2\pi} \int_{s_0\rho}^{\infty} dt \Im m \left[\frac{ds_x}{dt} \frac{1}{\sqrt{s_0^2 - s_x^2}} \right]_L e^{i\omega t}. \tag{4.2.12}$$

The above is just a Fourier transform integral. Therefore, we conclude that

$$g(t, \rho) = -\frac{1}{2\pi} \Im m \left[\frac{ds_x}{dt} \frac{1}{\sqrt{s_0^2 - s_x^2}} \right]_L H(t - s_0\rho), \tag{4.2.13}$$

where $H(t)$ is a Heaviside step function, and s_x is a function of t as given in (9). More specifically, for $t > s_0\rho$,

$$s_x = \frac{t}{\rho} \cos\phi \pm i \sin\phi \left(\frac{t^2}{\rho^2} - s_0^2 \right)^{\frac{1}{2}}, \tag{4.2.14a}$$

$$\sqrt{s_0^2 - s_x^2} = \frac{t}{\rho} \sin\phi \mp i \cos\phi \left(\frac{t^2}{\rho^2} - s_0^2 \right)^{\frac{1}{2}}, \tag{4.2.14b}$$

and

$$\frac{ds_x}{dt} = \frac{\cos\phi}{\rho} \pm i \frac{t \sin\phi}{\rho^2(\frac{t^2}{\rho^2} - s_0^2)^{\frac{1}{2}}} = \frac{\pm i \sqrt{s_0^2 - s_x^2}}{(t^2 - s_0^2\rho^2)^{\frac{1}{2}}}. \tag{4.2.14c}$$

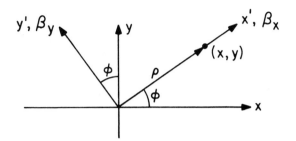

Figure 4.2.2 The Cagniard-de Hoop transformation.

The $+$ and $-$ signs are for P_U and P_L respectively. Therefore,

$$g(t, \rho) = \frac{1}{2\pi(t^2 - s_0^2\rho^2)^{\frac{1}{2}}} H(t - s_0\rho). \qquad (4.2.15)$$

But since $g(t, \rho)$ is cylindrically symmetric, the above can be more readily derived by setting $y = 0$, and $x = \rho$ in (6). The general derivation of (15), however, familiarizes us with the manipulations of the Cagniard-de Hoop method. Note that the contour deformation in (10a) to keep the argument of the exponent real has the same effect as setting $y = 0$, $x = \rho$ in (6).

Notice that the impulse response due to a line source in two dimensions has an infinite tail even though the exciting source is an impulse which has a finite time duration. The reason being that a line source is infinite in length. Even though the source is excited by an impulse $\delta(t)$ at $t = 0$, and switched off for $t > 0$, the observation point at ρ, $z = 0$ receives radiation of the line source continuously from the nearest point on the line source to the farthest point on the line source, which is infinity. Moreover, the field from the farther points will arrive at the observation point at a later time, resulting in an infinite tail for the impulse response (see Exercise 4.6).

§§4.2.2 Point Source in Free-Space
—Three-Dimensional Green's Function

Even though the Fourier inverse transform of the three-dimensional scalar Green's function, $e^{ik_0r}/4\pi r$, is easily sought, it is useful to go through the mechanics of the Cagniard-de Hoop method applied to a point source. Furthermore, by understanding the procedure for this simpler problem, we will better understand the solution of a point source on top of a layered medium. First, consider the identity

$$\tilde{g}(\omega, r) = \frac{e^{ik_0r}}{4\pi r} = \frac{i}{8\pi^2} \int\!\!\!\int\limits_{-\infty}^{\infty} dk_x dk_y \frac{e^{ik_xx+ik_yy+ik_z|z|}}{k_z}. \qquad (4.2.16)$$

A coordinate rotation can be performed (see Figure 4.2.2) so that the observation point (x, y) lies on the x' axis of the rotated coordinates. By letting

$$k_x = \beta_x \cos \phi - \beta_y \sin \phi, \qquad (4.2.17a)$$

$$k_y = \beta_x \sin \phi + \beta_y \cos \phi, \qquad (4.2.17b)$$

then,

$$\tilde{g}(\omega, r) = \frac{i}{8\pi^2} \int_{-\infty}^{\infty} d\beta_y \int_{-\infty}^{\infty} d\beta_x \frac{e^{i\beta_x \rho + i\beta_z |z|}}{\beta_z}, \qquad (4.2.18)$$

where $\beta_z = (k_0^2 - \beta_x^2 - \beta_y^2)^{\frac{1}{2}}$. Note that the inner integral resembles that of a line source given by (5). Furthermore, by letting

$$\beta_x = \omega s_x, \quad \beta_y = \omega s_y, \qquad (4.2.19)$$

one obtains

$$\tilde{g}(\omega, r) = \frac{i\omega}{8\pi^2} \int_{-\infty}^{\infty} ds_y \int_{-\infty}^{\infty} ds_x \frac{e^{i\omega(s_x \rho + s_z |z|)}}{s_z}, \qquad (4.2.20)$$

where $s_z = \sqrt{s_0^2 - s_x^2 - s_y^2}$, $s_0 = 1/c_0$. At this point, it is more expedient to deform the ds_y integration to the imaginary axis, which is permitted by Jordan's lemma and Cauchy's theorem (Exercise 4.7). Consequently, after the contour deformation, and with a change of variable $s_y = -iq$, we have

$$\tilde{g}(\omega, r) = \frac{\omega}{8\pi^2} \int_{-\infty}^{\infty} dq \int_{-\infty}^{\infty} ds_x \frac{e^{i\omega(s_x \rho + s_z |z|)}}{s_z}, \qquad (4.2.21)$$

where $s_z = \sqrt{s_0^2 + q^2 - s_x^2}$. Observe that the inner integral is now identical in form to (6).

From this point onward, the procedure is the same as the line source case. Hence, the inner integral can be deformed to a path similar to that of Figure 4.2.1 to get

$$\tilde{g}(\omega, r) = -\frac{i\omega}{4\pi^2} \int_{-\infty}^{\infty} dq \int_{(s_0^2 + q^2)^{\frac{1}{2}} r}^{\infty} \frac{dt\, e^{i\omega t}}{\sqrt{t^2 - (s_0^2 + q^2) r^2}}, \qquad (4.2.22)$$

where $r = \sqrt{\rho^2 + z^2}$ and we have made use of the result of (15). Furthermore, after exchanging the order of integrations, we can show that

$$\tilde{g}(\omega, r) = -\frac{i\omega}{4\pi^2} \int_{s_0 r}^{\infty} dt\, e^{i\omega t} \int_{-\left(\frac{t^2}{r^2} - s_0^2\right)^{\frac{1}{2}}}^{\left(\frac{t^2}{r^2} - s_0^2\right)^{\frac{1}{2}}} dq \frac{1}{r\sqrt{\left(\frac{t^2}{r^2} - s_0^2\right) - q^2}}. \qquad (4.2.23)$$

Notice that the limits of integration in (23) have to be changed accordingly because the lower limit of integration in the inner integral in (22) is dependent on q. Moreover, the inner integral in (23) evaluates to π, and the above becomes

$$\tilde{g}(\omega, r) = -\frac{i\omega}{4\pi r} \int\limits_{s_0 r}^{\infty} dt\, e^{i\omega t}. \qquad (4.2.24)$$

Therefore, the Fourier inverse transform of $-\tilde{g}(\omega, r)/i\omega$ is

$$\frac{H(t - s_0 r)}{4\pi r}. \qquad (4.2.25)$$

Then, the Fourier inverse transform of $\tilde{g}(\omega, r)$ is just the derivative of the above, giving

$$g(t, r) = \frac{\delta(t - s_0 r)}{4\pi r}. \qquad (4.2.26)$$

Of course, this is also easily obtained by Fourier inverse transforming $e^{ik_0 r}/4\pi r$. Notice that, unlike the line source, the impulse response in three dimensions does not have an infinitely long tail when the exciting source has a finite time duration. Alternatively, Equation (26) can also be convolved with a line source to yield (15) (Exercise 4.8).

§§4.2.3 Line Source Over Half-Space—Transient Response

A time harmonic electric line source $\mathbf{J} = \hat{z} I\, \delta(x)\, \delta(y)$ generates a field that satisfies the equation

$$\left(\frac{\partial^2}{\partial x^2} + \frac{\partial^2}{\partial y^2} + \omega^2 \mu_1 \epsilon_1 \right) E_{1z} = -i\omega \mu_1 I\, \delta(x)\, \delta(y). \qquad (4.2.27)$$

The time harmonic field is then

$$E_{1z} = -\frac{\omega \mu_1 I}{4\pi} \int\limits_{-\infty}^{\infty} dk_x \frac{e^{ik_x x + ik_{1y}|y|}}{k_{1y}}, \qquad (4.2.28)$$

where $k_{1y} = \sqrt{k_1^2 - k_x^2}$. Moreover, when such a line source is placed over a half-space at $y = -d_1$, the solution is

$$E_{1z} = -\frac{\omega \mu_1 I}{4\pi} \int\limits_{-\infty}^{\infty} dk_x \frac{e^{ik_x x}}{k_{1y}} \left[e^{ik_{1y}|y|} + R_{12}^{TE} e^{ik_{1y}(y + 2d_1)} \right]. \qquad (4.2.29)$$

The transient response of the first term has been found previously. To find the transient response of the second term, we let $k_x = \omega s_x$, $k_i = \omega s_i$, where $s_i = 1/c_i = \sqrt{\mu_i \epsilon_i}$. Then the second term, corresponding to the reflected field, becomes

$$E_{1z}^R = -\frac{\omega \mu_1 I}{4\pi} \int\limits_{-\infty}^{\infty} ds_x \frac{R_{12}^{TE}}{s_{1y}} e^{i\omega[s_x x + s_{1y}(y + 2d_1)]}, \qquad (4.2.30)$$

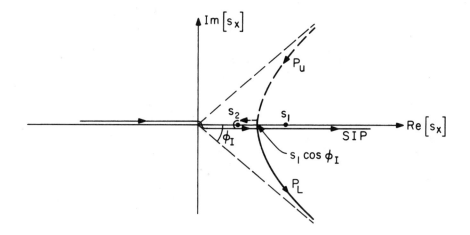

Figure 4.2.3 The contours on the complex s_x plane.

where $s_{iy} = \sqrt{s_i^2 - s_x^2}$, and

$$R_{12}^{TE} = \frac{\mu_2 s_{1y} - \mu_1 s_{2y}}{\mu_2 s_{1y} + \mu_1 s_{2y}}, \tag{4.2.31}$$

which is frequency independent. Notice that there are branch points at $s_x = \pm s_1$ and $s_x = \pm s_2$. Moreover, by letting

$$t = s_x x + s_{1y}(y + 2d_1), \tag{4.2.32}$$

we can solve for s_x in terms of t as in (9), yielding

$$s_x = \frac{t}{\rho_I} \cos \phi_I \pm i \sin \phi_I \left(\frac{t^2}{\rho_I^2} - s_1^2 \right)^{\frac{1}{2}}, \tag{4.2.33}$$

where $\rho_I = \sqrt{x^2 + (y + 2d_1)^2}$, $\phi_I = \cos^{-1}(x/\rho_I)$. Observe that if $s_2 < s_1$, then the branch points will be as shown in Figure 4.2.3. Moreover, if $s_1 \cos \phi_I > s_2$, the branch point s_2 will be encountered in deforming the contour to P from SIP. Furthermore, when $s_x = s_2$, $t = \tau = s_2 x + \sqrt{s_1^2 - s_2^2}(y + 2d_1)$. Then, similar to (10b), (30) becomes

$$E_{1z}^R = -\frac{\omega \mu_1 I}{4\pi} \int_{\tau}^{\infty} dt \left[\left(\frac{ds_x}{dt} \frac{R_{12}^{TE}}{s_{1y}} \right)_L - \left(\frac{ds_x}{dt} \frac{R_{12}^{TE}}{s_{1y}} \right)_U \right] e^{i\omega t}, \tag{4.2.34}$$

because now, the contour defined by s_2, $s_1 \cos \phi_I$, and P_U, or P_L on the complex s_x plane (see Figure 4.2.3) maps to $\tau < t < \infty$ on the real t axis on the complex t plane.

As in (14c), we have

$$\frac{1}{s_{1y}} \frac{ds_x}{dt} = \pm \frac{i}{(t^2 - s_1^2 \rho_I^2)^{\frac{1}{2}}}. \tag{4.2.35}$$

Indeed, for $t < s_1 \rho_I$ or $s_x < s_1 \cos \phi_I$,

$$\frac{1}{s_{1y}} \frac{ds_x}{dt}$$

is pure real. Furthermore, R_{12}^{TE} is the complex conjugate of each other above and below the real s_x axis, as s_{2y} is on two different Riemann sheets. Therefore,

$$\left(\frac{ds_x}{dt} \frac{R_{12}^{TE}}{s_{1y}} \right)_L - \left(\frac{ds_x}{dt} \frac{R_{12}^{TE}}{s_{1y}} \right)_U = \frac{2i}{(s_1^2 \rho_I^2 - t^2)^{1/2}} \Im m \left[R_{12}^{TE} \right]_L , \quad t < s_1 \rho_I. \tag{4.2.36}$$

But for $t > s_1 \rho_I$,

$$\frac{1}{s_{1y}} \frac{ds_x}{dt}$$

is pure imaginary, so that

$$\left(\frac{ds_x}{dt} \frac{R_{12}^{TE}}{s_{1y}} \right)_L - \left(\frac{ds_x}{dt} \frac{R_{12}^{TE}}{s_{1y}} \right)_U = \frac{-2i}{(t^2 - s_1^2 \rho_I^2)^{\frac{1}{2}}} \Re e \left[R_{12}^{TE} \right], \quad t > s_1 \rho_I. \tag{4.2.37}$$

By inspection, it follows that the Fourier inverse transform of $E_{1z}^R / [i\omega I(\omega)]$ is

$$f(t) = -\frac{\mu_1}{2\pi} \frac{\Im m \left[R_{12}^{TE} \right]_L}{(s_1^2 \rho_I^2 - t^2)^{\frac{1}{2}}} [H(t - \tau) - H(t - s_1 \rho_I)]$$

$$+ \frac{\mu_1}{2\pi} \frac{\Re e \left[R_{12}^{TE} \right]}{(t^2 - s_1^2 \rho_I^2)^{\frac{1}{2}}} H(t - s_1 \rho_I). \tag{4.2.38}$$

Then,

$$E_{1z}^R(t) = -\frac{\partial}{\partial t} I(t) * f(t), \tag{4.2.39}$$

where the asterisk implies a convolution. The first term in (38) is the contribution of the lateral wave or the head wave (see Figure 4.2.4). It would not be there if s_2 is not less than s_1. In fact, when $s_2 < s_1$, the wave in region 2 is faster than the wave in region 1. Hence, a wave through a refracted path, or the head wave as shown in Figure 4.2.4, reaches the receiver before the reflected wave. As a result, the first term in (38) exists when $\tau < s_1 \rho_I$, where $\tau = s_2 x + \sqrt{s_1^2 - s_2^2}\,(y + 2d_1)$ is the time taken by the head wave to traverse

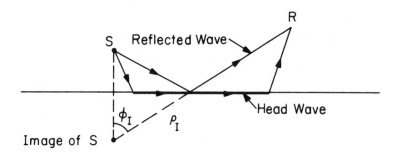

Figure 4.2.4 The paths for the arrivals of the direct wave and the head wave.

the refracted path shown in Figure 4.2.4. But when $\tau > s_1\rho_I$, only the direct reflection in the second term of (38) is observed.

§§4.2.4 Dipole Over Half-Space—Transient Response

The time harmonic response of a vertical electric dipole (VED) on top of a half-space is given by[1]

$$E_{1z} = \frac{I(\omega)\ell}{4\pi\omega\epsilon_1} \int_0^\infty dk_\rho \frac{k_\rho^3}{k_{1z}} J_0(k_\rho\rho) \left[e^{ik_{1z}|z|} + R_{12}^{TM} e^{ik_{1z}z + 2ik_{1z}d_1} \right], \qquad (4.2.40)$$

where R_{12}^{TM} is the Fresnel reflection coefficient. Here, the integral has been folded so that it ranges from 0 to ∞. Since a VED produces only an H_ϕ component of the magnetic field, we will use (2.3.17) to arrive at

$$H_{1\phi} = \frac{iI(\omega)\ell}{4\pi} \int_0^\infty dk_\rho \frac{k_\rho^2}{k_{1z}} J_1(k_\rho\rho) \left[e^{ik_{1z}|z|} + R_{12}^{TM} e^{ik_{1z}z + 2ik_{1z}d_1} \right]. \qquad (4.2.41)$$

In addition, the above could be written as

$$H_{1\phi} = -\frac{\partial}{\partial\rho} \frac{iI(\omega)\ell}{4\pi} \int_0^\infty dk_\rho \frac{k_\rho}{k_{1z}} J_0(k_\rho\rho) \left[e^{ik_{1z}|z|} + R_{12}^{TM} e^{ik_{1z}z + 2ik_{1z}d_1} \right].$$

$$(4.2.42)$$

In order to apply the Cagniard-de Hoop method, we write the integral in (42) in Cartesian coordinates using the Weyl identity given in Chapter 2

---

[1] Here, we have adapted a coordinate system where the interface is at $z = -d_1$, instead of at $y = -d_1$ as in the previous subsection. A similar solution to this problem has also been given by de Hoop and Frankena (1960).

[Equation (2.2.27)],

$$P = \frac{iI(\omega)\ell}{8\pi^2} \iint\limits_{-\infty}^{\infty} dk_y dk_x \frac{e^{ik_x x + ik_y y}}{k_{1z}} \left[e^{ik_{1z}|z|} + R_{12}^{TM} e^{ik_{1z}z + 2ik_{1z}d_1} \right]. \quad (4.2.43)$$

Note that the first integral is similar to (16), and the general solution is then

$$P_D = \frac{I(t - s_1 r)\ell}{4\pi r}, \quad (4.2.44)$$

where $s_1 = 1/c_1 = \sqrt{\mu_1\epsilon_1}$, and $I(t)$ is the Fourier inverse transform of $I(\omega)$.

For the reflected wave term in (43), a coordinate rotation gives

$$P_R = \frac{iI(\omega)\ell}{8\pi^2} \iint\limits_{-\infty}^{\infty} d\beta_y d\beta_x \frac{e^{i\beta_x \rho}}{\beta_{1z}} R_{12}^{TM} e^{i\beta_{1z}(z + 2d_1)}, \quad (4.2.45)$$

where $\beta_{1z} = \sqrt{k_1^2 - \beta_x^2 - \beta_y^2}$ and

$$R_{12}^{TM} = \frac{\epsilon_2\beta_{1z} - \epsilon_1\beta_{2z}}{\epsilon_2\beta_{1z} + \epsilon_1\beta_{2z}},$$

$$\beta_{2z} = \sqrt{k_2^2 - \beta_x^2 - \beta_y^2}.$$

By letting $\beta_x = \omega s_x$, $\beta_y = \omega s_y$, one obtains

$$P_R = \frac{i\omega I(\omega)\ell}{8\pi^2} \int\limits_{-\infty}^{\infty} ds_y \int\limits_{-\infty}^{\infty} ds_x \frac{e^{i\omega[s_x\rho + s_{1z}(z + 2d_1)]}}{s_{1z}} R_{12}^{TM}, \quad (4.2.45a)$$

where $R_{12}^{TM} = (\epsilon_2 s_{1z} - \epsilon_1 s_{2z})/(\epsilon_2 s_{1z} + \epsilon_1 s_{2z})$, $s_{iz} = \sqrt{s_i^2 - s_x^2 - s_y^2}$, $s_i = \sqrt{\mu_i\epsilon_i}$. Moreover, by virtue of Jordan's lemma and Cauchy's theorem, one can deform the ds_y integral to the imaginary axis, and with a change of variable $s_y = -iq$, as in (21), one obtains (see Exercise 4.9)

$$P_R = \frac{\omega I(\omega)\ell}{8\pi^2} \int\limits_{-\infty}^{\infty} dq \int\limits_{-\infty}^{\infty} ds_x \frac{R_{12}^{TM}}{s_{1z}} e^{i\omega[s_x\rho + s_{1z}(z + 2d_1)]}, \quad (4.2.46)$$

where $s_{iz} = \sqrt{\alpha_i^2 - s_x^2}$, $\alpha_i^2 = s_i^2 + q^2$. After applying the transformation

$$t = s_x\rho + s_{1z}(z + 2d_1) \quad (4.2.47)$$

as in (7), and inverting (47) as in (9), one obtains

$$s_x = \frac{t}{r_I}\sin\theta_I \pm i\cos\theta_I \left(\frac{t^2}{r_I^2} - s_1^2 - q^2\right)^{\frac{1}{2}}. \quad (4.2.48)$$

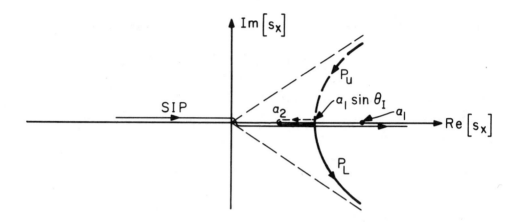

Figure 4.2.5 Contours on the complex s_x plane for the point source case.

where $r_I = \sqrt{\rho^2 + (z + 2d_1)^2}$, $\theta_I = \sin^{-1}(\rho/r_I)$. Notice now the branch-point singularities are at $s_x = \pm\alpha_1$ and $s_x = \pm\alpha_2$. Moreover, if $s_2 < s_1$, the branch-point singularities are as shown in Figure 4.2.5 with the contour P, which is the image of the real-t axis on the complex s_x plane.

When the Sommerfeld integration path is deformed to the P contour, we encounter the α_2 singularity. Furthermore, the point $s_x = \alpha_2$ corresponds to $t = \tau = \alpha_2\rho + \sqrt{s_1^2 - s_2^2}\,(z + 2d_1)$. Therefore, (46), whose inner integral is similar to (30), becomes

$$P_R = \frac{\omega I(\omega)\ell}{8\pi^2} \int\limits_{-\infty}^{\infty} dq \left[\int\limits_{\tau}^{\infty} dt \left(\frac{ds_x}{dt} \frac{R_{12}^{TM}}{s_{1z}} \right)_L e^{i\omega t} - \int\limits_{\tau}^{\infty} dt \left(\frac{ds_x}{dt} \frac{R_{12}^{TM}}{s_{1z}} \right)_U e^{i\omega t} \right].$$

$$(4.2.49)$$

The above can be shown to be the same as

$$P_R = \frac{-i\omega I(\omega)\ell}{4\pi^2} \int\limits_{-\infty}^{\infty} dq \int\limits_{\tau}^{\infty} dt\, \Im m \left(\frac{ds_x}{dt} \frac{R_{12}^{TM}}{s_{1z}} \right)_L e^{i\omega t}. \qquad (4.2.50)$$

Exchanging the order of integrations as in (23), one obtains

$$P_R = \frac{-i\omega I(\omega)\ell}{4\pi^2} \int\limits_{t_0}^{\infty} dt\, e^{i\omega t} \int\limits_{-q_0(t)}^{q_0(t)} dq\, \Im m \left(\frac{ds_x}{dt} \frac{R_{12}^{TM}}{s_{1z}} \right)_L, \qquad (4.2.51)$$

where $t_0 = s_2\rho + \sqrt{s_1^2 - s_2^2}\,(z + 2d_1)$, and $q_0(t) = \frac{1}{\rho}\{[t - \sqrt{s_1^2 - s_2^2}\,(z + 2d_1)]^2 - s_2^2\rho^2\}^{\frac{1}{2}}$. Therefore, the Fourier inverse transform of $P_R(\omega)/[-i\omega I(\omega)]$ is

$$f(t) = \frac{\ell}{4\pi^2} \int_{-q_0(t)}^{q_0(t)} dq\, \Im m \left(\frac{ds_x}{dt} \frac{R_{12}^{TM}}{s_{1z}} \right)_L H(t - t_0). \qquad (4.2.52)$$

From the above, one derives that

$$P_R(t) = \frac{\partial}{\partial t} I(t) * f(t), \qquad (4.2.53)$$

$$H_{1\phi R}(t) = -\frac{\partial}{\partial \rho} P_R(t). \qquad (4.2.54)$$

The integrand of (52) could be simplified as in (38). Equation (52) involves only a single integral, and it results in a great savings in computation time for calculating the transient response of a point source over a dispersionless medium. Such a technique can be applied to dipoles of other orientations on top of a half-space (Exercise 4.10).

§4.3 Multi-interface Problems

The Cagniard-de Hoop method cannot be used directly for a layered medium, because the generalized reflection coefficient for a layered medium is frequency dispersive. For example, for a three-layer medium,

$$\tilde{R}_{12} = R_{12} + \frac{T_{12}R_{23}T_{21}e^{2ik_{2y}(d_2-d_1)}}{1 - R_{21}R_{23}e^{2ik_{2y}(d_2-d_1)}}, \qquad (4.3.1)$$

where d_1 is the position of the first interface and d_2 is the position of the second interface, $R_{ij} = (p_j k_{iy} - p_i k_{jy})/(p_j k_{iy} + p_i k_{jy})$, and $p_j = \mu_j$ for TE waves and $p_j = \epsilon_j$ for TM waves. Even when each of the layers is dispersionless, a transformation of the type $k_{iy} = \omega s_{iy}$ where $s_{iy} = \sqrt{s_i^2 - s_x^2}$, does not make the generalized reflection coefficient frequency independent. Hence, this is unlike the dispersionless half-space where the reflection coefficient becomes frequency independent under a similar transformation. The success of the Cagniard-de Hoop method is contingent on the frequency independence of these coefficients, and hence, cannot be applied to a layered medium directly.

As an alternative, if (1) is expanded in terms of a ray series, then

$$\tilde{R}_{12} = R_{12} + \sum_{m=1}^{\infty} T_{12}R_{23}^m R_{21}^{m-1} T_{21} e^{2im\omega s_{2y}(d_2-d_1)}. \qquad (4.3.2)$$

Note that the frequency dependence of each of the terms is now factorizable into the exponential term. Now, if (2) is substituted into (4.2.30), the m-th term of the series becomes

$$E_{1z}^{Rm} = -\frac{\omega\mu_1 I}{4\pi} \int\limits_{-\infty}^{\infty} ds_x \frac{1}{s_{1y}} T_{12} R_{23}^m R_{21}^{m-1} T_{21} e^{i\omega[s_x x + s_{1y}(y+2d_1)+2ms_{2y}(d_2-d_1)]}.$$

$$(4.3.3)$$

To make this look like a Fourier inverse transform, we let

$$t = s_x x + s_{1y}(y + 2d_1) + 2ms_{2y}(d_2 - d_1). \qquad (4.3.4)$$

Subsequently, the P contour can be found on the complex s_x plane by keeping t real and solving this equation. Moreover, if $y = 2d_1 = 0$, then P is found easily as before, and s_x can be solved in terms of t (Exercise 4.11). But if y and d_1 are not zeroes, then (4) is an implicit equation for s_x in terms of t and it needs to be solved numerically. Once the P contour is found, however, the Cagniard-de Hoop method proceeds as before. The same idea could be extended to the N interface problem by expanding the reflection coefficients in terms of general ray series. But the P contour in this case could be quite complex (see Helmberger 1968; Aki and Richards 1980).

§4.4 Direct Inversion

Even though the Cagniard-de Hoop method is applicable to a point source by casting the point source into an integral summation of line sources, there is a more direct way of obtaining the point source solution. This method was introduced by Strick (1959), and extended by Helmberger (1968), and Gilbert and Helmberger (1972). It was also described by Aki and Richards (1980).

Before proceeding, let us study the integral identity

$$\frac{i}{4} \int\limits_{-\infty}^{\infty} d\omega \, e^{-i\omega t} H_0^{(1)}(\omega s_\rho \rho) = \frac{H(t - s_\rho \rho)}{[t^2 - s_\rho^2 \rho^2]^{\frac{1}{2}}}. \qquad (4.4.1)$$

Such an integral identity follows from the result of the Cagniard-de Hoop method in (4.2.15), which is the Fourier inverse transform of (4.2.1). At $\omega = 0$, however, $H_0^{(1)}(\omega s_\rho \rho)$ has a logarithmic singularity. Therefore, the integral in (1) is undefined if the integration path is on the real axis. Consequently, causality has to be invoked to properly define the integration path on the complex ω plane.

Notice that when $\omega \to +i\infty$,

$$H_0^{(1)}(\omega s_\rho \rho) \sim \sqrt{\frac{2}{i\omega s_\rho \rho}} e^{i\omega s_\rho \rho}, \qquad (4.4.2)$$

becomes exponentially small, and $e^{-i\omega t}$ becomes exponentially large. Nevertheless, if $s_\rho \rho > t$, the integrand is still exponentially small, and we can

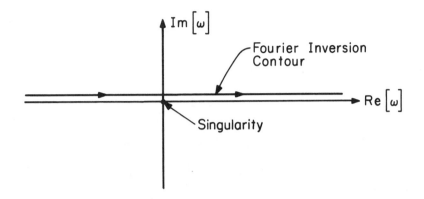

Figure 4.4.1 The Fourier inversion path has to be above all singularities in order to satisfy causality.

deform the path of integration to the upper-half plane. Then, by virtue of
Jordan's lemma, this integral evaluates to zero. But for $s_\rho\rho > t$, (1) must be
zero due to causality, implying that no singularities should be enclosed in the
course of this deformation. Consequently, in order to satisfy causality, the
path of integration in (1) must always be above the singularity at the origin,
which is the only singularity in the integrand (see Figure 4.4.1).

When $t > s_\rho\rho$, we can deform the contour downward to obtain the right-
hand side of (1). The integral identity in (1) holds true even for complex t
and complex $s_\rho\rho$. In this case,

$$\frac{i}{4} \int_{-\infty}^{\infty} d\omega\, e^{-i\omega t} H_0^{(1)}(\omega s_\rho\rho) = \frac{H[\Re e(t - s_\rho\rho)]}{[t^2 - s_\rho^2\rho^2]^{\frac{1}{2}}}. \tag{4.4.3}$$

Next, we shall see how the above integral identity can be applied to the
direct inversion of the frequency domain solution

$$I(\omega) = \frac{i}{2} \int_{-\infty}^{\infty} dk_\rho \frac{k_\rho}{k_{1z}} H_0^{(1)}(k_\rho\rho) e^{ik_{1z}z}. \tag{4.4.4}$$

If now, we change the variable by letting $k_\rho = \omega s_\rho$, $k_{1z} = \omega s_{1z}$ where
$s_{1z} = \sqrt{s_1^2 - s_\rho^2}$, $s_1 = 1/c_1$, then,

$$I(\omega) = \frac{i\omega}{2} \int_{-\infty}^{\infty} ds_\rho \frac{s_\rho}{s_{1z}} H_0^{(1)}(\omega s_\rho\rho) e^{i\omega s_{1z}z}. \tag{4.4.5}$$

Moreover, the Sommerfeld integration path can first be deformed to the P contour as in the Cagniard-de Hoop method, or

$$F(\omega) = \frac{I(\omega)}{i\omega} = \frac{1}{2} \int_P ds_\rho \frac{s_\rho}{s_{1z}} H_0^{(1)}(\omega s_\rho \rho) e^{i\omega s_{1z} z}. \qquad (4.4.6)$$

The P contour in this case is described by the equation

$$s_\rho \rho + s_{1z} z = \tau, \qquad (4.4.7)$$

where τ is real. Equation (7) can be inverted to yield

$$s_\rho = \frac{\tau}{r} \sin\theta \pm i \cos\theta \left(\frac{\tau^2}{r^2} - s_1^2 \right)^{1/2}, \qquad (4.4.8a)$$

$$s_{1z} = \frac{\tau}{r} \cos\theta \mp i \sin\theta \left(\frac{\tau^2}{r^2} - s_1^2 \right)^{1/2}. \qquad (4.4.8b)$$

Consequently, we can Fourier inverse transform (6) to obtain

$$F(t) = \frac{1}{4\pi} \int_P ds_\rho \frac{s_\rho}{s_{1z}} \int_{-\infty}^{\infty} d\omega \, e^{-i\omega(t - s_{1z} z)} H_0^{(1)}(\omega s_\rho \rho), \qquad (4.4.9)$$

where the order of integration has been exchanged. By virtue of (3), one then obtains

$$F(t) = \frac{1}{i\pi} \int_P ds_\rho \frac{s_\rho}{s_{1z}} \frac{H(t - s_{1z} z - s_\rho \rho)}{[(t - s_{1z} z)^2 - s_\rho^2 \rho^2]^{\frac{1}{2}}}. \qquad (4.4.10)$$

Notice that $s_{1z} z + s_\rho \rho$ is real on the P contour, so that we can eliminate the "$\Re e$" sign used in (3). Furthermore, the contour in (10) can be folded by combining the parts associated with P_U and P_L. Then, after changing the variable of integration to τ, we finally have

$$F(t) = \frac{2}{\pi} \Im m \int_{s_1 r}^{t} d\tau \frac{ds_\rho}{d\tau} \frac{s_\rho}{s_{1z}} \frac{1}{[(t - s_{1z} z)^2 - s_\rho^2 \rho^2]^{1/2}}. \qquad (4.4.11)$$

The upper limit in the integration is due to the Heaviside step function. Notice that the above is a finite-range integration. Moreover, it can be performed in closed form when the medium is homogeneous (see Exercise 4.12). When the point source is placed on top of a half-space, the analysis proceeds in a similar manner (Exercise 4.12); however, the finite-range integral equivalent to (11) will have to be evaluated numerically.

The idea of the Cagniard-de Hoop method has been adapted to lossy media (Kuester 1984) to expedite transient field computation albeit with the retention of double integrals. In addition, the idea of double deformation has also been used to expedite the numerical computation of such Fourier integrals (Tsang and Kong 1979; Ezzedine et al. 1982; Poh and Kong 1987).

§4.5 Numerical Integration of Fourier Integrals

In problems involving dispersions or loss, the Cagniard-de Hoop method cannot be used. One way of obtaining a transient solution then is to integrate (4.1.10) numerically. In this case, the function $\phi(\omega)$ is a frequency domain solution that may be obtained via the methods described in Chapter 2 or Chapter 3. Moreover, numerical integration has the advantage that a transient solution can be obtained once the frequency domain solution is obtained.

An integral like (4.1.10) may be subject to the contour deformation technique to accelerate its convergence. Even though it accelerates the convergence of the integral, numerical integration via contour deformation usually has to be tailored accordingly, depending on the parameter variations in the geometry. The computer code thus generated may not be robust. A robust way of generating a computer code is to just numerically integrate the integral on an integration path parallel to or along the real axis. This can usually withstand much parameter variation in a code.

Equation (4.1.10) implies that

$$\phi(\omega) = \int_{-\infty}^{\infty} dt\, e^{i\omega t}\phi(t). \tag{4.5.1}$$

But since $\phi(t)$ is real valued, it implies that $\phi(-\omega^*) = \phi^*(\omega)$ (in particular, $\phi(-\omega) = \phi^*(\omega)$ if ω is real). Furthermore, since there could be singularities on the real axis for lossless media (or even above the real axis for active media), the path of integration would have to be slightly above the real axis. Consequently, with this path of integration, (4.1.10) can be written as

$$\phi(t) = \frac{1}{\pi}\Re e \int_{i\Delta}^{i\Delta+\infty} d\omega\, e^{-i\omega t}\phi(\omega). \tag{4.5.2}$$

The above follows from folding the $d\omega$ integral using the fact $\phi(-\omega^*) = \phi^*(\omega)$. Hence, the infinite integral can be converted into a semi-infinite integral. Next, by letting $\omega = i\Delta + \omega'$, (2) becomes

$$\phi(t) = \frac{1}{\pi}e^{\Delta t}\int_{0}^{\infty} d\omega'\, \Re e[e^{-i\omega' t}\phi(\omega' + i\Delta)]. \tag{4.5.3}$$

But when the integration path is close to a singularity, the integrand will be sharply peaked in the vicinity of the singularity. This sharp peak will exacerbate the error in the numerical integration of the integral. Thus, to avoid this sharp peak, it is expedient to choose an integration path that is not too close to the singularities of the integrand. Consequently, with $\Delta > 0$, the path of integration can be chosen to be above any singularities that may lie on the real axis. By doing so, the integrand in (3) becomes smoother and the numerical evaluation of (3) becomes simpler and converges faster, because the error committed in a numerical scheme is proportional to the derivatives of the integrand. In addition, adaptive integration techniques mentioned in Subsection 2.7.3 of Chapter 2 can also be used to evaluate such integrals. For instance, when t is large, the integrand in (3) is rapidly oscillating. Methods like Filon quadrature (Hildebrand 1974) may then be used to accelerate convergence in this case. If, however, the value of $\phi(t)$ is required at many t's, fast Fourier transform (FFT; see Oppenheim and Schafer 1975) can be used to evaluate the integral (3). Furthermore, the cost of high sampling rate required for large t to reduce aliasing can be avoided using interpolation techniques (Anderson and Chew 1989).

Very often, the field at a receiver can be decomposed into a direct wave (primary field) contribution plus a reflected wave contribution. Because of the singular nature of its field, a direct wave term usually has slow numerical convergence. One way of avoiding the numerical evaluation of the direct wave term is then to evaluate it in closed form. In the case of a dispersionless medium, this is obviated by the Cagniard-de Hoop method. When a medium is lossy, however, a closed-form solution still exists despite the inapplicability of the Cagniard-de Hoop method. The following subsection describes how the closed-form solution for the direct field can be derived in a lossy medium. Similar solutions are also described by Courant and Hilbert (1962).

§§4.5.1 *Direct Field in a Lossy Medium—Two-Dimensional Case*

Consider first a lossy wave equation in two dimensions,

$$\left(\nabla_s^2 - \frac{1}{c^2}\frac{\partial^2}{\partial t^2} - \frac{1}{c^2\tau}\frac{\partial}{\partial t}\right) g(x,y,t) = -\delta(x)\,\delta(y)\,\delta(t) \qquad (4.5.4)$$

where $\tau = \epsilon/\sigma$, and

$$\nabla_s^2 = \frac{\partial^2}{\partial x^2} + \frac{\partial^2}{\partial y^2}.$$

With the transformation $g(x,y,t) = e^{-t/2\tau}\hat{g}(x,y,t)$, the above becomes

$$\left(\nabla_s^2 - \frac{1}{c^2}\frac{\partial^2}{\partial t^2} + \frac{1}{4c^2\tau^2}\right) \hat{g}(x,y,t) = -\delta(x)\,\delta(y)\,\delta(t). \qquad (4.5.5)$$

By letting $z = ict$, $k^2 = 1/4c^2\tau^2$, the left-hand side of (5) is similar to the left-hand side of the Helmholtz equation

$$\left(\nabla_s^2 + \frac{\partial^2}{\partial z^2} + k^2\right) \phi(x,y,z) = 0. \qquad (4.5.6)$$

Since the homogeneous solution to (6) is of the form $e^{\pm ikr}/r$, where $r = \sqrt{x^2 + y^2 + z^2}$, a possible homogeneous solution to (5) is (for $\rho \neq 0$, $ct \neq 0$)

$$\hat{g}(x,y,t) = \left[Ae^{-\frac{i}{2c\tau}\sqrt{\rho^2 - c^2t^2}} + Be^{\frac{i}{2c\tau}\sqrt{\rho^2 - c^2t^2}}\right]/\sqrt{\rho^2 - c^2t^2}, \qquad (4.5.7)$$

where $\rho = \sqrt{x^2 + y^2}$. However, the solution to (5) must be real valued, since (5) is an equation in the time domain. Note that this can be obtained by taking either the real part or the imaginary part of (7), since either of these is still a solution to (5). Hence, after taking the imaginary part of (7) and requiring that the solution be zero for $ct < \rho$ to satisfy causality, we find that $A = B$ and that they have to be pure-real. Consequently, a possible real-valued solution to (5) is given by

$$\hat{g}(x,y,t) = \begin{cases} 0, & ct < \rho, \\ 2A\dfrac{\cosh\left(\frac{1}{2c\tau}\sqrt{c^2t^2 - \rho^2}\right)}{\sqrt{c^2t^2 - \rho^2}}, & ct > \rho. \end{cases} \qquad (4.5.8)$$

Therefore,

$$\hat{g}(x,y,t) = H(ct - \rho)2A\frac{\cosh\left(\sqrt{c^2t^2 - \rho^2}/2c\tau\right)}{\sqrt{c^2t^2 - \rho^2}}, \qquad (4.5.9)$$

where $H(t)$ is the Heaviside step function. Next, A can be obtained by substituting (9) into (5) and matching the singularities on both sides of Equation (5) (see Exercise 4.13). By doing so, we find that $A = c/4\pi$. Consequently,

$$g(x,y,t) = cH(ct - \rho)e^{-t/2\tau}\frac{\cosh\left(\sqrt{c^2t^2 - \rho^2}/2c\tau\right)}{2\pi\sqrt{c^2t^2 - \rho^2}}. \qquad (4.5.10)$$

Notice that the above reduces to the Cagniard-de Hoop solution [see Equation (4.2.15)] when $\tau \to \infty$ for the lossless case. Moreover, only an axisymmetric solution is considered in (7) because the source in (5) is axially symmetric. Even though the $\partial^n/\partial t^n$ derivative of (7) is also a solution to (5), they are not considered because these solutions give rise to higher order singularities at the origin.

§§4.5.2 Direct Field in a Lossy Medium—Three-Dimensional Case

The lossy wave equation in three dimensions is

$$\left(\nabla^2 - \frac{1}{c^2}\frac{\partial^2}{\partial t^2} - \frac{1}{c^2\tau}\frac{\partial}{\partial t}\right)g(x,y,z,t) = -\delta(x)\,\delta(y)\,\delta(z)\,\delta(t), \qquad (4.5.11)$$

where $\nabla^2 = \frac{\partial^2}{\partial x^2} + \frac{\partial^2}{\partial y^2} + \frac{\partial^2}{\partial z^2}$. Letting $g(x,y,z,t) = e^{-t/2\tau}\hat{g}(x,y,z,t)$, we arrive at

$$\left(\nabla^2 - \frac{1}{c^2}\frac{\partial^2}{\partial t^2} + \frac{1}{4c^2\tau^2}\right)\hat{g}(x,y,z,t) = -\delta(x)\,\delta(y)\,\delta(z)\,\delta(t). \qquad (4.5.12)$$

But the previous equation can be reduced to two dimensions by letting

$$\hat{g}(x, y, z, t) = \frac{1}{2\pi} \int\limits_{-\infty}^{\infty} dk_z e^{ik_z z} \tilde{g}(x, y, k_z, t), \qquad (4.5.13a)$$

and

$$\delta(z) = \frac{1}{2\pi} \int\limits_{-\infty}^{\infty} dk_z e^{ik_z z}. \qquad (4.5.13b)$$

As a result, (12) becomes

$$\left(\nabla_s^2 - \frac{1}{c^2} \frac{\partial^2}{\partial t^2} + \frac{1}{4c^2\tau^2} - k_z^2 \right) \tilde{g}(x, y, k_z, t) = -\delta(x)\,\delta(y)\,\delta(t). \qquad (4.5.14)$$

Now, if we compare the above with (5), it is apparent that a possible homogeneous solution to (14) for $\rho \neq 0$ and $ct \neq 0$ is

$$\tilde{g}(x, y, k_z, t) = \frac{A}{\sqrt{\rho^2 - c^2 t^2}} e^{i\sqrt{\frac{1}{4c^2\tau^2} - k_z^2} \left(\rho^2 - c^2 t^2 \right)^{1/2}}. \qquad (4.5.15)$$

Next, by using (13), a homogeneous solution to (12) is then

$$\hat{g}(x, y, z, t) = \frac{A}{2\pi\sqrt{\rho^2 - c^2 t^2}} \int\limits_{-\infty}^{\infty} dk_z e^{ik_z z + i\sqrt{\frac{1}{4c^2\tau^2} - k_z^2} \left(\rho^2 - c^2 t^2 \right)^{1/2}}$$

$$= -\frac{A}{2\pi ict} \frac{\partial}{\partial ct} \int\limits_{-\infty}^{\infty} dk_z \frac{e^{ik_z z + i\sqrt{\frac{1}{4c^2\tau^2} - k_z^2} \left(\rho^2 - c^2 t^2 \right)^{1/2}}}{\sqrt{\frac{1}{4c^2\tau^2} - k_z^2}}. \qquad (4.5.16)$$

On invoking the identity (4.2.1), we then have

$$\hat{g}(x, y, z, t) = -\frac{A}{2ict} \frac{\partial}{\partial ct} H_0^{(1)} \left(\frac{1}{2c\tau} \sqrt{r^2 - c^2 t^2} \right), \qquad (4.5.17)$$

where $r = \sqrt{\rho^2 + z^2}$.

The above yields only one homogeneous solution of (12), but (12) should have two linearly independent homogeneous solutions. By taking a linear combination of the homogeneous solution as in (7), it is easy to show that the other linearly independent homogeneous solution is obtained by replacing the Hankel function in (17) with a Bessel function. Consequently, the most general homogeneous solution to (12) is

$$\hat{g}(x, y, z, t) = \frac{1}{ct} \frac{\partial}{\partial ct} \left[A' H_0^{(1)} \left(\frac{1}{2c\tau} \sqrt{r^2 - c^2 t^2} \right) + B' J_0 \left(\frac{1}{2c\tau} \sqrt{r^2 - c^2 t^2} \right) \right]. \qquad (4.5.18)$$

However, $\hat{g}(x, y, z, t)$ has to be real valued because (12) is a time domain equation. Furthermore, the real or the imaginary part of (18) is still a solution to (12). Therefore, by taking the imaginary part of (18) and demanding that it should be zero for $ct < r$ to satisfy causality, we conclude that $A' = 0$, and B' is pure-real. Consequently, we have

$$\hat{g}(x, y, z, t) = \frac{1}{ct}\frac{\partial}{\partial ct}\begin{cases} 0, & ct < r. \\ B'I_0\left(\frac{1}{2c\tau}\sqrt{c^2t^2 - r^2}\right), & ct > r, \end{cases} \qquad (4.5.19)$$

Notice that in (18), when $ct > r$, the arguments of J_0 and $H_0^{(1)}$ are pure imaginary. Therefore, they can be replaced by the modified Bessel functions I_0 and K_0 (Abramowitz and Stegun 1965, reference for Chapter 3). As a result,

$$\hat{g}(x, y, z, t) = \frac{B'}{ct}\frac{\partial}{\partial ct}H(ct - r)I_0\left(\frac{1}{2c\tau}\sqrt{c^2t^2 - r^2}\right). \qquad (4.5.20)$$

B' can be found by substituting (20) into (12) and matching the singularities on both sides of the equation. Alternatively, it can be determined by requiring (20) to reduce to (4.2.26) when $\tau \to \infty$ for the lossless case. Hence, $B' = c/4\pi$. Consequently,

$$g(x, y, z, t) = \frac{e^{-t/2\tau}}{4\pi t}\frac{\partial}{\partial ct}H(ct - r)I_0\left(\frac{1}{2c\tau}\sqrt{c^2t^2 - r^2}\right). \qquad (4.5.21)$$

Note that we have derived the transient solution in a two-dimensional lossy medium in Equation (7) by using the solution of the Helmholtz wave equation in three dimensions. Analogously, we could have derived (21) using the solutions of the Helmholtz wave equation in four dimensions instead of using the above procedure. In fact, it can be shown that solutions of the Helmholtz wave equation in four dimensions are expressible in terms of Bessel functions. Using them will lead directly to (21) (see Exercise 4.14).

§4.6 Finite-Difference Method

The methods discussed in the previous sections allow one to derive the transient solution in a layered medium of the type discussed in Chapters 2 and 3. To obtain the transient solution of the wave equation for a more general, inhomogeneous medium, a numerical method has to be used. The finite-difference method, a numerical method, is particularly suitable for solving transient problems. Moreover, it is quite versatile, and given the present computer technology, it has been used with great success in solving many practical problems.

In the finite-difference method, continuous space-time is replaced with a discrete space-time. Then, in the discrete space-time, partial differential equations are replaced with difference equations. These difference equations are readily implemented on a digital computer. Furthermore, an iterative scheme can be implemented without having to solve large matrices, resulting in a great savings in computer time. More recently, the development of parallel processor architectures in computers has also further enhanced the efficiency of the finite-difference scheme. This method is also described in numerous works (see, for example, Potter 1973; Richtmeyer and Morton 1967; Ames 1977; Taflove 1988).

§§4.6.1 The Finite-Difference Approximation

Consider first a scalar wave equation of the form

$$\frac{1}{c^2(\mathbf{r})}\frac{\partial^2}{\partial t^2}\phi(\mathbf{r},t) = \mu(\mathbf{r})\nabla\cdot\mu^{-1}(\mathbf{r})\nabla\phi(\mathbf{r},t). \tag{4.6.1}$$

Then, the time derivative can be approximated in many ways. For example,

Forward difference: $\dfrac{\partial\phi(\mathbf{r},t)}{\partial t} = \dfrac{\phi(\mathbf{r},t+\Delta t)-\phi(\mathbf{r},t)}{\Delta t},$ (4.6.2a)

Backward difference: $\dfrac{\partial\phi(\mathbf{r},t)}{\partial t} = \dfrac{\phi(\mathbf{r},t)-\phi(\mathbf{r},t-\Delta t)}{\Delta t},$ (4.6.2b)

Central difference: $\dfrac{\partial\phi(\mathbf{r},t)}{\partial t} = \dfrac{\phi(\mathbf{r},t+\frac{\Delta t}{2})-\phi(\mathbf{r},t-\frac{\Delta t}{2})}{\Delta t},$

(4.6.2c)

where Δt is a small number. Of the three methods of approximating the time derivative, the central-difference scheme is the best approximation, as is evident in Figure 4.6.1. The errors in the forward and backward differences are $O(\Delta t)$ (first-order error) while the central-difference approximation has an error $O[(\Delta t)^2]$ (second-order error). This can be easily illustrated by Taylor series expanding the right-hand side of (2) (Exercise 4.15; also see Abramowitz and Stegun 1965, reference for Chapter 3).

Consequently, using the central-difference formula twice, we arrive at

$$\frac{\partial^2}{\partial t^2}\phi(\mathbf{r},t) = \frac{\partial}{\partial t}\left[\frac{\phi(\mathbf{r},t+\frac{\Delta t}{2})-\phi(\mathbf{r},t-\frac{\Delta t}{2})}{\Delta t}\right]$$
$$= \frac{\phi(\mathbf{r},t+\Delta t)-2\phi(\mathbf{r},t)+\phi(\mathbf{r},t-\Delta t)}{(\Delta t)^2}. \tag{4.6.3}$$

Next, if the function $\phi(\mathbf{r},t)$ is put on discrete time steps on the t axis, such that $\phi(\mathbf{r},t) = \phi(\mathbf{r},l\Delta t) = \phi^l(\mathbf{r})$, where l is an integer, Equation (3) then becomes

$$\frac{\partial^2}{\partial t^2}\phi(\mathbf{r},t) = \frac{\phi^{l+1}(\mathbf{r})-2\phi^l(\mathbf{r})+\phi^{l-1}(\mathbf{r})}{(\Delta t)^2}. \tag{4.6.4}$$

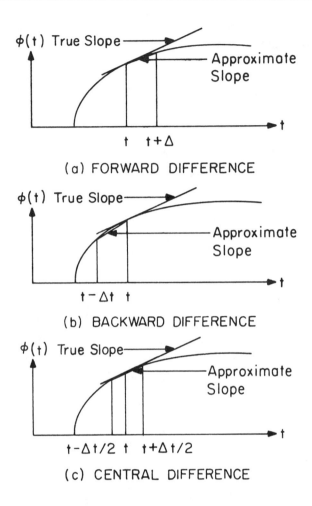

Figure 4.6.1 Different finite-difference approximations.

With this notation, Equation (1) becomes a time-stepping formula,

$$\phi^{l+1}(\mathbf{r}) = c^2(\mathbf{r})(\Delta t)^2 \mu(\mathbf{r}) \nabla \cdot \mu^{-1}(\mathbf{r}) \nabla \phi^l(\mathbf{r}) + 2\phi^l(\mathbf{r}) - \phi^{l-1}(\mathbf{r}). \qquad (4.6.5)$$

Therefore, given the knowledge of $\phi(\mathbf{r}, t)$ at $t = l\Delta t$ and $t = (l-1)\Delta t$, one can deduce $\phi(\mathbf{r}, t)$ at $t = (l+1)\Delta t$. In other words, given the initial values of $\phi(\mathbf{r}, t)$ at, for example, $t = 0$ and $t = \Delta t$, $\phi(\mathbf{r}, t)$ can be deduced for all subsequent times, provided that the time-stepping formula is stable.

At this point, the right-hand side of (5) involves the space derivatives. There exist a plethora of ways to approximate the right-hand side of (5) numerically, examples of which are the finite-element method, Galerkin's method, the method of weighted residuals, the point-matching method (also known as the method of collocation), and the finite-difference method. Here, we shall illustrate the use of the finite-difference approximation for the right-hand side of (5). Before proceeding further, note that the space derivatives on the right-hand side in Cartesian coordinates are

$$\mu(\mathbf{r})\nabla \cdot \mu^{-1}(\mathbf{r})\nabla\phi(\mathbf{r}) = \mu\frac{\partial}{\partial x}\mu^{-1}\frac{\partial}{\partial x}\phi + \mu\frac{\partial}{\partial y}\mu^{-1}\frac{\partial}{\partial y}\phi + \mu\frac{\partial}{\partial z}\mu^{-1}\frac{\partial}{\partial z}\phi. \quad (4.6.6)$$

Then, one can approximate

$$\frac{\partial}{\partial z}\phi(x,y,z) = \frac{1}{\Delta z}\left[\phi\left(x,y,z+\frac{\Delta z}{2}\right) - \phi\left(x,y,z-\frac{\Delta z}{2}\right)\right], \quad (4.6.7)$$

using central differencing. Consequently,

$$\frac{\partial}{\partial z}\mu^{-1}\frac{\partial}{\partial z}\phi(x,y,z) = \frac{1}{(\Delta z)^2}\left\{\mu^{-1}\left(z+\frac{\Delta z}{2}\right)\phi(x,y,z+\Delta z)\right.$$
$$- \left[\mu^{-1}\left(z+\frac{\Delta z}{2}\right) + \mu^{-1}\left(z-\frac{\Delta z}{2}\right)\right]\phi(x,y,z)$$
$$\left.\mu^{-1}\left(z-\frac{\Delta z}{2}\right)\phi(x,y,z-\Delta z)\right\}. \quad (4.6.8)$$

Furthermore, after letting $\phi(x,y,z) = \phi_{m,n,p}$, $\mu(x,y,z) = \mu_{m,n,p}$, on a discretized grid point at $x = m\Delta x$, $y = n\Delta y$, $z = p\Delta z$, we have

$$\frac{\partial}{\partial z}\mu^{-1}\frac{\partial}{\partial z}\phi(x,y,z) = \frac{1}{(\Delta z)^2}\left[\mu^{-1}_{m,n,p+\frac{1}{2}}\phi_{m,n,p+1}\right.$$
$$\left. - \left(\mu^{-1}_{m,n,p+\frac{1}{2}} + \mu^{-1}_{m,n,p-\frac{1}{2}}\right)\phi_{m,n,p} + \mu^{-1}_{m,n,p-\frac{1}{2}}\phi_{m,n,p-1}\right]. \quad (4.6.9)$$

With similar approximations to the other terms in (6), Equation (5) becomes

$$\phi^{l+1}_{m,n,p} = (\Delta t)^2 c^2_{m,n,p}\mu_{m,n,p}\left\{\frac{1}{(\Delta x)^2}\left[\mu^{-1}_{m+\frac{1}{2},n,p}\phi^l_{m+1,n,p}\right.\right.$$
$$\left. - \left(\mu^{-1}_{m+\frac{1}{2},n,p} + \mu^{-1}_{m-\frac{1}{2},n,p}\right)\phi^l_{m,n,p} + \mu^{-1}_{m-\frac{1}{2},n,p}\phi^l_{m-1,n,p}\right]$$
$$+ \frac{1}{(\Delta y)^2}\left[\mu^{-1}_{m,n+\frac{1}{2},p}\phi^l_{m,n+1,p} - \left(\mu^{-1}_{m,n+\frac{1}{2},p} + \mu^{-1}_{m,n-\frac{1}{2},p}\right)\phi^l_{m,n,p} + \mu^{-1}_{m,n-\frac{1}{2},p}\phi^l_{m,n-1,p}\right]$$
$$+ \frac{1}{(\Delta z)^2}\left[\mu^{-1}_{m,n,p+\frac{1}{2}}\phi^l_{m,n,p+1} - \left(\mu^{-1}_{m,n,p+\frac{1}{2}} + \mu^{-1}_{m,n,p-\frac{1}{2}}\right)\phi^l_{m,n,p}\right.$$
$$\left.\left. + \mu^{-1}_{m,n,p-\frac{1}{2}}\phi^l_{m,n,p-1}\right]\right\} + 2\phi^l_{m,n,p} - \phi^{l-1}_{m,n,p}. \quad (4.6.10)$$

The above can be readily implemented on a computer for time stepping. Notice however, that the use of central differencing results in the evaluation of medium property μ at half grid points. This is inconvenient, as the introduction of half grid points increases computer memory requirements. Hence, it is customary to approximate

$$\mu_{m+\frac{1}{2},n,p} \simeq \frac{1}{2}(\mu_{m+1,n,p} + \mu_{m,n,p}), \quad (4.6.11a)$$

$$\mu_{m+\frac{1}{2},n,p} + \mu_{m-\frac{1}{2},n,p} \simeq 2\mu_{m,n,p}, \qquad (4.6.11\text{b})$$

and so on. Moreover, it is easy to show that the errors in the above approximations are of second order by Taylor series expansions, if μ is a smooth function of space (Exercise 4.24).

For a homogeneous medium, with $\Delta x = \Delta y = \Delta z = \Delta s$, (10) becomes

$$\phi_{m,n,p}^{l+1} = \left(\frac{\Delta t}{\Delta s}\right)^2 c^2 \left[\phi_{m+1,n,p}^l + \phi_{m-1,n,p}^l + \phi_{m,n+1,p}^l + \phi_{m,n-1,p}^l + \phi_{m,n,p+1}^l\right.$$
$$\left. + \phi_{m,n,p-1}^l - 6\phi_{m,n,p}^l\right] + 2\phi_{m,n,p}^l - \phi_{m,n,p}^{l-1}. \quad (4.6.12)$$

Notice then that with the central-difference approximation, the value of $\phi_{m,n,p}^{l+1}$ is dependent only on $\phi_{m,n,p}^l$, and its nearest neighbors, $\phi_{m\pm1,n,p}^l$, $\phi_{m,n\pm1,p}^l$, $\phi_{m,n,p\pm1}^l$, and $\phi_{m,n,p}^{l-1}$. Moreover, in the finite-difference scheme outlined above, no matrix inversion is required at each time step. Such a scheme is also known as an explicit scheme. The use of an explicit scheme is a major advantage of the finite-difference method compared to the finite-element method. Consequently, in order to update N grid points using (10) or (12), $O(N)$ multiplications are required for each time step. In comparison, $O(N^3)$ multiplications are required to invert an $N \times N$ full matrix, e.g., in the method of moments (Chapter 5).

§§4.6.2 Stability Analysis

The implementation of the finite-difference scheme does not always lead to a stable scheme. Hence, in order for the solution to converge, the time-stepping scheme must at least be stable. Consequently, it is useful to find the condition under which a numerical finite-difference scheme is stable. To do this, one performs the von Neumann stability analysis (von Neumann 1943) on Equation (12).

As shown in Chapter 2, any wave can be expanded in terms of a spectrum of plane waves. So if a scheme is not stable for a plane wave, it would not be stable for any wave. Consequently, to perform the stability analysis, we let

$$\phi(x,y,z,t) = A(t)e^{ik_x x + ik_y y + ik_z z}, \qquad (4.6.13)$$

which denotes a plane wave. In discretized form, it is just

$$\phi_{m,n,p}^l = A^l e^{ik_x m\Delta s + ik_y n\Delta s + ik_z p\Delta s}. \qquad (4.6.13\text{a})$$

Using (13a), it is easy to show that

$$\phi_{m+1,n,p}^l - 2\phi_{m,n,p}^l + \phi_{m-1,n,p}^l = 2[\cos(k_x\Delta s) - 1]\phi_{m,n,p}^l$$
$$= -4\sin^2\left(\frac{k_x\Delta s}{2}\right)\phi_{m,n,p}^l.$$
$$(4.6.14)$$

Furthermore, one can assume that

$$A^{l+1} = gA^l. \tag{4.6.15}$$

For the solution (13a) to be stable, the requirement is for $|g| \leq 1$. Then, using (14) and (15) in (12), and repeating (14) for the n and p variables, one obtains

$$(g - 2 + g^{-1})\phi^l_{m,n,p} = -4 \left(\frac{\Delta t}{\Delta s}\right)^2 c^2 \left[\sin^2\left(\frac{k_x \Delta s}{2}\right) + \sin^2\left(\frac{k_y \Delta s}{2}\right)\right. $$
$$\left. + \sin^2\left(\frac{k_z \Delta s}{2}\right)\right] \phi^l_{m,n,p}$$
$$= -4r^2 s^2 \phi^l_{m,n,p}, \tag{4.6.16}$$

where

$$r = \left(\frac{\Delta t}{\Delta s}\right) c, \qquad s^2 = \sin^2\left(\frac{k_x \Delta s}{2}\right) + \sin^2\left(\frac{k_y \Delta s}{2}\right) + \sin^2\left(\frac{k_z \Delta s}{2}\right). $$
$$\tag{4.6.16a}$$

Equation (16) implies that, for nonzero $\phi^l_{m,n,p}$,

$$g^2 - 2g + 4r^2 s^2 g + 1 = 0, \tag{4.6.17}$$

or that

$$g = (1 - 2r^2 s^2) \pm 2rs\sqrt{(r^2 s^2 - 1)}. \tag{4.6.18}$$

But if

$$r^2 s^2 < 1, \tag{4.6.19}$$

the second term in (18) is pure imaginary, and

$$|g|^2 = (1 - 2r^2 s^2)^2 + 4r^2 s^2(1 - r^2 s^2) = 1, \tag{4.6.20}$$

or stability is ensured. Since $s^2 \leq 3$ for all k_x, k_y, and k_z, from (19), one concludes that the general condition for stability is

$$r < \frac{1}{\sqrt{3}}, \qquad \text{or} \qquad \Delta t < \frac{\Delta s}{c\sqrt{3}}. \tag{4.6.21}$$

It is clear from the above analysis that for an n-dimensional problem,

$$\Delta t < \frac{\Delta s}{c\sqrt{n}}. \tag{4.6.22}$$

One may ponder on the meaning of this inequality further: but it is only natural that the time step Δt has to be bounded from above. Otherwise, one arrives at the ridiculous notion that the time step can be arbitrarily large thus violating causality. Moreover, if the grid points of the finite-difference scheme are regarded as a simple cubic lattice, then the distance $\Delta s/\sqrt{n}$ is also the distance between the closest lattice planes through the simple cubic lattice. Notice that the time for the wave to travel between these two lattice planes is $\Delta s/(c\sqrt{n})$. Consequently, the stability criterion (22) implies that the time step Δt has to be less than the shortest travel time for the wave between the lattice planes in order to satisfy causality. In other words, if the wave is time-stepped ahead of the time on the right-hand side of (22), instability ensues. The above is also known as the CFL (Courant, Friedrichs, and Lewy 1928) stability criterion. It could be easily modified for $\Delta x \neq \Delta y \neq \Delta z$ (see Exercise 4.16).

We can also ascertain how the computation time of a finite-difference scheme grows as the number of unknowns. Consider a box of N grid points on each side used to represent a region. Then, the total number of grid points is N^n where n is the dimension of the box. Since at each time step, one has to calculate (12) for all the grid points in the box, the number of multiplications needed is proportional to N^n at each time step. Moreover, to reach the solution at $t = T$, the total number of time steps is equal to $T/\Delta t$. From (22), $1/\Delta t$ is proportional to $1/\Delta s = N/L$ where L is the length of the side of the box. Therefore, the total time steps needed to reach T is proportional to N, and consequently, the total number of computations required is proportional to N^{n+1}. Next, we define $M = N^n$, where M is the total number of grid points in the box; then the total number of computations required is proportional to $M^{(n+1)/n}$. Hence, this ascertains how the computation time grows with the number of grid points in different dimensions n. For example, to arrive at the solution at time T, the number of computations is proportional to M^2 in one dimension, $M^{3/2}$ in two dimensions, and $M^{4/3}$ in three dimensions, where M is the number of grid points which is proportional to N^n.

If the wave equation is lossy, for instance, due to conductive losses, then Equation (1) is modified accordingly to become

$$\frac{1}{c^2}\frac{\partial^2}{\partial t^2}\phi(\mathbf{r}, t) + \mu\sigma\frac{\partial}{\partial t}\phi(\mathbf{r}, t) = \mu\nabla \cdot \mu^{-1}\nabla\phi(\mathbf{r}, t). \tag{4.6.23}$$

Using a central-differencing scheme for the above, it follows that the CFL stability criterion still holds true (Exercise 4.17; also see Richtmeyer and Morton 1967).

When propagation effects or the displacement currents are ignored in (23), it becomes

$$\mu\sigma\frac{\partial}{\partial t}\phi(\mathbf{r}, t) = \mu\nabla \cdot \mu^{-1}\nabla\phi(\mathbf{r}, t). \tag{4.6.24}$$

The above is also known as the diffusion equation—the same equation that governs heat flow. But now, if a central-difference scheme is used for time and the same scheme is also used for space (the central-difference in time here is also known as leapfrogging), then the time-stepping formula that results is always unstable (see Exercise 4.18). On the other hand, if forward difference in time and central difference in space are used (also known as the

Euler scheme), the scheme is stable if (see Potter 1973)

$$\Delta t < \frac{\mu\sigma\Delta s^2}{2n}. \tag{4.6.25}$$

The time $\tau = \mu\sigma\Delta s^2/n$ is just the diffusion time for the field to diffuse from one lattice plane to another lattice plane. Stability requires that the time step be no larger than half this diffusion time.

A simple analysis of the Euler scheme shows that the number of computations grows as $M^{(n+2)/n}$, where M is the total number of grid points and n is the dimension of the problem. Hence, the computational effort in solving the diffusion equation grows faster than that in solving the wave equation. The reason being that Equation (25) implies that Δt is very small when Δs is small. One way to surmount this inefficiency in the Euler scheme is to use an implicit scheme known as the Crank-Nicholson (1947) method to solve (24) (see Exercise 4.18).

An alternative way of making the diffusion equation stable is to add a wavelike term to (24), resulting in

$$n\left(\frac{\Delta t}{\Delta s}\right)^2 \frac{\partial^2}{\partial t^2}\phi(\mathbf{r},t) + \mu\sigma\frac{\partial}{\partial t}\phi(\mathbf{r},t) = \mu\nabla\cdot\mu^{-1}\nabla\phi(\mathbf{r},t). \tag{4.6.26}$$

Now, when a central-difference scheme is applied to the above equation, the CFL stability criterion is always satisfied. Therefore, the above equation becomes unconditionally stable for all Δt (Exercise 4.19). If, however, Δt is made too large, the purpose is defeated because the diffusion equation changes into a wavelike equation, thereby altering the physics of the problem. This is especially true at an early time, when the propagation effect is more important. This method of stabilizing the diffusion equation is also known as the DuFort-Frankel (1953) scheme.

§§4.6.3 Grid-Dispersion Error

When a finite-difference scheme is stable, it still may not produce good results because of the errors in the scheme. Hence, it is useful to ascertain the errors in terms of the size of the grid and the time step. An easy error to analyze is the *grid-dispersion error*. In a homogeneous, dispersionless medium, all plane waves propagate with the same phase velocity. However, in the finite-difference approximation, all plane waves will not propagate at the same velocity. As a consequence, a pulse in the time domain, which is a linear superposition of plane waves, will be distorted if the dispersion introduced by the finite-difference scheme is intolerable.

To ascertain the grid-dispersion error, we assume that $A^l = e^{-i\omega l\Delta t}$ in (13a). In this case, the left-hand side of (16) becomes

$$\left(e^{-i\omega\Delta t} - 2 + e^{+i\omega\Delta t}\right)\phi_{m,n,p}^l = -4\sin^2\left(\frac{\omega\Delta t}{2}\right)\phi_{m,n,p}^l. \tag{4.6.27}$$

Then, from Equation (16), it follows that

$$\sin\left(\frac{\omega\Delta t}{2}\right) = rs, \tag{4.6.28}$$

where r and s are given in (16a). Now, Equation (28) governs the relationship between ω and k_x, k_y, and k_z in the finite-difference scheme, and hence, is a dispersion relation. But if a medium is homogeneous, it is well known that (1) has a plane-wave solution of the type given by (13) where

$$\omega = c\sqrt{k_x^2 + k_y^2 + k_z^2}. \tag{4.6.29}$$

Equation (28) departs from Equation (29) as a consequence of the finite-difference approximation. This departure gives rise to errors, which are the consequence of grid dispersion. For example, for a constant c, (29) states that the phase velocities of plane waves of different wavelengths are the same. However, this is not true for (28), as shall be shown.

Assuming s small, (28) can be written as

$$\frac{\omega\Delta t}{2} = \sin^{-1} rs \cong rs + \frac{r^3 s^3}{6}. \tag{4.6.30}$$

When Δs is small, using the small argument approximation for the sine function, one obtains from (16a)

$$s \simeq \frac{\Delta s}{2}(k_x^2 + k_y^2 + k_z^2)^{1/2}\left[1 - \frac{\Delta s^2}{24}\left(\frac{k_x^4 + k_y^4 + k_z^4}{k_x^2 + k_y^2 + k_z^2}\right)\right]. \tag{4.6.31}$$

Equation (30) then becomes

$$\frac{\omega\Delta t}{2} \simeq r\frac{\Delta s}{2}(k_x^2 + k_y^2 + k_z^2)^{1/2}\left[1 - \frac{\Delta s^2}{24}\frac{k_x^4 + k_y^4 + k_z^4}{k_x^2 + k_y^2 + k_z^2} + \frac{r^2\Delta s^2}{24}(k_x^2 + k_y^2 + k_z^2)\right]. \tag{4.6.32}$$

Since $r = c\Delta t/\Delta s$, on comparing (32) with (29), we note that they differ by a factor of

$$1 - \frac{\Delta s^2}{24}\left(\frac{k_x^4 + k_y^4 + k_z^4}{k_x^2 + k_y^2 + k_z^2}\right) + \frac{r^2\Delta s^2}{24}(k_x^2 + k_y^2 + k_z^2). \tag{4.6.32a}$$

This implies that the phase velocities for different plane waves are different. Moreover, since k_x, k_y, and k_z are proportional to $2\pi/\lambda$, this error is proportional to $(\Delta s/\lambda)^2$. Hence, waves with shorter wavelengths would suffer more errors compared to waves with longer wavelengths. In addition, when k_x, k_y, and k_z are written in terms of direction cosines, the error is also dependent on the direction of the plane wave. The reason is that since the grid points are rectilinear, and not spherically symmetric, they have preferred directions for the wave to propagate in. Hence, to reduce the grid dispersion error, it is necessary to have

$$\left(\frac{\Delta s}{\lambda}\right)^2 \ll 1. \qquad (4.6.33)$$

For a wave with frequency content at ω, the wavelength is related to k via

$$\frac{\omega}{c} = \frac{2\pi}{\lambda} = \sqrt{k_x^2 + k_y^2 + k_z^2}. \qquad (4.6.34)$$

Consequently, in order for the finite-difference scheme to propagate this frequency content accurately, the grid size must be much less than this wavelength. Furthermore, Δt must be chosen so that the CFL stability criterion is met. Hence, the rule of thumb is to choose about 10 to 20 grid points per wavelength.

§§4.6.4 The Yee Algorithm

The Yee algorithm (Yee 1966) is specially designed to solve vector electromagnetic field problems on a rectilinear grid. To derive it, Maxwell's equations are first written in Cartesian coordinates:

$$-\frac{\partial B_x}{\partial t} = \frac{\partial E_z}{\partial y} - \frac{\partial E_y}{\partial z}, \qquad (4.6.35\text{a})$$

$$-\frac{\partial B_y}{\partial t} = \frac{\partial E_x}{\partial z} - \frac{\partial E_z}{\partial x}, \qquad (4.6.35\text{b})$$

$$-\frac{\partial B_z}{\partial t} = \frac{\partial E_y}{\partial x} - \frac{\partial E_x}{\partial y}, \qquad (4.6.35\text{c})$$

$$\frac{\partial D_x}{\partial t} = \frac{\partial H_z}{\partial y} - \frac{\partial H_y}{\partial z} - J_x, \qquad (4.6.35\text{d})$$

$$\frac{\partial D_y}{\partial t} = \frac{\partial H_x}{\partial z} - \frac{\partial H_z}{\partial x} - J_y, \qquad (4.6.35\text{e})$$

$$\frac{\partial D_z}{\partial t} = \frac{\partial H_y}{\partial x} - \frac{\partial H_x}{\partial y} - J_z. \qquad (4.6.35\text{f})$$

After denoting $f(n\Delta x, m\Delta y, p\Delta z, l\Delta t) = f_{m,n,p}^l$, and replacing derivatives with central finite-differences in accordance with Figure 4.6.2, (35a) becomes

$$\frac{1}{\Delta t}\left[B_{x,m,n+\frac{1}{2},p+\frac{1}{2}}^{l+\frac{1}{2}} - B_{x,m,n+\frac{1}{2},p+\frac{1}{2}}^{l-\frac{1}{2}}\right] = \frac{1}{\Delta z}\left[E_{y,m,n+\frac{1}{2},p+1}^l - E_{y,m,n+\frac{1}{2},p}^l\right]$$
$$- \frac{1}{\Delta y}\left[E_{z,m,n+1,p+\frac{1}{2}}^l - E_{z,m,n,p+\frac{1}{2}}^l\right]. \qquad (4.6.36)$$

Moreover, the above can be repeated for (35b) and (35c). Notice that in Figure 4.6.2, the electric field is always assigned to the edge center of a cube, whereas the magnetic field is always assigned to the face center of a cube.

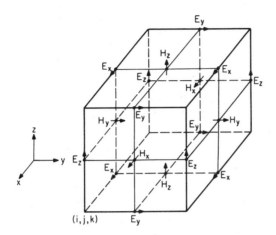

Figure 4.6.2 The assignment of fields on a grid in the Yee algorithm.

In fact, after multiplying (36) by $\Delta z \Delta y$, (36) is also the approximation of the integral forms of Maxwell's equations when applied at a face of a cube. By doing so, the left-hand side of (36) becomes

$$(\Delta y \Delta z / \Delta t) \left[B^{l+\frac{1}{2}}_{x,m,n+\frac{1}{2},p+\frac{1}{2}} - B^{l-\frac{1}{2}}_{x,m,n+\frac{1}{2},p+\frac{1}{2}} \right],$$

which is the time variation of the total flux through an elemental area $\Delta y \Delta z$. Moreover, by summing this flux on the six faces of the cube shown in Figure 4.6.2, and using the right-hand side of (36) and its equivalent, it can be shown that the magnetic flux adds up to zero. Hence, $\nabla \cdot \mathbf{B} = 0$ condition is satisfied within the numerical approximations of Yee's algorithm. Therefore, only two of the three finite-difference approximations of (35a) to (35c) are truly independent (Exercise 4.20).

Furthermore, a similar approximation of (35d) leads to

$$\frac{1}{\Delta t} \left[D^l_{x,m+\frac{1}{2},n,p} - D^{l-1}_{x,m+\frac{1}{2},n,p} \right] = \frac{1}{\Delta y} \left[H^{l-\frac{1}{2}}_{z,m+\frac{1}{2},n+\frac{1}{2},p} - H^{l-\frac{1}{2}}_{z,m+\frac{1}{2},n-\frac{1}{2},p} \right]$$

$$- \frac{1}{\Delta z} \left[H^{l-\frac{1}{2}}_{y,m+\frac{1}{2},n,p+\frac{1}{2}} - H^{l-\frac{1}{2}}_{y,m+\frac{1}{2},n,p-\frac{1}{2}} \right] - J^{l-\frac{1}{2}}_{x,m+\frac{1}{2},n,p}. \quad (4.6.37)$$

Also, similar approximations apply for (35e) and (35f). In addition, the above has an interpretation similar to (36) if one thinks in terms of a cube that is shifted by half a grid point in each direction. Hence, the approximations of (35d) to (35f) are consistent with the approximation of $\frac{\partial}{\partial t} \nabla \cdot \mathbf{D} = -\nabla \cdot \mathbf{J}$ (see Exercise 4.21). Again, if $\mathbf{J}$ is known in (35d) to (35f), then only two of the three equations are truly independent. This fact, together with the fact that (36) satisfies $\nabla \cdot \mathbf{B} = 0$, can be used to eliminate two of the six field components to economize on storage in the computer memory. In the above, $\mathbf{D} = \epsilon \mathbf{E}$ and $\mathbf{B} = \mu \mathbf{H}$. Since the magnetic field and the electric field are assigned on staggered grids, μ and ϵ may have to be assigned on staggered grids. This does not usually lead to serious problems if the grid size is small. Alternatively, (11) can be used to remove this problem.

By eliminating the $\mathbf{E}$ or the $\mathbf{H}$ field from the Yee algorithm, it can be shown that the Yee algorithm is equivalent to finite differencing the vector wave equation directly. Hence, the Yee algorithm is also constrained by the CFL stability criterion (see Exercise 4.22).

§4.7 Absorbing Boundary Conditions

In a finite-difference scheme, the grid that contains the points for finite differencing has to be of finite extent due to the finite size of computer memory and resource. For example, the finite-difference grid may lie in a box. In order to generate solutions that are unique, boundary conditions have to be imposed on the surface of this box. But in many applications, the media are of infinite extent. In this case, imposing boundary conditions on the surface of the box may give rise to reflections that are not reflective of the actual physical situation. Consequently, to simulate an infinite geometry, absorbing boundary conditions are imposed at the surface of the box to diminish these reflections. The use of absorbing boundary conditions reduces computer memory requirements since the size of the box can be made smaller.

An early viscous absorbing boundary condition was described by Lysmer and Kuhlemeyer (1969). We shall describe a few popular schemes below.

§§4.7.1 Engquist-Majda Absorbing Boundary Condition

This absorbing boundary condition was discussed by Clayton and Engquist (1977), Engquist and Majda (1977, 1979), Mur (1981), and Wagatha (1983). As shown in Chapter 2, an arbitrary wave can be expanded in terms of a spectrum of plane waves. Hence, absorbing boundary conditions can be derived first for plane waves of arbitrary incidence. If the absorbing boundary condition is made such that it is independent of the angle of incidence and wavelength of the plane wave, then it could be used for an arbitrary wave.

If a plane wave is normally incident on a plane surface, and the surface is perfectly absorbing, there will be no reflected wave. Next, notice that an equation that has a wave solution traveling only in the negative x direction is

$$\left[\frac{\partial}{\partial x} - \frac{1}{c} \frac{\partial}{\partial t} \right] \phi(x, t) = 0. \tag{4.7.1}$$

The above is also known as the advective equation, whose solution can be easily shown to be $f(x + ct)$, a left traveling wave. Hence, if we impose the condition (1) on a wave normally incident on the plane surface, the wave will not be reflected. Consequently, an absorbing boundary condition for a normally incident wave is

$$\frac{\partial}{\partial x}\phi(x,t)\Big|_{x=0} = \frac{1}{c}\frac{\partial}{\partial t}\phi(x,t)\Big|_{x=0}. \qquad (4.7.2)$$

One way to approximate the above is

$$\phi_1^l - \phi_0^l = \frac{\Delta x}{c\Delta t}\left(\phi_0^{l+1} - \phi_0^l\right), \qquad (4.7.3)$$

that is, using forward differencing for both space and time. Consequently,

$$\phi_0^{l+1} = \phi_0^l\left(1 - \frac{c\Delta t}{\Delta x}\right) + \frac{c\Delta t}{\Delta x}\phi_1^l. \qquad (4.7.4)$$

Now, if 0 is the grid point at $x = 0$, the above boundary condition allows us to find ϕ_0^l for all l. Note that if $\Delta x/c\Delta t = 1$, the above is just $\phi_0^{l+1} = \phi_1^l$, a simple formula denoting that the wave at 0 at time step $l+1$ is just the wave at 1 at time step l. In other words, the wave just shifts over one grid point to the left for each time step. Finally, one can show the above to be stable for $c\Delta t/\Delta x \leq 1$, which is also a causality requirement (see Exercise 4.23).

Another way to impose (1) to second-order accuracy is to impose (1) at $x = \frac{1}{2}\Delta x$. Then, at $x = \frac{1}{2}\Delta x$ and $t = \left(n + \frac{1}{2}\right)\Delta t$,

$$\frac{\partial}{\partial x}\phi(x,t)\Big|_{x=\frac{1}{2}\Delta x} \simeq \frac{1}{\Delta x}\left[\phi_1^{n+\frac{1}{2}} - \phi_0^{n+\frac{1}{2}}\right], \qquad (4.7.5a)$$

$$\frac{1}{c}\frac{\partial}{\partial t}\phi(x,t)\Big|_{x=\frac{1}{2}\Delta x} \simeq \frac{1}{c\Delta t}\left[\phi_{\frac{1}{2}}^{n+1} - \phi_{\frac{1}{2}}^n\right]. \qquad (4.7.5b)$$

In this manner, the finite-difference approximation is accurate to the second order. But since the values at the half grid points and half time steps are not available, it is expedient to approximate

$$\phi_m^{n+\frac{1}{2}} \simeq \frac{1}{2}(\phi_m^{n+1} + \phi_m^n), \qquad (4.7.6a)$$

$$\phi_{m+\frac{1}{2}}^n \simeq \frac{1}{2}(\phi_{m+1}^n + \phi_m^n). \qquad (4.7.6b)$$

The above approximation is also second-order accurate if $\phi(x,t)$ is a smooth function. This can be shown easily by Taylor series expansion (Exercise 4.24). Consequently, using (6) in (5) and (1), one obtains

$$\phi_0^{n+1} = \phi_1^n + \left(\frac{c\Delta t - \Delta x}{c\Delta t + \Delta x}\right)(\phi_1^{n+1} - \phi_0^n). \qquad (4.7.7)$$

Equation (7) allows one to update $\phi(x = 0)$ at a new time step given the knowledge of $\phi(x = 0)$ at a previous time step, and the knowledge of $\phi(x =$

Δx) at the old and the new time steps. Moreover, it can be shown to be unconditionally stable (Exercise 4.23).

The above boundary condition is correct only for a plane wave at normal incidence. Hence, the wave will still be reflected for an obliquely incident wave. Therefore, an absorbing boundary condition for which obliquely incident waves are also absorbed is needed.

For a plane wave at oblique incidence and traveling to the left, the equation that it satisfies in the frequency domain is

$$\frac{d}{dx}\tilde{\phi}(\mathbf{r},\omega) + ik_x(\omega)\tilde{\phi}(\mathbf{r},\omega) = 0, \tag{4.7.8}$$

where we have assumed $e^{-i\omega t}$ time dependence, and $k_x = \sqrt{\omega^2/c^2 - k_y^2 - k_z^2}$. When the wave is close to normal incidence, then $c^2/\omega^2(k_y^2 + k_z^2) \ll 1$, and we can approximate

$$k_x \simeq \frac{\omega}{c}\left[1 - \frac{1}{2}\frac{c^2}{\omega^2}\left(k_y^2 + k_z^2\right)\right]. \tag{4.7.9}$$

The above is also known as the **paraxial approximation**. On substituting (9) into (8) and multiplying the resultant equation throughout by $i\omega$, we have

$$i\omega\frac{d}{dx}\tilde{\phi}(\mathbf{r},\omega) - \frac{\omega^2}{c}\tilde{\phi}(\mathbf{r},\omega) + \frac{1}{2}c(k_y^2 + k_z^2)\tilde{\phi}(\mathbf{r},\omega) = 0. \tag{4.7.10}$$

Then, after rewriting the above in the time domain by replacing $i\omega$ with $-\partial/\partial t$, ik_y with $\partial/\partial y$ and ik_z with $\partial/\partial z$, one derives

$$\frac{\partial^2}{\partial x \partial t}\phi(\mathbf{r},t) - \frac{1}{c}\frac{\partial^2}{\partial t^2}\phi(\mathbf{r},t) + \frac{1}{2}c\left(\frac{\partial^2}{\partial y^2} + \frac{\partial^2}{\partial z^2}\right)\phi(\mathbf{r},t) = 0. \tag{4.7.11}$$

Note that if $\nabla_s^2\phi = (\frac{\partial^2}{\partial y^2} + \frac{\partial^2}{\partial z^2})\phi = 0$, then the above is just the time derivative of (1). Therefore, the finite-difference approximation of the above is related to the time derivative of (3), i.e.,

$$\frac{\partial}{\partial t}\left(\phi_{1,n,p}^l - \phi_{0,n,p}^l\right) = \frac{\Delta x}{c\Delta t}\frac{\partial}{\partial t}\left(\phi_{0,n,p}^{l+1} - \phi_{0,n,p}^l\right) - \frac{\Delta x}{2}c\tilde{\nabla}_s^2\phi_{0,n,p}^l, \tag{4.7.12}$$

where $\tilde{\nabla}_s^2$ is the finite-difference approximation of $\nabla_s^2 = \partial^2/\partial y^2 + \partial^2/\partial z^2$. Moreover, by approximating the time derivative once more, one obtains

$$\phi_{1,n,p}^l - \phi_{0,n,p}^l - \phi_{1,n,p}^{l-1} + \phi_{0,n,p}^{l-1} = \frac{\Delta x}{c\Delta t}\left(\phi_{0,n,p}^{l+1} - 2\phi_{0,n,p}^l + \phi_{0,n,p}^{l-1}\right)$$
$$- \frac{\Delta t \Delta x}{2}c\tilde{\nabla}_s^2\phi_{0,n,p}^l. \tag{4.7.13}$$

Therefore,

$$\phi_{0,n,p}^{l+1} = \phi_{0,n,p}^{l}\left(2 - \frac{c\Delta t}{\Delta x}\right) + \frac{c\Delta t}{\Delta x}(\phi_{1,n,p}^{l} - \phi_{1,n,p}^{l-1}) + \left(\frac{c\Delta t}{\Delta x} - 1\right)\phi_{0,n,p}^{l-1}$$
$$+ \frac{c^2(\Delta t)^2}{2}\tilde{\nabla}_s^2\phi_{0,n,p}^{l}. \quad (4.7.14)$$

The above can be used as an absorbing boundary condition to deduce the field on the surface of the box for all time. Note that if the wave is at normal incidence, then $\tilde{\nabla}_s^2 = 0$, and (14) can be deduced from (4). Moreover, the finite-difference approximation of (11), which is accurate to the second order, has also been derived by Mur (1981).

§§4.7.2 Lindman Absorbing Boundary Condition

The paraxial approximation of (8) breaks down when the angle of incidence is too large. In the Lindman absorbing boundary condition (Lindman 1975; Randall 1988), a better approximation is made for k_x in (8) and (9). Since $k_x = \frac{\omega}{c}\cos\theta$, Equation (8) can be rewritten as

$$\left[\frac{1}{\cos\theta}\frac{d}{dx} + i\frac{\omega}{c}\right]\tilde{\phi}(\mathbf{r}, \omega) = 0. \quad (4.7.15)$$

The above is only exactly valid for a plane wave incident at angle θ; however, we can approximate $[\cos\theta]^{-1}$ with a rational expression as

$$\frac{1}{\cos\theta} \simeq 1 + \sum_{m=1}^{M}\frac{\alpha_m \sin^2\theta}{1 - \beta_m \sin^2\theta} \quad (4.7.16)$$

where α_m and β_m are constants yet to be determined. Since $\sin^2\theta = (ck_y/\omega)^2$ (assume $k_z = 0$) for each plane wave, (15) can be written as

$$\left[\frac{d}{dx} + \frac{i\omega}{c}\right]\tilde{\phi}(\mathbf{r}, \omega) = -\sum_{m=1}^{M}\tilde{h}_m \quad (4.7.17)$$

where

$$\tilde{h}_m = \frac{\alpha_m c^2 k_y^2}{\omega^2 - \beta_m c^2 k_y^2}\frac{d}{dx}\tilde{\phi}(\mathbf{r}, \omega). \quad (4.7.18)$$

Consequently, Equation (17), when converted to the time and spatial domain representation, yields

$$\left[\frac{\partial}{\partial x} - \frac{1}{c}\frac{\partial}{\partial t}\right]\phi(\mathbf{r}, t) = -\sum_{m=1}^{M}h_m(\mathbf{r}, t). \quad (4.7.19)$$

Moreover, Equation (18) can be multiplied by $\omega^2 - \beta_m c^2 k_y^2$ and converted to the time and spatial domain, resulting in

$$\frac{\partial^2}{\partial t^2}h_m(\mathbf{r}, t) - \beta_m c^2\frac{\partial^2}{\partial y^2}h_m(\mathbf{r}, t) = \alpha_m c^2\frac{\partial^3}{\partial y^2 \partial x}\phi(\mathbf{r}, t). \quad (4.7.20)$$

Notice that $h_m(\mathbf{r},t)$ is now independent of θ. Furthermore, Equations (19) and (20) can be approximated by a finite-difference scheme and solved in tandem near a boundary (Exercise 4.25).

It has been found that with $M = 3$, this absorbing boundary condition gives a reflection coefficient of less than 1% for angles of incidence less than 89°. Here, α_m and β_m are chosen in (16) such that the reflected wave is minimized for angles of incidence less than 89°. They are given in Table 4.3.1.

Table 4.3.1. Coefficients α_m and β_m

	$m = 1$	$m = 2$	$m = 3$
α_m	0.3264	0.1272	0.0309
β_m	0.7375	0.98384	0.9996472

The above absorbing boundary condition is valid for propagating plane waves. Similar absorbing boundary conditions can be derived for evanescent waves if there exists a preponderance of evanescent waves in a spectrum (Lindman 1975).

§§4.7.3 Bayliss-Turkel Absorbing Boundary Condition

An alternative absorbing boundary condition written in the spherical coordinates is the Bayliss-Turkel absorbing boundary condition (Bayliss and Turkel 1980; Bayliss, Gunzburger, and Turkel 1982). In this absorbing boundary condition, it is assumed that the wave has the form

$$\phi(\mathbf{r},t) = \sum_{i=1}^{\infty} \frac{f_i(ct - r, \theta, \phi)}{r^i} \tag{4.7.21}$$

in the vicinity of the box boundary. Notice that the above representation is intuitively obvious since the far field from a source can be expanded in terms of spherical harmonics.

On defining an operator

$$L_i = \left(\frac{\partial}{\partial r} + \frac{1}{c}\frac{\partial}{\partial t} + \frac{i}{r} \right), \tag{4.7.22}$$

we can show that

$$L_i \frac{f_i(ct - r, \theta, \phi)}{r^i} = 0. \tag{4.7.23}$$

Hence, if we apply L_1 to (21), it will annihilate the first term in (21) and convert the other terms to higher order terms, namely,

$$L_1\phi(\mathbf{r},t) = \sum_{i=2}^{\infty} \frac{(1 - i)f_i(ct - r, \theta, \phi)}{r^{i+1}}. \tag{4.7.24}$$

Consequently, $L_1\phi(\mathbf{r}, t) \sim O(r^{-3})$. Similarly, it can be shown that $L_3 L_1 \phi(\mathbf{r}, t)$ $\sim O(r^{-5})$ and so on. Then, on defining

$$B_m = L_{2m-1} B_{m-1}, \qquad (4.7.25\text{a})$$

$$B_1 = L_1, \qquad (4.7.25\text{b})$$

we have

$$B_m = \prod_{l=m}^{1} \left(\frac{\partial}{\partial r} + \frac{1}{c} \frac{\partial}{\partial t} + \frac{2l-1}{r} \right). \qquad (4.7.26)$$

Furthermore, it is easy to show that

$$B_m \phi(\mathbf{r}, t) \sim O(1/r^{2m+1}). \qquad (4.7.27)$$

Hence, a boundary condition

$$B_m \phi(\mathbf{r}, t) = 0 \qquad (4.7.28)$$

can be used as an absorbing boundary condition. It is good up to $O(1/r^{2m+1})$. In particular, if we assume that $m = 2$ explicitly, the boundary condition is

$$\left(\frac{\partial}{\partial r} + \frac{1}{c} \frac{\partial}{\partial t} + \frac{3}{r} \right) \left(\frac{\partial}{\partial r} + \frac{1}{c} \frac{\partial}{\partial t} + \frac{1}{r} \right) \phi(\mathbf{r}, t)$$
$$= \left(\frac{\partial^2}{\partial r^2} + \frac{2}{c} \frac{\partial^2}{\partial r \partial t} + \frac{4}{r} \frac{\partial}{\partial r} + \frac{1}{c^2} \frac{\partial^2}{\partial t^2} + \frac{4}{rc} \frac{\partial}{\partial t} + \frac{3}{r^2} \right) \phi(\mathbf{r}, t) = 0. \quad (4.7.29)$$

This boundary condition is increasingly difficult to implement when m becomes large. Notice that the $\partial^2/\partial r^2$ term may be removed by using the fact that $\phi(\mathbf{r}, t)$ is a solution to the wave equation in spherical coordinates, converting it into $\partial^2/\partial\theta^2$ and $\partial^2/\partial\phi^2$ derivatives. By doing so, the boundary condition is easier to implement because it involves only first derivatives in r. In this case, the values on the boundary can be deduced from the values at grid points in the immediate neighborhood of the boundary (see Exercise 4.26).

§§4.7.4 Liao's Absorbing Boundary Condition

The Liao's absorbing boundary condition is based on the work of Liao, Wong, Yang, and Yuan (1984). Its advantage is its ease in implementation even at the corners of a box. This boundary condition is derived as follows.

A plane-wave solution to the wave equation in the time domain can be written as $u(ct - x\cos\theta)$ where the direction of the plane wave makes an angle θ with the x axis. Moreover, an arbitrary wave can be approximated by a summation of plane waves with different angles. Therefore,

$$\phi(x, t) = \sum_i u_i(ct - x\cos\theta_i). \qquad (4.7.30)$$

But if $\phi(x,t)$ represents only a single plane wave, then $\phi(x, t + \Delta t) = \phi(x - c\Delta t / \cos\theta, t)$, which is a primitive absorbing boundary condition to update the boundary point at x with the interior point if the angle of the wave θ is known. Motivated by this fact, we can write

$$\phi(x, t+\Delta t) = \sum_i u_i[c(t+\Delta t) - x\cos\theta_i]$$

$$= \sum_i u_i(ct + c\Delta t - x\cos\theta_i + \alpha c\Delta t\cos\theta_i - \alpha c\Delta t\cos\theta_i)$$

$$= \sum_i u_i[ct - (x - \alpha c\Delta t)\cos\theta_i + c\Delta t(1 - \alpha\cos\theta_i)]$$

$$= \sum_i u_i(\eta_i + \epsilon_i), \qquad (4.7.31)$$

where

$$\eta_i = ct - (x - \alpha c\Delta t)\cos\theta_i, \qquad (4.7.32a)$$
$$\epsilon_i = c\Delta t(1 - \alpha\cos\theta_i). \qquad (4.7.32b)$$

The above is useless at this point unless we can express $u_i(\eta_i + \epsilon_i)$, which are the values at the boundary point, in terms of $u_i(\eta_i - l\epsilon_i)$, $l \geq 0$, which are the values at the interior points. Note that if only one plane-wave component exists, α can be chosen to equal $1/\cos\theta$, rendering $\epsilon = 0$, and this problem disappears. But in general, a spectrum of plane waves exists for which an absorbing boundary condition is needed.

To this end, we define a difference operator

$$\Delta u(\eta + \epsilon) = u(\eta + \epsilon) - u(\eta). \qquad (4.7.33)$$

Then,

$$\Delta^m u(\eta + \epsilon) = \Delta^{m-1}u(\eta + \epsilon) - \Delta^{m-1}u(\eta), \quad m > 0,$$

and $\Delta^0 = 1$. Using this relationship, it can be shown that (see Exercise 4.27)

$$u(\eta + \epsilon) = \sum_{m=1}^{N} \Delta^{m-1}u(\eta) + \Delta^N u(\eta + \epsilon). \qquad (4.7.34)$$

The above equation is just the finite-difference analogue of the N-th order Taylor series expansion. Note that in the limit when $\epsilon \to 0$, we have

$$\Delta^N u(\eta + \epsilon) \sim \epsilon^N \frac{d^N}{d\eta^N}u(\eta). \qquad (4.7.35)$$

Therefore, for small ϵ, the last term in (34) is of order ϵ^N and can be neglected. As a result,

$$u(\eta + \epsilon) \simeq \sum_{m=1}^{N} \Delta^{m-1}u(\eta). \qquad (4.7.36)$$

On defining a shifting operator $\mathcal{Z}$ such that $\mathcal{Z}u(\eta) = u(\eta - \epsilon)$, the difference operator becomes

$$\Delta u(\eta) = (\mathcal{I} - \mathcal{Z})u(\eta), \tag{4.7.37a}$$

$$\Delta^{m-1} u(\eta) = (\mathcal{I} - \mathcal{Z})^{m-1} u(\eta), \tag{4.7.37b}$$

where $\mathcal{I}$ is the identity operator. Then,

$$
u(\eta + \epsilon) = \sum_{m=1}^{N} \Delta^{m-1} u(\eta) = \sum_{m=1}^{N} (\mathcal{I} - \mathcal{Z})^{m-1} u(\eta) = \mathcal{Z}^{-1} \left[\mathcal{I} - (\mathcal{I} - \mathcal{Z})^N \right] u(\eta)
$$

$$
= \sum_{j=1}^{N} (-1)^{j+1} C_j^N \mathcal{Z}^{j-1} u(\eta) = \sum_{j=1}^{N} (-1)^{j+1} C_j^N u[\eta - (j-1)\epsilon],
\tag{4.7.38}
$$

where $C_j^N = N!/[j!(N-j)!]$. Note that in the above, $u(\eta + \epsilon)$ is interpolated from $u[\eta - (j-1)\epsilon]$. Now, using (38) in (31), we have

$$
\phi(x, t + \Delta t) = \sum_{j=1}^{N} (-1)^{j+1} C_j^N \sum_i u_i[\eta_i - (j-1)\epsilon_i].
\tag{4.7.39}
$$

With the definition of η_i and ϵ_i given by (32), one obtains

$$
\eta_i - (j-1)\epsilon_i = c[t - (j-1)\Delta t] - (x - j\alpha c \Delta t) \cos \theta_i
\tag{4.7.40}
$$

and

$$
\sum_i u_i[\eta_i - (j-1)\epsilon_i] = \phi[x - j\alpha c \Delta t, t - (j-1)\Delta t].
\tag{4.7.41}
$$

Consequently, (39) becomes

$$
\phi(x, t + \Delta t) = \sum_{j=1}^{N} (-1)^{j+1} C_j^N \phi[x - j\alpha c \Delta t, t - (j-1)\Delta t].
\tag{4.7.42}
$$

The above is now an absorbing boundary condition expressing $\phi(x, t + \Delta t)$, where x is the boundary point, in terms of the values of ϕ at interior points on the x axis and at previous time steps (see Figure 4.7.1). Note that α is arbitrary at this point. Since α multiplies c in the above, we do not need to know the velocity c accurately to apply this absorbing boundary condition. In fact, it can be shown that this absorbing boundary condition holds even when the waves have different intrinsic velocities (Exercise 4.27). For all practical purposes, α can be chosen to be 1 in (42).

Note that when $\alpha = 1$, $\epsilon = 0$ in (32) for a normally incident wave. Moreover, ϵ becomes larger for larger angles of incident. Hence, the interpolation scheme in (42) is exact for a normally incident wave, and corrects for an obliquely incident wave via (36).

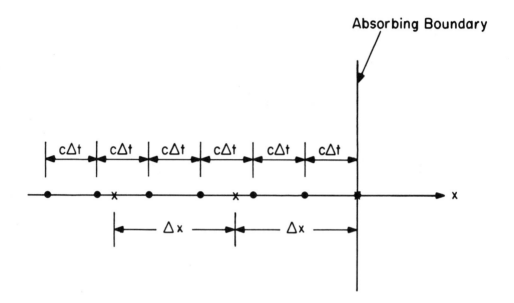

Figure 4.7.1 Implementation of the Liao's absorbing boundary condition. Dots are points needed by Equation (42) (with $\alpha = 1$), while crosses are the finite-difference grid.

Equation (42) expresses the boundary point values at x in terms of interior point values at $x - jc\Delta t$ (for $\alpha = 1$), but these points need not lie on the grid points. An interpolation scheme is further needed to infer the values required in (42) from the values available on the grid points. For example, it is easily shown via a quadratic interpolation that (see Exercise 4.28)

$$\phi(x - c\Delta t, t) = \frac{(2 - s)(1 - s)}{2}\phi_1^1(x) + s(2 - s)\phi_2^1(x) + \frac{s(s - 1)}{2}\phi_3^1(x),$$

(4.7.43)

where

$$\phi_i^m(x) = \phi[x - (i - 1)\Delta x, t - (m - 1)\Delta t],$$ (4.7.44a)

$$s = c\Delta t / \Delta x.$$ (4.7.44b)

Here, $\phi_i^m(x)$ are values available on the finite-difference grid only. Alternatively, Equation (43) can be expressed as

$$\phi(x - c\Delta t, t) = \mathbf{T}^1 \cdot \mathbf{\Phi}_3^1(x),$$ (4.7.45)

where

$$\mathbf{T}^1 = \left[\frac{(2-s)(1-s)}{2}, s(2-s), \frac{s(s-1)}{2} \right], \tag{4.7.46a}$$

$$\boldsymbol{\Phi}_3^1(x) = \left[\phi_1^1(x), \phi_2^1(x), \phi_3^1(x) \right]^t. \tag{4.7.46b}$$

The subscript 3 in $\boldsymbol{\Phi}_3^1(x)$ indicates that it is a three-component vector. The superscript 1 indicates that it corresponds to $m = 1$ in (44a). Similarly,

$$\begin{aligned}
\phi(x - 2c\Delta t, t - \Delta t) &= \phi(x - c\Delta t - c\Delta t, t - \Delta t) \\
&= \frac{(2-s)(1-s)}{2} \phi_1^2(x - c\Delta t) + s(2-s)\phi_2^2(x - c\Delta t) \\
&\quad + \frac{s(s-1)}{2} \phi_3^2(x - c\Delta t) \\
&= \mathbf{T}^1 \cdot \left[\phi_1^2(x - c\Delta t), \phi_2^2(x - c\Delta t), \phi_3^2(x - c\Delta t) \right]^t \\
&= \mathbf{T}^1 \cdot \boldsymbol{\Phi}_3^2(x - c\Delta t).
\end{aligned} \tag{4.7.47}$$

Since

$$\phi_1^2(x - c\Delta t) = \mathbf{T}^1 \cdot \boldsymbol{\Phi}_3^2(x), \tag{4.7.48a}$$

$$\phi_2^2(x - c\Delta t) = \phi_1^2(x - \Delta x - c\Delta t) = \mathbf{T}^1 \cdot \boldsymbol{\Phi}_3^2(x - \Delta x), \tag{4.7.48b}$$

$$\phi_3^2(x - c\Delta t) = \phi_1^2(x - 2\Delta x - c\Delta t) = \mathbf{T}^1 \cdot \boldsymbol{\Phi}_3^2(x - 2\Delta x), \tag{4.7.48c}$$

where $\boldsymbol{\Phi}_3^2(x) = [\phi_1^2(x), \phi_2^2(x), \phi_3^2(x)]^t$, $\boldsymbol{\Phi}_3^2(x - \Delta x) = [\phi_2^2(x), \phi_3^2(x), \phi_4^2(x)]^t$, and $\boldsymbol{\Phi}_3^2(x - 2\Delta x) = [\phi_3^2(x), \phi_4^2(x), \phi_5^2(x)]^t$, we can write the (48) as (Exercise 4.28)

$$\boldsymbol{\Phi}_3^2(x - c\Delta t) = \begin{bmatrix} \mathbf{T}^1 & 0 & 0 \\ 0 & \mathbf{T}^1 & 0 \\ 0 & 0 & \mathbf{T}^1 \end{bmatrix} \cdot \boldsymbol{\Phi}_5^2(x), \tag{4.7.49}$$

where

$$\boldsymbol{\Phi}_5^2(x) = \left[\phi_1^2(x), \phi_2^2(x), \phi_3^2(x), \phi_4^2(x), \phi_5^2(x) \right]^t. \tag{4.7.50}$$

Consequently, from (47) and (49), one obtains

$$\phi(x - 2c\Delta t, t - \Delta t) = \mathbf{T}^1 \cdot \begin{bmatrix} \mathbf{T}^1 & 0 & 0 \\ 0 & \mathbf{T}^1 & 0 \\ 0 & 0 & \mathbf{T}^1 \end{bmatrix} \cdot \boldsymbol{\Phi}_5^2(x) = \mathbf{T}^2 \cdot \boldsymbol{\Phi}_5^2(x). \tag{4.7.51}$$

After applying this idea recursively, one obtains

$$\phi(x - jc\Delta t, t - (j-1)\Delta t) = \mathbf{T}^j \cdot \boldsymbol{\Phi}_{2j+1}^j(x), \tag{4.7.52}$$

where

$$\mathbf{T}^j = \mathbf{T}^1 \cdot \begin{bmatrix} \mathbf{T}^{j-1} & 0 & 0 \\ 0 & \mathbf{T}^{j-1} & 0 \\ 0 & 0 & \mathbf{T}^{j-1} \end{bmatrix} \tag{4.7.53}$$

is a row vector of length $2j + 1$, and

$$\boldsymbol{\Phi}_{2j+1}^j(x) = \left[\phi_1^j(x), \phi_2^j(x), \cdots, \phi_{2j+1}^j(x) \right]^t. \tag{4.7.54}$$

In this manner, the absorbing boundary condition (42) becomes

$$\phi(x, t + \Delta t) = \sum_{j=1}^{N} (-1)^{j+1} C_j^N \mathbf{T}^j \cdot \mathbf{\Phi}_{2j+1}^j(x). \tag{4.7.55}$$

Notice that it is now expressed in terms of interior grid points at previous time steps.

Absorbing boundary conditions are fervently studied by a number of scientists (e.g. see Fang and Mei 1988) because of their potential in reducing the requirements for computer storage and computer time.

Exercises for Chapter 4

4.1 Show that for the Fourier transform of a causal function, the real part and the imaginary part are just the Hibert transforms of each other.

4.2 The Hilbert transform of a function $f(t)$ is

$$g(t) = \frac{1}{\pi} P.V. \int_{-\infty}^{\infty} dt' \frac{f(t')}{t' - t}.$$

(a) Show that

$$i \operatorname{sgn}(\omega) = \frac{1}{\pi} P.V. \int_{-\infty}^{\infty} dt \frac{e^{i\omega t}}{t}.$$

Therefore, $i \operatorname{sgn}(\omega)$ can be thought of as the Fourier transform of the function $\frac{1}{\pi} P.V. \frac{1}{t}$. Then, the Hilbert transform of a function $f(t)$ is the convolution of $f(t)$ with $-\frac{1}{\pi} P.V. \frac{1}{t}$, or

$$g(t) = f(t) * \left(-\frac{1}{\pi} P.V. \frac{1}{t} \right).$$

(b) From the above, show that

$$g(\omega) = -i \operatorname{sgn}(\omega) f(\omega).$$

Therefore, to obtain the Fourier transform $g(\omega)$ of the function $g(t)$, which is the Hilbert transform of $f(t)$, one must only multiply $f(\omega)$ with $-i \operatorname{sgn}(\omega)$ to engender an appropriate 90° phase shift to $f(\omega)$ for $\omega > 0$ and $\omega < 0$.

4.3 (a) Show that

$$-i \operatorname{sgn}(t) = \frac{1}{\pi} P.V. \int_{-\infty}^{\infty} d\omega \frac{e^{-i\omega t}}{\omega}.$$

Hence, the Hilbert transform of $f(\omega)$, which is the convolution of $f(\omega)$ with $-\frac{1}{\pi}P.V.\frac{1}{\omega}$, is equivalent to multiplying $f(t)$ by $i\,\mathrm{sgn}(t)$ in the time domain.

(b) Show that (4.1.8) is equivalent to

$$\epsilon'(t) - \epsilon(\infty)\,\delta(t) = i\,\mathrm{sgn}(t)\epsilon''(t),$$

and (4.1.9) is equivalent to

$$\epsilon''(t) = -i\,\mathrm{sgn}(t)[\epsilon'(t) - \epsilon(\infty)\,\delta(t)],$$

where

$$\epsilon'(t) = \frac{1}{2\pi}\int_{-\infty}^{\infty} d\omega\, e^{-i\omega t}\epsilon'(\omega),$$

$$\epsilon''(t) = \frac{1}{2\pi}\int_{-\infty}^{\infty} d\omega\, e^{-i\omega t}\epsilon''(\omega).$$

(c) Show from the above that

$$\epsilon(t) - \epsilon(\infty)\,\delta(t) = \epsilon'(t) + i\epsilon''(t) - \epsilon(\infty)\,\delta(t)$$
$$= i[\,\mathrm{sgn}(t)\epsilon''(t) + \epsilon''(t)] = 0, \quad t < 0.$$

Hence, $\epsilon(t) - \epsilon(\infty)\,\delta(t)$ is causal if $\epsilon''(\omega)$ and $\epsilon'(\omega)$ are related as in (4.1.8) and (4.1.9).

4.4 When a medium is active, the singularities in (4.1.10) may be above the real axis. Show that in order for causality not to be violated, the Fourier inversion contour would have to be above all the singularities. When $t \to \infty$, what are the salient features of the solution expressed by (4.1.13) when the medium is active?

4.5 Show that the real-t axis in Equation (4.2.10) maps to P_U and P_L which form a hyperbola on the complex s_x plane.

4.6 A fictitious, uniform current is turned on at $t = 0$ on the wire of length l and turned off completely on the wire at $t = T$. Find the time duration at $\rho = a$ over which the observed field is nonzero (see figure).

4.7 Show that Jordan's lemma and Cauchy's theorem permit the transformation from (4.2.20) to (4.2.21).

4.8 Since a line source is a linear superposition of point sources, integrate (4.2.26) over a line source (i.e., convolve it with a line source) to obtain (4.2.15).

4.9 By using Cauchy's theorem and Jordan's lemma, show that (4.2.45a) transforms to (4.2.46).

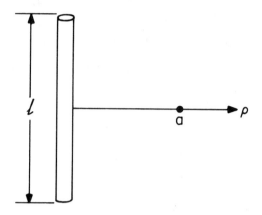

Figure for Exercise 4.6

4.10 A horizontal electric dipole is pointing in the x direction. It is located at a distance d above a lossless, dispersionless half-space. Assuming that the lower half-space is optically less dense than the upper half-space where the dipole is located, use the Cagniard-de Hoop method to find the transient reponse of the x component of the dipole by doing the following:

(a) Find the integral representation for the x component of the dipole field in Cartesian coordinates. Show that there is no pole at $k_x = ik_y$.

(b) Perform a coordinate rotation to simplify the integral.

(c) Find the P contour on the complex s_x plane which is the mapping of the real-t axis for the direct field term and the reflected field term.

(d) Go through the necessary procedure that gives the time domain solution for the direct field term and the reflected field term in terms of a single integral.

4.11 Assuming that $d_1 = y = 0$, find the Cagniard-de Hoop solution to (4.3.3) in the time domain.

4.12 (a) By writing (4.4.11) as

$$F(t) = -\frac{2}{\pi}\Im m \int_C ds_{1z} \frac{1}{[(t - s_{1z}z)^2 - s_\rho^2 \rho^2]^{1/2}},$$

where C is the image of the segment of the real τ axis, $s_1 r < \tau < t$ on the complex s_{1z} plane, show that $F(t)$ reduces to (4.2.25) after performing the integral in closed form. The relationship between τ and s_{1z} is given by (4.4.7).

(b) Use the method in Section 4.4 to derive the transient response of a vertical electric dipole on top of a half-space.

4.13 When $\tau \to \infty$, (4.5.9) reduces to

$$\hat{g}(x, y, t) = \frac{2AH(ct - \rho)}{\sqrt{c^2 t^2 - \rho^2}}.$$

(a) By substituting $\hat{g}(x, y, t)$ into (4.5.5) with $\tau \to \infty$, i.e.,

$$\left(\nabla_s^2 - \frac{1}{c^2} \frac{\partial^2}{\partial t^2}\right) \hat{g}(x, y, t) = -\frac{\delta(\rho)}{2\pi\rho} \delta(t),$$

where $\nabla_s^2 = \frac{1}{\rho} \frac{\partial}{\partial \rho} \rho \frac{\partial}{\partial \rho}$, show that the left-hand side of the above reduces to

$$-2A \frac{\delta(ct - \rho)}{\rho} \frac{\sqrt{ct - \rho}}{(ct + \rho)^{3/2}}.$$

(b) The last expression in (a) is only nonzero if $\rho = 0$ and $t = 0$. Otherwise, it is identically zero everywhere else. By rewriting it as

$$\lim_{\alpha \to 0} -2A \frac{\delta(ct - \rho)}{\rho} \frac{\sqrt{ct - \rho + \alpha}}{(ct + \rho + \alpha)^{3/2}},$$

show that

$$-2A \lim_{\alpha \to 0} \int_0^\infty d\rho \int_{-\infty}^\infty dt\, \delta(ct - \rho) \frac{\sqrt{ct - \rho + \alpha}}{(ct + \rho + \alpha)^{3/2}}$$

$$= -2A \lim_{\alpha \to 0} \frac{\sqrt{\alpha}}{c} \int_0^\infty d\rho \frac{1}{(2\rho + \alpha)^{3/2}} = -\frac{2A}{c}.$$

(c) Hence, one concludes that

$$-2A \frac{\delta(ct - \rho)\sqrt{ct - \rho}}{\rho(ct + \rho)^{3/2}} = -\frac{2A}{c} \frac{\delta(t)}{\rho} \frac{\delta(\rho)}{\rho},$$

or that the above functions are distributionally equivalent (see Appendix C for further discussions). By comparing the above with the right-hand side of (4.5.5), deduce that $A = c/4\pi$.

(d) Give an argument for why A cannot be a function of τ. Hence, A derived above is also valid for $\tau \neq 0$.

4.14 The Green's function satisfies $(\nabla^2 + k^2)g(r) = -\delta(\mathbf{r})$. In three dimensions, the Green's function is $e^{ikr}/4\pi r$. Using this fact, show that in four dimensions (i.e., $\nabla^2 = \partial^2/\partial x_1^2 + \partial^2/\partial x_2^2 + \partial^2/\partial x_3^2 + \partial^2/\partial x_4^2$), the Green's function $g(r)$ is related to a Hankel function.

Hint: Fourier transform the four-dimensional equation to reduce it to three dimensions, which has a closed-form solution. Then, the four-dimensional Green's function is expressible in terms of a Fourier integral, which can be related to a Hankel function.

4.15 (a) Using Taylor series expansions, show that the forward and backward difference approximations give rise to errors $O(\Delta t)$, while the central-difference approximation has an error $O(\Delta t^2)$.

   (b) Can you design a differencing scheme such that the error is $O[(\Delta t)^4]$ for $\partial\phi(\mathbf{r}, t)/\partial t$?

4.16 How would the stability criterion (4.6.22) be changed if $\Delta x \neq \Delta y \neq \Delta z$?

4.17 Show that the lossy wave Equation (4.6.23) has the same stability criterion as the lossless wave equation.

4.18 The diffusion equation is

$$\nabla^2 \phi(\mathbf{r}, t) - \mu\sigma \frac{\partial}{\partial t}\phi(\mathbf{r}, t) = 0.$$

   (a) Show that if we use central-difference approximation in time and space, the finite-difference scheme is always unstable.

   (b) Show that if the Euler scheme is employed where forward difference is used in time, but central difference is used in space, then the finite-difference scheme is conditionally stable, and that the condition is given by (4.6.25).

   (c) An unconditionally stable scheme, known as the Crank-Nicholson implicit method, can be used to solve the diffusion equation. This is obtained by averaging the space derivative over time in the Euler scheme. In one dimension, the diffusion equation then becomes

$$\phi_m^{l+1} - \phi_m^l = \frac{\Delta t}{2(\Delta x)^2 \mu\sigma} \left[\phi_{m+1}^{l+1} + \phi_{m+1}^l - 2(\phi_m^{l+1} + \phi_m^l) + \phi_{m-1}^{l+1} + \phi_{m-1}^l \right].$$

   This is not an explicit equation for time-stepping. Describe how you would use this equation to time-step forward in time.

   (d) By using a Fourier mode in the above, show that the Crank-Nicholson method is unconditionally stable.

4.19 Show that Equation (4.6.26) is unconditionally stable.

4.20 The Yee algorithm conserves flux, i.e.,

$$\int_S dS\, \hat{n} \cdot \mathbf{B} = 0.$$

   (a) Show that the $\mathbf{B}$ flux is conserved from the finite-difference approximations of (4.6.35a) to (4.6.35c).

(b) With this knowledge, B_z could be derived from B_x and B_y instead of time stepping (4.6.35c). Derive a relationship between B_z, B_x, and B_y in the Yee algorithm.

4.21 Show that (4.6.37) and its equivalences for (4.6.35e) and (4.6.35f) imply the same approximation for $(\partial/\partial t)\nabla \cdot \mathbf{D} = -\nabla \cdot \mathbf{J}$.

4.22 In a homogeneous medium, the magnetic field satisfies

$$\left(\nabla^2 - \frac{1}{c^2}\frac{\partial^2}{\partial t^2}\right)\mathbf{H}(\mathbf{r}, t) = 0.$$

(a) Using central differencing, derive a finite-difference approximation of the above wave equation in three dimensions.

(b) By eliminating the $\mathbf{E}$ field from the approximations of (4.6.35a) to (4.6.35f) in the Yee algorithm, show that the finite-difference approximation of H_x derived in (a) is also derivable from the Yee algorithm.

(c) Therefore, derive a stability criterion for the Yee algorithm.

4.23 (a) By substituting a Fourier mode in (4.7.4) as in the von Neumann stability analysis, show that it is stable when $c\Delta t/\Delta x \leq 1$.

(b) By similar analysis, show that (4.7.7) is always stable.

4.24 Using Taylor series expansions, show that (4.6.11a) and (4.7.6) are second-order accurate for smooth functions of x and t.

4.25 Using central differencing, derive a finite-difference approximation of the Lindman absorbing boundary condition.

4.26 (a) Use the wave equation to eliminate the $\partial^2/\partial r^2$ derivatives in the second-order Bayliss-Turkel absorbing boundary condition given in (4.7.29) and derive a boundary condition that involves only $\partial/\partial r$.

(b) Explain how you would implement this absorbing boundary condition in spherical coordinates.

4.27 (a) Derive Equation (4.7.34).

(b) Some waves like elastic waves in solids or electromagnetic waves in anisotropic media have more than one velocity for a given plane wave. In this case, Equation (4.7.30) may be written as

$$\phi(x, t) = \sum_i \sum_l u_{il}(c_l t - x\cos\theta_i),$$

where c_l indicates waves with different velocities. By assuming that α is α_l in (4.7.31), derive the equivalence of (4.7.42) for elastic waves.

4.28 (a) Derive the quadratic interpolation formula (4.7.43) and (4.7.47).

(b) Show that $\Phi_3^2(x - c\Delta t)$ can be written as (4.7.49), and hence, derive the recursive relation (4.7.52).

References for Chapter 4

Achenbach, J. D. 1973. *Wave Propagation in Elastic Solids.* Amsterdam: North-Holland.

Aki, K., and P. G. Richards. 1980. *Quantitative Seismology, Theory and Methods*, vol. I and II. New York: Freeman.

Ames, W. F. 1977. *Numerical Methods for Partial Differential Equations.* New York: Academic Press.

Anderson, B., and W. C. Chew. 1989. "Transient response of some electromagnetic logging tools." *Geophysics* 54(2): 216–24.

Bayliss, A., and E. Turkel. 1980. "Radiation boundary conditions for wavelike equations." *Communications on Pure and Applied Mathematics* 33: 707–25.

Bayliss, A., M. Gunzburger, and E. Turkel. 1982. "Boundary conditions for the numerical solution of elliptical equations in exterior regions." *SIAM J. Appl. Math.* 42: 430.

Cagniard, L. 1939. *Réflection et Réfraction des Ondes Séismiques Progressives.* Gauthier-Villars (Translated and revised by E. A. Flinn and C. H. Dix. 1962). *Reflection and Refraction of Progressive Seismic Waves.* New York: McGraw-Hill.

Clayton, R., and B. Engquist. 1977. "Absorbing boundary conditions for acoustic and elastic wave equations." *Bull. Seis. Soc. Am.* 67: 1529–40.

Courant, R., K. O. Friedrichs, and H. Lewy. 1928. "Über die partiallen differenzengleichungen der mathematischen physik." *Math. Ann.* 100: 32.

Courant, R., and D. Hilbert. 1962. *Methods of Mathematical Physics*, vol. II. New York: Interscience.

Crank, J., and P. Nicholson. 1947. "A practical method for numerical integration of solution of partial differential equations of heat-conductive type." *Proc. Camb. Philos. Soc.* 43: 50.

de Hoop, A. T. 1960. "A modification of Cagniard's method for solving seismic pulse problems." *Appl. Sci. Res.* B8: 349–56.

de Hoop, A. T., and H. J. Frankena. 1960. "Radiation of pulses generated by a vertical electric dipole above a plane non-conducting earth." *Appl. Sci. Res.* B8: 369–77.

Doak, P. E. 1952. "The reflexion of a spherical acoustic pulse by an absorbent infinite plane and related problems." *Proc. Roy. Soc. (London)* A215: 233–54.

Dufort E. C., and S. P. Frankel. 1953. "Stability condition for the numerical treatment of parabolic differential equations." *Math. Tables and Other Aids to Comp.* 7: 135.

Engquist, B., and A. Majda. 1977. "Absorbing boundary conditions for the numerical simulation of waves." *Math. Comp.* 31: 629–51.

Engquist, B., and A. Majda. 1979. "Radiation boundary conditions for acoustic and elastic wave calculations." *Comm. Pure Appl. Math.* 32: 314–57.

Ezzedine, A., J. A. Kong, and L. Tsang. 1982. "Time response of a vertical electric dipole over a two-layer medium by the double deformation technique." *J. Appl. Phys.* 53: 813–22.

Fang, J., and K. K. Mei. 1988. "A super-absorbing boundary algorithm for solving electromagnetic problems by time-domain finite-difference method." *IEEE AP-S International Symposium Digest* 2: 472-74.

Gilbert, F., and D. V. Helmberger. 1972. "Generalized ray theory for a layered sphere." *Geo. J. Roy. Astr. Soc.* 27: 57–80.

Helmberger, D. V. 1968. "The crust-mantle transition in the Bering Sea." *Bull. Seism. Soc. Am.* 58: 179–214.

Hildebrand, F. B. 1974. *Introduction to Numerical Analysis.* New York: McGraw-Hill.

Kramers, H. A. 1927. *Atti Congr. Internaz. Fisici*, Sept. 1927, Como, Italy.

Kuester, E. F. 1984. "The transient electromagnetic field of a pulsed line source located above a dispersively reflecting surface." *IEEE Trans. Antennas Propagat.* AP-32: 1154–62.

Liao, Z. P., H. L. Wong, B. P. Yang, and Y. F. Yuan. 1984. "A transmitting boundary for transient wave analysis." *Scientia Sinica. (series A.)* 27(10): 1063–76.

Lindman, E. L. 1975. "Free space boundary conditions for the time dependent wave equation." *J. Comp. Phys.* 18: 66–78.

Lysmer, J., and R. L. Kuhlemeyer. 1969. "Finite dynamic model for infinite media." *J. Eng. Mech. Div., ASCE* 95: 859–77.

Mur, G. 1981. "Absorbing boundary conditions for the finite-difference approximation of the time-domain electromagnetic-field equations." *IEEE Trans. Electromagn. Compat.* EMC-23: 377–82.

Oppenheim, A. V., and R. W. Schafer. 1975. *Digital Signal Processing.* Englewood Cliffs, N.J.: Prentice-Hall.

Poh, S. Y., and J. A. Kong. 1987. "Transient response of a vertical electric dipole on a two-layer medium." *J. Elect. Waves Appl.* 1(2): 133–58.

Potter, D. 1973. *Computational Physics.* New York: Wiley-Interscience.

Randall, C. L. 1988. "Absorbing boundary conditions for the elastic wave equation." *Geophysics.* 53(5): 611–24.

Richtmeyer, R. D., and K. W. Morton. 1967. *Difference Methods for Initial Value Problems.* New York: Wiley-Interscience.

Strick, E. 1959. "Propagation of elastic wave motion from an impulsive source along a fluid/solid interface. II. Theoretical pressure pulse." *Phil. Trans. Roy. Soc.*, London, A251: 465–523.

Taflove, A. 1988. "Review of the formulation and applications of the finite-difference time-domain method for numerical modeling of the electromagnetic wave interactions with arbitrary structures." *Wave Motions* 10: 547–82.

Titchmarsh, E. C. 1948. *Introduction to the Theory of Fourier Integrals.* Oxford: Clarendon Press.

Tsang, L., and J. A. Kong. 1979. "Asymptotic methods for the first compressional head wave arrival in a fluid-filled borehole." *J. Acoust. Soc. Am.* 65: 647–54.

Tygel, M., and P. Hubral. 1987. *Transient Waves in Layered Media.* New York: Elsevier.

von Neumann, J. 1943. *Mathematische Grundlagen der Quantenmechanik.* New York: Dover.

Wagatha, L. 1983. "Approximation of pseudodifferential operators in absorbing boundary conditions for hyperbolic equations." *Numer. Math.* 42: 1–64.

Yee, K. S. 1966. "Numerical solution of initial boundary value problems involving Maxwell's equations in isotropic media." *IEEE Trans. Antennas Propagat.* AP-14: 302–07.

Further Readings for Chapter 4

Alford, R. M., K. R. Kelly, and D. M. Boore. 1974. "Accuracy of finite-difference modeling of the acoustic wave equation." *Geophysics* 39: 834–42.

Alterman, Z., and F. C. Karal, Jr. 1968. "Propagation of elastic waves in layered media by finite difference methods." *Bulletin of the Seismological Society of America* 58: 3C7–98.

Baum, C. E. 1976. "The singularity expansion method." In *Transient electromagnetic fields*, ed. L. B. Felsen, pp. 129–79. Berlin: Springer-Verlag.

Bhattacharyya, B. K. 1957. "Propagation of an electric pulse through a homogeneous and isotropic medium." *Geophysics* 22: 905–21.

Birtwistle, G. M. 1968. "The explicit solution of the equation of heat conduction." *Comput. J.* 11: 317.

Boerner, W. M., and Y. M. Antar. 1979. "Aspects of electromagnetic pulse scattering from a grounded dielectric slab." *Arch. Elek. Ubertragungstech.* 26: 14–21.

Boore, D. M. 1972. "Finite-difference methods for seismic wave propagation in heterogeneous materials." In *Methods in Computational Physics*, eds. B. Alder, S. Fernback, and M. Rotenberg, vol. 11, pp. 1–37. New York: Academic Press.

Botros, A. Z., and S. F. Mahmoud. 1978. "The transient fields of simple radiators from the point of view of remote sensing of the ground subsurface." *Radio Sci.* 13: 379–89.

Chapman, C. H. 1978. "A new method for computing seismograms." *Geophys. J. Roy. Astr. Soc.* 54: 481–518.

Chew, W. C., S. Gianzero, and K. J. Kaplin. 1981. "Transient response of an induction logging tool in a borehole." *Geophysics* 46: 1291–1300.

Davis, J. L. 1986. *Finite Difference Methods in Dynamics of Continuous Media*. New York: MacMillan.

de Hoop, A. T. 1979. "Pulsed electromagnetic radiation from a line source in a two-media configuration." *Radio Sci.* 14: 253–68.

de Hoop, A. T. 1988. "Large-offset approximations in the modified Cagniard method for computing synthetic seismograms: a survey." *Geophysical Prospecting* 36: 465–77.

du Cloux, R. 1984. "Pulsed electromagnetic radiation from a line source in the presence of a semi-infinite screen in the plane interface of two different media." *Wave Motion* 6: 459–76.

Dudley, D. G., T. M. Papazoglou, and R. C. White. 1974. "On the interaction of a transient electromagnetic plane wave and a lossy half space." *J. Appl. Phys.* 45: 1171–75.

Emerman, S. H., W. Schmidt, and R. A. Stephen. 1982. "An implicit finite-difference formulation of the elastic wave equation." *Geophysics* 47: 1521–26.

Felsen, L. B., ed. 1976. *Transient Electromagnetic Fields.* New York: Springer-Verlag.

Felsen, L. B., and N. Marcuvitz. 1973. *Radiation and Scattering of Electromagnetic Waves.* New Jersey: Prentice-Hall.

Forsythe, G. E., and W. R. Wasow. 1960. *Finite-Difference Methods for Partial Differential Equations.* New York: John Wiley & Sons.

Frankena, H. J. 1960. "Transient phenomena associated with Sommerfeld's horizontal dipole problem." *Appl. Sci. Res.* B8: 349.

Fuller, J. A., and J. R. Wait. 1976. "A pulsed dipole in the earth." *Topics in Appl. Phys.* 10: 237–69.

Gilbert, F., and S. J. Laster. 1962. "Excitation and propagation of pulses on an interface." *Bull. Scism. Soc. Am.* 52: 299–319.

Goldman, M. M., and C. H. Stoyer. 1983. "Finite-difference calculations of the transient field of an axially symmetric earth for vertical magnetic dipole excitation." *Geophysics* 48: 953–63.

Gray, K. G., and S. A. Bowhill. 1974. "Transient response of stratified media: multiple scattering integral and differential equation for an impulsive incident plane wave." *Radio Sci.* 9: 57–62.

Hermance, J. F. 1982. "Refined finite-difference simulations using local integral forms: application to telluric fields in two dimensions." *Geophysics* 47: 825–31.

Hjelt, S. E. 1971. "The transient electromagnetic field of a two layer sphere." *Geoexploration* 9: 213–30.

Holland, R., L. Simpson, and K. S. Kunz. 1980. "Finite-difference analysis of EMP coupling to lossy dielectric structures." *IEEE Trans. Electromagn. Compat.* EMC-22: 203–09.

Kamenetski, F. M. 1968. "Transient processes using combined loops for a two layer section with a non-conducting base." *Izv. Vozov, Sect., Geology Prosp.* 6.

Kaufman, A. 1979. "Harmonic and transient fields on the surface of a two layered medium." *Geophysics* 44: 1208–17.

Kooij, B. J., and D. Quak. 1988. "Three-dimensional scattering of impulsive acoustic waves by a semi-infinite crack in the plane interface of a half-space and a layer." *J. Math. Phys.* 29: 1712–21.

Kunz, K. S., and K.-M. Lee. 1978. "A three-dimensional finite-difference solution of the external response of an aircraft to a complex transient EM environment: Part I—The method and its implementation." *IEEE Trans.*

Electromagn. Compat. EMC-20: 328–33.

Kuo, J. T., and D. Cho. 1980. "Transient time-domain electromagnetics." *Geophysics* 45: 271–91.

Lamb, H. 1904. "On the propagation of tremors over the surface of an elastic solid." *Phil. Trans. Roy. Soc. (London), Ser. A* 203: 1–42.

Lee, T. 1975. "Transient electromagnetic response of a sphere in a layered medium." *Geophysical Prospecting* 23: 492–512.

Lee, T. 1982. "Asymptotic expansions for transient electomagnetic fields." *Geophysics* 47: 38–46.

Lee, T., and R. J. G. Lewis. 1974. "Transient EM response of a large loop on a layered ground." *Geophysical Prospecting* 22: 430–44.

Lewis, R. M. 1967. "Asymptotic theory of transients." *Electromagnetic Wave Theory*, Part 2, ed. J. Brown, pp. 845–69. New York: Pergamon Press.

Lindemuth, I., and J. Killen. 1973. "Alternating-direction implicit techniques for two-dimensional magnetohydrodynamic calculations." *J. Comp. Physics* 13: 181–208.

Mahmoud, S. F., A. Z. Batras, and J. R. Wait. 1979. "Transient electromagnetic fields of a vertical magnetic dipole on a two layered earth." *Proc. IEEE* 67: 1022–29.

Merewether, D. E. 1971. "Transient currents induced on a metallic body of revolution by an electromagnetic pulse." *IEEE Trans. Electromagn. Compat.* EMC-13: 41–44.

Metwally, A. D., and S. F. Mahmoud. "Transient response of a dissipative earth by use of image theory." *IEEE Trans. Antennas Propagat.* AP-32(3): 287–91.

Mitchell, A. R. 1969. *Computational Methods in Partial Differential Equations.* New York: John Wiley & Sons.

Morrison, H. F., P. J. Phillips, and D. P. O'Brien. 1969. "Quantitative interpretation of transient electromagnetic fields over a layered half space." *Geophysical Prospecting* 17: 82–101.

Mur, G. 1981. "The modeling of singularities in the finite-difference approximation of the time-domain electromagnetic-field equations." *IEEE Trans. Microwave Theory Tech.* MTT-29: 1073–77.

Nabighian, M. N. 1970. "Quasistatic response of a conducting sphere in a dipolar field." *Geophysics* 35: 303–09.

Nabighian, M. N. 1971. "Quasistatic response of a conducting permeable two layered sphere in a dipolar field." *Geophysics* 36: 25–37.

Oristaglio, M. L., and G. W. Hohmann. 1984. "Diffusion of electromagnetic fields into a two-dimensional earth: a finite difference approach." *Geo-*

physics 49: 870–94.

Peaceman, D. W., and H. H. Rachford. 1955. "The numerical solution of parabolic and elliptic differential equations." *J. Industr. Math. Soc.* 3: 28–41.

Peterson, E. W. 1974. "Acoustic wave propagation along a fluid-filled cylinder." *J. Appl. Phys.* 45: 3340–50.

Raiche, A. P., and B. R. Spies. 1981. "Coincident loop transient electromagnetic master curves for intepretation of two-layer earths." *Geophysics* 46: 53–64.

Roever, W. L., J. H. Rosenbaum, and T. F. Vining. 1974. "Acoustic waves from an impulsive source in a fluid-filled borehole." *J. Acoust. Soc. Am.* 55: 1144–57.

Sezginer, A., and W. C. Chew. 1984. "Closed-form expression of the Green's function for the time-domain wave equation for a lossy two-dimensional medium." *IEEE Trans. Antennas Propagat.* AP-32: 527–28.

Sezginer, A., and J. A. Kong. 1984. "Transient response of line source excitation in cylindrical geometry." *Electromagnetics* 4: 35–54.

Singh, R. N. 1972. "Transient electromagnetic screening in a two concentric spherical shell model." *Pure and Applied Geophysics* 94: 226–32.

Singh, R. N. 1973. "Electromagnetic transient response of a conducting sphere in a conductive medium." *Geophysics* 38: 864–93.

Singh, S. K. 1973. "Electromagnetic response of a conducting sphere embedded in a conductive medium." *Geophysics* 38: 384–93.

Stoyer, C. H., and R. J. Greenfield. "Numerical solutions of the response of a two-dimensional earth to an oscillating magnetic dipole source." *Geophysics* 41: 519–30.

Tabarovsky, L. A., and V. S. Krioputsky. 1978. "Solution of the transient electromagnetic problems for axially-symmetric models by the grid method." *Geol. and Geophys.*, no. 7: 68–78.

Taflove, A. 1980. "Application of the finite-difference time-domain method to sinusoidal steady-state electromagnetic-penetration problems." *IEEE Trans. Electromagn. Compat.* EMC-22: 191–202.

Taflove, A., and M. E. Brodwin. 1975. "Numerical solution of steady-state electromagnetic scattering problems using the time-dependent Maxwell's equations." *IEEE Trans. Microwave Theory Tech.* MTT-23: 623–30.

Taylor, C. D., D.-H. Lam, and T. H. Shumpert. 1969. "Electromagnetic pulse scattering in time-varying inhomogeneous media." *IEEE Trans. Antennas Propagat.* AP-17: 585–89.

Thacker, W. C. 1978a. "Comparison of finite-element and finite-difference schemes. Part I: One-dimensional gravity wave motion." *J. Phys. Ocean.*

8: 676–79.

Thacker, W. C. 1978b. "Comparison of finite-element and finite-difference schemes. Part II: Two-dimensional gravity wave motion." *J. Phys. Ocean.* 8: 680–89, 1978.

Tijhuis, A. G., and H. Blok. 1984a. "SEM approach to the transient scattering by an inhomogeneous, lossy dielectric slab; Part 1: the homogeneous case." *Wave Motion* 6: 61–78.

Tijhuis, A. G., and H. Blok. 1984b. "SEM approach to the transient scattering by an inhomogeneous, lossy dielectric slab; Part 2: the inhomogeneous case." *Wave Motion* 6: 167–82.

Tsubota. K., and T. R. Wait. 1980. "The frequency- and the time-domain reponse of a buried axial conductor." *Geophysics* 45: 941–51.

van der Hijden, J. H. M. T. 1987. *Propagation of Transient Elastic Waves in Stratified Anisotropic Media.* Amsterdam: North-Holland.

Velekin, A. B., and J. U. Bulgakov. 1967. "Transient method of electrical prospecting (one-loop version). *Int. Sem. on Geophy. Meth.* Moscow.

Verma, S. K. 1972. "Transient electromagnetic response of a conducting sphere excited by different types of input pulses." *Geophysical Prospecting* 20: 752–71.

Verma, S. K., and R. N. Singh. 1970. "Transient electromagnetic response of an inhomogeneous conducting sphere." *Geophysics* 35: 331–36.

Wait, J. R. 1951. "Transient electromagnetic propagation in a conducting medium." *Geophysics* 16: 213–21.

Wait, J. R., and D. A. Hill. 1972. "Electromagnetic surface fields produced by a pulse-excited loop buried in the earth." *J. Appl. Physics* 43: 3988–91.

Wait, J. R., and K. P. Spies. 1969. "On the image representation of the quasi-static fields of a line current source above the ground." *Can. J. Phys.* 47(23): 2731–33.

Wait, J. R., L. Thrane, and R. J. King. 1975. "The transient electric field response of an array of parallel wires on the Earth's surface." *IEEE Trans. Antennas Propagat.* AP-23: 261.

Weidelt, P. 1975. "Electromagnetic induction in three-dimensional structures." *J. Geophys.* 41: 85–109.

Wu, T. T., and H. Lehmann. 1985. "Spreading of electromagnetic pulses." *J. Appl. Phys.* 55: 2064–65.

CHAPTER 5

VARIATIONAL METHODS

Variational methods have been used historically to solve many physical problems. In a variational method, a functional, which has a stationary property at the exact solution of a physical problem, is first derived. Trial functions are then used to approximate the solution in order to minimize or maximize the functional. In this manner, a systematic method can be developed to find an optimal solution to the physical problem. Such a systematic way of obtaining an optimal solution is also known as the Rayleigh-Ritz procedure. Moreover, because of the advent of high-speed digital computers, the Rayleigh-Ritz procedure can be implemented systematically and efficiently. In addition, variational methods are also the foundations of the finite-element method, Galerkin's method, and the method of moments (also known as the method of weighted residuals). These are numerical methods with marked impact on our present day science and technology because they can solve problems which have been previously thought unsolvable.

In this chapter, variational expressions are derived for several physical wave problems. Furthermore, the finite-element method or Galerkin's method will be shown to follow from applying the Rayleigh-Ritz procedure to obtain the optimal solution to a variational expression. At this juncture, a proper understanding of the infinite-dimensional linear vector space is useful. Hence, the chapter begins with a review of the linear vector space. It is assumed that the readers possess a basic knowledge of linear algebra.

§5.1 Review of Linear Vector Space

An understanding of the techniques used to solve the related problems of wave propagations in inhomogeneities varying in more than one dimension is enhanced by insight into the infinite-dimensional linear vector space. Hence, we shall present a brief review here. A number of books are available that contain a more extensive discussion of this topic (e.g., Berberian 1961; Stakgold 1967, 1979; Mikhlin 1970).

§§5.1.1 Inner Product Spaces

The three-dimensional vector space is a familiar linear vector space. For instance, if u and v are elements of a linear vector space, then $\alpha u + \beta v$ is also an element of the linear vector space embodying the concept of linearity. Moreover, algebraic properties exist for a three-dimensional linear vector space which are also true for higher-dimensional linear vector spaces. These algebraic properties can be summarized as follows: If two vectors u and v are elements of a linear vector space, then (Exercise 5.1)

$$u + v = v + u, \tag{5.1.1a}$$

$$u + (v + w) = (u + v) + w. \tag{5.1.1b}$$

For two numbers α and β,

$$\alpha(\beta u) = (\alpha\beta)u, \tag{5.1.1c}$$

$$(\alpha + \beta)u = \alpha u + \beta u, \tag{5.1.1d}$$

$$\alpha(u + v) = \alpha u + \alpha v. \tag{5.1.1e}$$

Note that these algebraic properties are the same as those of scalar numbers.

Similarly, many concepts in linear algebra for a finite-dimensional linear vector space can be extended for an infinite-dimensional linear vector space. For example, the concept of the inner product is familiar in a finite-dimensional space. This concept can be extended to an infinite-dimensional space. The inner product between two vectors whose dimensions are countably infinite is

$$\mathbf{f}^t \cdot \mathbf{g} = \sum_{i=1}^{\infty} f_i\, g_i. \tag{5.1.2}$$

Only vectors whose "energies," $\mathbf{f}^t \cdot \mathbf{f}$, are finite are considered here. Moreover, the inner product (2) could be extended to two infinite-dimensional vectors whose indices are a continuum and uncountably infinite. For example, the function $f(x)$, defined over $a < x < b$, can be considered an infinite-dimensional vector whose index x is a continuum. Hence, $f(x)$ is also called a vector. Moreover, when the vector $f(x)$ is used to describe the state of an equation or a system, it is also called a *state vector* (see Merzbacher 1970). By a straightforward extension of (2), the inner product between vectors $f(x)$ and $g(x)$ over the domain $a < x < b$ can be defined as

$$\langle f, g \rangle = \int_a^b dx\, f(x)\, g(x). \tag{5.1.3}$$

The "energy" $\mathbf{f}^t \cdot \mathbf{f}$ or $\langle f, f \rangle$ is a real number if $\mathbf{f}$ and $f(x)$ are real vectors and real functions respectively. But when $\mathbf{f}$ and $f(x)$ are complex, it is customary to define the inner product in (2) as

$$\mathbf{f}^\dagger \cdot \mathbf{g} = \sum_{i=1}^{\infty} f_i^*\, g_i \tag{5.1.4}$$

and that in (3) as

$$\langle f^*, g \rangle = \int_a^b dx\, f^*(x)\, g(x). \tag{5.1.5}$$

The bracket notation $\langle f^*, g \rangle$ is often used to denote the inner product between two vectors, regardless of whether their indices are countably or uncountably infinite. In some books, a complex conjugate now shown explicitly in (5) is implicitly implied in all inner products. In this book, however, it is more convenient to denote a complex conjugate explicitly when it is needed in an inner product, because the inner product (3) is used in electromagnetics to denote the reaction between two field quantities (Rumsey 1954).

A linear vector space with an inner product defined as (5) is also known as an *inner product space*. [Other definitions of inner products are also possible (see Axelsson and Barker 1984).] The norm of a vector can be defined as

$$\|f\| = \langle f^*, f \rangle^{\frac{1}{2}}, \qquad (5.1.6)$$

which is always a positive number. This is also known as the L_2 norm. Moreover, a linear vector space with a defined norm is also known as a *norm space*. The distance between two vectors can be defined as

$$d(f, g) = \|f - g\|, \qquad (5.1.7)$$

which is always nonzero if $f \neq g$. Here, $d(f, g)$ is the *metric* of the inner product space. A space with a defined metric is also known as a *metric space*.

An inner product space which is *complete* is called a *Hilbert space* because it was extensively studied by Hilbert. A space is complete if it has no "holes" in it. For example, the set of rational numbers is not complete over the real number line because there are irrational numbers on the real number line that leave "holes" which cannot be filled in by the rational numbers. However, the union of the sets of rational and irrational numbers is complete over the real number line because there are no "holes" left. This concept of completeness can be extended to higher-dimensional linear vector spaces.

A complete Hilbert space with an inner product defined in (5) and a norm defined in (6) is also known as the L_2 space. It is the space of square integrable functions in the interval a to b. This space includes functions with step discontinuities but not the Dirac delta function.

Equations (2) and (3) are similar in concept except that (2) is a discrete summation, while (3) consists of a continuous summation. If (2) or (3) is implemented on a computer, however, the infinite summation in (2) will have to be truncated somewhere, and (3) must be discretized. For example, (2) would become

$$\mathbf{f}^t \cdot \mathbf{g} \simeq \sum_{i=1}^{N} f_i \, g_i, \qquad (5.1.8)$$

while (3) becomes

$$\langle f, g \rangle = \sum_{i=1}^{N} \Delta x \, f(i\Delta x) \, g(i\Delta x). \qquad (5.1.9)$$

Hence, in numerical implementations, they are quite similar.

§§5.1.2 *Linear Operators*

The extension of the matrix concept in a finite-dimensional linear vector space to an infinite-dimensional linear vector space allows us to write

$$\int_0^a G(x, x')\, f(x')\, dx' = h(x). \qquad (5.1.10)$$

The above is just the extension of the concept

$$\sum_{j=1}^N G_{ij} f_j = h_i, \qquad (5.1.11)$$

where G_{ij} is a matrix element. Either (10) or (11) can be written symbolically as

$$\mathcal{G} f = h. \qquad (5.1.12)$$

$\mathcal{G}$ is known varyingly as a transformation, a mapping, or an operator. In the case of (10), $\mathcal{G}$ is an integral operator.

Similarly, a linear differential equation

$$\left[\frac{d^2}{dx^2} + k^2(x) \right] f(x) = h(x), \quad a < x < b \qquad (5.1.13)$$

can be written as

$$\mathcal{D} f = h, \qquad (5.1.14)$$

where $\mathcal{D}$ is the differential operator. Moreover, Equations (13) and (10) can be paralleled by writing (13) as

$$\int_a^b dx'\, \delta(x' - x) \left[\frac{d^2}{dx'^2} + k^2(x') \right] f(x') = h(x). \qquad (5.1.15)$$

Note that the above is the analogue of

$$\sum_{j=1}^N \delta_{ij} D_{ij} f_j = h_i. \qquad (5.1.16)$$

Now, by comparing (15) and (16) with (10) and (11), we note that a differential operator is diagonal in nature: A differential operator is *local* while an integral operator is *global*. In other words, a differential operator relates x and x' that are close together, while an integral operator relates x and x' that are close by as well as far apart.

A function f maps the number x into the number $f(x)$. The set of numbers x for which f is defined is known as the **domain** of the function f. The set of numbers $f(x)$ generated by f for x in the domain is known as the **range** of the function f. On the other hand, an operator maps a vector from one space to a vector in another space. In a similar manner, the set of vectors on which an operator $\mathcal{G}$ operates is known as the **domain** of the operator $\mathcal{G}$. Then, the set of vectors into which $\mathcal{G}$ maps, i.e., the resultant vectors after $\mathcal{G}$ operates on all possible vectors in its domain, is known as the **range** of the operator.

A number g is a **functional** of a function f if g is a number that depends on f. For example, $g = \langle h, f \rangle$ and $g = \langle f, f \rangle$ are both functionals of f. Moreover, a functional $g(f)$ is linear if $\alpha g_1 + \beta g_2 = g(\alpha f_1 + \beta f_2)$, provided that $g_1 = g(f_1)$ and $g_2 = g(f_2)$. Note that a functional is a mapping from a vector space to a scalar space (Exercise 5.2). It can be proven that all bounded linear functionals of f, where $f \in \mathcal{H}$ and $\mathcal{H}$ is a Hilbert space, can be written as $\langle h_1^*, f \rangle$, where $h_1 \in \mathcal{H}$ also. This follows from the **Riesz representation theorem** (see Stakgold 1967, 1979).

§§5.1.3 Basis Functions

A function $f(x)$, defined in an interval $0 < x < a$, may be expanded in terms of a complete set of functions $\{f_n(x)\}$, namely,

$$f(x) = \sum_{n=1}^{\infty} a_n f_n(x). \tag{5.1.17}$$

A set $\{f_n(x)\}$ is complete over an interval $0 < x < a$ if the expansion (17) exists when its left-hand side is an arbitrary, square integrable function[1] in the same interval. Then the set $\{f_n(x)\}$ forms a **basis** in the function space, or the set $\{f_n(x)\}$ spans the Hilbert space. Moreover, a Hilbert space is **separable** if it is spanned by a countably infinite basis set. Furthermore, if $f_n(x)$ is orthogonal so that $\langle f_n, f_m \rangle = 0$ when $n \neq m$, then f_n can be normalized so that it is orthonormal, that is,

$$\langle f_n, f_m \rangle = \delta_{nm}. \tag{5.1.18}$$

Consequently, from (17), and using the orthonormality property of f_n, one obtains

$$a_n = \langle f, f_n \rangle, \tag{5.1.19}$$

and (17) could be rewritten as

$$f(x) = \sum_{n=1}^{\infty} f_n(x) \langle f_n, f \rangle. \tag{5.1.20}$$

---

[1] A function $f(x)$ is square integrable over the interval $0 < x < a$ if $\int_0^a dx\,|f(x)|^2 < \infty$. In this case, $f(x)$ is in L_2 space which is a Hilbert space. The set $\{f_n(x)\}$ spans the L_2 space.

If $f_n(x)$ is not orthogonal, however, then finding the inner product of (17) with $f_m(x)$, we have

$$\langle f_m, f \rangle = \sum_{n=1}^{\infty} \langle f_m, f_n \rangle a_n. \tag{5.1.21}$$

After defining $B_{mn} = \langle f_m, f_n \rangle$, $c_m = \langle f_m, f \rangle$, the above equation is of the form

$$c_m = \sum_{n=1}^{\infty} B_{mn} a_n, \tag{5.1.22}$$

where $B_{mn} = B_{nm}$. An iterative method may be used to solve (22) for a_n. Yet another way to find a_n is to truncate the sum in (22) and solve (22) as a finite-dimensional matrix equation.

If $g_n(x)$ forms a complete and orthonormal set, then $f(x)$ can be expanded as

$$f(x) = \sum_{n=1}^{\infty} b_n g_n(x) = \sum_{n=1}^{\infty} \langle f, g_n \rangle g_n(x). \tag{5.1.23}$$

But $g_n(x)$ can be expanded in terms of $f_m(x)$ if $f_m(x)$ is orthonormal yielding

$$g_n(x) = \sum_{m=1}^{\infty} \langle g_n, f_m \rangle f_m(x). \tag{5.1.24}$$

Therefore, (23) becomes

$$f(x) = \sum_{n,m} \langle f, g_n \rangle \langle g_n, f_m \rangle f_m(x). \tag{5.1.25}$$

Then, on comparing (20) and (25), we deduce formally that

$$\mathcal{I} = \sum_{n} g_n \rangle \langle g_n \tag{5.1.26}$$

is an identity operator. In other words, it can always be inserted in an inner product and the value of the inner product will remain unchanged (Exercise 5.3). For example, if g_n is orthonormal and complete, then

$$\sum_{n} \langle f, g_n \rangle \langle g_n, g \rangle = \langle f, g \rangle. \tag{5.1.27}$$

Moreover, a similar statement applies to f_n. Equation (26) is also the sum of the outer product between the orthonormal vectors. Hence, the sum of the outer product of a complete set of orthonormal vectors forms an identity operator.

If f is expanded in terms of f_n as in (20), then an equivalent representation of the vector f is the numbers a_n, $n = 1, 2, 3, ..., \infty$ in (17) in a space spanned by $f_n(x)$ or, equivalently, the numbers b_n, $n = 1, 2, 3, ..., \infty$ in (23) in a space spanned by $g_n(x)$. Moreover, these two bases are related by (24) via a matrix $\langle g_n, f_m \rangle$. This is analogous to the representation of a vector in a three-dimensional space. For instance, a vector $\mathbf{A}$ can be written as

$$\mathbf{A} = \hat{x} A_x + \hat{y} A_y + \hat{z} A_z = \hat{x}' A_{x'} + \hat{y}' A_{y'} + \hat{z}' A_{z'}. \tag{5.1.28}$$

Here, $(\hat{x}', \hat{y}', \hat{z}')$ is related to $(\hat{x}, \hat{y}, \hat{z})$ via a coordinate-transformation matrix. Moreover, the two representations of $\mathbf{A}$ are completely equivalent. Furthermore, the above could be written analogous to (20) or (23) as

$$\mathbf{A} = \sum_{i=1}^{3} \hat{a}_i \hat{a}_i \cdot \mathbf{A} = \sum_{i=1}^{3} \hat{a}_i' \hat{a}_i' \cdot \mathbf{A}, \tag{5.1.29}$$

where the $\hat{a}_i$ or $\hat{a}_i'$ are orthogonal unit vectors.

Just as it is convenient to write a vector $\mathbf{A}$ without explicitly displaying its constituent components or how it is represented, it is also convenient to adopt a similar notation for a vector $f(x)$. A vector $f(x)$ can then be denoted analogous to $\mathbf{A}$ as $f\rangle$ (or simply f), and its transpose as $\langle f$ (or simply f^t). Equations (20) and (23) then become

$$f\rangle = \sum_{n=1}^{\infty} f_n \rangle \langle f_n, f \rangle = \sum_{n=1}^{\infty} g_n \rangle \langle g_n, f \rangle, \tag{5.1.30}$$

analogous to (29). Since the representation of the vector $f(x)$ is comparable to explicitly stating its constituent components at each position x, one may define an orthonormal coordinate vector $x\rangle$ such that[2] $\langle x, f \rangle = f(x)$ (see Merzbacher 1970; Sakurai 1985). Hence, $f(x)$ is also called the coordinate-space representation of the vector $f\rangle$.

Assuming that $x\rangle$ is complete and orthonormal for $0 < x < a$, analogous to (30), we can write

$$f\rangle = \int_0^a dx' \, x' \rangle \langle x', f \rangle, \tag{5.1.31}$$

where $\langle x', f \rangle = f(x')$ is the component of the vector $f\rangle$ at position x'. Note that an integral in (31) replaces the summation in (30) because $x\rangle$ is continuously indexed. On finding the inner product of (31) with x, we have

$$\langle x, f \rangle = f(x) = \int_0^a dx' \, \langle x, x' \rangle f(x'), \quad 0 < x < a. \tag{5.1.32}$$

---

[2] This notation, often used by physicists, can be confusing, for one may think that $f(x)$ is a linear functional of x. But $x_1 \rangle + x_2 \rangle \neq (x_1 + x_2) \rangle$, and hence, $f(x_1 + x_2) \neq f(x_1) + f(x_2)$. It will be less confusing to think of $x\rangle$ as some $p_x\rangle$ which has the property that $\langle p_x, f \rangle = f(x)$. For the perplexed readers, $x\rangle$ could be replaced with $p_x\rangle$ for clarity.

Therefore,

$$\langle x, x' \rangle = \delta(x - x').$$ (5.1.33)

The above is just the orthonormality condition similar to (18) for a continuously indexed vector. Hence, the coordinate-space representation of the coordinate vector $x'\rangle$ is $\delta(x - x')$.

Analogous to (26), we have

$$\mathcal{I} = \int\limits_0^a dx\, x\rangle\langle x,$$ (5.1.34)

which is an identity operator. It can be inserted into an inner product, leaving its value unchanged. For example, inserting this into an inner product $\langle f, g \rangle$, we have

$$\langle f, g \rangle = \int\limits_0^a dx\, \langle f, x \rangle\langle x, g \rangle = \int\limits_0^a dx\, f(x)\, g(x).$$ (5.1.35)

The left-hand side of (35) is the inner product between two vectors whose representation is not explicit. The right-hand side of (35) is the explicit representation of this inner product in coordinate space. Alternatively, we can insert (26) into $\langle f, g \rangle$ to obtain

$$\langle f, g \rangle = \sum_n \langle f, g_n \rangle\langle g_n, g \rangle.$$ (5.1.36)

The above is closely related to the Parseval's theorem, as we shall show next.

§§5.1.4 Parseval's Theorem

Note that the right-hand side of (36) is the explicit representation of the inner product between two vectors f and g in the space spanned by the basis vector $g_n\rangle$. On comparing (35) and (36), we notice that the inner product $\langle f, g \rangle$ has different explicit representations for different basis vectors. In fact, Parseval's theorem can be derived easily using (35) and (36).

For example, if

$$g_n(x) = \sqrt{\frac{2}{a}}\, \sin\frac{n\pi x}{a}, \quad n = 1, 2, 3, 4, \ldots, \infty,$$ (5.1.37)

then $g_n(x)$ is orthonormal and complete in the domain $0 < x < a$. Then, we can write

$$g(x) = \sum_{n=1}^{\infty} \langle g_n, g \rangle \sqrt{\frac{2}{a}}\, \sin\frac{n\pi x}{a}, \quad 0 < x < a,$$ (5.1.38)

where $\langle g_n, g \rangle$ is the Fourier coefficient of the Fourier series. By the same token, we can write

$$f(x) = \sum_{n=1}^{\infty} \langle g_n, f \rangle \sqrt{\frac{2}{a}}\, \sin\frac{n\pi x}{a}.$$ (5.1.39)

But from (35) and (36), we have

$$\int_0^a dx\, f(x)\, g(x) = \sum_{n=1}^\infty \langle f, g_n\rangle\langle g_n, g\rangle. \tag{5.1.40}$$

The above is just the Parseval's theorem for the Fourier series. Hence, Parseval's theorem is just the statement that the inner product between two vectors is invariant with respect to their representations (Exercise 5.4).

§§5.1.5 Parseval's Theorem for Complex Vectors

It is convenient to define the inner product between two complex vectors as $\langle f^*, g\rangle$, so that $\langle f^*, f\rangle^{\frac{1}{2}}$ defines the length or the norm of a complex vector. According to this definition, a basis is orthonormal if

$$\langle f_n^*, f_m\rangle = \delta_{nm}. \tag{5.1.41}$$

Consequently, expansions in terms of orthonormal basis will take the form

$$f(x) = \sum_{n=1}^\infty f_n(x)\,\langle f_n^*, f\rangle \tag{5.1.42a}$$

or

$$f\rangle = \sum_{n=1}^\infty f_n\rangle\langle f_n^*, f\rangle, \tag{5.1.42b}$$

and the identity operator becomes

$$\mathcal{I} = \sum_{n=0}^\infty f_n\rangle\langle f_n^*. \tag{5.1.43}$$

It is easy to show, as in (33), that $\langle x^*, x'\rangle = \delta(x - x')$. Furthermore, the identity operator in this case is $\mathcal{I} = \int dx\, x\rangle\langle x^*$. Moreover, Parseval's theorem for complex vectors can be similarly derived. For example, if $f(x)$ is defined over an infinite domain, using Fourier transform, we can write

$$f(x) = \frac{1}{\sqrt{2\pi}} \int_{-\infty}^\infty dk\, e^{ikx} f(k). \tag{5.1.44}$$

Similarly, we can define a coordinate vector $x\rangle$ such that $f(x) = \langle x^*, f\rangle$, and $x\rangle$ is complete and orthonormal. Analogous to the coordinate vector $x\rangle$, we can define a spectral vector $k\rangle$ which is complete and orthonormal such that $\langle k^*, f\rangle = f(k),$[3] $\langle k^*, k'\rangle = \delta(k - k')$, and $\mathcal{I} = \int dk\, k\rangle\langle k^*$. Then,

$$f(x) = \langle x^*, f\rangle = \int_{-\infty}^\infty dk\, \langle x^*, k\rangle\langle k^*, f\rangle. \tag{5.1.45}$$

---

[3] Note again that the vector $k\rangle$ is actually a vector $s_k\rangle$ with the property that $\langle s_k^*, f\rangle = f(k)$.

Consequently, on comparing (44) and (45), we conclude that

$$\langle x^*, k \rangle = \frac{1}{\sqrt{2\pi}} \, e^{ikx}. \tag{5.1.46}$$

Note that the above is consistent with the idea that $\langle k^*, k' \rangle = \delta(k - k')$ because

$$\langle k^*, k' \rangle = \int_{-\infty}^{\infty} dx \, \langle k^*, x \rangle \langle x^*, k' \rangle$$

$$= \frac{1}{2\pi} \int_{-\infty}^{\infty} e^{i(k'-k)x} \, dx = \delta(k' - k). \tag{5.1.47}$$

As a result, Parseval's theorem follows from

$$\langle f^*, g \rangle = \int dx \, \langle f^*, x \rangle \langle x^*, g \rangle = \int dx \, f^*(x) \, g(x)$$

$$= \int dk \, \langle f^*, k \rangle \langle k^*, g \rangle = \int dk \, f^*(k) \, g(k). \tag{5.1.48}$$

Moreover, a similar Parseval's theorem can be established for the Fourier series (see Exercise 5.5).

§§5.1.6 Solutions to Operator Equations—A Preview

A linear operator defined in Equation (10) can be written symbolically as

$$\mathcal{G}f \rangle = h \rangle. \tag{5.1.49}$$

If $f \rangle$ is unknown, but $h \rangle$ is known, the above is known as an operator equation. It is analogous to the matrix equation $\overline{\mathbf{G}} \cdot \mathbf{f} = \mathbf{h}$. Then, with a countably infinite, orthonormal complete set $\{g_n(x)\}$, or $g_n \rangle$, and using the identity operator defined in (26), we can rewrite (49) as

$$\sum_{n=1}^{\infty} \mathcal{G} g_n \rangle \langle g_n, f \rangle = h \rangle. \tag{5.1.50}$$

On taking the inner product of the above with $\langle w_m$, one obtains

$$\sum_{n=1}^{\infty} \langle w_m, \mathcal{G} g_n \rangle \langle g_n, f \rangle = \langle w_m, h \rangle. \tag{5.1.51}$$

In the above, $w_m(x)$ is called the **weighting function** or the **testing function**. The above is an infinite-dimensional matrix equation. Moreover, note that the unknown vector $f\rangle$ is now expanded in terms of $g_n\rangle$ but with unknown coefficients $\langle g_n, f\rangle$. That is,

$$f\rangle = \sum_n g_n\rangle\langle g_n, f\rangle. \tag{5.1.52}$$

Consequently, Equation (51) can be truncated and inverted as a matrix equation to yield $\langle g_n, f\rangle$. Once $\langle g_n, f\rangle$ is known, $f\rangle$ is found from (52). The matrix $\langle w_m, \mathcal{G}g_n\rangle$ is known as the **matrix representation** of the operator $\mathcal{G}$ in the space spanned by $w_m\rangle$ and $g_n\rangle$. But for complex vectors where the inner product is defined as $\langle f^*, g\rangle$, the matrix representation of $\mathcal{G}$ is defined as $\langle w_m^*, \mathcal{G}g_n\rangle$ in the space spanned by $w_m\rangle$ and $g_n\rangle$.

The **transpose** of an operator $\mathcal{G}$, denoted $\mathcal{G}^t$, is an operator such that

$$\langle f, \mathcal{G}g\rangle = \langle g, \mathcal{G}^t f\rangle.$$

Hence, an operator $\mathcal{G}$ is **symmetric** if

$$\langle f, \mathcal{G}g\rangle = \langle g, \mathcal{G}f\rangle. \tag{5.1.53}$$

In other words, its transpose is itself. Then, its matrix representation is also symmetric if $w_m = g_m$ in (51). The **adjoint**[4] of an operator $\mathcal{G}$, denoted $\mathcal{G}^a$, is an operator such that $\langle f^*, \mathcal{G}g\rangle = \langle g^*, \mathcal{G}^a f\rangle^*$. An operator $\mathcal{G}$ is **self-adjoint** or **Hermitian** if

$$\langle f^*, \mathcal{G}g\rangle = \langle g^*, \mathcal{G}f\rangle^*. \tag{5.1.54}$$

Consequently, its matrix representation in a complex inner product space is also Hermitian if $w_m = g_m$. Notice that the above definitions are similar to those for matrix operators (Exercise 5.6).

This method of solving an operator equation like (49) can be extended to using a nonorthogonal but complete basis set. In this case, we expand the unknown $f\rangle$ in terms of $g_n\rangle$ with the unknown coefficients a_n as

$$f\rangle = \sum_n g_n\rangle a_n. \tag{5.1.55}$$

Then, after substituting (55) into (49), one obtains

$$\sum_n \mathcal{G}g_n\rangle a_n = h\rangle. \tag{5.1.56}$$

Moreover, by finding the inner product of the above with $\langle w_m$, one obtains

$$\sum_n \langle w_m, \mathcal{G}g_n\rangle a_n = \langle w_m, h\rangle. \tag{5.1.57}$$

---

[4] Some books do not distinguish between a transpose operator and an adjoint operator. This distinction is important in electromagnetics because both inner products $\langle f, g\rangle$ and $\langle f^*, g\rangle$ are used (see Cairo and Kahan 1965).

Consequently, a_n can be solved by inverting a truncated version of the matrix representation of $\mathcal{G}$. Then, $f\rangle$ can be found from (55). This method of solving an operator equation is also known as the **Petrov-Galerkin's method** (Petrov 1940), the **method of weighted residuals** (Finlayson and Scriven 1967, and references therein), or the **method of moments** (Kravchuk 1932, 1936; Kantorovich and Krylov 1958; Harrington 1968).

If instead, w_m are chosen to be Dirac delta functions, the above method is known as the **point-matching method** (Harrington 1968), or the **method of collocation.**[5] Notice that if $\mathcal{G}$ is an integral operator, then to find its matrix representation $\langle w_m, \mathcal{G}g_n \rangle$ requires two-fold integration, which could be numerically quite intensive. But the method of collocation allows one of these integrals to be evaluated in closed form. On the other hand, if $\mathcal{G}$ is a differential operator, finding its matrix representation requires only a single integration (Exercise 5.7). If, however, the method of collocation is used, no integration is involved.

When the basis sets used are subdomain basis functions,[6] it is known as the **finite-element method.**[7] Sometimes, it may be convenient to find the inner product of (56) with a same set of functions, thereby deriving

$$\sum_n \langle g_m, \mathcal{G}g_n \rangle \, a_n = \langle g_m, h \rangle. \qquad (5.1.58)$$

For example, this yields a symmetric matrix when $\mathcal{G}$ is a symmetric operator. This is then known as **Galerkin's method** (1915).

There is no reason to suppose that (49) has a unique solution. For example, if $\mathcal{G}$ is a differential operator, the solution to (49) is nonunique without stipulated boundary conditions. In this case, the solution to (57) or (58) will be nonunique unless $g_n(x)$ satisfies some boundary conditions (Exercise 5.7).

As a result of their versatility, subdomain basis functions are widely used to approximate functions defined over a complex-shaped domain. Also, because of their simplicity, many integrals in the matrix representations can be evaluated in closed forms, avoiding the need for actual numerical integrations. Furthermore, their use in the finite-element method for solving partial differential equations yields sparse matrices which are banded. This reduces the memory and computational requirements of finite-element codes (Axelsson and Barker 1984).

---

[5] These are all classified as Galerkin-type methods by some authors.

[6] A subdomain basis function $g_n(x)$ is a function defined only over a subset of the domain of the functions $f(x)$ on which $\mathcal{G}$ operates.

[7] The finite-element method is sometimes used as a generic name for methods wherein subdomain basis functions are used. Many books on finite-element method are listed in the references.

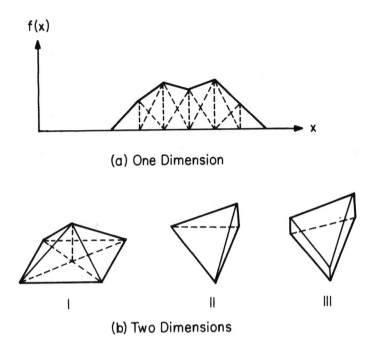

Figure 5.1.1 Some subdomain basis functions for piecewise-linear approximations of a smooth function. (a) A one-dimensional function can be piecewise-linearly approximated by a linear superposition of triangular or hat (chapeau) functions. (b) In two dimensions, a piecewise-linear function can be approximated by pyramidal functions as in (i), or it can be approximated by the subdomain basis functions as shown in (ii) and (iii).

Examples of subdomain basis functions are shown in Figure 5.1.1. A one-dimensional domain is a union of line segments. A two-dimensional domain can be approximated by a union of triangles, while a three-dimensional domain can be approximated by a union of tetrahedrons. Hence, in one dimension, a smooth function can be approximated by a piecewise-linear function which in turn is a linear superposition of triangular functions. In two dimensions, a function can be approximated by a piecewise-linear function defined over a triangle. This in turn can be approximated by a linear superposition of pyramidal functions as shown in Figure 5.1.1(b.i). Moreover, a piecewise-linear function is also a linear superposition of the functions shown in Figure 5.1.1(b.ii) and (b.iii). Furthermore, this can be generalized to tetrahedrons in three dimensions.

In addition to linear interpolations between the nodes in the finite-element method, more sophisticated quadratic or spline interpolations can be devised (e.g., see Strang and Fix 1973; Axelsson and Barker 1984; Wait and Mitchell 1985). Moreover, because of their popularity, the matrix representations of some commonly used operators like the Laplacian ∇^2 are tabulated for such subdomain basis functions (see Silvester and Ferrari 1983). A whole body of literature has been written about the finite-element method, some of which are listed in the references.

§§5.1.7 The Eigenvalue Problem

The eigenvalue problem surfaces when one calculates the resonant or guided mode of a structure. We shall discuss some salient features of the eigenvalue problem.

If

$$\mathcal{G}f_i\rangle = \lambda_i f_i\rangle, \tag{5.1.59}$$

$f_i\rangle$ is known as the eigenvector of $\mathcal{G}$, and λ_i is the corresponding eigenvalue. If $\mathcal{G}$ is symmetric, however, then eigenvectors corresponding to distinct eigenvalues are orthogonal. For if

$$\mathcal{G}f_j\rangle = \lambda_j f_j\rangle, \tag{5.1.60}$$

then by finding the products of (59) with $\langle f_j$ and (60) with $\langle f_i$, and using the definition of symmetric operator, one obtains

$$\lambda_i\langle f_j, f_i\rangle = \lambda_j\langle f_i, f_j\rangle. \tag{5.1.61}$$

Therefore, if $\lambda_i \neq \lambda_j$, then $\langle f_j, f_i\rangle = 0$ when $i \neq j$, implying mutual orthogonality.

If $\mathcal{G}$ is Hermitian, however, the eigenvalues are always real, and the eigenvectors are orthogonal with a complex inner product. We can show this by finding the product of (59) with $\langle f_j^*$ and (60) with $\langle f_i^*$. Then, using the definition of Hermitian operators, one obtains

$$\lambda_i\langle f_j^*, f_i\rangle = \lambda_j^*\langle f_j^*, f_i\rangle. \tag{5.1.62}$$

If $i = j$, the above implies that $\lambda_i = \lambda_i^*$ or that λ_i is always real. Furthermore, if λ_i and λ_j are distinct, then $\langle f_j^*, f_i\rangle = 0$ implying orthogonality. Moreover, the eigenvectors of a Hermitian operator form a complete basis, but the eigenvectors of a symmetric operator may not form a complete basis (see Exercise 5.8).

In electromagnetics, if a medium is reciprocal and linear, the linear operators representing the vector wave equations are always symmetric (see Exercise 5.9). Moreover, if the medium is lossless, the operators are also Hermitian. But if the medium is lossy, the operators are symmetric but not Hermitian. The symmetric and Hermitian properties of these operators save computation time in finding the matrix representation of an operator. For example, in Equation (57), if we use N basis functions for our matrix representation, we need to compute only $\frac{N(N+1)}{2}$ elements instead of N^2 elements. The computation of the matrix element $\langle g_m, \mathcal{G}g_n \rangle$ often involves double integrals. Hence, if these integrals cannot be computed in closed form, the computation time needed to compute $\langle g_m, \mathcal{G}g_n \rangle$ can be extensive. Consequently, the time needed to fill the matrix elements of the matrix representation can be substantial.

§5.2 Variational Expressions for Self-Adjoint Problems

The solutions of the operator equations have been discussed in a cursory manner in the previous section. Many of the methods previously mentioned, however, such as the Galerkin's method, moment method, and the finite-element method, can be associated with a variational method. In order to understand the convergence properties of these methods, it is helpful to study their variational properties. It can be shown that these methods follow from the application of the Rayleigh-Ritz[8] procedure to their corresponding variational expressions. As discussed previously, operators can be classified as self-adjoint or non-self-adjoint. The corresponding variational expressions for these two classes of operators are quite different. Hence, we shall discuss first the case of the self-adjoint operators. The case of non-self-adjoint operators shall be considered in the next section. Moreover, this topic is usually discussed in books on finite-element method listed in the references, and is also covered in Cairo and Kahan (1965), and Harrington (1968).

§§5.2.1 General Concepts

The derivation of variational expressions for self-adjoint or symmetric problems can be considered in the context of linear integral or differential equations. First, consider the operator equation

$$\mathcal{L}f(\mathbf{r}) = g(\mathbf{r}), \tag{5.2.1}$$

where $\mathcal{L}$ is a linear operator. As mentioned in (5.1.54), $\mathcal{L}$ is self-adjoint if

$$\langle f_2^*, \mathcal{L}f_1 \rangle = \langle (\mathcal{L}f_2)^*, f_1 \rangle = \langle f_1^*, \mathcal{L}f_2 \rangle^*. \tag{5.2.2}$$

But if the complex conjugate is not used in the inner product, and

$$\langle f_2, \mathcal{L}f_1 \rangle = \langle f_1, \mathcal{L}f_2 \rangle, \tag{5.2.3}$$

then $\mathcal{L}$ is symmetric. In the above, the inner product in its coordinate-space representation is

$$\langle f_1, f_2 \rangle = \int d\mathbf{r} \, f_1(\mathbf{r}), f_2(\mathbf{r}). \tag{5.2.4}$$

---

[8] Named after Lord Rayleigh, a prodigious English scientist, and Walter Ritz, a Swiss mathematician.

Consequently, a variational expression for Equation (1) when $\mathcal{L}$ is self-adjoint is

$$I = \langle f^*, \mathcal{L}f \rangle - \langle f^*, g \rangle - \langle g^*, f \rangle. \tag{5.2.5}$$

Note that I is real valued in (5) (see Exercise 5.10). Moreover, it is a nonlinear functional of f. The stationary point of the functional I with respect to variations in f in an infinite-dimensional space is a solution to (1). To show this, we take the first variation of (5) about the solution of (1) and show that indeed $\delta I = 0$. So, we let

$$f = f_e + \delta f, \tag{5.2.6}$$

where f_e is the exact solution to (1). Next, by substituting (6) into (5) and equating the first-order terms, we have

$$\delta I = \langle f_e^*, \mathcal{L}\delta f \rangle + \langle \delta f^*, \mathcal{L}f_e \rangle - \langle \delta f^*, g \rangle - \langle g^*, \delta f \rangle. \tag{5.2.7}$$

But from the self-adjoint property of $\mathcal{L}$, $\langle f_e^*, \mathcal{L}\delta f \rangle = \langle \delta f^*, \mathcal{L}f_e \rangle^*$. This together with (1) implies that the right-hand side of (7) equals zero. Therefore, δI or the first variation of the functional about f_e, the exact solution, is zero implying the stationary property of I about f_e.

Moreover, if $\mathcal{L}$ is a symmetric operator, then it can be shown that the corresponding variational expression is

$$I = \langle f, \mathcal{L}f \rangle - 2\langle f, g \rangle. \tag{5.2.8}$$

The first variation in I about the exact solution f_e in the above can again be shown to be zero. Notice that if f is a complex function, then the functional I need not be real valued in this case.

The advantage of the variational expressions (5) and (8) is that they give an alternate formulation of the solution to (1). Here, the solution to (1) is obtained by finding the f such that the first variation of I vanishes with a first variation of f in either (5) or (8), depending on whether $\mathcal{L}$ is self-adjoint or symmetric. Because of the quadratic nature of expression (5) or (8), it only has one stationary point. Therefore, if an f that will make I stationary is found, this f must be the solution to (1).

One may wonder as to where the motivation for the expressions (5) and (8) has come from. Actually, both the variational expressions (5) and (8) are extensions of the concept of the quadratic expression

$$I = ax^2 + bx \tag{5.2.9}$$

to an infinite-dimensional space. For instance, in a finite-dimensional space, (5) and (8) correspond to

$$I = \mathbf{a}^\dagger \cdot \overline{\mathbf{L}} \cdot \mathbf{a} - \mathbf{a}^\dagger \cdot \mathbf{g} - \mathbf{g}^\dagger \cdot \mathbf{a} \tag{5.2.10}$$

and

$$I = \mathbf{a}^t \cdot \overline{\mathbf{L}} \cdot \mathbf{a} - 2\mathbf{a}^t \cdot \overline{\mathbf{g}}. \tag{5.2.11}$$

Now, if $\overline{\mathbf{L}}$ is Hermitian, then I in (10) is stationary about $\mathbf{a}_e$, which is the solution to the matrix equation

$$\overline{\mathbf{L}} \cdot \mathbf{a} = \mathbf{g}. \tag{5.2.12}$$

But if $\overline{\mathbf{L}}$ is symmetric, then I in (11) is stationary about $\mathbf{a}_e$, which is the solution to (12). Moreover, the stationary point will be a minimum or a maximum depending on the positive or negative definiteness of $\overline{\mathbf{L}}$ (Exercise 5.11).

It has been shown that the functional I stated in the above has a second order error, if the error in f is of the first order. We may not, however, be directly interested in I. Hence, it may be preferable to derive a variational expression that relates directly to a measured quantity. For example, a measured quantity u may be related to f via a linear functional

$$u = \langle h^*, f \rangle, \tag{5.2.13}$$

where f is a solution to (1). To find a variational expression for u, we define an auxiliary equation as

$$\mathcal{L} f_a = h. \tag{5.2.14}$$

Then, when $\mathcal{L}$ is self-adjoint, a variational expression for u, the measured quantity, is (Jones 1956)

$$u = \frac{\langle f_a^*, g \rangle \langle h^*, f \rangle}{\langle f_a^*, \mathcal{L}f \rangle}. \tag{5.2.15}$$

To show that (15) is in fact variational, we cross-multiply (15) and take its first variation about the exact solution, yielding

$$u_e \left[\langle f_{ae}^*, \mathcal{L}\delta f \rangle + \langle \delta f_a^*, \mathcal{L}f_e \rangle \right] + \delta u \langle f_{ae}^*, \mathcal{L}f_e \rangle$$
$$= \langle \delta f_a^*, g \rangle \langle h^*, f_e \rangle + \langle f_{ae}^*, g \rangle \langle h^*, \delta f \rangle, \tag{5.2.16}$$

where the subscript e denotes exact values of the functions f, f_a and the functional u. By making use of (1), (13), (14), the self-adjointness of $\mathcal{L}$, and the fact that

$$\langle f_{ae}^*, g \rangle = \langle f_{ae}^*, \mathcal{L}f_e \rangle = \langle f_e^*, \mathcal{L}f_{ae} \rangle^* = \langle h^*, f_e \rangle, \tag{5.2.17}$$

we can show that $\delta u = 0$. This analysis implies that, given rough estimates of f and f_a which depart from the exact solutions by δf and δf_a respectively, the error in u estimated via (15) is of higher order. Hence, to obtain a good estimate of u, we need a good approximation of the solution to the auxiliary problem in addition to the original problem.

When $\mathcal{L}$ is symmetric but not self-adjoint, an expression corresponding to Equation (15) becomes

$$u = \frac{\langle f_a, g \rangle \langle f, h \rangle}{\langle f_a, \mathcal{L}f \rangle}, \tag{5.2.18}$$

which can easily be shown to be a variational expression (Exercise 5.12).

§§5.2.2 Rayleigh-Ritz Procedure—Self-Adjoint Problems

Given a variational expression, there exists a systematic way to obtain the best estimate of I in (5) leading to the solution of (1). This is the Rayleigh-Ritz procedure. Note that the meaning of the variational expression in (5) is that when an estimate of f with an error of δf is substituted into (5), an even better estimate of I with a higher order error of order $(\delta f)^2$ is obtained. Hence, we can substitute trial functions admissible by $\mathcal{L}$ into (5) to obtain a better estimate of I. (A function f is admissible by $\mathcal{L}$ if $\mathcal{L}f$ is not singular.) Another meaning of (5) is that I achieves a stationary value only for the value of f that satisfies Equation (1). This knowledge can be used to develop a systematic way to obtain an approximate solution to (1).

The trial field can be given N degrees of freedom by expanding it in terms of N known basis functions. That is, let

$$f = \sum_{n=1}^{N} a_n f_n, \qquad (5.2.19)$$

where f_n's are known basis functions, and a_n's are unknowns yet to be determined. Now, if f_n is from a complete set, then there exist values of a_n's which will form an increasingly good approximation of f via (19) when $N \to \infty$. Moreover, if the a_n's are chosen to make I stationary, then in the limit when $N \to \infty$, f in (19) approaches the value f_e, which is the exact solution to (1). This is the mark of the Rayleigh-Ritz procedure. It can be placed on a more systematic ground as follows.

On using (19) in (5), we have

$$I = \sum_{n=1}^{N} \sum_{n'=1}^{N} a_{n'}^* a_n \langle f_{n'}^*, \mathcal{L} f_n \rangle - \sum_{n=1}^{N} a_n^* \langle f_n^*, g \rangle - \sum_{n=1}^{N} a_n \langle g^*, f_n \rangle. \qquad (5.2.20)$$

The above is the same as

$$I = \mathbf{a}^\dagger \cdot \overline{\mathbf{L}} \cdot \mathbf{a} - 2\Re e \, [\mathbf{a}^\dagger \cdot \mathbf{g}], \qquad (5.2.21)$$

where

$$[\overline{\mathbf{L}}]_{mn} = \langle f_m^*, \mathcal{L} f_n \rangle, \qquad (5.2.22)$$

$$[\mathbf{g}]_n = \langle f_n^*, g \rangle, \qquad (5.2.23)$$

and $\mathbf{a}$ is a column vector that contains a_n's. Since I is quadratic as given by (21), it has only one stationary point in the space spanned by a_n's. Then, the optimal $\mathbf{a}$ is the $\mathbf{a}$ that will make I in (21) stationary. Hence, around the optimal values of a_n's, the first variation of I should vanish. That is, by letting $\mathbf{a} = \mathbf{a}_0 + \delta\mathbf{a}$ in (21), where $\mathbf{a}_0$ is the optimal $\mathbf{a}_0$, and collecting the first order terms, we have

$$\delta I = \delta\mathbf{a}^\dagger \cdot \overline{\mathbf{L}}\mathbf{a}_0 + \mathbf{a}_0^\dagger \cdot \overline{\mathbf{L}} \cdot \delta\mathbf{a} - 2\Re e\,[\delta\mathbf{a}^\dagger \cdot \mathbf{g}] = 0. \qquad (5.2.24)$$

The last equality follows because $\delta I = 0$ for variation about the optimal $\mathbf{a}_0$.

Since $\mathcal{L}$ is a self-adjoint operator, $\overline{\mathbf{L}}$ is a Hermitian matrix. Therefore, in order for $\delta I = 0$, one requires

$$2\Re e\,[\delta\mathbf{a}^\dagger \cdot \overline{\mathbf{L}} \cdot \mathbf{a}_0] = 2\Re e\,[\delta\mathbf{a}^\dagger \cdot \mathbf{g}]. \qquad (5.2.25)$$

But since $\delta\mathbf{a}^\dagger$ is arbitrary, one obtains

$$\overline{\mathbf{L}} \cdot \mathbf{a}_0 = \mathbf{g}. \qquad (5.2.26)$$

Consequently, the optimal value of $\mathbf{a}$ to be used in (19) is the solution to Equation (26). Note that Equation (26) can also be obtained from (1) directly by applying Galerkin's method described in Section 5.1. However, the above analysis indicates the link between Galerkin's method and the variational method. Furthermore, it also shows the underpinning principle behind Galerkin's method.

By the same token, the Rayleigh-Ritz procedure can also be applied to Equation (15). In this case, we let

$$f = \sum_{n=1}^{N} a_n f_n, \qquad (5.2.27)$$

$$f_a = \sum_{m=1}^{N} b_m f_{am}. \qquad (5.2.28)$$

On substituting them into (15), we have

$$u = \frac{\displaystyle\sum_{m=1}^{N} \sum_{n=1}^{N} b_m^* a_n \langle f_{am}^*, g\rangle \langle h^*, f_n\rangle}{\displaystyle\sum_{m=1}^{N} \sum_{n=1}^{N} b_m^* a_n \langle f_{am}^*, \mathcal{L}f_n\rangle}. \qquad (5.2.29)$$

The above can be written more compactly as

$$u = \frac{\mathbf{b}^\dagger \cdot \mathbf{g}\mathbf{h}^\dagger \cdot \mathbf{a}}{\mathbf{b}^\dagger \cdot \overline{\mathbf{L}} \cdot \mathbf{a}}, \qquad (5.2.30)$$

where $[\mathbf{g}]_m = \langle f_{am}^*, g\rangle$, $[\mathbf{h}]_n = \langle f_n^*, h\rangle$, $[\overline{\mathbf{L}}]_{mn} = \langle f_{am}^*, \mathcal{L}f_n\rangle$, and $\mathbf{a}$ and $\mathbf{b}$ are column vectors containing a_n's and b_n's. The optimal values of $\mathbf{a}$ and $\mathbf{b}$ are then obtained by requiring that they make (30) stationary. Hence, the first variation of u due to the first variations of $\mathbf{a}$ and $\mathbf{b}$ about their optimal values, $\mathbf{a}_0$ and $\mathbf{b}_0$, must vanish. Note that both (15) and (30) have only one stationary point. Hence, by demanding that the optimal values of $\mathbf{a}$ and $\mathbf{b}$ in (30) be at a stationary point, f, f_a, and u thus obtained will converge to the exact values when $N \to \infty$.

Consequently, after cross-multiplying (30) and taking its first variation by letting $\mathbf{a} = \mathbf{a}_0 + \delta \mathbf{a}$ and $\mathbf{b} = \mathbf{b}_0 + \delta \mathbf{b}$, we have

$$\delta u\, \mathbf{b}_0^\dagger \cdot \overline{\mathbf{L}} \cdot \mathbf{a}_0 + u_0 \left(\delta \mathbf{b}^\dagger \cdot \overline{\mathbf{L}} \cdot \mathbf{a}_0 + \mathbf{b}_0^\dagger \cdot \overline{\mathbf{L}} \cdot \delta \mathbf{a} \right)$$
$$= \delta \mathbf{b}^\dagger \cdot \mathbf{g}\mathbf{h}^\dagger \cdot \mathbf{a}_0 + \mathbf{b}_0^\dagger \cdot \mathbf{g}\mathbf{h}^\dagger \cdot \delta \mathbf{a}. \quad (5.2.31)$$

In order for $\delta u = 0$ for arbitrary $\delta \mathbf{a}$ and $\delta \mathbf{b}$ so that $\mathbf{a}_0$ and $\mathbf{b}_0$ are at the stationary point of (30), we require that

$$u_0\, \delta \mathbf{b}^\dagger \cdot \overline{\mathbf{L}} \cdot \mathbf{a}_0 = \delta \mathbf{b}^\dagger \cdot \mathbf{g}\mathbf{h}^\dagger \cdot \mathbf{a}_0, \quad (5.2.32a)$$

$$u_0\, \mathbf{b}_0^\dagger \cdot \overline{\mathbf{L}} \cdot \delta \mathbf{a} = \mathbf{b}_0^\dagger \cdot \mathbf{g}\mathbf{h}^\dagger \cdot \delta \mathbf{a}. \quad (5.2.32b)$$

But since $\delta \mathbf{a}$ and $\delta \mathbf{b}$ are arbitrary, the above implies that

$$u_0 \overline{\mathbf{L}} \cdot \mathbf{a}_0 = \mathbf{g}\mathbf{h}^\dagger \cdot \mathbf{a}_0, \quad (5.2.33a)$$
$$u_0 \mathbf{b}_0^\dagger \cdot \overline{\mathbf{L}} = \mathbf{b}_0^\dagger \cdot \mathbf{g}\mathbf{h}^\dagger, \quad (5.2.33b)$$

where

$$u_0 = \frac{\mathbf{b}_0^\dagger \cdot \mathbf{g}\mathbf{h}^\dagger \cdot \mathbf{a}_0}{\mathbf{b}_0^\dagger \cdot \overline{\mathbf{L}} \cdot \mathbf{a}_0}. \quad (5.2.34)$$

Note that Equation (33a) is actually an eigenvalue problem where u_0 is the eigenvalue, and $\mathbf{a}_0$ is the eigenvector. But the matrix $\mathbf{g}\mathbf{h}^\dagger$, formed by the outer product of $\mathbf{g}$ and $\mathbf{h}^*$, has a nullspace of rank $N - 1$, because there are $N - 1$ independent vectors $\mathbf{v}$ in an N-dimensional space that are orthogonal to $\mathbf{h}$ that could make $\mathbf{g}\mathbf{h}^\dagger \cdot \mathbf{v} = 0$. Consequently, there is only one nontrivial eigenvalue u_0. By inspection, if

$$\overline{\mathbf{L}} \cdot \mathbf{a}_0 = \mathbf{g}, \quad (5.2.35a)$$

$$\overline{\mathbf{L}} \cdot \mathbf{b}_0 = \mathbf{h}, \quad (5.2.35b)$$

then Equations (33a) and (33b) are satisfied, and δu would be zero. Furthermore, from (34),

$$u_0 = \mathbf{h}^\dagger \cdot \mathbf{a}_0 = \mathbf{b}_0^\dagger \cdot \mathbf{g}. \quad (5.2.36)$$

Therefore, the optimal $\mathbf{a}_0$ and $\mathbf{b}_0$ are obtained by solving (35).

Note that in (35), the solutions to the actual problem and the auxiliary problem are actually decoupled. Consequently, we need only solve one of the two equations in (35) and obtain u_0 via (36). Therefore, if we were to solve (35a), the only involvement with the auxiliary problem is via the matrix $\left[\overline{\mathbf{L}} \right]_{mn} = \langle f_{am}^*, \mathcal{L} f_n \rangle$. The connection with the auxiliary problem seems tenuous at this point. How does the auxiliary problem then play a role in the convergence of the functional u to its exact value? But note that the preceding analysis and the analysis in Equation (16) indicate that the error in u is small only if both the errors in the original and the auxiliary problems are small. Hence, in computing the matrix $\overline{\mathbf{L}}$, f_{am}, the basis function used for solving the auxiliary problem, should be chosen so that it can best approximate the solution to the auxiliary equation.

Equations (35a) and (35b) can also be obtained by the direct application of the method of weighted residuals or method of moments (described in Section 5.1) to (1) and (14). But the above analysis shows the connection of (35) with a variational expression and an auxiliary problem. For instance, in the method of weighted residuals, it is not obvious how the weighting functions should be chosen. But now the above analysis provides a guideline for the choice of the expansion and weighting functions.

§§5.2.3 *Applications to Scalar Wave Equations*

Given a scalar wave equation for an inhomogeneous medium

$$\nabla \cdot p(\mathbf{r})\nabla \phi(\mathbf{r}) + k^2(\mathbf{r})\, p(\mathbf{r})\, \phi(\mathbf{r}) = s(\mathbf{r}), \tag{5.2.37}$$

we can identify the linear operator as

$$\mathcal{L} = \nabla \cdot p\nabla + k^2 p. \tag{5.2.38}$$

Moreover, when p and k^2 are real, corresponding to a lossless medium, it can be shown that $\mathcal{L}$ is self-adjoint for ϕ satisfying appropriate boundary conditions, because

$$\langle \phi_1^*, \nabla \cdot p\nabla \phi_2 \rangle = \langle \phi_2^*, \nabla \cdot p\nabla \phi_1 \rangle^* \tag{5.2.39a}$$

and

$$\langle \phi_1^*, k^2 p\, \phi_2 \rangle = \langle \phi_2^*, k^2 p\, \phi_1 \rangle^*, \tag{5.2.39b}$$

where the inner product

$$\langle f, g \rangle = \int_V d\mathbf{r}\, f(\mathbf{r})\, g(\mathbf{r}), \tag{5.2.40}$$

and V is the domain over which Equation (37) is defined.

To show (39a), we use integration by parts, or Gauss' theorem. Then, the left-hand side of (39a) is just

$$\int_V d\mathbf{r}\, \phi_1^*(\mathbf{r})\nabla \cdot p\nabla \phi_2(\mathbf{r}) = \int_S dS\, \hat{n} \cdot \phi_1^*(\mathbf{r})\, p\nabla \phi_2(\mathbf{r}) - \int_V d\mathbf{r}\, p\nabla \phi_1^*(\mathbf{r}) \cdot \nabla \phi_2(\mathbf{r}).$$

$$\tag{5.2.41}$$

If the ϕ_i's satisfy either the homogeneous Dirichlet or homogeneous Neumann boundary condition on S, then the first integral on the right-hand side of (41) vanishes. With these constraints on ϕ_1 and ϕ_2, we have

$$\int_V d\mathbf{r}\, \phi_1^*(\mathbf{r}) \nabla \cdot p \nabla \phi_2(\mathbf{r}) = - \int_V d\mathbf{r}\, p \nabla \phi_1^*(\mathbf{r}) \cdot \nabla \phi_2(\mathbf{r}). \qquad (5.2.42)$$

Note that the right-hand side of (42) is symmetrical about ϕ_1^* and ϕ_2. Clearly, (39a) is satisfied. Moreover, Equation (39b) follows easily from the definition of the inner product (40). Hence, $\mathcal{L}$ is a self-adjoint operator when operating on a class of functions satisfying the requisite boundary condition. Consequently, a variational expression for (37), according to Equation (5), is

$$I = - \int_V d\mathbf{r}\, p |\nabla \phi(\mathbf{r})|^2 + \int_V d\mathbf{r}\, k^2 p |\phi(\mathbf{r})|^2 - 2\Re e \int_V d\mathbf{r}\, \phi^*(\mathbf{r})\, s(\mathbf{r}). \qquad (5.2.43)$$

If p and k^2 are complex, however, corresponding to lossy or active media, then the operator $\mathcal{L}$ in (38) is not self-adjoint, but it is still symmetric. In this case, applying the rule of Equation (8), a variational expression is

$$I = - \int_V d\mathbf{r}\, p \left(\nabla \phi(\mathbf{r})\right)^2 + \int_V d\mathbf{r}\, k^2 p \phi^2(\mathbf{r}) - 2 \int_V d\mathbf{r}\, \phi(\mathbf{r})\, s(\mathbf{r}). \qquad (5.2.44)$$

The expression I in Equation (43) is real valued and can be associated with the energy of the system. The expression I in Equation (44) cannot be directly associated with energy because it is a complex number in general. But since Equations (43) and (44) are variational expressions, we can use the Rayleigh-Ritz procedure to systematically obtain an approximate solution to (37) (Exercise 5.13).

Conversely, if the value of interest is the field at a point, then the variational expression to use is (15). For example, if the value of interest in the solution to (37) is $\phi(\mathbf{r})$ at $\mathbf{r} = \mathbf{r}_0$, then the value of interest, according to Equation (13), is

$$u = \phi(\mathbf{r}_0) = \langle h^*, \phi(\mathbf{r}) \rangle, \qquad (5.2.45)$$

where $h^* = \delta(\mathbf{r} - \mathbf{r}_0)$. Accordingly, from (14), the auxiliary problem is

$$\nabla \cdot p(\mathbf{r}) \nabla \phi_a(\mathbf{r}) + k^2(\mathbf{r}) p(\mathbf{r})\, \phi_a(\mathbf{r}) = \delta(\mathbf{r} - \mathbf{r}_0). \qquad (5.2.46)$$

In accordance with (15), the variational expression for $u = \hat{\phi}(\mathbf{r}_0)$ is[9]

$$u = \hat{\phi}(\mathbf{r}_0) = \frac{\phi(\mathbf{r}_0) \int_V d\mathbf{r}\, \phi_a^* s}{- \int_V d\mathbf{r}\, p \nabla \phi_a^* \cdot \nabla \phi + \int_V d\mathbf{r}\, k^2 p\, \phi_a^* \phi}. \qquad (5.2.47)$$

---

[9] We use $\hat{\phi}(\mathbf{r}_0)$ to denote the desired value for which the variational expression is written and $\phi(\mathbf{r}_0)$ to denote the trial function.

Hence, given a rough estimate of ϕ_a and ϕ, they can be substituted in (47) to obtain an estimate of $\hat{\phi}(\mathbf{r}_0)$, whose error will be of the second order. Moreover, the Rayleigh-Ritz procedure can be used to find an optimal approximation to (47) (Exercise 5.13). As previously discussed, the optimal ϕ and ϕ_a in (47) are obtained by solving (37) and (46) by the method of weighted residuals. According to (46) then, $\phi_a(\mathbf{r})$ is the field due to a point source at $\mathbf{r} = \mathbf{r}_0$, which is singular at $\mathbf{r} = \mathbf{r}_0$. Hence, the expansion functions for ϕ_a [which is the weighting function for solving (37)] should be chosen to closely approximate the singular field ϕ_a. As a result, a better approximation for u in (47) will be obtained.

§§5.2.4 Applications to Vector Wave Equations

Given the vector wave equation

$$\nabla \times \mu^{-1}\nabla \times \mathbf{E}(\mathbf{r}) - \mu^{-1}k^2\,\mathbf{E}(\mathbf{r}) = i\omega\mathbf{J}(\mathbf{r}), \qquad (5.2.48)$$

the linear operator can be identified as

$$\mathcal{L} = \left(\nabla \times \mu^{-1}\nabla\times\right) - \mu^{-1}k^2. \qquad (5.2.49)$$

If μ and k^2 are real, corresponding to a lossless medium, then it can be shown that $\mathcal{L}$ is self-adjoint with appropriate boundary conditions on $\mathbf{E}$. Consequently,

$$\left\langle \mathbf{E}_1^*(\mathbf{r}), \nabla \times \mu^{-1}\nabla \times \mathbf{E}_2(\mathbf{r})\right\rangle = \left\langle \mathbf{E}_2^*(\mathbf{r}), \nabla \times \mu^{-1}\nabla \times \mathbf{E}_1(\mathbf{r})\right\rangle^* \qquad (5.2.50a)$$

and

$$\left\langle \mathbf{E}_1^*(\mathbf{r}), \mu^{-1}k^2\,\mathbf{E}_2(\mathbf{r})\right\rangle = \left\langle \mathbf{E}_2^*(\mathbf{r}), \mu^{-1}k^2\,\mathbf{E}_1(\mathbf{r})\right\rangle^*, \qquad (5.2.50b)$$

where the inner product is defined as

$$\langle \mathbf{A}, \mathbf{B}\rangle = \int_V d\mathbf{r}\,\mathbf{A}(\mathbf{r}) \cdot \mathbf{B}(\mathbf{r}), \qquad (5.2.51)$$

and V is the domain over which Equation (48) is defined.

To prove (50a), we use Gauss' theorem and the vector identity

$$\nabla \cdot (\mathbf{A} \times \mathbf{B}) = \mathbf{B} \cdot \nabla \times \mathbf{A} - \mathbf{A} \cdot \nabla \times \mathbf{B}$$

to deduce that

$$\int_V d\mathbf{r}\,\mathbf{E}_1^*(\mathbf{r}) \cdot \nabla \times \mu^{-1}\nabla \times \mathbf{E}_2(\mathbf{r}) = -\int_S dS\,\hat{n} \cdot \left[\mu^{-1}\mathbf{E}_1^*(\mathbf{r}) \times \nabla \times \mathbf{E}_2(\mathbf{r})\right]$$

$$+ \int_V d\mathbf{r}\,\mu^{-1}\nabla \times \mathbf{E}_1^*(\mathbf{r}) \cdot \nabla \times \mathbf{E}_2(\mathbf{r}). \qquad (5.2.52)$$

Now, if $\mathbf{E}(\mathbf{r})$ satisfies the electric wall boundary condition ($\hat{n} \times \mathbf{E} = 0$) or the magnetic wall boundary condition ($\hat{n} \times \nabla \times \mathbf{E} = i\omega\mu\hat{n} \times \mathbf{H} = 0$) on S, then the surface integral on the right-hand side of (52) vanishes, and

$$\int_V d\mathbf{r}\, \mathbf{E}_1^*(\mathbf{r}) \cdot \nabla \times \mu^{-1}\nabla \times \mathbf{E}_2(\mathbf{r}) = \int_V d\mathbf{r}\, \mu^{-1}\nabla \times \mathbf{E}_1^*(\mathbf{r}) \cdot \nabla \times \mathbf{E}_2(\mathbf{r}).$$
(5.2.53)

The right-hand side of the above is symmetric about $\mathbf{E}_1^*$ and $\mathbf{E}_2$. Clearly, (50a) is satisfied. Moreover, (50b) follows from the definition of the inner product (51). Consequently, a variational expression using Equation (5) is

$$I = \int_V d\mathbf{r}\, \mu^{-1}|\nabla \times \mathbf{E}(\mathbf{r})|^2 - \int_V d\mathbf{r}\, \mu^{-1}k^2\, |\mathbf{E}(\mathbf{r})|^2 - 2\Re\mathrm{e}\left[i\omega \int_V d\mathbf{r}\, \mathbf{E}^*(\mathbf{r}) \cdot \mathbf{J}(\mathbf{r})\right].$$
(5.2.54)

However, if μ and k^2 are complex, corresponding to lossy or active media, then the operator $\mathcal{L}$ in (47) is not self-adjoint but still symmetric. In this case, applying the rule of Equation (8), a variational expression is

$$I = \int_V d\mathbf{r}\, \mu^{-1}[\nabla \times \mathbf{E}(\mathbf{r})]^2 - \int_V d\mathbf{r}\, \mu^{-1}k^2\, [\mathbf{E}(\mathbf{r})]^2 - 2i\omega \int_V d\mathbf{r}\, \mathbf{E}(\mathbf{r}) \cdot \mathbf{J}(\mathbf{r}).$$
(5.2.55)

The expression (54) is real and can be related to the energy of the system. But the expression (55) is complex in general. The inner product

$$\int_V \mathbf{E} \cdot \mathbf{J}\, d\mathbf{r}$$

has the dimension of power; however, it is not related to complex power. It is the reaction between $\mathbf{E}$ and $\mathbf{J}$. Consequently, (55) can be related to the reaction between the field and the source. Then, the variational expressions (54) and (55) can be used to obtain an approximate solution to (48) via the use of the Rayleigh-Ritz procedure.

But if the $\hat{a}$ component of the field at $\mathbf{r} = \mathbf{r}_0$ is the quantity of interest, then according to Equation (13), it is

$$u = \hat{a} \cdot \mathbf{E}(\mathbf{r}_0) = \langle \mathbf{h}^*, \mathbf{E}(\mathbf{r})\rangle,$$
(5.2.56)

where $\mathbf{h}^* = \hat{a}\delta(\mathbf{r} - \mathbf{r}_0)$. Then, an auxiliary problem can be defined such that

$$\nabla \times \mu^{-1}\nabla \times \mathbf{E}_a(\mathbf{r}) - \mu^{-1}k^2\, \mathbf{E}_a(\mathbf{r}) = \hat{a}\delta(\mathbf{r} - \mathbf{r}_0).$$
(5.2.57)

In accordance with Equation (15), a variational expression for $u = \hat{a} \cdot \hat{\mathbf{E}}(\mathbf{r}_0)$ is

$$\hat{a} \cdot \hat{\mathbf{E}}(\mathbf{r}_0) = \frac{i\omega\hat{a} \cdot \mathbf{E}(\mathbf{r}_0) \int_V d\mathbf{r}\, \mathbf{E}_a^*(\mathbf{r}) \cdot \mathbf{J}(\mathbf{r})}{\int_V d\mathbf{r}\, \mu^{-1}\nabla \times \mathbf{E}_a^*(\mathbf{r}) \cdot \nabla \times \mathbf{E}(\mathbf{r}) + \int_V d\mathbf{r}\, \mu^{-1}k^2\, \mathbf{E}_a^*(\mathbf{r}) \cdot \mathbf{E}(\mathbf{r})}.$$
(5.2.58)

In Equation (58), with the first order errors in the estimates of $\mathbf{E}(\mathbf{r})$ and $\mathbf{E}_a(\mathbf{r})$, the estimate of $\hat{a} \cdot \hat{\mathbf{E}}(\mathbf{r}_0)$ will have second order error.

Note that in both (47) and (58), when the first order errors are large, the corresponding second order errors will be large. Hence, in order to obtain a good estimate of $\hat{a} \cdot \hat{\mathbf{E}}(\mathbf{r}_0)$, good approximations of $\mathbf{E}(\mathbf{r})$ and $\mathbf{E}_a(\mathbf{r})$ are needed. That is, a good estimate of the solution to the auxiliary equation in (57) is necessary as well as that of the original equation. The discussion at the end of Subsection 5.2.3 also applies here.

§5.3 Variational Expressions for Non-Self-Adjoint Problems

Many physical problems are nonsymmetric and non-self-adjoint. Despite this, we can still write down a variational expression from which a solution can be derived. For example, for a reciprocal medium, the vector wave equation is both symmetric and self-adjoint when the medium is lossless. If the medium is lossy, however, the linear operator corresponding to the vector wave equation is symmetric but not self-adjoint. Moreover, if the medium is nonreciprocal, then the corresponding linear operator is neither self-adjoint nor symmetric. For such a class of problems, an auxiliary problem has to be defined in order to derive a variational expression (see Chen and Lien 1980).

§§5.3.1 General Concepts

When a linear operator is non-self-adjoint but still symmetric, the concepts of the previous section can be used to derive a variational expression. However, if a linear operator is non-self-adjoint and nonsymmetric, a different way of deriving a variational expression is needed.

First, let us define the adjoint operator $\mathcal{L}^a$ of $\mathcal{L}$ to satisfy

$$\langle f_1^*, \mathcal{L} f_2 \rangle = \langle (\mathcal{L}^a f_1)^*, f_2 \rangle. \tag{5.3.1}$$

In addition, a transpose operator $\mathcal{L}^t$ of $\mathcal{L}$ is defined as

$$\langle f_1, \mathcal{L} f_2 \rangle = \langle f_2, \mathcal{L}^t f_1 \rangle. \tag{5.3.2}$$

If an operator is non-self-adjoint and nonsymmetric, then $\mathcal{L}^a \neq \mathcal{L}$ and $\mathcal{L}^t \neq \mathcal{L}$. Note that from the definitions (1) and (2), $(\mathcal{L}^t)^* = \mathcal{L}^a$. For the case when the complex conjugate is used in the inner product, a variational expression is then

$$I = \langle f_a^*, \mathcal{L} f \rangle - \langle f_a^*, g \rangle - \langle g_a^*, f \rangle, \tag{5.3.3}$$

where

$$\mathcal{L} f = g, \quad \mathcal{L}^a f_a = g_a. \tag{5.3.4}$$

Note that now I is a complex functional.

To prove the variational property of (3), we let $f = f_e + \delta f$, $f_a = f_{ae} + \delta f_a$, where the subscripts e stand for exact values. Then

$$\delta I = \langle f_{ae}^*, \mathcal{L} \delta f \rangle + \langle \delta f_a^*, \mathcal{L} f_e \rangle - \langle \delta f_a^*, g \rangle - \langle g_a^*, \delta f \rangle. \tag{5.3.5}$$

Moreover, using the definition of the adjoint operator and Equation (4), we can readily show that $\delta I = 0$, implying the stationarity of (3) about the exact solutions of (4). The variational expression (3) offers an alternate way of deriving solutions to Equation (4). Consequently, the Rayleigh-Ritz procedure can be systematically used to obtain approximate solutions to (4) via (3).

If the inner product is defined without complex conjugation, then a variational expression is

$$I = \langle f_a, \mathcal{L}f \rangle - \langle f_a, g \rangle - \langle g_a, f \rangle, \tag{5.3.6}$$

where

$$\mathcal{L}f = g, \quad \mathcal{L}^t f_a = g_a. \tag{5.3.7}$$

Furthermore, it is quite straightforward to show that $\delta I = 0$ for first variations in f_a and f in (6). But since $(\mathcal{L}^t)^* = \mathcal{L}^a$, Equation (6) is actually equivalent to (3) if g_a in (7) is the complex conjugate of g_a in (4). In addition, when $\mathcal{L}$ is symmetric and $g_a = g$, then from (7), $f_a = f$. In this case, Equation (6) reduces to (5.2.8). Hence, (5.2.8) can be considered a special case of (6) where $g_a = g$ and when $\mathcal{L}$ is symmetric.

On the other hand, if a quantity

$$u = \langle h^*, f \rangle \tag{5.3.8}$$

is desired, a variational expression can be derived by defining an auxiliary equation as

$$\mathcal{L}^a f_a = h. \tag{5.3.9}$$

The desired variational expression is then

$$u = \frac{\langle f_a^*, g \rangle \langle h^*, f \rangle}{\langle f_a^*, \mathcal{L}f \rangle} \tag{5.3.10}$$

similar to (5.2.15). The variational property of the above expression can be easily established similar to (5.2.16).

If a complex conjugate is not used in the inner product definition, and the desired quantity is

$$u = \langle \theta, f \rangle, \tag{5.3.11}$$

then a variational expression can be derived by defining the auxiliary equation as

$$\mathcal{L}^t f_a = \theta. \tag{5.3.12}$$

Consequently, the variational expression for u is

$$u = \frac{\langle f_a, g \rangle \langle \theta, f \rangle}{\langle f_a, \mathcal{L}f \rangle}. \tag{5.3.13}$$

Notice that since $(\mathcal{L}^t)^* = \mathcal{L}^a$, if we assume $\theta = h^*$, then f_a in (12) is precisely the complex conjugate of f_a in (8). Moreover, (13) is indeed equivalent to (10).

From the above discussions, it is evident that for non-self-adjoint problems, it is immaterial if a complex conjugate is used in an inner product. Equation (6) is essentially equivalent to (3), while (13) is essentially equivalent to (10). Furthermore, the case of symmetric but non-self-adjoint problem expounded by (5.2.8) is just a special case of (6) where the symmetry property of $\mathcal{L}$ is used to simplify (6).

We shall show how these concepts can be used to derive variational expressions for non-self-adjoint and nonsymmetric problems.

§§5.3.2 Rayleigh-Ritz Procedure—Non-Self-Adjoint Problems

When an operator is non-self-adjoint, a variational expression is given by (3) where the original and auxiliary equations are given by (4). To invoke the Rayleigh-Ritz procedure, one first lets

$$f = \sum_{n=1}^{N} a_n f_n, \quad f_a = \sum_{m=1}^{N} b_m f_{am} \qquad (5.3.14)$$

for both the original and the auxiliary problem. Then, after substituting them into (3), one obtains

$$I = \sum_{n=1}^{N}\sum_{m=1}^{N} a_n b_m^* \langle f_{am}^*, \mathcal{L} f_n \rangle - \sum_{m=1}^{N} b_m^* \langle f_{am}^*, g \rangle - \sum_{n=1}^{N} a_n \langle g_a^*, f_n \rangle, \qquad (5.3.15)$$

which can be rewritten more compactly as

$$I = \mathbf{b}^\dagger \cdot \overline{\mathbf{L}} \cdot \mathbf{a} - \mathbf{b}^\dagger \cdot \mathbf{g} - \mathbf{g}_a^\dagger \cdot \mathbf{a}, \qquad (5.3.16)$$

where $\left[\overline{\mathbf{L}}\right]_{mn} = \langle f_{am}^*, \mathcal{L} f_n \rangle$, $[\mathbf{g}]_m = \langle f_{am}^*, g \rangle$, $[\mathbf{g}_a]_n = \langle f_n^*, g_a \rangle$, and $\mathbf{a}$ and $\mathbf{b}$ are column vectors containing a_n's and b_m's. Note that $\overline{\mathbf{L}}$ is a non-Hermitian matrix now because $\mathcal{L}$ is non-self-adjoint. Subsequently, the optimal $\mathbf{a}$ and $\mathbf{b}$ are obtained by demanding that the first variation of I due to the first variation of $\mathbf{a}$ and $\mathbf{b}$ about the optimal solutions, $\mathbf{a}_0$ and $\mathbf{b}_0$, be zero. By letting $\mathbf{a} = \mathbf{a}_0 + \delta\mathbf{a}$, $\mathbf{b} = \mathbf{b}_0 + \delta\mathbf{b}$, and taking the first variation of (16), we have

$$\delta I = \mathbf{b}_0^\dagger \cdot \overline{\mathbf{L}} \cdot \delta\mathbf{a} + \delta\mathbf{b}^\dagger \cdot \overline{\mathbf{L}} \cdot \mathbf{a}_0 - \delta\mathbf{b}^\dagger \cdot \mathbf{g} - \mathbf{g}_a^\dagger \cdot \delta\mathbf{a}. \qquad (5.3.17)$$

For δI to be zero, we require that

$$\overline{\mathbf{L}} \cdot \mathbf{a}_0 = \mathbf{g}, \qquad (5.3.18a)$$

$$\overline{\mathbf{L}}^\dagger \cdot \mathbf{b}_0 = \mathbf{g}_a. \qquad (5.3.18b)$$

Hence, the optimal solutions, $\mathbf{a}_0$ and $\mathbf{b}_0$, are obtained by solving (18a) and (18b). Note that the original problem and the auxiliary problem are decoupled. Therefore, if we need only the solution to the original problem, we need only solve (18a). The only involvement with the auxiliary problem is in $\overline{\mathbf{L}}$. Furthermore, the basis functions f_{am}'s should be chosen so that they can best approximate the auxiliary problem.

In the variational approach to a non-self-adjoint problem, we first derive a variational expression for the equation

$$\mathcal{L}f = g \tag{5.3.19}$$

by defining an auxiliary problem. Then, the optimal solution to (19) is obtained by applying the Rayleigh-Ritz procedure. Consequently, the resultant equations are the matrix equations for the original and the auxiliary problem, Equations (18a) and (18b). Moreover, these two equations are decoupled. Alternatively, we may arrive at (18a) more directly using the method of weighted residuals on (4). In spite of this, the above analysis establishes a relationship of (18a) and (18b) with the variational expression and the auxiliary problem. For faster convergence, we want the weighting functions in the method of weighted residuals to approximate the solutions of the adjoint equation well.

The Rayleigh-Ritz procedure can also be applied to (10) and (13), yielding equations similar to (5.2.35). They also correspond to solving $\mathcal{L}f = g$ by the method of weighted residuals where the weighting functions are chosen to approximate the auxiliary solution effectively.

§§5.3.3 Applications to Scalar Wave Equations

Consider the scalar wave equation

$$\nabla \cdot p\nabla\phi(\mathbf{r}) + k^2 p\,\phi(\mathbf{r}) = s(\mathbf{r}), \tag{5.3.20}$$

where p and k^2 are complex corresponding to lossy media. Then, the operator

$$\mathcal{L} = \nabla \cdot p\nabla + k^2 p \tag{5.3.21}$$

is non-self-adjoint even though it is still symmetric. The adjoint operator to $\mathcal{L}$, using the definition of adjoint operators, is

$$\mathcal{L}^a = \nabla \cdot p^*\nabla + k^{*2}p^*. \tag{5.3.22}$$

This can be proved using manipulations similar to those for Equations (5.2.41) and (5.2.42). Hence, (22) is the adjoint of (21) if ϕ satisfies either the homogeneous Dirichlet or Neumann boundary condition on S, the surface bounding V. Consequently, and the auxiliary equation is

$$\nabla \cdot p^*\nabla\phi_a(\mathbf{r}) + k^{*2}p^*\phi_a(\mathbf{r}) = s_a(\mathbf{r}). \tag{5.3.23}$$

Note that $\mathcal{L}^a$ is an adjoint operator to $\mathcal{L}$ for a class of functions defined in a volume V, satisfying either the homogeneous Dirichlet or Neumann boundary condition on the boundary of V.

Notice that the auxiliary equation is for an active medium if the original equation is for a lossy medium. Consequently, using Equation (3), a variational expression for this non-self-adjoint problem is

$$I = -\int_V d\mathbf{r}\, p\, (\nabla \phi_a)^* \cdot \nabla \phi + \int_V d\mathbf{r}\, k^2 p \phi_a^* \phi - \int_V d\mathbf{r}\, \phi_a^* s - \int_V d\mathbf{r}\, s_a^* \phi, \qquad (5.3.24)$$

where the appropriate boundary conditions are imposed on ϕ and ϕ_a on S, the surface bounding V.

But since $\mathcal{L}$ is symmetric, it is preferable to use the symmetric property of $\mathcal{L}$ to derive a simpler variational expression for this example, as is given in (5.2.44). Note that if we choose $s_a = s^*$, then $\phi_a = \phi^*$ by comparing (20) and (23). In this case, (24) reduces to (5.2.44). Hence, (5.2.44) is a special case of (24). Also, the Rayleigh-Ritz procedure can be applied to (24) to obtain optimal solutions given by (18).

§§5.3.4 Applications to Vector Wave Equations

Many electromagnetic problems are related to operators which are neither self-adjoint nor symmetric. In the previous section, we notice that when a medium is lossy, the electromagnetic problem is not self-adjoint; however, it is still symmetric. The symmetric property of an operator is related to the reciprocal nature of many wave problems. As seen in the previous subsection, when the operator is non-self-adjoint, but symmetric, there is no real advantage of using the formulation of Subsection 5.3.2. The use of Equation (5.2.8) actually yields a simpler expression. But when a medium is non-reciprocal, the corresponding electromagnetic operators are not self-adjoint and not symmetric, as we shall see. Then, the formulation of Subsection 5.3.2 comes in useful.

Consider an inhomogeneous, anisotropic medium. The general vector wave equation describing the electric field in a domain V is

$$\nabla \times \overline{\mu}^{-1} \cdot \nabla \times \mathbf{E}(\mathbf{r}) - \omega^2 \overline{\epsilon} \cdot \mathbf{E}(\mathbf{r}) = i\omega \mathbf{J}(\mathbf{r}). \qquad (5.3.25)$$

The linear operator in this equation can be identified as

$$\mathcal{L} = \left(\nabla \times \overline{\mu}^{-1} \cdot \nabla \times \right) - \omega^2 \overline{\epsilon}. \qquad (5.3.26)$$

Next, let us examine the inner product

$$\langle \mathbf{E}_1^*(\mathbf{r}), \mathcal{L}\mathbf{E}_2(\mathbf{r}) \rangle = \langle \mathbf{E}_1^*(\mathbf{r}), \nabla \times \overline{\mu}^{-1} \cdot \nabla \times \mathbf{E}_2(\mathbf{r}) \rangle - \omega^2 \langle \mathbf{E}_1^*(\mathbf{r}), \overline{\epsilon} \cdot \mathbf{E}_2(\mathbf{r}) \rangle. \qquad (5.3.27)$$

It can be shown, with the necessary boundary condition that $\hat{n} \times \mathbf{E} = 0$ or $\hat{n} \times \mathbf{H} = 0$ on S (a surface enclosing V), that (Exercise 5.14)

$$\langle \mathbf{E}_1^*(\mathbf{r}), \nabla \times \overline{\mu}^{-1} \cdot \nabla \times \mathbf{E}_2(\mathbf{r}) \rangle = \int_V d\mathbf{r}\, \nabla \times \mathbf{E}_1^*(\mathbf{r}) \cdot \overline{\mu}^{-1} \cdot \nabla \times \mathbf{E}_2(\mathbf{r}). \qquad (5.3.28)$$

Hence, the operator $(\nabla \times \overline{\mu}^{-1} \cdot \nabla \times)$ is not self-adjoint unless $\overline{\mu}$ is Hermitian. Furthermore, the operator $\omega^2 \overline{\epsilon}$ in (26) is not self-adjoint unless $\overline{\epsilon}$ is Hermitian. But $\overline{\mu}$ and $\overline{\epsilon}$ are Hermitian only for a lossless medium (see Chapter 1). Therefore, for a general, lossy, anisotropic medium, $\mathcal{L}$ is non-self-adjoint.

If we examine the inner product

$$\langle \mathbf{E}_1(\mathbf{r}), \mathcal{L}\mathbf{E}_2(\mathbf{r}) \rangle = \langle \mathbf{E}_1(\mathbf{r}), \nabla \times \overline{\mu}^{-1} \cdot \nabla \times \mathbf{E}_2(\mathbf{r}) \rangle - \omega^2 \langle \mathbf{E}_1(\mathbf{r}), \overline{\epsilon} \cdot \mathbf{E}_2(\mathbf{r}) \rangle \tag{5.3.29}$$

with the necessary boundary condition on $\mathbf{E}$, it can be shown that

$$\langle \mathbf{E}_1(\mathbf{r}), \nabla \times \overline{\mu}^{-1} \cdot \nabla \times \mathbf{E}_2(\mathbf{r}) \rangle = \int_V d\mathbf{r} \, \nabla \times \mathbf{E}_1(\mathbf{r}) \cdot \overline{\mu}^{-1} \cdot \nabla \times \mathbf{E}_2(\mathbf{r}). \tag{5.3.30}$$

Hence, the operator $(\nabla \times \overline{\mu}^{-1} \cdot \nabla \times)$ will not be symmetric unless $\overline{\mu}$ is symmetric. Furthermore, the operator $\omega^2 \overline{\epsilon}$ in (26) is not symmetric unless $\overline{\epsilon}$ is symmetric. But the condition for $\overline{\mu}$ and $\overline{\epsilon}$ to be symmetric is for the medium to be reciprocal (see Chapter 1). Consequently, for a nonreciprocal medium, $\mathcal{L}$ is nonsymmetric.

From the above discussions, we note that the self-adjointness property of an electromagnetic operator is closely related to the lossless property of the medium, while the symmetric property of an electromagnetic operator is intimately related to the reciprocal nature of the medium. In a lossless medium, $\overline{\mu}$ and $\overline{\epsilon}$ are always Hermitian as shown in Chapter 1, and the corresponding vector wave equation is always self-adjoint as indicated by (28), even if the medium is nonreciprocal. If a medium is nonreciprocal, then the corresponding vector wave equation is always nonsymmetric as (30) shows. But it can still be self-adjoint if it is lossless!

If a medium is lossy but reciprocal, we could exploit the symmetric property of the operator and derive a simpler variational expression using the method given in Section 5.2. However, for a general lossy, nonreciprocal, and anisotropic medium, a variational expression which follows from (3) is

$$I = \int_V d\mathbf{r} \, \nabla \times \mathbf{E}_a^*(\mathbf{r}) \cdot \overline{\mu}^{-1} \cdot \nabla \times \mathbf{E}(\mathbf{r}) - \omega^2 \int_V d\mathbf{r} \, \mathbf{E}_a^*(\mathbf{r}) \cdot \overline{\epsilon} \cdot \mathbf{E}(\mathbf{r})$$

$$- i\omega \int_V d\mathbf{r} \, \mathbf{E}_a^*(\mathbf{r}) \cdot \mathbf{J}(\mathbf{r}) + i\omega \int_V d\mathbf{r} \, \mathbf{J}_a^*(\mathbf{r}) \cdot \mathbf{E}(\mathbf{r}), \tag{5.3.31}$$

where $\mathbf{E}_a(\mathbf{r})$ is a solution to the auxiliary equation $\mathcal{L}^a f_a = g_a$. For this case, it is (Exercise 5.15)

$$\nabla \times \left(\overline{\mu}^\dagger\right)^{-1} \cdot \nabla \times \mathbf{E}_a(\mathbf{r}) - \omega^2 \overline{\epsilon}^\dagger \cdot \mathbf{E}_a(\mathbf{r}) = i\omega \mathbf{J}_a(\mathbf{r}), \tag{5.3.32}$$

and $\mathbf{E}$ and $\mathbf{E}_a$ satisfy the appropriate boundary conditions on the surface S bounding V. Note that if $\overline{\mu}$ and $\overline{\epsilon}$ are for a lossy medium, then Equation (32) is for an active medium.

If instead, the $\hat{a}$ component of the field $\mathbf{E}(\mathbf{r})$ is desired at $\mathbf{r} = \mathbf{r}_0$, then a variational expression can be written for

$$u = \hat{a} \cdot \mathbf{E}(\mathbf{r}_0) = \langle \hat{a}\delta(\mathbf{r} - \mathbf{r}_0), \mathbf{E}(\mathbf{r}) \rangle, \qquad (5.3.33)$$

assuming that $\hat{a}$ is real. When a complex conjugate is used in the inner product, the auxiliary equation is then

$$\nabla \times \left(\overline{\mu}^{\dagger}\right)^{-1} \cdot \nabla \times \mathbf{E}_a(\mathbf{r}) - \omega^2 \overline{\epsilon}^{\dagger} \cdot \mathbf{E}_a(\mathbf{r}) = \hat{a}\delta(\mathbf{r} - \mathbf{r}_0). \qquad (5.3.34)$$

Consequently, the variational expression for $\hat{a} \cdot \mathbf{E}(\mathbf{r}_0)$ is (Exercise 5.15)

$$\hat{a} \cdot \hat{\mathbf{E}}(\mathbf{r}_0) = \frac{i\omega \hat{a} \cdot \mathbf{E}(\mathbf{r}_0) \int\limits_V d\mathbf{r} \, \mathbf{E}_a^*(\mathbf{r}) \cdot \mathbf{J}(\mathbf{r})}{\int\limits_V d\mathbf{r} \nabla \times \mathbf{E}_a^*(\mathbf{r}) \cdot \overline{\mu}^{-1} \cdot \nabla \times \mathbf{E}(\mathbf{r}) - \omega^2 \int\limits_V d\mathbf{r} \, \mathbf{E}_a^*(\mathbf{r}) \cdot \overline{\epsilon} \cdot \mathbf{E}(\mathbf{r})}, \qquad (5.3.35)$$

where the appropriate boundary conditions are imposed on $\mathbf{E}$ and $\mathbf{E}_a$ on S.

Applications of the above method in deriving a variational expression for other non-self-adjoint problems are given in Exercises 5.16 and 5.17. Moreover, the Rayleigh-Ritz procedure can be applied to (31) and (35) to yield optimal solutions.

§5.4 Variational Expressions for Eigenvalue Problems

Many physical problems are expressible as eigenvalue problems. Examples of these are the cavity resonance problem and the wave guidance problem. For the cavity resonance problem, the eigenvalues correspond to the resonant frequencies of the cavity. For the wave guidance problem, the eigenvalues correspond to the wavenumbers of the guided modes of the waveguiding structure. In order to obtain good estimates for the eigenvalues of the problem, the variational expression for the eigenvalues of the problem is needed. Once such a variational expression is obtained, the Rayleigh-Ritz method provides a systematic approach for finding good estimates of these eigenvalues.

§§5.4.1 General Concepts

Consider the general eigenvalue problem

$$\mathcal{L}f = \lambda \mathcal{B}f \qquad (5.4.1)$$

where $\mathcal{L}$ and $\mathcal{B}$ are operators. If $\mathcal{L}$ and $\mathcal{B}$ are self-adjoint operators, a variational expression from which λ can be deduced is

$$\langle f^*, \mathcal{L}f \rangle - \lambda \langle f^*, \mathcal{B}f \rangle = 0, \qquad (5.4.2)$$

or

$$\lambda = \frac{\langle f^*, \mathcal{L}f \rangle}{\langle f^*, \mathcal{B}f \rangle}. \qquad (5.4.2a)$$

This indeed has a stationary value when f is a solution to (1). The stationary value for λ is the corresponding eigenvalue of (1). To show this, we take the first variation of Equation (2) about the exact values to yield

$$\langle \delta f^*, \mathcal{L} f_e \rangle + \langle f_e^*, \mathcal{L} \delta f \rangle - \lambda_e (\langle \delta f^*, \mathcal{B} f_e \rangle + \langle f_e^*, \mathcal{B} \delta f \rangle) - \delta \lambda \langle f_e^*, \mathcal{B} f_e \rangle = 0, \quad (5.4.3)$$

where the subscripts e stand for exact values. Furthermore, using the self-adjoint properties of $\mathcal{L}$ and $\mathcal{B}$, and that f_e is the solution to (1) with eigenvalue λ_e, we can show that $\delta \lambda = 0$ from (3), implying the stationarity of (2a). Note that λ in (5.4.2) is always real.

For the case when $\mathcal{L}$ and $\mathcal{B}$ are symmetric but not self-adjoint operators, a variational expression from which λ can be deduced is

$$\langle f, \mathcal{L} f \rangle - \lambda \langle f, \mathcal{B} f \rangle = 0, \quad (5.4.4)$$

or

$$\lambda = \frac{\langle f, \mathcal{L} f \rangle}{\langle f, \mathcal{B} f \rangle}. \quad (5.4.4a)$$

It can be readily shown that (4) or (4a) is stationary about the exact eigenvalue as in (3). Notice that in this case, λ need not be real.

The Rayleigh-Ritz procedure can also be applied to (2) and (4) to obtain matrix eigenvalue equations. These matrix eigenvalue equations are the same as those obtained by applying Galerkin's method directly to (1). In applying Galerkin's method to (1), we use $\langle f^*, g \rangle$ inner product when $\mathcal{L}$ and $\mathcal{B}$ are self-adjoint operators, and $\langle f, g \rangle$ inner product when $\mathcal{L}$ and $\mathcal{B}$ are symmetric.

When $\mathcal{L}$ and $\mathcal{B}$ are non-self-adjoint, we define an auxiliary equation such that

$$\mathcal{L}^a f_a = \lambda^* \mathcal{B}^a f_a, \quad (5.4.5)$$

where $\mathcal{L}^a$ and $\mathcal{B}^a$ are the adjoint operators of $\mathcal{L}$ and $\mathcal{B}$ respectively. A variational expression for λ is then

$$\lambda = \frac{\langle f_a^*, \mathcal{L} f \rangle}{\langle f_a^*, \mathcal{B} f \rangle}. \quad (5.4.6)$$

Then, cross-multiplying the above and taking the first variation yields

$$\delta \lambda \langle f_{ae}^*, \mathcal{B} f_e \rangle + \lambda_e (\langle \delta f_a^*, \mathcal{B} f_e \rangle + \langle f_{ae}^*, \mathcal{B} \delta f \rangle) = \langle \delta f_a^*, \mathcal{L} f_e \rangle + \langle f_{ae}^*, \mathcal{L} \delta f \rangle. \quad (5.4.7)$$

By making use of Equations (1) and (5), and the definition of adjoint operators, we can show that $\delta \lambda = 0$, implying the stationarity of (6).

If $\mathcal{L}$ and $\mathcal{B}$ are nonsymmetric, we define an auxiliary equation such that

$$\mathcal{L}^t f_a = \lambda \mathcal{B}^t f_a, \quad (5.4.8)$$

where $\mathcal{L}^t$ and $\mathcal{B}^t$ are the transpose operators of $\mathcal{L}$ and $\mathcal{B}$ respectively. A variational expression for λ is then

$$\lambda = \frac{\langle f_a, \mathcal{L} f \rangle}{\langle f_a, \mathcal{B} f \rangle}. \quad (5.4.9)$$

Since $(\mathcal{L}^t)^* = \mathcal{L}^a$, if $\tilde{f}_a$ is a solution to (5), then $\tilde{f}_a^*$ is a solution to (8). Therefore, (6) and (9) are actually equivalent to each other. That is to say, the conjugation in the inner product is immaterial when the operator in non-self-adjoint. Furthermore, (4a) is a special case of (6) or (9) where $\mathcal{L}$ and $\mathcal{B}$ are symmetric. Note that in (6) and (9), λ need not be real.

The Rayleigh-Ritz procedure can be applied to (6) or (9) to obtain matrix eigenvalue equations. These matrix eigenvalue equations are also obtained by applying the method of weighted residuals to (1) and (5) directly as in (5.2.35) (see Exercise 5.18). But the analysis here ties together the actual problem and the auxiliary problem.

§§5.4.2 Applications to Scalar Wave Equations

Consider the scalar wave equation

$$\nabla \cdot p(\mathbf{r})\nabla\phi(\mathbf{r}) + \omega^2 p(r)/c^2(\mathbf{r})\,\phi(\mathbf{r}) = 0 \qquad (5.4.10)$$

with either the homogeneous Dirichlet or Neumann boundary condition on S, the surface bounding the domain V over which Equation (10) is defined. Since (10) does not have an excitation source, the nontrivial solutions of (10) are the resonance solutions or the natural solutions. Moreover, these solutions can exist only at certain values of ω, which are the resonant frequencies of the system.

In the above $\mathcal{L}$, $\mathcal{B}$, and λ in Equation (1) can be identified as

$$\mathcal{L} = \nabla \cdot p\nabla, \quad \mathcal{B} = -\frac{p(\mathbf{r})}{c^2(\mathbf{r})}, \quad \lambda = \omega^2. \qquad (5.4.11)$$

But if $p(\mathbf{r})$ and $c^2(\mathbf{r})$ are real, then $\mathcal{L}$ and $\mathcal{B}$ are self-adjoint with the appropriate boundary conditions for ϕ on S [see (5.2.42)]. Then, a variational expression for ω^2 is

$$\omega^2 = \frac{\int\limits_V d\mathbf{r}\, p(\mathbf{r})|\nabla\phi(\mathbf{r})|^2}{\int\limits_V d\mathbf{r}\, p(\mathbf{r})|\phi(\mathbf{r})|^2/c^2(\mathbf{r})}. \qquad (5.4.12)$$

On the other hand, if $p(\mathbf{r})$ and $c^2(\mathbf{r})$ are complex, $\mathcal{L}$ and $\mathcal{B}$ are not self-adjoint but symmetric, then a variational expression using (4a) is

$$\omega^2 = \frac{\int\limits_V d\mathbf{r}\, p(\mathbf{r})[\nabla\phi(\mathbf{r})]^2}{\int\limits_V d\mathbf{r}\, p(\mathbf{r})[\phi(\mathbf{r})]^2/c^2(\mathbf{r})}. \qquad (5.4.13)$$

Note that when $\mathcal{L}$ and $\mathcal{B}$ are self-adjoint, all the eigenvalues are real. This is evident from Equation (12) since it is a real-valued functional for all possible $\phi(\mathbf{r})$. Conversely, when $\mathcal{L}$ and $\mathcal{B}$ are symmetric but not self-adjoint, the eigenvalues need not be real valued. This is reflected in the functional expressed by Equation (13) which need not be real valued.

The Rayleigh-Ritz procedure can also be used on (13) to yield a matrix eigenvalue equation.

§§5.4.3 Applications to Electromagnetic Problems

(a) Resonator Cavity Problem

Consider the vector wave equation for anisotropic medium,

$$\nabla \times \overline{\mu}^{-1} \cdot \nabla \times \mathbf{E}(\mathbf{r}) - \omega^2 \overline{\epsilon} \cdot \mathbf{E}(\mathbf{r}) = 0 \qquad (5.4.14)$$

defined in a domain V bounded by a surface S. If the medium is lossless, then $\overline{\mu}$ and $\overline{\epsilon}$ are Hermitian. In this case, the operators $\nabla \times \overline{\mu}^{-1} \cdot \nabla \times$ and $\overline{\epsilon}$ are self-adjoint operators on $\mathbf{E}(\mathbf{r})$ with appropriate boundary conditions [see (5.2.53)]. A variational expression for the eigenvalue ω^2 from (2a) is then

$$\omega^2 = \frac{\int\limits_V d\mathbf{r}\, \nabla \times \mathbf{E}^*(\mathbf{r}) \cdot \overline{\mu}^{-1} \cdot \nabla \times \mathbf{E}(\mathbf{r})}{\int\limits_V d\mathbf{r}\, \mathbf{E}^*(\mathbf{r}) \cdot \overline{\epsilon} \cdot \mathbf{E}(\mathbf{r})}, \qquad (5.4.15)$$

where the appropriate electric wall or magnetic wall boundary condition is imposed for $\mathbf{E}$ on S; here, ω is the resonant frequency of the system. Notice that because of the Hermitian properties of $\overline{\mu}$ and $\overline{\epsilon}$, the above is always a real-valued functional; hence, ω^2 is always real.

When the medium is lossy, $\overline{\mu}$ and $\overline{\epsilon}$ are not Hermitian. But when the medium is reciprocal, $\overline{\mu}$ and $\overline{\epsilon}$ are symmetric matrices. In this case, the operators $\nabla \times \overline{\mu}^{-1} \cdot \nabla \times$ and $\overline{\epsilon}$ are symmetric operators. Then, a variational expression for the resonant frequency from (4a) is

$$\omega^2 = \frac{\int\limits_V d\mathbf{r}\, \nabla \times \mathbf{E}(\mathbf{r}) \cdot \overline{\mu}^{-1} \cdot \nabla \times \mathbf{E}(\mathbf{r})}{\int\limits_V d\mathbf{r}\, \mathbf{E}(\mathbf{r}) \cdot \overline{\epsilon} \cdot \mathbf{E}(\mathbf{r})}, \qquad (5.4.16)$$

where the appropriate boundary conditions have been imposed on $\mathbf{E}$. Note that ω need not be real in the above.

If $\overline{\mu}$ and $\overline{\epsilon}$ are neither Hermitian nor symmetric, which is the case of lossy, nonreciprocal media, then the linear operators associated with Equation (1) are neither self-adjoint nor symmetric. Consequently, an auxiliary equation to (14) can be defined as

$$\nabla \times \left(\overline{\mu}^t\right)^{-1} \cdot \nabla \times \mathbf{E}_a(\mathbf{r}) - \omega^2 \overline{\epsilon}^t \cdot \mathbf{E}_a(\mathbf{r}) = 0. \qquad (5.4.17)$$

A variational expression for ω^2 using (9) is then

$$\omega^2 = \frac{\int\limits_V d\mathbf{r}\, \nabla \times \mathbf{E}_a(\mathbf{r}) \cdot \overline{\mu}^{-1} \cdot \nabla \times \mathbf{E}(\mathbf{r})}{\int\limits_V d\mathbf{r}\, \mathbf{E}_a(\mathbf{r}) \cdot \overline{\epsilon} \cdot \mathbf{E}(\mathbf{r})}. \qquad (5.4.18)$$

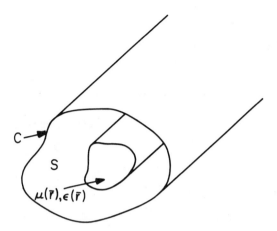

Figure 5.4.1 The cross section of a uniform waveguide loaded with inhomogeneity.

Moreover, an expression using (6) can be derived, but as previously discussed, (6) and (9) are equivalent to each other. Another variational expression for the resonant frequencies involving both $\mathbf{E}$ and $\mathbf{H}$ fields may be derived directly from Maxwell's equations [see Exercise 5.17(d)] rather than from (14).

The Rayleigh-Ritz procedure, when applied to (15) or (16), will yield a matrix eigenvalue equation. But when it is applied to (18), two matrix eigenvalue equations ensue: one for the actual problem and one for the auxiliary problem, as in (5.2.35). Previous discussions on the auxiliary problem also apply here.

(b) Uniform, Inhomogeneous Waveguide Problem

A uniform, inhomogeneous waveguide is filled with an inhomogeneous medium uniform in the direction of wave guidance (see Figure 5.4.1). The field of the guided modes in the waveguide can be assumed to have $e^{ik_z z}$ dependence where z is the axis of the waveguide. As a result of this, and the $\nabla \cdot \epsilon \mathbf{E} = 0$ condition, we have $E_z = -(ik_z \epsilon)^{-1} \nabla_s \cdot \epsilon \mathbf{E}_s$, implying that the z component of the electric field is directly dependent on the transverse components. Therefore, only two components of the electric field are truly independent in a waveguide.[10] Consequently, the equation governing the

---

[10] If the $\nabla \cdot \epsilon \mathbf{E} = 0$ condition is not imposed in the waveguide problem, spurious guided modes will occur due to the presence of fictitious charges.

electric field in the waveguide can be written as (Exercise 5.19; also see Chew and Nasir 1989)[11]

$$\mu \hat{z} \times \nabla_s \times \mu^{-1} \nabla_s \times \mathbf{E}_s - \hat{z} \times \nabla_s \epsilon^{-1} \nabla_s \cdot \epsilon \, \mathbf{E}_s - k^2 \hat{z} \times \mathbf{E}_s + k_z^2 \, \hat{z} \times \mathbf{E}_s = 0, \tag{5.4.19}$$

where $k^2 = \omega^2 \mu \epsilon$. Next, if we identify

$$\mathcal{L} = \left(\mu \, \hat{z} \times \nabla_s \times \mu^{-1} \nabla_s \times \right) - \hat{z} \times \nabla_s \epsilon^{-1} \nabla_s \cdot \epsilon - (k^2 \, \hat{z} \times), \tag{5.4.20}$$

$\mathcal{B} = \hat{z} \times$, and $-\lambda = k_z^2$, then $\mathcal{L}$ is neither symmetric nor self-adjoint, because $\langle \mathbf{A}_s, \mathcal{L} \mathbf{E}_s \rangle \neq \langle \mathbf{E}_s, \mathcal{L} \mathbf{A}_s \rangle$, and $\langle \mathbf{A}_s^*, \mathcal{L} \mathbf{E}_s \rangle \neq \langle (\mathcal{L} \mathbf{A}_s)^*, \mathbf{E}_s \rangle$.

More explicitly,

$$\langle \mathbf{A}_s, \mathcal{L} \mathbf{E}_s \rangle = \langle \mathbf{A}_s, \mu \, \hat{z} \times \nabla_s \times \mu^{-1} \nabla_s \times \mathbf{E}_s \rangle$$
$$- \langle \mathbf{A}_s, \hat{z} \times \nabla_s \epsilon^{-1} \nabla_s \cdot \epsilon \, \mathbf{E}_s \rangle - \langle \mathbf{A}_s, k^2 \, \hat{z} \times \mathbf{E}_s \rangle, \tag{5.4.21}$$

where $\langle \mathbf{A}_s, \mathbf{B}_s \rangle = \int_S d\mathbf{r} \, \mathbf{A}_s \cdot \mathbf{B}_s$ is a two-dimensional integral over the waveguide cross section. The first term on the right-hand side of (21) is

$$\langle \mathbf{A}_s, \hat{z} \times \mu \nabla_s \times \mu^{-1} \nabla_s \times \mathbf{E}_s \rangle = - \int_S d\mathbf{r} \, \mu \, \hat{z} \times \mathbf{A}_s \cdot \nabla_s \times \mu^{-1} \nabla_s \times \mathbf{E}_s$$

$$= \int_S d\mathbf{r} \, \nabla_s \cdot \left(\mu \, \hat{z} \times \mathbf{A}_s \times \mu^{-1} \nabla_s \times \mathbf{E}_s \right)$$

$$- \int_S d\mathbf{r} (\nabla_s \times \hat{z} \times \mu \mathbf{A}_s) \cdot \mu^{-1} \nabla_s \times \mathbf{E}_s \tag{5.4.22}$$

after using the identity $\nabla_s \cdot (\mathbf{B}_s \times \mathbf{C}_s) = \mathbf{C}_s \cdot \nabla_s \times \mathbf{B}_s - \mathbf{B}_s \cdot \nabla_s \times \mathbf{C}_s$. The first integral on the right-hand side of (22) can be converted into a line integral after using the two-dimensional Gauss' theorem and made to vanish with the appropriate boundary conditions on $\mathbf{E}_s$ or $\mathbf{A}_s$, which is either $\hat{n} \times \mathbf{E}_s = 0$ on the boundary or $\hat{n} \times \mathbf{A}_s = 0$ on the boundary. Moreover, using the fact that $\nabla_s \times \hat{z} \times \mu \mathbf{A}_s = \hat{z} \nabla_s \cdot \mu \mathbf{A}_s$, and $\hat{z} \cdot \nabla_s \times \mathbf{E}_s = -\nabla_s \cdot (\hat{z} \times \mathbf{E}_s)$, (22) becomes

$$\langle \mathbf{A}_s, \mu \hat{z} \times \nabla_s \times \mu^{-1} \nabla_s \times \mathbf{E}_s \rangle = - \int_S d\mathbf{r} \, (\nabla_s \cdot \mu \mathbf{A}_s) \mu^{-1} \hat{z} \cdot \nabla_s \times \mathbf{E}_s$$

$$= \int_S d\mathbf{r} \, (\nabla_s \cdot \mu \mathbf{A}_s) \mu^{-1} \nabla_s \cdot (\hat{z} \times \mathbf{E}_s). \tag{5.4.23}$$

Clearly, this part of the operator is neither self-adjoint nor symmetric. Nevertheless, its transpose can be found. Then, after using the identity

---

[11] We have to multiply the equation with $\hat{z} \times$ so that its transpose equation is easily obtained, as shall be shown.

$$\nabla_s \cdot (\mathbf{A}_s \phi) = (\nabla_s \cdot \mathbf{A}_s) \phi + \mathbf{A}_s \cdot \nabla_s \phi,$$

the right-hand side of (23) becomes

$$\langle \mathbf{A}_s, \mu \hat{z} \times \nabla_s \times \mu^{-1} \nabla_s \times \mathbf{E}_s \rangle = \int_S d\mathbf{r}\, \nabla_s \cdot (\hat{z} \times \mathbf{E}_s \mu^{-1} \nabla_s \cdot \mu \mathbf{A}_s)$$

$$- \int_S d\mathbf{r}\, (\hat{z} \times \mathbf{E}_s) \cdot \nabla_s \mu^{-1} \nabla_s \cdot \mu \mathbf{A}_s. \quad (5.4.24)$$

Note that the first term on the right-hand side of (24) can again be converted to a line integral using Gauss' theorem and made to vanish with the appropriate boundary condition that $\hat{n} \times \mathbf{E}_s$ or $\hat{n} \times \mathbf{A}_s$ is zero on the boundary. Consequently, we conclude that

$$\langle \mathbf{A}_s, \mu \hat{z} \times \nabla_s \times \mu^{-1} \nabla_s \times \mathbf{E}_s \rangle = \int_S d\mathbf{r}\, \mathbf{E}_s \cdot \hat{z} \times \nabla_s \mu^{-1} \nabla_s \cdot \mu \mathbf{A}_s$$

$$= \langle \mathbf{E}_s, \hat{z} \times \nabla_s \mu^{-1} \nabla_s \cdot \mu \mathbf{A}_s \rangle. \quad (5.4.25)$$

This demonstrates that $(\mu \hat{z} \times \nabla_s \times \mu^{-1} \nabla_s \times)^t = (\hat{z} \times \nabla_s \mu^{-1} \nabla_s \cdot \mu)$, with appropriate boundary conditions on $\mathbf{A}_s$ or $\mathbf{E}_s$. Similarly, the transpose of the second term in (20) is easily obtained, i.e., $(\hat{z} \times \nabla_s \epsilon^{-1} \nabla_s \cdot \epsilon)^t = (\epsilon \hat{z} \times \nabla_s \times \epsilon^{-1} \nabla_s \times)$. Moreover, the last term in (21) is

$$- \langle \mathbf{A}_s, k^2 \hat{z} \times \mathbf{E}_s \rangle = \int_S d\mathbf{r}\, k^2 \mathbf{E}_s \cdot \hat{z} \times \mathbf{A}_s = \langle \mathbf{E}_s, k^2 \hat{z} \times \mathbf{A}_s \rangle \quad (5.4.26)$$

which corresponds to a skew-symmetric operator.

From the above, we conclude that the operator given in Equation (20) is neither self-adjoint nor symmetric, but its transpose is easily found to be

$$\mathcal{L}^t = - \left(\epsilon \hat{z} \times \nabla_s \times \epsilon^{-1} \nabla_s \times \right) + \hat{z} \times \nabla_s \mu^{-1} \nabla_s \cdot \mu + (k^2\, \hat{z} \times) \quad (5.4.27a)$$

and

$$\mathcal{B}^t = -\hat{z} \times. \quad (5.4.27b)$$

Except for a minus sign, (27a) is also the electromagnetic dual of (20) (for duality, see Chapter 1). Consequently, if we assume that $\mathbf{A}_s = \mathbf{H}_s$, then the transpose or auxiliary equation to Equation (19) becomes

$$-\epsilon \hat{z} \times \nabla_s \times \epsilon^{-1} \nabla_s \times \mathbf{H}_s + \hat{z} \times \nabla_s \mu^{-1} \nabla_s \cdot \mu \mathbf{H}_s + k^2\, \hat{z} \times \mathbf{H}_s - k_z^2\, \hat{z} \times \mathbf{H}_s = 0, \quad (5.4.28)$$

which is actually the equation satisfied by $\mathbf{H}_s$ in the inhomogeneously filled waveguide. Furthermore, if μ and ϵ are real representing a lossless medium, then (28) is also the adjoint equation of (19). As previously mentioned,

Equation (28) is also the dual of Equation (19). Consequently, from (9), a variational expression for the propagation constant k_z^2 of a uniform waveguide is

$$k_z^2 = \frac{\langle \hat{z}\nabla_s \cdot \mu \, \mathbf{H}_s, \mu^{-1}\nabla_s \times \mathbf{E}_s \rangle - \langle \epsilon^{-1}\nabla_s \times \mathbf{H}_s, \hat{z}\nabla_s \cdot \epsilon \, \mathbf{E}_s \rangle + \langle \mathbf{H}_s, k^2 \, \hat{z} \times \mathbf{E}_s \rangle}{\langle \mathbf{H}_s, \hat{z} \times \mathbf{E}_s \rangle}.$$

(5.4.29)

Here, the fields $\mathbf{E}_s$ and $\mathbf{H}_s$ in (29) must satisfy the appropriate boundary conditions on the perimeter C of the waveguide.

The Rayleigh-Ritz procedure, when applied to (29), yields matrix eigenvalue equations for the actual and the auxiliary problems.

§5.5 Essential and Natural Boundary Conditions

When we solve the equation

$$\mathcal{L}f = g,$$

(5.5.1)

where $\mathcal{L}$ is a differential operator, the solution is nonunique unless boundary conditions are imposed on f. To formulate (1) as a variational problem, we first derive a variational expression (e.g., for the case where $\mathcal{L}$ is symmetric) as

$$I = \langle f, \mathcal{L}f \rangle - 2\langle f, g \rangle.$$

(5.5.2)

The solution to (1) is the f that will make (2) stationary. Moreover, in searching for the right f, we look for f in a function space where f satisfies the requisite boundary conditions. Such a boundary condition imposed on the f to be used in (2) is known as an ***essential boundary condition***. Frequently, $\mathcal{L}$ is symmetric (or self-adjoint) only if f satisfies these requisite boundary conditions.

It will be interesting to ask: "If we were to look for f in a function space larger than that originally defined, i.e., with unspecified boundary conditions, is there a unique f that would make I stationary, and if so, what is the eventual boundary condition it would satisfy?" It turns out that an f in an extended space can still make (2) stationary. In addition, this f would satisfy a boundary condition known as the ***natural boundary condition***, i.e., it is the boundary condition ***naturally*** satisfied by the f that would make I stationary. It need not be imposed on f like the essential boundary condition.

§§5.5.1 The Scalar Wave Equation Case

As an example, we first look at the partial differential equation

$$\nabla \cdot p(\mathbf{r})\nabla\phi(\mathbf{r}) + k^2 p(\mathbf{r})\,\phi(\mathbf{r}) = s(\mathbf{r}), \quad \mathbf{r} \in V,$$

(5.5.3)

where $\mathcal{L} = \nabla \cdot p\nabla + k^2 p$. Here, the linear operator $\mathcal{L}$ is symmetric for a class of functions ϕ satisfying either the homogeneous Dirichlet or Neumann boundary condition on the boundary of V, and V is the region over which the partial differential equation in (3) is defined. Hence, a variational expression, from (5.2.8) is

$$I = -\int_V d\mathbf{r}\, p\,(\nabla\phi)^2 + \int_V d\mathbf{r}\, k^2 p\,\phi^2 - 2\int_V d\mathbf{r}\,\phi\, s. \qquad (5.5.4)$$

Even though (4) is derived for a class of ϕ satisfying the requisite boundary condition, we can still search for a ϕ that will minimize I outside the above class. What is then the value of ϕ that will minimize (4) without the boundary condition specified for ϕ on S (the surface surrounding V), and what is the corresponding natural boundary condition satisfied by ϕ on S? To answer these questions, we take the first variation of (4) to yield

$$\delta I = -2\int_V d\mathbf{r}\, p\,(\nabla\phi)\cdot\nabla\delta\phi + 2\int_V d\mathbf{r}\, k^2 p\,\phi\,\delta\phi - 2\int_V d\mathbf{r}\,\delta\phi\, s. \qquad (5.5.5)$$

Then, using Gauss' theorem, we deduce that

$$\int_V d\mathbf{r}\, p\,(\nabla\phi)\cdot\nabla\delta\phi = \int_S dS\,\hat{n}\cdot(p\nabla\phi\,\delta\phi) - \int_V d\mathbf{r}\,(\nabla\cdot p\nabla\phi)\,\delta\phi. \qquad (5.5.6)$$

Hence, (5) becomes

$$\delta I = 2\int_V d\mathbf{r}\,(\nabla\cdot p\nabla\phi)\,\delta\phi + 2\int_V d\mathbf{r}\, k^2 p\,\phi\,\delta\phi - 2\int_V d\mathbf{r}\,\delta\phi\, s - 2\int_S dS\,\hat{n}\cdot(p\nabla\phi\,\delta\phi). \qquad (5.5.7)$$

In order for δI to vanish, we require that Equation (3) be satisfied; in addition,

$$\hat{n}\cdot\nabla\phi = 0, \quad \text{on} \quad S, \qquad (5.5.8)$$

and ϕ need not be zero on S. Therefore, if we were to search for a ϕ that minimizes (4) and in a function space such that $\phi \neq 0$ on S, then the ϕ required satisfies (3) plus the natural boundary condition (8) on S. It is the homogeneous Neumann boundary condition. Furthermore, this result can also be explained from another perspective, as shall be discussed next.

The operator $\nabla\cdot p\nabla$ is a nonsymmetric operator unless it operates on a class of functions in V that satisfies the homogeneous Dirichlet or Neumann boundary condition on S. Hence, (4) is derived for a class of functions satisfying this boundary condition. We can, however, remove this restriction in the derivation of (4) by defining an **extended operator**

$$\mathcal{L} = \nabla\cdot p\nabla - \Delta(S)\,p\,\hat{n}\cdot\nabla + k^2 p. \qquad (5.5.9)$$

In the above, $\Delta(S)$ implies that a surface integral on the surface S is to be invoked when a volume integral is performed with $\mathcal{L}$, namely,

$$\langle\phi_1, \Delta(S)\,p\,\hat{n}\cdot\nabla\phi_2\rangle = \int_S dS\,\phi_1\, p\,\hat{n}\cdot\nabla\phi_2,$$

where $\hat{n}$ is an outward pointing unit vector normal to the surface S. Hence, $\Delta(S)$ can be considered a delta function shell on S [this notation is adapted from Chen and Lien (1980)]. Moreover, this shell is contained in V. As a consequence, it can be shown easily that (Exercise 5.20)

$$\langle \phi_1, \mathcal{L}\phi_2 \rangle = -\int_V d\mathbf{r}\, (\nabla \phi_1) \cdot p\, (\nabla \phi_2) + \int_V d\mathbf{r}\, k^2 p\, \phi_1 \phi_2 \qquad (5.5.10)$$

without imposing any boundary conditions on ϕ_1 or ϕ_2. In other words, $\mathcal{L}$ is now a symmetric operator for an extended class of functions ϕ. Furthermore, when (9) is used in (2) to derive a variational expression, the same expression (4) ensues. But the same expression (4) now admits a larger class of functions ϕ, which need not satisfy the homogeneous Dirichlet or Neumann boundary condition on S. From the analysis of (5) to (8), however, the resultant solution ϕ in the extended class that minimizes I satisfies a homogeneous Neumann boundary condition on S. This can be understood as follows.

Since

$$\mathcal{L}\phi = \nabla \cdot p \nabla \phi - \Delta(S) p\, \hat{n} \cdot \nabla \phi + k^2 p\, \phi, \qquad (5.5.11)$$

the second term can be viewed as "charges" on S that are negative of those induced by ϕ. Moreover, these "charges" induced by ϕ are proportional to $p\, \hat{n} \cdot \nabla \phi$. Hence, the second term always produces "charges" to annihilate those induced by ϕ. As a consequence, $\hat{n} \cdot \nabla \phi = 0$ on S, if we were to solve the following equation:

$$\nabla \cdot p \nabla \phi - \Delta(S) p\, \hat{n} \cdot \nabla \phi + k^2 p\, \phi = s \qquad (5.5.12)$$

even without imposing the boundary condition. Note that these "charges" are true charges if $k^2 = 0$, $p = \epsilon$, and (12) becomes Poisson's equation. Moreover, the Neumann boundary condition naturally follows from solving (12).

The original operator in (3) can be extended in another way and yet remains symmetric. To do this, we define

$$\mathcal{L} = \nabla \cdot p \nabla - \nabla \cdot p\, \hat{n} \Delta(S) + k^2 p, \qquad (5.5.13)$$

where the divergence operators are to act on everything to their right. In finding $\langle \phi_1, \mathcal{L}\phi_2 \rangle$, we perform integration by parts on the first and second terms in (13) to yield (Exercise 5.20)

$$\langle \phi_1, \mathcal{L}\phi_2 \rangle = -\int_V d\mathbf{r}\, (\nabla \phi_1) \cdot p\, (\nabla \phi_2) + \int_S dS\, \hat{n} \cdot (\phi_1 p \nabla \phi_2)$$

$$+ \int_S dS\, (\nabla \phi_1) \cdot p\, \hat{n} \phi_2 + \int_V d\mathbf{r}\, k^2 p\, \phi_1 \phi_2. \qquad (5.5.14)$$

Clearly, $\mathcal{L}$ is symmetric with no requisite boundary conditions on ϕ_1 and ϕ_2. Consequently, using $\mathcal{L}$ in (2), we derive a variational expression

$$I = -\int_V dr\, p\,(\nabla\phi)^2 + 2\int_S dS\,\hat{n}\cdot(\phi\,p\nabla\phi) + \int_V dr\, k^2 p\,\phi^2 - 2\int_V dr\,\phi\, s. \tag{5.5.15}$$

Now, taking the first variation of (15), we have

$$\delta I = 2\int_V dr\,(\nabla\cdot p\nabla\phi)\,\delta\phi + 2\int_S dS\,\hat{n}\cdot(\phi p\nabla\delta\phi) + 2\int_V dr\, k^2 p\,\phi\,\delta\phi - 2\int_V dr\,\delta\phi\, s. \tag{5.5.16}$$

In order for δI to vanish, ϕ has to satisfy (3) plus the boundary condition that $\phi = 0$ on S, i.e., ϕ satisfies the homogeneous Dirichlet boundary condition on S.

The second term in (13) can be thought of as a surface source placed on S. Here, the surface source is of a "double-layer" type (Stratton 1941, reference for Chapter 3), and it cancels any "double-layer" source induced by a nonzero ϕ on S. Consequently, the boundary condition $\phi = 0$ follows naturally from the stationary solution to (15). It also follows naturally from the solution of the equation $\mathcal{L}\phi = s$, where $\mathcal{L}$ is as defined in (13).

Using a similar idea, if an inhomogeneous Neumann boundary condition is desired on S, a modification of (2) yields a variational expression, which is (Exercise 5.21a)

$$I = -\int_V dr\, p\,(\nabla\phi)^2 + 2\int_S dS\,\beta\,\phi + \int_V dr\, k^2 p\,\phi^2 - 2\int_V dr\,\phi\, s. \tag{5.5.17}$$

Moreover, it can be shown that the stationary solution of the above variational expression satisfies the inhomogeneous Neumann boundary condition $\hat{n}\cdot p\nabla\phi = \beta$ on S. The second term in (17) may be thought of as "surface charges" (independent of ϕ) impressed on S to force an inhomogeneous Neumann boundary condition for ϕ on S. Furthermore, it may be grouped with the last term in (17) if we so wish.

Similarly, for an inhomogeneous Dirichlet boundary condition, by modifying (15), the corresponding variational expression is (Exercise 5.21b)

$$I = -\int_V dr\, p\,(\nabla\phi)^2 + 2\int_S dS\,\hat{n}\cdot(\phi-\gamma)\,p\nabla\phi + \int_V dr\, k^2 p\,\phi^2 - 2\int_V dr\,\phi\, s. \tag{5.5.18}$$

It can be readily shown that the stationary solution to the above equation satisfies the Dirichlet boundary condition $\phi = \gamma$ on S.

If an impedance boundary condition is desired, the variational expression is (Exercise 5.21c)

$$I = -\int_V dr\, p\,(\nabla\phi)^2 + \int_S dS\,\alpha\phi^2 + \int_V dr\, k^2 p\,\phi^2 - 2\int_V dr\,\phi\, s. \tag{5.5.19}$$

The stationary solution to the above satisfies the boundary condition that $\hat{n} \cdot p \nabla \phi = \alpha \phi$ on S. The impedance boundary may be thought of as an impenetrable impedance layer arising from the values of $k^2 p$ being infinite on S.

§§5.5.2 The Electromagnetic Case

The vector wave equation that governs the **E** field in a volume V for a general inhomogeneous, anisotropic medium is

$$\nabla \times \overline{\mu}^{-1} \cdot \nabla \times \mathbf{E} - \omega^2 \overline{\epsilon} \cdot \mathbf{E} = i\omega \mathbf{J}. \tag{5.5.20}$$

Then, the corresponding linear operator

$$\mathcal{L} = \left(\nabla \times \overline{\mu}^{-1} \cdot \nabla \times \right) - \omega^2 \overline{\epsilon} \tag{5.5.21}$$

is in general non-self-adjoint and nonsymmetric. If, however, **E** satisfies an electric wall or a magnetic wall boundary condition on S (the surface bounding V), then it can be shown that the adjoint operator of $\mathcal{L}$ is

$$\mathcal{L}^a = \left[\nabla \times \left(\overline{\mu}^\dagger \right)^{-1} \cdot \nabla \times \right] - \omega^2 \overline{\epsilon}^\dagger, \tag{5.5.22}$$

while the transpose operator is

$$\mathcal{L}^t = \left[\nabla \times \left(\overline{\mu}^t \right)^{-1} \cdot \nabla \times \right] - \omega^2 \overline{\epsilon}^t. \tag{5.5.23}$$

Note that (22) is simply the complex conjugate of (23). Consequently, a variational expression, in accordance with (5.3.3), is

$$I = \left\langle \nabla \times \mathbf{E}_a^*, \overline{\mu}^{-1} \cdot \nabla \times \mathbf{E} \right\rangle - \omega^2 \langle \mathbf{E}_a^*, \overline{\epsilon} \cdot \mathbf{E} \rangle$$
$$- i\omega \langle \mathbf{E}_a^*, \mathbf{J} \rangle + i\omega \langle \mathbf{J}_a^*, \mathbf{E} \rangle, \tag{5.5.24}$$

where $\mathbf{E}_a$ satisfies the auxiliary equation

$$\mathcal{L}^a \mathbf{E}_a = i\omega \mathbf{J}_a, \tag{5.5.25}$$

and both **E** and $\mathbf{E}_a$ are to satisfy the electric wall or the magnetic wall boundary condition on S.

The requirement on **E** and $\mathbf{E}_a$ to satisfy the aforementioned boundary condition could be lifted if we extend the operators (21) and (22) just as in the scalar wave case (also see Chen and Lien 1980). For example, if we want the boundary condition $\hat{n} \times \mathbf{H} = 0$ on S, it could be achieved, in the spirit of the second term in (9), by adding current sources on the surface S in order to cancel the current induced on S by the field. This required surface current, negative of the surface current induced on S, is (note that $\hat{n}$ is pointing outward to the surface S)

$$i\omega \mathbf{J}_s = i\omega \hat{n} \times \mathbf{H} = \hat{n} \times \overline{\mu}^{-1} \cdot \nabla \times \mathbf{E}. \tag{5.5.26}$$

Consequently, the operator (21) can be extended as

$$\mathcal{L} = \left(\nabla \times \overline{\boldsymbol{\mu}}^{-1} \cdot \nabla \times\right) - \left[\triangle(S)\,\hat{n} \times \overline{\boldsymbol{\mu}}^{-1} \cdot \nabla \times\right] - \omega^2 \overline{\boldsymbol{\epsilon}}, \qquad (5.5.27)$$

with (20) becoming

$$\mathcal{L}\mathbf{E} = i\omega \mathbf{J}. \qquad (5.5.27a)$$

Then,

$$\langle \mathbf{E}_1^*, \mathcal{L}\mathbf{E}_2 \rangle = \langle \nabla \times \mathbf{E}_1^*, \overline{\boldsymbol{\mu}}^{-1} \cdot \nabla \times \mathbf{E}_2 \rangle - \omega^2 \langle \mathbf{E}_1^*, \overline{\boldsymbol{\epsilon}} \cdot \mathbf{E}_2 \rangle. \qquad (5.5.28)$$

Now, if $\mathcal{L}^a$ is extended as

$$\mathcal{L}^a = \left[\nabla \times \left(\overline{\boldsymbol{\mu}}^\dagger\right)^{-1} \cdot \nabla \times\right] - \left[\triangle(S)\,\hat{n} \times \left(\overline{\boldsymbol{\mu}}^\dagger\right)^{-1} \cdot \nabla \times\right] - \omega^2 \overline{\boldsymbol{\epsilon}}^\dagger, \qquad (5.5.29)$$

and used in (25), then the resultant equation is the adjoint equation of (27a). As a result,

$$\langle \mathbf{E}_2^*, \mathcal{L}^a \mathbf{E}_1 \rangle = \left\langle \nabla \times \mathbf{E}_2^*, \left(\overline{\boldsymbol{\mu}}^\dagger\right)^{-1} \cdot \nabla \times \mathbf{E}_1 \right\rangle - \omega^2 \langle \mathbf{E}_2^*, \overline{\boldsymbol{\epsilon}}^\dagger \cdot \mathbf{E}_1 \rangle. \qquad (5.5.30)$$

It is clear that (30) is the complex conjugate of (28). Hence, (29) is the adjoint operator of (27). With the operators extended in this manner, the variational expression is still the same as (24). But now, $\mathbf{E}$ and $\mathbf{E}_a$ in (24) need not be chosen from a class of functions that satisfies the electric wall or the magnetic wall boundary condition. In addition, it can be shown that the stationary solution to (24) now satisfies (20) and (25) plus the natural boundary conditions (Exercise 5.22a)

$$\hat{n} \times \overline{\boldsymbol{\mu}}^{-1} \cdot \nabla \times \mathbf{E} = 0, \quad \text{on} \quad S, \qquad (5.5.31a)$$

$$\hat{n} \times \left(\overline{\boldsymbol{\mu}}^\dagger\right)^{-1} \cdot \nabla \times \mathbf{E}_a = 0, \quad \text{on} \quad S, \qquad (5.5.31b)$$

which are the magnetic wall boundary conditions.

Consequently, with the operator extended as in (27), the equation that we are solving is

$$\nabla \times \overline{\boldsymbol{\mu}}^{-1} \cdot \nabla \times \mathbf{E} - \triangle(S)\,\hat{n} \times \overline{\boldsymbol{\mu}}^{-1} \cdot \nabla \times \mathbf{E} - \omega^2 \overline{\boldsymbol{\epsilon}} \cdot \mathbf{E} = i\omega\mu\,\mathbf{J}. \qquad (5.5.32)$$

The second term in (32) is a surface current that will cancel the induced surface current on S. Moreover, a similar statement can be made with regard to the adjoint equations.

On the contrary, if the boundary condition that $\hat{n} \times \mathbf{E} = 0$ on S is desired, it can be achieved by introducing a surface magnetic current to cancel the surface magnetic current induced by the field on S. Such a current is

$$\mathbf{M}_s = -\hat{n} \times \mathbf{E} \qquad (5.5.33)$$

(note that $\hat{n}$ points outward from the surface). Then, the corresponding equation is

$$\nabla \times \overline{\boldsymbol{\mu}}^{-1} \cdot \nabla \times \mathbf{E} - \nabla \times \left[\overline{\boldsymbol{\mu}}^{-1} \cdot \triangle(S)(\hat{n} \times \mathbf{E})\right] - \omega^2 \overline{\boldsymbol{\epsilon}} \cdot \mathbf{E} = i\omega \mathbf{J} \qquad (5.5.34)$$

with the corresponding linear operator

$$\mathcal{L} = \left(\nabla \times \overline{\mu}^{-1} \cdot \nabla\times\right) - \left[\nabla \times \overline{\mu}^{-1} \cdot \triangle(S)\,\hat{n}\times\right] - \omega^2\,\overline{\epsilon}. \qquad (5.5.35)$$

Consequently, using integration by parts,

$$\langle \mathbf{E}_1^*, \mathcal{L}\mathbf{E}_2 \rangle = \left\langle \nabla \times \mathbf{E}_1^*, \overline{\mu}^{-1} \cdot \nabla \times \mathbf{E}_2 \right\rangle - \int_S dS\,\hat{n} \cdot \left(\mathbf{E}_1^* \times \overline{\mu}^{-1} \cdot \nabla \times \mathbf{E}_2\right)$$

$$+ \int_S dS\,\left(\nabla \times \mathbf{E}_1^* \cdot \overline{\mu}^{-1}\right) \times \mathbf{E}_2 \cdot \hat{n} - \omega^2\langle\mathbf{E}_1^*, \overline{\epsilon}\cdot\mathbf{E}_2\rangle. \qquad (5.5.36)$$

Therefore, if the adjoint operator in (22) is extended as

$$\mathcal{L}^a = \left(\nabla \times \left(\overline{\mu}^\dagger\right)^{-1} \cdot \nabla\times\right) - \left[\nabla \times \left(\overline{\mu}^\dagger\right)^{-1} \cdot \triangle(S)\,\hat{n}\times\right] - \omega^2\,\overline{\epsilon}^\dagger, \qquad (5.5.37)$$

then

$$\langle \mathbf{E}_2^*, \mathcal{L}^a\mathbf{E}_1 \rangle = \left\langle \nabla \times \mathbf{E}_2^*, \left(\overline{\mu}^\dagger\right)^{-1} \cdot \nabla \times \mathbf{E}_1 \right\rangle - \int_S dS\,\hat{n} \cdot \left(\mathbf{E}_2^* \times \left(\overline{\mu}^\dagger\right)^{-1} \cdot \nabla \times \mathbf{E}_1\right)$$

$$+ \int_S dS\,\left[\nabla \times \mathbf{E}_2^* \cdot \left(\overline{\mu}^\dagger\right)^{-1}\right] \times \mathbf{E}_1 \cdot \hat{n} - \omega^2\langle\mathbf{E}_2^*, \overline{\epsilon}^\dagger\cdot\mathbf{E}_1\rangle. \qquad (5.5.38)$$

Clearly, (38) is the complex conjugate of (36). Hence, the extended operator (37) is the adjoint of the extended operator (35). Consequently, a variational expression is (Exercise 5.22b)

$$I = \langle \mathbf{E}_a^*, \mathcal{L}\mathbf{E} \rangle - i\omega\langle\mathbf{E}_a^*, \mathbf{J}\rangle + i\omega\langle\mathbf{J}_a^*, \mathbf{E}\rangle$$
$$= \left\langle \nabla \times \mathbf{E}_a^*, \overline{\mu}^{-1} \cdot \nabla \times \mathbf{E} \right\rangle + \left\langle \mathbf{E}_a^*, \triangle(S)\,\hat{n}\times\overline{\mu}^{-1}\cdot\nabla\times\mathbf{E}\right\rangle$$
$$+ \left\langle \hat{n}\times\nabla\times\mathbf{E}_a^*\cdot\overline{\mu}^{-1}, \triangle(S)\mathbf{E}\right\rangle - \omega^2\langle\mathbf{E}_a^*, \overline{\epsilon}\cdot\mathbf{E}\rangle - i\omega\langle\mathbf{E}_a^*, \mathbf{J}\rangle + i\omega\langle\mathbf{J}_a^*, \mathbf{E}\rangle.$$
$$(5.5.39)$$

Moreover, if the boundary conditions that $\hat{n} \times \mathbf{H} = \boldsymbol{\alpha}$ and $\hat{n} \times \mathbf{H}_a = \boldsymbol{\alpha}_a$ on S are desired, then surface currents $\mathbf{J}_s = -\boldsymbol{\alpha}$ and $\mathbf{J}_{sa} = -\boldsymbol{\alpha}_a$ can be induced in the source term of the equations corresponding to (27) and (29), i.e.,

$$\nabla \times \overline{\mu}^{-1} \cdot \nabla \times \mathbf{E} - \triangle(S)\,\hat{n}\times\overline{\mu}^{-1}\cdot\nabla\times\mathbf{E} - \omega^2\,\overline{\epsilon}\cdot\mathbf{E} = i\omega\mathbf{J} - i\omega\triangle(S)\,\boldsymbol{\alpha},$$
$$(5.5.40a)$$
$$\nabla \times \left(\overline{\mu}^\dagger\right)^{-1} \cdot \nabla \times \mathbf{E}_a - \triangle(S)\,\hat{n}\times\left(\overline{\mu}^\dagger\right)^{-1}\cdot\nabla\times\mathbf{E}_a - \omega^2\,\overline{\epsilon}^\dagger\cdot\mathbf{E}_a = i\omega\mathbf{J}_a - i\omega\triangle(S)\,\boldsymbol{\alpha}_a.$$
$$(5.5.40b)$$

As a result, the corresponding variational expression, which is a modification of (24), is (Exercise 5.23a)

$$I = \left\langle \nabla \times \mathbf{E}_a^*, \overline{\mu}^{-1} \cdot \nabla \times \mathbf{E} \right\rangle - \omega^2\langle\mathbf{E}_a^*, \overline{\epsilon}\cdot\mathbf{E}\rangle$$
$$- i\omega\langle\mathbf{E}_a^*, [\mathbf{J} - \triangle(S)\boldsymbol{\alpha}]\rangle + i\omega\langle[\mathbf{J}_a^* - \triangle(S)\boldsymbol{\alpha}_a^*], \mathbf{E}\rangle. \qquad (5.5.41)$$

But if we require that $\hat{n} \times \mathbf{E} = \boldsymbol{\beta}$ and $\hat{n} \times \mathbf{E}_a = \boldsymbol{\beta}_a$ on S, then magnetic surface currents $\mathbf{M}_s = \boldsymbol{\beta}$ and $\mathbf{M}_{sa} = \boldsymbol{\beta}_a$ can be added to the surface S. Consequently, Equation (34) becomes

$$\nabla \times \overline{\boldsymbol{\mu}}^{-1} \cdot \nabla \times \mathbf{E} - \nabla \times \overline{\boldsymbol{\mu}}^{-1} \cdot \Delta(S)\,\hat{n} \times \mathbf{E} - \omega^2 \overline{\boldsymbol{\epsilon}} \cdot \mathbf{E}$$
$$= i\omega\mathbf{J} - \nabla \times \Delta(S)\,\overline{\boldsymbol{\mu}}^{-1} \cdot \boldsymbol{\beta}, \quad (5.5.42a)$$

and the equation corresponding to (37) is

$$\nabla \times \left(\overline{\boldsymbol{\mu}}^\dagger\right)^{-1} \cdot \nabla \times \mathbf{E}_a - \nabla \times \left(\overline{\boldsymbol{\mu}}^\dagger\right)^{-1} \cdot \Delta(S)\,\hat{n} \times \mathbf{E}_a - \omega^2 \overline{\boldsymbol{\epsilon}}^\dagger \cdot \mathbf{E}_a$$
$$= i\omega\mathbf{J}_a - \nabla \times \Delta(S) \left(\overline{\boldsymbol{\mu}}^\dagger\right)^{-1} \cdot \boldsymbol{\beta}_a. \quad (5.5.42b)$$

The variational expression, which is a modification of (39), then becomes (Exercise 5.23b)

$$I = \langle \mathbf{E}_a^*, \mathcal{L}\mathbf{E} \rangle - i\omega\langle \mathbf{E}_a^*, \mathbf{J} \rangle + \left\langle \nabla \times \mathbf{E}_a^*, \Delta(S)\,\overline{\boldsymbol{\mu}}^{-1} \cdot \boldsymbol{\beta} \right\rangle$$
$$+ i\omega\langle \mathbf{J}_a^*, \mathbf{E} \rangle + \left\langle \Delta(S) \left(\overline{\boldsymbol{\mu}}^t\right)^{-1} \cdot \boldsymbol{\beta}_a^*, \nabla \times \mathbf{E} \right\rangle. \quad (5.5.43)$$

Notice that in (41) and (43), by including surface sources on S, the natural boundary conditions for the tangential components of the electric field and the magnetic field are inhomogeneous.

If an impedance boundary condition is required on S, it may be considered the result of a thin sheet of material with $\overline{\boldsymbol{\epsilon}}$ infinite. Hence, a variational expression is (Exercise 5.23c)

$$I = \left\langle \nabla \times \mathbf{E}_a^*, \overline{\boldsymbol{\mu}}^{-1} \cdot \nabla \times \mathbf{E} \right\rangle - \omega^2 \langle \mathbf{E}_a^*, [\overline{\boldsymbol{\epsilon}} - \Delta(S)\,\overline{\mathbf{r}}] \cdot \mathbf{E} \rangle$$
$$- i\omega\langle \mathbf{E}_a^*, \mathbf{J} \rangle + i\omega\langle \mathbf{J}_a^*, \mathbf{E} \rangle. \quad (5.5.44)$$

Consequently, the stationary solution to the above has the following impedance boundary conditions:

$$\hat{n} \times \mathbf{H} = -i\omega\overline{\mathbf{r}} \cdot \mathbf{E}, \quad \hat{n} \times \mathbf{H}_a = -i\omega\overline{\mathbf{r}}^\dagger \cdot \mathbf{E}_a, \quad (5.5.45)$$

where $i\omega\overline{\mathbf{r}}$ is an impedance tensor. Mixed type natural boundary conditions can be derived by a combination of the above techniques.

In the preceding sections, we have only discussed the use of the extended operators in deriving the variational expressions I. Similar extended operators can be used in deriving variational expressions as in (5.2.15).

Exercises for Chapter 5

5.1 Show that any two vectors from the three-dimensional vector space satisfy properties (5.1.1a) to (5.1.1e). Are these algebraic properties satisfied by real and complex numbers?

5.2 Show that the norm of f, i.e., $\|f\|$ defined in (5.1.6), is a nonlinear, real functional of f.

5.3 Using $\hat{x}$, $\hat{y}$, and $\hat{z}$ as an orthonormal basis set in a three-dimensional linear vector space, derive the corresponding identity operator given by (5.1.26). Show that this identity operator could be inserted between an inner product and yet leave its value unchanged.

5.4 If we have two different orthonormal basis sets for the three-dimensional linear vector space, $\{\hat{x}, \hat{y}, \hat{z}\}$ and $\{\hat{x}', \hat{y}', \hat{z}'\}$, write the analogue of Parseval's theorem for transformations between these two basis sets. Show that this is equivalent to the statement that an inner product between two vectors is invariant with respect to their representations.

5.5 An arbitrary, square integrable function $f(x)$ defined for $-a < x < a$ can be expanded as a Fourier series

$$f(x) = \sum_{n=-\infty}^{\infty} b_n e^{\frac{in\pi x}{a}}.$$

(a) Show that the basis $e^{in\pi x/a}$ is orthogonal with a complex inner product, and derive a new orthonormal basis.

(b) Derive the equivalence of (5.1.42a), and hence, the identity operator (5.1.43) for this basis. Give the explicit expression for $\langle x^*, \mathcal{I}x' \rangle$, the coordinate-space representation of this identity operator.

(c) Derive Parseval's theorem for the Fourier series as in (5.1.48).

5.6 Show that the definitions of symmetric and Hermitian operators given by (5.1.53) and (5.1.54) are similar to those for matrix operators.

5.7 For a differential equation

$$\left[\frac{d^2}{dx^2} + k^2(x)\right] f(x) = g(x),$$

or $\mathcal{D}f\rangle = g\rangle$ where $0 < x < a$, and $f(0) = f(a) = 0$, we can approximate $f(x) = \sum_{n=1}^{N-1} a_n f_n(x)$ where (see figure)

$$f_n(x) = \begin{cases} \dfrac{[x - (n-1)\Delta]}{\Delta}, & (n-1)\Delta < x < n\Delta, \\ \dfrac{[(n+1)\Delta - x]}{\Delta}, & n\Delta < x < (n+1)\Delta. \end{cases}$$

In the above, $\Delta = a/N$ and $f_n(x)$ is a triangular basis function, also known as a chapeau function.

(a) Find the matrix representation of $\mathcal{D}$ in a space spanned by the chapeau functions. Also, find the representation of the state vector $g\rangle$ in this space, and hence, find the matrix equation equivalence of the above differential equation.

Hint: Use integration by parts to simplify your answer.

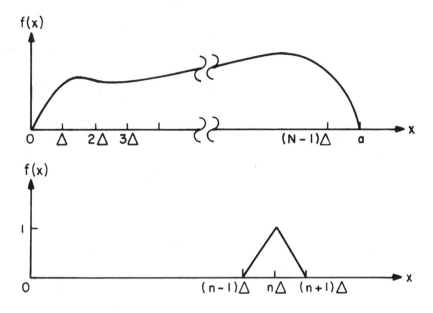

Figure for Exercise 5.7

(b) Assuming that $k(x)$ is a constant, use finite difference with central differencing to solve the above differential equation. Compare this with the result in part (a) when $k(x)$ is a constant. What are the differences and the similarities?

(c) Explain how the local nature of the differential operator expressed by (5.1.15) is manifested in the numerical approximations.

(d) Show that if half triangular elements are added at $x = 0$ and $x = a$ so that $f(0) \neq 0$, and $f(a) \neq 0$, then the matrix equation will be different, implying a different solution. Hence, the solution depends on the specified boundary conditions.

5.8 A set of vectors $f_i\rangle$ in an N-dimensional space is linearly independent if $\sum_{i=1}^{N} a_i f_i\rangle = 0$ implies that $a_i = 0$, $i = 1, \dots, N$.

(a) Show that if $f_i\rangle$ are mutually orthogonal and that $|\langle f_i, f_i \rangle| > 0$, then $f_i\rangle$ must be linearly independent.

(b) Show that for the vector $\mathbf{a} = \hat{x} + i\hat{y}$, $\mathbf{a}^\dagger \cdot \mathbf{a} \neq 0$ while $\mathbf{a}^t \cdot \mathbf{a} = 0$. Therefore, for a noncomplex inner product, a nonzero length vector may have zero self inner product.

(c) For a symmetric operator that has distinct eigenvalues, we have proven that $\langle f_i, f_j \rangle = 0$, $i \neq j$ as in (5.1.61). But this does not imply that $|\langle f_i, f_i \rangle| > 0$, because of (b). Hence, the eigenvectors may not be linearly independent. Show that this problem does not exist for a Hermitian operator.

(d) Show that a set of N mutually orthogonal and linearly independent vectors forms a complete basis in an N-dimensional space, i.e., there is no state vector in this N-dimensional space that is orthogonal to all the N state vectors. Hence, show that the eigenvectors of a Hermitian operator form a complete set. If the eigenvalues are not all distinct, show how the Gram-Schmidt orthogonalization procedure can be used to obtain orthogonal vectors.

5.9 If a medium is reciprocal, the electromagnetic operators are often symmetric. The equation governing an electromagnetic field in an inhomogeneous, bounded medium is

$$\nabla \times \mu^{-1} \nabla \times \mathbf{E} - \omega^2 \epsilon \, \mathbf{E} = i\omega \mathbf{J},$$

or $\mathcal{L} \mathbf{E} = i\omega \mathbf{J}$.

(a) Show that $\mathcal{L}$ is a symmetric operator in the sense that $\langle \mathbf{E}_1, \mathcal{L} \mathbf{E}_2 \rangle = \langle \mathbf{E}_2, \mathcal{L} \mathbf{E}_1 \rangle$, and hence, its matrix representation is symmetric in a space spanned by independent vectors $\mathbf{E}_n$, $n = 1, 2, \ldots$. (Three-dimensional integration is implied here in the inner product, and the dot product is taken between two vector functions in a multiplication. Assume that the medium is bounded by a perfectly conducting cavity.)

(b) Explain why the above is related to reciprocity.

(c) If the medium is lossless, show that $\mathcal{L}$ is Hermitian, but when the medium is lossy, it is no longer Hermitian.

5.10 Prove that the functional I given by (5.2.5) is a real-valued functional if $\mathcal{L}$ is a self-adjoint operator.

5.11 By taking the first variation of (5.2.10) about $\mathbf{a}_e$, which is the solution of (5.2.12), show that

(a) I has a minimum at $\mathbf{a}_e$ if $\overline{\mathbf{L}}$ is a positive definite matrix.

(b) I has a maximum at $\mathbf{a}_e$ if $\overline{\mathbf{L}}$ is a negative definite matrix.

(c) I has a saddle point at $\mathbf{a}_e$ if $\overline{\mathbf{L}}$ is an indefinite matrix.

5.12 Show that Equation (5.2.18) is a variational expression when $\mathcal{L}$ is a symmetric operator.

5.13 (a) Using the Rayleigh-Ritz procedure and the variational expression in (5.2.43), derive a matrix equation from which an approximate solution to (5.2.37) can be derived.

(b) Using the Rayleigh-Ritz procedure, derive an optimal approximation for u in (5.2.47).

5.14 Show that Equation (5.3.28) follows from (5.3.27) with the boundary condition $\hat{n} \times \mathbf{E} = 0$ or $\hat{n} \times \mathbf{H} = 0$ on S, the surface enclosing V.

5.15 (a) Prove that (5.3.32) is the adjoint equation of (5.3.25).

(b) Derive the variational expression (5.3.35).

5.16 A partial differential equation is described by

$$\nabla^2 \phi + \mathbf{a} \cdot \nabla \phi + k^2 \phi = s(\mathbf{r}),$$

where $\mathbf{a}$ is a constant real vector and k^2 is a constant real scalar.

(a) Is the equation self-adjoint?

(b) Find the adjoint equation if it is not self-adjoint.

(c) Derive a variational functional I for the equation.

5.17 (a) Show that Maxwell's equations imply

$$\begin{bmatrix} 0 & \nabla \times \\ \nabla \times & 0 \end{bmatrix} \begin{bmatrix} \mathbf{E} \\ \mathbf{H} \end{bmatrix} + i\omega \begin{bmatrix} \bar{\epsilon} & 0 \\ 0 & -\bar{\mu} \end{bmatrix} \begin{bmatrix} \mathbf{E} \\ \mathbf{H} \end{bmatrix} = \begin{bmatrix} \mathbf{J} \\ 0 \end{bmatrix},$$

which is of the form $\mathcal{L}f = s$.

(b) Deduce the conditions under which the linear operator $\mathcal{L}$ is self-adjoint and the conditions under which it is symmetric.

(c) Derive a variational expression I for the above equation when it is non-self-adjoint.

(d) Assuming that $\mathbf{J} = 0$ and the equation is now defined inside a cavity, derive a variational expression for the resonant frequencies of the cavity when the linear operator is symmetric.

5.18 (a) Apply the Rayleigh-Ritz procedure to (5.4.2a) and (5.4.4a) and show that the ensuing matrix eigenvalue equation is of the form

$$\overline{\mathbf{L}} \cdot \mathbf{a} = \lambda \overline{\mathbf{B}} \cdot \mathbf{a}.$$

How are $\overline{\mathbf{L}}$ and $\overline{\mathbf{B}}$ related to $\mathcal{L}$ and $\mathcal{B}$?

(b) Apply the Rayleigh-Ritz procedure to (5.4.6) and (5.4.9) and show that the ensuing matrix eigenvalue equations are

$$\overline{\mathbf{L}} \cdot \mathbf{a} = \lambda \overline{\mathbf{B}} \cdot \mathbf{a}$$

and

$$\overline{\mathbf{L}}^a \cdot \mathbf{b} = \lambda^a \overline{\mathbf{B}}^a \cdot \mathbf{b},$$

where a is † for (5.4.6), and a is t for (5.4.9). How are $\overline{\mathbf{L}}$ and $\overline{\mathbf{B}}$ related to $\mathcal{L}$ and $\mathcal{B}$?

(c) Could the above matrix equation be more directly obtained using the method of weighted residuals?

5.19 Derive Equation (5.4.19) from Maxwell's equations.

5.20 (a) Show that (5.5.9) implies (5.5.10) for an extended class of functions.

(b) Show that (5.5.14) follows from (5.5.13) for an extended class of functions.

5.21 (a) Show that the optimal solution to (5.5.17) satisfies the Neumann boundary condition that $\hat{n} \cdot p\nabla\phi = \beta$ on S, the surface of V.

(b) Show that the optimal solution to (5.5.18) satisfies the Dirichlet boundary condition that $\phi = \gamma$ on S.

(c) Show that the optimal solution to (5.5.19) satisfies the impedance boundary condition that $\hat{n} \cdot p\nabla\phi = \alpha\phi$ on S.

5.22 (a) Show that the optimal solution to (5.5.24), with the operator $\mathcal{L}$ and $\mathcal{L}^a$ extended as in (5.5.27) and (5.5.29), satisfies the boundary condition (5.5.31).

(b) Show that the optimal solution to (5.5.39) satisfies

$$\hat{n} \times \mathbf{E} = \hat{n} \times \mathbf{E}_a = 0$$

on S.

5.23 (a) Show that the optimal solution to (5.5.41) satisfies $\hat{n} \times \mathbf{H} = \boldsymbol{\alpha}$ and $\hat{n} \times \mathbf{H}_a = \boldsymbol{\alpha}_a$.

(b) Show that the optimal solution to (5.5.43) satisfies

$$\hat{n} \times \mathbf{E} = \boldsymbol{\beta}$$

and

$$\hat{n} \times \mathbf{E}_a = \boldsymbol{\beta}_a.$$

(c) Show that the optimal solution to (5.5.44) satisfies the impedance boundary condition $\hat{n} \times \mathbf{H} = -i\omega\bar{\mathbf{r}} \cdot \mathbf{E}$ and $\hat{n} \times \mathbf{H}_a = -i\omega\bar{\mathbf{r}}^\dagger \cdot \mathbf{E}_a$.

References for Chapter 5

Axelsson, O., and V. A. Barker. 1984. *Finite Element Solution of Boundary Value Problems: Theory and Computation.* New York: Academic Press.

Berberian, S. K. 1961. *Introduction to Hilbert Space.* New York: Oxford University Press.

Cairo, L., and T. Kahan. 1965. *Variational Techniques in Electromagnetism.* New York: Gordon and Breach.

Chen, C. H., and C. D. Lien. 1980. "The variational formulation for non-self-adjoint electromagnetic problems." *IEEE Trans. Microwave Theory Tech.* 28: 878–86.

Chew, W. C., and M. Nasir. 1989. "A variational analysis of anisotropic, inhomogeneous dielectric waveguides." *IEEE Trans. Microwave Theory Tech.* MTT-37: 661–68.

Finlayson, B. A., and L. E. Scriven. 1967. *Int. J. Heat Mass Trans.* 10: 799.

Galerkin, B. G. 1915. "Series occurring in various questions concerning the elastic equilibrium of plates." (Russian). *Eng. Bull. (Vestn. Inzh. Tech.)* 19: 897–908.

Harrington, R. 1983. *Field Computation by Moment Method.* Malabar, FL: Krieger. (First printing 1968.)

Jones, D. S. 1956. "A critique of the variational method in scattering problems." *IRE Trans.* AP-4(3): 297–301.

Kantorovich, L. V., and V. I. Krylov. 1958. *Approximate Method of Higher Analysis.* Gronigen: Noordhoff Ltd., pp. 336.

Kravchuk, M. F. 1932. "Application of the method of moments to the solution of linear differential and integral equations." (Ukrainian). *Kiev. Soobshch. AN USSR* 1: 168.

Kravchuk, M. F. 1936. "On the convergence of the method of moments for partial differential equations." (Ukrainian). *Zh. in-ta matem. AN USSR* 1: 23–27.

Merzbacher, E. 1970. *Quantum Mechanics.* New York: John Wiley & Sons.

Mikhlin, S. G. 1970. *Mathematices Physics.* Amsterdam: North-Holland.

Petrov, G. 1940. "Application of Galerkin's method to a problem of the stability of the flow of a viscous liquid." (Russian). *Prikad. Matem. i Mekh.* 4(3): 3–12.

Rumsey, V. H. 1954. "Reaction concepts in electromagnetic theory." *Phys. Rev.* 94: 1483–91; 95: 1706.

Sakurai, J. J. 1985. *Modern Quantum Mechanics.* New York: Wiley-Interscience.

Silvester, P. P., and R. L. Ferrari. 1983. *Finite Elements for Electrical*

Engineers. Cambridge: Cambridge University Press.

Stakgold, I. 1967. *Boundary Value Problems of Mathematical Physics*, vol. I. London: Macmillan Co.

Stakgold, I. 1979. *Green's Functions and Boundary Value Problems.* New York: John Wiley & Sons.

Strang, G., and G. J. Fix. 1973. *Analysis of the Finite Element Method.* Englewood Cliffs, NJ: Prentice Hall.

Wait, R., and A. R. Mitchell. 1985. *Finite Element Analysis and Applications.* New York: John Wiley & Sons.

Further Readings for Chapter 5

Akiba, S., and H. A. Haus. 1982. "Variational analysis of optical waveguides with rectangular cross sections." *Appl. Opt.* 21: 804–808.

Anderson, B., and S. K. Chang. 1982. "Synthetic induction logs by the finite element method." *Log Analyst* 23: 17–20.

Angkaew, T., M. Matsuhara, and N. Kumagai. 1987. "Finite-element analysis of waveguide modes: a novel approach that eliminates spurious modes." *IEEE Trans. Microwave Theory Tech.* MTT-35: 117–23.

Angulo, C. M., and W. S. C. Chang. 1959. "A variational expression for the terminal admittance of a semi-infinite dielectric rod." *IRE Trans. Antennas Propagat.* AP-7: 207–12.

Bathe, K.-J., and E. L. Wilson. 1976. *Numerical Methods in Finite-Element Analysis.* Englewood Cliffs, NJ: Prentice Hall.

Bierwirth, K., N. Schulz, and F. Arndt. 1986. "Finite-difference analysis of rectangular dielectric waveguide structures." *IEEE Trans. Microwave Theory Tech.* MTT-34: 1104–14.

Bossavit, A., and J.-C. Vérité. 1982. "A mixed FEM-BIEM method to solve 3-D eddy-current problems." *IEEE Trans. Magnetics* MAG-18: 431–35.

Brandt, A. 1977. "Multi-level adaptive solutions to boundary-value problems." *Math. Comp.* 31: 333–90.

Chari, M. V. K., and P. P. Silvester. 1980. *Finite Elements in Electric and Magnetic Field Problems.* New York: John Wiley & Sons.

Chen, C. H., and Y. W. Kiang. 1980. "A variational theory for wave propagation in a one-dimensional inhomogeneous medium." *IEEE Trans. Antennas Propagat.* AP-28: 762–69.

Chew, W. C., and R. L. Kleinberg. 1988. "Theory of microinduction measurements." *IEEE Trans. Geosci. Remote Sensing* GE-26: 707–19.

Csenges, Z. J. 1976. "A note on the finite-element solution of exterior–field problems." *IEEE Trans. Microwave Theory Tech.* MTT-24: 468–73.

English, W. J. 1971. "Vector variational solutions of inhomogeneously loaded cylindrical waveguide structures." *IEEE Trans. Microwave Theory Tech.* MTT-19: 9–18.

English, W. J., and F. J. Young. 1971. "An E vector variational formulation of the Maxwell equations for cylindrical waveguide problems." *IEEE Trans. Microwave Theory Tech.* MTT-19: 40–46.

Gurtin, M. E. 1964. "Variational problems for linear initial-value problems." *Quart. J. Appl. Mech.* 22: 252–56.

Gurtin, M. E. 1964. "Variational principles for linear elasto-dynamics." *Archive for Rational Mech. and Anal.* 16: 34–50.

Hayata, K., M. Koshiba, M. Eguchi, and M. Suzuki. 1986. "Vectorial finite element method without any spurious solutions for dielectric waveguiding problems using transverse magnetic field component." *IEEE Trans. Microwave Theory Tech.* MTT-34: 1120–24.

Hinton, E., and D. R. J. Owen. 1977. *Finite Element Programming.* Orlando, FL: Academic.

Ikeuchi, M., H. Sawami, and H. Niki. 1981. "Analysis of open-type dielectric waveguides by the finite-element iterative method." *IEEE Trans. Microwave Theory Tech.* MTT-29: 234–39.

Irons, B. M. 1970. "A frontal solution program for finite element analysis," *International J. for Numerical Method in Engineering.* 2: 5–32.

Jeng, S. K., and C. H. Chen. 1984. "Variational finite element solution of electromagnetic wave propagation in a one-dimensional inhomogeneous anisotropic medium." *J. Appl. Phys.* 55: 630–36.

Jeng, S. K., and C. H. Chen. 1984. "On variational electromagnetics: Theory and applications." *IEEE Trans. Antennas Propagat.* AP-32: 902–07.

Jeng, G., and A. Wexler. 1978. "Self-adjoint variational formulation of problems having non-self-adjoint operators." *IEEE Trans. Microwave Theory Tech.* MTT-26: 91–94.

Jeng, S.-K., R.-B. Wu, and C. H. Chen. 1986. "Waves obliquely incident upon a stratified anisotropic slab: A variational reaction approach." *Radio Sci.* 21: 681.

Kaplan, S. 1964. "The use of the Rayleigh-Ritz method in non-self-adjoint problems." *IEEE Trans. Microwave Theory Tech.* MTT-12: 254–55.

Katz, J. 1982. "Novel solution of 2-D waveguide using the finite element method." *Appl. Opt.* 21: 2747.

Konrad, A. 1976. "Vector variational formulation of electromagnetic fields in anisotropic media." *IEEE Trans. Microwave Theory Tech.* MTT-24: 553–59.

Koshiba, M., K. Hayata, and M. Suzuki. 1984. "Approximate scalar finite element analysis of anisotropic optical waveguides with off-diagonal elements in a permittivity tensor." *IEEE Trans. Microwave Theory Tech.* MTT-32: 587–93.

Koshiba, M., K. Hayata, and M. Suzuki. 1984. "Vectorial finite element formulation without spurious modes for dielectric waveguides." *Trans. IECE Japan* E 67: 191–96.

Kotiuga, P. R., and P. P. Silvester. 1982. "Vector potential formulation for three-dimensional magnetostatics." *J. Appl. Phys.* 53: 8399–8401.

Lanczos, C. 1970. *The Variational Principles of Mechanics*, 4th ed. Toronto: University of Toronto Press.

Mabaya, N., P. E. Lagasse, and P. Vanderbulcke. 1981. "Finite element analysis of optical waveguides." *IEEE Trans. Microwave Theory Tech.* MTT-29: 600–05.

McAulay, A. D. 1977. "Variational finite-element solution for dissipative waveguides and transportation application." *IEEE Trans. Microwave Theory Tech.* MTT-25: 382–392.

Morishita, K., and N. Kumagai. 1977. "Unified approach to the derivation of variational expression for electromagnetic fields." *IEEE Trans. Microwave Theory Tech.* MTT-25: 34–40.

Morishita, K., and N. Kumagai. 1978. "Systematic derivation of variational expressions for electromagnetic and/or acoustic waves." *IEEE Trans. Microwave Theory Tech.* MTT-26: 684–89.

Mur, G. 1988. "Optimum choice of finite elements for computing three-dimensional electromagnetic fields in inhomogeneous media." *IEEE Trans. Magnetics* 24: 330–32.

Mur, G., and A. T. de Hoop. 1985. "A finite-element method for computing three-dimensional electromagnetic fields in inhomogeneous media." *IEEE Trans. Magnetics* MAG-21: 2188–91.

Nédelec, J. C. 1980. "Mixed finite elements in R3." *Numer. Math.* 35: 315–41.

Pridmore, D. F., G. W. Hohmann, S. W. Ward, and W. R. Sill. 1981. "An investigation of the finite-element modeling for electrical and electromagnetic data in three dimensions." *Geophysics* 46: 1009–24.

Rahman, B. M. A., and J. B. Davies. 1984. "Finite-element analysis of optical and microwave waveguide problems." *IEEE Trans. Microwave Theory Tech.* MTT-32: 20–28.

Rozzi, T. E., and G. H. Veld. 1980. "Variational treatment of the diffraction at the facet of d.h. lasers and of dielectric millimeter wave antennas." *IEEE Trans. Microwave Theory Tech.* MTT-28: 61–73.

Smith, W. D. 1975. "The application of finite-element analysis to body wave propagation problems." *Geophys. J. Roy. Astr. Soc.* 42: 747–68.

Smith, W. D., and B. A. Bolt. 1976. "Rayleigh's principle in finite-element calculations for seismic wave response." *Geophys. J. R. Astr. Soc.* 45: 647–755.

Su, C. C. 1986. "A combined method for dielectric waveguides using the finite-element technique and the surface integral equations method." *IEEE Trans. Microwave Theory Tech.* MTT-34: 1140–46.

Thacker, W. C. 1978a. "Comparison of finite-element and finite-difference schemes. Part I: One-dimensional gravity wave motion." *J. Phys. Ocean.* 8: 676–79.

Thacker, W. C. 1978b. "Comparison of finite-element and finite-difference schemes. Part II: Two-dimensional gravity wave motion." *J. Phys. Ocean.* 8: 680–89.

Washizu, K. 1962. "Variational principles in continuum mechanics." Dept. Aero. Eng., Seattle: University of Washington.

Wu, R. B., and C. H. Chen. 1985. "On the variational reaction theory for dielectric waveguides." *IEEE Trans. Microwave Theory Tech.* MTT-33: 477–83.

Wu, R. B., and C. H. Chen. 1985. "A variational analysis of dielectric waveguides by the conformal mapping technique." *IEEE Trans. Microwave Theory Tech.* MTT-33: 681–85.

Wu, R. B., and C. H. Chen. 1986. "A scalar variational conformal mapping technique by weakly guiding dielectric waveguide." *IEEE J. Quantum Electron.* QE-22: 595–608.

Yeh, C., K. Ha, S. B. Dong, and W. P. Brown. 1979. "Single mode optical waveguides." *Appl. Optics* 1810: 1490–1504.

Zienkiewicz, O. C. 1977. *The Finite Element Method in Engineering Science.* New York: McGraw-Hill.

CHAPTER 6

MODE MATCHING METHOD

When a two-dimensional inhomogeneity is formed by a junction of two layered regions, the solution to this problem may be sought by the mode matching method. In this method, the eigenmodes of each region are found first and matched to each other at the junction discontinuity to satisfy the boundary conditions.

If a source is in a layered structure, the field due to the source can be expanded in terms of the eigenmodes of the structure. When the structure is enclosed by impenetrable walls, these eigenmodes have a discrete spectra. However, when the structure is open, extending to infinity, some of these modes form a continuum, known as the radiation modes. In this chapter, the eigenmodes of a layered medium will be studied first. Assuming discrete-mode spectra only, we shall show how these modes are used to match boundary conditions at a junction of two dissimilar layered regions. Subsequently, this is generalized to continuum-mode spectra. Later, we shall show how the canonical solution of one junction discontinuity can be used to find the solution of the problem involving multiple junction discontinuities. In this manner, fairly complex two-dimensional problems are solved as a concatenation of one-dimensional problems. Furthermore, the method can be shown to be more efficient than a direct use of the finite-element method to solve such a problem.[1]

§6.1 Eigenmodes of a Planarly Layered Medium

Before introducing the mode matching method, it is imperative to understand the eigenmodes of a layered medium.[2] The field of an open, layered medium extends to infinity. As a result, an open, layered medium has a continuum of modes which makes its understanding less transparent. Therefore, it is more appropriate to study first the eigenmodes of a layered medium enclosed by two impenetrable walls (see Figure 6.1.1). In this case, all the eigenmodes of the layered medium are discrete. The modes of an open medium can then be obtained by allowing the locations of the impenetrable walls to tend to infinity.

---

[1] A knowledge of linear vector space as discussed in Chapter 5 is required to understand the subject matter in this chapter.

[2] See also Marcuse (1972) for a discussion.

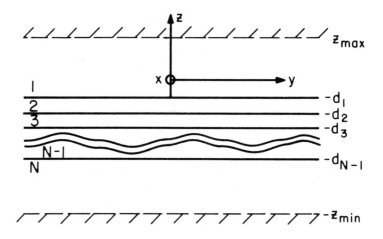

Figure 6.1.1 A layered medium bounded by two impenetrable walls.

§§6.1.1 *Orthogonality of Eigenmodes in a Layered Medium*

When a layered medium is bounded by two impenetrable walls at z_{max} and $-z_{min}$ (Figure 6.1.1), the natural solutions to the boundary value problem are the eigenmodes of the structure. The walls in this case may be electric, magnetic, or even reactive. The governing equation for such a problem, as shown in Chapter 2, is

$$\left[p\frac{d}{dz}p^{-1}\frac{d}{dz} + k^2(z) - k_s^2 \right]\phi(z) = 0, \qquad k_s^2 = k_x^2 + k_y^2, \tag{6.1.1}$$

where $p = \epsilon$ for TM waves and $p = \mu$ for TE waves. The natural solutions or the eigenmodes can only exist for certain discrete values of k_s^2, found by imposing the guidance conditions (similar to finding the modes of a parallel plate waveguide). These solutions are the eigenfunctions or the eigenvectors of the operator equation

$$\left[p\frac{d}{dz}p^{-1}\frac{d}{dz} + k^2(z) \right]\phi(z) = k_s^2\phi(z), \tag{6.1.2}$$

with the appropriate boundary conditions at z_{max} and $-z_{min}$. An alternative way to write Equation (2) is

$$\left[\frac{d}{dz}p^{-1}\frac{d}{dz} + k^2 p^{-1} \right]\phi(z) = k_s^2 p^{-1}\phi(z) \tag{6.1.3}$$

making it self-adjoint.

The above is an operator equation of the form

$$\mathcal{D}\phi\rangle = k_s^2\mathcal{B}\phi\rangle, \tag{6.1.4}$$

where k_s^2 is the eigenvalue, $\phi\rangle$ the eigenvector, and $\mathcal{D}$ and $\mathcal{B}$ are operators. The coordinate-space representation of Equation (4) is retrieved by projecting (4) onto coordinate-space $\langle z'$, and inserting the identity operator $\int dz\, z\rangle\langle z$ into (4).[3] For example,

$$\langle z', \mathcal{D}\phi \rangle = \int \langle z', \mathcal{D}z \rangle \langle z, \phi \rangle\, dz = \int dz\, \mathcal{D}(z', z)\phi(z), \qquad (6.1.5a)$$

$$\langle z', \mathcal{B}\phi \rangle = \int \langle z', \mathcal{B}z \rangle \langle z, \phi \rangle\, dz = \int dz\, \mathcal{B}(z', z)\phi(z). \qquad (6.1.5b)$$

On comparing the above with (3), we deduce that

$$\mathcal{D}(z', z) = \delta(z' - z) \left[\frac{d}{dz} p^{-1} \frac{d}{dz} + k^2 p^{-1} \right], \qquad (6.1.5)$$

$$\mathcal{B}(z', z) = \delta(z' - z) p^{-1}(z). \qquad (6.1.6)$$

With some boundary conditions on $\phi(z)$, it can be shown easily that $\mathcal{D}$ and $\mathcal{B}$ are Hermitian operators when the medium is lossless (Exercise 6.1). In this case, Equation (4) is equivalent to a matrix equation of the form

$$\overline{\mathbf{D}} \cdot \boldsymbol{\phi} = \lambda \overline{\mathbf{B}} \cdot \boldsymbol{\phi}, \qquad (6.1.7)$$

where $\overline{\mathbf{D}}$ and $\overline{\mathbf{B}}$ are Hermitian matrices. Moreover, Equation (7) gives rise to only real eigenvalues because if

$$\overline{\mathbf{D}} \cdot \boldsymbol{\phi}_1 = \lambda_1 \overline{\mathbf{B}} \cdot \boldsymbol{\phi}_1, \qquad (6.1.8a)$$

$$\overline{\mathbf{D}} \cdot \boldsymbol{\phi}_2 = \lambda_2 \overline{\mathbf{B}} \cdot \boldsymbol{\phi}_2, \qquad (6.1.8b)$$

then multiplying (8a) by $\boldsymbol{\phi}_2^\dagger$ and (8b) by $\boldsymbol{\phi}_1^\dagger$, one obtains

$$\boldsymbol{\phi}_2^\dagger \cdot \overline{\mathbf{D}} \cdot \boldsymbol{\phi}_1 = \lambda_1 \boldsymbol{\phi}_2^\dagger \cdot \overline{\mathbf{B}} \cdot \boldsymbol{\phi}_1, \qquad (6.1.9a)$$

$$\boldsymbol{\phi}_1^\dagger \cdot \overline{\mathbf{D}} \cdot \boldsymbol{\phi}_2 = \lambda_2 \boldsymbol{\phi}_1^\dagger \cdot \overline{\mathbf{B}} \cdot \boldsymbol{\phi}_2. \qquad (6.1.9b)$$

Then, after taking the conjugate transpose of (9b) and subtracting it from (9a), one obtains

$$(\lambda_1 - \lambda_2^*) \boldsymbol{\phi}_2^\dagger \cdot \overline{\mathbf{B}} \cdot \boldsymbol{\phi}_1 = 0. \qquad (6.1.10)$$

But if $\lambda_1 = \lambda_2$ or $\boldsymbol{\phi}_1 = \boldsymbol{\phi}_2$ in (10), then (10) could be satisfied only if $\lambda_1 = \lambda_1^*$. That is, λ_1 is real, since $\boldsymbol{\phi}_1^\dagger \cdot \overline{\mathbf{B}} \cdot \boldsymbol{\phi}_1$ is not zero in general. Furthermore, if $\lambda_1 \neq \lambda_2$, then $\boldsymbol{\phi}_2^\dagger \cdot \overline{\mathbf{B}} \cdot \boldsymbol{\phi}_1 = 0$, i.e., $\boldsymbol{\phi}_1$ and $\boldsymbol{\phi}_2$ are $\overline{\mathbf{B}}$ orthogonal. A similar proof can be applied to the operator equation. Therefore, in a lossless medium, Equation (4) possesses only real eigenvalues, and the eigenvectors are $\mathcal{B}$ orthogonal, i.e.,

$$\langle \phi_1^*, \mathcal{B}\phi_2 \rangle = 0, \qquad (6.1.11)$$

---

[3] See Chapter 5 for a review of linear vector space.

for two eigenvectors $\phi_1\rangle$ and $\phi_2\rangle$ corresponding to two distinct eigenvalues.

If the medium is lossy, however, $\mathcal{D}$ and $\mathcal{B}$ are no longer Hermitian but are still symmetric. In this case, the eigenvalues k_s^2 are complex. But for distinct eigenvalues, it can be shown that (Exercise 6.2)

$$\langle \phi_1, \mathcal{B}\phi_2 \rangle = 0. \tag{6.1.12}$$

§§6.1.2 Guided Modes and Radiation Modes of a Layered Medium

Let us elicit some salient features of the eigenfunctions of (2). These eigenfunctions are actually the eigenmodes of the layered medium described by Equation (2). There are two kinds of eigenmodes in Equation (2):

(i) one that corresponds to guided modes in the layered slab,

(ii) and one that corresponds to exterior modes not guided by the layered slab, but guided by the impenetrable walls.

(We define the layered slab to be the regions excluding region 1 and region N.) For the slab-guided modes, the field is evanescent in both the open regions 1 and N (both k_{1z} and k_{Nz} are imaginary); hence, the energy is trapped mainly in the layered slab. These modes are in **bound states**. An exterior mode is propagatory in region 1 or N, or both (k_{1z} and k_{Nz} are not imaginary). Hence, the energy of the exterior mode is not bound to the layered slab but is flowing in the open region; therefore, it is in a **free state** with respect to the layered slab.

Now, let us discuss how these modes can be found. Assume first that the impenetrable walls are at z_{max} and $-z_{min}$. Then, the field in each region can be derived as in Chapter 2. For example, if the inhomogeneity consists of piecewise constant layers, the field of a mode in the i-th layer can be expressed as

$$\phi_i(z) = A_i \left[e^{-ik_{iz}z} + \tilde{R}_{i,i+1} e^{ik_{iz}(2d_i + z)} \right], \tag{6.1.13}$$

where $k_{iz} = \sqrt{k_i^2 - k_s^2}$. In region 1, the boundary condition at the impenetrable wall may be a homogeneous Dirichlet boundary condition, which is

$$\phi_1(z = z_{max}) = 0. \tag{6.1.14}$$

This boundary condition corresponds to an electric wall for TE waves, while it corresponds to a magnetic wall for TM waves (the converse will be true for a homogeneous Neumann boundary condition). As we shall show, it is from this boundary condition (14) that the eigenvalue k_s^2 in (2) is determined.

The boundary condition (14) at $z = z_{max}$ implies that the solution can exist only if

$$1 + \tilde{R}_{12} e^{2ik_{1z}(d_1 + z_{max})} = 0. \tag{6.1.15}$$

$\tilde{R}_{12}$ can be defined in terms of k_{iz}'s of all the regions as shown in Chapter 2. Since $k_{iz} = \sqrt{k_i^2 - k_s^2}$, the above is an implicit equation for k_s^2, which is the eigenvalue of (2). It can be solved to yield eigenvalues corresponding to modes which are either:

(i) guided by the layered slab with energy bound to the slab, or

(ii) guided by the impenetrable walls with energy outside the slab.

It can be shown that (Exercises 6.3, 6.4) Equation (15) produces roots for k_s^2 only at discrete values, implying that all the eigenvalues are discrete. Once the eigenvalue is found, the field of the eigenmode can be derived from (13). Moreover, $\tilde{R}_{i,i+1}$ and A_i can be found as in Chapter 2—A_i is determined within a multiplicative factor.

The left-hand side of (15) at this point is a multivalued function because of the branch point due to $k_{1z} = \sqrt{k_1^2 - k_s^2}$. But a meromorphic function (a single-valued function which has only poles and zeroes) can be constructed to have the same zeroes as (15). It is much easier to look for the roots of such a function on the complex plane with a numerical equation root solver because of its single-valued property. Such a function is

$$f(k_s) = \frac{1 + \tilde{R}_{12}e^{2ik_{1z}(d_1+z_{max})}}{\tilde{R}_{12} + e^{2ik_{1z}(d_1+z_{max})}}. \qquad (6.1.16)$$

It can be shown that the above is a single-valued function (see Exercises 6.3, 6.4). Incidentally, the above can be derived by putting a source in region 1 of Figure 6.1.1. Then, from the uniqueness principle, one can prove that the solution is branch-point free. Since (16) is branch-point free, it is a meromorphic function with discrete poles and zeroes.

When z_{max} and z_{min} are made large, we expect the number of exterior modes (modes not guided by the slab) to increase but the number of guided modes (or bound states) in the slab to remain essentially the same. This becomes clearer if we study the modes of a parallel plate waveguide, namely, the case when the layered slab is absent. In this case, it is easy to show that for the n-th mode, $k_s = [k^2 - \left(\frac{n\pi}{d}\right)^2]^{1/2}$, where d is the separation of the parallel plates. Then, if the separation of a parallel plate waveguide is increased, the number of guided modes between the parallel plates becomes denser on the complex k_s plane. As a result, when the modes for a layered medium are sketched on the complex k_s plane for large z_{max} and z_{min} for a lossless medium, they appear as shown in Figure 6.1.2 for the case when $k_N < k_1$.

For a lossless medium, k_s^2 is pure real (positive real as well as negative real), as shown in Subsection 6.1.1. Therefore, k_s is either pure real or pure imaginary. The modes corresponding to $k_s > k_1 > k_N$ are modes guided by the slab since these modes are evanescent in both regions 1 and N because, in this case, k_{1z} and k_{Nz} are pure imaginary.[4] On the other hand, modes with

---

[4] Note that $k_{iz} = \sqrt{k_i^2 - k_s^2}$.

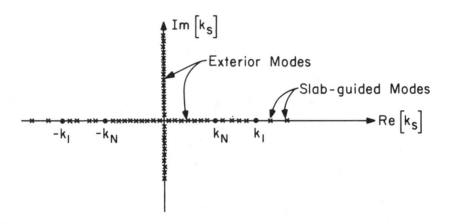

Figure 6.1.2 Slab-guided modes and exterior modes on the complex k_s plane.

$k_N < k_s < k_1$ are evanescent in region N but propagatory in region 1. Hence, if z_{max} becomes larger, the number of these modes becomes denser as region 1 can accommodate more guided modes. Moreover, when either $k_s < k_N$ or k_s is purely imaginary, both k_{1z} and k_{Nz} are purely real. Consequently, these modes are propagatory in the z direction in both regions 1 and N. Again, their number becomes denser for larger z_{max} and z_{min}. Note that a mode whose k_s is purely imaginary is evanescent in the s direction, and hence is cut off. These modes have $k_{1z} > k_1$ and $k_{Nz} > k_N$.

Moreover, when region 1 and region N become open regions, $f(k_s)$ in (16) is no longer meromorphic. This is corroborated by the analysis of Section 2.7 in Chapter 2, where it has been shown that branch points exist at $k_s = k_1$ and $k_s = k_N$. Hence, branch cuts have to be defined for these branch points. The radiation modes then fall on the Sommerfeld branch cuts defined for k_{1z} and k_{Nz}. Surprisingly, this is even the case when the medium is lossy, as shall be shown next.

When z_{max} and z_{min} tend to infinity, the exterior modes, i.e., modes not guided by the slab, become a continuum. Then, the distribution of modes is as shown in Figure 6.1.3. In other words, the continuum modes lie on the Sommerfeld branch cut.[5] These continuum modes are called the *radiation modes*—so called because they carry energy to infinity.[6]

---

[5] See Chapter 2, Subsection 2.2.3, for a definition of the Sommerfeld branch cut.

[6] It is well-known that an operator can have continuous as well as discrete spectra (e.g., see Stakgold 1967; Marcuse 1972; Felsen and Marcuvitz 1973).

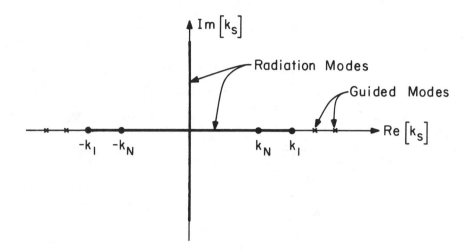

Figure 6.1.3 Distribution of modes in an open layered slab. The radiation modes are distributed on the Sommerfeld branch cuts.

When the medium is lossy so that k_1 and k_N are complex, the Sommerfeld branch cuts are not as shown in Figure 6.1.3, and they do not hug the real and imaginary axes anymore [see Chapter 2, Equation (2.2.33)]. Even though this is not obvious, it can be proven that the radiation modes still lie along the Sommerfeld branch cut as follows: If (15) is solved for the exterior modes in region 1 when z_{max} becomes large, k_{1z} can only have a small imaginary part, otherwise $e^{ik_{1z}z_{max}}$ will become inordinately large or small, and Equation (15) cannot be satisfied unless $\tilde{R}_{12}$ becomes small or large respectively. For a mode guided in region 1, however, $\tilde{R}_{12}$ is finite.[7] Therefore, when z_{max} tends to infinity, k_{1z} corresponding to the exterior modes in region 1 has to be real. Since k_{1z} is real on the Sommerfeld branch cut (defined to be $\Im m[k_{1z}] = 0$), these exterior modes are again distributed along the Sommerfeld branch cut! Similar argument applies to region N; the exterior modes corresponding to region N will lie on the Sommerfeld branch cut corresponding to k_{Nz}, which is defined to be $\Im m[k_{Nz}] = 0$. Consequently, the distribution of the radiation modes (which evolve from the exterior modes) is as shown in Figure 6.1.4 (also see Exercise 6.5).

The continuum modes plus the discrete modes of a layered slab may be

---

[7] $\tilde{R}_{12}$ can become infinite only at values of k_s corresponding to guided modes in the subsurface region with respect to region 1. This can be understood by studying the expression for $\tilde{R}_{12}$ given in Chapter 2, Equation (2.1.24).

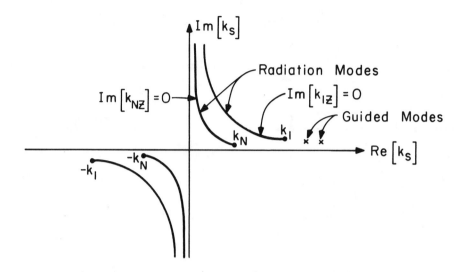

Figure 6.1.4 Distribution of modes in a lossy medium case. The radiation modes are distributed on the Sommerfeld branch cuts.

assumed complete.[8] Hence, an arbitrary field $\phi(z)$ may be expanded as

$$\phi(z) = \sum_{n=1}^{N} a_n \tilde{\phi}(n, z) + \int_0^\infty dk_{1z}\, a(k_{1z}) \tilde{\phi}(k_{1z}, z)$$

$$+ \int_0^\infty dk_{Nz}\, a(k_{Nz}) \tilde{\phi}(k_{Nz}, z). \quad (6.1.17)$$

The first term corresponds to discrete modes, and the second term corresponds to radiation modes in region 1, while the third term corresponds to radiation modes in region N. From (12), however, we deduce that the eigenmodes are p^{-1} orthogonal. Hence, these eigenmodes can always be normalized so that they are p^{-1} orthonormal (Exercise 6.6). Then, for discrete modes, we have

$$\int_{-\infty}^\infty dz\, \tilde{\phi}(n, z) \tilde{\phi}(m, z) p^{-1}(z) = \delta_{nm}. \quad (6.1.18)$$

---

[8] This is only strictly true for a lossless slab because only the eigenvectors of a Hermitian operator can be shown to be complete (see Chapter 5, Exercise 5.8). For a lossy slab, the modes cannot be proved to be complete, but they cannot be proved to be incomplete either.

For the continuum modes, we have[9]

$$\int_{-\infty}^{\infty} dz \, \tilde{\phi}(k_{1z}, z) \tilde{\phi}(k'_{1z}, z) p^{-1}(z) = \delta(k_{1z} - k'_{1z}), \qquad (6.1.19)$$

$$\int_{-\infty}^{\infty} dz \, \tilde{\phi}(k_{Nz}, z) \tilde{\phi}(k'_{Nz}, z) p^{-1}(z) = \delta(k_{Nz} - k'_{Nz}). \qquad (6.1.20)$$

Note that a discrete mode and a continuum mode are also mutually orthogonal.

§6.2 Eigenfunction Expansion of a Field

The eigenmodes of a layered medium are also the eigenfunctions of the medium. Since the set of eigenmodes of a layered slab is complete, a solution due to an arbitrary source may be expanded in terms of the eigenmodes of the slab. Moreover, these eigenmodes are orthogonal, making the solution simpler (Chew 1985). Even though this method of deriving the field due to a source in a layered medium is different from that in Chapter 2, they can be shown to be equivalent to each other (Exercise 6.8).

§§6.2.1 Excitation of Modes due to a Line Source

If we have a line source with $e^{ik_x x}$ dependence, all the fields it generates will have $e^{ik_x x}$ dependence. Now, if this source is embedded in a medium as shown in Figure 6.1.1, from Maxwell's equations, it can be shown that the z component of the electric flux D_z satisfies the equation (Exercise 6.7)

$$\left[\frac{\partial^2}{\partial y^2} + \epsilon \frac{\partial}{\partial z} \epsilon^{-1} \frac{\partial}{\partial z} + k^2(z) - k_x^2 \right] D_z(y, z) = -i\omega\mu\epsilon J_z + \epsilon \frac{\partial}{\partial z} \epsilon^{-1} \varrho,$$

$$(6.2.1)$$

where J_z is the z component of the current source and ϱ is the charge density. $\mathbf{J}$ and ϱ are related by the continuity equation. Furthermore, the z component of the magnetic flux B_z satisfies the equation

$$\left[\frac{\partial^2}{\partial y^2} + \mu \frac{\partial}{\partial z} \mu^{-1} \frac{\partial}{\partial z} + k^2(z) - k_x^2 \right] B_z(y, z) = -\mu \left(\nabla_s \times \mathbf{J}_s \right)_z. \qquad (6.2.2)$$

Equations (1) and (2) are for the TM and TE waves in the layered medium respectively.

In the above equations, the $e^{ik_x x}$ dependence has been suppressed from both sides of the equations. Now, if the source is a line source of horizontal dipoles pointing in the y direction,

$$\mathbf{J} = \hat{y} I \ell \, \delta(y) \, \delta(z), \qquad (6.2.3)$$

---

[9] These orthogonal relationships can be obtained by first considering the discrete mode case and, finally, allowing the locations of the impenetrable walls to tend to infinity.

with the implied $e^{ik_x x}$ dependence, then

$$\varrho = \frac{\nabla \cdot \mathbf{J}}{i\omega} = \frac{I\ell}{i\omega}\,\delta'(y)\,\delta(z) \qquad (6.2.4)$$

and

$$\nabla_s \times \mathbf{J}_s = \hat{z}\, i I \ell\, k_x\, \delta(y)\,\delta(z). \qquad (6.2.5)$$

In this case, (1) and (2) become

$$\left[\frac{\partial^2}{\partial y^2} + \epsilon\frac{\partial}{\partial z}\epsilon^{-1}\frac{\partial}{\partial z} + k^2 - k_x^2\right] D_z(y, z) = \frac{I\ell}{i\omega}\,\delta'(y)\,\delta'(z), \qquad (6.2.6)$$

$$\left[\frac{\partial^2}{\partial y^2} + \mu\frac{\partial}{\partial z}\mu^{-1}\frac{\partial}{\partial z} + k^2 - k_x^2\right] B_z(y, z) = -i\mu I\ell\, k_x\, \delta(y)\,\delta(z). \qquad (6.2.7)$$

To solve (6), $D_z(y, z)$ is expanded in terms of the eigenmodes of (6.1.2) that correspond to TM waves. Here, we denote the n-th TM eigenmode as $\phi_\epsilon(n, z)$. We shall assume that only discrete modes exist, which is the case, for example, when the inhomogeneous slab is placed in between a parallel plate waveguide. Consequently,

$$D_z(y, z) = \sum_{n=1}^{\infty} a_n(y)\phi_\epsilon(n, z). \qquad (6.2.8)$$

The continuum modes can be thought of as a limiting case of the discrete modes when the walls of the parallel plate waveguide tend to infinity. Next, on substituting (8) into (6), one obtains

$$\sum_{n=1}^{\infty}\left[\frac{d^2}{dy^2} + k_{ns\epsilon}^2 - k_x^2\right] a_n(y)\phi_\epsilon(n, z) = \frac{I\ell}{i\omega}\,\delta'(y)\,\delta'(z), \qquad (6.2.9)$$

after making use of the fact that

$$\left(\epsilon\frac{d}{dz}\epsilon^{-1}\frac{d}{dz} + k^2\right)\phi_\epsilon(n, z) = k_{ns\epsilon}^2\phi_\epsilon(n, z), \qquad (6.2.10)$$

corresponding to (6.1.2) where $k_{ns\epsilon}^2$ is the n-th eigenvalue for the TM modes.

Assuming that $\phi_\epsilon(n, z)$ are $\epsilon^{-1}(z)$ orthonormal, on multiplying (9) by

$$\epsilon^{-1}\phi_\epsilon(m, z)$$

and integrating, we have

$$\left[\frac{d^2}{dy^2} + k_{ms\epsilon}^2 - k_x^2\right] a_m(y) = \frac{-I\ell}{i\omega\epsilon(0)}\,\phi_\epsilon'(m, 0)\,\delta'(y). \qquad (6.2.11)$$

Furthermore, by matching the singularity at the origin, Equation (11) can be solved to yield

$$a_m(y) = \frac{\mp I\ell}{2i\omega\epsilon(0)}\phi'_\epsilon(m,0)e^{ik_{my\epsilon}|y|}, \qquad (6.2.12)$$

where $k_{my\epsilon}^2 = k_{ms\epsilon}^2 - k_x^2$; the upper sign in $\pm$ is taken when $y > 0$, and the lower sign is taken when $y < 0$. Finally, after substituting the above into (8), we have the eigenfunction expansion solution

$$D_z(y,z) = \frac{\mp I\ell}{2i\omega\epsilon(0)}\sum_{n=1}^\infty e^{ik_{ny\epsilon}|y|}\phi_\epsilon(n,z)\phi'_\epsilon(n,0). \qquad (6.2.13)$$

Similarly, B_z can be expanded in terms of the eigenmodes of (6.1.2) corresponding to TE modes, $\phi_\mu(n,z)$, giving

$$B_z(y,z) = -\frac{I\ell k_x}{2}\sum_{n=1}^\infty \frac{e^{ik_{ny\mu}|y|}}{k_{ny\mu}}\phi_\mu(n,z)\phi_\mu(n,0). \qquad (6.2.14)$$

These fields D_z and B_z represent the TM and TE fields respectively, excited in the layered slab by the source. Similar expressions could be derived for other kinds of source excitation (Exercise 6.8). As a final note, the results above can also be obtained by the Fourier integral technique of Chapter 2 via a contour deformation (Exercise 6.8).

§§6.2.2 The Use of Vector Notation

Equations (13) and (14) are cumbersome to write. To make them more compact, we introduce vector notation (Chew et al. 1984; Chew 1985). By doing so, they can be written more succinctly as

$$D_z = \frac{\mp I\ell}{2i\omega\epsilon(0)}\boldsymbol{\phi}_\epsilon^t(z) \cdot e^{i\overline{K}_\epsilon|y|} \cdot \boldsymbol{\phi}'_\epsilon(0), \qquad (6.2.15)$$

$$B_z = -\frac{I\ell k_x}{2}\boldsymbol{\phi}_\mu^t(z) \cdot e^{i\overline{K}_\mu|y|} \cdot \overline{\mathbf{K}}_\mu^{-1} \cdot \boldsymbol{\phi}_\mu(0), \qquad (6.2.16)$$

where $\phi(z)$ is a column vector containing $\phi(n,z)$ and $\overline{\mathbf{K}}$ is a diagonal matrix containing $k_{ny}\delta_{nn'}$ (Exercise 6.9). The above can be further compressed by defining

$$\mathbf{A}_z = \begin{bmatrix} D_z \\ B_z \end{bmatrix} = \overline{\boldsymbol{\Phi}}^t(z) \cdot e^{i\overline{\mathbf{K}}|y|} \cdot \boldsymbol{\phi}_{s\pm}(0), \qquad (6.2.17)$$

where

$$\overline{\boldsymbol{\Phi}}(z) = \begin{bmatrix} \phi_\epsilon(z) & 0 \\ 0 & \phi_\mu(z) \end{bmatrix}, \qquad (6.2.18a)$$

and

$$\overline{\mathbf{K}} = \begin{bmatrix} \overline{\mathbf{K}}_\epsilon & 0 \\ 0 & \overline{\mathbf{K}}_\mu \end{bmatrix}, \qquad \boldsymbol{\phi}_{s\pm}(0) = \frac{I\ell}{2}\begin{bmatrix} \mp\frac{1}{i\omega\epsilon(0)}\phi'_\epsilon(0) \\ -k_x\overline{\mathbf{K}}_\mu^{-1} \cdot \phi_\mu(0) \end{bmatrix}. \qquad (6.2.18b)$$

In the preceding equations, $\phi_{s\pm}(0)$ contains the excitation coefficients, $e^{i\overline{\mathbf{K}}y}$ is a propagator that propagates the eigenmodes through a distance y, and $\overline{\mathbf{\Phi}}^t(z)$ contains the eigenmodes of the structure.

The transverse to z components of the fields can be found via the equations derivable from Maxwell's equations:

$$\mathbf{E}_{ns} = \frac{1}{k_{ns\epsilon}^2}\frac{\partial}{\partial z}\nabla_s E_{nz} + \frac{i\omega\mu}{k_{ns\mu}^2}\nabla_s \times \mathbf{H}_{nz}, \qquad (6.2.19a)$$

$$\mathbf{H}_{ns} = \frac{1}{k_{ns\mu}^2}\frac{\partial}{\partial z}\nabla_s H_{nz} - \frac{i\omega\epsilon}{k_{ns\epsilon}^2}\nabla_s \times \mathbf{E}_{nz}. \qquad (6.2.19b)$$

The above applies only to the summands in (13) and (14), or only to the eigenmodes.[10] Hence, E_{nz} and H_{nz} correspond to the electric field and the magnetic field respectively derived from the n-th eigenmode of (13) and (14). The above formulas are valid over a piecewise homogeneous layer only, because across a discontinuity $\frac{\partial}{\partial z}E_{nz}$ and $\frac{\partial}{\partial z}H_{nz}$ are singular, since E_{nz} and H_{nz} are discontinuous. We can, however, by a simple trick rewrite (19a) and (19b) so that they are valid for all z in the following manner:

$$\mathbf{E}_{ns} = \frac{1}{k_{ns\epsilon}^2}\epsilon^{-1}\frac{\partial}{\partial z}\nabla_s D_{nz} + \frac{i\omega}{k_{ns\mu}^2}\nabla_s \times \mathbf{B}_{nz}, \qquad (6.2.20a)$$

$$\mathbf{H}_{ns} = \frac{1}{k_{ns\mu}^2}\mu^{-1}\frac{\partial}{\partial z}\nabla_s B_{nz} - \frac{i\omega}{k_{ns\epsilon}^2}\nabla_s \times \mathbf{D}_{nz}. \qquad (6.2.20b)$$

Now, the $\partial/\partial z$ operates on functions D_{nz} and B_{nz} which are continuous across discontinuities in ϵ and μ respectively. Hence, the right-hand sides of (20) are bounded and continuous across an interface, which is also required of $\mathbf{E}_{ns}$ and $\mathbf{H}_{ns}$ on the left-hand side. In other words, (20a) and (20b) are valid for all z. Consequently, they can be used for layers which are piecewise homogeneous, and also in the limit when these layers become infinitely thin, reducing to arbitrary, inhomogeneous layers. The above equations could also be established more rigorously from Maxwell's equations (Exercise 6.10; also see Chew 1985, 1988a, b).

Using the above, we find that

$$E_{nx} = \frac{1}{k_{ns\epsilon}^2}\epsilon^{-1}ik_x\frac{\partial}{\partial z}D_{nz} + \frac{i\omega}{k_{ns\mu}^2}\frac{\partial}{\partial y}B_{nz}, \qquad (6.2.21a)$$

$$H_{nx} = \frac{1}{k_{ns\mu}^2}\mu^{-1}ik_x\frac{\partial}{\partial z}B_{nz} - \frac{i\omega}{k_{ns\epsilon}^2}\frac{\partial}{\partial y}D_{nz}, \qquad (6.2.21b)$$

where the $e^{ik_x x}$ dependence of the field has been used to replace $\frac{\partial}{\partial x}$ with ik_x. To write (21) more compactly, a vector can be defined as follows:

$$\mathbf{A}_x = \begin{bmatrix} H_x \\ E_x \end{bmatrix} = \sum_{n=1}^{\infty}\begin{bmatrix} H_{nx} \\ E_{nx} \end{bmatrix} = \begin{bmatrix} -i\omega\frac{\partial}{\partial y} & \frac{ik_x}{\mu}\frac{\partial}{\partial z} \\ \frac{ik_x}{\epsilon}\frac{\partial}{\partial z} & i\omega\frac{\partial}{\partial y} \end{bmatrix} \cdot \sum_{n=1}^{\infty}\begin{bmatrix} k_{ns\epsilon}^{-2}D_{nz} \\ k_{ns\mu}^{-2}B_{nz} \end{bmatrix}. \qquad (6.2.22)$$

---

[10] These equations are similar to those of (2.3.17) in Chapter 2.

Moreover, if

$$\sum a_n b_n = \mathbf{a}^t \cdot \mathbf{b}, \tag{6.2.23a}$$

then

$$\sum a_n \lambda_n b_n = \mathbf{a}^t \cdot \overline{\boldsymbol{\lambda}} \cdot \mathbf{b}, \tag{6.2.23b}$$

where $\overline{\boldsymbol{\lambda}}$ is a diagonal matrix containing λ_n. Consequently, with the above, we deduce that

$$\sum_{n=1}^{\infty} k_{ns\epsilon}^{-2} D_{nz} = \frac{\pm I\ell}{2i\omega\epsilon(0)} \boldsymbol{\phi}_\epsilon^t(z) \cdot \overline{\mathbf{K}}_{s\epsilon}^{-2} \cdot e^{i\overline{\mathbf{K}}_\epsilon |y|} \cdot \boldsymbol{\phi}_\epsilon'(0) \tag{6.2.24}$$

after making use of (13) and its more compact equivalence (15). A similar statement could be made of the summation over $k_{ns\mu}^{-2} B_{nz}$ in (22). Consequently, (22) becomes

$$\mathbf{A}_x = \begin{bmatrix} -i\omega\frac{\partial}{\partial y} & \frac{ik_x}{\mu}\frac{\partial}{\partial z} \\ \frac{ik_x}{\epsilon}\frac{\partial}{\partial z} & i\omega\frac{\partial}{\partial y} \end{bmatrix} \cdot \begin{bmatrix} \frac{\pm I\ell}{2i\omega\epsilon(0)}\boldsymbol{\phi}_\epsilon^t(z) \cdot \overline{\mathbf{K}}_{s\epsilon}^{-2} \cdot e^{i\overline{\mathbf{K}}_\epsilon|y|} \cdot \boldsymbol{\phi}_\epsilon'(0) \\ -\frac{I\ell k_x}{2}\boldsymbol{\phi}_\mu^t(z) \cdot \overline{\mathbf{K}}_{s\mu}^{-2} \cdot e^{i\overline{\mathbf{K}}_\mu|y|} \cdot \overline{\mathbf{K}}_\mu^{-1} \cdot \boldsymbol{\phi}_\mu(0) \end{bmatrix}, \tag{6.2.25}$$

where $\overline{\mathbf{K}}_s$ is a diagonal matrix containing k_{ns}. The above can be factorized to yield

$$\mathbf{A}_x = \begin{bmatrix} -i\omega\frac{\partial}{\partial y} & \frac{ik_x}{\mu}\frac{\partial}{\partial z} \\ \frac{ik_x}{\epsilon}\frac{\partial}{\partial z} & i\omega\frac{\partial}{\partial y} \end{bmatrix} \cdot \begin{bmatrix} \boldsymbol{\phi}_\epsilon^t(z) \cdot \overline{\mathbf{K}}_{s\epsilon}^{-2} & 0 \\ 0 & \boldsymbol{\phi}_\mu^t(z) \cdot \overline{\mathbf{K}}_{s\mu}^{-2} \end{bmatrix}$$
$$\cdot \begin{bmatrix} e^{i\overline{\mathbf{K}}_\epsilon|y|} & 0 \\ 0 & e^{i\overline{\mathbf{K}}_\mu|y|} \end{bmatrix} \cdot \begin{bmatrix} \frac{\pm I\ell}{2i\omega\epsilon(0)}\boldsymbol{\phi}_\epsilon'(0) \\ -\frac{I\ell k_x}{2}\overline{\mathbf{K}}_\mu^{-1} \cdot \boldsymbol{\phi}_\mu(0) \end{bmatrix}. \tag{6.2.26}$$

Moreover, the derivatives can be carried out to give

$$\mathbf{A}_x = \begin{bmatrix} \pm\omega\boldsymbol{\phi}_\epsilon^t(z) \cdot \overline{\mathbf{K}}_\epsilon \cdot \overline{\mathbf{K}}_{s\epsilon}^{-2} & \frac{ik_x}{\mu}\boldsymbol{\phi}_\mu^{t'}(z) \cdot \overline{\mathbf{K}}_{s\mu}^{-2} \\ \frac{ik_x}{\epsilon}\boldsymbol{\phi}_\epsilon^{t'}(z) \cdot \overline{\mathbf{K}}_{s\epsilon}^{-2} & \mp\omega\boldsymbol{\phi}_\mu^t(z) \cdot \overline{\mathbf{K}}_\mu \cdot \overline{\mathbf{K}}_{s\mu}^{-2} \end{bmatrix}$$
$$\cdot \begin{bmatrix} e^{i\overline{\mathbf{K}}_\epsilon|y|} & 0 \\ 0 & e^{i\overline{\mathbf{K}}_\mu|y|} \end{bmatrix} \cdot \begin{bmatrix} \frac{\pm I\ell}{2i\omega\epsilon(0)}\boldsymbol{\phi}_\epsilon'(0) \\ -\frac{I\ell k_x}{2}\overline{\mathbf{K}}_\mu^{-1} \cdot \boldsymbol{\phi}_\mu(0) \end{bmatrix}. \tag{6.2.27}$$

Note that since the $\overline{\mathbf{K}}$ matrices are diagonal, they commute with each other. Consequently, we can write (27) more compactly as

$$\mathbf{A}_x = \overline{\mathbf{N}}_\pm(z) \cdot e^{i\overline{\mathbf{K}}|y|} \cdot \boldsymbol{\phi}_{s\pm}(0), \tag{6.2.28}$$

where $\overline{\mathbf{N}}_\pm(z)$ is the first matrix in (27).

In summary, given a horizontal electric dipole at the origin, the fields can be written compactly as

$$\mathbf{A}_z = \begin{bmatrix} D_z \\ B_z \end{bmatrix} = \overline{\boldsymbol{\Phi}}^t(z) \cdot e^{i\overline{\mathbf{K}}|y|} \cdot \boldsymbol{\phi}_{s\pm}(0), \tag{6.2.29a}$$

$$\mathbf{A}_x = \begin{bmatrix} H_x \\ E_x \end{bmatrix} = \overline{\mathbf{N}}_\pm(z) \cdot e^{i\overline{\mathbf{K}}|y|} \cdot \boldsymbol{\phi}_{s\pm}(0). \tag{6.2.29b}$$

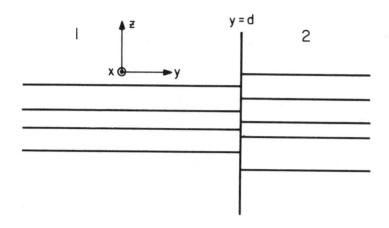

Figure 6.3.1 A junction between two dissimilar layered slabs.

We can derive a similar expression for the y components of the field. Note that in the preceding equations the solutions have been factored into their y and z dependent parts. The $\pm$ sign in $\overline{\mathbf{N}}_\pm$ is for $y > 0$ and $y < 0$ as it assumes different forms for these two regions. When the source is changed in this factorized notation, we only need to change $\phi_s(0)$ (Exercise 6.8). Moreover, Equation (29) resembles the separation of variables, but is done in an infinite-dimensiona linear vector space. It exhibits the physical clarity of propagating waves despite its compact notation.

The above derivation is for a line source modulated with a frequency of $e^{ik_x x}$. To retrieve the point source solution, we only need to multiply (29) by $\frac{1}{2\pi} e^{ik_x x}$ and integrate over all k_x.

§6.3 Reflection and Transmission at a Junction Discontinuity

Having derived the field due to a source in a layered medium, we can study how this field, expressed in terms of the eigenmodes of the structure, is reflected and transmitted at a junction discontinuity between two layered media shown in Figure 6.3.1. At such a junction discontinuity, a single eigenmode incident onto it will be reflected and transmitted into all the other eigenmodes of the two layered structures. This phenomenon is also known as **mode conversion**. It has been studied by many scientists whose works are listed in the references. We shall see how such a phenomenon can be described in terms of reflection and transmission operators following Chew (1985).

§§6.3.1 Derivation of Reflection and Transmission Operators

When two regions of layered inhomogeneities meet at $y = d$ (see Figure 6.3.1), the discontinuity at $y = d$ will scatter any eigenmodes incident on it. But if a source is at the origin, it will excite all the eigenmodes. Furthermore, the presence of the discontinuity at $y = d$ results in the reflection of the eigenmodes in region 1. Therefore, the field in region 1 can be written as

$$\mathbf{A}_{1z} = \overline{\mathbf{\Phi}}_1^t(z) \cdot e^{i\overline{\mathbf{K}}_1|y|} \cdot \boldsymbol{\phi}_{s\pm}(0) + \overline{\mathbf{\Phi}}_1^t(z) \cdot e^{-i\overline{\mathbf{K}}_1 y} \cdot \mathbf{a}. \tag{6.3.1}$$

Physically, the second term in the above equation denotes a collection of eigenmodes propagating in the negative y direction representing the reflected modes from the junction discontinuity. Here, $\mathbf{a}$ is a vector denoting the amplitudes of the reflected eigenmodes: it is an unknown yet to be determined. The second term in (1), however, can be written in a physically more suggestive manner by writing $\mathbf{a}$ as

$$\mathbf{a} = e^{i\overline{\mathbf{K}}_1 d} \cdot \overline{\mathbf{R}}_{12} \cdot e^{i\overline{\mathbf{K}}_1 d} \cdot \boldsymbol{\phi}_{s+}(0).$$

The above suggests that $\mathbf{a}$ is linearly related to $\boldsymbol{\phi}_{s+}(0)$, the excitation coefficient vector of the eigenmodes at $y = 0$, via a matrix operator $\overline{\mathbf{R}}_{12}$ and the phase factors $e^{i\overline{\mathbf{K}}_1 d}$. Consequently, (1) becomes

$$\mathbf{A}_{1z} = \overline{\mathbf{\Phi}}_1^t(z) \cdot \left[e^{i\overline{\mathbf{K}}_1|y|} \cdot \boldsymbol{\phi}_{s\pm}(0) + e^{-i\overline{\mathbf{K}}_1(y-d)} \cdot \overline{\mathbf{R}}_{12} \cdot e^{i\overline{\mathbf{K}}_1 d} \cdot \boldsymbol{\phi}_{s+}(0) \right]. \tag{6.3.1a}$$

Then, applying (6.2.28) for positively and negatively traveling waves in the y direction, we have

$$\mathbf{A}_{1x} = \left[\overline{\mathbf{N}}_{1\pm}(z) \cdot e^{i\overline{\mathbf{K}}_1|y|} \cdot \boldsymbol{\phi}_{s\pm}(0) + \overline{\mathbf{N}}_{1-}(z) \cdot e^{-i\overline{\mathbf{K}}_1(y-d)} \cdot \overline{\mathbf{R}}_{12} \cdot e^{i\overline{\mathbf{K}}_1 d} \cdot \boldsymbol{\phi}_{s+}(0) \right]. \tag{6.3.2}$$

The terms associated with $\overline{\mathbf{R}}_{12}$ correspond to waves traveling in the negative y direction. Physically, $\overline{\mathbf{R}}_{12}$ is a reflection operator that is the "ratio" of the reflected modes to the incident modes (this is obviated by setting the expression (1a) at $y = d$). Furthermore, it accounts for mode conversions at the discontinuity, i.e., an n mode incident can give rise to an m mode reflected, and a TE mode can give rise to a TM mode and vice versa.

In region 2, only transmitted modes exist, and the field can be written in a similar manner as

$$\mathbf{A}_{2z} = \overline{\mathbf{\Phi}}_2^t(z) \cdot e^{i\overline{\mathbf{K}}_2(y-d)} \cdot \overline{\mathbf{T}}_{12} \cdot e^{i\overline{\mathbf{K}}_1 d} \cdot \boldsymbol{\phi}_{s+}(0), \tag{6.3.3}$$

$$\mathbf{A}_{2x} = \overline{\mathbf{N}}_{2+}(z) \cdot e^{i\overline{\mathbf{K}}_2(y-d)} \cdot \overline{\mathbf{T}}_{12} \cdot e^{i\overline{\mathbf{K}}_1 d} \cdot \boldsymbol{\phi}_{s+}(0), \tag{6.3.4}$$

where $\overline{\mathbf{T}}_{12}$ is a transmission operator. Note that (E_z, H_z) and (E_x, H_x) are continuous at $y = d$. Consequently, equating (E_z, H_z) from (1a) and (3), and (E_x, H_x) from (2) and (4) at $y = d$ results in

$$\overline{\mathbf{E}}_1^{-1}(z) \cdot \overline{\mathbf{\Phi}}_1^t(z) \cdot \left[\overline{\mathbf{I}} + \overline{\mathbf{R}}_{12} \right] = \overline{\mathbf{E}}_2^{-1}(z) \cdot \overline{\mathbf{\Phi}}_2^t(z) \cdot \overline{\mathbf{T}}_{12}, \tag{6.3.5}$$

$$\overline{\mathbf{N}}_{1+}(z) + \overline{\mathbf{N}}_{1-}(z) \cdot \overline{\mathbf{R}}_{12} = \overline{\mathbf{N}}_{2+}(z) \cdot \overline{\mathbf{T}}_{12}, \tag{6.3.6}$$

where

$$\overline{\mathbf{E}}_i(z) = \begin{bmatrix} \epsilon_i(z) & 0 \\ 0 & \mu_i(z) \end{bmatrix}.$$

The above are two infinite-dimensional matrix equations with two unknown operators $\overline{\mathbf{R}}_{12}$ and $\overline{\mathbf{T}}_{12}$. They can be solved in many ways, for example, by projecting them onto an orthonormal and complete set. To do this, we multiply (5) by a set of weighting functions and integrate over z. If the set is chosen to be

$$\overline{\mathbf{\Phi}}_1(z), \tag{6.3.7}$$

then from the fact that the eigenmodes $\phi_\epsilon(n, z)$ and $\phi_\mu(n, z)$ are ϵ and μ orthonormal respectively, we have

$$\left\langle \overline{\mathbf{\Phi}}_1(z) \cdot \overline{\mathbf{E}}_1^{-1}(z),\ \overline{\mathbf{\Phi}}_1^t(z) \right\rangle = \overline{\mathbf{I}}. \tag{6.3.8}$$

As a result, (5), after projecting onto $\overline{\mathbf{\Phi}}_1(z)$, becomes

$$(\overline{\mathbf{I}} + \overline{\mathbf{R}}_{12}) = \left\langle \overline{\mathbf{\Phi}}_1(z),\ \overline{\mathbf{E}}_2^{-1}(z) \cdot \overline{\mathbf{\Phi}}_2^t(z) \right\rangle \cdot \overline{\mathbf{T}}_{12}. \tag{6.3.9}$$

Furthermore, after multiplying (6) by $\overline{\mathbf{\Phi}}_1(z) \cdot \overline{\mathbf{E}}_1^{-1}(z)$ and integrating, one obtains

$$\left\langle \overline{\mathbf{\Phi}}_1(z), \overline{\mathbf{E}}_1^{-1}(z) \cdot \overline{\mathbf{N}}_{1+}(z) \right\rangle + \left\langle \overline{\mathbf{\Phi}}_1(z), \overline{\mathbf{E}}_1^{-1}(z) \cdot \overline{\mathbf{N}}_{1-}(z) \right\rangle \cdot \overline{\mathbf{R}}_{12}$$
$$= \left\langle \overline{\mathbf{\Phi}}_1(z), \overline{\mathbf{E}}_1^{-1}(z) \cdot \overline{\mathbf{N}}_{2+}(z) \right\rangle \cdot \overline{\mathbf{T}}_{12}. \tag{6.3.10}$$

The above equations are of the form

$$\overline{\mathbf{I}} + \overline{\mathbf{R}}_{12} = \overline{\mathbf{B}}_{12} \cdot \overline{\mathbf{T}}_{12}, \tag{6.3.11}$$
$$\overline{\mathbf{H}}_{1+} + \overline{\mathbf{H}}_{1-} \cdot \overline{\mathbf{R}}_{12} = \overline{\mathbf{H}}_{2+} \cdot \overline{\mathbf{T}}_{12}, \tag{6.3.12}$$

where $\overline{\mathbf{B}}_{12} = \left\langle \overline{\mathbf{\Phi}}_1(z),\ \overline{\mathbf{E}}_2^{-1}(z) \cdot \overline{\mathbf{\Phi}}_2^t(z) \right\rangle$ and $\overline{\mathbf{H}}_{i\pm} = \left\langle \overline{\mathbf{\Phi}}_1(z),\ \overline{\mathbf{E}}_1^{-1}(z) \cdot \overline{\mathbf{N}}_{i\pm}(z) \right\rangle$. The eigenmodes making up $\overline{\mathbf{\Phi}}_1(z)$ and $\overline{\mathbf{\Phi}}_2(z)$ can be found from Equation (6.1.13) after the eigenvalues are found from (6.1.15). Since the eigenmodes consist of exponential functions, and $\overline{\mathbf{E}}_1^{-1}(z)$ and $\overline{\mathbf{E}}_2^{-1}(z)$ are piecewise-constant functions, these inner products can be obtained in closed forms. (An alternative way of calculating these matrices via a numerical method will be discussed in Section 6.4.) The matrices are, in theory, infinite-dimensional matrices. In practice, however, they must be truncated and solved as matrix equations, giving

$$\overline{\mathbf{R}}_{12} = -\left(\overline{\mathbf{B}}_{12}^{-1} - \overline{\mathbf{H}}_{2+}^{-1} \cdot \overline{\mathbf{H}}_{1-} \right)^{-1} \cdot \left(\overline{\mathbf{B}}_{12}^{-1} - \overline{\mathbf{H}}_{2+}^{-1} \cdot \overline{\mathbf{H}}_{1+} \right), \tag{6.3.13}$$

$$\overline{\mathbf{T}}_{12} = \overline{\mathbf{B}}_{12}^{-1} \cdot \left(\overline{\mathbf{I}} + \overline{\mathbf{R}}_{12} \right). \tag{6.3.14}$$

By truncating the infinite-dimensional matrices, we are implying that only N eigenmodes are used in the calculation, even though an infinite number of eigenmodes are excited by the source and generated by mode conversion at the junction discontinuity. As seen from Figure 6.1.2, many of these eigenmodes have large imaginary k_s, and hence, large imaginary $k_y(= \sqrt{k_s^2 - k_x^2})$. Therefore, these modes are highly evanescent in the y direction and can be partially ignored. In general, $\overline{\mathbf{R}}_{12}$ and $\overline{\mathbf{T}}_{12}$ are dependent on the relative location of the source to the discontinuity (Exercise 6.11; also see Chew 1985, 1988a, b, 1989).

In general, $\overline{\mathbf{R}}_{12}$ and $\overline{\mathbf{T}}_{12}$ are full matrices. They account for the mode conversions within TE and TM modes and also between TE and TM modes. More explicitly,

$$
\overline{\mathbf{R}}_{12} = \left[
\begin{array}{c|c}
\begin{array}{ccc} \text{TM} & \text{to} & \text{TM} \\ \text{mode} & \text{conversion} \end{array} & \begin{array}{ccc} \text{TE} & \text{to} & \text{TM} \\ \text{mode} & \text{conversion} \end{array} \\
\hline
\begin{array}{ccc} \text{TM} & \text{to} & \text{TE} \\ \text{mode} & \text{conversion} \end{array} & \begin{array}{ccc} \text{TE} & \text{to} & \text{TE} \\ \text{mode} & \text{conversion} \end{array}
\end{array}
\right]. \qquad (6.3.15)
$$

The same could be said of the $\overline{\mathbf{T}}_{12}$ operator in its mode conversion properties. Note that if $k_x = 0$, then $\overline{\mathbf{N}}_{\pm}(z)$ in Equation (6) is block diagonal. Then, the corresponding $\overline{\mathbf{H}}_{i\pm}$ in (12) is block diagonal, and subsequently, $\overline{\mathbf{R}}_{12}$ and $\overline{\mathbf{T}}_{12}$ in (13) and (14) are also block diagonal. In this case, there are no inter- TE and TM mode conversions.

Having obtained the line source solution, we can Fourier transform in the x direction to retrieve the point source solution. This type of problem, where the field varies in three dimensions but the inhomogeneity varies only in two dimensions, is known as a two-and-a-half-dimensional problem.

§§6.3.2 The Continuum Limit Case

The previous discussion is valid for when there are discrete modes only. However, when z_{max} and z_{min} tend to infinity, continuum modes as well as discrete modes constitute the set of eigenmodes. In such a case, the dot products in the previous section, which imply infinite summations, also involve integral summations. For clarity, we will use double-dot products to indicate products where a continuum may be involved. In this case, Equations (5) and (6) become

$$
\overline{\mathbf{E}}_1^{-1}(z) \cdot \overline{\boldsymbol{\Phi}}_1^t(z) : (\overline{\mathbf{I}} + \overline{\mathbf{R}}_{12}) = \overline{\mathbf{E}}_2^{-1}(z) \cdot \overline{\boldsymbol{\Phi}}_2^t(z) : \overline{\mathbf{T}}_{12}, \qquad (6.3.16)
$$

$$
\overline{\mathbf{N}}_{1+}(z) + \overline{\mathbf{N}}_{1-}(z) : \overline{\mathbf{R}}_{12} = \overline{\mathbf{N}}_{2+}(z) : \overline{\mathbf{T}}_{12}. \qquad (6.3.17)
$$

Now, if the same operations were applied to the above equations as in the previous section, the resulting equivalence of Equations (11) and (12) still involves integrals, and may not be easily solvable.

One way to overcome this problem is to find an orthonormal basis set which is countable and yet complete over the infinite domain $-\infty < z < \infty$, e.g., Laguerre polynomials or Hermite polynomials. The eigenmodes can then be expanded in terms of this basis set to give[11]

$$\phi_p(k_z, z) = \sum_{n=1}^{\infty} b_{np}(k_z) S_n(z), \qquad (6.3.18)$$

where $S_n(z)$ is an orthonormal complete set over $-\infty < z < \infty$ and k_z implies either the discrete modes or the continuum modes. Therefore, $\phi_p(k_z, z)$ can be thought of as a vector of continuum dimension indexed by k_z, i.e.,

$$\phi_p(z) = \overline{\mathbf{b}}_p \cdot \mathbf{S}(z), \qquad (6.3.19)$$

where $\phi_p(z)$ is a column vector of continuum dimension indexed by k_z, $\overline{\mathbf{b}}_p$ is a matrix of continuum dimension by discrete dimension, and $\mathbf{S}(z)$ is a column vector of discrete dimension. Hence, we write

$$\overline{\boldsymbol{\Phi}}_1^t(z) = \begin{bmatrix} \phi_{1\epsilon}^t(z) & 0 \\ 0 & \phi_{1\mu}^t(z) \end{bmatrix} = \begin{bmatrix} \mathbf{S}^t(z) & 0 \\ 0 & \mathbf{S}^t(z) \end{bmatrix} \cdot \begin{bmatrix} \overline{\mathbf{b}}_{1\epsilon}^t & 0 \\ 0 & \overline{\mathbf{b}}_{1\mu}^t \end{bmatrix} = \overline{\mathbf{S}}^t(z) \cdot \overline{\mathbf{B}}_1^t,$$

$$\qquad (6.3.20)$$

where $\overline{\mathbf{S}}(z)$ and $\overline{\mathbf{B}}_1$ are defined accordingly as the matrices above. On using this in (16), one obtains

$$\overline{\mathbf{E}}_1^{-1}(z) \cdot \overline{\mathbf{S}}^t(z) \cdot \overline{\mathbf{B}}_1^t : (\overline{\mathbf{I}} + \overline{\mathbf{R}}_{12}) = \overline{\mathbf{E}}_2^{-1}(z) \cdot \overline{\mathbf{S}}^t(z) \cdot \overline{\mathbf{B}}_2^t : \overline{\mathbf{T}}_{12}. \qquad (6.3.21a)$$

Then, after multiplying the above by $\overline{\mathbf{S}}(z)$ and integrating, one obtains

$$\overline{\mathbf{P}}_1^I \cdot \overline{\mathbf{B}}_1^t : (\overline{\mathbf{I}} + \overline{\mathbf{R}}_{12}) = \overline{\mathbf{P}}_2^I \cdot \overline{\mathbf{B}}_2^t : \overline{\mathbf{T}}_{12}, \qquad (6.3.21b)$$

where $\overline{\mathbf{P}}_i^I = \left\langle \overline{\mathbf{S}}(z), \overline{\mathbf{E}}_i^{-1}(z) \cdot \overline{\mathbf{S}}^t(z) \right\rangle$ is the matrix representation of $\overline{\mathbf{E}}_i^{-1}(z)$ in a space spanned by $S_n(z)$.

From (6.2.27), after using (19), then we have

$$\overline{\mathbf{N}}_{\pm}(z) = \begin{bmatrix} \pm\omega\mathbf{S}^t(z) \cdot \overline{\mathbf{b}}_\epsilon^t : \overline{\mathbf{K}}_\epsilon : \overline{\mathbf{K}}_{s\epsilon}^{-2} & ik_x\mu^{-1}\frac{\partial}{\partial z}\mathbf{S}^t(z) \cdot \overline{\mathbf{b}}_\mu^t : \overline{\mathbf{K}}_{s\mu}^{-2} \\ ik_x\epsilon^{-1}\frac{\partial}{\partial z}\mathbf{S}^t(z) \cdot \overline{\mathbf{b}}_\epsilon^t : \overline{\mathbf{K}}_{s\epsilon}^{-2} & \mp\omega\mathbf{S}^t(z) \cdot \overline{\mathbf{b}}_\mu^t : \overline{\mathbf{K}}_\mu : \overline{\mathbf{K}}_{s\mu}^{-2} \end{bmatrix}. \qquad (6.3.22)$$

Moreover, using the properties of orthonormal basis, we can deduce that (Exercise 6.12)

$$p^{-1}(z)\frac{\partial}{\partial z}\mathbf{S}^t(z) = \mathbf{S}^t(z) \cdot \left\langle \mathbf{S}(z), p^{-1}\frac{\partial}{\partial z}\mathbf{S}^t(z) \right\rangle = \mathbf{S}^t(z) \cdot \overline{\mathbf{D}}_p, \qquad (6.3.23)$$

---

[11] It is debatable whether a nonseparable Hilbert with undenumerable (continuum) eigenmodes can be expanded in terms of a denumerable (discrete) basis function. Nevertheless, this method has been adopted by a number of scientists to solve this problem (e.g., see Rozzi 1978; Rozzi and Veld 1980).

where $\overline{\mathbf{D}}_p = \langle \mathbf{S}(z), p^{-1}\frac{\partial}{\partial z}\mathbf{S}^t(z) \rangle$, the matrix representation of the operator $p^{-1}\frac{\partial}{\partial z}$ in a space spanned by $S_n(z)$. Subsequently, (22) becomes

$$\overline{\mathbf{N}}_{\pm}(z) = \begin{bmatrix} \mathbf{S}^t(z) & 0 \\ 0 & \mathbf{S}^t(z) \end{bmatrix} \cdot \begin{bmatrix} \pm\omega\overline{\mathbf{b}}_{\epsilon}^t : \overline{\mathbf{K}}_{\epsilon} : \overline{\mathbf{K}}_{s\epsilon}^{-2} & ik_x\overline{\mathbf{D}}_{\mu} \cdot \overline{\mathbf{b}}_{\mu}^t : \overline{\mathbf{K}}_{s\mu}^{-2} \\ ik_x\overline{\mathbf{D}}_{\epsilon} \cdot \overline{\mathbf{b}}_{\epsilon}^t : \overline{\mathbf{K}}_{s\epsilon}^{-2} & \mp\omega\overline{\mathbf{b}}_{\mu}^t : \overline{\mathbf{K}}_{\mu} : \overline{\mathbf{K}}_{s\mu}^{-2} \end{bmatrix}$$

$$= \overline{\mathbf{S}}^t(z) \cdot \overline{\mathbf{M}}_{\pm}. \tag{6.3.24}$$

Using the above in (17), one derives

$$\overline{\mathbf{M}}_{1+} + \overline{\mathbf{M}}_{1-} : \overline{\mathbf{R}}_{12} = \overline{\mathbf{M}}_{2+} : \overline{\mathbf{T}}_{12}. \tag{6.3.25}$$

Because of the meaning of the double-dot products, the above is an integral equation. Therefore, it needs to be manipulated into a matrix equation so that its solution can be computed.

Using the orthogonality property that [see (6.1.18) to (6.1.20)]

$$\langle \phi_p(z), \; p^{-1}\phi_p^t(z) \rangle = \delta(k_z - k_z'), \tag{6.3.26}$$

we deduce that

$$\overline{\mathbf{b}}_p \cdot \langle \mathbf{S}(z), \; p^{-1}\mathbf{S}^t(z) \rangle \cdot \overline{\mathbf{b}}_p^t = \delta(k_z - k_z'). \tag{6.3.27}$$

Calling $\overline{\mathbf{p}}^I = \langle \mathbf{S}(z), \; p^{-1}\mathbf{S}^t(z) \rangle$ the matrix representation of p^{-1}, (27) can be rewritten as

$$\overline{\mathbf{b}}_p \cdot \overline{\mathbf{p}}^I \cdot \overline{\mathbf{b}}_p^t = \delta(k_z - k_z'). \tag{6.3.28}$$

The above delta function becomes a Kronecker delta function when discrete modes are involved, which is implied in (26) to (28). Hence, $\overline{\mathbf{b}}_p \cdot \overline{\mathbf{p}}^I \cdot \overline{\mathbf{b}}_p^t$ is an identity operator, or using the definition of $\overline{\mathbf{B}}$ given by (20),

$$\overline{\mathbf{B}} \cdot \overline{\mathbf{P}}^I \cdot \overline{\mathbf{B}}^t = \delta(k_z - k_z'). \tag{6.3.29}$$

Finally, on inserting this piece into the double-dot product in (25), one obtains

$$\overline{\mathbf{M}}_{1+} + \overline{\mathbf{M}}_{1-} : \overline{\mathbf{B}}_1 \cdot \overline{\mathbf{P}}_1^I \cdot \overline{\mathbf{B}}_1^t : \overline{\mathbf{R}}_{12} = \overline{\mathbf{M}}_{2+} : \overline{\mathbf{B}}_2 \cdot \overline{\mathbf{P}}_2^I \cdot \overline{\mathbf{B}}_2^t : \overline{\mathbf{T}}_{12}. \tag{6.3.30}$$

Then, after multiplying the above by $\overline{\mathbf{B}}_1$ from the right, one derives

$$\overline{\mathbf{M}}_{1+} : \overline{\mathbf{B}}_1 + \overline{\mathbf{M}}_{1-} : \overline{\mathbf{B}}_1 \cdot \overline{\mathbf{P}}_1^I \cdot \left(\overline{\mathbf{B}}_1^t : \overline{\mathbf{R}}_{12} : \overline{\mathbf{B}}_1 \right)$$

$$= \overline{\mathbf{M}}_{2+} : \overline{\mathbf{B}}_2 \cdot \overline{\mathbf{P}}_2^I \cdot \overline{\mathbf{B}}_2^t : \overline{\mathbf{T}}_{12} : \overline{\mathbf{B}}_1. \tag{6.3.31}$$

Consequently, the above becomes

$$\overline{\mathbf{m}}_{1+} + \overline{\mathbf{m}}_{1-} \cdot \overline{\mathbf{P}}_1^I \cdot \overline{\mathbf{r}}_{12} = \overline{\mathbf{m}}_{2+} \cdot \overline{\mathbf{P}}_2^I \cdot \overline{\mathbf{t}}_{12}, \tag{6.3.32}$$

where

$$\overline{m}_{i\pm} = \overline{M}_{i\pm} : \overline{B}_i, \qquad \overline{r}_{12} = \overline{B}_1^t : \overline{R}_{12} : \overline{B}_1, \qquad \overline{t}_{12} = \overline{B}_2^t : \overline{T}_{12} : \overline{B}_1. \qquad (6.3.33)$$

Similarly, after multiplying (21b) from the right by $\overline{B}_1$, one obtains

$$\overline{P}_1^I \cdot \left(\overline{B}_1^t : \overline{B}_1 + \overline{r}_{12} \right) = \cdot \overline{P}_2^I \cdot \overline{t}_{12}. \qquad (6.3.34)$$

Equations (32) and (34) are now matrices with discrete, infinite dimensions. But they can be truncated and solved as matrix equations, yielding $\overline{r}_{12}$ and $\overline{t}_{12}$. Then, the relationship between $\overline{r}_{12}$ and $\overline{t}_{12}$, and $\overline{R}_{12}$ and $\overline{T}_{12}$ can be found from (33). For example,

$$\overline{r}_{12} = \overline{B}_1^t : \overline{R}_{12} : \overline{B}_1. \qquad (6.3.35)$$

Moreover, using (29), it can be shown that (Exercise 6.13)

$$\overline{R}_{12} = \overline{B}_1 \cdot \overline{P}_1^I \cdot \overline{r}_{12} \cdot \overline{P}_1^I \cdot \overline{B}_1^t, \qquad (6.3.36)$$

and similarly,

$$\overline{T}_{12} = \overline{B}_2 \cdot \overline{P}_2^I \cdot \overline{t}_{12} \cdot \overline{P}_2^I \cdot \overline{B}_2^t. \qquad (6.3.37)$$

Therefore, the continuum reflection and transmission operators are derivable from the discrete reflection and transmission operators.

§6.4 A Numerical Method to Find the Eigenmodes

Finding the eigenmodes of an inhomogeneous slab is often a rather laborious procedure, especially if the inhomogeneity is arbitrary and general. To do so, one has to solve the implicit equations (6.1.15) or (6.1.16) for the eigenvalues. Then, the eigenmodes are found using (6.1.13). Since the eigenmodes satisfy Equation (6.1.3) where p is either μ or ϵ for TE and TM waves respectively, a systematic way of finding the eigenvalues k_s^2 and the eigenfunction $\phi(z)$ is via a numerical method. In this method, Equation (6.1.3) is converted into a matrix eigenvalue problem whose eigenvalues and eigenvectors are found efficiently using matrix eigenvalue solvers (Pudensi and Ferreira 1982; Chew et al. 1984; Chew 1985).

To solve (6.1.3) numerically, we let

$$\phi(z) = \sum_{n=1}^{N} b_n S_n(z), \qquad (6.4.1)$$

where $S_n(z)$ is a basis set that can approximate $\phi(z)$ fairly well over its support $-z_{min} < z < z_{max}$. For example, $S_n(z)$ may be the Fourier series basis, some orthogonal polynomials, or subdomain functions like the triangular functions. If the support of $\phi(z)$ is infinite, we may choose $S_n(z)$ to be Laguerre polynomials or Hermite polynomials. Consequently, using (1) in (6.1.3), we have

$$\sum_{n=1}^{N} b_n \left[\frac{d}{dz} p^{-1} \frac{d}{dz} + k^2(z) p^{-1} \right] S_n(z) = k_s^2 \sum_{n=1}^{N} b_n p^{-1} S_n(z). \tag{6.4.2}$$

The above could then be multiplied by $S_m(z)$ and integrated over the support of $\phi(z)$, or

$$\sum_{n=1}^{N} b_n \left[\left\langle S_m(z), \frac{d}{dz} p^{-1} \frac{d}{dz} S_n(z) \right\rangle + \left\langle S_m(z), k^2(z) p^{-1} S_n(z) \right\rangle \right]$$

$$= k_s^2 \sum_{n=1}^{N} b_n \langle S_m(z), p^{-1} S_n(z) \rangle, \quad m = 1, \ldots, N, \tag{6.4.3}$$

where $\langle a, b \rangle = \int\limits_{-z_{min}}^{z_{max}} a(z)b(z)\, dz$. Next, using integration by parts, we have

$$\left\langle S_m(z), \frac{d}{dz} p^{-1} \frac{d}{dz} S_n(z) \right\rangle = \int\limits_{-z_{min}}^{z_{max}} S_m(z) \frac{d}{dz} p^{-1} \frac{d}{dz} S_n(z)\, dz$$

$$= S_m(z) p^{-1} \frac{d}{dz} S_n(z) \Big|_{-z_{min}}^{z_{max}} - \int\limits_{-z_{min}}^{z_{max}} \left(\frac{d}{dz} S_m(z) \right) p^{-1} \left(\frac{d}{dz} S_n(z) \right) dz. \tag{6.4.4}$$

However, if the basis function $S_n(z)$ is chosen such that either $S_m(z)$ or $\frac{d}{dz} S_n(z)$ is zero at $-z_{min}$ and z_{max},[12] then the first term on the right-hand side of (4) vanishes, and we have

$$\left\langle S_m(z), \frac{d}{dz} p^{-1} \frac{d}{dz} S_n(z) \right\rangle = - \left\langle S_m'(z), p^{-1} S_n'(z) \right\rangle. \tag{6.4.5}$$

The choice of $S_m(z) = 0$ at the end points corresponds to a homogeneous Dirichlet boundary condition. In this case, the walls at $-z_{min}$ and z_{max} correspond to electric walls and magnetic walls for TE and TM waves respectively. On the other hand, the choice of $S_m'(z) = 0$ at $-z_{min}$ and z_{max} implies a homogeneous Neumann boundary condition. This corresponds to a magnetic wall and an electric wall for TE and TM waves respectively.[13]

As a result, (3) becomes

$$\sum_{n=1}^{N} b_n \left[- \left\langle S_m'(z), p^{-1} S_n'(z) \right\rangle + \left\langle S_m(z), k^2(z) p^{-1} S_n(z) \right\rangle \right]$$

$$= k_s^2 \sum_{n=1}^{N} b_n \left\langle S_m(z), p^{-1} s_n(z) \right\rangle, \quad m = 1, \ldots, N. \tag{6.4.6}$$

---

[12] These are the essential boundary conditions described in Chapter 5.

[13] If $S_m(z)$'s values at $-z_{min}$ and z_{max} are allowed to float, then the solution satisfies the homogeneous Neumann boundary condition naturally as described in Chapter 5.

Equation (6) is now a matrix equation of the form

$$\sum_{n=1}^{N} L_{mn} b_n = k_s^2 \sum_{n=1}^{N} p_{mn}^{-1} b_n, \quad m = 1, \ldots, N, \qquad (6.4.7)$$

or

$$\overline{\mathbf{L}} \cdot \mathbf{b} = k_s^2 \overline{\mathbf{p}}^I \cdot \mathbf{b}, \qquad (6.4.7a)$$

where

$$L_{mn} = - \left\langle S_m'(z), p^{-1} S_n'(z) \right\rangle + \left\langle S_m(z), k^2 p^{-1} S_n(z) \right\rangle, \qquad (6.4.7b)$$

$$p_{mn}^I = \left\langle S_m(z), p^{-1} S_n(z) \right\rangle. \qquad (6.4.7c)$$

Therefore, (6.1.3) has finally been converted to a matrix Equation (7). It corresponds to a conventional eigenvalue problem where the eigenvalue is k_s^2 and the eigenvector is $\mathbf{b}$. Moreover, it is easy to show that L_{mn} and p_{mn}^I are symmetric matrices due to reciprocity. In fact, L_{mn} is also the matrix representation of the operator $\frac{d}{dz} p^{-1} \frac{d}{dz} + k^2 p^{-1}$ in the space spanned by $S_n(z)$, and likewise, p_{mn}^I is a matrix representation of p^{-1} (Exercise 6.14).

Now, we can solve (7a) for the eigenvector $\mathbf{b}$ and the eigenvalue k_s^2. And if N basis functions are used in (1), there would be N eigenvalues and N eigenvectors. Hence, the general solution to (6.1.3) corresponding to the i-th eigenfunction is

$$\tilde{\phi}(i, z) = \sum_{n=1}^{N} b_{in} S_n(z), \quad i = 1, \ldots, N, \qquad (6.4.8)$$

where b_{in} is the i-th eigenvector of (7a). The tilde is used here to denote a numerical approximation of the actual eigenfunction. Furthermore, since (7a) is a conventional eigenvalue problem, b_{in} can be found efficiently and systematically. In addition, since $\overline{\mathbf{L}}$ and $\overline{\mathbf{p}}^I$ are symmetric, it can be readily shown that the eigenvectors are $\overline{\mathbf{p}}^I$ orthogonal, i.e.,

$$\mathbf{b}_i^t \cdot \overline{\mathbf{p}}^I \cdot \mathbf{b}_j = 0, \quad i \neq j. \qquad (6.4.9)$$

From (9), it also follows that (Exercise 6.15)

$$\left\langle \tilde{\phi}(i, z), p^{-1}(z) \tilde{\phi}(j, z) \right\rangle = 0, \quad i \neq j. \qquad (6.4.10)$$

Hence, these approximate eigenfunctions are still p^{-1} orthogonal.

The numerically approximate eigenfunction $\tilde{\phi}(i, z)$ given in (8) can be used in place of that in (6.2.8), yielding

$$D_z(y, z) = \sum_{n=1}^{N} a_n(y) \tilde{\phi}_\epsilon(n, z). \qquad (6.4.11)$$

Moreover, the eigenvectors and eigenfunctions can be orthonormalized for convenience. Then, after substituting the (11) into (6.2.6), one obtains

$$\sum_{n=1}^{N} \left[\frac{\partial^2}{\partial y^2} + \epsilon \frac{\partial}{\partial z} \epsilon^{-1} \frac{\partial}{\partial z} + k^2 - k_x^2 \right] a_n(y) \tilde{\phi}_\epsilon(n, z) = \frac{I\ell}{i\omega} \delta'(y) \, \delta'(z).$$

(6.4.12)

Next, on multiplying the above by $\tilde{\phi}_\epsilon(m, z)\epsilon^{-1}$ and integrating over z, one obtains

$$\frac{\partial^2}{\partial y^2} a_m(y) + \sum_{n=1}^{N} \left\langle \tilde{\phi}_\epsilon(m, z), \left(\frac{\partial}{\partial z} \epsilon^{-1} \frac{\partial}{\partial z} + \frac{k^2}{\epsilon} \right) \tilde{\phi}_\epsilon(n, z) \right\rangle a_n(y)$$

$$- k_x^2 a_m(y) = \frac{-I\ell}{i\omega\epsilon(0)} \tilde{\phi}'_\epsilon(m, 0) \, \delta'(y). \quad (6.4.13)$$

On using (8), it follows that (Exercise 6.16)

$$\left\langle \tilde{\phi}_\epsilon(m, z), \left(\frac{\partial}{\partial z} \epsilon^{-1} \frac{\partial}{\partial z} + \frac{k^2}{\epsilon} \right) \tilde{\phi}_\epsilon(n, z) \right\rangle = \sum_{j=1}^{N} \sum_{i=1}^{N} b_{mi} b_{nj}$$

$$\left\langle S_i(z), \left(\frac{\partial}{\partial z} \epsilon^{-1} \frac{\partial}{\partial z} + \frac{k^2}{\epsilon} \right) S_j(z) \right\rangle$$

$$= \sum_{j=1}^{N} \sum_{i=1}^{N} b_{mi} L_{ij} b_{nj}$$

$$= \mathbf{b}_m^t \cdot \overline{\mathbf{L}} \cdot \mathbf{b}_n. \quad (6.4.14)$$

But since $\overline{\mathbf{L}} \cdot \mathbf{b}_n = k_{ns}^2 \overline{\mathbf{p}}^I \cdot \mathbf{b}_n$, the above equals

$$k_{ns}^2 \mathbf{b}_m^t \cdot \overline{\mathbf{p}}^I \cdot \mathbf{b}_n = k_{ns}^2 \, \delta_{mn}. \quad (6.4.15)$$

Therefore, (13) becomes

$$\left[\frac{d^2}{dy^2} + k_{ns}^2 - k_x^2 \right] a_m(y) = \frac{-I\ell}{i\omega\epsilon(0)} \tilde{\phi}_\epsilon(m, 0) \, \delta'(y). \quad (6.4.16)$$

The above is similar to (6.2.11) but for quite a different reason.

From this point onward, the development of the theory is the same as that after (6.2.11) before. Moreover, Equation (8) can be written as

$$\tilde{\phi}(z) = \overline{\mathbf{b}} \cdot \mathbf{S}(z), \quad (6.4.17)$$

where $\overline{\mathbf{b}}$ is the matrix containing b_{in}. As a matter of fact, Equation (7a) can be solved for both TE and TM waves. Therefore, in general, the eigenfunction $\overline{\boldsymbol{\Phi}}(z)$ defined in (6.2.18a) can be written as

$$\overline{\boldsymbol{\Phi}}_i(z') = \begin{bmatrix} \tilde{\phi}_{i\epsilon}(z) & 0 \\ 0 & \tilde{\phi}_{i\mu}(z) \end{bmatrix} = \begin{bmatrix} \overline{\mathbf{b}}_{i\epsilon} & 0 \\ 0 & \overline{\mathbf{b}}_{i\mu} \end{bmatrix} \cdot \begin{bmatrix} \mathbf{S}(z) & 0 \\ 0 & \mathbf{S}(z) \end{bmatrix} = \overline{\mathbf{B}}_i \cdot \overline{\mathbf{S}}(z),$$

(6.4.18)

where the subscript i represents the region, and ϵ and μ represent TM and TE wave eigenvectors obtained from (7a) by letting p be ϵ and μ respectively. Now, Equation (18) is similar to Equation (6.3.20). Note that the basis $\overline{\mathbf{S}}(z)$ is assumed to be the same for all regions. With the above definition, $\overline{\mathbf{B}}_{12}$ in (6.3.11) can then be written as

$$\overline{\mathbf{B}}_{12} = \left\langle \overline{\boldsymbol{\Phi}}_1(z), \overline{\mathbf{E}}_2^{-1}(z) \cdot \overline{\boldsymbol{\Phi}}_2^t(z) \right\rangle = \overline{\mathbf{B}}_1 \cdot \left\langle \overline{\mathbf{S}}(z), \overline{\mathbf{E}}_2^{-1}(z) \cdot \overline{\mathbf{S}}^t(z) \right\rangle \cdot \overline{\mathbf{B}}_2^t$$

$$= \overline{\mathbf{B}}_1 \cdot \overline{\mathbf{P}}_2^I \cdot \overline{\mathbf{B}}_2^t, \qquad (6.4.19)$$

where $\overline{\mathbf{P}}_2^I = \left\langle \overline{\mathbf{S}}(z), \overline{\mathbf{E}}_2^{-1} \cdot \overline{\mathbf{S}}^t(z) \right\rangle$. After using (18) in the definition of $\overline{\mathbf{N}}_\pm(z)$ in (6.2.28), $\mathbf{N}_\pm(z)$ becomes

$$\overline{\mathbf{N}}_\pm(z) = \begin{bmatrix} \pm\omega\mathbf{S}^t(z) \cdot \overline{\mathbf{b}}_\epsilon^t \cdot \overline{\mathbf{K}}_\epsilon \cdot \overline{\mathbf{K}}_{s\epsilon}^{-2} & ik_x\mu^{-1}\frac{\partial}{\partial z}\mathbf{S}^t(z) \cdot \overline{\mathbf{b}}_\mu^t \cdot \overline{\mathbf{K}}_{s\mu}^{-2} \\ ik_x\epsilon^{-1}\frac{\partial}{\partial z}\mathbf{S}^t(z) \cdot \overline{\mathbf{b}}_\epsilon^t \cdot \overline{\mathbf{K}}_{s\epsilon}^{-2} & \mp\omega\mathbf{S}^t(z) \cdot \overline{\mathbf{b}}_\mu^t \cdot \overline{\mathbf{K}}_\mu \cdot \overline{\mathbf{K}}_{s\mu}^{-2} \end{bmatrix}, \qquad (6.4.20)$$

similar to (6.3.22), except that the double-dot products have been replaced with dot products here because of the absence of continuum modes. Furthermore, the basis $S_n(z)$ can always be orthonormalized if it is a linearly independent basis. Consequently, we can approximate

$$p^{-1}(z)\frac{\partial}{\partial z}\mathbf{S}^t(z) \simeq \mathbf{S}^t(z) \cdot \left\langle \mathbf{S}(z), p^{-1}(z)\frac{\partial}{\partial z}\mathbf{S}^t(z) \right\rangle = \mathbf{S}^t(z) \cdot \overline{\mathbf{D}}_p, \qquad (6.4.21)$$

where $\overline{\mathbf{D}}_p$ is the approximate matrix representation of the operator $p^{-1}(z)\frac{\partial}{\partial z}$. Then, (20) becomes

$$\overline{\mathbf{N}}_\pm = \begin{bmatrix} \mathbf{S}^t(z) & 0 \\ 0 & \mathbf{S}^t(z) \end{bmatrix} \cdot \begin{bmatrix} \pm\omega\overline{\mathbf{b}}_\epsilon^t \cdot \overline{\mathbf{K}}_\epsilon \cdot \overline{\mathbf{K}}_{s\epsilon}^{-2} & ik_x\overline{\mathbf{D}}_\mu \cdot \overline{\mathbf{b}}_\mu^t \cdot \overline{\mathbf{K}}_{s\mu}^{-2} \\ ik_x\overline{\mathbf{D}}_\epsilon \cdot \overline{\mathbf{b}}_\epsilon^t \cdot \overline{\mathbf{K}}_{s\epsilon}^{-2} & \mp\omega\overline{\mathbf{b}}_\mu^t \cdot \overline{\mathbf{K}}_\mu \cdot \overline{\mathbf{K}}_{s\mu}^{-2} \end{bmatrix}$$

$$= \overline{\mathbf{S}}^t(z) \cdot \overline{\mathbf{M}}_\pm. \qquad (6.4.22)$$

As a result, $\overline{\mathbf{H}}_{i\pm}$ defined in (6.3.12) becomes

$$\overline{\mathbf{H}}_{i\pm} = \left\langle \overline{\boldsymbol{\Phi}}_1(z), \overline{\mathbf{E}}_1^{-1}(z) \cdot \overline{\mathbf{N}}_{i\pm}^t(z) \right\rangle = \overline{\mathbf{B}}_1 \cdot \overline{\mathbf{P}}_1^I \cdot \overline{\mathbf{M}}_{i\pm}, \qquad (6.4.23)$$

where $\overline{\mathbf{P}}_1^I = \left\langle \overline{\mathbf{S}}(z), \overline{\mathbf{E}}_1^{-1} \cdot \overline{\mathbf{S}}^t(z) \right\rangle$.

In this manner, $\overline{\mathbf{B}}_{12}$ and $\overline{\mathbf{H}}_{i\pm}$ in (6.3.13) and (6.3.14) can be computed. Moreover, if $\overline{\mathbf{E}}_1^{-1}(z)$ is a piecewise-constant function, and if the basis functions are subdomain functions, e.g., triangular functions, then all the integrals needed to compute $\overline{\mathbf{P}}^I$ and $\overline{\mathbf{D}}_p$ can be found in closed forms. Consequently, the reflection operator $\overline{\mathbf{R}}_{12}$ and the transmission operator $\overline{\mathbf{T}}_{12}$ can be found in a systematic manner for quite a versatile range of $\epsilon_i(z)$ and $\mu_i(z)$. This manner of performing the mode matching is known as the ***numerical mode matching*** method. The numerical mode matching method exploits the versatility of a numerical method like the finite-element method in solving (6.1.3).

In addition, it makes use of well-developed software in solving the matrix eigenvalue problem in (7a) to find the eigenfunctions of the structure. More importantly, when eigenmodes of different regions at a junction discontinuity are expressed in terms of the same basis set as in (18), the matrix elements of $\overline{\mathbf{B}}_{12}$ and $\overline{\mathbf{H}}_{i\pm}$ can be found in closed forms. Furthermore, the approximation (21) allows $\overline{\mathbf{H}}_{i\pm}$ to be expressed in a simple form as in (23) (Chew 1985). As a result, this method of solving a two-dimensional, inhomogeneous problem is a lot more efficient than the direct use of the finite-element method.

§6.5 The Cylindrically Layered Medium Case

The scattering of waves by a junction of two cylindrically layered half-spaces finds applications in geophysical prospecting and in the study of optical fibers and dielectric resonators. This problem can be solved by the mode matching method, as will be shown in the following. Unlike a planarly layered medium, a cylindrically layered medium does not support a pure TE or TM wave except for the axisymmetric waves (see Chapter 3). Therefore, the problem is inherently vector in nature even before a junction discontinuity is introduced. However, when the source is axially symmetric so that the fields generated are also axially symmetric, then the scattering problem can be decomposed into two decoupled scalar problems of the TE and TM types.

The scattering of waves by junction discontinuities has been reported in the literature (Chew et al. 1983, 1984; Tsang et al. 1984; Pai and Huang 1988; Moghaddam and Chew 1988). Such junction discontinuities have also been studied in the context of dielectric resonators (Hong and Jansen 1982, 1984; Zaki and Atia 1983; Zaki and Chen 1986).

§§6.5.1 Eigenmodes of a Cylindrically Layered Medium

The eigenmodes of a cylindrically layered medium can be found in a manner similar to that of the eigenmodes of a planarly layered structure. We shall consider first the case of the cylindrically layered medium enclosed in an impenetrable, outermost cylindrical wall. In this case, the eigenmodes of the structure are discrete in nature. The field in region i, in the manner of Equation (3.2.20) of Chapter 3, can then be written as

$$\begin{bmatrix} E_{iz} \\ H_{iz} \end{bmatrix} = \left[H_n^{(1)}(k_{i\rho}\rho)\widetilde{\overline{\mathbf{R}}}_{i,i-1} + J_n(k_{i\rho}\rho)\,\overline{\mathbf{I}} \right] \cdot \mathbf{a}_i, \qquad (6.5.1)$$

where E_{iz} represents the TM to z waves while H_{iz} represents the TE to z waves. Referring to Figure 3.2.2 of Chapter 3, $\widetilde{\overline{\mathbf{R}}}_{10} = 0$ in the above. Now, if an electric wall is placed at $\rho = \rho_{max}$, then the field in region N can be written alternatively as

$$\begin{bmatrix} E_{Nz} \\ H_{Nz} \end{bmatrix} = \left[H_n^{(1)}(k_{N\rho}\rho)\,\overline{\mathbf{I}} + J_n(k_{N\rho}\rho)\,\overline{\mathbf{R}}_{N,N+1} \right] \cdot \mathbf{b}_N, \qquad (6.5.2)$$

where (Exercise 6.17)

$$\overline{\mathbf{R}}_{N,N+1} = \begin{bmatrix} -H_n^{(1)}(k_{N\rho}\rho_{max})/J_n(k_{N\rho}\rho_{max}) & 0 \\ 0 & -H_n^{(1)'}(k_{N\rho}\rho_{max})/J_n'(k_{N\rho}\rho_{max}) \end{bmatrix},$$
$$(6.5.2a)$$

is the reflection matrix for the wave due to a perfect electric conducting wall at $\rho = \rho_{max}$. Using (1) for region N, (1) and (2) imply that

$$\widetilde{\overline{\mathbf{R}}}_{N,N-1} \cdot \mathbf{a}_N = \mathbf{b}_N, \quad \overline{\mathbf{R}}_{N,N+1} \cdot \mathbf{b}_N = \mathbf{a}_N. \qquad (6.5.3)$$

The above is only possible if

$$\det\left(\overline{\mathbf{I}} - \overline{\mathbf{R}}_{N,N+1} \cdot \widetilde{\overline{\mathbf{R}}}_{N,N-1}\right) = 0 \qquad (6.5.4)$$

which is the guidance condition for the modes of the structure. Here, $\widetilde{\overline{\mathbf{R}}}_{N,N-1}$ embodies the physics of the layered medium. Moreover, the above yields discrete modes when ρ_{max} is a finite number. When $\rho_{max} \to \infty$, a portion of the discrete modes becomes a continuum of modes just as in the planarly layered medium case. Furthermore, it can also be proven that the radiation modes lie on the Sommerfeld branch cut.

Equation (4) is a multivalued function. Hence, its direct use to search for the eigenvalues of the structure may not work well. It can be shown from the uniqueness principle that the solution of a problem involving only a closed region should be single-valued. This fact can be used to derive an expression which is branch-point free, and hence, single-valued as in the case of (6.1.16). This will expedite the solution for the eigenvalues using a numerical root solver. Once the eigenvalues are found, they can be used in (1) to derive the eigenfunctions within a multiplicative constant. The coefficients $\mathbf{a}_i$ and $\widetilde{\overline{\mathbf{R}}}_{i,i-1}$ in (1) can be derived as shown in Chapter 3.

An orthogonality relationship exists between the eigenfunctions of a cylindrically layered structure. Because of the vector nature of the problem, the orthogonality relationship exists as (Collin 1960)

$$\int_S (\mathbf{E}_{is} \times \mathbf{H}_{js}) \cdot \hat{z} \, dS = 0, \quad i \neq j, \qquad (6.5.5)$$

where $\mathbf{E}_{is}$ is the transverse to the z component of the electric field of the i-th mode and $\mathbf{H}_{js}$ is the transverse to the z component of the magnetic field of the j-th mode; S is the cross-sectional area of the whole structure.

Consequently, when a source is placed inside a cylindrically layered medium, the transverse to z fields have to be found from (1). Then, the eigenfunction expansion method can be used to deduce the excitation coefficients of all the modes via the use of the orthogonality relationship (5).

Because of the complexity of the cylindrically layered problem compared to the planarly layered problem, it may be more expedient to find the eigen-

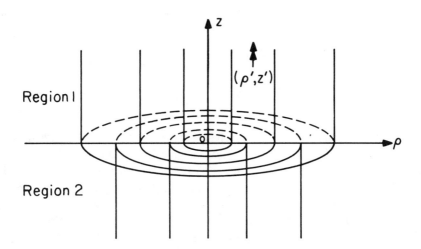

Figure 6.5.1 A junction consisting of two cylindrically layered half-spaces.

modes via a numerical approach. In the next subsection, we shall discuss the use of such a numerical approach to find the eigenmodes. But before doing so, we need to derive the pertinent differential equations.

§§6.5.2 Differential Equations of a Cylindrical Structure

Prior to applying the mode matching method across a junction discontinuity as shown in Figure 6.5.1, the modes in each region need to be found. These modes are found by solving the implicit equation discussed in the previous section, or they can be found by numerically solving the differential equation governing the modes directly. Therefore, it is useful first to derive these equations satisfied by the electromagnetic field in a cylindrical structure.

The vector wave equation satisfied by the electric field in an inhomogeneous medium with electric and magnetic current sources is

$$\mu \nabla \times \mu^{-1} \nabla \times \mathbf{E}(\mathbf{r}) - k^2 \mathbf{E}(\mathbf{r}) = i\omega\mu \mathbf{J}(\mathbf{r}) - \mu \nabla \times \mu^{-1} \mathbf{M}(\mathbf{r}). \qquad (6.5.6)$$

Since $\nabla \cdot \mathbf{D} = \varrho$, only two of the three components of the electric field are independent given the knowledge of ϱ. Hence, by letting $\mathbf{E} = \mathbf{E}_s + \mathbf{E}_z$ and $\nabla = \nabla_s + \hat{z}\frac{\partial}{\partial z}$, the transverse components of (6) can be extracted to yield (Exercise 6.18)

$$\mu \nabla_s \times \mu^{-1} \nabla_s \times \mathbf{E}_s(\mathbf{r}) - \nabla_s \epsilon^{-1} \nabla_s \cdot \epsilon \mathbf{E}_s - \left(k^2 + \frac{\partial^2}{\partial z^2} \right) \mathbf{E}_s$$

$$= i\omega\mu \mathbf{J}_s - \nabla_s \frac{\varrho}{\epsilon} - \mu(\nabla \times \mu^{-1}\mathbf{M})_s, \qquad (6.5.7)$$

where we have made use of the $\nabla \cdot \epsilon \mathbf{E} = \varrho$ condition to remove the $\mathbf{E}_z$ component in Equation (7). Furthermore, by duality, the equation for the transverse magnetic field is

$$\epsilon \nabla_s \times \epsilon^{-1} \nabla_s \times \mathbf{H}_s(\mathbf{r}) - \nabla_s \mu^{-1} \nabla_s \cdot \mu \mathbf{H}_s - \left(k^2 + \frac{\partial^2}{\partial z^2} \right) \mathbf{H}_s$$
$$= i\omega\epsilon \, \mathbf{M}_s - \nabla_s \frac{\varrho_m}{\mu} + \epsilon(\nabla \times \epsilon^{-1}\mathbf{J})_s. \quad (6.5.8)$$

In an inhomogeneous cylindrical structure, the field in general cannot be decomposed into TE and TM fields as in (6.2.1) and (6.2.2). Hence, at least two components of the field have to be considered simultaneously, as is implicit in (7) and (8) above. The eigenfunctions of the above equations are their solutions when the sources are absent. Note that when the source terms on the right-hand sides of (7) and (8) are set to zero, (8) is the transpose equation of (7), as was shown in Chapter 5, Subsection 5.4.3. Furthermore, if we assume the field to have $e^{ik_z z}$ dependence, then (7) and (8) become

$$\mu \nabla_s \times \mu^{-1} \nabla_s \times \mathbf{E}_s - \nabla_s \epsilon^{-1} \nabla_s \cdot \epsilon \mathbf{E}_s - k^2 \mathbf{E}_s + k_z^2 \mathbf{E}_s = 0, \quad (6.5.9)$$

$$\epsilon \nabla_s \times \epsilon^{-1} \nabla_s \times \mathbf{H}_s - \nabla_s \mu^{-1} \nabla_s \cdot \mu \mathbf{H}_s - k^2 \mathbf{H}_s + k_z^2 \mathbf{H}_s = 0. \quad (6.5.10)$$

These are the eigenequations in differential forms that yield the eigenvalues k_z^2. Moreover, the orthogonality relationship expressed by (5) can be easily proven from (9) and (10). Since (9) and (10) are the transpose of each other, they share the same set of eigenvalues (Exercise 6.19). For a general $\mu(x,y)$ and $\epsilon(x,y)$, these eigenmodes have to be found numerically. However, when $\mu = \mu(\rho)$ and $\epsilon = \epsilon(\rho)$, i.e., the inhomogeneity has a cylindrical symmetry, the eigenmodes can be found via the method in Chapter 3, just as the eigenmodes of a planar slab can be found via the method in Chapter 2. For more versatility, a numerical method for finding these eigenmodes may be preferred. We shall discuss such a numerical method next.

§§6.5.3 Numerical Solution of the Eigenmodes

When Equations (7) and (8) are source-free, the solutions can be assumed to have $e^{in\phi}$ dependence where n is an integer. By doing so, $\partial/\partial\phi = in$, and the source-free equation for the electric field simplifies to (Exercise 6.18)

$$-\frac{1}{\rho\epsilon_r} \left[\rho\epsilon_r \frac{\partial}{\partial\rho} \frac{1}{\rho\epsilon_r} \frac{\partial}{\partial\rho} - \frac{n^2}{\rho^2} + k^2 + \frac{\partial^2}{\partial z^2} \right] \rho\epsilon_r E_\rho + \frac{2in}{\rho^2} E_\phi = 0, \quad (6.5.11a)$$

$$in \left(\frac{\partial}{\partial\rho} \frac{1}{\rho^2\mu_r\epsilon_r} \right) \rho\epsilon_r E_\rho - \frac{1}{\rho\mu_r} \left[\rho\mu_r \frac{\partial}{\partial\rho} \frac{1}{\rho\mu_r} \frac{\partial}{\partial\rho} - \frac{n^2}{\rho^2} + k^2 + \frac{\partial^2}{\partial z^2} \right] \rho E_\phi = 0. \quad (6.5.11b)$$

The magnetic field equations are just the dual of the above equations:

$$-\frac{1}{\rho\epsilon_r} \left[\rho\epsilon_r \frac{\partial}{\partial\rho} \frac{1}{\rho\epsilon_r} \frac{\partial}{\partial\rho} - \frac{n^2}{\rho^2} + k^2 + \frac{\partial^2}{\partial z^2} \right] \rho H_\phi + in \left(\frac{\partial}{\partial\rho} \frac{1}{\rho^2\mu_r\epsilon_r} \right) \rho\mu_r H_\rho = 0, \quad (6.5.12a)$$

$$\frac{2in}{\rho^2}H_\phi - \frac{1}{\rho\mu_r}\left[\rho\mu_r\frac{\partial}{\partial\rho}\frac{1}{\rho\mu_r}\frac{\partial}{\partial\rho} - \frac{n^2}{\rho^2} + k^2 + \frac{\partial^2}{\partial z^2}\right]\rho\mu_r H_\rho = 0. \qquad (6.5.12b)$$

For compactness, (12a) and (12b) can be written in a matrix form

$$\overline{\mathcal{L}}(\rho)\cdot\mathbf{E}(\rho,z) + \frac{\partial^2}{\partial z^2}\overline{\mathcal{R}}(\rho)\cdot\mathbf{E}(\rho,z) = 0, \qquad (6.5.13)$$

where

$$\overline{\mathcal{L}}(\rho) = \begin{bmatrix} \frac{1}{\rho\epsilon_r}\left(\rho\epsilon_r\frac{\partial}{\partial\rho}\frac{1}{\rho\epsilon_r}\frac{\partial}{\partial\rho} - \frac{n^2}{\rho^2} + k^2\right) & -\frac{2in}{\rho^3} \\ -in\left(\frac{\partial}{\partial\rho}\frac{1}{\rho^2\mu_r\epsilon_r}\right) & \frac{1}{\rho\mu_r}\left(\rho\mu_r\frac{\partial}{\partial\rho}\frac{1}{\rho\mu_r}\frac{\partial}{\partial\rho} - \frac{n^2}{\rho^2} + k^2\right) \end{bmatrix},$$
$$(6.5.13a)$$

$$\overline{\mathcal{R}}(\rho) = \begin{bmatrix} \frac{1}{\rho\epsilon_r} & 0 \\ 0 & \frac{1}{\rho\mu_r} \end{bmatrix}, \qquad \mathbf{E}(\rho,z) = \begin{bmatrix} \rho\epsilon_r E_\rho \\ \rho E_\phi \end{bmatrix}. \qquad (6.5.13b)$$

Similarly, Equations (12a) and (12b) become

$$\overline{\mathcal{L}}^t(\rho)\cdot\mathbf{H}(\rho,z) + \frac{\partial^2}{\partial z^2}\overline{\mathcal{R}}(\rho)\cdot\mathbf{H}(\rho,z) = 0, \qquad (6.5.14)$$

where

$$\mathbf{H}(\rho,z) = \begin{bmatrix} \rho H_\phi \\ \rho\mu_r H_\rho \end{bmatrix}. \qquad (6.5.14a)$$

Note that (14) is just the transpose of (13). Hence, once the solution to (13) is found, the solution to (14) can be obtained easily. Moreover, Equations (13) and (14) are inherently vector in nature. But when $n = 0$, (11a) and (11b), and (12a) and (12b) are decoupled. For this case, the problem reduces to two sets of scalar problems representing the TE and the TM waves in the geometry.

Because of the simple $\partial^2/\partial z^2$ operator, the general solutions to (13) and (14) have $e^{ik_z z}$ dependences. Hence, (13) becomes

$$\overline{\mathcal{L}}(\rho)\cdot\mathbf{E}(\rho) - k_z^2\overline{\mathcal{R}}(\rho)\cdot\mathbf{E}(\rho) = 0. \qquad (6.5.15)$$

To solve the above numerically, we let

$$\mathbf{E}(\rho) = \begin{bmatrix} \rho\epsilon_r E_\rho \\ \rho E_\phi \end{bmatrix} = \sum_{i=1}^N \begin{bmatrix} S_{i1}(\rho)a_{i1} \\ S_{i2}(\rho)a_{i2} \end{bmatrix} = \sum_{i=1}^N \begin{bmatrix} S_{i1}(\rho) & 0 \\ 0 & S_{i2}(\rho) \end{bmatrix}\begin{bmatrix} a_{i1} \\ a_{i2} \end{bmatrix} = \overline{\mathbf{S}}^t(\rho)\cdot\mathbf{a},$$
$$(6.5.16)$$

where $S_{i1}(\rho)$ and $S_{i2}(\rho)$ are basis functions that can approximate $\rho\epsilon_r E_\rho$ and ρE_ϕ respectively, over the support $0 < \rho < \rho_{max}$. For example, the basis functions could be triangular functions, Bessel functions, Laguerre or Hermite polynomials. In the above,

$$\overline{\mathbf{S}}(\rho) = \begin{bmatrix} \mathbf{S}_1(\rho) & 0 \\ 0 & \mathbf{S}_2(\rho) \end{bmatrix}, \qquad \mathbf{a} = \begin{bmatrix} \mathbf{a}_1 \\ \mathbf{a}_2 \end{bmatrix}, \qquad (6.5.17)$$

where $\mathbf{S}_j(\rho)$ is a column vector containing $S_{ij}(\rho)$, $i = 1, \ldots, N$, and $\mathbf{a}_j$ is a column vector containing a_{ij}, $i = 1, \ldots, N$, as elements. Consequently, using (16) in (15), we have

$$\overline{\mathcal{L}}(\rho) \cdot \overline{\mathbf{S}}^t(\rho) \cdot \mathbf{a} - k_z^2 \overline{\mathcal{R}}(\rho) \cdot \overline{\mathbf{S}}^t(\rho) \cdot \mathbf{a} = 0. \tag{6.5.18}$$

The above can be left-multiplied by $\overline{\mathbf{S}}(\rho)$ and integrated over ρ to yield

$$\overline{\mathbf{L}} \cdot \mathbf{a} - k_z^2 \overline{\mathbf{R}} \cdot \mathbf{a} = 0, \tag{6.5.19}$$

where

$$\overline{\mathbf{L}} = \left\langle \overline{\mathbf{S}}(\rho), \overline{\mathcal{L}}(\rho) \cdot \overline{\mathbf{S}}^t(\rho) \right\rangle, \tag{6.5.20a}$$

$$\overline{\mathbf{R}} = \left\langle \overline{\mathbf{S}}(\rho), \overline{\mathcal{R}}(\rho) \cdot \overline{\mathbf{S}}^t(\rho) \right\rangle. \tag{6.5.20b}$$

In the above,

$$\langle f(\rho), g(\rho) \rangle = \int\limits_0^{\rho_{max}} f(\rho) g(\rho) d\rho,$$

and $\overline{\mathbf{L}}$ and $\overline{\mathbf{R}}$ are $2N \times 2N$ matrices. $\overline{\mathbf{L}}$ is asymmetric while $\overline{\mathbf{R}}$ is symmetric (Exercise 6.20). A similar procedure applied to (14) then yields

$$\overline{\mathbf{L}}^t \cdot \mathbf{b} - k_z^2 \overline{\mathbf{R}} \cdot \mathbf{b} = 0 \tag{6.5.21}$$

from which $2N$ eigenvalues $k_{\alpha z}^2$ with $2N$ eigenvectors $\mathbf{a}_\alpha$ or $\mathbf{b}_\alpha$ can be found. It is easy to show that (Exercise 6.20)

$$\mathbf{a}_\alpha^t \cdot \overline{\mathbf{R}} \cdot \mathbf{b}_\beta = \delta_{\alpha\beta} D_\alpha, \tag{6.5.22}$$

where D_α is a constant. Hence, $\mathbf{a}_\alpha$ and $\mathbf{b}_\beta$, i.e., eigenvectors to (19) and (21), for different eigenvalues $k_{\alpha z}^2$ and $k_{\beta z}^2$, are $\overline{\mathbf{R}}$ orthogonal. Furthermore, it can be shown that

$$\left\langle \mathbf{E}_\alpha^t(\rho), \overline{\mathcal{R}}(\rho) \cdot \mathbf{H}_\beta(\rho) \right\rangle = \delta_{\alpha\beta} D_\alpha, \tag{6.5.23}$$

where $\mathbf{E}_\alpha(\rho) = \overline{\mathbf{S}}^t(\rho) \cdot \mathbf{a}_\alpha$ and $\mathbf{H}_\beta(\rho) = \overline{\mathbf{S}}^t(\rho) \cdot \mathbf{b}_\beta$. The eigenvectors and the eigenfunctions can be normalized for convenience setting $D_\alpha = 1$. Note that Equations (22) and (23) are just different ways of expressing the orthogonality relationship (5).

§§6.5.4 Eigenfunction Expansion of a Field

Given the eigenfunctions of Equations (7) and (8), we can use them to solve for the fields when the source is present by the eigenfunction expansion method. Consider the case when the source is a vertical magnetic dipole such that

$$\mathbf{M} = \hat{z} I_m \ell \, \delta(\mathbf{r}) = \hat{z} I_m \ell \, \frac{\delta(\rho - \rho')}{\rho'} \delta(\phi - \phi') \, \delta(z - z'). \tag{6.5.24}$$

Then, $(\nabla \times \mathbf{M})_s$ becomes

$$(\nabla \times \mathbf{M})_s = -\hat{\phi}I_m\ell\frac{\delta'(\rho - \rho')}{\rho'}\delta(\phi - \phi')\delta(z - z') + \hat{\rho}\frac{I_m\ell}{\rho'^2}\delta(\rho - \rho')\delta'(\phi - \phi')\delta(z - z').$$
(6.5.25)

The ϕ variation of the source could be expanded using Fourier series as (see Appendix C)

$$\delta(\phi - \phi') = \frac{1}{2\pi}\sum_{n=-\infty}^{\infty}e^{in(\phi - \phi')}.$$
(6.5.26)

Hence, the source (25) can be expanded in terms of $e^{in\phi}$ harmonics. Furthermore, each of these harmonics will excite a field with $e^{in\phi}$ dependence. For the n-th harmonic, Equation (7) can be written as

$$\overline{\mathcal{L}}(\rho) \cdot \mathbf{E}(\rho, z) + \frac{\partial^2}{\partial z^2}\overline{\mathcal{R}}(\rho) \cdot \mathbf{E}(\rho, z) = \mathbf{S}(\rho, z) = \begin{bmatrix} S_1(\rho, z) \\ S_2(\rho, z) \end{bmatrix},$$
(6.5.27)

where

$$S_1(\rho, z) = \frac{iI_m\ell ne^{-in\phi'}}{2\pi\rho'^2}\delta(\rho - \rho')\,\delta(z - z'),$$
(6.5.27a)

$$S_2(\rho, z) = -\frac{I_m\ell e^{-in\phi'}}{2\pi\rho'}\delta'(\rho - \rho')\,\delta(z - z').$$
(6.5.27b)

The unknown field, $\mathbf{E}(\rho, z)$, can be expanded in terms of the eigenfunctions $\mathbf{E}_\alpha(\rho)$ such that

$$\mathbf{E}(\rho, z) = \sum_{\alpha=1}^{2N}\mathbf{E}_\alpha(\rho)a_\alpha(z).$$
(6.5.28)

Then, (27) becomes

$$\sum_{\alpha=1}^{2N}\overline{\mathcal{L}}(\rho) \cdot \mathbf{E}_\alpha(\rho)a_\alpha(z) + \frac{\partial^2}{\partial z^2}\sum_{\alpha=1}^{2N}\overline{\mathcal{R}}(\rho) \cdot \mathbf{E}_\alpha(\rho)a_\alpha(z) = \begin{bmatrix} S_1 \\ S_2 \end{bmatrix}.$$
(6.5.29)

Consequently, after multiplying the above by $\mathbf{H}_\beta^t(\rho)$ and integrating over ρ, we have

$$\sum_{\alpha=1}^{2N}\left\langle\mathbf{H}_\beta^t(\rho), \overline{\mathcal{L}}(\rho) \cdot \mathbf{E}_\alpha(\rho)\right\rangle a_\alpha(z) + \frac{\partial^2}{\partial z^2}\sum_{\alpha=1}^{2N}\left\langle\mathbf{H}_\beta^t(\rho), \overline{\mathcal{R}}(\rho) \cdot \mathbf{E}_\alpha(\rho)\right\rangle a_\alpha(z)$$

$$= A_\beta\delta(z - z'),$$
(6.5.30)

where $A_\beta = \frac{I_m\ell e^{-in\phi'}}{2\pi\rho'}\left[\frac{in}{\rho'}H_{\alpha1}(\rho') + H'_{\alpha2}(\rho')\right]$. Then, using $\mathbf{E}_\alpha(\rho) = \overline{\mathbf{S}}^t(\rho) \cdot \mathbf{a}_\alpha$ and $\mathbf{H}_\beta(\rho) = \overline{\mathbf{S}}^t(\rho) \cdot \mathbf{b}_\beta$, it can be shown using (21) and (23) that

$$\left\langle\mathbf{H}_\beta^t(\rho), \overline{\mathcal{L}}(\rho) \cdot \mathbf{E}_\alpha(\rho)\right\rangle = \mathbf{b}_\beta^t \cdot \overline{\mathbf{L}} \cdot \mathbf{a}_\alpha = k_{\alpha z}^2\mathbf{b}_\beta^t \cdot \overline{\mathbf{R}} \cdot \mathbf{a}_\alpha = k_{\beta z}^2\delta_{\alpha\beta},$$
(6.5.31)

where we have assumed an orthonormal set so that $D_\alpha = 1$ in (23). Equation (31), together with the orthogonality relationship (23), transforms (30) to

$$\frac{d^2}{dz^2}a_\beta(z) + k_{\beta z}^2 a_\beta(z) = A_\beta \delta(z - z'). \tag{6.5.32}$$

Solving (32) yields

$$a_\beta(z) = A_\beta \frac{e^{ik_{\beta z}|z-z'|}}{2ik_{\beta z}}. \tag{6.5.33}$$

Hence, the solution (28) becomes

$$\mathbf{E}(\rho, z) = \sum_{\alpha=1}^{2N} \mathbf{E}_\alpha(\rho) A_\alpha \frac{e^{ik_{\alpha z}|z-z'|}}{2ik_{\alpha z}}. \tag{6.5.34}$$

Moreover, using vector notation, the above becomes

$$\mathbf{E}(\rho, z) = \overline{\mathbf{E}}^t(\rho) \cdot e^{i\overline{\mathbf{K}}_z|z-z'|} \cdot \mathbf{S}(\rho'), \tag{6.5.35}$$

where $\overline{\mathbf{E}}^t(\rho) = [\mathbf{E}_1(\rho), \mathbf{E}_2(\rho), \ldots, \mathbf{E}_{2N}(\rho)]$ is a $2 \times 2N$ matrix, $\overline{\mathbf{K}}_z$ is a diagonal matrix containing $k_{\alpha z}$, and $\mathbf{S}(\rho')$ is a column vector containing A_α which embodies the characteristics of the source alone. For some sources, $\mathbf{S}(\rho')$ may assume different values for $z > z'$ and $z < z'$. Hence, more generally,

$$\mathbf{E}(\rho, z) = \overline{\mathbf{E}}^t(\rho) \cdot e^{i\overline{\mathbf{K}}_z|z-z'|} \cdot \mathbf{S}_\pm(\rho'), \tag{6.5.36}$$

where the $\pm$ refers to different $\mathbf{S}(\rho')$ for $z > z'$ for the upper sign and $z < z'$ for the lower sign. In the above, $\mathbf{S}_\pm(\rho')$ is the excitation coefficient vector, $e^{i\overline{\mathbf{K}}_z|z-z'|}$ is a propagator that propagates the field from z to z', and $\overline{\mathbf{E}}^t(\rho)$ is a matrix that contains the eigenmodes of the structure. A similar expression can also be derived for other source excitations (Exercise 6.21).

§§6.5.5 Reflection from a Junction Discontinuity

The expression (36) represents the field when the source is in an infinitely long, cylindrically layered medium. But when a junction discontinuity is present at $z = 0$, the field will be reflected by the discontinuity. For example, if the source is in region 1 at $z = z'$, in the manner of (6.3.1), the field in region 1 can be written as

$$\mathbf{E}_1(\rho, z) = \overline{\mathbf{E}}_1^t(\rho) \cdot \left[e^{i\overline{\mathbf{K}}_{1z}|z-z'|} \cdot \mathbf{S}_{1\pm}(\rho') + e^{i\overline{\mathbf{K}}_{1z}z} \cdot \overline{\mathbf{R}}_{12} \cdot e^{i\overline{\mathbf{K}}_{1z}|z'|} \cdot \mathbf{S}_{1-}(\rho') \right]. \tag{6.5.37}$$

Here, the subscript 1 denotes quantities associated with region 1. Furthermore, the term additional to (36) is due to reflection from the discontinuity at $z = 0$. In region 2, a transmitted wave will exist and

$$\mathbf{E}_2(\rho, z) = \overline{\mathbf{E}}_2^t(\rho) e^{-i\overline{\mathbf{K}}_{2z}z} \cdot \overline{\mathbf{T}}_{12} \cdot e^{i\overline{\mathbf{K}}_{1z}|z'|} \cdot \mathbf{S}_{1-}(\rho'). \tag{6.5.38}$$

In (37) and (38), $\overline{\mathbf{R}}_{12}$ is the reflection operator while $\overline{\mathbf{T}}_{12}$ is the transmission operator, which are unknowns yet to be sought. To find them, the boundary conditions for the continuity of tangential electric and tangential magnetic fields have to be imposed at $z = 0$. Hence, we need to derive an expression for the transverse component of the magnetic field from the transverse component of the electric field.

From Maxwell's equations, we can show that (Exercise 6.22)

$$-i\omega\hat{z} \times \frac{\partial}{\partial z}\mathbf{H}_s = \nabla_s \times \mu^{-1}\nabla_s \times \mathbf{E}_s - \omega^2\epsilon\,\mathbf{E}_s. \qquad (6.5.39)$$

The above can be written in a matrix form relating $\mathbf{H}(\rho, z)$ to $\mathbf{E}(\rho, z)$, yielding

$$\frac{\partial}{\partial z}\mathcal{R}(\rho) \cdot \mathbf{H}(\rho, z) = \mathcal{M}(\rho) \cdot \mathbf{E}(\rho, z), \qquad (6.5.40)$$

where

$$\mathcal{M}(\rho) = \frac{1}{i\omega\mu_0}\begin{bmatrix} \frac{n^2}{\rho^3\mu_r\epsilon_r^2} - \frac{k_0^2}{\rho\epsilon_r} & \frac{in}{\rho^2\mu_r\epsilon_r}\frac{\partial}{\partial\rho} \\ -in\frac{\partial}{\partial\rho}\frac{1}{\rho^2\mu_r\epsilon_r} & \frac{\partial}{\partial\rho}\frac{1}{\rho\mu_r}\frac{\partial}{\partial\rho} + \frac{k^2}{\rho\mu_r} \end{bmatrix}. \qquad (6.5.40a)$$

Then, on applying (40) to (37) and (38), we have

$$\mathbf{H}_1(\rho, z) = \mathcal{R}_1^{-1} \cdot \mathcal{M}_1 \cdot \mathbf{E}_1^t(\rho) \cdot \left(i\overline{\mathbf{K}}_{1z}\right)^{-1}$$
$$\cdot \left[\pm e^{i\overline{\mathbf{K}}_{1z}|z-z'|} \cdot \mathbf{S}_{1\pm}(\rho') + e^{i\overline{\mathbf{K}}_{1z}z} \cdot \overline{\mathbf{R}}_{12} \cdot e^{i\overline{\mathbf{K}}_{1z}|z'|} \cdot \mathbf{S}_{1-}(\rho')\right], \qquad (6.5.41)$$

$$\mathbf{H}_2(\rho, z) = \mathcal{R}_2^{-1} \cdot \mathcal{M}_2 \cdot \mathbf{E}_2^t(\rho) \cdot \left(-i\overline{\mathbf{K}}_{2z}\right)^{-1} \cdot e^{-i\overline{\mathbf{K}}_{2z}z} \cdot \overline{\mathbf{T}}_{12} \cdot e^{i\overline{\mathbf{K}}_{1z}|z'|} \cdot \mathbf{S}_{1-}(\rho'). \qquad (6.5.42)$$

Moreover, by matching the continuity of the tangential $\mathbf{E}$ and tangential $\mathbf{H}$ fields at $z = 0$, one obtains

$$\overline{\epsilon}_1^{-1} \cdot \mathbf{E}_1^t(\rho) \cdot \left[\overline{\mathbf{I}} + \overline{\mathbf{R}}_{12}\right] = \overline{\epsilon}_2^{-1} \cdot \mathbf{E}_2^t(\rho) \cdot \overline{\mathbf{T}}_{12}, \qquad (6.5.43a)$$

$$\overline{\mu}_1^{-1} \cdot \mathcal{R}_1^{-1} \cdot \mathcal{M}_1 \cdot \mathbf{E}_1^t(\rho) \cdot \overline{\mathbf{K}}_{1z}^{-1} \cdot \left[\overline{\mathbf{I}} - \overline{\mathbf{R}}_{12}\right]$$
$$= \overline{\mu}_2^{-1} \cdot \mathcal{R}_2^{-1} \cdot \mathcal{M}_2 \cdot \mathbf{E}_2^t(\rho) \cdot \overline{\mathbf{K}}_{2z}^{-1} \cdot \overline{\mathbf{T}}_{12}, \qquad (6.5.43b)$$

where

$$\overline{\epsilon}_i^{-1} = \begin{bmatrix} \epsilon_{ir}^{-1} & 0 \\ 0 & 1 \end{bmatrix}, \qquad \overline{\mu}_i^{-1} = \begin{bmatrix} 1 & 0 \\ 0 & \mu_{ir}^{-1} \end{bmatrix}. \qquad (6.5.43c)$$

Next, after multiplying (43a) by $\frac{1}{\rho}\overline{\mu}_1^{-1}$ and $\overline{\mathbf{H}}_1(\rho)$, and integrating over ρ, one derives

$$\left\langle \overline{\mathbf{H}}_1(\rho), \frac{1}{\rho}\overline{\mu}_1^{-1} \cdot \overline{\epsilon}_1^{-1} \cdot \mathbf{E}_1^t(\rho)\right\rangle \cdot \left[\overline{\mathbf{I}} + \overline{\mathbf{R}}_{12}\right] = \left\langle \overline{\mathbf{H}}_1(\rho), \frac{1}{\rho}\overline{\mu}_1^{-1} \cdot \overline{\epsilon}_2^{-1} \cdot \mathbf{E}_2^t(\rho)\right\rangle \cdot \overline{\mathbf{T}}_{12}. \qquad (6.5.44)$$

Since $\frac{1}{\rho}\overline{\mu}_1^{-1}\cdot\overline{\epsilon}_1^{-1} = \mathcal{R}(\rho)$, the inner product on the left-hand side of (44) results in an identity matrix from Equation (16) if the eigenvectors are orthonormal. Hence, the above becomes

$$\left[\overline{\mathbf{I}} + \overline{\mathbf{R}}_{12}\right] = \overline{\mathbf{L}}_{12} \cdot \overline{\mathbf{T}}_{12}, \qquad (6.5.45)$$

where $\overline{\mathbf{L}}_{12} = \left\langle \overline{\mathbf{H}}_1(\rho), \frac{1}{\rho}\overline{\mu}_1^{-1} \cdot \overline{\epsilon}_2^{-1} \cdot \mathbf{E}_2^t(\rho)\right\rangle$.

Similarly, we can multiply (43b) by $\frac{1}{\rho}\overline{\epsilon}_1^{-1}$ and $\overline{\mathbf{E}}_1(\rho)$, and integrate over ρ to yield

$$\left\langle \overline{\mathbf{E}}_1(\rho), \overline{\mathcal{M}}_1 \cdot \overline{\mathbf{E}}_1^t(\rho) \right\rangle \cdot \overline{\mathbf{K}}_{1z}^{-1} \cdot \left[\overline{\mathbf{I}} - \overline{\mathbf{R}}_{12} \right]$$
$$= \left\langle \overline{\mathbf{E}}_1(\rho), \frac{1}{\rho}\overline{\epsilon}_1^{-1} \cdot \overline{\mu}_2^{-1} \cdot \overline{\mathcal{R}}_2^{-1} \cdot \overline{\mathcal{M}}_2 \cdot \overline{\mathbf{E}}_2^t(\rho) \right\rangle \cdot \overline{\mathbf{K}}_{2z}^{-1} \cdot \overline{\mathbf{T}}_{12}. \quad (6.5.46)$$

The above is just of the form
$$\overline{\mathbf{A}}_{11} \cdot \left[\overline{\mathbf{I}} - \overline{\mathbf{R}}_{12} \right] = \overline{\mathbf{A}}_{12} \cdot \overline{\mathbf{T}}_{12}. \quad (6.5.47)$$

Hence, solving (45) and (47) yields

$$\overline{\mathbf{R}}_{12} = \left(\overline{\mathbf{A}}_{12}^{-1} \cdot \overline{\mathbf{A}}_{11} + \overline{\mathbf{L}}_{12}^{-1} \right)^{-1} \cdot \left(\overline{\mathbf{A}}_{12}^{-1} \cdot \overline{\mathbf{A}}_{11} - \overline{\mathbf{L}}_{12}^{-1} \right), \quad (6.5.48a)$$

$$\overline{\mathbf{T}}_{12} = 2\overline{\mathbf{L}}_{12}^{-1} \cdot \left(\overline{\mathbf{A}}_{12}^{-1} \cdot \overline{\mathbf{A}}_{11} + \overline{\mathbf{L}}_{12}^{-1} \right)^{-1} \cdot \overline{\mathbf{A}}_{12}^{-1} \cdot \overline{\mathbf{A}}_{11}. \quad (6.5.48b)$$

Once $\overline{\mathbf{R}}_{12}$ and $\overline{\mathbf{T}}_{12}$ are found, the field can be found everywhere.

§6.6 The Multiregion Problem

In the previous sections, we have developed the theory for the radiation and propagation of waves in a two-region inhomogeneous region. The two-region problem is a canonical problem; once the reflection operator and transmission operator for a single discontinuity, two-region problem are found, they can be used to construct the reflection and transmission operators for a multiregion problem shown in Figure 6.6.1. The theory developed in this section, though using the results of Section 6.5 as examples, can be easily adapted to the results of Sections 6.3 and 6.4.[14]

§§6.6.1 The Three-Region Problem

For the three-region problem as shown in Figure 6.6.2, the solution in region 1, using the notation of Section 6.5, is

$$\mathbf{E}_1(\rho, z) = \overline{\mathbf{E}}_1^t(\rho) \cdot \left[e^{i\overline{\mathbf{K}}_{1z}|z-z'|} \cdot \mathbf{S}_{1\pm}(\rho') + e^{i\overline{\mathbf{K}}_{1z}(z-d_1)} \cdot \widetilde{\overline{\mathbf{R}}}_{12} \cdot e^{i\overline{\mathbf{K}}_{1z}|d_1-z'|} \cdot \mathbf{S}_{1-}(\rho') \right], \quad (6.6.1)$$

where $\widetilde{\overline{\mathbf{R}}}_{12}$ is now a reflection operator that incorporates subsurface reflections. Furthermore, in region 2, the solution is

$$\mathbf{E}_2(\rho, z) = \overline{\mathbf{E}}_2^t(\rho) \cdot \left[e^{-i\overline{\mathbf{K}}_{2z}(z-d_2)} + e^{i\overline{\mathbf{K}}_{2z}(z-d_2)} \cdot \overline{\mathbf{R}}_{23} \right] \cdot \mathbf{A}_2, \quad (6.6.2)$$

since the upgoing wave and the downgoing wave have to be related by $\overline{\mathbf{R}}_{23}$ at $z = d_2$. In addition, in region 3, the solution is

$$\mathbf{E}_3(\rho, z) = \overline{\mathbf{E}}_3^t(\rho) \cdot e^{-i\overline{\mathbf{K}}_{3z}(z-d_2)} \cdot \mathbf{A}_3. \quad (6.6.3)$$

---

[14] For example, see Chew and Anderson (1985); Chew (1988a, b); Liu (1989).

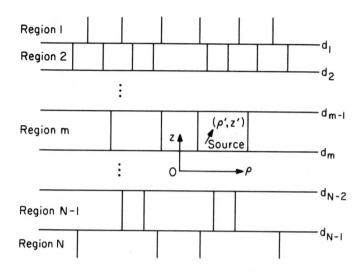

Figure 6.6.1 The multiregion geometry.

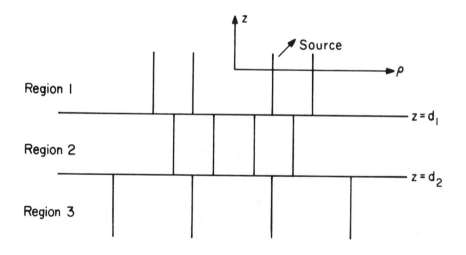

Figure 6.6.2 The three-region geometry.

To find $\mathbf{A}_2$, we make use of constraint conditions similar to those adopted for Chapter 2, Subsection 2.1.3, or Chapter 3, Section 3.2. Consequently, we use the fact that the downgoing wave in region 2 is a consequence of the transmission of the downgoing wave in region 1 plus the reflection of the upgoing wave in region 2. In other words, at $z = d_1$,

$$e^{-i\overline{\mathbf{K}}_{2z}(d_1-d_2)} \cdot \mathbf{A}_2 = \overline{\mathbf{T}}_{12} \cdot e^{i\overline{\mathbf{K}}_{1z}|d_1-z'|} \cdot \mathbf{S}_{1-}(\rho')$$
$$+ \overline{\mathbf{R}}_{21} \cdot e^{i\overline{\mathbf{K}}_{2z}(d_1-d_2)} \cdot \overline{\mathbf{R}}_{23} \cdot \mathbf{A}_2. \quad (6.6.4)$$

Then, solving the above yields

$$\mathbf{A}_2 = e^{i\overline{\mathbf{K}}_{2z}(d_1-d_2)} \cdot \overline{\mathbf{D}}_{2-} \cdot \overline{\mathbf{T}}_{12} \cdot e^{i\overline{\mathbf{K}}_{1z}|d_1-z'|} \cdot \mathbf{S}_{1-}(\rho'), \quad (6.6.5)$$

where

$$\overline{\mathbf{D}}_{2-} = \left[\overline{\mathbf{I}} - \overline{\mathbf{R}}_{21} \cdot e^{i\overline{\mathbf{K}}_{2z}(d_1-d_2)} \cdot \overline{\mathbf{R}}_{23} \cdot e^{i\overline{\mathbf{K}}_{2z}(d_1-d_2)}\right]^{-1}, \quad (6.6.5a)$$

and it accounts for multiple reflections in a layered region (Exercise 6.23).

Similarly, $\mathbf{A}_3$ is just the transmission of the downgoing wave in region 2, which is

$$\mathbf{A}_3 = \overline{\mathbf{T}}_{23} \cdot \mathbf{A}_2. \quad (6.6.6)$$

But in region 1, the reflected wave, which is described by the second term in (1), is the reflection of the downgoing wave in region 1 due to the discontinuity at $z = d_1$ plus a transmission of the upgoing wave in region 2. Hence, at $z = d_1$,

$$\widetilde{\overline{\mathbf{R}}}_{12} \cdot e^{i\overline{\mathbf{K}}_{1z}|d_1-z'|} \cdot \mathbf{S}_{1-}(\rho') = \overline{\mathbf{R}}_{12} \cdot e^{i\overline{\mathbf{K}}_{1z}|d_1-z'|} \cdot \mathbf{S}_{1-}(\rho')$$
$$+ \overline{\mathbf{T}}_{21} \cdot e^{i\overline{\mathbf{K}}_{2z}(d_1-d_2)} \cdot \overline{\mathbf{R}}_{23} \cdot \mathbf{A}_2. \quad (6.6.7)$$

Consequently, using $\mathbf{A}_2$ from (5), we deduce that

$$\widetilde{\overline{\mathbf{R}}}_{12} = \overline{\mathbf{R}}_{12} + \overline{\mathbf{T}}_{21} \cdot e^{i\overline{\mathbf{K}}_{2z}(d_1-d_2)} \cdot \overline{\mathbf{R}}_{23} \cdot e^{i\overline{\mathbf{K}}_{2z}(d_1-d_2)} \cdot \overline{\mathbf{D}}_{2-} \cdot \overline{\mathbf{T}}_{12}. \quad (6.6.8)$$

Hence, the generalized reflection operator $\widetilde{\overline{\mathbf{R}}}_{12}$ is found. Now, if a subsurface layer exists below region 3, we need only change $\overline{\mathbf{R}}_{23}$ to $\widetilde{\overline{\mathbf{R}}}_{23}$ in the above and in $\overline{\mathbf{D}}_{2-}$. Moreover, the above analysis can be easily repeated for planarly layered regions of Section 6.3 (Exercise 6.24).

§§6.6.2 The N-Region Problem

In general, when we have N regions below the source, the field in region m can be written in the manner of (2) as

$$\mathbf{E}_m(\rho, z) = \overline{\mathbf{E}}_m^t(\rho) \cdot \left[e^{-i\overline{\mathbf{K}}_{mz}(z-d_m)} + e^{i\overline{\mathbf{K}}_{mz}(z-d_m)} \cdot \widetilde{\overline{\mathbf{R}}}_{m,m+1}\right] \cdot \mathbf{A}_m, \quad (6.6.9)$$

where $\widetilde{\overline{\mathbf{R}}}_{m,m+1}$ is a generalized reflection operator. As shown previously, it can be found recursively as

$$\widetilde{\overline{\mathbf{R}}}_{m,m+1} = \overline{\mathbf{R}}_{m,m+1} + \overline{\mathbf{T}}_{m+1,m} \cdot e^{i\overline{\mathbf{K}}_{m+1,z}(d_m-d_{m+1})}$$
$$\cdot \widetilde{\overline{\mathbf{R}}}_{m+1,m+2} \cdot e^{i\overline{\mathbf{K}}_{m+1,z}(d_m-d_{m+1})} \cdot \overline{\mathbf{D}}_{m+1,-} \cdot \overline{\mathbf{T}}_{m,m+1}, \quad (6.6.10)$$

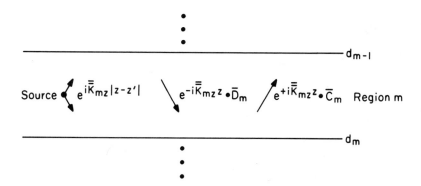

Figure 6.6.3 In region m where the source is, the field consists of that due to the source plus an upgoing wave and a downgoing wave.

where $\overline{\mathbf{D}}_{m+1,-}$ is the generalization of (5a) given by

$$\overline{\mathbf{D}}_{m+1,-} = \left[\overline{\mathbf{I}} - \overline{\mathbf{R}}_{m+1,m} \cdot e^{i\overline{\mathbf{K}}_{m+1,z}(d_m-d_{m+1})}\widetilde{\overline{\mathbf{R}}}_{m+1,m+2} \cdot e^{i\overline{\mathbf{K}}_{m+1,z}(d_m-d_{m+1})}\right]^{-1}.$$
(6.6.10a)

Moreover, a generalization of Equation (5) can be used to find $\mathbf{A}_m$, i.e.,

$$\mathbf{A}_m = e^{i\overline{\mathbf{K}}_{mz}(d_{m-1}-d_m)} \cdot \overline{\mathbf{D}}_{m,-} \cdot \overline{\mathbf{T}}_{m-1,m} \cdot \mathbf{A}_{m-1}.$$
(6.6.11)

Therefore, the field in all the regions can be found in this manner. Furthermore, if there are N regions above the source, the problem can be treated in a similar manner.

Now, if the source is embedded in region m as shown in Figure 6.6.3, the field in region m becomes

$$\mathbf{E}_m(\rho, z) = \overline{\mathbf{E}}_m^t(\rho) \cdot \left[e^{i\overline{\mathbf{K}}_{mz}|z-z'|} \cdot \mathbf{S}_{m\pm}(\rho') + e^{i\overline{\mathbf{K}}_{mz}z} \cdot \mathbf{C}_m + e^{-i\overline{\mathbf{K}}_{mz}z} \cdot \mathbf{D}_m\right].$$
(6.6.12)

At $z = d_m$, the upgoing wave must be related to the downgoing wave by the generalized reflection operator $\widetilde{\overline{\mathbf{R}}}_{m,m+1}$. Hence,

$$e^{i\overline{\mathbf{K}}_{mz}d_m} \cdot \mathbf{C}_m = \widetilde{\overline{\mathbf{R}}}_{m,m+1} \cdot \left[e^{i\overline{\mathbf{K}}_{mz}|d_m-z'|} \cdot \mathbf{S}_{m-}(\rho') + e^{-i\overline{\mathbf{K}}_{mz}d_m} \cdot \mathbf{D}_m\right].$$
(6.6.13)

At $z = d_{m-1}$, the downgoing wave must be related to the upgoing wave by the generalized reflection operator $\widetilde{\overline{\mathbf{R}}}_{m,m-1}$. Hence,

$$e^{-i\overline{\mathbf{K}}_{mz}d_{m-1}} \cdot \mathbf{D}_m = \widetilde{\overline{\mathbf{R}}}_{m,m-1} \left[e^{i\overline{\mathbf{K}}_{mz}|d_{m-1}-z'|} \cdot \mathbf{S}_{m+}(\rho') + e^{i\overline{\mathbf{K}}_{mz}d_{m-1}} \cdot \mathbf{C}_m\right].$$
(6.6.14)

In (14), $\widetilde{\overline{\mathbf{R}}}_{m,m+1}$ can be found recursively via the relation (10). Note that a similar recursive relation for $\widetilde{\overline{\mathbf{R}}}_{m,m-1}$ can be written as follows:

$$\widetilde{\overline{\mathbf{R}}}_{m,m-1} = \overline{\mathbf{R}}_{m,m-1} + \overline{\mathbf{T}}_{m-1,m} e^{i\overline{\mathbf{K}}_{m-1,z}(d_{m-2}-d_{m-1})}$$
$$\cdot \widetilde{\overline{\mathbf{R}}}_{m-1,m-2} \cdot e^{i\overline{\mathbf{K}}_{m-1,z}(d_{m-2}-d_{m-1})} \cdot \overline{\mathbf{D}}_{m-1,+} \cdot \overline{\mathbf{T}}_{m,m-1}, \quad (6.6.15)$$

where

$$\overline{\mathbf{D}}_{m-1,+} = \left[\overline{\mathbf{I}} - \overline{\mathbf{R}}_{m-1,m} \cdot e^{i\overline{\mathbf{K}}_{m-1,z}(d_{m-2}-d_{m-1})} \right.$$
$$\left. \cdot \widetilde{\overline{\mathbf{R}}}_{m-1,m-2} \cdot e^{i\overline{\mathbf{K}}_{m-1,z}(d_{m-2}-d_{m-1})} \right]^{-1}. \quad (6.6.15a)$$

Consequently, we can solve (13) and (14) for $\mathbf{C}_m$ and $\mathbf{D}_m$ to yield (Exercise 6.25)

$$e^{i\overline{\mathbf{K}}_{mz}d_m} \mathbf{C}_m = \widetilde{\overline{\mathbf{M}}}_{m+} \cdot \widetilde{\overline{\mathbf{R}}}_{m,m+1} \cdot \left[e^{i\overline{\mathbf{K}}_{mz}|d_m-z'|} \cdot \mathbf{S}_{m-} \right.$$
$$\left. + e^{i\overline{\mathbf{K}}_{mz}(d_{m-1}-d_m)} \cdot \widetilde{\overline{\mathbf{R}}}_{m,m-1} \cdot e^{i\overline{\mathbf{K}}_{mz}|d_{m-1}-z'|} \cdot \mathbf{S}_{m+} \right],$$
$$(6.6.16a)$$

$$e^{-i\overline{\mathbf{K}}_{mz}d_{m-1}} \mathbf{D}_m = \widetilde{\overline{\mathbf{M}}}_{m-} \cdot \widetilde{\overline{\mathbf{R}}}_{m,m-1} \cdot \left[e^{i\overline{\mathbf{K}}_{mz}|d_{m-1}-z'|} \cdot \mathbf{S}_{m+} \right.$$
$$\left. + e^{i\overline{\mathbf{K}}_{mz}(d_{m-1}-d_m)} \cdot \widetilde{\overline{\mathbf{R}}}_{m,m+1} \cdot e^{i\overline{\mathbf{K}}_{mz}|d_m-z'|} \cdot \mathbf{S}_{m-} \right],$$
$$(6.6.16b)$$

where

$$\widetilde{\overline{\mathbf{M}}}_{m+} = \left[\overline{\mathbf{I}} - \widetilde{\overline{\mathbf{R}}}_{m,m+1} \cdot e^{i\overline{\mathbf{K}}_{mz}(d_{m-1}-d_m)} \cdot \widetilde{\overline{\mathbf{R}}}_{m,m-1} \cdot e^{i\overline{\mathbf{K}}_{mz}(d_{m-1}-d_m)} \right]^{-1},$$
$$(6.6.17a)$$
$$\widetilde{\overline{\mathbf{M}}}_{m-} = \left[\overline{\mathbf{I}} - \widetilde{\overline{\mathbf{R}}}_{m,m-1} \cdot e^{i\overline{\mathbf{K}}_{mz}(d_{m-1}-d_m)} \cdot \widetilde{\overline{\mathbf{R}}}_{m,m+1} \cdot e^{i\overline{\mathbf{K}}_{mz}(d_{m-1}-d_m)} \right]^{-1}.$$
$$(6.6.17b)$$

Hence, the field in region m is derived.

When (16) is substituted into (12), it can be shown after some algebraic manipulation that for $z > z'$ (Exercise 6.25),

$$\mathbf{E}_m(\rho, z) = \overline{\mathbf{E}}_m^t(\rho) \cdot \left[e^{i\overline{\mathbf{K}}_{mz}(z-d_m)} + e^{-i\overline{\mathbf{K}}_{mz}(z-d_{m-1})} \cdot \widetilde{\overline{\mathbf{R}}}_{m,m-1} \cdot e^{i\overline{\mathbf{K}}_{mz}(d_{m-1}-d_m)} \right]$$
$$\cdot \widetilde{\overline{\mathbf{M}}}_{m+} \cdot \left[e^{-i\overline{\mathbf{K}}_{mz}(z'-d_m)} \cdot \mathbf{S}_{m+} + \widetilde{\overline{\mathbf{R}}}_{m,m+1} \cdot e^{i\overline{\mathbf{K}}_{mz}(z'-d_m)} \cdot \mathbf{S}_{m-} \right]. \quad (6.6.18a)$$

For $z < z'$,

$$\mathbf{E}_m(\rho, z) = \overline{\mathbf{E}}_m^t(\rho) \cdot \left[e^{-i\overline{\mathbf{K}}_{mz}(z-d_{m-1})} + e^{i\overline{\mathbf{K}}_{mz}(z-d_m)} \cdot \widetilde{\overline{\mathbf{R}}}_{m,m+1} \cdot e^{i\overline{\mathbf{K}}_{mz}(d_{m-1}-d_m)} \right]$$
$$\cdot \widetilde{\overline{\mathbf{M}}}_{m-} \cdot \left[e^{i\overline{\mathbf{K}}_{mz}(z'-d_{m-1})} \cdot \mathbf{S}_{m-} + \widetilde{\overline{\mathbf{R}}}_{m,m-1} \cdot e^{i\overline{\mathbf{K}}_{mz}(d_{m-1}-z')} \cdot \mathbf{S}_{m+} \right]. \quad (6.6.18b)$$

The preceding equation has the advantage of being more symmetrical.

In region n where $n > m$, the field can be written as

$$\mathbf{E}_n(\rho, z) = \mathbf{E}_n^t(\rho) \cdot \left[e^{-i\overline{\mathbf{K}}_{nz}(z-d_n)} + e^{i\overline{\mathbf{K}}_{nz}(z-d_n)} \cdot \widetilde{\overline{\mathbf{R}}}_{n,n+1} \right] \cdot \mathbf{A}_n, \qquad (6.6.19)$$

where $\widetilde{\overline{\mathbf{R}}}_{N,N+1} = 0$, and $\widetilde{\overline{\mathbf{R}}}_{n,n+1}$ can be found recursively using (10). Then, in the manner of (11), $\mathbf{A}_n$ can be found in terms of $\mathbf{A}_{n-1}$ recursively, until region m is reached. In region m, $\mathbf{A}_m$, the amplitude of the downgoing wave at $z = d_m$, is given by

$$\mathbf{A}_m = e^{i\overline{\mathbf{K}}_{mz}|d_m - z'|} \cdot \mathbf{S}_{m-}(\rho') + e^{-i\overline{\mathbf{K}}_{mz}d_m} \cdot \mathbf{D}_m \qquad (6.6.20)$$

from (12). Hence, the field in all the regions n where $n > m$ can be found.

In region n where $n < m$, the field can be written as

$$\mathbf{E}_n(\rho, z) = \overline{\mathbf{E}}_n^t(\rho) \cdot \left[e^{-i\overline{\mathbf{K}}_{nz}(z-d_{n-1})} \cdot \widetilde{\overline{\mathbf{R}}}_{n,n-1} + e^{i\overline{\mathbf{K}}_{nz}(z-d_{n-1})} \right] \cdot \mathbf{B}_n, \qquad (6.6.21)$$

where $\widetilde{\overline{\mathbf{R}}}_{10} = 0$. Here, $\widetilde{\overline{\mathbf{R}}}_{n,n-1}$ can be found recursively using (15). Moreover, a recursive relation for $\mathbf{B}_n$ analogous to (11) can be derived:

$$\mathbf{B}_n = e^{i\overline{\mathbf{K}}_{nz}(d_{n-1}-d_n)} \cdot \overline{\mathbf{D}}_{n,+} \cdot \overline{\mathbf{T}}_{n+1,n} \cdot \mathbf{B}_{n+1}, \qquad (6.6.22)$$

where $\overline{\mathbf{D}}_{n,+}$ is given by (15a). The above lets one find $\mathbf{B}_n$ in terms of $\mathbf{B}_{n+1}$ recursively until $\mathbf{B}_m$ is reached where $\mathbf{B}_m$ is the amplitude of the upgoing wave in region m at $z = d_{m-1}$. Then, from (12),

$$\mathbf{B}_m = e^{i\overline{\mathbf{K}}_{mz}|d_{m-1}-z'|} \cdot \mathbf{S}_{m+}(\rho') + e^{i\overline{\mathbf{K}}_{mz}d_{m-1}} \cdot \mathbf{C}_m. \qquad (6.6.23)$$

In this manner, the field in all regions can be found when the source is embedded in region m. Note that in this algorithm of solving the inhomogeneous medium problem, the computational time grows linearly with N, where N is the number of regions. The eigenfunctions of each region can be found analytically or numerically. And if a numerical method is used where M basis functions are used to expand the field, it will yield M eigenfunctions for each region. But the computational effort needed to find the eigenvalues and eigenfunctions is proportional to M^3. Therefore, in total, the computational effort grows as NM^3 using this approach. For many application problems, like waveguide junctions, dielectric resonators, or geophysical prospecting problems, N need not be large. Hence, the mode matching approach can be much more efficient than the direct use of the finite-element method in solving such problems.

Exercises for Chapter 6

6.1 Find the conditions on the class of functions $\phi(z)$ and the operators $\mathcal{D}$ and $\mathcal{B}$ in (6.1.4) so that $\mathcal{D}$ and $\mathcal{B}$ are (a) Hermitian and (b) symmetric.

6.2 Given the operator equation (6.1.4), prove that two eigensolutions with different eigenvalues are $\mathcal{B}$ orthogonal if the operators are symmetric.

6.3 (a) By putting a source field $e^{ik_{1z}|z-z'|}/k_{1z}$ in region 1 of Figure 6.1.1, derive the expression for the total field in the manner of (2.4.6) of Section 2.4, Chapter 2.

   (b) Using the uniqueness theorem, show that such an expression is branch-point free in the manner of Section 2.7, Chapter 2.

   (c) Show that such an expression can be used to derive (6.1.16) which is branch-point free.

6.4 (a) Write the guidance condition (6.1.15) for the geometry (see figure).

   (b) Show that the function

$$f(k_s) = \frac{1 + \tilde{R}_{12}e^{2ik_{1z}d_1}}{\tilde{R}_{12} + e^{2ik_{1z}d_1}}$$

   is a single-valued function. Therefore, it is a meromorphic function with discrete zeroes and poles.

   (c) Show that the zeroes of $f(k_s)$ are the zeroes of the guidance condition derived in (a). This implies that the modes of geometry (see figure) exist only at discrete values of k_s^2. This proof can be generalized to an enclosed layered slab.

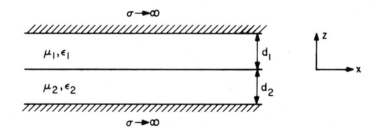

Figure for Exercise 6.4

6.5 For the geometry (see figure for Exercise 6.4):

   (a) Write down the guidance conditions for the TE and TM modes. Simplify where possible.

   (b) Solve for the locations of the modes on the complex k_x plane when d_1 and $d_2 \to \infty$. Sketch the locations approximately on the complex k_x plane.

(c) Repeat part (b) by introducing some loss into the medium. Again, sketch the locations of the modes approximately on the complex k_x plane.

6.6 For the geometry shown in Exercise 6.4,

    (a) Write the closed-form expressions for a TE and a TM eigensolution.

    (b) Normalize this solution such that (6.1.18) is satisfied.

    (c) Assuming $\mu_1 = \mu_2$, $\epsilon_1 = \epsilon_2$, explain what happens to this normalization when d_1 and $d_2 \to \infty$.

6.7 From Maxwell's equations, assuming that μ and ϵ are functions of z only, derive Equations (6.2.1) and (6.2.2).

6.8 (a) If $\mathbf{J} = \hat{z} I \ell \, \delta(y) \, \delta(z)$, i.e., a line source of dipoles pointing in the $\hat{z}$ direction, using the eigenfunction expansion method, derive the solution for D_z and B_z in a layered slab.

    (b) Now, using the Fourier transform method described in Chapter 2, derive a solution for the same source in terms of a Fourier inverse transform.

    (c) By deforming the Fourier inversion contour in part (b), show that the results in parts (a) and (b) are equivalent.

6.9 If $\overline{\mathbf{K}}$ is a diagonal matrix with elements $k_1, k_2, k_3, k_4, \ldots$, show that $e^{i\overline{\mathbf{K}}z}$ is a diagonal matrix with elements $e^{ik_1 z}, e^{ik_2 z}, e^{ik_3 z}, e^{ik_4 z}, \ldots$.

6.10 Derive Equations (6.2.20a) and (6.2.20b) rigorously from Maxwell's equations.

6.11 With the source now located to the right of a junction discontinuity as shown in the figure:

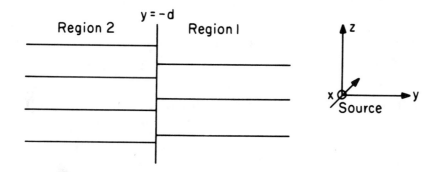

Figure for Exercise 6.11

(a) Find the reflection and transmission operators.

(b) Explain why the reflection and transmission operators are not identical to those given in (6.3.13) and (6.3.14).

6.12 For a complete, orthonormal basis set $\{S_i(z)\}$, show that $f(z) = \mathbf{S}^t(z) \cdot \langle \mathbf{S}(z), f(z) \rangle$, where $\mathbf{S}(z)$ is a vector containing $S_i(z)$ and $f(z)$ has the same support as $S_i(z)$. Therefore, derive the result in (6.3.23).

6.13 Derive the results in Equations (6.3.36) and (6.3.37).

6.14 Show that

(a) Equation (6.4.1) can be written as $\phi(z) = \mathbf{S}^t(z) \cdot \mathbf{a}$, where $\mathbf{S}(z)$ is a vector that contains $S_n(z)$ and $\mathbf{a}$ is a vector that contains a_n.

(b) Therefore, show that in Equation (6.4.7a),

$$\overline{\mathbf{L}} = \langle \mathbf{S}(z), \mathcal{L}(z)\mathbf{S}^t(z) \rangle, \quad \overline{\mathbf{p}}^I = \langle \mathbf{S}(z), p^{-1}(z)\mathbf{S}^t(z) \rangle,$$

where $\mathcal{L}(z) = \frac{d}{dz}p^{-1}\frac{d}{dz} + k^2(z)$.

(c) Show that $\overline{\mathbf{L}}$ and $\overline{\mathbf{p}}^I$ are symmetric matrices.

6.15 (a) Show that the eigenvectors of (6.4.7a) are $\overline{\mathbf{p}}^I$ orthogonal, i.e., prove Equation (6.4.9).

(b) From (a), deduce that the eigenfunctions $\tilde{\phi}(i, z)$ and $\tilde{\phi}(j, z)$ are $p^{-1}(z)$ orthogonal, i.e., prove Equation (6.4.10).

6.16 Derive the results of (6.4.14) and (6.4.15).

6.17 Derive (6.5.2a) which is valid for a metallic cylinder.

6.18 From the vector wave equation for inhomogeneous media, where $\epsilon(\mathbf{r}_s)$ and $\mu(\mathbf{r}_s)$ are not constants:

(a) Derive (6.5.7) and (6.5.8).

(b) When $\epsilon(\rho)$ and $\mu(\rho)$ are functions of ρ only, derive (6.5.11) and (6.5.12).

6.19 (a) Show that $\left(\overline{\mathbf{A}} + \lambda\overline{\mathbf{B}}\right) \cdot \mathbf{a} = 0$ and $\left(\overline{\mathbf{A}}^t + \lambda\overline{\mathbf{B}}^t\right) \cdot \mathbf{b} = 0$ share the same set of eigenvalues.

(b) Show that $\mathbf{b}_i^t \cdot \overline{\mathbf{B}} \cdot \mathbf{a}_j = 0$, $i \neq j$ for two different eigenvalues.

(c) Show that the eigensolutions of (6.5.9) and (6.5.10) are orthogonal in the manner of (6.5.5).

6.20 (a) Give expressions for the mn elements of the matrices $\overline{\mathbf{L}}$ and $\overline{\mathbf{R}}$ given in (6.5.20). Show that $\overline{\mathbf{L}}$ is asymmetrical while $\overline{\mathbf{R}}$ is symmetrical.

(b) Show that Equation (6.5.21) follows from (6.5.14), and it is just the transpose equation of (6.5.19). Hence, derive (6.5.22).

6.21 Derive the equivalence of (6.5.35) or (6.5.36) for a vertical electric dipole.

6.22 From Maxwell's equations, derive Equation (6.5.39), and hence, Equation (6.5.40).

6.23 Expand the term $\overline{\mathbf{D}}_{2-}$ in (6.6.5) into a geometric series. Use this in (6.6.8) to derive a series expansion for $\widetilde{\overline{\mathbf{R}}}_{12}$ and give a physical interpretation to the series.

6.24 Repeat the three-region problem in Subsection 6.6.1 for the case when each region is planarly layered rather than cylindrically layered.

6.25 (a) Derive Equations (6.6.16a) and (6.6.16b) from Equations (6.6.13) and (6.6.14).

(b) Derive Equations (6.6.18a) and (6.6.18b).

References for Chapter 6

Chew, W. C. 1985. "Response of a source on top of a vertically stratified half-space." *IEEE Trans. Antennas Propagat.* AP-33: 649–54.

Chew, W. C. 1988a. "Modeling of the high frequency dielectric logging tool: theory." *IEEE Trans. Geosci. Remote Sensing* GE-26(4): 382.

Chew, W. C. 1988b. "Modeling of the high frequency dielectric logging tool: applications and results." *IEEE Trans. Geosci. Remote Sensing* GE-26(4): 388.

Chew, W. C. 1989. "Analysis of optical and millimeter wave dielectric waveguide." *J. Electromagnetic Waves and Applications* 3(4): 359–77.

Chew, W. C., and B. Anderson. 1985. "Propagation of electromagnetic waves through geological beds in a geophysical probing environment." *Radio Sci.* 20: 611–21.

Chew, W. C., S. Barone, and C. H. Hennessy. 1983. "Diffraction of waves by discontinuities in open, cylindrical structures." *Proceedings, Int. Symp. IEEE Ant. Propag.* Houston, 503–06.

Chew, W. C., S. Barone, B. Anderson, and C. Hennessy. 1984. "Diffraction of axisymmetric waves in a borehole by bed boundary discontinuities." *Geophysics* 49: 1586.

Collin, R. E. 1960. *Field Theory of Guided Waves.* New York: McGraw-Hill.

Felsen, L. P., and N. Marcuvitz. 1973. *Radiation and Scattering of Waves.* Englewood Cliffs, NJ: Prentice Hall.

Hong, U. S., and R. H. Jansen. Nov. 1982. "Numerical analysis of shielded dielectric resonators including substrates, support disk, and tuning post." *Electron. Lett.* 18: 1000–02.

Hong, U. S., and R. H. Jansen. 1984. "Veranderung der Resonanz frequenzen dielektrischer Resonatoren in Mikrostripschaltungen durch Umgebungs-parameter." *Arch. Elek. Ubertragung* (AEU) 38: 106–12.

Liu, Q. 1989. *Some composite boundary value problems in electromagnetics and their applications.* Ph.D. thesis, Dept. of Elect. Comp. Eng., University of Illinois at Urbana-Champaign.

Marcuse, D. 1972. *Light Transmission Optics.* New York: Van Nostrand Reinhold.

Moghaddam, M., and W. C. Chew. 1988. "Response of a point source near the interface of multicylindrically layered half-spaces." *Intern. Union Radio Science Meeting Digest*, p. 174, June 1988, Syracuse.

Pai, D. M., and M. Huang. 1988. "A generalized Haskell matrix method for borehole electromagnetics: theory and applications." *Geophysics* 53(12): 1577–86.

Pudensi, M., and L. Ferreira. 1982. "Method to calculate the reflection and transmission of guided waves." *J. Opt. Soc. Am.* 72: 126–30.

Rozzi, T. 1978. "Rigorous analysis of the step discontinuities in planar dielectric waveguides." *IEEE Trans. Microwave Theory Tech.* MTT-26: 738–46.

Rozzi, T. E., and G. H. Veld. 1980. "Variational treatment of the diffraction at the facet of d.h. lasers and of dielectric millimeter wave antennas." *IEEE Trans. Microwave Theory Tech.* MTT-28: 61–73.

Stakgold, I. 1967. *Boundary Value Problems in Mathematical Physics*, vol. I. London: Macmillan.

Tsang, L., A. K. Chan, and S. Gianzero. 1984. "Solution of the fundamental problem in resistivity logging with a hybrid method." *Geophysics* 49 (10): 1596.

Zaki, K., and A. E. Atia. 1983. "Modes in dielectric-loaded waveguides and resonators." *IEEE Trans. Microwave Theory Tech.* MTT-31(12): 1039–45.

Zaki, K., and C. Chen. 1986. "New results in dielectric-loaded resonators." *IEEE Trans. Microwave Theory Tech.* MTT-34(7): 815–24.

Further Readings for Chapter 6

Anderson, B., and W. C. Chew. 1984. "A new high speed technique for calculating synthetic induction and DPT logs." *SPWLA Symp. Trans.*, paper HH.

Angulo, C. M., and W. S. C. Chang. 1953. "The excitation of a dielectric rod by a cylindrical waveguide." *IEEE Trans. Microwave Theory Tech.* MTT-6: 389–93.

Angulo, C. M., and W. S. C. Chang. 1959a. "The launching of surface waves by a parallel plate waveguide." *IEEE Trans. Antennas Propagat.* AP-7: 359–68.

Angulo, C. M., and W. S. C. Chang. 1959b. "A variational expression for the terminal admittance of a semi-infinite dielectric rod." *IRE Trans. Antennas Propagat.* AP-7: 207–12.

Aoki, K., and T. Miyazaki. 1982. "On junction of two semi-infinite dielectric guides." *Radio Sci.* 17: 11–19.

Bahar, E. 1970. "Propagation of radio waves over a nonuniform layered medium" *Radio Sci.* 5: 1069–76.

Bahar, E. 1973a. "Electromagnetic wave propagation in inhomogeneous multilayered structures of arbitrary thickness—generalized field transforms." *J. Math. Phys.* 14(8): 1024–29.

Bahar, E. 1973b. "Electromagnetic wave propagation in inhomogeneous multilayered structures of arbitrary thickness—full wave solutions." *J. Math. Phys.* 14(8): 1030–36.

Bahar, E. 1976. "Generalized characteristic functions for simultaneous linear differential equations with variable coefficients applied to propagation in inhomogeneous anisotropic media." *Can. J. Phys.* 54: 301–16.

Bahar, E. 1986. "Full wave solutions for electromagnetic scattering and depolarization in irregular stratified media." *Radio Sci.* 21: 543.

Bresler, A. D. 1959. "On the discontinuity problem at the input to an anisotropic waveguide." *IRE Trans. Antennas Propagat.* AP-7: 261–74.

Brooke, G. H., and M. M. Z. Kharadly. 1982. "Scattering by abrupt discontinuities on planar dielectric waveguides." *IEEE Trans. Microwave Theory Tech.* MTT-30: 760–70.

Budden, K. G. 1961. *The Waveguide Mode Theory of Wave Propagation.* London: Logos Press.

Chew, W. C., and L. Gurel. 1988. "Reflection and transmission operators for strips or disks embedded in homogeneous and layered media." *IEEE Trans. Microwave Theory Tech.* MTT-36(11): 1488–97.

Chew, W. C., and T. M. Habashy. 1986. "The use of vector transforms in solving some electromagnetic scattering problems." *IEEE Trans. Anten-*

nas Propagat. AP-34(7): 871.

Clarricoats, P. J. B., and A. B. Sharpe. 1972. "Modal matching applied to a discontinuity in a planar surface waveguide." *Electron. Lett.* 8: 28–29.

Crombach, U. 1981. "Analysis of single and coupled rectangular dielectric waveguides." *IEEE Trans. Microwave Theory Tech.* MTT-29: 870–74.

Friedman, B. 1956. *Principles and Techniques of Applied Mathematics.* New York: John Wiley & Sons.

Galejs, J. 1971. "VLF propagation across discontinuous daytime to nighttime transitions in anisotropic terrestrial waveguides." *IEEE Trans. Antennas Propagat.* AP-19(6): 756–62.

Gelin, P., M. Petenzi, and J. Citerne. 1979. "New rigorous analysis of the step discontinuity in a slab dielectric waveguide." *Electron. Lett.* 15: 355–65.

Gelin, P., M. Petenzi, and J. Citerne. 1981. "Rigorous analysis of the scattering of surface waves in an abruptly ended slab dielectric waveguide" *IEEE Trans. Microwave Theory Tech.* MTT-29: 107–14.

Gurel, L., and W. C. Chew. 1988. "Guidance and resonance conditions for strips or disks embedded in homogeneous and layered media." *IEEE Trans. Microwave Theory Tech.* MTT-36(11): 1498–1506.

Hockman, G. A., and A. B. Sharpe. 1972. "Dielectric waveguide discontinuities." *Electron. Lett.* 8: 230–31.

Ikegami, T. 1972. "Reflectivity of mode at facet and oscillation mode in double-heterostructure injection lasers." *IEEE J. Quantum Electron.* QE-6: 470–76.

Inoue, Y., and S. Horowitz. 1966. "Numerical solution of full-wave equation with mode coupling." *Radio Sci.* 2: 5–32.

Kay, A. F. 1959. "Scattering of a surface wave by a discontinuity in reactance." *IRE Trans. Antennas Propagat.* AP-7: 22–31.

Lewin, L. 1951. *Advanced Theory of Waveguides.* London: Illiffe and Sons.

Mahmoud, S., and J. Beal. 1975. "Scattering of surface waves at a dielectric discontinuity on a planar waveguide." *IEEE Trans. Microwave Theory Tech.* MTT-23: 193–98.

Marcuse, D. 1973. *Integrated Optics.* New York: IEEE Press.

Marcuvitz, N. 1951. *Waveguide Handbook, MIT Radiation Laboratory Series,* vol. 10. New York: McGraw-Hill.

Marcuvitz, N., and J. Schwinger. 1951. "On the representation of the electric and magnetic fields produced by currents and discontinuities in waveguides. I," *J. Appl. Phys.* 22: 806–19.

Mittra, R., and S. W. Lee. 1971. *Analytical Techniques in the Theory of*

Guided Waves. New York: Macmillan.

Mittra, R., Y. L. Hou, and V. Jamnejad. 1980. "Analysis of open dielectric waveguides using mode-matching technique and variational methods." *IEEE Trans. Microwave Theory Tech.* MTT-28: 36–43.

Morishita, K., S. E. Inagaki, and N. Kumagai. 1979. "Analysis of discontinuities in dielectric waveguides by means of the least squares boundary residual method." *IEEE Trans. Microwave Theory Tech.* MTT-27: 310–15.

Pappert, R. A., and J. A. Ferguson. 1986. "VLF/LF mode conversion model calculations for air to air transmissions in the earth-ionosphere waveguide." *Radio Sci.* 21: 551.

Peng, S. T., and T. Tamir. 1974. "Directional blazing of waves guided by asymmetrical dielectric gratings." *Opt. Comm.* 11: 405–09.

Reinhart, F. K., I. Hayashi, and M. Panish. 1971. "Mode reflectivity and waveguide properties of double-heterostructure injection lasers." *J. Appl. Phys.* 42: 4466–79.

Schlak, G. A., and J. R. Wait. 1967. "Electromagnetic wave propagation over a nonparallel stratified conducting medium." *Can. J. Phys.* 45: 3697–3720.

Schlak, G. A., and J. R. Wait. 1968. "Attenuation function for propagation over a nonparallel stratified ground." *Can. J. Phys.* 46: 1135–36.

Schlosser, W., and H. G. Unger. 1966. "Partially filled waveguides and surface waveguides of rectangular cross-section." *Advances in Microwaves.* New York: Academic Press.

Tamir, T., and A. A. Oliner. 1963. "Guided complex waves. I. Fields at an interface. II. Relation to radiation patterns." *Proc. IEE (London)* 110: 310–34.

Wait, J. R. 1968a. "Mode conversion and refraction effects in the earth-ionosphere waveguide for VLF radio waves." *J. Geophy. Res.* 73(11): 3537–48.

Wait, J. R. 1968b. "On the theory of VLF propagation for a step model of the nonuniform earth-ionosphere waveguide." *Can. J. Phys.* 46(17): 1979–83.

Wait, J. R. 1970. "Factorization method applied to electromagnetic wave propagation in a curved waveguide with nonuniform walls." *Radio Sci.* 5(7): 1059–68.

CHAPTER 7

DYADIC GREEN'S FUNCTIONS

A dyadic Green's function is a dyad[1] that relates a vector field to a vector current source. It was introduced briefly in Chapter 1. In this chapter, we will study it in greater detail. The use of dyadic Green's functions makes the formulations and the solutions of some electromagnetic problems more compact. Even though most problems can be solved without using dyadic Green's functions, the symbolic simplicity they offer makes their use attractive. Only in a few simple geometries can the dyadic Green's functions be solved in closed form. In many instances, the dyadic Green's functions are obtained in terms of the vector wave functions of the geometry considered.

We shall first derive the various representations of the dyadic Green's function for a homogeneous medium. Later, the dyadic Green's functions for planarly, cylindrically, and spherically layered media will be derived. Only the electric-type dyadic Green's function that relates the electric field to the current density will be discussed, since the magnetic-type dyadic Green's function can be found by invoking duality. Once the electric field is obtained, the magnetic field is derivable by taking the curl of the electric field, and vice versa. Much has been written on the topic of dyadic Green's functions, and some of these works are found in the references for this chapter. Yaghjian (1980) presents a historical review of the works that predate van Bladel's work (1961).

§7.1 Dyadic Green's Function in a Homogeneous Medium

The dyadic Green's function in a homogeneous, isotropic medium can be represented in many forms. First, it can be represented in terms of simple algebraic functions which are the derivatives of the scalar Green's function (Exercise 7.1),

$$g(\mathbf{r}, \mathbf{r}') = \frac{e^{ik_0|\mathbf{r}-\mathbf{r}'|}}{4\pi|\mathbf{r}-\mathbf{r}'|}, \tag{7.1.1}$$

as shown in Chapter 1, where $k_0 = \omega\sqrt{\mu\epsilon}$. This is the spatial representation of the dyadic Green's function. Furthermore, it can be represented in terms of the eigenfunctions or the vector wave functions of the geometry. This is then the eigenfunction representation of the dyadic Green's function. The spatial representation of the dyadic Green's function shall be examined first.

---

[1] See Appendix B for a review of dyads.

§§7.1.1 The Spatial Representation

The dyadic Green's function relates the electric field $\mathbf{E}(\mathbf{r})$ to the current density $\mathbf{J}(\mathbf{r})$ via

$$\mathbf{E}(\mathbf{r}) = i\omega\mu \int_V \overline{\mathbf{G}}(\mathbf{r}, \mathbf{r}') \cdot \mathbf{J}(\mathbf{r}') \, d\mathbf{r}'. \tag{7.1.2}$$

Since $\mathbf{E}(\mathbf{r})$ is a solution of

$$\nabla \times \nabla \times \mathbf{E}(\mathbf{r}) - k_0^2 \mathbf{E}(\mathbf{r}) = i\omega\mu \, \mathbf{J}(\mathbf{r}), \tag{7.1.3}$$

using (2) in (3), we have

$$\nabla \times \nabla \times \overline{\mathbf{G}}(\mathbf{r}, \mathbf{r}') - k_0^2 \overline{\mathbf{G}}(\mathbf{r}, \mathbf{r}') = \overline{\mathbf{I}} \, \delta(\mathbf{r} - \mathbf{r}'). \tag{7.1.4}$$

One way to obtain $\overline{\mathbf{G}}(\mathbf{r}, \mathbf{r}')$ is described in Chapter 1. Yet another way to obtain $\overline{\mathbf{G}}(\mathbf{r}, \mathbf{r}')$ is via an indirect means using scalar and vector potentials.

The electric field $\mathbf{E}(\mathbf{r})$ is a solution of Maxwell's equations in the presence of a source $\mathbf{J}(\mathbf{r})$, i.e.,

$$\nabla \times \mathbf{E}(\mathbf{r}) = i\omega\mu \, \mathbf{H}(\mathbf{r}), \tag{7.1.5a}$$

$$\nabla \times \mathbf{H}(\mathbf{r}) = -i\omega\epsilon \, \mathbf{E}(\mathbf{r}) + \mathbf{J}(\mathbf{r}). \tag{7.1.5b}$$

Since $\nabla \cdot \mu \mathbf{H} = 0$, $\mu \mathbf{H}$ must be of the form $\mu \mathbf{H} = \nabla \times \mathbf{A}$, where $\mathbf{A}$ is a vector potential. This ensures the zero divergence of $\mu \mathbf{H}$. Consequently, (5a) becomes

$$\nabla \times \mathbf{E}(\mathbf{r}) = i\omega\nabla \times \mathbf{A}(\mathbf{r}). \tag{7.1.6}$$

Since $\nabla \times \nabla\phi = 0$, the above implies that

$$\mathbf{E}(\mathbf{r}) = i\omega\mathbf{A}(\mathbf{r}) - \nabla\phi(\mathbf{r}), \tag{7.1.7}$$

where $\phi(\mathbf{r})$ is a scalar potential. Also, (5b) becomes

$$\nabla \times \nabla \times \mathbf{A}(\mathbf{r}) = \omega^2\mu\epsilon \, \mathbf{A}(\mathbf{r}) + i\omega\mu\epsilon\nabla\phi(\mathbf{r}) + \mu \, \mathbf{J}(\mathbf{r}), \tag{7.1.8}$$

which, after using the identity $\nabla \times \nabla \times \mathbf{A} = \nabla(\nabla \cdot \mathbf{A}) - \nabla^2\mathbf{A}$, is the same as

$$\nabla^2\mathbf{A}(\mathbf{r}) + k_0^2\mathbf{A}(\mathbf{r}) - \nabla[\nabla \cdot \mathbf{A}(\mathbf{r})] = -i\omega\mu\epsilon\nabla\phi(\mathbf{r}) - \mu \, \mathbf{J}(\mathbf{r}), \tag{7.1.9}$$

where $k_0^2 = \omega^2\mu\epsilon$. In the above, $\mathbf{A}(\mathbf{r})$ is nonunique because we can define a new $\mathbf{A}' = \mathbf{A} - \nabla\psi$, that will yield the same $\mathbf{H}$ field. To uniquely define $\mathbf{A}$, we need to specify $\nabla \cdot \mathbf{A}$ in addition to specifying $\nabla \times \mathbf{A} = \mu\mathbf{H}$. From (9), it is apparent that letting

$$\nabla \cdot \mathbf{A} = +i\omega\mu\epsilon\phi(\mathbf{r}) \tag{7.1.10}$$

greatly simplifies (9), reducing it to

$$(\nabla^2 + k_0^2) \, \mathbf{A}(\mathbf{r}) = -\mu \, \mathbf{J}(\mathbf{r}). \tag{7.1.11}$$

The method of pinning the value of $\mathbf{A}$ by Equation (10) is known as the *Lorentz gauge*.[2] Since $\nabla \cdot \mathbf{E} = \varrho/\epsilon$ in a homogeneous medium, taking the divergence of (7) and using the *Lorentz gauge* yields[3]

$$(\nabla^2 + k_0^2)\phi(\mathbf{r}) = -\varrho(\mathbf{r})/\epsilon. \qquad (7.1.12)$$

Equation (11) consists of three scalar wave equations. Hence, using the scalar Green's function (1), its solution becomes

$$\mathbf{A}(\mathbf{r}) = \mu \int_V d\mathbf{r}' \, g(\mathbf{r}, \mathbf{r}') \, \mathbf{J}(\mathbf{r}'). \qquad (7.1.13)$$

Similarly, the solution to (12) is

$$\phi(\mathbf{r}) = \frac{1}{\epsilon} \int_V d\mathbf{r}' g(\mathbf{r}, \mathbf{r}') \, \varrho(\mathbf{r}'). \qquad (7.1.14)$$

Consequently, the electric field given by (7) is

$$\mathbf{E}(\mathbf{r}) = i\omega\mu \int_V d\mathbf{r}' \, g(\mathbf{r}, \mathbf{r}') \, \mathbf{J}(\mathbf{r}') - \frac{\nabla}{\epsilon} \int_V d\mathbf{r}' \, g(\mathbf{r}, \mathbf{r}') \, \varrho(\mathbf{r}'). \qquad (7.1.15)$$

From the continuity equation that $\nabla \cdot \mathbf{J}(\mathbf{r}) = i\omega\varrho(\mathbf{r})$, a relationship between the electric field $\mathbf{E}(\mathbf{r})$ and current source $\mathbf{J}(\mathbf{r})$ is

$$\mathbf{E}(\mathbf{r}) = i\omega\mu \int_V d\mathbf{r}' \, g(\mathbf{r}, \mathbf{r}) \, \mathbf{J}(\mathbf{r}') - \frac{\nabla}{i\omega\epsilon} \int_V d\mathbf{r}' \, g(\mathbf{r}, \mathbf{r}') \nabla' \cdot \mathbf{J}(\mathbf{r}'). \qquad (7.1.16)$$

Using integration by parts, and the fact that $\nabla g(\mathbf{r}, \mathbf{r}') = -\nabla' g(\mathbf{r}, \mathbf{r}')$ and that $\mathbf{J}(\mathbf{r})$ is only supported in V, (16) becomes

$$\mathbf{E}(\mathbf{r}) = i\omega\mu \int_V d\mathbf{r}' \, g(\mathbf{r}, \mathbf{r}') \, \mathbf{J}(\mathbf{r}') - \frac{\nabla\nabla\cdot}{i\omega\epsilon} \int_V d\mathbf{r}' \, g(\mathbf{r}, \mathbf{r}') \, \mathbf{J}(\mathbf{r}'). \qquad (7.1.17)$$

It seems natural at this point to write

$$\mathbf{E}(\mathbf{r}) = i\omega\mu \int_V d\mathbf{r}' \, \overline{\mathbf{G}}(\mathbf{r}, \mathbf{r}') \cdot \mathbf{J}(\mathbf{r}'), \qquad (7.1.18)$$

---

[2] Other ways of pinning the value of $\mathbf{A}$ are possible; e.g., $\nabla \cdot \mathbf{A} = 0$ is known as the Coulomb gauge (Exercise 7.2).

[3] These manipulations are found in standard books on electromagnetics available in the references for Chapter 1.

where

$$\overline{\mathbf{G}}(\mathbf{r},\mathbf{r}') = \left[\overline{\mathbf{I}} + \frac{\nabla\nabla}{k_0^2}\right] g(\mathbf{r},\mathbf{r}'). \tag{7.1.19}$$

But if $\mathbf{r}$, the point where the field $\mathbf{E}(\mathbf{r})$ is observed, is in the source region V, $|\mathbf{r} - \mathbf{r}'|$ could be zero. When $|\mathbf{r} - \mathbf{r}'| \to 0$, $g(\mathbf{r},\mathbf{r}')$ becomes singular, and the term $\nabla\nabla$, operating on $g(\mathbf{r},\mathbf{r}')$, gives rise to terms $\sim O(1/|\mathbf{r} - \mathbf{r}'|^3)$. Hence, the resultant integral is nonuniformly convergent.

§§7.1.2 The Singularity of the Dyadic Green's Function

When the vector and scalar potentials are calculated from the source $\mathbf{J}(\mathbf{r})$ and $\varrho(\mathbf{r})$ as in (13) and (14), the integrands of the integrals involve singularities due to $g(\mathbf{r},\mathbf{r}')$. Therefore, to be strictly correct, when $\mathbf{r} \in V$, (14) should be written as

$$\phi(\mathbf{r}) = \frac{1}{\epsilon} \lim_{V_\delta \to 0} \int_{V - V_\delta} d\mathbf{r}' \, g(\mathbf{r},\mathbf{r}') \, \varrho(\mathbf{r}'), \tag{7.1.20}$$

where V_δ is an **exclusion volume** surrounding $\mathbf{r}$. In this way, the integral with the singularity is properly defined. An integral defined this way to circumvent integrating over the singularity in the integrand is known as an improper integral in the classical function theory (Kellogg 1953; Fikioris 1965). The improper integral converges if the integral tends to a fixed limit regardless of the shape of V_δ. The condition for (20) to be convergent is for $\varrho(\mathbf{r})$ to be piecewise continuous. Hence, if $\mathbf{J}(\mathbf{r})$ is continuous, then the improper integrals in (13) and the first terms of (16) and (17) are convergent. However, this is not true for the second terms because derivatives are involved as we shall show.

In order for the second terms of (16) and (17) to converge as an improper integral, $\mathbf{J}(\mathbf{r})$ has to satisfy the Holder's condition (Kellogg 1953) at $\mathbf{r}$, i.e., there exist positive constants c, A, and α such that $|\mathbf{J}(\mathbf{r}) - \mathbf{J}(\mathbf{r}')| \le A|\mathbf{r}' - \mathbf{r}|^\alpha$ for $|\mathbf{r}' - \mathbf{r}| < c$. This condition is slightly stronger than the continuity condition (Exercise 7.3). The improper integral in (18), however, does not converge in the classical sense because of the invalid exchange of the integral and the differential operators. In particular,

$$\lim_{V_\delta \to 0} \int_{V - V_\delta} d\mathbf{r}' \, \overline{\mathbf{G}}(\mathbf{r},\mathbf{r}') \cdot \mathbf{J}(\mathbf{r}') \tag{7.1.21}$$

converges but to a value dependent on the shape of V_δ. Hence, the principal value of this integral exists, and its value depends on the shape of V_δ. Therefore, Equation (18), as it stands, is nonunique and undefined in a classical sense, if $\mathbf{r} \in V$. The nonconvergence comes from the $\nabla\nabla$ part of the integrand because it introduces an $O(1/|\mathbf{r} - \mathbf{r}'|^3)$ term. Since the field generated by a current source has to be unique, there is a way of uniquely defining this integral to relate the electric field to the current source.

To do this, we write the second term in (17) as

$$
\nabla\nabla \cdot \int_V d\mathbf{r}'\, g(\mathbf{r},\mathbf{r}')\, \mathbf{J}(\mathbf{r}') = \lim_{V_\delta \to 0} \left[\nabla\nabla \cdot \int_{V-V_\delta} d\mathbf{r}'\, g(\mathbf{r},\mathbf{r}')\, \mathbf{J}(\mathbf{r}') \right.
$$

$$
\left. + \nabla\nabla \cdot \int_{V_\delta} d\mathbf{r}'\, g(\mathbf{r},\mathbf{r}')\, \mathbf{J}(\mathbf{r}') \right]
$$

$$
= \lim_{V_\delta \to 0} \left[\int_{V-V_\delta} d\mathbf{r}'\, \nabla\nabla g(\mathbf{r},\mathbf{r}') \cdot \mathbf{J}(\mathbf{r}') \right.
$$

$$
\left. - \nabla \int_{V_\delta} d\mathbf{r}'\, \nabla' g(\mathbf{r},\mathbf{r}') \cdot \mathbf{J}(\mathbf{r}') \right], \tag{7.1.22}
$$

when $\mathbf{r} \in V_\delta \in V$. Since the first integral is excluded from the singularity at $\mathbf{r}$, we can bring the $\nabla\nabla$ operator into the integrand. The second integral contains $\mathbf{r}$, and it converges only if one ∇ operator is brought into the integrand. Hence, when $V_\delta \to 0$, both the first and the second integrals on the right-hand side of (22) converge, but to values dependent on the shape of V_δ. Nevertheless, the sum of the two integrals equals the left-hand side, which should be independent of V_δ. More specifically, using integration by parts, the second integral reduces to

$$
\int_{V_\delta} d\mathbf{r}'\, \nabla' g(\mathbf{r},\mathbf{r}') \cdot \mathbf{J}(\mathbf{r}') = \int_{S_\delta} dS'\, \hat{n} \cdot \mathbf{J}(\mathbf{r}') g(\mathbf{r},\mathbf{r}') - i\omega \int_{V_\delta} d\mathbf{r}'\, g(\mathbf{r},\mathbf{r}')\, \varrho(\mathbf{r}'), \tag{7.1.23}
$$

where we have made use of $\nabla \cdot \mathbf{J} = i\omega\varrho$. For a volume current, $\varrho(\mathbf{r})$ is continuous; hence, the second integral in (23) vanishes when $V_\delta \to 0$. In the first integral, $\hat{n} \cdot \mathbf{J}(\mathbf{r}')$ is the surface charge on S_δ (the surface surrounding V_δ). Hence, the first integral is proportional to the field at $\mathbf{r}$ due to a surface charge on S_δ, which is scale invariant[4]; i.e., it does not vanish when $V_\delta \to 0$, but it very much depends on the shape of V_δ. In the limit when $V_\delta \to 0$, (23) should be linearly proportional to $\mathbf{J}(\mathbf{r})$. Therefore, (22) can be written as

$$
\nabla\nabla \cdot \int_V d\mathbf{r}'\, g(\mathbf{r},\mathbf{r}')\, \mathbf{J}(\mathbf{r}') = \lim_{V_\delta \to 0} \int_{V-V_\delta} d\mathbf{r}'\, \nabla\nabla g(\mathbf{r},\mathbf{r}') \cdot \mathbf{J}(\mathbf{r}') - \overline{\mathbf{L}} \cdot \mathbf{J}(\mathbf{r}), \tag{7.1.24}
$$

where $\overline{\mathbf{L}}$ is a dyad dependent on the shape of V_δ. Consequently, (18) becomes

$$
\mathbf{E}(\mathbf{r}) = i\omega\mu \lim_{V_\delta \to 0} \int_{V-V_\delta} d\mathbf{r}'\, \overline{\mathbf{G}}(\mathbf{r},\mathbf{r}') \cdot \mathbf{J}(\mathbf{r}') + \frac{\overline{\mathbf{L}} \cdot \mathbf{J}(\mathbf{r})}{i\omega\epsilon}, \tag{7.1.25}
$$

---

[4] This follows from the concept that the solutions to Laplace's equation only depend on the shape of the geometry but not on the size of the geometry because Laplace's equation is scale invariant.

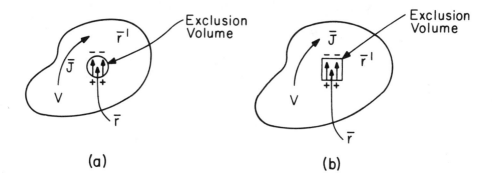

Figure 7.1.1 A cavity in a current source region. The discontinuity in the current generates a local field inside the cavity.

or in short,

$$\mathbf{E}(\mathbf{r}) = i\omega\mu P.V. \int_V d\mathbf{r}' \, \overline{\mathbf{G}}(\mathbf{r},\mathbf{r}') \cdot \mathbf{J}(\mathbf{r}') + \frac{\overline{\mathbf{L}} \cdot \mathbf{J}(\mathbf{r})}{i\omega\epsilon}, \qquad (7.1.26)$$

when $\mathbf{r} \in V$, where

$$P.V. \int_V$$

stands for a shape dependent principal value integral. In this manner, $\mathbf{E}(\mathbf{r})$ is unique, but the two terms in (26) are both dependent on the shape of V_δ, the exclusion volume or the principal volume. The above method of treating the singular nature of electric dyadic Green's function is known as the **principal volume** method (van Bladel 1961).

Physically, the first integral in (26) corresponds to putting the observation point $\mathbf{r}$ inside a cavity excavated in the current source region V. Since the current is discontinuous on the surface of this cavity, charges accumulate on the surface of the cavity (see Figure 7.1.1). When the size of the cavity is very small, the field due to these charges is essentially electrostatic in nature in the cavity. Since the electrostatic field satisfies Laplace's equation which is scale invariant, this field persists even in the limit when the exclusion volume tends to zero. This electrostatic field is also a function of the shape of the cavity, irrespective of how small it is.

These charges give rise to a field that should not have been there since the exclusion volume is absent in the first place. Therefore, to obtain a correct answer, the expression (26) is augmented by a second term which removes the effect of the surface charges around the exclusion volume. The second term could also be thought of as due to a current element in the shape of the exclusion volume containing the observation point $\mathbf{r}$.

As a consequence of the preceding discussion, a symbolic way to write the dyadic Green's function which is valid for all observation points $\mathbf{r}$ is

$$\overline{\mathbf{G}}(\mathbf{r}, \mathbf{r}') = P.V.\overline{\mathbf{G}}(\mathbf{r}, \mathbf{r}') - \frac{\overline{\mathbf{L}}\,\delta(\mathbf{r} - \mathbf{r}')}{k_0^2}. \tag{7.1.27}$$

When this dyadic Green's function operates on a current, the first term yields a principal value integral to be integrated over a specific exclusion volume, and the second term depends on the shape of the exclusion volume chosen for the first term. In this way, the dyadic Green's function is uniquely defined.

In light of the above discussions, even though the representation (18) is not valid in the classical function sense, it is valid in the distributional function sense if $\overline{\mathbf{G}}(\mathbf{r}, \mathbf{r}')$ is regarded as a generalized function. A generalized function is a function that does not have a meaning by itself, but whose properties are defined by its action on other functions, i.e., by the values of the integrals when it is multiplied by other functions and integrated over space.[5]

Hence, the right-hand side of (27) can be considered as generalized functions. A volume integral over $P.V.\overline{\mathbf{G}}(\mathbf{r}, \mathbf{r}')$ acting on a current $\mathbf{J}(\mathbf{r}')$ implies that a principal volume integral has to be invoked, and the result is a function of the shape of the exclusion volume in the principal volume integral. The second term is just the well-known generalized function, which is the singular distribution of the Dirac delta function.

The values of $\overline{\mathbf{L}}$ in (27) for various exclusion volumes shown in Figure 7.1.2 are (Exercise 7.4; see Yaghjian 1980; Lee et al. 1980)

$$\overline{\mathbf{L}} = \frac{\overline{\mathbf{I}}}{3}, \quad \text{for spheres or cubes,} \tag{7.1.28a}$$

$$\overline{\mathbf{L}} = \hat{z}\hat{z}, \quad \text{for disks,} \tag{7.1.28b}$$

$$\overline{\mathbf{L}} = \frac{\hat{x}\hat{x} + \hat{y}\hat{y}}{2}, \quad \text{for needles,} \tag{7.1.28c}$$

$$\overline{\mathbf{L}} = L_1\hat{x}\hat{x} + L_2\hat{y}\hat{y} + L_3\hat{z}\hat{z}, \quad \text{for upright ellipsoid.} \tag{7.1.28d}$$

$\overline{\mathbf{L}}$ is also related to the depolarization factors discussed by Stratton (1941). In general, the trace $[\overline{\mathbf{L}}] = 1$ (Yaghjian 1980).

§§7.1.3 The Spectral Representation

Sometimes, it is preferable to express the dyadic Green's function in terms of a Fourier transform or in terms of vector wave functions (Stratton 1941).

---

[5] For a review of generalized functions, see Lighthill (1958), Stakgold (1967), and Appendix C.

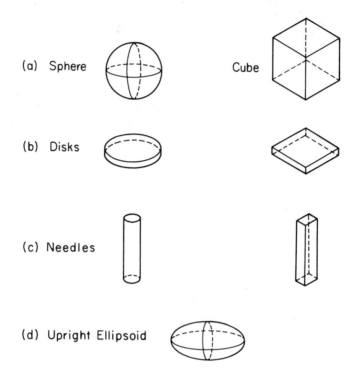

Figure 7.1.2 Various types of exclusion volumes for the principal value integral.

When the dyadic Green's function is expressed as a Fourier integral, it is in its *spectral representation*. When the dyadic Green's function is expressed in terms of the vector wave functions, it is in its *vector-wave-function representation*. Since the Fourier transform method is simpler than the vector-wave-function method, and its manipulations more transparent, it will be presented first. The vector-wave-function method will be presented in the next section.

With the understanding that the Fourier space can be expanded to accommodate generalized functions described in (27), the dyadic Green's function in its spectral representation can also be derived by first Fourier transforming Equation (4) in the **r** variable giving

$$-\mathbf{k} \times \mathbf{k} \times \widetilde{\overline{\mathbf{G}}}(\mathbf{k}, \mathbf{r}') - k_0^2 \widetilde{\overline{\mathbf{G}}}(\mathbf{k}, \mathbf{r}') = \overline{\mathbf{I}}\, e^{-i\mathbf{k}\cdot\mathbf{r}'}, \qquad (7.1.29)$$

where $\mathbf{k} = \hat{x}k_x + \hat{y}k_y + \hat{z}k_z$. The above can be inverted to yield (Exercise 7.5;

also see Tsang et al. 1975)

$$\tilde{\overline{\mathbf{G}}}(\mathbf{k}, \mathbf{r}') = \frac{\overline{\mathbf{I}} k_0^2 - \mathbf{kk}}{k_0^2 (k^2 - k_0^2)} e^{-i\mathbf{k} \cdot \mathbf{r}'}, \tag{7.1.30}$$

where $k^2 = k_x^2 + k_y^2 + k_z^2$. Equation (30) can be shown to be a solution of (29) by backsubstituting (30) into (29).[6] Consequently, we can write

$$\overline{\mathbf{G}}(\mathbf{r}, \mathbf{r}') = \frac{1}{(2\pi)^3} \iiint\limits_{-\infty}^{\infty} d\mathbf{k} \, e^{i\mathbf{k} \cdot (\mathbf{r} - \mathbf{r}')} \frac{\overline{\mathbf{I}} k_0^2 - \mathbf{kk}}{k_0^2 (k^2 - k_0^2)}, \tag{7.1.31}$$

where $d\mathbf{k} = dk_x dk_y dk_z$.

The above Fourier representation of $\overline{\mathbf{G}}(\mathbf{r}, \mathbf{r}')$ is a nonconvergent integral in the classical sense because the magnitude of the integrand tends to a constant when $|\mathbf{k}| \to \infty$. This Fourier representation, however, includes generalized functions or distributions such as the Dirac delta function. Such Fourier representations are usually nonconvergent integrals as in (31). Hence, in deriving (31), the principal volume integral need not be specified as in (25) or (26), since the Fourier space representation of $\overline{\mathbf{G}}(\mathbf{r}, \mathbf{r}')$ in (31) can accommodate the generalized functions defined in (27).

To extract the delta function singularity from Equation (31), note that the integrand tends to a constant when $|\mathbf{k}| \to \infty$. For example, if $k_z \to \infty$ while k_x and k_y are finite, then

$$\frac{\overline{\mathbf{I}} k_0^2 - \mathbf{kk}}{k^2 - k_0^2} \sim -\hat{z}\hat{z}, \qquad k_z \to \infty. \tag{7.1.32}$$

Hence, contour integration techniques used in Chapter 2 cannot be used to evaluate (31) directly, as Jordan's lemma is violated. To remedy this, (31) is rewritten as

$$\overline{\mathbf{G}}(\mathbf{r}, \mathbf{r}') = \frac{1}{(2\pi)^3} \iiint\limits_{-\infty}^{\infty} d\mathbf{k} \, e^{i\mathbf{k} \cdot (\mathbf{r} - \mathbf{r}')} \left[\frac{\overline{\mathbf{I}} k_0^2 - \mathbf{kk}}{k_0^2 (k^2 - k_0^2)} + \frac{\hat{z}\hat{z}}{k_0^2} \right]$$
$$- \frac{\hat{z}\hat{z}}{(2\pi)^3 k_0^2} \iiint\limits_{-\infty}^{\infty} d\mathbf{k} \, e^{i\mathbf{k} \cdot (\mathbf{r} - \mathbf{r}')}. \tag{7.1.33}$$

The first integrand now vanishes when $k_z \to \infty$, while k_x and k_y are held finite. By virtue of Jordan's lemma and Cauchy's theorem, the dk_z integral can be evaluated first using residue calculus for $(z - z') > 0$, by picking up the

---

[6] Note that in (30), the $1/r^3$ singularity in (19) is mapped to the high-frequency spectral component in (30), causing $\tilde{\overline{\mathbf{G}}}(k, \mathbf{r}') \sim$ const. as $|\mathbf{k}| \to \infty$ (Exercise 7.6; also see Doetsch 1974).

residue contributions at the pole locations given by $k_{0z} = \pm(k_0^2 - k_x^2 - k_y^2)^{\frac{1}{2}}$, as was done in Chapter 2.

The second integral in (33) is just the Fourier representation of the distribution $\delta(\mathbf{r} - \mathbf{r}')$. Consequently, we arrive at

$$\overline{\mathbf{G}}(\mathbf{r}, \mathbf{r}') = \frac{i}{8\pi^2} \iint\limits_{-\infty}^{\infty} d\mathbf{k}_s \, e^{i\mathbf{k}_s \cdot (\mathbf{r}_s - \mathbf{r}_s') + ik_{0z}|z - z'|} \left[\frac{\overline{\mathbf{I}}k_0^2 - \mathbf{k}_0\mathbf{k}_0}{k_0^2 k_{0z}} \right] - \frac{\hat{z}\hat{z}}{k_0^2} \delta(\mathbf{r} - \mathbf{r}'), \tag{7.1.34}$$

where $\mathbf{k}_s = \hat{x}k_x + \hat{y}k_y$, $d\mathbf{k}_s = dk_x dk_y$, $k_{0z} = \sqrt{k_0^2 - k_x^2 - k_y^2}$, $\mathbf{r}_s = \hat{x}x + \hat{y}y$, and

$$\mathbf{k}_0 = \begin{cases} \hat{x}k_x + \hat{y}k_y + \hat{z}k_{0z}, & z - z' > 0, \\ \hat{x}k_x + \hat{y}k_y - \hat{z}k_{0z}, & z - z' < 0. \end{cases} \tag{7.1.34a}$$

It is understood that the integration path is chosen such that $\Im m[k_{0z}] > 0$ in order to satisfy the radiation condition. It should be noted that when $|z - z'| = 0$, the above integral is not uniformly convergent (Exercise 7.7). Furthermore, it should be emphasized that when (31) or (34) is applied to a current source, they are to be used in the manner of Equation (18) since they are distributions.

Note that a delta function singularity has surfaced in Equation (34). This may be surprising since an exclusion volume is not specified. Furthermore, Equation (31) is symmetrical about k_x, k_y, and k_z. If the dk_x or the dk_y integral is evaluated first, one would obtain an $\hat{x}\hat{x}$ or a $\hat{y}\hat{y}$ dyad for the singular term respectively. Therefore, it seems that the singularity in the last term of Equation (34) is nonunique. However, if the dk_x or the dk_y integral is performed first, the first term in (34) would also have a different form. Hence, a combination of the first term and the second term in (34) still yields a unique value. Similar representations of the dyadic Green's function have also been presented by Johnson et al. (1979), Kerns (1981), and Kong (1986).

§§7.1.4 *Equivalence of Spectral and Spatial Representations*[7]

The reason for the apparent nonuniqueness of $\overline{\mathbf{G}}(\mathbf{r}, \mathbf{r}')$ in its spectral representation and its many forms of singularities is intimately related to the nonuniqueness of the principal volume integral and the corresponding Dirac delta function term in the spatial representation in (27). In the principal volume integral in (26), for example, the integral approaches different values depending on the shape of the exclusion volume that is approaching zero. Since a singularity has high spectral components, this singularity in the spatial domain is mapped to infinity in the spectral domain. In Equation (30), note that $\widetilde{\overline{\mathbf{G}}}(\mathbf{k}, \mathbf{r}')$ tends to different values depending on how one approaches infinity in the $\mathbf{k}$ space, very much the way $\overline{\mathbf{G}}(\mathbf{r}, \mathbf{r}')$ approaches infinity depending on how $|\mathbf{r} - \mathbf{r}'| \to 0$.

---

[7] Also, see Chew 1989.

To see this more clearly, let us assume a current distribution in the shape of a cuboid of the form

$$\mathbf{J}(\mathbf{r}) = \hat{\alpha} I \ell \frac{B(\frac{x-x_0}{a})B(\frac{y-y_0}{b})B(\frac{z-z_0}{c})}{abc}, \tag{7.1.35}$$

where $\hat{\alpha}$ is an arbitrary unit vector, and

$$B(s) = \begin{cases} 1, & |s| \leq \frac{1}{2}, \\ 0, & |s| > \frac{1}{2}. \end{cases} \tag{7.1.35a}$$

The electric field due to this current, using Equation (31) in Equation (18) and after performing the spatial integration, is

$$\mathbf{E}(\mathbf{r}) = \frac{i\omega\mu_0 I \ell}{(2\pi)^3} \int\!\!\!\int\!\!\!\int\limits_{-\infty}^{\infty} d\mathbf{k}\, e^{i\mathbf{k}\cdot(\mathbf{r}-\mathbf{r}_0)} \frac{\overline{\mathbf{I}}k_0^2 - \mathbf{k}\mathbf{k}}{k_0^2(k^2 - k_0^2)} \cdot \tilde{\mathbf{J}}(\mathbf{k}), \tag{7.1.36}$$

where $\tilde{\mathbf{J}}(\mathbf{k})$ is related to the Fourier transform of $\mathbf{J}(\mathbf{r})$ and is given by

$$\tilde{\mathbf{J}}(\mathbf{k}) = \hat{\alpha} I \ell \frac{2}{k_x a} \sin\left(\frac{k_x a}{2}\right) \frac{2}{k_y b} \sin\left(\frac{k_y b}{2}\right) \frac{2}{k_z c} \sin\left(\frac{k_z c}{2}\right). \tag{7.1.37}$$

Since $\tilde{\mathbf{J}}(\mathbf{k}) \to 0$ when $\mathbf{k} \to \infty$, the integral in (36) is well defined. We may consider Equation (36) as consisting of an integral summation of the field due to distributed current sources. If a, b, and $c \to 0$, then

$$\mathbf{J}(\mathbf{r}) = \hat{\alpha} I \ell\, \delta(x - x_0)\, \delta(y - y_0)\, \delta(z - z_0) = \hat{\alpha} I \ell\, \delta(\mathbf{r} - \mathbf{r}_0), \tag{7.1.38}$$

which is a point source whose Fourier transform is $\hat{\alpha} I \ell e^{-i\mathbf{k}\cdot\mathbf{r}_0}$. The integral (36) can be manipulated to different representations, however, depending on the order in which a, b, and c tend to zero. For instance, if $c \to 0$ first, then Equation (37) becomes

$$\tilde{\mathbf{J}}_1(\mathbf{k}) = \hat{\alpha} I \ell \frac{2}{k_x a} \sin\left(\frac{k_x a}{2}\right) \frac{2}{k_y b} \sin\left(\frac{k_y b}{2}\right). \tag{7.1.39}$$

Also, note that $\tilde{\mathbf{J}}_1(\mathbf{k})$ now contains more high-frequency components of k_z than before. After substituting (39) into (36), applying Jordan's lemma and Cauchy's theorem as in (33) by subtracting a constant term corresponding to the constant when $k_z \to \infty$ first, we have

$$\mathbf{E}(\mathbf{r}) = -\frac{\omega\mu_0}{8\pi^2} \int\!\!\!\int\limits_{-\infty}^{\infty} d\mathbf{k}_s\, e^{i\mathbf{k}_s\cdot(\mathbf{r}_s-\mathbf{r}_{0s})+ik_z|z-z_0|} \left[\frac{\overline{\mathbf{I}}k_0^2 - \mathbf{k}_0\mathbf{k}_0}{k_0^2 k_z}\right] \cdot \tilde{\mathbf{J}}_1(\mathbf{k}_s)$$

$$- \frac{i\omega\mu_0}{(2\pi)^3 k_0^2} \int\!\!\!\int\!\!\!\int\limits_{-\infty}^{\infty} d\mathbf{k}\, \hat{z}\hat{z} \cdot e^{i\mathbf{k}\cdot(\mathbf{r}-\mathbf{r}_0)} \tilde{\mathbf{J}}_1(\mathbf{k}_s). \tag{7.1.40}$$

Now, if a and $b \to 0$, then $\tilde{\mathbf{J}}_1(\mathbf{k}_s) \to \hat{a}I\ell$, and

$$\mathbf{E}(\mathbf{r}) = -\frac{\omega\mu_0}{8\pi^2} \int\limits_{-\infty}^{\infty}\!\!\int d\mathbf{k}_s\, e^{i\mathbf{k}_s \cdot (\mathbf{r}_s - \mathbf{r}_{0s}) + ik_z|z - z_0|} \left[\frac{\overline{\mathbf{I}}k_0^2 - \mathbf{k}_0\mathbf{k}_0}{k_0^2 k_z}\right] \cdot \hat{a}I\ell$$

$$-\frac{i\omega\mu_0}{k_0^2}\hat{z}\hat{z} \cdot \hat{a}I\ell\,\delta(\mathbf{r} - \mathbf{r}_0). \quad (7.1.41)$$

The above can be written as

$$\mathbf{E}(\mathbf{r}) = i\omega\mu_0 \int d\mathbf{r}'\,\overline{\mathbf{G}}(\mathbf{r}, \mathbf{r}') \cdot \hat{a}I\ell\,\delta(\mathbf{r}' - \mathbf{r}_0), \quad (7.1.42)$$

where $\overline{\mathbf{G}}$ is as defined in (34).[8] Of course, if a or b tends to zero first, (41) will have different representations corresponding to different representations of $\overline{\mathbf{G}}(\mathbf{r}, \mathbf{r}')$.

Therefore, the representation (34) is equivalent to a point source response where the point source is the limit of a disk-shaped source; that is, the z dimension of the disk vanishes before its transverse dimensions. Equation (34) can be viewed as due to a linear superposition of current sheet sources; hence, the z dimension becomes more highly resolved before the transverse dimensions.

A more succinct but less formal approach for deriving (34) is to substitute the Weyl identity (Chapter 2)

$$\frac{e^{ik_0 r}}{r} = \frac{i}{2\pi} \int\limits_{-\infty}^{\infty}\!\!\int d\mathbf{k}_s\, \frac{e^{i\mathbf{k}_s \cdot \mathbf{r}_s + ik_z|z|}}{k_z}, \quad (7.1.43)$$

for $g(\mathbf{r}, \mathbf{r}')$ in (17) and interchange the order of integration and differentiation (Exercise 7.8).

There is a close relationship between the spectral representation in the Fourier domain and the spatial representation of the dyadic Green's function. The last term in (34) is precisely the effect due to the charges on the disk-shaped box [see Equation (28b)] if the observation point is inside this box. It is nonvanishing even when the size of the box vanishes. Hence, in order for (34) to represent the dyadic Green's function uniquely, the first term in (34) must be a representation of $P.V.\overline{\mathbf{G}}(\mathbf{r}, \mathbf{r}')$ with a disk-shaped principal volume. Therefore, the action of the first term in (34) on a current source is the same as the action of $P.V.\overline{\mathbf{G}}(\mathbf{r}, \mathbf{r}')$ (with a disk-shaped principal volume) on a current source. In other words, the first term in (34) is the spectral representation of $P.V.\overline{\mathbf{G}}(\mathbf{r}, \mathbf{r}')$ with a disk-shaped exclusion volume. This is

---

[8] Here, we assume that $\int \delta(\mathbf{r} - \mathbf{r}')\,\delta(\mathbf{r}' - \mathbf{r}_0)d\mathbf{r}' = \delta(\mathbf{r} - \mathbf{r}_0)$.

very much like the distribution $P.V.\frac{1}{y}$ in one dimension, which has a spectral representation (see Exercise 7.9)

$$P.V.\frac{1}{y} = \frac{-i}{2} \int\limits_{-\infty}^{\infty} dx \, \text{sgn}(x) \, e^{ixy}. \qquad (7.1.44)$$

The distribution on the left-hand side, when integrated with another function, is defined to yield the Cauchy principal value (see Stakgold 1967).

§7.2 Vector Wave Functions

The dyadic Green's function in a homogeneous medium can also be expressed in terms of vector wave functions. In this section, the vector wave functions for an unbounded medium will be derived. Once these vector wave functions are known, the dyadic Green's function for an unbounded medium can be expressed in terms of these vector wave functions. These vector wave functions can be derived in the Cartesian, cylindrical, and spherical coordinates (Stratton 1941).

§§7.2.1 Derivation of Vector Wave Functions

In a source-free region V filled with a homogeneous medium, the electric field and the magnetic field satisfy the following vector wave equation

$$\nabla \times \nabla \times \mathbf{F}(\mathbf{r}) - k^2 \mathbf{F}(\mathbf{r}) = 0. \qquad (7.2.1)$$

Moreover, for a homogeneous medium, $\mathbf{F}(\mathbf{r})$ is derivable from the scalar potential $\psi(\mathbf{r})$. For example, if $\psi(\mathbf{r})$ satisfies

$$(\nabla^2 + k^2)\psi(\mathbf{r}) = 0, \qquad (7.2.2)$$

and a vector wave function is defined as

$$\mathbf{M}(\mathbf{r}) = \nabla \times \mathbf{c}\psi(\mathbf{r}), \qquad (7.2.3)$$

where $\mathbf{c}$ is a constant vector known as the ***pilot vector***, then $\mathbf{M}(\mathbf{r})$ is easily shown to be a solution to (1) (see Exercise 7.10). Furthermore, if another vector wave function is defined as

$$\mathbf{N}(\mathbf{r}) = \frac{1}{k}\nabla \times \mathbf{M}(\mathbf{r}), \qquad (7.2.4)$$

then $\mathbf{N}(\mathbf{r})$ can be shown to be a solution to (1). Also, it follows that $\nabla \times \mathbf{N}(\mathbf{r}) = k\mathbf{M}(\mathbf{r})$ from (1). Notice that $\mathbf{M}(\mathbf{r})$ is the equivalence of an electric (or a magnetic field) and $\mathbf{N}(\mathbf{r})$ is the equivalence of a magnetic field (or an electric field) respectively.

It is clear that the $\mathbf{M}$ and $\mathbf{N}$ vector wave functions are divergence-free but not curl-free: They are also known as the ***solenoidal vector wave functions***. For a bounded volume V, they are related to the solenoidal modes of the cavity formed by V. Consequently, they cannot be used to represent all electromagnetic fields, as not all fields are divergence-free.[9] As a result, a third vector wave function $\mathbf{L}(\mathbf{r})$, defined as

$$\mathbf{L}(\mathbf{r}) = \nabla\psi(\mathbf{r}), \qquad (7.2.5)$$

is needed to represent fields with nonzero divergence. Clearly, $\nabla \cdot \mathbf{L}(\mathbf{r}) = \nabla^2\psi = -k^2\psi(\mathbf{r})$ is nonzero. Also, $\nabla \times \mathbf{L}(\mathbf{r}) = 0$, implying that it is curl-free or that it has no rotation. Hence, $\mathbf{L}(\mathbf{r})$ is also known as the *irrotational vector wave function*. Furthermore, for a bounded region V, $\mathbf{L}(\mathbf{r})$ is related to the irrotational modes of a cavity formed by V (also see Kurokawa 1958; Collin 1966, 1973).

Alternatively, one knows that the eigenvectors of the eigenequation $\mathcal{L}f = \lambda f$ are complete if $\mathcal{L}$ is Hermitian (Exercise 5.8). In (1), $\mathcal{L} = \nabla \times \nabla\times$ is Hermitian if $\mathbf{F}$ satisfies some prescribed boundary conditions on S, the surface bounding V (see Chapter 5), and the eigenvalue is k^2. However, $\mathcal{L}$ has an infinite-dimensional nullspace formed by all solutions of $\mathcal{L}f = 0$, corresponding to $\lambda = 0$. Also, this nullspace is spanned by functions $\mathbf{L}(\mathbf{r})$ given by (5), where $\psi(\mathbf{r})$ is all the possible solutions of the eigenequation $(\nabla^2 + k^2)\psi = 0$ for all possible eigenvalues k^2. But if ∇^2 is Hermitian, $\psi(\mathbf{r})$ again forms a complete set. Note that $\nabla \times \nabla\times$ and ∇^2 operators are Hermitian only under prescribed boundary conditions (Chapter 5). Hence, the eigenvectors in (1) and (2) span the space of functions satisfying the same boundary conditions. In other words, the set of eigenvectors $\mathbf{F}$ and the set of eigenvectors ψ are complete in their respective space of functions satisfying the same boundary conditions.

§§7.2.2 *Orthogonality Relationships of Vector Wave Functions*

The orthogonality relationship between vector wave functions defined in a bounded V can be shown. To do this, we need the orthogonality between the scalar functions $\psi(\mathbf{r})$. For a bounded V, the pilot vector $\mathbf{c}$ produces orthogonal vector wave functions only for a V with a shape shown in Figure 7.2.1: It consists of a section of a uniform, arbitrarily shaped waveguide terminated with two end-caps.

If two scalar potentials $\psi_a(\mathbf{r})$ and $\psi_b(\mathbf{r})$ satisfy the equations

$$(\nabla^2 + k_a^2)\psi_a(\mathbf{r}) = 0, \qquad (7.2.6a)$$

$$(\nabla^2 + k_b^2)\psi_b(\mathbf{r}) = 0, \qquad (7.2.6b)$$

with prescribed boundary conditions, then $\psi_a(\mathbf{r})$ and $\psi_b(\mathbf{r})$ are orthogonal if $k_a^2 \neq k_b^2$. To show this, we multiply (6a) by ψ_b and (6b) by ψ_a and subtract the result. Consequently,

$$\psi_b(\mathbf{r})\nabla^2\psi_a(\mathbf{r}) - \psi_a(\mathbf{r})\nabla^2\psi_b(\mathbf{r}) = (k_b^2 - k_a^2)\psi_a(\mathbf{r})\psi_b(\mathbf{r}). \qquad (7.2.7)$$

---

[9] It follows from the Helmholtz theorem (Morse and Feshbach 1953; Plonsey and Collin 1961) that an arbitrary vector field can be decomposed into a divergence-free part and a curl-free part.

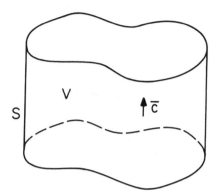

Figure 7.2.1 The surface S bounding the volume V and the pilot vector **c** necessary for the derivation of (15).

Furthermore, one can show that

$$\nabla \cdot [\psi_b \nabla \psi_a - \psi_a \nabla \psi_b] = \psi_b \nabla^2 \psi_a - \psi_a \nabla^2 \psi_b. \qquad (7.2.7a)$$

Hence, integrating (7) over volume and invoking Gauss' theorem for the left-hand side yields

$$\int_S dS\, \hat{n} \cdot (\psi_b \nabla \psi_a - \psi_a \nabla \psi_b) = (k_b^2 - k_a^2) \int_V d\mathbf{r}\, \psi_a(\mathbf{r})\psi_b(\mathbf{r}), \qquad (7.2.8)$$

where S is the surface bounding V. Consequently, if the homogeneous Dirichlet boundary condition $\psi = 0$ on S or the homogeneous Neumann boundary condition $\hat{n} \cdot \nabla \psi = 0$ on S is satisfied, then the orthogonality relation that

$$\int_V d\mathbf{r}\, \psi_a(\mathbf{r})\psi_b(\mathbf{r}) = 0 \qquad (7.2.9)$$

follows when $k_a^2 \neq k_b^2$.

Note that with the homogeneous Dirichlet boundary condition $\psi = 0$ on S, then $\hat{n} \cdot \mathbf{M}(\mathbf{r}) = 0$ and $\hat{n} \times \mathbf{N}(\mathbf{r}) = 0$ on the side wall, while $\hat{n} \times \mathbf{M}(\mathbf{r}) = 0$ and $\hat{n} \cdot \mathbf{N}(\mathbf{r}) = 0$ on the end-caps. Furthermore, with the homogeneous Neumann boundary condition $\hat{n} \cdot \nabla \psi = 0$ on S, then $\hat{n} \cdot \mathbf{N}(\mathbf{r}) = 0$ and $\hat{n} \times \mathbf{M}(\mathbf{r}) = 0$ on the side wall and $\hat{n} \times \mathbf{N}(\mathbf{r}) = 0$ and $\hat{n} \cdot \mathbf{M}(\mathbf{r}) = 0$ on the end-caps (Exercise 7.11).

From the above, similar orthogonality relationships can be inferred about the vector wave functions. Consequently, one can show that

$$\int_V d\mathbf{r}\, \mathbf{M}_a(\mathbf{r}) \cdot \mathbf{M}_b(\mathbf{r}) = \int_V d\mathbf{r}\, [\nabla \times \mathbf{c}\psi_a(\mathbf{r})] \cdot [\nabla \times \mathbf{c}\psi_b(\mathbf{r})]$$

$$= \int_S dS\, \hat{n} \cdot [\psi_a(\mathbf{r})\, \mathbf{c} \times \nabla \times \mathbf{c}\psi_b(\mathbf{r})] + \int_V d\mathbf{r}\, \psi_a(\mathbf{r})\, \mathbf{c} \cdot \nabla \times \nabla \times \mathbf{c}\psi_b(\mathbf{r}),$$
$$(7.2.10)$$

where the identity that $\nabla \cdot (\mathbf{A} \times \mathbf{B}) = \mathbf{B} \cdot \nabla \times \mathbf{A} - \mathbf{A} \cdot \nabla \times \mathbf{B}$ and Gauss' theorem have been used. But it can be shown that

$$\mathbf{c} \times \nabla \times \mathbf{c}\psi_b(\mathbf{r}) = [c^2\nabla - \mathbf{cc} \cdot \nabla]\psi_b(\mathbf{r}) = c^2\nabla_s\psi_b(\mathbf{r}), \qquad (7.2.11)$$

where $\nabla_s = \nabla - \hat{c}\hat{c} \cdot \nabla$ is a del operator transverse to $\hat{c} = \mathbf{c}/c$. Furthermore, one can show that

$$\mathbf{c} \cdot \nabla \times \nabla \times \mathbf{c}\psi_b(\mathbf{r}) = -c^2\nabla_s^2\psi_b(\mathbf{r}) = c^2k_{sb}^2\psi_b(\mathbf{r}), \qquad (7.2.12)$$

where ∇_s^2 is the Laplacian operator transverse to $\hat{c}$, and $k_{sb}^2 = k_b^2 - k_c^2$; for if $\psi_b(\mathbf{r})$ is assumed to have $e^{\pm ik_c c}$ dependence, then

$$(\nabla^2 + k_b^2)\psi_b(\mathbf{r}) = \left(\nabla_s^2 + \frac{\partial^2}{\partial c^2} + k_b^2\right)\psi_b(\mathbf{r}) = (\nabla_s^2 + k_{sb}^2)\psi_b(\mathbf{r}). \qquad (7.2.13)$$

Therefore, (10) becomes

$$\int_V d\mathbf{r}\, \mathbf{M}_a(\mathbf{r}) \cdot \mathbf{M}_b(\mathbf{r}) = c^2\int_S dS\, \psi_a(\mathbf{r})\hat{n} \cdot \nabla_s\psi_b(\mathbf{r}) + c^2k_{sb}^2\int_V d\mathbf{r}\, \psi_a(\mathbf{r})\psi_b(\mathbf{r}).$$
$$(7.2.14)$$

Note that with the surface S and the pilot vector $\mathbf{c}$ chosen such that $\hat{n} \cdot \nabla_s\psi_b(\mathbf{r}) = \hat{n} \cdot \nabla\psi_b(\mathbf{r})$ on the side wall as shown in Figure 7.2.1, then the homogeneous Dirichlet or the homogeneous Neumann boundary condition on $\psi(\mathbf{r})$ will make the first term vanish on the side wall as well as the end walls. Hence,

$$\int_V d\mathbf{r}\, \mathbf{M}_a(\mathbf{r}) \cdot \mathbf{M}_b(\mathbf{r}) = c^2k_{sb}^2\int_V d\mathbf{r}\, \psi_a(\mathbf{r})\psi_b(\mathbf{r}) = 0, \quad \text{if } k_a^2 \neq k_b^2. \qquad (7.2.15)$$

Furthermore, it can be shown that

$$\int_V d\mathbf{r}\, \mathbf{N}_a(\mathbf{r}) \cdot \mathbf{N}_b(\mathbf{r}) = \frac{1}{k_ak_b}\int_V d\mathbf{r}\, \nabla \times \mathbf{M}_a(\mathbf{r}) \cdot \nabla \times \mathbf{M}_b(\mathbf{r})$$

$$= \frac{1}{k_ak_b}\int_S dS\, \hat{n} \cdot [\mathbf{M}_a(\mathbf{r}) \times \nabla \times \mathbf{M}_b(\mathbf{r})]$$

$$+ \frac{k_b}{k_a}\int_V d\mathbf{r}\, \mathbf{M}_a(\mathbf{r}) \cdot \mathbf{M}_b(\mathbf{r}),$$
$$(7.2.16)$$

where the identities that $\nabla \cdot (\mathbf{A} \times \mathbf{B}) = \mathbf{B} \cdot \nabla \times \mathbf{A} - \mathbf{A} \cdot \nabla \times \mathbf{B}$, $\nabla \times \nabla \times \mathbf{M}_b(\mathbf{r}) = k_b^2 \mathbf{M}_b(\mathbf{r})$, and Gauss' theorem have been used. Since $\psi(\mathbf{r})$ satisfies either the homogeneous Dirichlet or the homogeneous Neumann boundary condition on S, the first integral on the right-hand side of (16) can be shown to vanish. As a result,

$$
\int_V d\mathbf{r}\, \mathbf{N}_a(\mathbf{r}) \cdot \mathbf{N}_b(\mathbf{r}) = \frac{k_b}{k_a} \int d\mathbf{r}\, \mathbf{M}_a(\mathbf{r}) \cdot \mathbf{M}_b(\mathbf{r})
$$

$$
= c^2 k_{sb}^2 \frac{k_b}{k_a} \int_V d\mathbf{r}\, \psi_a(\mathbf{r}) \psi_b(\mathbf{r}) = 0, \quad \text{if } k_a^2 \neq k_b^2.
$$
$$(7.2.17)$$

Note that the second equality follows from (15).

Similarly,

$$
\int_V d\mathbf{r}\, \mathbf{L}_a(\mathbf{r}) \cdot \mathbf{L}_b(\mathbf{r}) = \int_V d\mathbf{r}\, \nabla \psi_a(\mathbf{r}) \cdot \nabla \psi_b(\mathbf{r})
$$

$$
= \int_S dS\, \hat{n} \cdot [\psi_a(\mathbf{r}) \nabla \psi_b(\mathbf{r})] + k_b^2 \int_V d\mathbf{r}\, \psi_a(\mathbf{r}) \psi_b(\mathbf{r})
$$
$$(7.2.18)$$

by using the identity that $\nabla \cdot (\psi \mathbf{A}) = \nabla \psi \cdot \mathbf{A} + \psi \nabla \cdot \mathbf{A}$, Gauss' theorem, and that $\nabla^2 \psi_b = -k_b^2 \psi_b$. By the same token, the first integral on the right-hand side of (18) again vanishes by virtue of the boundary condition on ψ. Hence,

$$
\int_V d\mathbf{r}\, \mathbf{L}_b(\mathbf{r}) \cdot \mathbf{L}_b(\mathbf{r}) = k_b^2 \int_V d\mathbf{r}\, \psi_a(\mathbf{r}) \psi_b(\mathbf{r}) = 0 \quad \text{if } k_a^2 \neq k_b^2.
$$
$$(7.2.19)$$

The $\mathbf{M}$, $\mathbf{N}$, and $\mathbf{L}$ functions are also mutually orthogonal under one condition: if all of them represent the electric field, or all of them represent the magnetic field. This distinction is unimportant for the unbounded medium case, but for a bounded V, this ensures that $\mathbf{M}$, $\mathbf{N}$, and $\mathbf{L}$ satisfy the same boundary condition on S. Note that for a bounded medium, if

$$
\mathbf{M}_a(\mathbf{r}) = \nabla \times \mathbf{c}\psi_a(\mathbf{r}),
$$

then

$$
\mathbf{N}_b(\mathbf{r}) = \frac{1}{k} \nabla \times \nabla \times \mathbf{c}\psi_b(\mathbf{r}),
$$

where $\psi_a(\mathbf{r})$ satisfies the homogeneous Neumann boundary condition, while $\psi_b(\mathbf{r})$ satisfies the homogeneous Dirichlet boundary condition, or vice versa. In this manner, $\mathbf{M}_a$ and $\mathbf{N}_b$ satisfy the same boundary condition on S. Otherwise, they cannot be used to represent the electric field or the magnetic field simultaneously, and yet satisfy a physical boundary condition on S, the wall of V. For example, if $\psi_a(\mathbf{r})$ satisfies the homogeneous Neumann boundary condition on S, then $\mathbf{M}_a(\mathbf{r})$ satisfies the boundary condition $\hat{n} \times \mathbf{M}_a(\mathbf{r}) = 0$ on the side wall of V while it satisfies $\hat{n} \cdot \mathbf{M}_a(\mathbf{r}) = 0$ on the end walls of V as shown in Figure 7.2.1. In other words, if $\mathbf{M}_a(\mathbf{r})$ represents an electric field, the side wall has an electric wall boundary condition, while the end walls have a magnetic wall boundary condition. Thus, in order for $\mathbf{N}_b(\mathbf{r})$ to satisfy the same physical boundary conditions on S, it is necessary that $\psi_b(\mathbf{r})$ satisfy the homogeneous Dirichlet boundary condition on S. Because of the different boundary condition on ψ_a and ψ_b, $\mathbf{N}_b(\mathbf{r})$ cannot be derived from $\mathbf{M}_a(\mathbf{r})$ using (4).

To show the mutual orthogonality of $\mathbf{M}$, $\mathbf{N}$, and $\mathbf{L}$ functions under the above condition, a similar procedure is used. For example,

$$
\int_V d\mathbf{r}\, \mathbf{M}_a(\mathbf{r}) \cdot \mathbf{N}_b(\mathbf{r}) = \int_V d\mathbf{r}\, [\nabla \times \mathbf{c}\psi_a(\mathbf{r})] \cdot \mathbf{N}_b(\mathbf{r})
$$

$$
= \int_S dS\, \hat{n} \cdot [\mathbf{c}\psi_a(\mathbf{r}) \times \mathbf{N}_b(\mathbf{r})]
$$

$$
+ \int_V d\mathbf{r}\, \psi_a \mathbf{c} \cdot \nabla \times \mathbf{N}_b(\mathbf{r}) = 0. \tag{7.2.20}
$$

The last equality follows because $\mathbf{c} \cdot \nabla \times \mathbf{N}_b(\mathbf{r}) = k_b \mathbf{c} \cdot \mathbf{M}_b(\mathbf{r}) = k_b \mathbf{c} \cdot \nabla \times \mathbf{c}\psi_b(\mathbf{r}) = 0$, and $\hat{n} \cdot \mathbf{c}\psi_a(\mathbf{r}) \times \mathbf{N}_b(\mathbf{r}) = -\mathbf{c}\psi_a(\mathbf{r}) \cdot \hat{n} \times \mathbf{N}_b(\mathbf{r})$, which is zero if $\psi_a(\mathbf{r})$ or $\psi_b(\mathbf{r})$ satisfies the homogeneous Dirichlet boundary condition on S.

Similarly,

$$
\int_V d\mathbf{r}\, \mathbf{M}_a(\mathbf{r}) \cdot \mathbf{L}_b(\mathbf{r}) = \int_V d\mathbf{r}\, [\nabla \times \mathbf{c}\psi_a(\mathbf{r})] \cdot \nabla\psi_b(\mathbf{r})
$$

$$
= \int_S dS\, \hat{n} \cdot (\psi_b \nabla \times \mathbf{c}\psi_a) = 0, \tag{7.2.21}
$$

if either $\psi_a(\mathbf{r})$ or $\psi_b(\mathbf{r})$ satisfies the homogeneous Dirichlet boundary condition on S. Notice that in the above, we have used the fact that $\nabla \cdot (\psi_b \nabla \times \mathbf{c}\psi_a) = \nabla \times \mathbf{c}\psi_a \cdot \nabla\psi_b$.

Furthermore,

$$
\int_V d\mathbf{r}\, \mathbf{N}_a(\mathbf{r}) \cdot \mathbf{L}_b(\mathbf{r}) = \frac{1}{k_a} \int_V d\mathbf{r}\, (\nabla \times \nabla \times \mathbf{c}\psi_a) \cdot \nabla\psi_b
$$

$$
= \frac{1}{k_a} \int_S dS\, \hat{n} \cdot (\mathbf{c} \times \nabla\psi_a) \times \nabla\psi_b = 0 \tag{7.2.22}
$$

for the same condition on ψ_a and ψ_b as before.

When the medium is unbounded, so that V is unbounded, it could be viewed as the limiting case of the above when V and $S \to \infty$ in all directions.

Consequently, in the next subsection, we will give examples of the vector wave functions for unbounded media in Cartesian, cylindrical, and spherical coordinates.

§§7.2.3 Vector Wave Functions for Unbounded Media

The vector wave functions for unbounded media can be derived for a number of coordinate systems. Therefore, we shall present the derivations for Cartesian, cylindrical, and spherical coordinate systems. As shown previously, when the region V is bounded, the vector wave functions have a discrete spectrum k. But when the region V is unbounded, this spectrum becomes a continuum. In spite of this, the unbounded region vector wave functions can be thought of as a limiting case of the bounded region vector wave functions. [These results are also presented by Stratton (1941).]

(a) Cartesian Coordinates

In Cartesian coordinates, the solution to (2) is

$$\psi(\mathbf{k}, \mathbf{r}) = e^{ik_x x + ik_y y + ik_z z} = e^{i\mathbf{k}\cdot\mathbf{r}}, \tag{7.2.23}$$

where $k_x^2 + k_y^2 + k_z^2 = k^2$ and k is a continuous variable. To use (3), we may choose the pilot vector $\mathbf{c}$ as $\hat{z}$. Then,

$$\mathbf{M}(\mathbf{k}, \mathbf{r}) = \nabla \times \hat{z} e^{i\mathbf{k}\cdot\mathbf{r}} = i\mathbf{k} \times \hat{z} e^{i\mathbf{k}\cdot\mathbf{r}}. \tag{7.2.24}$$

Hence, $\mathbf{M}(\mathbf{k}, \mathbf{r})$ is a vector wave function transverse to $\hat{z}$. Similarly,

$$\mathbf{N}(\mathbf{k}, \mathbf{r}) = \frac{1}{k}\nabla \times \mathbf{M}(\mathbf{k}, \mathbf{r}) = -\frac{1}{k}\mathbf{k} \times \mathbf{k} \times \hat{z} e^{i\mathbf{k}\cdot\mathbf{r}} \tag{7.2.25}$$

and

$$\mathbf{L}(\mathbf{k}, \mathbf{r}) = \nabla e^{i\mathbf{k}\cdot\mathbf{r}} = i\mathbf{k} e^{i\mathbf{k}\cdot\mathbf{r}}. \tag{7.2.26}$$

Consequently, an arbitrary field $\mathbf{E}(\mathbf{r})$ can be expanded as

$$\mathbf{E}(\mathbf{r}) = \int\!\!\!\int\!\!\!\int_{-\infty}^{\infty} dk_x dk_y dk_z \left[a(\mathbf{k})\,\mathbf{M}(\mathbf{k}, \mathbf{r}) + b(\mathbf{k})\,\mathbf{N}(\mathbf{k}, \mathbf{r}) + c(\mathbf{k})\,\mathbf{L}(\mathbf{k}, \mathbf{r}) \right]. \tag{7.2.27}$$

We can show readily that (see Exercise 7.12)

$$\int\!\!\!\int\!\!\!\int_{-\infty}^{\infty} dx\,dy\,dz\; \psi(\mathbf{k}, \mathbf{r})\psi(-\mathbf{k}', \mathbf{r}) = (2\pi)^3\, \delta(\mathbf{k} - \mathbf{k}'). \tag{7.2.28}$$

Then, from (15), (17), and (19), it follows that

$$\iiint\limits_{-\infty}^{\infty} d\mathbf{r}\, \mathbf{M}(\mathbf{k},\mathbf{r})\cdot \mathbf{M}(-\mathbf{k}',\mathbf{r}) = k_s^2(2\pi)^3\,\delta(\mathbf{k}-\mathbf{k}'), \tag{7.2.29}$$

$$\iiint\limits_{-\infty}^{\infty} d\mathbf{r}\, \mathbf{N}(\mathbf{k},\mathbf{r})\cdot \mathbf{N}(-\mathbf{k}',\mathbf{r}) = k_s^2(2\pi)^3\,\delta(\mathbf{k}-\mathbf{k}'), \tag{7.2.30}$$

$$\iiint\limits_{-\infty}^{\infty} d\mathbf{r}\, \mathbf{L}(\mathbf{k},\mathbf{r})\cdot \mathbf{L}(-\mathbf{k}',\mathbf{r}) = k^2(2\pi)^3\,\delta(\mathbf{k}-\mathbf{k}'), \tag{7.2.31}$$

where $k_s^2 = k_x^2 + k_y^2$. Also, $\mathbf{M}$, $\mathbf{N}$, and $\mathbf{L}$ are mutually orthogonal. Thus, the above orthogonality relationships can be used to find the unknowns $a(\mathbf{k})$, $b(\mathbf{k})$, and $c(\mathbf{k})$ in (27).

(b) Cylindrical Coordinates

In cylindrical coordinates, the solution to (2) that is regular at the origin is

$$\psi_n(k_\rho, k_z, \mathbf{r}) = J_n(k_\rho\rho)e^{ik_z z + in\phi}, \tag{7.2.32}$$

where $k_\rho^2 + k_z^2 = k^2$ and k is a continuous variable. Choosing the pilot vector $\mathbf{c} = \hat{z}$, we have accordingly,

$$\mathbf{M}_n(k_\rho, k_z, \mathbf{r}) = \nabla \times \hat{z} J_n(k_\rho\rho)e^{ik_z z + in\phi}, \tag{7.2.33}$$

$$\mathbf{N}_n(k_\rho, k_z, \mathbf{r}) = \frac{1}{k}\nabla \times \nabla \times \hat{z} J_n(k_\rho\rho)e^{ik_z z + in\phi}, \tag{7.2.34}$$

$$\mathbf{L}_n(k_\rho, k_z, \mathbf{r}) = \nabla J_n(k_\rho\rho)e^{ik_z z + in\phi}. \tag{7.2.35}$$

As a result, an arbitrary field $\mathbf{E}(\mathbf{r})$ can be expanded as

$$\mathbf{E}(\mathbf{r}) = \sum_{n=-\infty}^{\infty} \int_{-\infty}^{\infty} dk_z \int_0^{\infty} dk_\rho \,[a_n(k_\rho, k_z)\,\mathbf{M}_n(k_\rho, k_z, \mathbf{r})$$
$$+ b_n(k_\rho, k_z)\,\mathbf{N}_n(k_\rho, k_z, \mathbf{r}) + c_n(k_\rho, k_z)\,\mathbf{L}_n(k_\rho, k_z, \mathbf{r})]. \tag{7.2.36}$$

But it can be shown that (see Exercise 7.13)

$$\int_0^{2\pi} d\phi \int_{-\infty}^{\infty} dz \int_0^{\infty} d\rho\, \rho\psi_n(k_\rho, k_z, \mathbf{r})\psi_{-n'}(-k_\rho', -k_z', \mathbf{r})$$

$$= (2\pi)^2\,\delta_{nn'}\,\delta(k_z-k_z')\frac{\delta(k_\rho - k_\rho')}{k_\rho}. \tag{7.2.37}$$

Hence, from (15), (17), and (19), it follows that

$$\int\limits_0^{2\pi} d\phi \int\limits_{-\infty}^{\infty} dz \int\limits_0^{\infty} d\rho\, \rho \mathbf{M}_n(k_\rho, k_z, \mathbf{r}) \cdot \mathbf{M}_{-n'}(-k'_\rho, -k'_z, \mathbf{r})$$

$$= (2\pi)^2 k_\rho\, \delta_{nn'}\, \delta(k_z - k'_z)\, \delta(k_\rho - k'_\rho), \quad (7.2.38)$$

$$\int\limits_0^{2\pi} d\phi \int\limits_{-\infty}^{\infty} dz \int\limits_0^{\infty} d\rho\, \rho \mathbf{N}_n(k_\rho, k_z, \mathbf{r}) \cdot \mathbf{N}_{-n'}(-k'_\rho, -k'_z, \mathbf{r})$$

$$= (2\pi)^2 k_\rho\, \delta_{nn'}\, \delta(k_z - k'_z)\, \delta(k_\rho - k'_\rho), \quad (7.2.39)$$

$$\int\limits_0^{2\pi} d\phi \int\limits_{-\infty}^{\infty} dz \int\limits_0^{\infty} d\rho\, \rho \mathbf{L}_n(k_\rho, k_z, \mathbf{r}) \cdot \mathbf{L}_{-n'}(-k'_\rho, -k'_z, \mathbf{r})$$

$$= (2\pi)^2 k^2\, \delta_{nn'}\, \delta(k_z - k'_z)\, \frac{\delta(k_\rho - k'_\rho)}{k_\rho}. \quad (7.2.40)$$

Furthermore, the **M**, **N**, and **L** functions are mutually orthogonal. Consequently, the above orthogonality relationships can be used to find the unknowns in (36).

(c) Spherical Coordinates

In spherical coordinates, the solution to (2) that is regular about the origin is

$$\psi_{nm}(k, \mathbf{r}) = j_n(kr)Y_{nm}(\theta, \phi), \quad (7.2.41)$$

where $j_n(kr)$ is a spherical Bessel function,

$$Y_{nm}(\theta, \phi) = \sqrt{\frac{(n - m)!(2n + 1)}{(n + m)!4\pi}}\, P_n^m(\cos\theta)e^{im\phi}, \quad (7.2.41a)$$

and $P_n^m(\cos\theta)$ is a Legendre's polynomial (see p. 193). As a result, it can be shown easily that $Y_{nm}(\theta, \phi)$ is orthonormal and $Y_{n,-m}(\theta, \phi) = (-1)^m Y_{nm}^*(\theta, \phi)$ [see Exercise 7.14 and Equation (3.7.3)].

In this case, the pilot vector $\mathbf{c} = \mathbf{r}$, and we have

$$\mathbf{M}_{nm}(k, \mathbf{r}) = \nabla \times \mathbf{r} j_n(kr)Y_{nm}(\theta, \phi), \quad (7.2.42)$$

$$\mathbf{N}_{nm}(k, \mathbf{r}) = \frac{1}{k}\nabla \times \nabla \times \mathbf{r} j_n(kr)Y_{nm}(\theta, \phi), \quad (7.2.43)$$

$$\mathbf{L}_{nm}(k, \mathbf{r}) = \frac{1}{k}\nabla j_n(kr)Y_{nm}(\theta, \phi). \quad (7.2.44)$$

Note that $1/k$ is included in (44) to keep the dimensions of all the vector wave functions the same. Now, the pilot vector $\mathbf{r}$ is not a constant vector as in the other coordinate systems. Hence, the orthogonality relationships have to be rederived.

It can be shown readily that $[\hat{\psi}_{nm}(k, \mathbf{r})$ here implies the complex conjugation of the angular part of $\psi_{nm}(k, \mathbf{r})$ in (41) only]

$$\int d\mathbf{r}\, \psi_{nm}(k, \mathbf{r})\hat{\psi}_{n'm'}(k', \mathbf{r})$$

$$= \int_0^{2\pi} d\phi \int_0^{\pi} d\theta \sin\theta Y_{nm}(\theta, \phi)Y_{n'm'}^*(\theta, \phi) \int_0^{\infty} dr\, r^2 j_n(kr)j_{n'}(k'r)$$

$$= \delta_{mm'}\,\delta_{nn'} \int_0^{\infty} dr\, r^2 j_n(kr)j_n(k'r). \tag{7.2.45}$$

Hence, using the identity that [see Exercise 2.10 and Equation (1.2.40a)]

$$\frac{\pi\,\delta(k - k')}{2k^2} = \int_0^{\infty} dr\, r^2 j_n(kr)j_n(k'r), \tag{7.2.46}$$

we have (see Exercise 7.14)

$$\int d\mathbf{r}\, \psi_{nm}(k, \mathbf{r})\hat{\psi}_{n'm'}(k', \mathbf{r}) = \pi\,\delta_{mm'}\,\delta_{nn'}\frac{\delta(k - k')}{2k^2}. \tag{7.2.47}$$

Moreover, it can be shown that

$$\int_V d\mathbf{r}\, \mathbf{M}_a(\mathbf{r}) \cdot \mathbf{M}_b(\mathbf{r}) = \int_V d\mathbf{r}\, \nabla \times \mathbf{r}\psi_a(\mathbf{r}) \cdot \nabla \times \mathbf{r}\psi_b(\mathbf{r})$$

$$= \int_S dS\, \hat{n} \cdot [\mathbf{r}\psi_a(\mathbf{r}) \times \nabla \times \mathbf{r}\psi_b(\mathbf{r})] + \int_V d\mathbf{r}\, \psi_a(\mathbf{r})\,\mathbf{r} \cdot \nabla \times \nabla \times \mathbf{r}\psi_b(\mathbf{r}). \tag{7.2.48}$$

In addition, using the definition of $\nabla\times$ operator in spherical coordinates, it follows that (see Subsection 3.5.1)

$$\mathbf{r} \cdot \nabla \times \nabla \times \mathbf{r}\psi = -\left(\frac{1}{\sin\theta}\frac{\partial}{\partial\theta}\sin\theta\frac{\partial}{\partial\theta} + \frac{1}{\sin^2\theta}\frac{\partial^2}{\partial\phi^2}\right)\psi$$

$$= n(n + 1)\psi \tag{7.2.49}$$

if ψ is of the form (41). Note that the second equality follows from Equation (1.2.39). Furthermore, the first integral on the right-hand side of (48) vanishes if S is a spherical surface. Therefore,

$$\int_V d\mathbf{r}\, \mathbf{M}_{nm}(k, \mathbf{r}) \cdot \hat{\mathbf{M}}_{n'm'}(k', \mathbf{r}) = n(n + 1) \int_V d\mathbf{r}\, \psi_{nm}(k, \mathbf{r})\hat{\psi}_{n'm'}(k', \mathbf{r})$$

$$= n(n + 1)\pi\,\delta_{mm'}\,\delta_{nn'}\frac{\delta(k - k')}{2k^2}. \tag{7.2.50}$$

Similar to (16) and (17), it can be shown that

$$
\int_V d\mathbf{r}\, \mathbf{N}_{nm}(k,\mathbf{r}) \cdot \hat{\mathbf{N}}_{n'm'}(k',\mathbf{r}) = n(n+1) \int_V d\mathbf{r}\, \psi_{nm}(k,\mathbf{r})\hat{\psi}_{n'm'}(k,\mathbf{r})
$$

$$
= n(n+1)\pi\, \delta_{mm'}\, \delta_{nn'} \frac{\delta(k-k')}{2k^2}. \tag{7.2.51}
$$

By the same token as (18), we have

$$
\int_V d\mathbf{r}\, \mathbf{L}_{nm}(k,\mathbf{r}) \cdot \hat{\mathbf{L}}_{n'm'}(k',\mathbf{r}) = \int d\mathbf{r}\, \psi_{nm}(k,\mathbf{r})\hat{\psi}_{n'm'}(k,\mathbf{r})
$$

$$
= \pi\, \delta_{mm'}\, \delta_{nn'} \frac{\delta(k-k')}{2k^2}. \tag{7.2.52}
$$

Furthermore, $\mathbf{M}$, $\mathbf{N}$, and $\mathbf{L}$ are mutually orthogonal for reasons analogous to those given in (20), (21), and (22). Therefore, given an arbitrary field $\mathbf{E}(\mathbf{r})$, it can be expanded as

$$
\mathbf{E}(\mathbf{r}) = \sum_{n=1}^{\infty} \sum_{m=-n}^{n} \int_0^{\infty} dk\, k^2 [a_{nm}(k)\, \mathbf{M}_{nm}(k,\mathbf{r}) + b_{nm}(k)\, \mathbf{N}_{nm}(k,\mathbf{r})
$$

$$
+ c_{nm}(k)\, \mathbf{L}_{nm}(k,\mathbf{r})]. \tag{7.2.53}
$$

Moreover, the coefficients $a_{nm}(k)$, $b_{nm}(k)$, and $c_{nm}(k)$ can be found by using the orthogonality relationships of the $\mathbf{M}$, $\mathbf{N}$, and $\mathbf{L}$ vector wave functions.

§7.3 Dyadic Green's Function Using Vector Wave Functions

Given the vector wave functions $\mathbf{M}$, $\mathbf{N}$, and $\mathbf{L}$, which are complete, the dyadic Green's function in terms of these vector wave functions can be found easily. Accordingly, we shall discuss how this can be done in Cartesian coordinates, cylindrical coordinates, and spherical coordinates.[10]

§§7.3.1 The Integral Representations

As can be seen from the previous section, a straightforward use of the vector wave functions to expand the dyadic Green's function results in the representation of the dyadic Green's function in terms of at least a single integral. But these eigenfunction representations can be extended to include distributions (see Appendix C). Hence, a dyadic Green's function, which should actually be thought of as a distribution, is accommodated by such representations.

---

[10] The $\mathbf{L}$ functions are left out by some authors. In this case, the dyadic Green's functions are invalid in a source region (Tai 1971, 1973).

(a) Cartesian Coordinates

In Cartesian coordinates, the dyadic Green's function is expandable in terms of dyads using vector wave functions as (Johnson et al. 1979)

$$\overline{\mathbf{G}}(\mathbf{r}, \mathbf{r}') = \int\!\!\!\int\!\!\!\int_{-\infty}^{\infty} d\mathbf{k} \, [\mathbf{M}(\mathbf{k}, \mathbf{r})\,\mathbf{a}(\mathbf{k}) + \mathbf{N}(\mathbf{k}, \mathbf{r})\,\mathbf{b}(\mathbf{k}) + \mathbf{L}(\mathbf{k}, \mathbf{r})\,\mathbf{c}(\mathbf{k})].$$

$$(7.3.1)$$

But then, it is required that

$$\nabla \times \nabla \times \overline{\mathbf{G}}(\mathbf{r}, \mathbf{r}') - k_0^2 \overline{\mathbf{G}}(\mathbf{r}, \mathbf{r}') = \overline{\mathbf{I}}\,\delta(\mathbf{r} - \mathbf{r}'). \qquad (7.3.2)$$

First, the right-hand side of (2) can be easily expanded as

$$\overline{\mathbf{I}}\,\delta(\mathbf{r} - \mathbf{r}') = \frac{1}{(2\pi)^3} \int\!\!\!\int\!\!\!\int_{-\infty}^{\infty} d\mathbf{k} \, \left[\frac{\mathbf{M}(\mathbf{k}, \mathbf{r})\,\mathbf{M}(-\mathbf{k}, \mathbf{r}')}{k_s^2} + \frac{\mathbf{N}(\mathbf{k}, \mathbf{r})\,\mathbf{N}(-\mathbf{k}, \mathbf{r}')}{k_s^2} \right.$$
$$\left. + \frac{\mathbf{L}(\mathbf{k}, \mathbf{r})\,\mathbf{L}(-\mathbf{k}, \mathbf{r}')}{k^2} \right] \qquad (7.3.3)$$

using the orthogonality relationships given in Equations (7.2.29) to (7.2.31) in the previous section, where $k_s^2 = k_x^2 + k_y^2$, and $k^2 = k_x^2 + k_y^2 + k_z^2$. Then, on substituting (1) into the left-hand side of (2) and equating it with (3), we deduce that

$$\mathbf{a}(\mathbf{k}) = \frac{\mathbf{M}(-\mathbf{k}, \mathbf{r}')}{(2\pi)^3(k^2 - k_0^2)k_s^2}, \quad \mathbf{b}(\mathbf{k}) = \frac{\mathbf{N}(-\mathbf{k}, \mathbf{r}')}{(2\pi)^3(k^2 - k_0^2)k_s^2}, \quad \mathbf{c}(\mathbf{k}) = -\frac{\mathbf{L}(-\mathbf{k}, \mathbf{r}')}{(2\pi)^3 k_0^2 k^2}.$$

$$(7.3.4)$$

Therefore, the dyadic Green's function becomes

$$\overline{\mathbf{G}}(\mathbf{r}, \mathbf{r}') = \frac{1}{(2\pi)^3} \int\!\!\!\int\!\!\!\int_{-\infty}^{\infty} d\mathbf{k} \, \left[\frac{\mathbf{M}(\mathbf{k}, \mathbf{r})\,\mathbf{M}(-\mathbf{k}, \mathbf{r}')}{(k^2 - k_0^2)k_s^2} \right.$$
$$\left. + \frac{\mathbf{N}(\mathbf{k}, \mathbf{r})\,\mathbf{N}(-\mathbf{k}, \mathbf{r}')}{(k^2 - k_0^2)k_s^2} - \frac{\mathbf{L}(\mathbf{k}, \mathbf{r})\,\mathbf{L}(-\mathbf{k}, \mathbf{r}')}{k_0^2 k^2} \right]. \qquad (7.3.5)$$

(b) Cylindrical Coordinates

Using the vector wave functions for cylindrical coordinates, the dyadic Green's function can be expanded as (Howard 1974; Pearson 1983)

$$\overline{\mathbf{G}}(\mathbf{r}, \mathbf{r}') = \sum_{n=-\infty}^{\infty} \int_{-\infty}^{\infty} dk_z \int_{0}^{\infty} dk_\rho \, k_\rho [\mathbf{M}_n(k_\rho, k_z, \mathbf{r})\,\mathbf{a}_n(k_\rho, k_z)$$
$$+ \mathbf{N}_n(k_\rho, k_z, \mathbf{r})\,\mathbf{b}_n(k_\rho, k_z) + \mathbf{L}_n(k_\rho, k_z, \mathbf{r})\,\mathbf{c}_n(k_\rho, k_z)]. \qquad (7.3.6)$$

On substituting $\overline{\mathbf{G}}(\mathbf{r}, \mathbf{r}')$ into (2) and using the orthogonality relationships of $\mathbf{M}$, $\mathbf{N}$, and $\mathbf{L}$ functions in Equation (7.2.38) to (7.2.40), the dyadic Green's function is derived to be (Exercise 7.15; note that $k^2 = k_z^2 + k_\rho^2$)

$$
\overline{\mathbf{G}}(\mathbf{r}, \mathbf{r}') = \frac{1}{(2\pi)^2} \sum_{n=-\infty}^{\infty} \int_{-\infty}^{\infty} dk_z \int_{0}^{\infty} dk_\rho\, k_\rho \left[\frac{\mathbf{M}_n(k_\rho, k_z, \mathbf{r})\mathbf{M}_{-n}(-k_\rho, -k_z, \mathbf{r}')}{(k^2 - k_0^2)k_\rho^2} \right.
$$
$$
\left. + \frac{\mathbf{N}_n(k_\rho, k_z, \mathbf{r})\mathbf{N}_{-n}(-k_\rho, -k_z, \mathbf{r}')}{(k^2 - k_0^2)k_\rho^2} - \frac{\mathbf{L}_n(k_\rho, k_z, \mathbf{r})\mathbf{L}_{-n}(-k_\rho, -k_z, \mathbf{r}')}{k_0^2 k^2} \right]. \quad (7.3.7)
$$

(c) Spherical Coordinates

The dyadic Green's function can be derived similarly in spherical coordinates. Thus, using the orthogonality relationships (7.2.50) to (7.2.52), we deduce that (Exercise 7.15; also see Collin 1986a)

$$
\overline{\mathbf{G}}(\mathbf{r}, \mathbf{r}') = \frac{2}{\pi} \sum_{n=1}^{\infty} \sum_{m=-n}^{n} \int_{0}^{\infty} dk\, k^2 \left[\frac{\mathbf{M}_{nm}(k, \mathbf{r})\,\hat{\mathbf{M}}_{nm}(k, \mathbf{r}')}{(k^2 - k_0^2)n(n+1)} \right.
$$
$$
\left. + \frac{\mathbf{N}_{nm}(k, \mathbf{r})\,\hat{\mathbf{N}}_{nm}(k, \mathbf{r}')}{(k^2 - k_0^2)n(n+1)} - \frac{\mathbf{L}_{nm}(k, \mathbf{r})\,\hat{\mathbf{L}}_{nm}(k, \mathbf{r}')}{k_0^2} \right]. \quad (7.3.8)
$$

Note that $\mathbf{M}$ and $\mathbf{N}$ are also zero when $n = 0$.

As shown in Equation (1.3.52), reciprocity requires that $\overline{\mathbf{G}}(\mathbf{r}, \mathbf{r}') = \overline{\mathbf{G}}^t (\mathbf{r}', \mathbf{r})$. Accordingly, the above representations of the dyadic Green's function can be shown to satisfy this property quite easily (Exercise 7.15).

§§7.3.2 Singularity Extraction

A dyadic Green's function is highly singular. As was shown in Section 7.1, the dyadic Green's function can be written as a principal value part plus a part proportional to a Dirac delta distribution. Moreover, this decomposition is nonunique. Notice that the integral representations given in Subsection 7.3.1 do not have the Dirac delta singularity explicitly shown. But by evaluating one of the integrals in the representation, a singularity can be explicitly extracted from the vector wave function representations. By doing so, we can reduce the above integral representations by one fold of integration. Also, the dyadic Green's functions thus obtained will be of the same form as those in Section 7.1.

(a) Cartesian Coordinates

Take for example Equation (5), which could be simplified by residue calculus. Since $k^2 = k_x^2 + k_y^2 + k_z^2$ and $d\mathbf{k} = dk_x dk_y dk_z$, the dk_z integral can be performed first. Therefore, a typical integral in (5) is of the form

$$
I = \int_{-\infty}^{\infty} dk_z\, \frac{f(k_z)e^{ik_z(z-z')}}{k_z^2 - (k_0^2 - k_x^2 - k_y^2)}. \quad (7.3.9)
$$

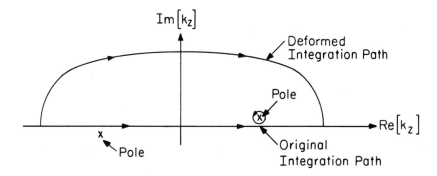

Figure 7.3.1 The original path of integration and the deformed path of integration.

At this point, the integrals in (5) and (9) are undefined because of the existence of poles at $k_z = \pm\sqrt{k_0^2 - k_x^2 - k_y^2}$. But by introducing some loss, the integrals are well defined (see Subsection 2.2.2 and Figure 7.3.1). Consequently, if $z > z'$ and $f(k_z) \to 0$ when $k_z \to \infty$, the contour of integration can be deformed from the real axis to the upper-half plane, thereby picking up a residue contribution of the pole located at $k_z = (k_0^2 - k_x^2 - k_y^2)^{1/2}$. Note that the integral on the deformed path (Figure 7.3.1) vanishes by virtue of Jordan's lemma. Therefore, (9) becomes

$$I = \pi i \frac{f(k_{0z})e^{ik_{0z}(z-z')}}{k_{0z}}, \qquad z > z', \qquad (7.3.10)$$

where $k_{0z} = \sqrt{k_0^2 - k_x^2 - k_y^2}$. By the same token, a similar operation can be performed when $z < z'$. Therefore,

$$I = \pi i \frac{f(\pm k_{0z})}{k_{0z}} e^{\pm ik_{0z}(z-z')}, \qquad \begin{matrix} z > z' \\ z < z' \end{matrix} . \qquad (7.3.11)$$

The above contour integration technique can be used to simplify the first term in (5). But in the second term in (5), because $\mathbf{N}(\mathbf{r}) = \frac{1}{k}\nabla \times \mathbf{M}(\mathbf{r})$, $\mathbf{N}(k,\mathbf{r})\,\mathbf{N}(-k,\mathbf{r}')$ has poles at $k = 0$ or at $k_z = \pm i\sqrt{k_x^2 + k_y^2}$ on the complex k_z plane. Furthermore, the residue contribution of this pole satisfies the dispersion relation $k_x^2 + k_y^2 + k_z^2 = 0$, rendering it static in nature. To one's disappointment, this static field is unphysical in view of this electrodynamic solution. Fortunately, the last term in (5) also has "static" poles at $k_z = \pm i\sqrt{k_x^2 + k_y^2}$ whose residue contribution should cancel that from the "static" pole in the second term.

From (7.2.26), $\mathbf{L}(\mathbf{k}, \mathbf{r})\,\mathbf{L}(-\mathbf{k}, \mathbf{r}) \sim \hat{z}\hat{z}k_z^2 e^{i\mathbf{k}\cdot(\mathbf{r}-\mathbf{r}')}$ when $k_z \to \infty$, violating Jordan's lemma. Therefore, the contour integration technique cannot be used directly on (5). But then, we can rewrite (5) as

$$\overline{\mathbf{G}}(\mathbf{r}, \mathbf{r}') = \frac{1}{(2\pi)^3} \iiint_{-\infty}^{\infty} d\mathbf{k} \left[\frac{\mathbf{M}(\mathbf{k}, \mathbf{r})\,\mathbf{M}(-\mathbf{k}, \mathbf{r}')}{(k^2 - k_0^2)k_s^2} + \frac{\mathbf{N}(\mathbf{k}, \mathbf{r})\,\mathbf{N}(-\mathbf{k}, \mathbf{r}')}{(k^2 - k_0^2)k_s^2} \right.$$

$$\left. - \left(\frac{\mathbf{L}(\mathbf{k}, \mathbf{r})\,\mathbf{L}(-\mathbf{k}, \mathbf{r}')}{k_0^2 k^2} - \hat{z}\hat{z}\frac{e^{i\mathbf{k}\cdot(\mathbf{r}-\mathbf{r}')}}{k_0^2} \right) \right] - \frac{1}{(2\pi)^3}\frac{\hat{z}\hat{z}}{k_0^2} \iiint_{-\infty}^{\infty} d\mathbf{k}\, e^{i\mathbf{k}\cdot(\mathbf{r}-\mathbf{r}')}. \quad (7.3.12)$$

Note that Jordan's lemma now holds true for the deformed integration path on the k_z plane for all the terms in the first integral. Furthermore, the pole contributions at $k_z = \pm i(k_x^2 + k_y^2)^{1/2}$ cancel each other from the second and the third terms in the first integral (see Exercise 7.16). Moreover, the second integral is just the Fourier integral representation of the Dirac delta function. Consequently, the dyadic Green's function becomes (also see Johnson et al. 1979)

$$\overline{\mathbf{G}}(\mathbf{r}, \mathbf{r}') = \frac{i}{8\pi^2} \iint_{-\infty}^{\infty} d\mathbf{k}_s \left[\frac{\mathbf{M}(\mathbf{k}_s, \pm k_{0z}, \mathbf{r})\,\mathbf{M}(-\mathbf{k}_s, \mp k_{0z}, \mathbf{r}')}{k_{0z}k_s^2} \right.$$

$$\left. + \frac{\mathbf{N}(\mathbf{k}_s, \pm k_{0z}, \mathbf{r})\,\mathbf{N}(-\mathbf{k}_s, \mp k_{0z}, \mathbf{r}')}{k_{0z}k_s^2} \right] - \frac{\hat{z}\hat{z}}{k_0^2}\delta(\mathbf{r} - \mathbf{r}'), \quad (7.3.13)$$

where $d\mathbf{k}_s = dk_x dk_y$. The above is the representation of the dyadic Green's function in the Cartesian coordinates using vector wave functions. Notice that it is very similar to the spectral representation of the dyadic Green's function given by (7.1.34).

Note that $\mathbf{M}$ and $\mathbf{N}$ functions in (13) are now discontinuous functions at $z = z'$. These discontinuities surface after the contour integration which is performed after the differentiations are performed in (7.2.24) to (7.2.26).

On first sight, it appears that $\nabla \cdot \overline{\mathbf{G}}(\mathbf{r}, \mathbf{r}') = -\frac{\hat{z}}{k_0^2}\frac{\partial}{\partial z}\delta(\mathbf{r} - \mathbf{r}')$ since $\nabla \cdot \mathbf{M} = \nabla \cdot \mathbf{N} = 0$. Apparently, this implies that

$$\nabla \cdot \mathbf{E} = \nabla \cdot i\omega\mu \int \overline{\mathbf{G}}(\mathbf{r}, \mathbf{r}') \cdot \mathbf{J}(\mathbf{r}')\, d\mathbf{r}' = \frac{1}{i\omega\epsilon}\frac{\partial}{\partial z}J_z(\mathbf{r}) \neq \frac{\varrho}{\epsilon},$$

and (13) seems to be in error. One must note, however, that $\mathbf{M}(\mathbf{k}_s, \pm k_{0z}, \mathbf{r})$ and $\mathbf{N}(\mathbf{k}_s, \pm k_{0z}, \mathbf{r})$ are now discontinuous functions across the plane $z = z'$ since (11) is a discontinuous function across such a plane. Fortunately, by taking this into account, it can be shown that in fact $\nabla \cdot \mathbf{E} = \varrho/\epsilon$ (Exercise 7.17).

(b) Cylindrical Coordinates

In the same manner, (7) can be simplified by evaluating either the dk_z or dk_ρ integral first. For instance, evaluating the dk_z integral first and using the technique described in (a) previously will give rise to an expression similar to (13), namely,

$$\overline{\mathbf{G}}(\mathbf{r}, \mathbf{r}') = \frac{i}{4\pi} \sum_{n=-\infty}^{\infty} \int_0^{\infty} dk_\rho \, k_\rho \left[\frac{\mathbf{M}_n(k_\rho, \pm k_{0z}, \mathbf{r}) \, \mathbf{M}_{-n}(-k_\rho, \mp k_{0z}, \mathbf{r}')}{k_{0z} k_\rho^2} \right.$$

$$\left. + \frac{\mathbf{N}_n(k_\rho, \pm k_{0z}, \mathbf{r}) \, \mathbf{N}_{-n}(-k_\rho, \mp k_{0z}, \mathbf{r}')}{k_{0z} k_\rho^2} \right] - \frac{\hat{z}\hat{z}}{k_0^2} \delta(\mathbf{r} - \mathbf{r}'). \quad (7.3.14)$$

Again, in (14), the derivatives in $\mathbf{M}$ and $\mathbf{N}$ are performed before the contour integration.

On the other hand, from Equation (7), another expression involving a dk_z integral only will be obtained if the dk_ρ integration is performed first. To this end, we write

$$\mathbf{M}_n(k_\rho, k_z, \mathbf{r}) \, \mathbf{M}_{-n}(-k_\rho, -k_z, \mathbf{r}') = (\nabla \times \hat{z})(\nabla' \times \hat{z})$$
$$\psi_n(k_\rho, k_z, \mathbf{r}) \psi_{-n}(-k_\rho, -k_z, \mathbf{r}'), \qquad (7.3.15a)$$

$$\mathbf{N}_n(k_\rho, k_z, \mathbf{r}) \, \mathbf{N}_{-n}(-k_\rho, -k_z, \mathbf{r}') = (\nabla \times \nabla \times \hat{z})(\nabla' \times \nabla' \times \hat{z}) \frac{1}{k^2}$$
$$\psi_n(k_\rho, k_z, \mathbf{r}) \psi_{-n}(-k_\rho, -k_z, \mathbf{r}'), \qquad (7.3.15b)$$

$$\mathbf{L}_n(k_\rho, k_z, \mathbf{r}) \, \mathbf{L}_{-n}(-k_\rho, -k_z, \mathbf{r}') = \nabla \nabla' \psi_n(k_\rho, k_z, \mathbf{r}) \psi_{-n}(-k_\rho, -k_z, \mathbf{r}'), \qquad (7.3.15c)$$

where $\psi_n(k_\rho, k_z, \mathbf{r})$ is given by (7.2.32). In actuality, the differentiations are performed before the integration. But in this case, it may be simpler to perform the dk_ρ integration before taking the derivative operations inside $\mathbf{M}_n$, $\mathbf{N}_n$, and $\mathbf{L}_n$. Hence, after exchanging the order of dk_ρ integration and differentiation, typical integrals involving $\mathbf{M}_n$, $\mathbf{N}_n$, and $\mathbf{L}_n$ terms in (7) are of the form

$$I_1 = \int_0^{\infty} dk_\rho \, \frac{J_n(k_\rho \rho) J_n(k_\rho \rho')}{k_\rho (k^2 - k_0^2)}, \qquad (7.3.16a)$$

$$I_2 = \int_0^{\infty} dk_\rho \, \frac{J_n(k_\rho \rho) J_n(k_\rho \rho')}{k_\rho k^2 (k^2 - k_0^2)}, \qquad (7.3.16b)$$

$$I_3 = \int_0^{\infty} dk_\rho \, k_\rho \frac{J_n(k_\rho \rho) J_n(k_\rho \rho')}{k^2 k_0^2}. \qquad (7.3.16c)$$

Let us consider first the evaluation of I_1 when $n \neq 0$. With $J_n(x) =$

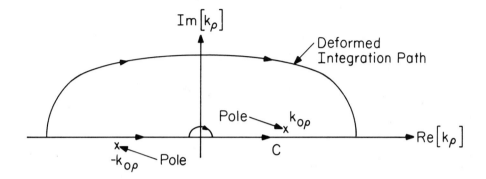

Figure 7.3.2 The contour C and the deformed path of integration on the complex k_ρ plane.

$$\tfrac{1}{2}[H_n^{(1)}(x) + H_n^{(2)}(x)],$$

$$I_1 = \lim_{\delta\to 0}\left[\frac{1}{2}\int\limits_\delta^\infty dk_\rho\,\frac{J_n(k_\rho\rho)H_n^{(1)}(k_\rho\rho')}{k_\rho(k^2 - k_0^2)} + \frac{1}{2}\int\limits_\delta^\infty dk_\rho\,\frac{J_n(k_\rho\rho)H_n^{(2)}(k_\rho\rho')}{k_\rho(k^2 - k_0^2)}\right].$$

(7.3.17)

Here, the limit is introduced because now, a pole at $k_\rho = 0$ exists in each of the integrands due to the Hankel functions. Furthermore, by letting $k_\rho = e^{-i\pi}k_\rho'$, and using the reflection formulas $H_n^{(2)}(e^{-i\pi}z) = -(-1)^n H_n^{(1)}(z)$ and $J_n(-z) = (-1)^n J_n(z)$ in the second integral in (17), we have

$$I_1 = \lim_{\delta\to 0}\left[\frac{1}{2}\int\limits_\delta^\infty dk_\rho\,\frac{J_n(k_\rho\rho)H_n^{(1)}(k_\rho\rho')}{k_\rho(k^2 - k_0^2)} + \frac{1}{2}\int\limits_{-\infty}^{-\delta} dk_\rho'\,\frac{J_n(k_\rho'\rho)H_n^{(1)}(k_\rho'\rho')}{k_\rho'(k^2 - k_0^2)}\right]$$

$$= P.V.\frac{1}{2}\int\limits_{-\infty}^\infty dk_\rho\,\frac{J_n(k_\rho\rho)H_n^{(1)}(k_\rho\rho')}{k_\rho(k^2 - k_0^2)},$$

(7.3.18)

where $P.V.$ represents a principal value integral. Notice that in (18), poles exist at $k_\rho = \pm\sqrt{k_0^2 - k_z^2} = \pm k_{0\rho}$ which may be on the real axis. But with the introduction of some loss, these poles are displaced from the real axis as shown in Figure 7.3.2 and the integral in (18) is well defined. Moreover, a residue contribution can be added to (18) at the origin to make it a complete contour integral. In other words,

$$I_1 = \frac{1}{2}\int\limits_C dk_\rho\,\frac{J_n(k_\rho\rho)H_n^{(1)}(k_\rho\rho')}{k_\rho(k^2 - k_0^2)} - \frac{1}{2nk_{0\rho}^2}\left(\frac{\rho}{\rho'}\right)^n,$$

(7.3.19)

where C is as shown in Figure 7.3.2. In (19), the last term is the residue contribution which has been included in the first term to make C a continuous contour (Exercise 7.18).

When $k_\rho \to \infty$, using the large argument approximations of $J_n(x)$ and $H_n^{(1)}(x)$, one can show that $J_n(k_\rho\rho)H_n^{(1)}(k_\rho\rho') \sim ae^{ik_\rho(\rho+\rho')} + be^{ik_\rho(\rho'-\rho)}$. Thus, if $\rho' > \rho$, the integrand becomes exponentially small in the upper half plane. Consequently, by virtue of Jordan's lemma and Cauchy's theorem, the integral along C can be deformed to the deformed path of integration, which vanishes, leaving a residue contribution due to the pole at $k_{0\rho}$ (Exercise 7.18). Hence,

$$I_1 = \pi i \frac{J_n(k_{0\rho}\rho)H_n^{(1)}(k_{0\rho}\rho')}{2k_{0\rho}^2} - \frac{1}{2nk_{0\rho}^2}\left(\frac{\rho}{\rho'}\right)^n, \quad \rho' > \rho. \tag{7.3.20}$$

When $\rho > \rho'$, we can use a similar technique to evaluate I_1. In general,

$$I_1 = \frac{\pi i}{2k_{0\rho}^2}J_n(k_{0\rho}\rho_<)H_n^{(1)}(k_{0\rho}\rho_>) - \frac{1}{2nk_{0\rho}^2}\left(\frac{\rho_<}{\rho_>}\right)^n, \quad n \neq 0, \tag{7.3.21}$$

where $\rho_>$ is the larger of ρ and ρ' and $\rho_<$ is the smaller of ρ and ρ'.

Analogously, we have (Exercise 7.18)

$$I_2 = \frac{\pi i}{2k_{0\rho}^2 k_0^2}J_n(k_{0\rho}\rho_<)H_n^{(1)}(k_{0\rho}\rho_>) + \frac{\pi i}{2k_z^2 k_0^2}J_n(ik_z\rho_<)H_n^{(1)}(ik_z\rho_>)$$

$$- \frac{1}{2nk_z^2 k_{0\rho}^2}\left(\frac{\rho_<}{\rho_>}\right)^n, \quad n \neq 0. \tag{7.3.22}$$

Note that the second term in (22) is derived because of the "static" pole at $k_\rho = \pm ik_z$ in the integrand of I_2 in (16b) due to the $1/k^2$ term. Indeed, this pole gives rise to a static-like field.

Similarly (Exercise 7.18),

$$I_3 = \frac{\pi i}{2k_0^2}J_n(ik_z\rho_<)H_n^{(1)}(ik_z\rho_>), \tag{7.3.23}$$

which is also static-like. As is expected, the static fields from (22) and (23) should cancel each other on physical grounds.

In simplifying (7), we are evaluating the dk_ρ integral first in I_1, I_2, and I_3 before performing the differential operation warranted by (15). As a result, Equations (21), (22), and (23) are apparent discontinuous functions of ρ and ρ'. Hence, it seems that if the differential operations are performed, Dirac delta functions would result. But instead of directly performing the derivatives to obtain the Dirac delta function, it is simpler to first anticipate the term from which the Dirac delta function would come.

A Dirac delta function is represented by (Exercise 2.10)

$$\frac{\delta(\rho - \rho')}{\rho} = \int\limits_0^\infty dk_\rho \, k_\rho J_n(k_\rho \rho) J_n(k_\rho \rho'). \tag{7.3.24}$$

With the large argument approximation for $J_n(x)$, notice that the amplitude of the integrand in (24) becomes a constant when $k_\rho \to \infty$.[11] Hence, a necessary condition for the dk_ρ integration to produce a Dirac delta function is that the amplitude of the integrand tends to a constant when $k_\rho \to \infty$. Consequently, to determine whether an integral has a Dirac delta function singularity, we study its integrand when $k_\rho \to \infty$.

It can be shown that when the following differential operator acts on a Bessel function,

$$(\nabla \times \hat{z})(\nabla' \times \hat{z}) \sim \hat{\phi}\hat{\phi}' \frac{\partial^2}{\partial\rho\partial\rho'} \sim O(k_\rho^2), \qquad k_\rho \to \infty, \tag{7.3.25a}$$

because $\partial/\partial\rho J_n(k_\rho\rho) = k_\rho J_n'(k_\rho\rho) \sim O(k_\rho)J_{n-1}(k_\rho\rho)$ when $k_\rho \to \infty$, and the $\partial/\partial\rho$ derivative dominates over all other derivatives when $k_\rho \to \infty$. (The Bessel functions $J_n(x)$ and $J_{n-1}(x)$ are of the same order of magnitude when $x \to \infty$.) Similarly, when the following operator acts on a Bessel function,

$$(\nabla \times \nabla \times \hat{z})(\nabla' \times \nabla' \times \hat{z}) = (\nabla ik_z - \hat{z}\nabla^2)(-\nabla'ik_z - \hat{z}\nabla'^2)$$

$$\sim \hat{z}\hat{z}\frac{\partial^2}{\partial\rho^2}\frac{\partial^2}{\partial\rho'^2} \sim O(k_\rho^4), \quad k_\rho \to \infty, \tag{7.3.25b}$$

$$\nabla\nabla' \sim \hat{\rho}\hat{\rho}'\frac{\partial^2}{\partial\rho\partial\rho'} \sim O(k_\rho^2), \qquad k_\rho \to \infty. \tag{7.3.25c}$$

Thus, after substituting these order-of-magnitude estimates into (15) and then into (7), it is evident that the first term and the second term in (7) do not produce a delta function after integration because their magnitudes vanish when $k_\rho \to \infty$. On the contrary, using (25c), the third term in (7) contains a delta function because it tends to a constant amplitude when $k_\rho \to \infty$. This can be verified in another way. For example, by applying $(\nabla \times \hat{z})(\nabla' \times \hat{z})$ to I_1 given by (21), the Dirac delta function terms from the first term in (21) cancel with those from the second term in (21), although this method of verifying the nonexistence of the Dirac delta function in the first two terms in (7) is rather tedious.

---

[11] Actually, the integrand becomes a sinusoidal function whose amplitude is a constant, because a Dirac delta function has a nondiminishing spectral content even in the very high spectrum region.

Consequently, after substituting I_1, I_2, and I_3 from (21), (22), and (23) into (7), we have (Exercise 7.19)[12]

$$\overline{\mathbf{G}}(\mathbf{r}, \mathbf{r}') = \frac{i}{8\pi} \sum_{n=-\infty}^{\infty} \int_{-\infty}^{\infty} dk_z \frac{1}{k_{0\rho}^2} [\mathbf{M}_n(k_{0\rho}, k_z, \mathbf{r}) \, \mathbf{M}_{-n}(k_{0\rho}, -k_z, \mathbf{r}')$$

$$+ \mathbf{N}_n(k_{0\rho}, k_z, \mathbf{r}) \, \mathbf{N}_{-n}(k_{0\rho}, -k_z, \mathbf{r}')] - \frac{\hat{\rho}\hat{\rho}}{k_0^2} \delta(\mathbf{r} - \mathbf{r}'). \quad (7.3.26)$$

Indeed, the static terms in I_2 and I_3 cancel each other as expected, except for the delta function term. This delta function term has to remain after substituting (23) into (7), as seen from the preceding argument. In addition, the last terms in I_1 and I_2 given in (21) and (22) also cancel each other. This is expected on physical grounds since this is an unphysical field with $k_\rho = 0$, $k_z \neq 0$. In fact, this field does not satisfy the dispersion relationship. More rigorously, they can be shown to cancel as a consequence of

$$(\nabla \times \hat{z})(\nabla' \times \hat{z}) \left(\frac{\rho_<}{\rho_>}\right)^n = -(\nabla \times \nabla \times \hat{z})(\nabla' \times \nabla' \times \hat{z}) \frac{1}{k_z^2} \left(\frac{\rho_<}{\rho_>}\right)^n. \quad (7.3.27)$$

To show this, first note that $\nabla \times \nabla \times \hat{z} = \nabla i k_z - \nabla^2 \hat{z} = \nabla_s i k_z - \nabla_s^2 \hat{z}$. Then,

$$(\nabla \times \nabla \times \hat{z})(\nabla' \times \nabla' \times \hat{z}) \frac{1}{k_z^2} \left(\frac{\rho_<}{\rho_>}\right)^n$$

$$= (\nabla_s i k_z - \nabla_s^2 \hat{z})(-\nabla'_s i k_z - \nabla_s'^2 \hat{z}) \frac{1}{k_z^2} \left(\frac{\rho_<}{\rho_>}\right)^n$$

$$= \nabla_s \nabla'_s \left(\frac{\rho_<}{\rho_>}\right)^n, \quad \rho \neq \rho'. \quad (7.3.28)$$

In the above, we assume that $\partial/\partial z = i k_z$ and $\partial/\partial z' = -i k_z$, and $\nabla_s^2 \left(\frac{\rho_<}{\rho_>}\right)^n = 0$ for $\rho \neq \rho'$. Moreover, assuming the field to have $e^{in\phi}$ and $e^{-in\phi'}$ dependences, we have

$$\nabla_s \left(\frac{\rho}{\rho'}\right)^n = \left(\hat{\rho} \frac{\partial}{\partial\rho} + \hat{\phi} \frac{in}{\rho}\right) \left(\frac{\rho}{\rho'}\right)^n = (\hat{\rho} + i\hat{\phi}) \frac{n}{\rho} \left(\frac{\rho}{\rho'}\right)^n, \quad (7.3.29a)$$

$$\nabla'_s \left(\frac{\rho}{\rho'}\right)^n = (-\hat{\rho}' - i\hat{\phi}') \frac{n}{\rho'} \left(\frac{\rho}{\rho'}\right)^n. \quad (7.3.29b)$$

Then,

$$\nabla_s \nabla'_s \left(\frac{\rho}{\rho'}\right)^n = (\hat{\rho} + i\hat{\phi})(-\hat{\rho}' - i\hat{\phi}') \frac{n^2}{\rho\rho'} \left(\frac{\rho}{\rho'}\right)^n. \quad (7.3.30)$$

---

[12] Different forms were obtained by Howard (1974) and Pearson (1983), but the errors were noted by Collin (private communication).

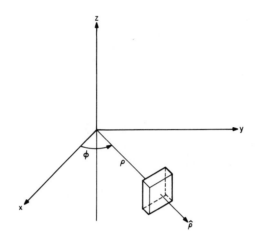

Figure 7.3.3 A pill box whose axis points in the ρ direction and whose ρ dimension shrinks before its $\rho\phi$ and z dimensions.

Furthermore, it can be shown that

$$
(\nabla_s \times \hat{z})(\nabla'_s \times \hat{z}) \left(\frac{\rho}{\rho'} \right)^n = (-\hat{\phi} + i\hat{\rho})(\hat{\phi}' - i\hat{\rho}') \frac{n^2}{\rho\rho'} \left(\frac{\rho}{\rho'} \right)^n
$$
$$
= -\nabla_s \nabla'_s \left(\frac{\rho}{\rho'} \right)^n , \tag{7.3.31}
$$

which verifies (27).

Since I_1 and I_2 in (16) have a nonintegrable pole at $k_\rho = 0$ when $n = 0$, the analyses leading to (21) and (22) for I_1 and I_2 are not strictly correct when $n = 0$. But when I_1 and I_2 are operated on by the differential operators in (15), it can be shown that the nonintegrable pole at $k_\rho = 0$ for $n = 0$ disappears. Furthermore, it can be shown that the $n = 0$ case is similar to ignoring the last terms in (21) and (22) (Exercise 7.19).

Note that in (26), the Bessel functions in $\mathbf{M}_n$ and $\mathbf{N}_n$ are to be replaced by Hankel functions if $\rho > \rho'$ and similarly for $\mathbf{M}_{-n}$ and $\mathbf{N}_{-n}$ when $\rho < \rho'$. Moreover, the form of the Dirac delta function term in (26) is also warranted by the physical interpretation. In this case, since the dk_ρ integral is performed first, it is equivalent to shrinking the ρ length scale first in the principal volume method, yielding a disk-shaped pill box whose axis is pointing in the ρ direction, as shown in Figure 7.3.3. Furthermore, the derivatives in the definitions of $\mathbf{M}$ and $\mathbf{N}$ are already carried out in the integrand of (26).

(c) Spherical Coordinates

In the vector wave function representation of the dyadic Green's function in spherical coordinates, a typical integral in the first term of (8), after exchanging the order of integration and differentiation, is of the form

$$I_1 = \int_0^\infty dk \, k^2 \frac{j_n(kr) j_n(kr')}{k^2 - k_0^2}. \tag{7.3.32}$$

Similarly, in the second term, a typical integral is

$$I_2 = \int_0^\infty dk \, \frac{j_n(kr) j_n(kr')}{k^2 - k_0^2}, \tag{7.3.33}$$

while in the third term, a typical integral is

$$I_3 = \int_0^\infty dk \, \frac{j_n(kr) j_n(kr')}{k_0^2}. \tag{7.3.34}$$

Consequently, on using a technique similar to that of the previous section, we can show that (Exercise 7.20)

$$I_1 = \frac{\pi i}{2} k_0 j_n(k_0 r_<) h_n^{(1)}(k_0 r_>), \tag{7.3.35}$$

where $r_<$ is the smaller of r and r', and $r_>$ is the larger of r and r'. Also, when using the technique on I_2, one has to be wary of a pole at $k = 0$, because $j_n(x) \sim x^n$ and $h_n^{(1)}(x) \sim x^{-n+1}$ when $x \to 0$. Therefore,

$$I_2 = \frac{\pi i}{2k_0} j_n(k_0 r_<) h_n^{(1)}(k_0 r_>) - \frac{\pi}{2(2n+1)k_0^2} \frac{r_<^n}{r_>^{n+1}}. \tag{7.3.36}$$

Similarly,

$$I_3 = \frac{\pi}{2(2n+1)k_0^2} \frac{r_<^n}{r_>^{n+1}}. \tag{7.3.37}$$

Notice that the second term in (36) and I_3 in (37) are the static-like terms. In addition, these static-like fields are expected to cancel each other on physical grounds since these fields should not be observed outside the source region. To illustrate this, we first show that

$$\nabla \times \nabla \times \mathbf{r} = \nabla \frac{\partial}{\partial r} r - \mathbf{r} \nabla^2. \tag{7.3.38}$$

First of all, the Laplacian becomes zero when it operates on the static-like term. Furthermore,

$$\nabla \frac{\partial}{\partial r} r \left(\frac{r^n}{r'^{n+1}} \right) = (n+1) \nabla \left(\frac{r^n}{r'^{n+1}} \right), \tag{7.3.39a}$$

$$\nabla' \frac{\partial}{\partial r'} r' \left(\frac{r^n}{r'^{n+1}} \right) = -n \nabla' \left(\frac{r^n}{r'^{n+1}} \right). \tag{7.3.39b}$$

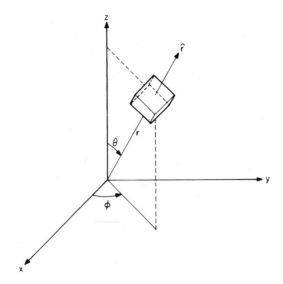

Figure 7.3.4 A pill box whose axis points in the r direction and whose r dimension shrinks before its transverse to r dimensions.

True to our expection, the above, together with (38), shows that the static-like terms in (36) and (37) cancel each other when they are substituted into (8).

By studying the spectral content of the dk integral when $k \to \infty$, only I_3, when substituted into (8), produces a Dirac delta function term in the dk integral. Consequently, (8) becomes (Exercise 7.20; also see Collin 1986b)

$$\overline{\mathbf{G}}(\mathbf{r}, \mathbf{r}') = ik_0 \sum_{n=1}^{\infty} \sum_{m=-n}^{n} \frac{1}{n(n+1)} [\mathbf{M}_{nm}(k_0, \mathbf{r}) \, \hat{\mathbf{M}}_{nm}(k_0, \mathbf{r}')$$

$$+ \mathbf{N}_{nm}(k_0, \mathbf{r}) \, \hat{\mathbf{N}}_{nm}(k_0, \mathbf{r}')] - \frac{\hat{r}\hat{r}}{k_0^2} \, \delta(\mathbf{r} - \mathbf{r}'). \quad (7.3.40)$$

The fact is that the Dirac delta function term in (40) is warranted by the physical interpretation, since we are evaluating the dk integral first, implying that the length scale in the r direction is being shrunk first. Therefore, this is similar to a principal volume integral where the length scale in the r direction is shrunk before length scales in other directions. Hence, the corresponding pill box has its axis pointing in the r direction, as shown in Figure 7.3.4.

In a manner corresponding to the result in the cylindrical-coordinate case, when $r > r'$, the spherical Bessel functions in $\mathbf{M}_{nm}(k_0, \mathbf{r})$ and $\mathbf{N}_{nm}(k_0, \mathbf{r})$ in (40) should be replaced by the spherical Hankel functions, and similarly, for $\hat{\mathbf{M}}_{nm}(k_0, \mathbf{r}')$ and $\hat{\mathbf{N}}_{nm}(k_0, \mathbf{r}')$ when $r < r'$. Furthermore, the derivatives in the definitions of the $\mathbf{M}$ and $\mathbf{N}$ functions are already carried out in the integrand of (40).

§7.4 Dyadic Green's Functions for Layered Media

In the previous section, we have illustrated how the dyadic Green's functions can be obtained in terms of the vector wave functions of the medium. In addition, these vector wave functions can be found for a homogeneous medium. But when the medium becomes inhomogeneous, they cannot be found easily. For example, in a planarly layered medium, the solution of Maxwell's equations is sought by decomposing the electromagnetic waves into TE and TM types. But for a cylindrically layered medium, the electromagnetic waves are in general hybrid because the TE and TM waves are usually coupled. On the other hand, for a spherically layered medium, again, the waves can be decomposed into TE and TM to r waves. Despite these complexities, we shall illustrate how to derive the dyadic Green's functions for layered media. Before doing so, a constraint on the dyadic Green's function due to the reciprocity requirement will be derived first for a general, inhomogeneous medium.

In passing, it should be mentioned that dyadic Green's functions for planarly layered media have also been presented by Tsang et al. (1975); Lee and Kong (1983); Kong (1986); Bagby and Nyquist (1987); and Viola and Nyquist (1988).

§§7.4.1 A General, Isotropic, Inhomogeneous Medium

A dyadic Green's function for a general, isotropic inhomogeneous medium can be defined even though its derivation may be difficult. Since for an inhomogeneous medium, the electric field satisfies

$$\nabla \times \mu^{-1}(\mathbf{r})\nabla \times \mathbf{E}(\mathbf{r}) - \omega^2 \epsilon(\mathbf{r})\, \mathbf{E}(\mathbf{r}) = i\omega \mathbf{J}(\mathbf{r}), \qquad (7.4.1)$$

a dyadic Green's function $\overline{\mathbf{G}}(\mathbf{r}, \mathbf{r}')$ for an inhomogeneous medium can be defined such that it is a solution of the following equation:

$$\nabla \times \mu^{-1}(\mathbf{r})\nabla \times \overline{\mathbf{G}}(\mathbf{r}, \mathbf{r}') - \omega^2 \epsilon(\mathbf{r})\, \overline{\mathbf{G}}(\mathbf{r}, \mathbf{r}') = \mu^{-1}(\mathbf{r})\, \overline{\mathbf{I}}\, \delta(\mathbf{r} - \mathbf{r}'). \qquad (7.4.2)$$

To aid in the thought process, (2) can be post-multiplied by an arbitrary constant vector $\mathbf{a}$ to yield

$$\nabla \times \mu^{-1}(\mathbf{r})\nabla \times \overline{\mathbf{G}}(\mathbf{r}, \mathbf{r}') \cdot \mathbf{a} - \omega^2 \epsilon(\mathbf{r})\, \overline{\mathbf{G}}(\mathbf{r}, \mathbf{r}') \cdot \mathbf{a} = \mu^{-1}(\mathbf{r})\, \overline{\mathbf{I}} \cdot \mathbf{a}\, \delta(\mathbf{r} - \mathbf{r}').$$
$$(7.4.2a)$$

Now, post-multiplying (1) by $\overline{\mathbf{G}}(\mathbf{r}, \mathbf{r}') \cdot \mathbf{a}$, pre-multiplying (2a) by $\mathbf{E}(\mathbf{r})$, and subtracting the results yield

$$\nabla \times \mu^{-1}(\mathbf{r})\nabla \times \mathbf{E}(\mathbf{r}) \cdot \overline{\mathbf{G}}(\mathbf{r}, \mathbf{r}') \cdot \mathbf{a} - \mathbf{E}(\mathbf{r}) \cdot \nabla \times \mu^{-1}(\mathbf{r})\nabla \times \overline{\mathbf{G}}(\mathbf{r}, \mathbf{r}') \cdot \mathbf{a}$$
$$= i\omega \mathbf{J}(\mathbf{r}) \cdot \overline{\mathbf{G}}(\mathbf{r}, \mathbf{r}') \cdot \mathbf{a} - \mu^{-1}(\mathbf{r})\, \mathbf{E}(\mathbf{r}) \cdot \mathbf{a}\, \delta(\mathbf{r} - \mathbf{r}'). \qquad (7.4.3)$$

On integrating the above over all $\mathbf{r}$, converting the left-hand side into a divergence with the necessary vector identity, invoking Gauss' theorem, and simplifying, one obtains (Exercise 7.21)

$$\mathbf{E}(\mathbf{r}') = i\omega\mu(\mathbf{r}') \int d\mathbf{r} \mathbf{J}(\mathbf{r}) \cdot \overline{\mathbf{G}}(\mathbf{r}, \mathbf{r}'). \tag{7.4.4}$$

As a result, the preceding equation is the relationship between the electric field and the current via the dyadic Green's function defined for the inhomogeneous medium. Because of reciprocity, we can easily show that (Exercise 7.21)

$$\overline{\mathbf{G}}(\mathbf{r}, \mathbf{r}')\mu(\mathbf{r}') = \overline{\mathbf{G}}^t(\mathbf{r}', \mathbf{r})\mu(\mathbf{r}). \tag{7.4.5}$$

Note that this property of the dyadic Green's function expressed by (5) should be a property satisfied by all dyadic Green's functions derived for an inhomogeneous medium. Thus, when deriving the dyadic Green's function for layered media, we shall keep property (5) in mind. Note further that Equation (1.3.52) of Chapter 1 is a special case of the above. With (5), (4) can be alternatively written as

$$\mathbf{E}(\mathbf{r}) = i\omega \int d\mathbf{r}' \, \overline{\mathbf{G}}(\mathbf{r}, \mathbf{r}')\mu(\mathbf{r}') \cdot \mathbf{J}(\mathbf{r}'), \tag{7.4.5a}$$

where we have swapped $\mathbf{r}$ and $\mathbf{r}'$. Without much difficulty, the corresponding magnetic field is derived from the above using Maxwell's equations as

$$\mathbf{H}(\mathbf{r}) = \int d\mathbf{r}' \, \mu^{-1}(\mathbf{r})\nabla \times \overline{\mathbf{G}}(\mathbf{r}, \mathbf{r}')\mu(\mathbf{r}') \cdot \mathbf{J}(\mathbf{r}'). \tag{7.4.5b}$$

The above can be extended to the anisotropic case (see Subsection 8.1.3, Chapter 8).

§§7.4.2 Planarly Layered Media

For a planarly layered medium, the dyadic Green's function can be found by starting with the dyadic Green's function in a homogeneous medium. Then, the inhomogeneity can be added later, giving rise to reflected wave terms in the dyadic Green's function.

As an example, for the source point in region m in a layered medium as shown in Figure 7.4.1, consider first the dyadic Green's function for a homogeneous medium consisting of the medium from region m. The dyadic Green's function in the absence of all the other regions is then

$$\overline{\mathbf{G}}(\mathbf{r}, \mathbf{r}') = \frac{i}{8\pi^2} \iint\limits_{-\infty}^{\infty} \frac{d\mathbf{k}_s}{k_{mz}k_s^2} [\mathbf{M}(\mathbf{k}_s, \mathbf{r})\mathbf{M}(-\mathbf{k}_s, \mathbf{r}')$$

$$+ \mathbf{N}(\mathbf{k}_s, \mathbf{r})\mathbf{N}(-\mathbf{k}_s, \mathbf{r}')] - \frac{\hat{z}\hat{z}}{k_m^2}\delta(\mathbf{r} - \mathbf{r}'), \tag{7.4.6}$$

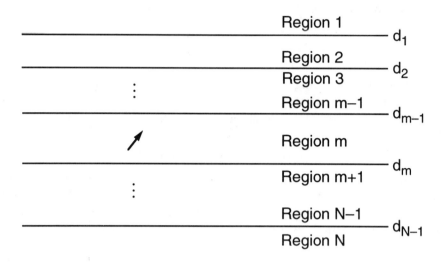

Figure 7.4.1 The dyadic Green's function for the case when the source point is in region m.

where

$$\mathbf{M}(\mathbf{k}_s, \mathbf{r})\,\mathbf{M}(-\mathbf{k}_s, \mathbf{r}') = (\mathbf{k}_s \times \hat{z})(\mathbf{k}_s \times \hat{z})e^{i\mathbf{k}_s \cdot (\mathbf{r}_s - \mathbf{r}'_s) + i k_{mz}|z - z'|},$$

$$(7.4.7a)$$

$$\mathbf{N}(\mathbf{k}_s, \mathbf{r})\,\mathbf{N}(-\mathbf{k}_s, \mathbf{r}') = \frac{1}{k_m^2}(\mathbf{k}_{m\pm} \times \mathbf{k}_s \times \hat{z})(\mathbf{k}_{m\mp} \times \mathbf{k}_s \times \hat{z})e^{i\mathbf{k}_s \cdot (\mathbf{r}_s - \mathbf{r}'_s) + i k_{mz}|z - z'|},$$

$$(7.4.7b)$$

$\mathbf{k}_{m\pm} = \mathbf{k}_s \pm \hat{z}k_{mz}$, and the upper sign is chosen when $z > z'$ and the lower sign is chosen when $z < z'$. Moreover, the $\mathbf{M}$ function represents the TE to z wave, whereas the $\mathbf{N}$ function, which is the curl of the TM to z magnetic field, represents the TM to z waves.

At this point, for bookkeeping purposes, it is more expedient to write (7a) and (7b) as

$$\mathbf{M}(\mathbf{k}_s, \mathbf{r})\mathbf{M}(-\mathbf{k}_s, \mathbf{r}') = \nabla \times \hat{z}\left[\nabla' \times \hat{z}e^{i\mathbf{k}_s \cdot (\mathbf{r}_s - \mathbf{r}'_s) + i k_{mz}|z - z'|}\right],$$

$$(7.4.8a)$$

$$\mathbf{N}(\mathbf{k}_s, \mathbf{r})\mathbf{N}(-\mathbf{k}_s, \mathbf{r}') = \frac{\nabla \times \nabla \times \hat{z}}{i\omega\epsilon_m}\left[\frac{\nabla' \times \nabla' \times \hat{z}}{-i\omega\mu_m}e^{i\mathbf{k}_s \cdot (\mathbf{r}_s - \mathbf{r}'_s) + i k_{mz}|z - z'|}\right]$$

$$(7.4.8b)$$

with the understanding that the Dirac delta function singularity from

$$\frac{\partial^2}{\partial z \partial z'}e^{i k_{mz}|z - z'|}$$

is ignored since it is already extracted as the last term in (6).

For plane waves, one can show that (Exercise 7.22)

$$\mathbf{E} = \frac{i\omega}{k_s^2}\left[\frac{1}{i\omega\epsilon}\nabla\times\nabla\times\hat{z}D_z + \nabla\times\hat{z}B_z\right]. \tag{7.4.9}$$

Since the dyadic Green's function (6) generates an electric field when it acts on an electric current source, the quantity in the square bracket in (8a) is proportional to B_z, while that in the square bracket in (8b) is proportional to D_z after comparing it with Equation (9). Therefore, in the presence of horizontal stratifications, these quantities transmit and reflect like B_z and D_z plane waves. But these reflections and transmissions were studied in Chapter 2, and those results can be used here.

In the presence of horizontal layers, when both z and z' are in region m of a planarly layered medium, (8a) is modified to become

$$\overline{\mathbf{M}}(\mathbf{k}_s,\mathbf{r},\mathbf{r}') = (\nabla\times\hat{z})(\nabla'\times\hat{z})e^{i\mathbf{k}_s\cdot(\mathbf{r}_s-\mathbf{r}_s')}\left[e^{ik_{mz}|z-z'|} + B_m^{TE}e^{-ik_{mz}z} + D_m^{TE}e^{ik_{mz}z}\right]. \tag{7.4.10}$$

Better still, B_m^{TE} and D_m^{TE} can be found as in Chapter 2, Equations (2.4.5a) and (2.4.5b). Note that $(\nabla\times\hat{z})(\nabla'\times\hat{z})$, which is $(\mathbf{k}_s\times\hat{z})(\mathbf{k}_s\times\hat{z})$, is just a constant dyad. Consequently, the use of (2.4.6a) and (2.4.6b) in (10) yields

$$\overline{\mathbf{M}}(\mathbf{k}_s,\mathbf{r},\mathbf{r}') = (\nabla\times\hat{z})(\nabla'\times\hat{z})e^{i\mathbf{k}_s\cdot(\mathbf{r}_s-\mathbf{r}_s')}\left[e^{-ik_{mz}z_<}\right.$$

$$\left. + e^{ik_{mz}(z_<+2d_m)}\tilde{R}_{m,m+1}^{TE}\right]\left[e^{ik_{mz}z_>} + e^{-ik_{mz}(z_>+2d_{m-1})}\tilde{R}_{m,m-1}^{TE}\right]\tilde{M}_m, \tag{7.4.11}$$

where $z_>$ is the larger of z and z' and $z_<$ is the lesser of z and z', and $\tilde{M}_m = \left[1 - \tilde{R}_{m,m-1}\tilde{R}_{m,m+1}e^{2ik_{mz}(d_m-d_{m-1})}\right]^{-1}$. Moreover, the generalized reflection coefficients $\tilde{R}$ can be derived as in Chapter 2. In general,

$$\overline{\mathbf{M}}(\mathbf{k}_s,\mathbf{r},\mathbf{r}') = (\nabla\times\hat{z})(\nabla'\times\hat{z})e^{i\mathbf{k}_s\cdot(\mathbf{r}_s-\mathbf{r}_s')}F_{\pm}^{TE}(z,z'), \tag{7.4.12}$$

where $F_{\pm}^{TE}(z,z')$ are given by (2.4.6a), (2.4.6b), (2.4.15), and (2.4.16) for TE waves, depending on where $\mathbf{r}$ is located with respect to $\mathbf{r}'$.

Similarly, in a layered medium, (8b) is modified to

$$\overline{\mathbf{N}}(\mathbf{k}_s,\mathbf{r},\mathbf{r}') = \left(\frac{\nabla\times\nabla\times\hat{z}}{i\omega\epsilon_n}\right)\left(\frac{\nabla'\times\nabla'\times\hat{z}}{-i\omega\mu_m}\right)e^{i\mathbf{k}_s\cdot(\mathbf{r}_s-\mathbf{r}_s')}F_{\pm}^{TM}(z,z'), \tag{7.4.13}$$

where $F_{\pm}^{TM}(z,z')$ are given by (2.4.6a), (2.4.6b), (2.4.15), and (2.4.16) for TM waves (Exercise 7.23), and $\mathbf{r}\in$ region n, $\mathbf{r}'\in$ region m. Note that $(\nabla\times\nabla\times\hat{z})(\nabla'\times\nabla'\times\hat{z}) = (\mathbf{k}_{n\pm}\times\mathbf{k}_s\times\hat{z})(\mathbf{k}_{m\mp}\times\mathbf{k}_s\times\hat{z})$, where $\mathbf{k}_{n\pm} = \mathbf{k}_s\pm\hat{z}k_{mz}$ and $\mathbf{k}_{m\mp} = \mathbf{k}_s\mp\hat{z}k_{mz}$, is a constant dyad.[13] Also, μ_m instead of μ_n is in

---

[13] The upper sign is chosen when $z > z'$ and the lower sign is chosen when $z < z'$.

the denominator of the second factor in (13) because the second factor is associated with $\mathbf{r}'$, the source point.

With these notations, the dyadic Green's function for a planarly layered medium becomes

$$\overline{\mathbf{G}}(\mathbf{r}, \mathbf{r}') = \frac{i}{8\pi^2} \iint\limits_{-\infty}^{\infty} \frac{d\mathbf{k}_s}{k_{mz}k_s^2} \left[\overline{\mathbf{M}}(\mathbf{k}_s, \mathbf{r}, \mathbf{r}') + \overline{\mathbf{N}}(\mathbf{k}_s, \mathbf{r}, \mathbf{r}') \right] - \frac{\hat{z}\hat{z}}{k_m^2} \delta(\mathbf{r} - \mathbf{r}'), \tag{7.4.14}$$

with $\overline{\mathbf{M}}(\mathbf{k}_s, \mathbf{r}, \mathbf{r}')$ and $\overline{\mathbf{N}}(\mathbf{k}_s, \mathbf{r}, \mathbf{r}')$ given by (12) and (13), and the source point $\mathbf{r}' \in$ region m while the field point $\mathbf{r} \in$ region n [see Equation (5a)].

§§7.4.3 Cylindrically Layered Media

When a medium is cylindrically layered, the TE and TM waves are, in general, coupled at the interfaces. Hence, the $\mathbf{M}$ and $\mathbf{N}$ vector wave functions in cylindrical coordinates, which represent the TE and TM to z waves, are coupled together. In this case, the TE and TM waves or the $\mathbf{M}$ and $\mathbf{N}$ functions should be treated together. Consequently, a function for the integrand of (7.3.26) can be defined as

$$\overline{\mathbf{C}}_n(k_z, \mathbf{r}, \mathbf{r}') = [\mathbf{N}_n(k_\rho, k_z, \mathbf{r}) \, \mathbf{N}_{-n}(-k_\rho, -k_z, \mathbf{r}')$$

$$+ \mathbf{M}_n(k_\rho, k_z, \mathbf{r}) \, \mathbf{M}_{-n}(-k_\rho, -k_z, \mathbf{r}')] \frac{1}{k_\rho^2}$$

$$= \frac{1}{k_\rho^2} [\mathbf{N}_n(k_\rho, k_z, \mathbf{r}), \mathbf{M}_n(k_\rho, k_z, \mathbf{r})] \begin{bmatrix} \mathbf{N}_{-n}(-k_\rho, -k_z, \mathbf{r}') \\ \mathbf{M}_{-n}(-k_\rho, -k_z, \mathbf{r}') \end{bmatrix}, \tag{7.4.15}$$

where $k_\rho^2 + k_z^2 = k^2 = \omega^2\mu\epsilon$. Alternatively, the above can be rewritten as

$$\overline{\mathbf{C}}_n(k_z, \mathbf{r}, \mathbf{r}') = \left[\frac{1}{k}\nabla \times \nabla \times \hat{z}, \nabla \times \hat{z} \right] \begin{bmatrix} \frac{1}{k}(\nabla' \times \nabla' \times \hat{z})^t \\ (\nabla' \times \hat{z})^t \end{bmatrix} \frac{1}{k_\rho^2}$$

$$\cdot J_n(k_\rho\rho_<)H_n^{(1)}(k_\rho\rho_>)e^{in(\phi-\phi')+ik_z(z-z')}. \tag{7.4.16}$$

In actuality, the derivatives are already carried out in (7.3.26), but in the form above, one avoids working with derivatives of Bessel functions explicitly. Equivalently, one may think that in the above, the action of differential operators produces a Dirac delta function which is to be ignored since it is already extracted in (7.3.26). Consequently, Equation (16) can be written as (Exercise 7.24)

$$\overline{\mathbf{C}}_n(k_z, \mathbf{r}, \mathbf{r}') = \frac{1}{(kk_\rho)^2} \overline{\mathbf{D}}_\mu J_n(k_\rho\rho_<)H_n^{(1)}(k_\rho\rho_>)e^{in(\phi-\phi')+ik_z(z-z')} \overset{\leftarrow}{\overline{\mathbf{D}}}{}'_\epsilon{}^t, \tag{7.4.17}$$

where $\overline{\mathbf{D}}_\mu = [\nabla \times \nabla \times \hat{z}, i\omega\mu\nabla \times \hat{z}]$, and $\overset{\leftarrow}{\overline{\mathbf{D}}}{}'_\epsilon = [\nabla' \times \nabla' \times \hat{z}, -i\omega\epsilon\nabla' \times \hat{z}]$ operates on the primed coordinates to its left. Using the fact that [see Equation (9)]

$$\mathbf{E} = \frac{1}{k_\rho^2} [\nabla \times \nabla \times \mathbf{E}_z + i\omega\mu\nabla \times \hat{z}H_z] \tag{7.4.18}$$

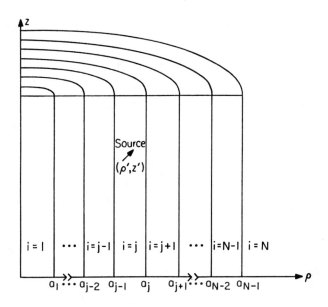

Figure 7.4.2 The dyadic Green's function for a source in region j of a cylindrically layered medium.

for fields with $e^{ik_z z}$ dependence, we note that the quantity to the right of $\overline{\mathbf{D}}_\mu$ in (17) is proportional to $[E_z, H_z]^t$. Hence, in the presence of a layered medium, this quantity will transmit and reflect like $[E_z, H_z]^t$, which was discussed in Chapter 3.

When the observation and the source points $\mathbf{r}$ and $\mathbf{r}'$ are in region j of a cylindrically layered medium as shown in Figure 7.4.2, the function $\overline{\mathbf{C}}_n(k_z, \mathbf{r}, \mathbf{r}')$ has to be modified to include reflected wave terms. As a result,

$$\overline{\mathbf{C}}_n(k_z, \mathbf{r}, \mathbf{r}') = \frac{1}{(k_j k_{j\rho})^2} \overline{\mathbf{D}}_{j\mu} \cdot \left[J_n(k_{j\rho}\rho_<)H_n^{(1)}(k_{j\rho}\rho_>)\,\overline{\mathbf{I}} + J_n(k_{j\rho}\rho)\,\overline{\mathbf{a}}_{jn} \right.$$
$$\left. + H_n^{(1)}(k_{j\rho}\rho)\,\overline{\mathbf{b}}_{jn} \right] f_n(z, \phi; z', \phi') \cdot \overleftarrow{\overline{\mathbf{D}}}'^{\,t}_{j\epsilon}, \quad (7.4.19)$$

where $k_{j\rho} = \sqrt{k_j^2 - k_z^2}$ and $f_n(z, \phi; z', \phi') = e^{in(\phi-\phi')+ik_z(z-z')}$. Since the quantity to the right of $\overline{\mathbf{D}}_{j\mu}$ in (19) is proportional to $[E_z, H_z]^t$, $\overline{\mathbf{a}}_{jn}$ and $\overline{\mathbf{b}}_{jn}$ are the same as that given in (3.3.11) and (3.3.12) of Chapter 3.

On the other hand, when $\mathbf{r} \in$ region $i > j$, the function $\overline{\mathbf{C}}_n(k_z, \mathbf{r}, \mathbf{r}')$ becomes

$$\overline{\mathbf{C}}_n(k_z, \mathbf{r}, \mathbf{r}') = \frac{1}{(k_j k_{i\rho})^2} \overline{\mathbf{D}}_{i\mu} \cdot \left[H_n^{(1)}(k_{i\rho}\rho)\,\overline{\mathbf{I}} + J_n(k_{i\rho}\rho)\widetilde{\overline{\mathbf{R}}}_{i,i+1} \right] \cdot \mathbf{a}_{in}.$$
$$(7.4.20)$$

$\mathbf{a}_{in}$ can be found via the methods of (3.3.16) and (3.3.17) of Chapter 3. Hence, analogous to (3.3.18),

$$\overline{\mathbf{C}}_n(k_z, \mathbf{r}, \mathbf{r}') = \frac{1}{(k_j k_{i\rho})^2} \overline{\mathbf{D}}_{i\mu} \cdot \left[H_n^{(1)}(k_{i\rho}\rho)\,\overline{\mathbf{I}} + J_n(k_{i\rho}\rho)\widetilde{\overline{\mathbf{R}}}_{i,i+1} \right] \cdot \widetilde{\overline{\mathbf{M}}}_{i+} \cdot \widetilde{\overline{\mathbf{T}}}_{ji} \cdot \widetilde{\overline{\mathbf{M}}}_{j+}$$

$$\cdot \left[J_n(k_{j\rho}\rho')\,\overline{\mathbf{I}} + H_n^{(1)}(k_{j\rho}\rho')\widetilde{\overline{\mathbf{R}}}_{j,j-1} \right] f_n(z, \phi; z', \phi') \cdot \overleftarrow{\overline{\mathbf{D}}}'^{\,t}_{j\epsilon}. \quad (7.4.21)$$

By the same token, when $\mathbf{r} \in$ region $i < j$, the function $\overline{\mathbf{C}}_n(k_z, \mathbf{r}, \mathbf{r}')$, analogous to (3.3.20), becomes

$$\overline{\mathbf{C}}_n(k_z, \mathbf{r}, \mathbf{r}') = \frac{1}{(k_j k_{i\rho})^2} \overline{\mathbf{D}}_{i\mu} \cdot \left[J_n(k_{i\rho}\rho)\,\overline{\mathbf{I}} + H_n^{(1)}(k_{i\rho}\rho)\widetilde{\overline{\mathbf{R}}}_{i,i-1} \right] \cdot \widetilde{\overline{\mathbf{M}}}_{i-} \cdot \widetilde{\overline{\mathbf{T}}}_{ji} \cdot \widetilde{\overline{\mathbf{M}}}_{j-}$$

$$\cdot \left[H_n^{(1)}(k_{j\rho}\rho')\,\overline{\mathbf{I}} + J_n(k_{j\rho}\rho')\widetilde{\overline{\mathbf{R}}}_{j,j+1} \right] f_n(z, \phi; z', \phi') \cdot \overleftarrow{\overline{\mathbf{D}}}'^{\,t}_{j\epsilon}. \quad (7.4.22)$$

In summary,

$$\overline{\mathbf{C}}_n(k_z, \mathbf{r}, \mathbf{r}') = \frac{1}{(k_j k_{i\rho})^2} \overline{\mathbf{D}}_{i\mu} \cdot \overline{\mathbf{F}}_n(\rho, \rho') f_n(z, \phi; z', \phi') \cdot \overleftarrow{\overline{\mathbf{D}}}'^{\,t}_{j\epsilon}, \quad (7.4.23)$$

where $\overline{\mathbf{F}}_n(\rho, \rho')$ is defined in (3.3.22) to (3.3.24) of Chapter 3. In the above, the subscript j denotes the region of the source point ρ', and the subscript i denotes the region of the observation point ρ.

In this manner, the dyadic Green's function for a cylindrically layered medium can be written as

$$\overline{\mathbf{G}}(\mathbf{r}, \mathbf{r}') = \frac{i}{8\pi} \sum_{n=-\infty}^{\infty} \int_{-\infty}^{\infty} dk_z \frac{1}{(k_j k_{i\rho})^2} \overline{\mathbf{D}}_{i\mu} \cdot \overline{\mathbf{F}}_n(\rho, \rho') f_n(z, \phi; z', \phi') \cdot \overleftarrow{\overline{\mathbf{D}}}'^{\,t}_{j\epsilon}$$

$$- \frac{\hat{\rho}\hat{\rho}}{k_j^2} \delta(\mathbf{r} - \mathbf{r}') \quad (7.4.24)$$

where the source point $\mathbf{r}' \in$ region j and the field point $\mathbf{r} \in$ region i [see Equation (5a)].

§§7.4.4 Spherically Layered Media

For a spherically layered medium, the waves can be decomposed into TE and TM to r waves. First, we note that for the homogeneous medium case with wave number k_j, the dyadic Green's function is given by Equation (7.3.40), where

$$\sum_{m=-n}^{n} \mathbf{M}_{nm}(k_j, \mathbf{r}) \hat{\mathbf{M}}_{nm}(k_j, \mathbf{r}') = (\nabla \times \mathbf{r})[\nabla' \times \mathbf{r}' j_n(k_j r_<) h_n^{(1)}(k_j r_>)$$

$$A_n(\theta, \phi; \theta', \phi')], \quad (7.4.25a)$$

$$\sum_{m=-n}^{n} \mathbf{N}_{nm}(k_j, \mathbf{r})\,\hat{\mathbf{N}}_{nm}(k_j, \mathbf{r}') =$$

$$\left(\frac{\nabla \times \nabla \times \mathbf{r}}{-i\omega\epsilon_j}\right)\left[\frac{\nabla' \times \nabla' \times \mathbf{r}'}{i\omega\mu_j}j_n(k_j r_<)h_n^{(1)}(k_j r_>)A_n(\theta, \phi; \theta', \phi')\right], \quad (7.4.25b)$$

where $r_<$ is the smaller of r and r', and conversely for $r_>$. Furthermore,

$$A_n(\theta, \phi; \theta', \phi') = \sum_{m=-n}^{n} Y_{nm}(\theta, \phi)Y_{nm}^*(\theta'\phi'). \quad (7.4.25c)$$

In actuality, the derivatives in (7.3.40) are already carried out; however, the notation in (25) avoids dealing with the derivatives of Bessel functions and spherical harmonics. Equivalently, the resultant Dirac delta function produced by the action of the differential operators on the discontinuous functions to their right is to be ignored.

In the presence of a layered medium, the above would have to be modified accordingly. By comparing with Equation (3.5.4) of Chapter 3, we note that the quantities in the square brackets of (25a) and (25b) are proportional to the Debye potentials. Hence, in the presence of spherical layers, these quantities will transmit and reflect like Debye potentials, which were studied in Chapter 3. Consequently, with the definition that $\overline{\mathbf{m}}_n(\mathbf{r}, \mathbf{r}')$ equals the summation on the left-hand side of (25a), (25a) in a layered medium becomes

$$\overline{\mathbf{m}}_n(\mathbf{r}, \mathbf{r}') = (\nabla \times \mathbf{r})(\nabla' \times \mathbf{r}')\left[j_n(k_j r_<)h_n^{(1)}(k_j r_>) + a_{jn}^{TE}h_n^{(1)}(k_j r)\right.$$
$$\left. + b_{jn}^{TE}j_n(k_j r)\right]A_n(\theta, \phi; \theta', \phi'), \quad (7.4.26)$$

where both $\mathbf{r}$ and $\mathbf{r}'$ are in region j. Moreover, the coefficients a_{jn}^{TE} and b_{jn}^{TE} can be found as in Equations (3.7.17a) and (3.7.17b) of Chapter 3. Analogous to (3.7.18), using them in (26) for TE waves yields

$$\overline{\mathbf{m}}_n(\mathbf{r}, \mathbf{r}') = (\nabla \times \mathbf{r})(\nabla' \times \mathbf{r}')\left[h_n^{(1)}(k_j r_>) + \tilde{R}_{j,j+1}^{TE}j_n(k_j r_>)\right]$$
$$\cdot\left[j_n(k_j r_<) + \tilde{R}_{j,j-1}^{TE}h_n^{(1)}(k_j r_<)\right]\tilde{M}_j^{TE}A_n(\theta, \phi; \theta', \phi'), \quad (7.4.27)$$

where $\tilde{M}_j^{TE} = (1 - \tilde{R}_{j,j+1}^{TE}\tilde{R}_{j,j-1}^{TE})^{-1}$ and $\tilde{R}^{TE}$ are TE generalized reflection coefficients derived in Chapter 3.

On the other hand, when $\mathbf{r} \in$ region $i > j$ and $\mathbf{r}' \in$ region j, using Equation (3.7.22) of Chapter 3 yields

$$\overline{\mathbf{m}}_n(\mathbf{r}, \mathbf{r}') = (\nabla \times \mathbf{r})(\nabla' \times \mathbf{r}')\left[h_n^{(1)}(k_i r) + \tilde{R}_{i,i+1}^{TE}j_n(k_i r)\right]$$
$$\cdot\left[j_n(k_j r') + \tilde{R}_{j,j-1}^{TE}h_n^{(1)}(k_j r')\right]\tilde{T}_{ji}^{TE}\tilde{M}_i^{TE}\tilde{M}_j^{TE}A_n(\theta, \phi; \theta', \phi'), \quad (7.4.28)$$

where $\tilde{T}_{ji}^{TE}$ is the TE generalized transmission coefficient for a layered slab between region j and region i.

By the same token, when $\mathbf{r} \in$ region $i < j$ and $\mathbf{r}' \in$ region j, using Equation (3.7.24) of Chapter 3, we have

$$\overline{\mathbf{m}}_n(\mathbf{r}, \mathbf{r}') = (\nabla \times \mathbf{r})(\nabla' \times \mathbf{r}') \left[j_n(k_i r) + \tilde{R}_{i,i-1}^{TE} h_n^{(1)}(k_i r) \right]$$
$$\cdot \left[h_n^{(1)}(k_j r') + \tilde{R}_{j,j+1}^{TE} j_n(k_j r') \right] \tilde{T}_{ji}^{TE} \tilde{M}_i^{TE} \tilde{M}_j^{TE} A_n(\theta, \phi; \theta', \phi'). \quad (7.4.29)$$

In summary,

$$\overline{\mathbf{m}}_n(\mathbf{r}, \mathbf{r}') = (\nabla \times \mathbf{r})(\nabla' \times \mathbf{r}') F_n^{TE}(r, r') A_n(\theta, \phi; \theta', \phi'), \quad (7.4.30)$$

where $F_n^{TE}(r, r')$ is given in (3.7.26) to (3.7.28) of Chapter 3, but adapted for TE waves. In (30), $\mathbf{r} \in$ region i, and $\mathbf{r}' \in$ region j.

Similarly, (25b) has to be modified for a spherically layered medium. Thus, with the definition that $\overline{\mathbf{n}}_n(\mathbf{r}, \mathbf{r}')$ equals the summation on the left-hand side of (25b), we have

$$\overline{\mathbf{n}}_n(\mathbf{r}, \mathbf{r}') = \left(\frac{\nabla \times \nabla \times \mathbf{r}}{-i\omega\epsilon_i} \right) \left(\frac{\nabla' \times \nabla' \times \mathbf{r}'}{i\omega\mu_j} \right) F_n^{TM}(r, r') A_n(\theta, \phi; \theta', \phi'),$$
$$(7.4.31)$$

where $F_n^{TM}(r, r')$ is given in (3.7.26) to (3.7.28) of Chapter 3, but adapted for TM waves. In (31), $\mathbf{r}' \in$ region j and $\mathbf{r} \in$ region i.

In this manner, the dyadic Green's function for a spherically layered medium is

$$\overline{\mathbf{G}}(\mathbf{r}, \mathbf{r}') = ik_j \sum_{n=1}^{\infty} \frac{1}{n(n+1)} [\overline{\mathbf{m}}_n(\mathbf{r}, \mathbf{r}') + \overline{\mathbf{n}}_n(\mathbf{r}, \mathbf{r}')] - \frac{\hat{r}\hat{r}}{k_j^2} \delta(\mathbf{r} - \mathbf{r}'), \quad (7.4.32)$$

where k_j is the wavenumber of the region in which the source point $\mathbf{r}'$ appears.

§§7.4.5 Reciprocity Considerations

The dyadic Green's functions derived in the previous subsections must satisfy reciprocity which, according to (5), is that $\overline{\mathbf{G}}(\mathbf{r}, \mathbf{r}')\mu(\mathbf{r}')$ must be equal to $\overline{\mathbf{G}}^t(\mathbf{r}', \mathbf{r})\mu(\mathbf{r})$. Therefore, we shall show that the dyadic Green's functions derived previously satisfy this condition.

(a) Planarly Layered Media Case

The dyadic Green's function for a planarly layered medium is given by (14). In order for reciprocity to be satisfied, we require that (Exercise 7.25)

$$\overline{\mathbf{M}}(\mathbf{k}_s, \mathbf{r}, \mathbf{r}') \frac{\mu_m}{k_{mz}} = \overline{\mathbf{M}}^t(\mathbf{k}_s, \mathbf{r}', \mathbf{r}) \frac{\mu_n}{k_{nz}}, \quad (7.4.33a)$$

$$\overline{\mathbf{N}}(\mathbf{k}_s, \mathbf{r}, \mathbf{r}') \frac{\mu_m}{k_{mz}} = \overline{\mathbf{N}}^t(\mathbf{k}_s, \mathbf{r}', \mathbf{r}) \frac{\mu_n}{k_{nz}}, \quad (7.4.33b)$$

where $\mathbf{r}' \in$ region m and $\mathbf{r} \in$ region n. From (12) and (13), Equations (33a) and (33b) imply that

$$\frac{\mu_m}{k_{mz}} F_\pm^{TE}(z, z') = \frac{\mu_n}{k_{nz}} F_\mp^{TE}(z', z), \qquad (7.4.34a)$$

$$\frac{\epsilon_m}{k_{mz}} F_\pm^{TM}(z, z') = \frac{\epsilon_n}{k_{nz}} F_\mp^{TM}(z', z). \qquad (7.4.34b)$$

Notice that when n and m denote the same region, or z and z' are in the same region, the above is clearly satisfied. But when n and m are of two different regions, from (2.4.15) and (2.4.16) of Chapter 2, (34a) and (34b) would be satisfied if

$$\frac{\mu_m}{k_{mz}} \tilde{T}_{mn}^{TE} = \frac{\mu_n}{k_{nz}} \tilde{T}_{nm}^{TE}, \qquad (7.4.35a)$$

$$\frac{\epsilon_m}{k_{mz}} \tilde{T}_{mn}^{TM} = \frac{\epsilon_n}{k_{nz}} \tilde{T}_{nm}^{TM}. \qquad (7.4.35b)$$

The above is proven in Exercise 2.4 of Chapter 2.

(b) Cylindrically Layered Media Case

The dyadic Green's function for a cylindrically layered medium is shown in (24). In order for reciprocity to be satisfied, it is necessary that (Exercise 7.25)

$$\frac{\mu_j}{k_j^2 k_{i\rho}^2} \overline{\mathbf{D}}_{i\mu} \cdot \mathbf{F}_n(\rho, \rho') \cdot \overleftarrow{\overline{\mathbf{D}}}{}'^{\,t}_{j\epsilon} = \frac{\mu_i}{k_i^2 k_{j\rho}^2} \overline{\mathbf{D}}_{i\epsilon} \cdot \mathbf{F}_n^t(\rho', \rho) \cdot \overleftarrow{\overline{\mathbf{D}}}{}'^{\,t}_{j\mu}, \qquad (7.4.36)$$

where $\rho' \in$ region j and $\rho \in$ region i. By inspecting the definition of $\overline{\mathbf{D}}_\mu$ and $\overline{\mathbf{D}}_\epsilon$, the above is the same as requiring

$$\frac{\mu_j}{k_j^2 k_{i\rho}^2} \overline{\mu}_i \cdot \mathbf{F}_n(\rho, \rho') \cdot \overline{\epsilon}_j = \frac{\mu_i}{k_i^2 k_{j\rho}^2} \overline{\epsilon}_i \cdot \mathbf{F}_n^t(\rho', \rho) \cdot \overline{\mu}_j, \qquad (7.4.37)$$

where

$$\overline{\mu}_i = \begin{bmatrix} 1 & 0 \\ 0 & \mu_i \end{bmatrix}, \qquad \overline{\epsilon}_j = \begin{bmatrix} 1 & 0 \\ 0 & -\epsilon_j \end{bmatrix}. \qquad (7.4.37a)$$

Before proceeding to show that the solution satisfies (37), we make use of the following observations. If new reflection and transmission matrices are defined as

$$\overline{\mathbf{r}}_{j,j\pm 1} = \overline{\mu}_j \cdot \overline{\mathbf{R}}_{j,j\pm 1} \cdot \overline{\epsilon}_j, \qquad (7.4.38a)$$

$$\overline{\mathbf{t}}_{j,j+1} = \overline{\mu}_{j+1} \cdot \overline{\mathbf{T}}_{j,j+1} \cdot \overline{\epsilon}_j, \qquad (7.4.38b)$$

$$\overline{\mathbf{t}}_{j+1,j} = \overline{\mu}_j \cdot \overline{\mathbf{T}}_{j+1,j} \cdot \overline{\epsilon}_{j+1}, \qquad (7.4.38c)$$

then it can be shown that $\overline{\mathbf{r}}_{j,j\pm 1}$ is symmetric (Exercise 7.26). Furthermore,

$$\frac{\epsilon_{j+1}}{k_{j+1,\rho}^2} \overline{\mathbf{t}}_{j,j+1} = \frac{\epsilon_j}{k_{j\rho}^2} \overline{\mathbf{t}}_{j+1,j}^t. \qquad (7.4.39)$$

Hence, using (39), it can be shown that the generalized reflection matrix is symmetric when transformed according to (38a) (Exercise 7.27), i.e.,

$$\widetilde{\mathbf{r}}_{j,j\pm1} = \overline{\boldsymbol{\mu}}_j \cdot \widetilde{\mathbf{R}}_{j,j\pm1} \cdot \overline{\boldsymbol{\epsilon}}_j = \widetilde{\mathbf{r}}_{j,j\pm1}^t. \tag{7.4.40}$$

Now, looking at Equation (3.3.22) of Chapter 3, when ρ and ρ' are both in region j, we can transform (3.3.22) as

$$\overline{\boldsymbol{\mu}}_j \cdot \overline{\mathbf{F}}_n(\rho, \rho') \cdot \overline{\boldsymbol{\epsilon}}_j = \left[H_n^{(1)}(k_{j\rho}\rho)\,\overline{\mathbf{I}} + J_n(k_{j\rho}\rho)\,\widetilde{\mathbf{r}}_{j,j+1} \cdot \overline{\mathbf{p}}_j^{-1} \right] \cdot \widetilde{\mathbf{m}}_{j+}$$
$$\cdot \left[J_n(k_{j\rho}\rho')\,\overline{\mathbf{I}} + H_n^{(1)}(k_{j\rho}\rho')\,\overline{\mathbf{p}}_j^{-1} \cdot \widetilde{\mathbf{r}}_{j,j-1} \right], \quad \rho > \rho', \tag{7.4.41a}$$

$$\overline{\boldsymbol{\mu}}_j \cdot \overline{\mathbf{F}}_n(\rho, \rho') \cdot \overline{\boldsymbol{\epsilon}}_j = \left[J_n(k_{j\rho}\rho')\,\overline{\mathbf{I}} + H_n^{(1)}(k_{j\rho}\rho')\,\widetilde{\mathbf{r}}_{j,j-1} \cdot \overline{\mathbf{p}}_j^{-1} \right] \cdot \widetilde{\mathbf{m}}_{j-}$$
$$\cdot \left[H_n^{(1)}(k_{j\rho}\rho)\,\overline{\mathbf{I}} + J_n(k_{j\rho}\rho)\,\widetilde{\overline{\mathbf{p}}}_j^{-1} \cdot \widetilde{\mathbf{r}}_{j,j+1} \right], \quad \rho < \rho', \tag{7.4.41b}$$

where

$$\widetilde{\mathbf{m}}_{j\pm} = \overline{\boldsymbol{\mu}}_j \cdot \widetilde{\overline{\mathbf{M}}}_{j\pm} \cdot \overline{\boldsymbol{\epsilon}}_j, \tag{7.4.42a}$$

$$\overline{\mathbf{p}}_j = \overline{\boldsymbol{\mu}}_j \cdot \overline{\boldsymbol{\epsilon}}_j. \tag{7.4.42b}$$

Then, in order for (37) to be satisfied, (41b) must be the transpose of (41a). Since the $\overline{\mathbf{r}}$ and $\overline{\mathbf{p}}$ matrices are symmetric, this requires that

$$\widetilde{\mathbf{m}}_{j+} = \widetilde{\mathbf{m}}_{j-}^t, \tag{7.4.43}$$

which can be proven easily using (40) (Exercise 7.28).

Furthermore, when $\rho \in$ region i and $\rho' \in$ region j, and $i > j$, using (3.3.23), we have

$$\overline{\boldsymbol{\mu}}_i \cdot \overline{\mathbf{F}}_n(\rho, \rho') \cdot \overline{\boldsymbol{\epsilon}}_j = \left[H_n^{(1)}(k_{i\rho}\rho)\,\overline{\mathbf{I}} + J_n(k_{i\rho}\rho)\,\widetilde{\mathbf{r}}_{i,i+1} \cdot \overline{\mathbf{p}}_i^{-1} \right] \cdot \widetilde{\mathbf{m}}_{i+} \cdot \overline{\mathbf{p}}_i^{-1} \cdot \widetilde{\mathbf{t}}_{ji}$$
$$\cdot \overline{\mathbf{p}}_j^{-1} \cdot \widetilde{\mathbf{m}}_{j+} \cdot \left[J_n(k_{j\rho}\rho')\,\overline{\mathbf{I}} + H_n^{(1)}(k_{j\rho}\rho')\,\overline{\mathbf{p}}_j^{-1} \cdot \widetilde{\mathbf{r}}_{j,j-1} \right]. \tag{7.4.44}$$

On the other hand, when ρ is the source point and ρ' is the observation point, but $\rho \in$ region i and $\rho' \in$ region j, using (3.3.24) with i replacing j, ρ replacing ρ' and vice versa, we have

$$\overline{\boldsymbol{\mu}}_j \cdot \overline{\mathbf{F}}_n(\rho', \rho) \cdot \overline{\boldsymbol{\epsilon}}_i = \left[J_n(k_{j\rho}\rho')\,\overline{\mathbf{I}} + H_n^{(1)}(k_{j\rho}\rho')\,\widetilde{\mathbf{r}}_{j,j-1} \cdot \overline{\mathbf{p}}_j^{-1} \right] \cdot \widetilde{\mathbf{m}}_{j-} \cdot \overline{\mathbf{p}}_j^{-1} \cdot \widetilde{\mathbf{t}}_{ij}$$
$$\cdot \overline{\mathbf{p}}_i^{-1} \cdot \widetilde{\mathbf{m}}_{i-} \cdot \left[H_n^{(1)}(k_{i\rho}\rho)\,\overline{\mathbf{I}} + J_n(k_{i\rho}\rho')\,\overline{\mathbf{p}}_i^{-1} \cdot \overline{\mathbf{r}}_{i,i+1} \right]. \tag{7.4.45}$$

Hence, in order for (37) to be satisfied, we require that

$$\frac{\epsilon_i}{k_{i\rho}^2}\widetilde{\mathbf{t}}_{ji} = \frac{\epsilon_j}{k_{j\rho}^2}\widetilde{\mathbf{t}}_{ij}^t. \tag{7.4.46}$$

The above is the generalization of (39) and it can be proven (Exercise 7.29).

(c) Spherically Layered Media Case

The dyadic Green's function for a spherically layered medium is given by (32). Therefore, in order for reciprocity to be satisfied, we require that (Exercise 7.25)

$$\overline{\mathbf{m}}_n(\mathbf{r}, \mathbf{r}') k_j \mu_j = \overline{\mathbf{m}}_n^t(\mathbf{r}', \mathbf{r}) k_i \mu_i, \tag{7.4.47a}$$

$$\overline{\mathbf{n}}_n(\mathbf{r}, \mathbf{r}') k_j \mu_j = \overline{\mathbf{n}}_n^t(\mathbf{r}', \mathbf{r}) k_i \mu_i, \tag{7.4.47b}$$

where $\mathbf{r} \in$ region i and $\mathbf{r}' \in$ region j. From (30) and (31), the above implies that

$$k_j \mu_j F_n^{TE}(r, r') = k_i \mu_i F_n^{TE}(r', r), \tag{7.4.48a}$$

$$k_j \epsilon_j F_n^{TM}(r, r') = k_i \epsilon_i F_n^{TM}(r', r). \tag{7.4.48b}$$

Notice that when i and j are the same, i.e., r and r' are in the same region, the above is clearly satisfied from the result of Chapter 3. On the contrary, when i and j are from two different regions, (47a) and (47b) will be met if

$$k_j \mu_j \tilde{T}_{ji}^{TE} = k_i \mu_i \tilde{T}_{ij}^{TE}, \tag{7.4.49a}$$

$$k_j \epsilon_j \tilde{T}_{ji}^{TM} = k_i \epsilon_i \tilde{T}_{ij}^{TM}. \tag{7.4.49b}$$

Again, the above results can be proven. The proof is very similar to that for Equations (35a) and (35b) (Exercise 7.30).

Exercises for Chapter 7

7.1 Show that the solution of the scalar wave equation

$$(\nabla^2 + k^2) g(\mathbf{r}, \mathbf{r}') = -\delta(\mathbf{r} - \mathbf{r}')$$

is given by Equation (7.1.1).

7.2 If we have specified that $\nabla \cdot \mathbf{A} = 0$ in (7.1.9), which is also known as *Coulomb's gauge*, what are the corresponding equations for (7.1.11) and (7.1.12)? Would the final electromagnetic field be changed with this gauge?

7.3 Show that the function

$$f(x) = \begin{cases} [\log(1/|x|)]^{-1}, & x \neq 0 \\ 0, & x = 0 \end{cases}$$

is continuous at $x = 0$ but it does not satisfy the Hölder condition.

7.4 For an exclusion volume which is a sphere (i.e., V_δ is a sphere), evaluate the second term in (7.1.22) and deduce the value of $\overline{\mathbf{L}}$ in Equation (7.1.25). Repeat the same for a cube, a disk, and a needle; and hence, derive the results of Equation (7.1.28).

7.5 Show that the inverse of (7.1.29) is in fact (7.1.30).

7.6 Using the method of stationary phase, derive the asymptotic approximation of

$$F(k_x) = \int\limits_0^\infty dx\, f(x) e^{-ik_x x},$$

where $f(x) \sim x^\alpha$, $x \to 0$, and $\alpha < 1$. Show that the more singular $f(x)$ is at $x = 0$, the more slowly $F(k_x)$ will decay when $k_x \to \infty$. Hence, a singular function has more high spectral components.

7.7 Show that when $z = z'$, the integral in (7.1.34) is divergent. Give an explanation for the divergence of this integral.

7.8 Derive (7.1.34) by substituting (7.1.43) into the definition of the dyadic Green's function given by (7.1.19).

7.9 Prove the identity presented by Equation (7.1.44).

7.10 Show that Equations (7.2.3) and (7.2.4) are solutions of (7.2.1) if (7.2.2) is satisfied.

7.11 (a) When the homogeneous Dirichlet boundary condition $\psi = 0$ on S is satisfied, show that $\hat{n} \cdot \mathbf{M} = \hat{n} \times \mathbf{N} = 0$ on the side wall, while $\hat{n} \times \mathbf{M} = \hat{n} \cdot \mathbf{N} = 0$ on the end-caps of Figure 7.2.1.

(b) When the homogeneous Neumann boundary condition $\hat{n} \cdot \nabla\psi = 0$ on S is satisfied, show that $\hat{n} \times \mathbf{M} = \hat{n} \cdot \mathbf{N} = 0$ on the side wall, while $\hat{n} \cdot \mathbf{M} = \hat{n} \times \mathbf{N} = 0$ on the end-caps of the same figure.

7.12 Derive the orthogonality relationships (7.2.28) to (7.2.31).

7.13 (a) Using the integral representation of Bessel functions [see (2.2.18) of Chapter 2], show that $J_{-n}(-x) = J_n(x)$.

(b) Derive the orthogonality relationships (7.2.37) to (7.2.40) (also see Exercise 2.10).

7.14 (a) Using the properties of Legendre polynomials, show that $Y_{n,-m}(\theta, \phi) = (-1)^m Y_{nm}^*(\theta, \phi)$ and that $Y_{nm}(\theta, \phi)$ is orthonormal.

(b) Derive the orthogonality relationships (7.2.47) to (7.2.52).

7.15 (a) Derive the dyadic Green's function shown in (7.3.7) and (7.3.8).

(b) Show that (7.3.5), (7.3.7), and (7.3.8) satisfy the reciprocity relation $\overline{\mathbf{G}}(\mathbf{r}, \mathbf{r}') = \overline{\mathbf{G}}^t(\mathbf{r}', \mathbf{r})$.

7.16 Show that the static-like pole contribution from the $\mathbf{N}$ and the $\mathbf{L}$ function terms in (7.3.12) cancel each other.

7.17 (a) Show that the form of $\overline{\mathbf{G}}(\mathbf{r}, \mathbf{r}')$ given by (7.3.13) can be manipulated to a form equivalent to (7.1.34).

(b) Show that the electric field $\mathbf{E}(\mathbf{r})$ generated by a current source $\mathbf{J}(\mathbf{r})$, which is

$$\mathbf{E}(\mathbf{r}) = i\omega\mu \int d\mathbf{r}' \, \overline{\mathbf{G}}(\mathbf{r}, \mathbf{r}') \cdot \mathbf{J}(\mathbf{r}')$$

satisfies $\nabla \cdot \mathbf{E} = \varrho/\epsilon$, where $\overline{\mathbf{G}}(\mathbf{r}, \mathbf{r}')$ is of the form (7.1.34).

7.18 (a) Determine that the residue contribution at the origin corresponds to the last term in (7.3.19).

    (b) Derive the expressions for I_1, I_2, and I_3 given by (7.3.21), (7.3.22), and (7.3.23).

7.19 (a) By a direct substitution of I_1, I_2, and I_3 given by (7.3.21), (7.3.22), and (7.3.23) into (7.3.7), show that (7.3.26) ensues.

    (b) Show that the derivation of (7.3.26) for the $n = 0$ term is analogous to ignoring the last terms in (7.3.21) and (7.3.22).

7.20 (a) Derive the expressions for I_1, I_2, and I_3 as given by (7.3.35), (7.3.36), and (7.3.37).

    (b) By a direct substitution of I_1, I_2, and I_3 given by (7.3.35), (7.3.36), and (7.3.37) into (7.3.8), show that (7.3.40) ensues.

7.21 (a) Derive Equation (7.4.4) from (7.4.3).

    (b) Derive Equation (7.4.5) by reciprocity consideration.

7.22 Derive Equation (7.4.9) from Maxwell's equations when the fields are plane waves.

7.23 Explain why the forms presented by (7.4.12) and (7.4.13) are indeed correct.

7.24 Show that (7.4.15) can be written as (7.4.17).

7.25 (a) Show that reciprocity implies (7.4.33) and consequently, (7.4.34) and (7.4.35).

    (b) Show that reciprocity implies (7.4.36) and consequently, (7.4.37).

    (c) Show that reciprocity implies (7.4.47) and consequently, (7.4.48) and (7.4.49).

7.26 (a) Prove that $\overline{\mathbf{r}}_{j,j\pm1}$ defined by (7.4.38a) is symmetric.

    (b) Prove that (7.4.39) is true.

7.27 Prove that (7.4.40) is true using the results of (7.4.38), (7.4.39), and the definition of generalized reflection matrices given in Chapter 3.

7.28 Show that (7.4.43) is true using the result of (7.4.40).

7.29 From Chapter 3,

$$\widetilde{\overline{\mathbf{T}}}_{1N} = \overline{\mathbf{T}}_{N-1,N} \cdot \left(\overline{\mathbf{I}} - \overline{\mathbf{R}}_{N-1,N-2} \cdot \overline{\mathbf{R}}_{N-1,N}\right)^{-1} \cdot \overline{\mathbf{T}}_{N-2,N-1} \cdots$$
$$\cdot \left(\overline{\mathbf{I}} - \overline{\mathbf{R}}_{21} \cdot \widetilde{\overline{\mathbf{R}}}_{23}\right)^{-1} \cdot \overline{\mathbf{T}}_{12},$$

$$\widetilde{\overline{\mathbf{T}}}_{N1} = \overline{\mathbf{T}}_{21} \cdot \left(\overline{\mathbf{I}} - \overline{\mathbf{R}}_{23} \cdot \overline{\mathbf{R}}_{21}\right)^{-1} \cdot \overline{\mathbf{T}}_{32} \cdots$$

$$\cdot \left(\overline{\mathbf{I}} - \overline{\mathbf{R}}_{N-1,N} \cdot \widetilde{\overline{\mathbf{R}}}_{N-1,N-2}\right)^{-1} \cdot \overline{\mathbf{T}}_{N,N-1}.$$

(a) Show that

$$\left(\overline{\mathbf{I}} - \overline{\mathbf{R}}_{32} \cdot \widetilde{\overline{\mathbf{R}}}_{34}\right)^{-1} \cdot \overline{\mathbf{T}}_{23} \cdot \left(\overline{\mathbf{I}} - \overline{\mathbf{R}}_{21} \cdot \widetilde{\overline{\mathbf{R}}}_{23}\right)^{-1} \cdot \overline{\mathbf{T}}_{12}$$

$$= \left(\overline{\mathbf{I}} - \widetilde{\overline{\mathbf{R}}}_{32} \cdot \widetilde{\overline{\mathbf{R}}}_{34}\right)^{-1} \cdot \overline{\mathbf{T}}_{23} \cdot \left(\overline{\mathbf{I}} - \overline{\mathbf{R}}_{21} \cdot \overline{\mathbf{R}}_{23}\right)^{-1} \cdot \overline{\mathbf{T}}_{12}.$$

In other words, the tilde sign can be moved from $\widetilde{\overline{\mathbf{R}}}_{23}$ to $\overline{\mathbf{R}}_{32}$ without violating the equality.

(b) Using the idea of (a), prove the equality given by (7.4.46).

7.30 Prove the identities given by (7.4.49) using a strategy given in Exercise 7.29.

References for Chapter 7

Bagby, J. S., and D. P. Nyquist. 1987. "Dyadic Green's functions for integrated electronic and optical circuits." *IEEE Trans. Microwave Theory Tech.* MTT-35: 206–10.

Chew, W. C. 1989. "Some observations on the spatial and eigenfunction representations of dyadic Green's function." *IEEE Trans. Antennas Propagat.* AP-37(10): 1322–27.

Collin, R. E. 1966. *Foundations for Microwave Engineering.* New York: McGraw-Hill.

Collin, R. E. 1973. "On the incompleteness of E and H modes in waveguides." *Can. J. Phys.* 51: 1135–40.

Collin, R. E. 1986a. "Dyadic Green's function expansions in spherical coordinates." *Electromagnetics* 6: 183–207.

Collin, R. E. 1986b. "The dyadic Green's function as an inverse operator." *Radio Sci.* 21(6): 883–90.

Doetsch, G. 1974. *Introduction to the Theory and Application of the Laplace Transformation.* New York: Springer-Verlag.

Fikioris, J. G. 1965. " Electromagnetic field inside a current-carrying region." *J. Math. Phys.* 6: 1671–20.

Howard, A. Q., Jr. 1974. "On the longitudinal component of the Green's functions in bounded media." *Proc. IEEE* 62: 1704–05.

Johnson, W. A., A. Q. Howard, and D. G. Dudley. 1979. "On the irrotational component of the electric Green's dyadic." *Radio Sci.* 14: 961–67.

Kellogg, O. D. 1953. *Foundations of Potential Theory.* New York: Dover Publications.

Kerns, D. M. 1981. *Plane Wave Scattering Matrix Theory of Antennas and Antenna-Antenna Interactions.* National Bureau of Standards Monograph. 126.

Kong, J. A. 1986. *Theory of Electromagnetic Waves.* New York: Wiley-Interscience.

Kurokawa, K. 1958. "The expansions of electromagnetic fields in cavities." *IRE Trans. Microwave Theory Tech.* MTT-6: 178–87.

Lee, J. K., and J. A. Kong. 1983. "Dyadic Green's functions for layered anisotropic medium." *Electromagnetics* 3: 111–30.

Lee, S. W., J. Boersma, C. L. Law, and G. A. Deschamps. 1980. "Singularity in Green's function and its numerical evaluation." *IEEE Trans. Antennas Propagat.* AP-28: 311–17.

Lighthill, M. J. 1958 *Introduction to Fourier Analysis and Generalized Functions.* Cambridge, England: Cambridge University Press.

Morse, P. M., and H. Feshbach. 1953. *Methods of Theoretical Physics.* New York: McGraw-Hill.

Pearson, L. W. 1983. "On the spectral expansion of electric and magnetic dyadic Green's functions in cylindrical harmonics." *Radio Science* 18: 166–74.

Plonsey, R., and R. E. Collin. 1961. *Principles and Applications of Electromagnetic Fields.* New York: McGraw-Hill.

Stakgold, I. 1967. *Boundary Value Problems of Mathematical Physics*, vol. I. London: Macmillan.

Stratton, J. A. 1941. *Electromagnetic Theory.* New York: McGraw-Hill.

Tai, C. T. 1971. *Dyadic Green's Functions in Electromagnetic Theory.* New York: Intext Publishers.

Tai, C. T. 1973. "On the eigenfunction expansion of dyadic Green's functions." *Proc. IEEE* 61: 480–81.

Tsang, L., E. Njoku, and J. A. Kong. 1975. "Microwave thermal emission from a stratified medium with nonuniform temperature distribution." *J. Appl. Phys.* 46: 5127–33.

van Bladel, J. 1961. "Some remarks on Green's dyadic for infinite space." *IEEE Trans. Antennas Propag.* 9: 563–66.

Viola, M. S., and D. P. Nyquist. 1988. "An observation on the Sommerfeld-integral representation of the electric dyadic Green's function for layered media." *IEEE Trans. Microwave Theory Tech.* MTT-36: 1289–92.

Yaghjian, A. D. 1980. "Electric dyadic Green's functions in the source region." *Proc. IEEE* 68: 248–63.

Further Readings for Chapter 7

Bressman, M., and G. Conciauro. 1985. "Singularity extraction from the electric Green's function for a spherical cavity." *IEEE Trans. Microwave Theory Tech.* MTT-33: 407–14.

Chen, H. C., and D. K. Cheng. 1965. "Dyadic Green's function for anisotropic and compressible media." *Proc. IEEE* 53: 1268–69.

Chen, K. M. 1977. "A simple physical picture of tensor Green's function in source region." *Proc. IEEE* 65: 1202–24.

Collin, R. E. 1960. *Field Theory of Guided Waves.* New York: McGraw-Hill.

Compton, R. T. 1966. "The time-dependent Green's function for electromagnetic waves in moving simple media." *J. Math. Phys.* 7: 2145–52.

Daniele, V. G., and M. Orefice. 1984. "Dyadic Green's functions in bounded media." *IEEE Trans. Antennas Propagat.* AP-32: 193–96.

Felsen, L. B., and N. Marcuvitz. 1973. *Radiation and Scattering of Electromagnetic Waves.* Englewood Cliffs, New Jersey: Prentice-Hall.

Hansen, W. W. 1935. "A new type of expansion in radiation problems." *Phys. Rev.* 47: 139–43.

Kisliuk, M. 1980. "The dyadic Green's functions for cylindrical waveguides and cavities." *IEEE Trans. Microwave Theory Tech.* MTT-28: 894–98.

Michalski, K. A. 1987. "On the scalar potential of a point charge associated with a time-harmonic dipole in a layered medium." *IEEE Trans. Antennas Propagat.* AP-35(11): 1299.

Pathak, P. H. 1983. "On the eigenfunction expansion of the electric and magnetic field dyadic Green's functions." *IEEE Trans. Antennas Propagat.* AP-31: 837–46.

Rahmat-Samii, Y. 1975. "On the question of computation of dyadic Green's function at a source region in waveguides and cavities." *IEEE Trans. Microwave Theory Tech.* 23: 762–65.

Su, C. C. 1987. "Principal value integrals for dyadic Green's function; evaluation using symmetry property." *IEEE Trans. Antennas Propagat.* AP-35: 1306–07.

Tai, C. T. 1981a. "Comments on 'Electric dyadic Green's functions in a source region.'" *Proc. IEEE* 69: 282–85.

Tai, C. T. 1981b. "Equivalent layers of surface charge, current sheet and polarization in the eigenfunction expansion of Green's functions in electromagnetic theory." *IEEE Trans. Antennas Propagat.* AP-29: 733–39.

Tai, C. T. 1983. "Dyadic Green's functions for a coaxial line." *IEEE Trans. Antennas Propagat.* AP-31: 355-58.

Tai, C. T. 1987. "Some essential formulas in dyadic analysis and their applications." *Radio Sci.* 22: 1283–88.

Tai, C. T. 1988. "Dyadic Green's functions for a rectangular waveguide filled with two dielectrics." *J. Electromagnetic Waves Appl.* 2: 245–53.

Tai, C. T., and P. Rozenfeld. 1976. "Different representations of dyadic Green's functions for a rectangular cavity." *IEEE Trans. Microwave Theory Tech.* MTT-24: 597–601.

Tsang, L., J. A. Kong, and R. T. Shin. 1985. *Theory of Microwave Remote Sensing.* New York: Wiley-Interscience.

Wang, J. J. H. 1982. "A unified consistent view on the singularities of the electric dyadic Green's function in the source region." *IEEE Trans. Antennas Propagat.* AP-30: 463–68.

Yaghjian, A. D. 1982. "A delta-distribution derivation of the electric field in the source region." *Electromagnetics* 2: 161–67.

CHAPTER 8

INTEGRAL EQUATIONS

Integral equations have arrested the interest of mathematicians for over a century (see Elliott 1980). Most integral equations do not have closed-form solutions; however, they can often be discretized and solved on a digital computer. Proof of the existence of the solution to an integral equation by discretization was first presented by Fredholm (1903). But such a discretization procedure was not feasible until the advent of the digital computer, which, in recent decades, created a "boom" in interest concerning integral equations.

When inhomogeneities are piecewise constant in each region, we may solve such problems using the surface integral equation technique. In this technique, the homogeneous-medium Green's functions are found for each region. Then, the field in each region is written in terms of the field due to any sources in the region plus the field due to surface sources at the interfaces between the regions, following Huygens' principle. Next, the boundary conditions at these interfaces are used to set up integral equations known as surface integral equations. From these integral equations, the unknown surface sources at the interfaces can be solved for.

The surface integral equation method is rather popular in a number of applications, because it employs a homogeneous-medium Green's function which is simple in form, and the unknowns are on a surface rather than in a volume. Moreover, the surface integral equation method is not limited to piecewise-constant inhomogeneities. For example, if each region is a layered medium whose Green's function is available as shown in Chapter 7, surface integral equations can be formulated for unknowns at interfaces between such regions.

For a bounded inhomogeneity, an alternative method is to view the scattered field as due to the induced currents flowing in the inhomogeneity. Now, the induced currents are proportional to the total field in the inhomogeneity. But in turn, the total field is the incident field plus the field due to induced currents in the inhomogeneity. So, this concept yields an equation called the volume integral equation from which the unknown field inside the inhomogeneity can be solved for.

In this chapter, the surface integral equations[1] both for scalar and vector fields will be studied first. Then, the volume integral equation will be discussed next.

§8.1 Surface Integral Equations

In an integral equation, the unknown to be sought is embedded in an integral. When the unknown of a linear integral equation is embedded inside the integral only, the integral equation is of the first kind. But when the unknown occurs both inside and outside the integral, the integral equation is of the second kind. An integral equation can be viewed as an operator equation. Thus, the matrix representation of such an operator equation (Chapter 5) can be obtained, and then the unknown is easily solved for with a computer. Later, we shall see how such integral equations with only surface integrals are derived, beginning with the scalar wave equation, followed by the vector wave equation case.

An early form of such a surface integral equation was first derived by Green (1818) through the use of Green's theorem. Since then, his idea has been adapted to solve different problems. Many of these works are listed in the references for this chapter. Poggio and Miller (1973), in addition, provide an extensive reference list for works in this area. The advantage of the surface integral equations is that they reduce the dimensionality of a problem by one. For example, a three-dimensional problem is reduced to a lower dimensional problem involving surface integrals.

§§8.1.1 Scalar Wave Equation

Consider a scalar wave equation for a two-region problem as shown in Figure 8.1.1. In region 1, the governing equation for the total field is

$$(\nabla^2 + k_1^2)\,\phi_1(\mathbf{r}) = Q(\mathbf{r}), \tag{8.1.1}$$

while in region 2, it is

$$(\nabla^2 + k_2^2)\,\phi_2(\mathbf{r}) = 0. \tag{8.1.2}$$

Therefore, we can define Green's functions for regions 1 and 2 respectively to satisfy the following equations:

$$(\nabla^2 + k_1^2)\,g_1(\mathbf{r}, \mathbf{r}') = -\delta(\mathbf{r} - \mathbf{r}'), \tag{8.1.3}$$

$$(\nabla^2 + k_2^2)\,g_2(\mathbf{r}, \mathbf{r}') = -\delta(\mathbf{r} - \mathbf{r}'). \tag{8.1.4}$$

On multiplying Equation (1) by $g_1(\mathbf{r}, \mathbf{r}')$ and Equation (3) by $\phi_1(\mathbf{r})$, subtracting the two resultant equations, and integrating over region 1, we have, for

---

[1] These are sometimes called boundary integral equations.

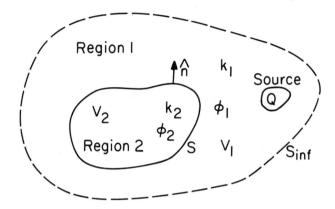

Figure 8.1.1 A two-region problem can be solved with a surface integral equation.

$\mathbf{r}' \in V_1$,

$$\int\limits_{V_1} dV \left[g_1(\mathbf{r}, \mathbf{r}') \nabla^2 \phi_1(\mathbf{r}) - \phi_1(\mathbf{r}) \nabla^2 g_1(\mathbf{r}, \mathbf{r}') \right]$$

$$= \int\limits_{V_1} dV \, g_1(\mathbf{r}, \mathbf{r}') Q(\mathbf{r}) + \phi_1(\mathbf{r}'), \quad \mathbf{r}' \in V_1. \quad (8.1.5)$$

Since $\nabla \cdot (g\nabla\phi - \phi\nabla g) = g\nabla^2\phi - \phi\nabla^2 g$, by applying Gauss' theorem, the volume integral on the left-hand side of (5) becomes a surface integral over the surface bounding V_1. Consequently,[2]

$$- \int\limits_{S+S_{inf}} dS \, \hat{n} \cdot \left[g_1(\mathbf{r}, \mathbf{r}') \nabla \phi_1(\mathbf{r}) - \phi_1(\mathbf{r}) \nabla g_1(\mathbf{r}, \mathbf{r}') \right]$$

$$= -\phi_{inc}(\mathbf{r}') + \phi_1(\mathbf{r}'), \quad \mathbf{r}' \in V_1. \quad (8.1.6)$$

In the above, we have let

$$\phi_{inc}(\mathbf{r}') = - \int\limits_{V_1} dV \, g_1(\mathbf{r}, \mathbf{r}') Q(\mathbf{r}), \quad\quad\quad (8.1.7)$$

since it is the incident field generated by the source $Q(\mathbf{r})$. Note that up to this point, $g_1(\mathbf{r}, \mathbf{r}')$ is not explicitly specified, and the manipulation up to (6)

---

[2] The equality of the volume integral on the left-hand side of (5) and the surface integral on the left-hand side of (6) is also known as Green's theorem.

is legitimate as long as $g_1(\mathbf{r}, \mathbf{r}')$ is a solution of (3). For example, a possible choice for $g_1(\mathbf{r}, \mathbf{r}')$ that satisfies the radiation condition is[3]

$$g_1(\mathbf{r}, \mathbf{r}') = \frac{e^{ik_1|\mathbf{r}-\mathbf{r}'|}}{4\pi|\mathbf{r} - \mathbf{r}'|}, \tag{8.1.8}$$

which is the unbounded, homogeneous medium scalar Green's function (see Subsection 1.3.4). In this case, $\phi_{inc}(\mathbf{r})$ is the incident field generated by the source $Q(\mathbf{r})$ in the absence of the scatterer. Moreover, the integral over S_{inf} vanishes when $S_{inf} \to \infty$ by virtue of the radiation condition (see Exercise 8.1). Then, after swapping $\mathbf{r}$ and $\mathbf{r}'$, we have

$$\phi_1(\mathbf{r}) = \phi_{inc}(\mathbf{r}) - \int_S dS' \, \hat{n}' \cdot [g_1(\mathbf{r},\mathbf{r}')\nabla'\phi_1(\mathbf{r}') - \phi_1(\mathbf{r}')\nabla'g_1(\mathbf{r},\mathbf{r}')], \quad \mathbf{r} \in V_1.$$
$$\tag{8.1.9}$$

But if $\mathbf{r}' \notin V_1$ in (5), the second term on the right-hand side of (5) would be zero, for $\mathbf{r}'$ would be in V_2 where the integration is not performed. Therefore, we can write (9) as

$$\left. \begin{array}{l} \mathbf{r} \in V_1, \ \phi_1(\mathbf{r}) \\ \mathbf{r} \in V_2, \ 0 \end{array} \right\} = \phi_{inc}(\mathbf{r}) - \int_S dS' \, \hat{n}' \cdot [g_1(\mathbf{r},\mathbf{r}')\nabla'\phi_1(\mathbf{r}') - \phi_1(\mathbf{r}')\nabla'g_1(\mathbf{r},\mathbf{r}')].$$
$$\tag{8.1.10}$$

The above equation is evocative of Huygens' principle. It says that when the observation point $\mathbf{r}$ is in V_1, then the total field $\phi_1(\mathbf{r})$ consists of the incident field, $\phi_{inc}(\mathbf{r})$, and the contribution of field due to surface sources on S. But if the observation point is in V_2, then the surface sources on S generate a field that exactly cancels the incident field $\phi_{inc}(\mathbf{r})$, making the total field in region 2 zero. This fact is the core of the **extinction theorem** (see Born and Wolf 1980).

In (10), $\hat{n} \cdot \nabla\phi_1(\mathbf{r})$ and $\phi_1(\mathbf{r})$ act as surface sources. Moreover, they are impressed on S, creating a field in region 2 that cancels exactly the incident field in region 2 (see Figure 8.1.2).

Applying the same derivation to region 2, we have (see Exercise 8.2)

$$\left. \begin{array}{l} \mathbf{r} \in V_2, \ \phi_2(\mathbf{r}) \\ \mathbf{r} \in V_1, \ 0 \end{array} \right\} = \int_S dS' \, \hat{n}' \cdot [g_2(\mathbf{r},\mathbf{r}')\nabla'\phi_2(\mathbf{r}') - \phi_2(\mathbf{r}')\nabla'g_2(\mathbf{r},\mathbf{r}')].$$
$$\tag{8.1.11}$$

The above states that the field in region 2 is due to some surface sources impressed on S. These surface sources generate $\phi_2(\mathbf{r})$ in V_2, but zero field in V_1. This is again evocative of Huygens' principle. [Note that $g_2(\mathbf{r}, \mathbf{r}')$ need not have the same form as $g_1(\mathbf{r}, \mathbf{r}')$ in (8) (Exercise 8.2).]

---

[3] Note that this is not the only form satisfying the radiation condition at infinity and satisfying (3) in V_1 (see Exercise 8.1).

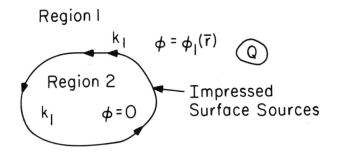

Figure 8.1.2 The illustration of the extinction theorem.

Applying the extinction theorem, integral equations can now be derived. So, using the lower parts of Equations (10) and (11), we have

$$\phi_{inc}(\mathbf{r}) = \int\limits_{S} dS' \, \hat{n}' \cdot [g_1(\mathbf{r}, \mathbf{r}')\nabla'\phi_1(\mathbf{r}') - \phi_1(\mathbf{r}')\nabla'g_1(\mathbf{r}, \mathbf{r}')], \quad \mathbf{r} \in V_2, \tag{8.1.12a}$$

$$0 = \int\limits_{S} dS' \, \hat{n}' \cdot [g_2(\mathbf{r}, \mathbf{r}')\nabla'\phi_2(\mathbf{r}') - \phi_2(\mathbf{r}')\nabla'g_2(\mathbf{r}, \mathbf{r}')], \quad \mathbf{r} \in V_1. \tag{8.1.12b}$$

Even though $g_1(\mathbf{r}, \mathbf{r}')$ and $g_2(\mathbf{r}, \mathbf{r}')$ need not be homogeneous-medium Green's functions (Exercise 8.2), homogeneous-medium Green's functions, for simplicity, are usually chosen. Then, the two integral equations above will have four independent unknowns, ϕ_1, ϕ_2, $\hat{n}\cdot\nabla\phi_1$, and $\hat{n}\cdot\nabla\phi_2$ on S. Next, boundary conditions can be used to eliminate two of these four unknowns. Exemplary boundary conditions are

$$\phi_1(\mathbf{r}) = \phi_2(\mathbf{r}), \quad \mathbf{r} \in S, \tag{8.1.13a}$$

$$p_1\hat{n} \cdot \nabla\phi_1(\mathbf{r}) = p_2\hat{n} \cdot \nabla\phi_2(\mathbf{r}), \quad \mathbf{r} \in S. \tag{8.1.13b}$$

Consequently, the integral equations in (12) can be treated as linear operator equations and solved with standard techniques (see Chapter 5). Here, the Green's function $g(\mathbf{r}, \mathbf{r}')$ and $\hat{n} \cdot \nabla'g(\mathbf{r}, \mathbf{r}')$ are the kernel of the integral equation.

§§8.1.2 Vector Wave Equation

Consider the vector electromagnetic wave equation for a two-region problem as shown in Figure 8.1.3. In region 1, the field satisfies the equation

$$\nabla \times \nabla \times \mathbf{E}_1(\mathbf{r}) - \omega^2\mu_1\epsilon_1\mathbf{E}_1(\mathbf{r}) = i\omega\mu_1\mathbf{J}(\mathbf{r}). \tag{8.1.14}$$

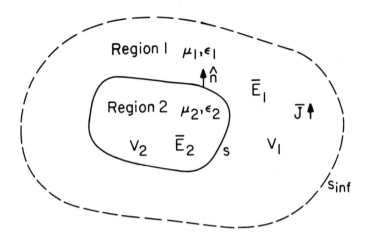

Figure 8.1.3 A two-region problem where a surface integral equation can be derived.

In region 2, the field satisfies

$$\nabla \times \nabla \times \mathbf{E}_2(\mathbf{r}) - \omega^2 \mu_2 \epsilon_2 \mathbf{E}_2(\mathbf{r}) = 0. \qquad (8.1.15)$$

Hence, the dyadic Green's functions for region 1 and region 2 respectively are defined by:

$$\nabla \times \nabla \times \overline{\mathbf{G}}_1(\mathbf{r}, \mathbf{r}') - \omega^2 \mu_1 \epsilon_1 \overline{\mathbf{G}}_1(\mathbf{r}, \mathbf{r}') = \overline{\mathbf{I}} \delta(\mathbf{r} - \mathbf{r}'), \qquad (8.1.16)$$

$$\nabla \times \nabla \times \overline{\mathbf{G}}_2(\mathbf{r}, \mathbf{r}') - \omega^2 \mu_2 \epsilon_2 \overline{\mathbf{G}}_2(\mathbf{r}, \mathbf{r}') = \overline{\mathbf{I}} \delta(\mathbf{r} - \mathbf{r}'). \qquad (8.1.17)$$

After post-multiplying Equation (14) with $\overline{\mathbf{G}}_1$ and pre-multiplying Equation (16) with $\mathbf{E}_1(\mathbf{r})$, subtracting the two equations, and integrating the result over V_1, we have

$$\int_{V_1} dV \left[\nabla \times \nabla \times \mathbf{E}_1(\mathbf{r}) \cdot \overline{\mathbf{G}}_1(\mathbf{r}, \mathbf{r}') - \mathbf{E}_1(\mathbf{r}) \cdot \nabla \times \nabla \times \overline{\mathbf{G}}_1(\mathbf{r}, \mathbf{r}') \right]$$

$$= i\omega\mu_1 \int_{V_1} dV \, \mathbf{J}(\mathbf{r}) \cdot \overline{\mathbf{G}}_1(\mathbf{r}, \mathbf{r}') - \mathbf{E}_1(\mathbf{r}'), \qquad \mathbf{r}' \in V_1. \quad (8.1.18)$$

The left-hand side of the above becomes a surface integral using the fact that[4]

$$\nabla \cdot \left\{ [\nabla \times \mathbf{E}_1(\mathbf{r})] \times \overline{\mathbf{G}}_1(\mathbf{r}, \mathbf{r}') + \mathbf{E}_1(\mathbf{r}) \times [\nabla \times \overline{\mathbf{G}}_1(\mathbf{r}, \mathbf{r}')] \right\}$$

$$= \nabla \times \nabla \times \mathbf{E}_1(\mathbf{r}) \cdot \overline{\mathbf{G}}_1(\mathbf{r}, \mathbf{r}') - \mathbf{E}_1(\mathbf{r}) \cdot \nabla \times \nabla \times \overline{\mathbf{G}}_1(\mathbf{r}, \mathbf{r}'). \quad (8.1.19)$$

---

[4] We can post-multiply (19) by an arbitrary constant vector $\mathbf{b}$ to aid in the derivation and cancel the constant vector later. The equality of the volume integral on the left-hand side of (18) to a surface integral is also known as the vector Green's theorem.

Moreover, the integral on the right-hand side of (18) corresponds to the incident wave. Hence, (18) becomes

$$\mathbf{E}_1(\mathbf{r}') = \mathbf{E}_{inc}(\mathbf{r}') + \int_{S+S_{inf}} dS\,\hat{n}\cdot\left\{[\nabla\times\mathbf{E}_1(\mathbf{r})]\times\overline{\mathbf{G}}_1(\mathbf{r},\mathbf{r}')\right.$$

$$\left.+\mathbf{E}_1(\mathbf{r})\times\nabla\times\overline{\mathbf{G}}_1(\mathbf{r},\mathbf{r}')\right\},\quad \mathbf{r}'\in V_1,\quad (8.1.20)$$

where

$$\mathbf{E}_{inc}(\mathbf{r}') = i\omega\mu_1\int_{V_1} dV\,\mathbf{J}(\mathbf{r})\cdot\overline{\mathbf{G}}_1(\mathbf{r},\mathbf{r}') = i\omega\mu_1\int_{V_1} dV\,\overline{\mathbf{G}}_1(\mathbf{r}',\mathbf{r})\cdot\mathbf{J}(\mathbf{r}) \tag{8.1.21}$$

is the incident field generated by the current $\mathbf{J}(\mathbf{r})$. In the above, we have made use of the reciprocity condition on the dyadic Green's function that [see (1.3.52b) of Chapter 1]

$$\overline{\mathbf{G}}_1^t(\mathbf{r},\mathbf{r}') = \overline{\mathbf{G}}_1(\mathbf{r}',\mathbf{r}). \tag{8.1.22}$$

Up to this point, $\overline{\mathbf{G}}_1(\mathbf{r},\mathbf{r}')$ need only satisfy (16). However, if $\overline{\mathbf{G}}_1(\mathbf{r},\mathbf{r}')$ is assumed to be the homogeneous-medium dyadic Green's function given in Chapter 1 and Chapter 7, then $\mathbf{E}_{inc}$ corresponds to the incident field generated by $\mathbf{J}(\mathbf{r})$ in the absence of the scatterer.

Using (22), we deduce that

$$\hat{n}\cdot[\nabla\times\mathbf{E}_1(\mathbf{r})]\times\overline{\mathbf{G}}_1(\mathbf{r},\mathbf{r}') = \hat{n}\times[\nabla\times\mathbf{E}_1(\mathbf{r})]\cdot\overline{\mathbf{G}}_1(\mathbf{r},\mathbf{r}')$$
$$= i\omega\mu_1\overline{\mathbf{G}}_1(\mathbf{r}',\mathbf{r})\cdot\hat{n}\times\mathbf{H}_1(\mathbf{r}). \tag{8.1.23}$$

Moreover, by assuming that $\overline{\mathbf{G}}_1(\mathbf{r},\mathbf{r}')$ is the unbounded homogeneous-medium dyadic Green's function, then[5]

$$\hat{n}\cdot\mathbf{E}_1(\mathbf{r})\times\nabla\times\overline{\mathbf{G}}_1(\mathbf{r},\mathbf{r}') = \hat{n}\times\mathbf{E}_1(\mathbf{r})\cdot\nabla\times\overline{\mathbf{G}}_1(\mathbf{r},\mathbf{r}')$$
$$= -\left[\nabla\times\overline{\mathbf{G}}_1(\mathbf{r}',\mathbf{r})\right]\cdot\hat{n}\times\mathbf{E}_1(\mathbf{r}), \tag{8.1.24}$$

Hence, Equation (20) becomes

$$\mathbf{E}_1(\mathbf{r}') = \mathbf{E}_{inc}(\mathbf{r}') + \int_S dS\left\{i\omega\mu_1\overline{\mathbf{G}}_1(\mathbf{r}',\mathbf{r})\cdot\hat{n}\times\mathbf{H}_1(\mathbf{r})\right.$$

$$\left.-\left[\nabla\times\overline{\mathbf{G}}_1(\mathbf{r}',\mathbf{r})\right]\cdot\hat{n}\times\mathbf{E}_1(\mathbf{r})\right\}. \tag{8.1.25}$$

Now, if $\overline{\mathbf{G}}_1(\mathbf{r},\mathbf{r}')$ satisfies the radiation condition, the integral over S_{inf} vanishes when $S_{inf}\to\infty$. Note that in (18), if $\mathbf{r}'\notin V_1$, the second term on

---

[5] The second equality follows from $[\nabla\times\overline{\mathbf{G}}_1(\mathbf{r},\mathbf{r}')]^t = -\nabla\times\overline{\mathbf{G}}_1(\mathbf{r}',\mathbf{r})$ [Exercise 8.3, and also (1.4.14) of Chapter 1].

the right-hand side of (18) vanishes. Consequently, the analogue of (10) for electromagnetic fields, after swapping $\mathbf{r}$ and $\mathbf{r}'$, is

$$\left.\begin{array}{l} \mathbf{r} \in V_1, \ \mathbf{E}_1(\mathbf{r}) \\ \mathbf{r} \in V_2, \ 0 \end{array}\right\} = \mathbf{E}_{inc}(\mathbf{r}) + \int_S dS' \left\{ i\omega\mu_1 \overline{\mathbf{G}}_1(\mathbf{r},\mathbf{r}') \cdot \hat{n}' \times \mathbf{H}_1(\mathbf{r}') \right.$$

$$\left. - \left[\nabla' \times \overline{\mathbf{G}}_1(\mathbf{r},\mathbf{r}')\right] \cdot \hat{n}' \times \mathbf{E}_1(\mathbf{r}') \right\}. \quad (8.1.26)$$

Again, the above is similar to Huygens' principle for electromagnetic fields. Furthermore, the lower part of the equation is the vector analogue of the extinction theorem.

In region 2, analogous to Equation (11), similar derivation yields (see Exercise 8.3)

$$\left.\begin{array}{l} \mathbf{r} \in V_2, \ \mathbf{E}_2(\mathbf{r}) \\ \mathbf{r} \in V_1, \ 0 \end{array}\right\} = - \int_S dS' \left\{ i\omega\mu_2 \overline{\mathbf{G}}_2(\mathbf{r},\mathbf{r}') \cdot \hat{n}' \times \mathbf{H}_2(\mathbf{r}') \right.$$

$$\left. - \left[\nabla' \times \overline{\mathbf{G}}_2(\mathbf{r},\mathbf{r}')\right] \cdot \hat{n}' \times \mathbf{E}_2(\mathbf{r}') \right\}, \quad (8.1.27)$$

where $\overline{\mathbf{G}}_2(\mathbf{r},\mathbf{r}')$ is assumed to be the unbounded homogeneous-medium Green's function.

The lower parts of (26) and (27) form integral equations which are[6]

$$\mathbf{E}_{inc}(\mathbf{r}) = \int_S dS' \left\{ i\omega\mu_1 \overline{\mathbf{G}}_1(\mathbf{r},\mathbf{r}') \cdot \hat{n}' \times \mathbf{H}_1(\mathbf{r}') \right.$$

$$\left. - \left[\nabla' \times \overline{\mathbf{G}}_1(\mathbf{r},\mathbf{r}')\right] \cdot \hat{n}' \times \mathbf{E}_1(\mathbf{r}') \right\}, \quad \mathbf{r} \in V_2,$$

$$(8.1.28a)$$

$$0 = \int_S dS' \left\{ i\omega\mu_2 \overline{\mathbf{G}}_2(\mathbf{r},\mathbf{r}') \cdot \hat{n}' \times \mathbf{H}_2(\mathbf{r}') \right.$$

$$\left. - \left[\nabla' \times \overline{\mathbf{G}}_2(\mathbf{r},\mathbf{r}')\right] \cdot \hat{n}' \times \mathbf{E}_2(\mathbf{r}') \right\}, \quad \mathbf{r} \in V_1.$$

$$(8.1.28b)$$

The above, together with the boundary conditions that

$$\hat{n} \times \mathbf{H}_1(\mathbf{r}) = \hat{n} \times \mathbf{H}_2(\mathbf{r}), \quad \hat{n} \times \mathbf{E}_1(\mathbf{r}) = \hat{n} \times \mathbf{E}_2(\mathbf{r}) \quad (8.1.29)$$

on S, can be solved for the surface unknowns $\hat{n} \times \mathbf{E}_1$ and $\hat{n} \times \mathbf{H}_1$. Once the surface fields are known, the field everywhere is derived from the upper parts of Equations (26) and (27). Note that the dyadic Green's functions in (28a) and (28b) need not be homogeneous-medium Green's functions given in Chapters 1 and 7, but homogeneous-medium Green's functions are chosen

---

[6] Variations of these integral equations are also given by Poggio and Miller (1973) and Ström (1975).

for simplicity (Exercise 8.3). If an arbitrary dyadic function satisfies (16) in region 1 but not the radiation condition at infinity, then the manipulation in (24) is not possible. In this case, the resultant integral equation will have the dyadic Green's functions to the right of the surface sources as in (20) (see next subsection).

Equations (28a) and (28b) are also known as the electric field integral equations (EFIE). By duality, the corresponding magnetic field integral equations (MFIE) can be derived.

§§8.1.3 The Anisotropic, Inhomogeneous Medium Case

Surface integral equations can be derived even when region 1 and region 2 in Figure 8.1.3 consist of anisotropic, inhomogeneous media. In this case, the electric field satisfies the vector wave equations (Section 1.3, Chapter 1)

$$\nabla \times \overline{\mu}_1^{-1} \cdot \nabla \times \mathbf{E}_1(\mathbf{r}) - \omega^2 \overline{\epsilon}_1 \cdot \mathbf{E}_1(\mathbf{r}) = i\omega \mathbf{J}(\mathbf{r}), \quad \mathbf{r} \in V_1, \qquad (8.1.30)$$

$$\nabla \times \overline{\mu}_2^{-1} \cdot \nabla \times \mathbf{E}_2(\mathbf{r}) - \omega^2 \overline{\epsilon}_2 \cdot \mathbf{E}_2(\mathbf{r}) = 0, \qquad \mathbf{r} \in V_2. \qquad (8.1.31)$$

Moreover, we can define dyadic Green's functions for regions 1 and 2 respectively as

$$\nabla \times (\overline{\mu}_1^t)^{-1} \cdot \nabla \times \overline{\mathbf{G}}_1(\mathbf{r}, \mathbf{r}') - \omega^2 \overline{\epsilon}_1^t \cdot \overline{\mathbf{G}}_1(\mathbf{r}, \mathbf{r}') = (\overline{\mu}_1^t)^{-1} \overline{\mathbf{I}} \delta(\mathbf{r} - \mathbf{r}'), \qquad (8.1.32)$$

$$\nabla \times (\overline{\mu}_2^t)^{-1} \cdot \nabla \times \overline{\mathbf{G}}_2(\mathbf{r}, \mathbf{r}') - \omega^2 \overline{\epsilon}_2^t \cdot \overline{\mathbf{G}}_2(\mathbf{r}, \mathbf{r}') = (\overline{\mu}_2^t)^{-1} \overline{\mathbf{I}} \delta(\mathbf{r} - \mathbf{r}'). \qquad (8.1.33)$$

On post-multiplying (30) by $\overline{\mathbf{G}}_1(\mathbf{r}, \mathbf{r}') \cdot \mathbf{b}$, we have[7]

$$\left[\nabla \times \overline{\mu}_1^{-1} \cdot \nabla \times \mathbf{E}_1(\mathbf{r})\right] \cdot \overline{\mathbf{G}}_1(\mathbf{r}, \mathbf{r}') \cdot \mathbf{b} - \omega^2 \mathbf{E}_1(\mathbf{r}) \cdot \overline{\epsilon}_1^t \cdot \overline{\mathbf{G}}_1(\mathbf{r}, \mathbf{r}') \cdot \mathbf{b}$$
$$= i\omega \mathbf{J}(\mathbf{r}) \cdot \overline{\mathbf{G}}_1(\mathbf{r}, \mathbf{r}') \cdot \mathbf{b}, \qquad (8.1.34)$$

where $\mathbf{b}$ is an arbitrary constant vector. Then, after pre-multiplying (32) by $\mathbf{E}_1(\mathbf{r})$, and post-multiplying it by $\mathbf{b}$, we have

$$\mathbf{E}_1(\mathbf{r}) \cdot \nabla \times (\overline{\mu}_1^t)^{-1} \cdot \nabla \times \overline{\mathbf{G}}_1(\mathbf{r}, \mathbf{r}') \cdot \mathbf{b} - \omega^2 \mathbf{E}_1(\mathbf{r}) \cdot \overline{\epsilon}_1^t \cdot \overline{\mathbf{G}}_1(\mathbf{r}, \mathbf{r}') \cdot \mathbf{b}$$
$$= \mathbf{E}_1(\mathbf{r}) \delta(\mathbf{r} - \mathbf{r}') \cdot (\overline{\mu}_1^t)^{-1} \cdot \mathbf{b}. \qquad (8.1.35)$$

Next, integrating the difference of (34) and (35) over V_1, for $\mathbf{r}' \in V_1$ yields

$$\int\limits_{V_1} dV \left[\nabla \times \overline{\mu}_1^{-1} \cdot \nabla \times \mathbf{E}_1(\mathbf{r}) \cdot \overline{\mathbf{G}}_1(\mathbf{r}, \mathbf{r}') \cdot \mathbf{b}\right.$$
$$\left. - \mathbf{E}_1(\mathbf{r}) \cdot \nabla \times (\overline{\mu}_1^t)^{-1} \cdot \nabla \times \overline{\mathbf{G}}_1(\mathbf{r}, \mathbf{r}') \cdot \mathbf{b}\right]$$
$$= i\omega \int\limits_{V_1} dV \mathbf{J}(\mathbf{r}) \cdot \overline{\mathbf{G}}_1(\mathbf{r}, \mathbf{r}') \cdot \mathbf{b} - \mathbf{E}_1(\mathbf{r}') \cdot (\overline{\mu}_1^t)^{-1} \cdot \mathbf{b}. \qquad (8.1.36)$$

---

[7] Note that the dot product between two vectors $\mathbf{A} \cdot \mathbf{B}$ is actually $\mathbf{A}^t \cdot \mathbf{B}$. Hence, the dot product between $\overline{\epsilon} \cdot \mathbf{E}$ and $\mathbf{B}$ is $\mathbf{E} \cdot \overline{\epsilon}^t \cdot \mathbf{B}$. The transpose sign t over the vector $\mathbf{E}$ is usually ignored.

With the identity that (Exercise 8.4)

$$\nabla \cdot \left\{ \left[\overline{\boldsymbol{\mu}}_1^{-1} \cdot \nabla \times \mathbf{E}_1(\mathbf{r}) \right] \times \overline{\mathbf{G}}_1(\mathbf{r}, \mathbf{r}') \cdot \mathbf{b} + \mathbf{E}_1(\mathbf{r}) \times \left[(\overline{\boldsymbol{\mu}}_1^t)^{-1} \cdot \nabla \times \overline{\mathbf{G}}_1(\mathbf{r}, \mathbf{r}') \cdot \mathbf{b} \right] \right\}$$
$$= \nabla \times \left[\overline{\boldsymbol{\mu}}_1^{-1} \cdot \nabla \times \mathbf{E}_1(\mathbf{r}) \right] \cdot \overline{\mathbf{G}}_1(\mathbf{r}, \mathbf{r}') \cdot \mathbf{b} - \mathbf{E}_1(\mathbf{r}) \cdot \nabla \times \left[(\overline{\boldsymbol{\mu}}_1^t)^{-1} \cdot \nabla \times \overline{\mathbf{G}}_1(\mathbf{r}, \mathbf{r}') \cdot \mathbf{b} \right],$$
$$(8.1.37)$$

Equation (36) becomes

$$\mathbf{E}_1(\mathbf{r}') = i\omega \int_{V_1} dV \mathbf{J}(\mathbf{r}) \cdot \overline{\mathbf{G}}_1(\mathbf{r}, \mathbf{r}') \cdot \overline{\boldsymbol{\mu}}_1^t(\mathbf{r}') + \int_S dS\, \hat{n} \cdot \left[\overline{\boldsymbol{\mu}}_1^{-1} \cdot \nabla \times \mathbf{E}_1(\mathbf{r}) \right.$$
$$\times \overline{\mathbf{G}}_1(\mathbf{r}, \mathbf{r}') \cdot \overline{\boldsymbol{\mu}}_1^t(\mathbf{r}') + \mathbf{E}_1(\mathbf{r}) \times (\overline{\boldsymbol{\mu}}_1^t)^{-1} \cdot \nabla \times \overline{\mathbf{G}}_1(\mathbf{r}, \mathbf{r}') \cdot \overline{\boldsymbol{\mu}}_1^t(\mathbf{r}') \right], \quad \mathbf{r}' \in V_1.$$
$$(8.1.38)$$

But if $\overline{\mathbf{G}}_1(\mathbf{r}, \mathbf{r}')$ satisfies the radiation condition at infinity, then the surface integral over S_{inf} vanishes by virtue of the radiation condition. This is if the anisotropic, inhomogeneous medium occupies only a finite region in space. Observe now, we have removed the constant vector $\mathbf{b}$ which, up to this point, has been used as a thinking aid. Then, using Maxwell's equations and the appropriate vector identity, Equation (38) becomes

$$\mathbf{E}_1(\mathbf{r}') = i\omega \int_{V_1} dV \mathbf{J}(\mathbf{r}) \cdot \overline{\mathbf{G}}_1(\mathbf{r}, \mathbf{r}') \cdot \overline{\boldsymbol{\mu}}_1^t(\mathbf{r}') + \int_S dS \left\{ i\omega \hat{n} \times \mathbf{H}_1(\mathbf{r}) \cdot \overline{\mathbf{G}}_1(\mathbf{r}, \mathbf{r}') \cdot \overline{\boldsymbol{\mu}}_1^t(\mathbf{r}') \right.$$
$$\left. + \hat{n} \times \mathbf{E}_1(\mathbf{r}) \cdot [\overline{\boldsymbol{\mu}}_1^t(\mathbf{r})]^{-1} \cdot \nabla \times \overline{\mathbf{G}}_1(\mathbf{r}, \mathbf{r}') \cdot \overline{\boldsymbol{\mu}}_1^t(\mathbf{r}') \right\}, \quad \mathbf{r}' \in V_1. \quad (8.1.39)$$

Now, the above is just the generalized Huygens' principle for a general anisotropic inhomogeneous medium.

For $\mathbf{r}' \in V_2$, the second term on the right-hand side of (36) vanishes, and (39) is modified accordingly. Consequently, the analogue of Equation (10) is

$$\left. \begin{array}{l} \mathbf{r} \in V_1, \ \mathbf{E}_1(\mathbf{r}) \\ \mathbf{r} \in V_2, \ 0 \end{array} \right\} = \mathbf{E}_{inc}(\mathbf{r}) + \int_S dS' \left\{ i\omega \hat{n}' \times \mathbf{H}_1(\mathbf{r}') \cdot \overline{\mathbf{G}}_1(\mathbf{r}', \mathbf{r}) \cdot \overline{\boldsymbol{\mu}}_1^t(\mathbf{r}) \right.$$
$$\left. + \hat{n}' \times \mathbf{E}_1(\mathbf{r}') \cdot [\overline{\boldsymbol{\mu}}_1^t(\mathbf{r}')]^{-1} \cdot \nabla' \times \overline{\mathbf{G}}_1(\mathbf{r}', \mathbf{r}) \cdot \overline{\boldsymbol{\mu}}_1^t(\mathbf{r}) \right\}, \quad (8.1.40)$$

where $\mathbf{E}_{inc}(\mathbf{r})$ is the first integral on the right-hand side of (39) which corresponds to the field generated by the current source $\mathbf{J}(\mathbf{r})$ in the inhomogeneous medium. Furthermore, the lower part of Equation (40) is just the generalized extinction theorem for anisotropic inhomogeneous media.

In region 2, analogous to Equation (11), a similar derivation yields (see Exercise 8.4)

$$\left. \begin{array}{l} \mathbf{r} \in V_2, \ \mathbf{E}_2(\mathbf{r}) \\ \mathbf{r} \in V_1, \ 0 \end{array} \right\} = -\int_S dS' \left\{ i\omega \hat{n}' \times \mathbf{H}_2(\mathbf{r}') \cdot \overline{\mathbf{G}}_2(\mathbf{r}', \mathbf{r}) \cdot \overline{\boldsymbol{\mu}}_2^t(\mathbf{r}) \right.$$
$$\left. + \hat{n}' \times \mathbf{E}_2(\mathbf{r}') \cdot [\overline{\boldsymbol{\mu}}_2^t(\mathbf{r}')]^{-1} \cdot \nabla' \times \overline{\mathbf{G}}_2(\mathbf{r}', \mathbf{r}) \cdot \overline{\boldsymbol{\mu}}_2^t(\mathbf{r}) \right\}. \quad (8.1.41)$$

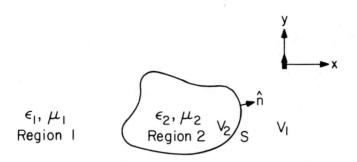

Figure 8.1.4 Inhomogeneity with two piecewise-constant regions for the two-dimensional problem.

Finally, the integral equations for an anisotropic, inhomogeneous medium are

$$-\mathbf{E}_{inc}(\mathbf{r}) = \int\limits_{S} dS' \ \left\{ i\omega\hat{n}' \times \mathbf{H}_1(\mathbf{r}') \cdot \overline{\mathbf{G}}_1(\mathbf{r}',\mathbf{r}) \cdot \overline{\mu}_1^t(\mathbf{r}) \right.$$

$$\left. +\hat{n}' \times \mathbf{E}_1(\mathbf{r}') \cdot [\overline{\mu}_1^t(\mathbf{r}')]^{-1} \cdot \nabla' \times \overline{\mathbf{G}}_1(\mathbf{r}',\mathbf{r}) \cdot \overline{\mu}_1^t(\mathbf{r}) \right\}, \quad \mathbf{r} \in V_2,$$
$$(8.1.42a)$$

$$0 = \int\limits_{S} dS' \ \left\{ i\omega\hat{n}' \times \mathbf{H}_2(\mathbf{r}') \cdot \overline{\mathbf{G}}_2(\mathbf{r}',\mathbf{r}) \cdot \overline{\mu}_2^t(\mathbf{r}) \right.$$

$$\left. +\hat{n}' \times \mathbf{E}_2(\mathbf{r}') \cdot [\overline{\mu}_2^t(\mathbf{r}')]^{-1} \cdot \nabla' \times \overline{\mathbf{G}}_2(\mathbf{r}',\mathbf{r}) \cdot \overline{\mu}_2^t(\mathbf{r}) \right\}, \quad \mathbf{r} \in V_1.$$
$$(8.1.42b)$$

§§8.1.4 *Two-Dimensional Electromagnetic Case*

If the fields are three-dimensional because of the point nature of the source, but the inhomogeneity is two dimensional and piecewise constant as shown in Figure 8.1.4, we can represent the fields and the source in terms of their Fourier transforms in the z direction. In other words, we can write

$$\mathbf{J}(\mathbf{r}) = \frac{1}{2\pi} \int\limits_{-\infty}^{\infty} dk_z \, e^{ik_z z} \mathbf{J}(k_z, \boldsymbol{\rho}), \qquad (8.1.43)$$

$$\mathbf{E}(\mathbf{r}) = \frac{1}{2\pi} \int\limits_{-\infty}^{\infty} dk_z \, e^{ik_z z} \mathbf{E}(k_z, \boldsymbol{\rho}), \qquad (8.1.44)$$

$$\mathbf{H}(\mathbf{r}) = \frac{1}{2\pi} \int\limits_{-\infty}^{\infty} dk_z \, e^{ik_z z} \mathbf{H}(k_z, \boldsymbol{\rho}), \qquad (8.1.45)$$

where $\rho = \hat{x}x + \hat{y}y$.

Physically, Equations (43) to (45) express the three-dimensional fields as linear superpositions of two-dimensional fields with $e^{ik_z z}$ dependence. But since only two components of the electromagnetic field are truly independent in this case, the field can be represented completely in terms of E_z and H_z. In other words, the field in such a geometry can be decomposed into TM (or E_z) and TE (or H_z) fields. These two-dimensional fields, in the integrands of (43) to (45) in a homogeneous region, satisfy the equations (see Exercise 8.5)

$$[\nabla_s^2 + k^2 - k_z^2]E_z(x,y) = -i\omega\mu J_z + ik_z\frac{\rho}{\epsilon}, \qquad (8.1.46a)$$

$$[\nabla_s^2 + k^2 - k_z^2]H_z(x,y) = -(\nabla_s \times \mathbf{J}_s)_z, \qquad (8.1.46b)$$

where $\nabla_s = \hat{x}\frac{\partial}{\partial x} + \hat{y}\frac{\partial}{\partial y}$. In the above derivations, the $e^{ik_z z}$ dependence of the transformed field has been assumed. Observe that the above equations are applicable in homogeneous region 1 and region 2. Thus, in this manner, a three-dimensional field problem with a two-dimensional inhomogeneity is reduced to a linear superposition of two-dimensional problems with varying k_z. Such a problem is also called a two-and-one-half-dimensional problem.[8] We shall discuss next the derivation of the corresponding surface integral equations.

A horizontal electric dipole pointing in the $\hat{y}$ direction has $\mathbf{J}(\mathbf{r}) = \hat{y}I\ell\delta(x)\delta(y)\delta(z)$. Therefore, (43) implies that

$$\mathbf{J}(\mathbf{r}) = \hat{y}I\ell\,\delta(x)\,\delta(y)\,\delta(z)$$

$$= \frac{\hat{y}}{2\pi}I\ell \int_{-\infty}^{\infty} dk_z\, e^{ik_z z}\delta(x)\,\delta(y). \qquad (8.1.47)$$

In other words, a point source is a linear superposition of line sources with different $e^{ik_z z}$ dependences. Furthermore, when such a source is embedded in region 1, (46a) and (46b) become

$$(\nabla_s^2 + k_{1s}^2)E_{1z}(\boldsymbol{\rho}) = \frac{I\ell k_z}{\omega\epsilon_1}\,\delta'(y)\,\delta(x), \qquad (8.1.48a)$$

$$(\nabla_s^2 + k_{1s}^2)H_{1z}(\boldsymbol{\rho}) = -I\ell\,\delta'(x)\,\delta(y), \qquad (8.1.48b)$$

where $k_{1s}^2 = k_1^2 - k_z^2$. The above are also known as the ***reduced wave equations*** because a three-dimensional field problem has been reduced to a two-dimensional problem. Next, we define a two-dimensional Green's function to satisfy the following equation:

$$(\nabla_s^2 + k_{is}^2)G_i(\boldsymbol{\rho}, \boldsymbol{\rho}') = -\delta(\boldsymbol{\rho} - \boldsymbol{\rho}'), \qquad (8.1.49)$$

---

[8] The reduction of such a three-dimensional problem to a two-dimensional problem is well known (see, e.g., Wait 1955 in references for Chapter 3; Chuang and Kong 1982).

where $\delta(\boldsymbol{\rho}) = \delta(x)\delta(y)$. On applying concepts similar to the three-dimensional scalar wave equation in arriving at (10), we derive

$$
\left.\begin{array}{cc} \boldsymbol{\rho} \in V_1, & E_{1z}(\boldsymbol{\rho}) \\ \boldsymbol{\rho} \in V_2, & 0 \end{array}\right\} = \frac{I\ell k_z}{\omega\epsilon_1}\frac{\partial}{\partial y'}G_1(\boldsymbol{\rho}, \boldsymbol{\rho}' = 0)
$$
$$
- \int_S dS'\,\hat{n}' \cdot [G_1(\boldsymbol{\rho}', \boldsymbol{\rho})\nabla'_s E_{1z}(\boldsymbol{\rho}') - E_{1z}(\boldsymbol{\rho}')\nabla'_s G_1(\boldsymbol{\rho}', \boldsymbol{\rho})]. \quad (8.1.50)
$$

In the above, $G_i(\boldsymbol{\rho}, \boldsymbol{\rho}')$ is the two-dimensional unbounded homogeneous-medium Green's function, which is [Equation (2.2.4), Chapter 2]

$$
G_i(\boldsymbol{\rho}, \boldsymbol{\rho}') = \frac{i}{4}H_0^{(1)}(k_{is}|\boldsymbol{\rho} - \boldsymbol{\rho}'|). \quad (8.1.51)
$$

By the same token, a similar equation can be derived for H_{1z}, giving

$$
\left.\begin{array}{cc} \boldsymbol{\rho} \in V_1, & H_{1z}(\boldsymbol{\rho}) \\ \boldsymbol{\rho} \in V_2, & 0 \end{array}\right\} = -I\ell\frac{\partial}{\partial x'}G_1(\boldsymbol{\rho}, \boldsymbol{\rho}' = 0)
$$
$$
- \int_S dS'\,\hat{n}' \cdot [G_1(\boldsymbol{\rho}, \boldsymbol{\rho}')\nabla'_s H_{1z}(\boldsymbol{\rho}') - H_{1z}(\boldsymbol{\rho}')\nabla'_s G_1(\boldsymbol{\rho}, \boldsymbol{\rho}')]. \quad (8.1.52)
$$

Moreover, applying the same concept to region 2 yields (Exercise 8.6)

$$
\left.\begin{array}{cc} \boldsymbol{\rho} \in V_2, & E_{2z}(\boldsymbol{\rho}) \\ \boldsymbol{\rho} \in V_1, & 0 \end{array}\right\} = \int_S dS'\,\hat{n}' \cdot [G_2(\boldsymbol{\rho}, \boldsymbol{\rho}')\nabla'_s E_{2z}(\boldsymbol{\rho}') - E_{2z}(\boldsymbol{\rho}')\nabla'_s G_2(\boldsymbol{\rho}, \boldsymbol{\rho}')],
$$
$$
(8.1.53)
$$
$$
\left.\begin{array}{cc} \boldsymbol{\rho} \in V_2, & H_{2z}(\boldsymbol{\rho}) \\ \boldsymbol{\rho} \in V_1, & 0 \end{array}\right\} = \int_S dS'\,\hat{n}' \cdot [G_2(\boldsymbol{\rho}, \boldsymbol{\rho}')\nabla'_s H_{2z}(\boldsymbol{\rho}') - H_{2z}(\boldsymbol{\rho}')\nabla'_s G_2(\boldsymbol{\rho}, \boldsymbol{\rho}')].
$$
$$
(8.1.54)
$$

In the above, $G_2(\boldsymbol{\rho}, \boldsymbol{\rho}')$ is the two-dimensional Green's function for region 2. It need only satisfy (49) and need not be of the form (51) (Exercise 8.6).

In the above, E_{iz} and H_{iz} and their normal derivatives are the unknowns. Subsequently, the extinction theorem can be used to write the integral equations as

$$
0 = \mathbf{S}_1(\boldsymbol{\rho}) - \int_S dS'\,\hat{n}' \cdot \left\{ G_1(\boldsymbol{\rho}, \boldsymbol{\rho}')\nabla'_s \begin{bmatrix} E_{1z}(\boldsymbol{\rho}') \\ H_{1z}(\boldsymbol{\rho}') \end{bmatrix} \right.
$$
$$
\left. - \begin{bmatrix} E_{1z}(\boldsymbol{\rho}') \\ H_{1z}(\boldsymbol{\rho}') \end{bmatrix} \nabla'_s G_1(\boldsymbol{\rho}, \boldsymbol{\rho}') \right\}, \quad \boldsymbol{\rho} \in V_2, \quad (8.1.55)
$$

and

$$
0 = \int_S dS'\,\hat{n}' \cdot \left\{ G_2(\boldsymbol{\rho}, \boldsymbol{\rho}')\nabla'_s \begin{bmatrix} E_{2z}(\boldsymbol{\rho}') \\ H_{2z}(\boldsymbol{\rho}') \end{bmatrix} \right.
$$
$$
\left. - \begin{bmatrix} E_{2z}(\boldsymbol{\rho}') \\ H_{2z}(\boldsymbol{\rho}') \end{bmatrix} \nabla'_s G_2(\boldsymbol{\rho}, \boldsymbol{\rho}') \right\}, \quad \boldsymbol{\rho} \in V_1, \quad (8.1.56)
$$

where $\mathbf{S}_1(\boldsymbol{\rho}) = \left[\frac{I\ell k_z}{\omega\epsilon_1}\frac{\partial}{\partial y'}G_1(\boldsymbol{\rho}, \boldsymbol{\rho}' = 0), -I\ell\frac{\partial}{\partial x'}G_1(\boldsymbol{\rho}, \boldsymbol{\rho}' = 0)\right]^t$ is the source field in region 1.

Notice that the above consists of four integral equations with eight unknowns, E_{iz}, H_{iz} and their normal derivatives on S. Therefore, the boundary conditions have to be imposed on S to eliminate four of the unknowns. To this end, we require that the tangential fields be continuous. Then,

$$E_{1z} = E_{2z}, \qquad H_{1z} = H_{2z}, \quad \text{on} \quad S. \tag{8.1.57}$$

Furthermore, it is necessary that

$$\hat{n} \times \mathbf{E}_{1s} = \hat{n} \times \mathbf{E}_{2s}, \qquad \hat{n} \times \mathbf{H}_{1s} = \hat{n} \times \mathbf{H}_{2s}, \quad \text{on} \quad S. \tag{8.1.58}$$

Hence, from the equation [see Equation (2.3.17), Chapter 2]

$$\mathbf{E}_s = \frac{i}{k_s^2}[k_z\nabla_s E_z + \omega\mu\nabla_s \times \mathbf{H}_z], \tag{8.1.59}$$

and $\hat{n} \times (\mathbf{E}_{1s} - \mathbf{E}_{2s}) = 0$, we have (see Exercise 8.7)

$$\hat{n} \cdot \nabla_s H_{2z} = -\frac{k_z}{\omega\mu_2}\left(\frac{k_{2s}^2}{k_{1s}^2} - 1\right)(\hat{z} \cdot \hat{n} \times \nabla_s)E_{1z} + \frac{\mu_1 k_{2s}^2}{\mu_2 k_{1s}^2}\hat{n} \cdot \nabla_s H_{1z}. \tag{8.1.60}$$

Furthermore, from the duality principle,

$$\hat{n} \cdot \nabla_s E_{2z} = \frac{k_z}{\omega\epsilon_2}\left(\frac{k_{2s}^2}{k_{1s}^2} - 1\right)(\hat{z} \cdot \hat{n} \times \nabla_s)H_{1z} + \frac{\epsilon_1 k_{2s}^2}{\epsilon_2 k_{1s}^2}\hat{n} \cdot \nabla_s E_{1z}. \tag{8.1.61}$$

Equations (60) and (61) can be combined as

$$\begin{bmatrix} \hat{n} \cdot \nabla_s E_{2z} \\ \hat{n} \cdot \nabla_s H_{2z} \end{bmatrix} = \overline{\mathbf{M}} \cdot \begin{bmatrix} E_{1z} \\ H_{1z} \end{bmatrix} + \overline{\mathbf{N}} \cdot \begin{bmatrix} \hat{n} \cdot \nabla_s E_{1z} \\ \hat{n} \cdot \nabla_s H_{1z} \end{bmatrix}, \tag{8.1.62}$$

where $M_{11} = M_{22} = N_{12} = N_{21} = 0$,

$$M_{12} = \frac{k_z}{\omega\epsilon_2}\left(\frac{k_{2s}^2}{k_{1s}^2} - 1\right)\hat{z} \cdot \hat{n} \times \nabla_s, \qquad N_{11} = \frac{\epsilon_1 k_{2s}^2}{\epsilon_2 k_{1s}^2}, \tag{8.1.63a}$$

$$M_{21} = -\frac{k_z}{\omega\mu_2}\left(\frac{k_{2s}^2}{k_{1s}^2} - 1\right)\hat{z} \cdot \hat{n} \times \nabla_s, \qquad N_{22} = \frac{\mu_1 k_{2s}^2}{\mu_2 k_{1s}^2}. \tag{8.1.63b}$$

By doing so, and making use of (57) and (62) in (56), then

$$\int_S dS' \left\{ G_2(\boldsymbol{\rho}, \boldsymbol{\rho}')\overline{\mathbf{N}} \cdot \begin{bmatrix} \hat{n}' \cdot \nabla'_s E_{1z}(\boldsymbol{\rho}') \\ \hat{n}' \cdot \nabla'_s H_{1z}(\boldsymbol{\rho}') \end{bmatrix} - [\hat{n}' \cdot \nabla'_s G_2(\boldsymbol{\rho}, \boldsymbol{\rho}') \right.$$

$$\left. - G_2(\boldsymbol{\rho}, \boldsymbol{\rho}')\overline{\mathbf{M}}] \cdot \begin{bmatrix} E_{1z}(\boldsymbol{\rho}') \\ H_{1z}(\boldsymbol{\rho}') \end{bmatrix} \right\} = 0, \quad \boldsymbol{\rho} \in V_1. \tag{8.1.64}$$

In addition, from (55),

$$\int_S dS' \left\{ G_1(\boldsymbol{\rho}, \boldsymbol{\rho}') \begin{bmatrix} \hat{n}' \cdot \nabla'_s E_{1z}(\boldsymbol{\rho}') \\ \hat{n}' \cdot \nabla'_s H_{1z}(\boldsymbol{\rho}') \end{bmatrix} \right.$$
$$\left. - \hat{n}' \cdot \nabla'_s G_1(\boldsymbol{\rho}, \boldsymbol{\rho}') \begin{bmatrix} E_{1z}(\boldsymbol{\rho}') \\ H_{1z}(\boldsymbol{\rho}') \end{bmatrix} \right\} = \mathbf{S}_1(\boldsymbol{\rho}), \quad \boldsymbol{\rho} \in V_2. \quad (8.1.65)$$

Equation (64) and Equation (65) together constitute two vector integral equations with two vector unknowns. In (64), $\overline{\mathbf{M}}$, being nondiagonal, couples the TE and TM fields together. But if $k_z = 0$, the TE and TM fields are decoupled again, and the problem reduces to two scalar problems (also see Chuang and Kong 1982; Wang and Chew 1989).

§8.2 Solutions by the Method of Moments

Given the integral equations and the boundary conditions, we can solve for the unknown surface field. Then, with the surface field known, the field everywhere can be calculated. The solutions of the integral equation, as such, are pertinent to many scattering problems. Unless the surfaces coincide with some curvilinear coordinate system, the integral equations in general do not have closed-form solutions, and more often than not, the unknowns have to be solved for numerically. Therefore, we shall illustrate the use of two methods, the **method of moments** (Harrington 1968) (also known as the method of weighted residuals; see Chapter 5 for references), and the **extended-boundary-condition method** (Waterman 1969, 1971) (also known as the **null-field approach**) to solve such integral equations.

§§8.2.1 Scalar Wave Case

The integral equations in (8.1.12a) and (8.1.12b) can be written symbolically as

$$\mathcal{L}_{11}(\mathbf{r}, \mathbf{r}') \, \hat{n}' \cdot \nabla' \phi_1(\mathbf{r}') + \mathcal{L}_{12}(\mathbf{r}, \mathbf{r}') \, \phi_1(\mathbf{r}') = \phi_{inc}(\mathbf{r}), \qquad \mathbf{r} \in V_2, \qquad (8.2.1a)$$

$$\mathcal{L}_{21}(\mathbf{r}, \mathbf{r}') \, \hat{n}' \cdot \nabla' \phi_2(\mathbf{r}') + \mathcal{L}_{22}(\mathbf{r}, \mathbf{r}') \, \phi_2(\mathbf{r}') = 0, \qquad \mathbf{r} \in V_1, \qquad (8.2.1b)$$

where the integral operators are

$$\mathcal{L}_{11}(\mathbf{r}, \mathbf{r}') = \int_S dS' \, g_1(\mathbf{r}, \mathbf{r}'), \quad \mathcal{L}_{12} = -\int_S dS' \, \hat{n}' \cdot \nabla' g_1(\mathbf{r}, \mathbf{r}'), \qquad (8.2.2a)$$

$$\mathcal{L}_{21}(\mathbf{r}, \mathbf{r}') = \int_S dS' \, g_2(\mathbf{r}, \mathbf{r}'), \quad \mathcal{L}_{22} = -\int_S dS' \, \hat{n}' \cdot \nabla' g_2(\mathbf{r}, \mathbf{r}'). \qquad (8.2.2b)$$

$\hat{n} \cdot \nabla \phi$ and ϕ are independent unknowns. But the boundary conditions given by (8.1.13), imply that (1a) and (1b) become

$$\mathcal{L}_{11}(\mathbf{r}, \mathbf{r}') \, \hat{n}' \cdot \nabla' \phi_1(\mathbf{r}') + \mathcal{L}_{12}(\mathbf{r}, \mathbf{r}') \, \phi_1(\mathbf{r}') = \phi_{inc}(\mathbf{r}), \qquad \mathbf{r} \in V_2, \qquad (8.2.3a)$$

$$\frac{p_1}{p_2} \mathcal{L}_{21}(\mathbf{r}, \mathbf{r}') \, \hat{n}' \cdot \nabla' \phi_1(\mathbf{r}') + \mathcal{L}_{22}(\mathbf{r}, \mathbf{r}') \, \phi_1(\mathbf{r}') = 0, \qquad \mathbf{r} \in V_1. \qquad (8.2.3b)$$

To solve (3a) and (3b) with the method of moments (Chapter 5), we let

$$\hat{n}' \cdot \nabla' \phi_1(\mathbf{r}') = \sum_{n=1}^{N} a_n f_{1n}(\mathbf{r}'), \quad \phi_1(\mathbf{r}') = \sum_{n=1}^{N} b_n f_{2n}(\mathbf{r}'), \qquad (8.2.4)$$

where $f_{1n}(\mathbf{r}')$ and $f_{2n}(\mathbf{r}')$ are known basis functions. Then, (3a) and (3b) become

$$\sum_{n=1}^{N} a_n \mathcal{L}_{11}(\mathbf{r}, \mathbf{r}') f_{1n}(\mathbf{r}') + \sum_{n=1}^{N} b_n \mathcal{L}_{12}(\mathbf{r}, \mathbf{r}') f_{2n}(\mathbf{r}') = \phi_{inc}(\mathbf{r}), \quad \mathbf{r} \in V_2, \tag{8.2.5a}$$

$$\sum_{n=1}^{N} a_n \frac{p_1}{p_2} \mathcal{L}_{21}(\mathbf{r}, \mathbf{r}') f_{1n}(\mathbf{r}') + \sum_{n=1}^{N} b_n \mathcal{L}_{22}(\mathbf{r}, \mathbf{r}') f_{2n}(\mathbf{r}') = 0, \qquad \mathbf{r} \in V_1. \tag{8.2.5b}$$

In (4), we assume the unknown $\hat{n} \cdot \nabla \phi_1(\mathbf{r}')$ is approximated well by $f_{1n}(\mathbf{r}')$ and $\phi_1(\mathbf{r}')$ by $f_{2n}(\mathbf{r}')$. Then, after multiplying (5a) by $w_{1m}(\mathbf{r})$ and (5b) by $w_{2m}(\mathbf{r})$, where $m = 1, \ldots, N$, and integrating over $\mathbf{r}$, we obtain

$$\sum_{n=1}^{N} a_n \langle w_{1m}(\mathbf{r}), \mathcal{L}_{11}(\mathbf{r}, \mathbf{r}') f_{1n}(\mathbf{r}') \rangle + \sum_{n=1}^{N} b_n \langle w_{1m}(\mathbf{r}), \mathcal{L}_{12}(\mathbf{r}, \mathbf{r}') f_{2n}(\mathbf{r}) \rangle$$
$$= \langle w_{1m}(\mathbf{r}), \phi_{inc}(\mathbf{r}) \rangle, \quad m = 1, \ldots, N, \quad (8.2.6a)$$

$$\sum_{n=1}^{N} a_n \left\langle w_{2m}(\mathbf{r}), \frac{p_1}{p_2} \mathcal{L}_{21}(\mathbf{r}, \mathbf{r}') f_{1n}(\mathbf{r}') \right\rangle + \sum_{n=1}^{N} b_n \langle w_{2m}(\mathbf{r}), \mathcal{L}_{22}(\mathbf{r}, \mathbf{r}') f_{2n}(\mathbf{r}) \rangle$$
$$= 0, \quad m = 1, \ldots, N. \quad (8.2.6b)$$

The above forms $2N$ linear algebraic equations which yield the $2N$ unknowns, a_n's, $n = 1, \ldots, N$ and b_n's, $n = 1, \ldots, N$. In Equation (6a), we can define $w_{1m}(\mathbf{r})$ anywhere in V_2 and $w_{2m}(\mathbf{r})$ anywhere in V_1. Often, the solution is most stable if $\mathbf{r}$ is chosen close to S. But if $\mathbf{r}$ is chosen to tend to S from either V_1 or V_2, then the singularity of the Green's function has to be properly accounted for.

To see this, notice that if $\mathbf{r} \in S$, the integral arising from the operator $\mathcal{L}_{12}$ and $\mathcal{L}_{22}$, i.e.,

$$\int_S dS' \, \phi(\mathbf{r}') \, \hat{n}' \cdot \nabla' g(\mathbf{r}, \mathbf{r}'), \quad \mathbf{r} \in S \tag{8.2.7}$$

does not converge, because the kernel

$$\hat{n}' \cdot \nabla' g(\mathbf{r}, \mathbf{r}') \sim O(1/|\mathbf{r} - \mathbf{r}'|^2), \quad |\mathbf{r} - \mathbf{r}'| \to 0 \tag{8.2.8}$$

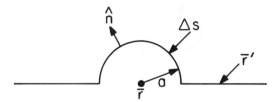

Figure 8.2.1 Diagram for evaluating the residue of a divergent integral.

gives rise to a divergent integral. Such integral equations are also called singular integral equations (see Baker 1977). However, the principal value of the integral and its residue exist, for if $\mathbf{r} \in S$, we can deform the $\mathbf{r}'$ integral around $\mathbf{r}$ and evaluate the integral in the limit when $a \to 0$ as shown in Figure 8.2.1. Therefore,

$$I(\mathbf{r}) = \int_S dS' \, \phi(\mathbf{r}') \, \hat{n}' \cdot \nabla' g(\mathbf{r}, \mathbf{r}') = \fint_S dS' \, \phi(\mathbf{r}') \, \hat{n}' \cdot \nabla' g(\mathbf{r}, \mathbf{r}') + \text{Res}, \qquad (8.2.9)$$

where

$$\fint_S dS'$$

denotes a principal value integral, while "Res" denotes the residue:

$$\text{Res} = \lim_{a \to 0} \int_{\Delta S} dS' \, \phi(\mathbf{r}') \, \hat{n}' \cdot \nabla' g(\mathbf{r}, \mathbf{r}') = \phi(\mathbf{r}) \lim_{a \to 0} \int_{\Delta S} dS' \, \hat{n}' \cdot \nabla' g(\mathbf{r}, \mathbf{r}').$$
$$(8.2.10)$$

Evaluating the last integral in (10) with $\mathbf{r}$ as the origin yields

$$\lim_{a \to 0} \int_{\Delta S} dS' \, \hat{n}' \cdot \nabla' g(\mathbf{r}, \mathbf{r}') = - \int_0^{\pi/2} \int_0^{2\pi} a^2 \sin\theta \, d\theta \, d\phi \, \frac{1}{4\pi a^2} = -\frac{1}{2}. \qquad (8.2.11)$$

Therefore,

$$I(\mathbf{r}) = -\frac{1}{2}\phi(\mathbf{r}) + \fint_S dS' \, \phi(\mathbf{r}') \, \hat{n}' \cdot \nabla' g(\mathbf{r}, \mathbf{r}'). \qquad (8.2.12)$$

We can use such principal value integrals to derive integral equations alternative to (8.1.12) (see Exercise 8.8). Note that the residue will have different signs if $\mathbf{r}$ approaches the surface S from different sides. In other words, $I(\mathbf{r})$ is a discontinuous function of $\mathbf{r}$ when $\mathbf{r}$ moves from one side of surface S to the other.

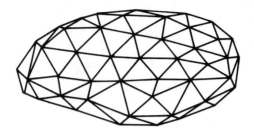

Figure 8.2.2 In the boundary-element method, a smooth surface is approximated by a union of triangles. In two dimensions, a line contour is replaced by a union of line segments.

In (2a) and (2b), if $\mathbf{r}$ and $\mathbf{r}'$ are both defined on the surface S, it can be shown that $\mathcal{L}_{11}$ and $\mathcal{L}_{21}$ are symmetric operators, while $\mathcal{L}_{12}$ and $\mathcal{L}_{22}$ are skew symmetric operators. Therefore, if one chooses $w_{1m} = w_{2m} = f_{1m} = f_{2m}$, as in Galerkin's method (Chapter 5), then the resultant matrix representations of the operators in (6a) and (6b) are either symmetric or skew symmetric (see Exercise 8.9). On the contrary, if w_{1m} and w_{2m} are chosen to be Dirac delta functions, then the solution corresponds to that obtained through the point-matching method or the method of collocation (see Chapter 5 for references).

When the surface S is approximated by a union of triangles or polygons (see Figure 8.2.2), and the expansion functions f_{1n} and f_{2n} are defined over a finite domain (subdomain), e.g., only over the triangles, the method is also known as the ***boundary-element method*** (Brebbia 1978). The boundary-element method is particularly suitable for arbitrarily shaped objects. The subdomain of each element is locally plane, and the choice and construction of basis functions for a subdomain are a lot easier.

§§8.2.2 The Electromagnetic Case

Having described a solution technique for the scalar integral equation, we ponder next on the electromagnetic (vector) case. Fortunately, the same idea is easily extended to the vector case for Equations (8.1.28a) and (8.1.28b).

To do this, we let

$$\hat{n}' \times \mathbf{E}_1(\mathbf{r}') = \sum_{n=1}^{N} a_n \mathbf{e}_n(\mathbf{r}'), \quad \hat{n}' \times \mathbf{H}_1(\mathbf{r}') = \sum_{n=1}^{N} b_n \mathbf{h}_n(\mathbf{r}'), \qquad (8.2.13)$$

where $\mathbf{e}_n(\mathbf{r}')$ and $\mathbf{h}_n(\mathbf{r}')$ are basis functions that can approximate the vector fields $\hat{n} \times \mathbf{E}_1(\mathbf{r}')$ and $\hat{n} \times \mathbf{H}_1(\mathbf{r}')$ on S fairly well. Testing or weighting functions $\mathbf{w}_{1m}(\mathbf{r})$ and $\mathbf{w}_{2m}(\mathbf{r})$ can be used as in the scalar case.

When the weighting functions are defined over S, as the expansion functions have been, the singularity of the dyadic Green's function must be properly accounted for. Even though the integral

$$\mathbf{I}_1 = \int_S dS' \, \overline{\mathbf{G}}(\mathbf{r}, \mathbf{r}') \cdot \hat{n}' \times \mathbf{H}(\mathbf{r}'), \quad \mathbf{r} \in S \qquad (8.2.14)$$

in (8.1.28) seemingly does not converge, it always yields a unique value either via a principal value integral or a vector potential approach (see Exercise 8.10), because $\hat{n} \times \mathbf{H}(\mathbf{r}')$ is an equivalent electric current sheet which produces an electric field that is continuous across this current sheet. However, the integral

$$\mathbf{I}_2 = \int_S dS' \, \nabla' \times \overline{\mathbf{G}}(\mathbf{r}, \mathbf{r}') \cdot \hat{n}' \times \mathbf{E}(\mathbf{r}'), \quad \mathbf{r} \in S \qquad (8.2.15)$$

is undefined, because $\hat{n}' \times \mathbf{E}(\mathbf{r}')$ is an equivalent magnetic current sheet that produces an $\mathbf{E}$ field whose tangential component is discontinuous across the surface S. We shall elaborate this further.

For an unbounded homogeneous medium,

$$\overline{\mathbf{G}}(\mathbf{r}, \mathbf{r}') = \left(\overline{\mathbf{I}} + \frac{\nabla \nabla}{k^2} \right) g(\mathbf{r}, \mathbf{r}') = \left(\overline{\mathbf{I}} + \frac{\nabla' \nabla'}{k^2} \right) g(\mathbf{r}, \mathbf{r}'), \qquad (8.2.16a)$$

and

$$\nabla' \times \overline{\mathbf{G}}(\mathbf{r}, \mathbf{r}') = \nabla' \times \overline{\mathbf{I}} \, g(\mathbf{r}, \mathbf{r}') = \nabla' g(\mathbf{r}, \mathbf{r}') \times \overline{\mathbf{I}}. \qquad (8.2.16b)$$

Therefore, Equation (15) becomes

$$\mathbf{I}_2 = \int_S dS' \, [\nabla' g(\mathbf{r}, \mathbf{r}')] \times [\hat{n}' \times \mathbf{E}(\mathbf{r}')]. \qquad (8.2.17)$$

On multiplying the above by $\hat{n} \times$, and after using the appropriate vector identity, we obtain

$$\hat{n} \times \mathbf{I}_2 = \int_S dS' \, \hat{n} \times \{[\nabla' g(\mathbf{r}, \mathbf{r}')] \times [\hat{n}' \times \mathbf{E}(\mathbf{r}')]\}$$

$$= -\int_S dS' \, \{\hat{n}' \times \mathbf{E}(\mathbf{r}') \, \hat{n} \cdot \nabla' g(\mathbf{r}, \mathbf{r}') - [\hat{n} \cdot \hat{n}' \times \mathbf{E}(\mathbf{r}')] \, \nabla' g(\mathbf{r}, \mathbf{r}')\} \qquad (8.2.18)$$

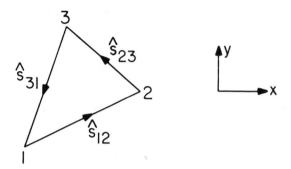

Figure 8.2.3 A triangle with $\hat{s}_{ij}$ defined on the edges.

Then, using the procedure of Equations (9) to (12) (see Exercise 8.10),

$$\hat{n} \times \mathbf{I}_2 = \frac{1}{2}\hat{n} \times \mathbf{E}(\mathbf{r}) - \oint\limits_S dS' \, \hat{n}' \times \mathbf{E}(\mathbf{r}') \, \hat{n} \cdot \nabla' g(\mathbf{r}, \mathbf{r}')$$

$$+ \int\limits_S dS' \, [\hat{n} \cdot \hat{n}' \times \mathbf{E}(\mathbf{r}')] \, \nabla' g(\mathbf{r}, \mathbf{r}') \qquad (8.2.19)$$

The first term in (19) is the singularity effect of the integral in (19). The second term in (19) is well defined after the extraction of this singularity. In addition, since the singularity of a dyadic Green's function is a local effect, this technique can also be applied to the integrals in Equations (8.1.42a) and (8.1.42b).

When the boundary-element method is applied to the vector problem, the surface unknowns $\hat{n} \times \mathbf{E}$ and $\hat{n} \times \mathbf{H}$ are expanded over each triangle of a boundary element. But since $\hat{n} \times \mathbf{E}$ and $\hat{n} \times \mathbf{H}$ represent currents, these currents should be continuous across contiguous elements on a surface approximated by a union of triangles (see Figure 8.2.3). For example, we may want the current components normal to the edges of the triangle to be continuous so that no charge accumulates at the edge of the triangle. More specifically, $\hat{n} \times \mathbf{H}$, which is the electric current on a triangle, can be expanded as

$$\mathbf{J}_s = \sum_{i=1}^{3} \mathbf{J}_{si} N_i(x, y), \qquad (8.2.20)$$

where xy is the plane of a local coordinate system that contains the triangular patch, and $N_i(x, y)$ is a shape function with value 1 at the i-th node and zero at the other nodes (see Figure 8.2.4). Also, $\mathbf{J}_{si}$ is the value of $\mathbf{J}_s$ at the i-th node, where i ranges from 1 to 3. Hence, $\mathbf{J}_{si}$ can be decomposed into

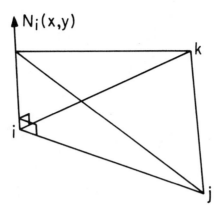

Figure 8.2.4 The shape function $N_i(x, y)$ defined as 1 at one node, zero at the other nodes.

components $\hat{s}_{ij}$ which are unit vectors along the edges of the triangle, as shown in Figure 8.2.3. As a result, the current at the i-th corner of the triangle can be expanded as (see Angkaew et al. 1987)

$$\mathbf{J}_{si} = a_i \hat{s}_{ij} + b_i \hat{s}_{ki}, \tag{8.2.21}$$

where, by simple vector algebra,

$$a_i = \frac{\hat{z} \cdot (\hat{s}_{ki} \times \mathbf{J}_{si})}{\hat{z} \cdot (\hat{s}_{ki} \times \hat{s}_{ij})}, \quad b_i = \frac{\hat{z} \cdot (\hat{s}_{ij} \times \mathbf{J}_{si})}{\hat{z} \cdot (\hat{s}_{ij} \times \hat{s}_{ki})}, \tag{8.2.22}$$

and $\hat{z}$ is normal to the patch surface. With (21) and (22) in mind, (20) is rewritten as

$$\mathbf{J}_s = \sum_{i=1}^{3} [\phi_{i1} \mathbf{N}_{i1}(x, y) + \phi_{i2} \mathbf{N}_{i2}(x, y)], \tag{8.2.23}$$

where

$$\mathbf{N}_{i1} = \frac{\hat{s}_{ij} N_i(x, y)}{\hat{z} \cdot (\hat{s}_{ki} \times \hat{s}_{ij})}, \quad \mathbf{N}_{i2} = \frac{\hat{s}_{ki} N_i(x, y)}{\hat{z} \cdot (\hat{s}_{ij} \cdot x \hat{s}_{ki})}, \tag{8.2.23a}$$

$$\phi_{i1} = \hat{z} \cdot \hat{s}_{ki} \times \mathbf{J}_{si}, \quad \phi_{i2} = \hat{z} \cdot \hat{s}_{ij} \times \mathbf{J}_{si}. \tag{8.2.23b}$$

Note that ϕ_{i1} and ϕ_{i2} are the normal components of $\mathbf{J}_{si}$ (the value of $\mathbf{J}_s$ on the i-th node) on the $\hat{s}_{ki}$ and $\hat{s}_{ij}$ edges respectively. Furthermore, these unknowns are shared by contiguous elements sharing the same edges of the triangles. In this manner, the currents normal to the edges of the triangle are easily rendered continuous from one triangular element to another triangular element, and this method can be used to expand the unknown currents $\hat{n} \times \mathbf{E}$ and $\hat{n} \times \mathbf{H}$ on a triangular patch on S. Subsequently, the matrix representation of the integral operator for the vector electromagnetic field can be obtained (see Exercise 8.11).

For the two-dimensional case, the integral Equations (8.1.64) and (8.1.65) may be written as

$$\overline{\mathcal{L}}_{11}(\rho, \rho') \cdot \psi(\rho') + \overline{\mathcal{L}}_{12}(\rho, \rho') \cdot \phi(\rho') = \mathbf{S}_1(\rho), \quad \rho' \in S, \quad \rho \in V_2, \tag{8.2.24}$$

$$\overline{\mathcal{L}}_{21}(\rho, \rho') \cdot \psi(\rho') + \overline{\mathcal{L}}_{22}(\rho, \rho') \cdot \phi(\rho') = 0, \qquad \rho' \in S, \quad \rho \in V_1. \tag{8.2.25}$$

In order to convert the above into matrix equations, we expand $\psi(\rho')$ and $\phi(\rho')$ in terms of basis functions and find the matrix representations of the integral operators. For example, $\psi(\rho')$ and $\phi(\rho')$ can be expanded as

$$\psi(\rho') = \sum_{n=1}^{N} \overline{\mathbf{g}}_n(\rho') \cdot \mathbf{a}_n, \quad \phi(\rho') = \sum_{n=1}^{N} \overline{\mathbf{g}}_n(\rho') \cdot \mathbf{b}_n, \tag{8.2.26}$$

where

$$\overline{\mathbf{g}}_n(\rho') = \begin{bmatrix} g_{1n}(\rho') & 0 \\ 0 & g_{2n}(\rho') \end{bmatrix}. \tag{8.2.26a}$$

Note that in general, we have to expand a vector in terms of a basis set of matrices in order to ensure completeness. For example, if $E_{1z} = \sum_n \alpha_n g_{1n}(\rho')$ and $H_{1z} = \sum_n \beta_n g_{2n}(\rho')$, the proper expansion for $[E_{1z}, H_{1z}]^t$ is (also see Exercise 8.12)

$$\begin{bmatrix} E_{1z} \\ H_{1z} \end{bmatrix} = \sum_n \begin{bmatrix} g_{1n}(\rho') & 0 \\ 0 & g_{2n}(\rho') \end{bmatrix} \cdot \begin{bmatrix} \alpha_n \\ \beta_n \end{bmatrix} = \sum_n \overline{\mathbf{g}}_n \cdot \boldsymbol{\gamma}_n. \tag{8.2.27}$$

On substituting (27) into (24) and (25) and weighting the equations by $\overline{\mathbf{w}}_m^t(\rho)$, where

$$\overline{\mathbf{w}}_m^t(\rho) = \begin{bmatrix} w_{1m}(\rho) & 0 \\ 0 & w_{2m}(\rho) \end{bmatrix}, \quad \rho \in S,$$

we obtain

$$\sum_{n=1}^{N} \overline{\mathbf{L}}_{11mn} \cdot \mathbf{a}_n + \sum_{n=1}^{N} \overline{\mathbf{L}}_{12mn} \cdot \mathbf{b}_n = \mathbf{S}_{1m}, \quad m = 1, \ldots, N, \tag{8.2.28a}$$

$$\sum_{n=1}^{N} \overline{\mathbf{L}}_{21mn} \cdot \mathbf{a}_n + \sum_{n=1}^{N} \overline{\mathbf{L}}_{22mn} \cdot \mathbf{b}_n = 0, \qquad m = 1, \ldots, N, \tag{8.2.28b}$$

where

$$\overline{\mathbf{L}}_{ijmn} = \langle \overline{\mathbf{w}}_m^t(\rho), \overline{\mathbf{L}}_{ij}(\rho, \rho'), \overline{\mathbf{g}}_n(\rho') \rangle \tag{8.2.29}$$

is the matrix representation of the operator $\overline{\mathbf{L}}_{ij}$ and

$$\mathbf{S}_{1m} = \langle \overline{\mathbf{w}}_m^t(\rho), \mathbf{S}_1(\rho) \rangle. \tag{8.2.30}$$

The system of linear equations in (28) can be solved for $\mathbf{a}_n$ and $\mathbf{b}_n$. Once they are found, the surface fields follow from (26). Then, knowing the surface fields, we can find the field everywhere via the use of Equations (8.1.50) to (8.1.54) in the previous section.

As a final note, firstly, the inner product in (29) usually involves double integrals. Moreover, if ρ and ρ' are both on the surface S, the singularities in $\hat{n} \cdot \nabla_s G_i(\rho - \rho')$ have to be properly accounted for, as has been discussed in Equations (7) to (12). Secondly, if $\overline{\mathbf{w}}_m^t(\rho)$ are chosen to comprise Dirac delta functions, then the method corresponds to the point-matching method.

§§8.2.3 Problem with Internal Resonances

The surface integral equations discussed previously are easily specialized to impenetrable scatterers. For instance, if an impenetrable scatterer has a Dirichlet boundary condition of $\phi_1(\mathbf{r}) = 0$ when $\mathbf{r} \in S$, then Equation (8.1.12a) becomes

$$\phi_{inc}(\mathbf{r}) = \int_S dS' \, g_1(\mathbf{r}, \mathbf{r}') \, \hat{n}' \cdot \nabla' \phi_1(\mathbf{r}'), \quad \mathbf{r} \in V_2. \tag{8.2.31}$$

Note that Equation (8.1.12b) is irrelevant now because $\phi_2(\mathbf{r}) = 0$ inside an impenetrable scatterer. Now, if (31) is imposed on $\mathbf{r} \in S$, severe errors could occur because (31) imposed on S may have a homogeneous solution. In other words,

$$0 = \int_S dS' \, g_1(\mathbf{r}, \mathbf{r}') \, \hat{n}' \cdot \nabla' \phi_1(\mathbf{r}'), \quad \mathbf{r} \in S \tag{8.2.32}$$

could have nontrivial solutions for $\phi_1(\mathbf{r})$, namely, at the internal resonant frequencies of the cavity formed by V_2 (Werner 1963; Schenck 1968; Mitzner 1968; Burton and Miller 1971; Bolomey and Tabbara 1973; Jones 1974; Mittra and Klein 1975; Mautz and Harrington 1978, 1979; Morita 1979; for a review, see Peterson 1990). At these internal resonances, the integral operator defined in (32) has a nonzero nullspace. Hence, its conversion to a matrix form via the method previously described yields an ill-conditioned matrix (see Exercise 8.13).

In addition, the surface source $\hat{n} \cdot \nabla \phi_1(\mathbf{r})$ on S generates no field outside V_2 at resonance [see Exercise 8.13(e)]. Also, from reciprocity (see Chapter 1, Exercise 1.13), the reaction of this surface source with $\phi_{inc}(\mathbf{r})$ is zero since $\phi_{inc}(\mathbf{r})$ is generated by some sources outside V_2. In other words,

$$\int_S dS' \, \phi_{inc}(\mathbf{r}') \, \hat{n}' \cdot \nabla' \phi_1(\mathbf{r}') = 0 \tag{8.2.33}$$

at the resonant frequencies of V_2 and $\hat{n} \cdot \nabla \phi_1(\mathbf{r})$ is the resonant surface source on S. Because of (33), the incident field is orthogonal to the resonant surface source. Therefore, in principle, this field does not excite the resonant surface source (see Exercise 8.13).

So, at resonances, the resonant surface sources $\hat{n} \cdot \nabla\phi_1(\mathbf{r})$ constitute a nullspace of the integral operator defined in (32). Moreover, the excitation coefficient of this resonant current is zero because it is orthogonal to the incident field as given by (33). What then is the amplitude of the resonant current of the scattering problem at the resonant frequency of the cavity formed by V_2? As it turns out, its amplitude is usually nonzero. In fact, it is similar to the case of finding the value of $\sin(x)/x$ at $x = 0$ (which has a removable singularity at $x = 0$) [see Exercise 8.13(d)]. Consequently, a numerical approximation of (31) is difficult to solve because the "poles" and "zeroes" do not cancel precisely.

A plethora of techniques have been proposed to remove this resonance effect in seeking the solution to (31). For instance, one way is to use the combined-field integral equation approach (see Exercise 8.14; also Mitzner 1968). Yet another way is to avoid imposing (31) only on S, but to impose (31) for all $\mathbf{r} \in V_2$. Then, the equation

$$0 = \int_S dS' \, \hat{n}' \cdot g_1(\mathbf{r}, \mathbf{r}') \nabla' \phi_1(\mathbf{r}'), \quad \text{all } \mathbf{r} \in V_2 \qquad (8.2.34)$$

will not have a nontrivial solution, since the field generated by the surface source $\hat{n} \cdot \nabla' \phi_1(\mathbf{r}')$ is forced to be zero everywhere in V_2, precluding an internal resonance. This is precisely the spirit of the extended-boundary-condition method to be discuss in the next section. In this manner, Equation (31) becomes

$$\phi_{inc}(\mathbf{r}) = \int_S dS' \, \hat{n}' \cdot g_1(\mathbf{r}, \mathbf{r}') \nabla' \phi_1(\mathbf{r}'), \quad \mathbf{r} \in V_2, \qquad (8.2.35)$$

which will avoid ill-conditioning. One way of achieving (35), then, is to impose it in a neighborhood interior to S. For example, if the point-matching method is used to solve (35), we will point-match it on S as well as on points just slightly interior to S. This will rid (35) of internal resonances for most practical purposes (see Exercise 8.15).

The scattering solution of a penetrable scatterer governed by (8.1.12) is nontrivial when $\phi_{inc}(\mathbf{r}) = 0$, implying a nonzero nullspace. Here, the physical meaning of the nontrivial solutions corresponds to the resonant modes of the open resonator, formed by two regions with different wave numbers, analogous to a dielectric resonator. Because of radiation damping, however, the resonant frequencies of this open resonator are complex. In other words, the poles of the structure are not on the real ω axis of the complex ω plane. Therefore, for a time harmonic solution where ω is always assumed purely real, the integral operator corresponding to (8.1.12) has no nullspace. Hence,

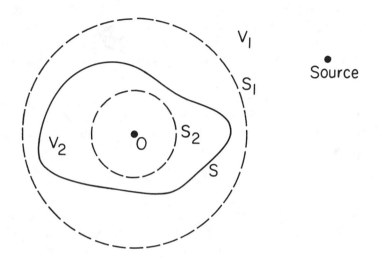

Figure 8.3.1 The surfaces for the extended-boundary-condition method or the null-field approach. V_1 is the volume outside S, while V_2 is the volume inside S.

the problems that plague the solution of (31) do not exist, unless the Q of the resonance mode is so very high that the poles of the structure lie very close to the real ω axis. (Note that the above discussions apply to the vector electromagnetic case too.)

§8.3 Extended-Boundary-Condition Method

The extended-boundary-condition (EBC) method, developed by Waterman (1969, 1971), is also known as the null-field approach. It is an alternative to solve the surface integral equation. In this method, the integral equations are imposed not on the surface S, but on some surfaces S_1 and S_2 away from S as shown in Figure 8.3.1, in order to simplify the solutions. In this section, we shall discuss this method for solving the surface integral equation for the scalar wave case first and discuss the electromagnetics case later.

§§8.3.1 The Scalar Wave Case

The scalar integral equations are

$$\phi_{inc}(\mathbf{r}) = \int_S dS' \left[g_1(\mathbf{r}, \mathbf{r}') \, \hat{n}' \cdot \nabla' \phi_1(\mathbf{r}') - \phi_1(\mathbf{r}') \, \hat{n}' \cdot \nabla' g_1(\mathbf{r}, \mathbf{r}') \right], \quad \mathbf{r} \in V_2,$$
(8.3.1)

$$0 = \int_S dS' \left[g_2(\mathbf{r}, \mathbf{r}') \, \hat{n}' \cdot \nabla' \phi_2(\mathbf{r}') - \phi_2(\mathbf{r}') \, \hat{n}' \cdot \nabla' g_2(\mathbf{r}, \mathbf{r}') \right], \quad \mathbf{r} \in V_1,$$
(8.3.2)

where Equation (1) is valid for $\mathbf{r}$ anywhere in V_2 bounded by S and Equation (2) for $\mathbf{r}$ anywhere in V_1 outside S. As such, it is often convenient to impose the integral equation in (1) on S_2, a spherical surface in V_2, and to impose integral equation in (2) on S_1, a spherical surface in V_1. This is further motivated by a subsequent simplification in the solutions of the integral equation. In this method, the integral equations are imposed on surfaces away from the actual surface of the object, and hence, its name, the extended-boundary-condition method.

The integral equations on the spherical surfaces simplify if spherical harmonics are used in the expansion of the field. For instance, via the addition theorem, the Green's function in a homogeneous medium can be expanded as (see Exercise 8.16)[9]

$$g(\mathbf{r}, \mathbf{r}') = ik \sum_n \psi_n(k, \mathbf{r}_>) \Re e \psi_n(k, \mathbf{r}_<), \qquad (8.3.3)$$

where $\psi_n(k, \mathbf{r})$ represents an outgoing wave spherical harmonic, and $\Re e \psi_n(k, \mathbf{r})$ is the *regular* part of $\psi_n(k, \mathbf{r})$.[10] $\mathbf{r}_>$ represents the larger of the $\mathbf{r}$ and $\mathbf{r}'$, and $\mathbf{r}_<$ is the smaller of the $\mathbf{r}$ and $\mathbf{r}'$ in magnitudes. In addition, since the incident wave is regular about the origin, it can be expanded as the regular spherical wave functions, i.e.,

$$\phi_{inc}(\mathbf{r}) = \sum_n a_n \Re e \psi_n(k_1, \mathbf{r}). \qquad (8.3.4)$$

As such, Equations (1) and (2) become

$$\sum_n a_n \Re e \psi_n(k_1, \mathbf{r}) = ik_1 \sum_n \Re e \psi_n(k_1, \mathbf{r}) \int_S dS' \left[\psi_n(k_1, \mathbf{r}') \, \hat{n}' \cdot \nabla' \phi_1(\mathbf{r}') \right.$$
$$\left. - \phi_1(\mathbf{r}') \, \hat{n}' \cdot \nabla' \psi_n(k_1, \mathbf{r}') \right], \quad \mathbf{r} \in S_2, \qquad (8.3.5)$$

$$0 = ik_2 \sum_n \psi_n(k_2, \mathbf{r}) \int_S dS' \left[\Re e \psi_n(k_2, \mathbf{r}') \, \hat{n}' \cdot \nabla' \phi_2(\mathbf{r}') \right.$$
$$\left. - \phi_2(\mathbf{r}') \, \hat{n}' \cdot \nabla' \Re e \psi_n(k_2, \mathbf{r}') \right], \quad \mathbf{r} \in S_1. \qquad (8.3.6)$$

In the above, (5) is actually valid for $\mathbf{r}$ anywhere in the volume bounded by S_2, and (6) is actually valid for $\mathbf{r}$ anywhere in the volume outside S_1. Consequently, from the orthogonality of the spherical harmonics, we deduce that

$$a_n = ik_1 \int_S dS' \left[\psi_n(k_1, \mathbf{r}') \, \hat{n}' \cdot \nabla' \phi_1(\mathbf{r}') - \phi_1(\mathbf{r}') \, \hat{n}' \cdot \nabla' \psi_n(k_1, \mathbf{r}') \right], \forall n,$$
$$(8.3.7)$$

$$0 = ik_2 \int_S dS' \left[\Re e \psi_n(k_2, \mathbf{r}') \, \hat{n}' \cdot \nabla' \phi_2(\mathbf{r}') - \phi_2(\mathbf{r}') \, \hat{n}' \cdot \nabla' \Re e \psi_n(k_2, \mathbf{r}') \right], \forall n.$$
$$(8.3.8)$$

---

[9] The factor k in (3) is replaced by $1/4$ in two dimensions.

[10] If $\psi_n(k, \mathbf{r})$ consists of a Hankel function, which is singular at $\mathbf{r} = 0$, then $\Re e \psi_n(k, \mathbf{r})$ consists of a Bessel function, which is regular at $\mathbf{r} = 0$.

Observe that the integral equations are greatly simplified now and do not involve $\mathbf{r}$ at all. To solve this new integral equations, we expand the surface unknowns $\hat{n} \cdot \nabla'\phi(\mathbf{r}')$ and $\phi(\mathbf{r}')$ in terms of a basis set to convert (7) and (8) into matrix equations. A clever way of expanding the surface unknowns is to let

$$\phi_2(\mathbf{r}') = \sum_m \alpha_m \Re e \psi_m(k_2, \mathbf{r}'), \qquad (8.3.9)$$

$$\hat{n}' \cdot \nabla'\phi_2(\mathbf{r}') = \sum_m \beta_m \hat{n}' \cdot \nabla'\Re e \psi_m(k_2, \mathbf{r}'). \qquad (8.3.10)$$

But this is only rigorously valid if $\Re e \psi_m(k_2, \mathbf{r}')$ and $\hat{n} \cdot \nabla'\Re e \psi_m(k_2, \mathbf{r}')$ are complete on S. As it turns out, they are complete except at the internal resonances of the cavity formed by V_2 with wave number k_2 (see Exercise 8.17; also see Waterman 1969). In this case, it may seem obvious that $\alpha_m = \beta_m$, but this may be inconsistent with (8). To check this consistency, we use (9) and (10) in (8) to yield

$$0 = \sum_m \beta_m \int_S dS' \left[\Re e \psi_n(k_2, \mathbf{r}') \, \hat{n}' \cdot \nabla'\Re e \psi_m(k_2, \mathbf{r}') \right]$$

$$- \sum_m \alpha_m \int_S dS' \left[\Re e \psi_m(k_2, \mathbf{r}') \, \hat{n}' \cdot \nabla'\Re e \psi_n(k_2, \mathbf{r}') \right]. \qquad (8.3.11)$$

At this point, it is not obvious at all that α_m should be β_m. However, by integrating

$$\nabla' \cdot \left[\Re e \psi_n(k_2, \mathbf{r}') \nabla'\Re e \psi_m(k_2, \mathbf{r}') - \Re e \psi_m(k_2, \mathbf{r}') \nabla'\Re e \psi_n(k_2, \mathbf{r}') \right]$$
$$(8.3.12)$$

over a volume V bounded by S and S_2, and applying Gauss' theorem and the fact that (12) is zero and that the integral over S_2 is zero due to the orthogonality of spherical harmonics on a spherical surface, we conclude that (see Exercise 8.18)

$$\int_S dS' \left[\Re e \psi_n(k_2, \mathbf{r}') \, \hat{n} \cdot \nabla'\Re e \psi_m(k_2, \mathbf{r}') \right] = \int_S dS' \, \Re e \psi_m(k_2, \mathbf{r}') \, \hat{n}' \cdot \nabla'\Re e \psi_n(k_2, \mathbf{r}').$$
$$(8.3.13)$$

This, when used in (11), indeed implies that $\alpha_m = \beta_m$. Hence, the reward for the clever choice of expansion functions in (9) and (10) is that it solves (8) immediately. As a result, Equation (7), with the use of (9) and (10), and the boundary conditions (8.1.13), becomes

$$a_n = ik_1 \sum_m \alpha_m \int_S dS' \left[\psi_n(k_1, \mathbf{r}') \, \hat{n}' \cdot \nabla'\Re e \psi_m(k_2, \mathbf{r}') \frac{p_2}{p_1} \right.$$

$$\left. - \Re e \psi_m(k_2, \mathbf{r}') \, \hat{n}' \cdot \nabla'\psi_n(k_1, \mathbf{r}') \right]. \qquad (8.3.14)$$

Then, Equation (14) is of the form

$$a_n = i \sum_m \alpha_m Q_{nm}, \qquad (8.3.15)$$

where

$$Q_{nm} = k_1 \int_S dS' \left[\psi_n(k_1, \mathbf{r}') \, \hat{n}' \cdot \nabla' \Re e \psi_m(k_2, \mathbf{r}') \frac{p_2}{p_1} \right.$$
$$\left. - \Re e \psi_m(k_2, \mathbf{r}') \, \hat{n}' \cdot \nabla' \psi_n(k_1, \mathbf{r}') \right]. \quad (8.3.16)$$

Note that Equation (15) is in theory an infinite-dimensional matrix equation. But in practice, α_m is solvable from (15) in terms of a_n, the amplitudes of the incident field, by truncating the infinite-dimensional matrix equation.

In the extended-boundary-condition method, the fields are extinct only inside the circle S_2 and outside the circle S_1. In actual fact, the surface sources impressed on S have to extinct the pertinent field everywhere inside S or everywhere outside S. As such, the extended-boundary-condition method provides a solution which sometimes is an approximation of the actual solution, albeit such an approximation is very good in a number of cases.

Because the testing surfaces S_1 and S_2 are away from the surfaces in the extended-boundary-condition method, the matrix equation (15) becomes very ill-conditioned if the maximum distances of S_1 and S_2 from S are large. For instance, this is the case for an elongated object or an object with high corrugations. Physically, the ill-conditioning arises because the surface sources on S generate fields which are localized in the vicinity of S for such surfaces (in the case of a planar, corrugated surface, this would be an evanescent type wave). Accordingly, information on S within such fields is greatly diminished on S_1 and S_2. Therefore, imposing the integral equations (1) and (2) on S_1 and S_2 results in a set of ill-conditioned equations. [The convergence of the Waterman EBC method has been studied by Bolomey and Wirgin (1974) and Bates and Wall (1977).] Despite this, the EBC method is attractive for many applications because it gives a simpler set of equations. It is particularly convenient for a scatterer where the fields around it are expandable in spherical harmonics, cylindrical harmonics, or Floquet modes (e.g., periodic rough surfaces).

The EBC method is also easily adaptable to impenetrable objects. In this case, we need to solve only Equation (7) with either the homogeneous Dirichlet or Neumann boundary condition, or the impedance boundary condition. By the same token as in (9) and (10), one can expand $\hat{n} \cdot \nabla \phi(\mathbf{r})$ in terms of $\hat{n} \cdot \nabla \Re e \psi_n(k_2, \mathbf{r})$, and $\phi(\mathbf{r})$ in terms of $\Re e \psi_n(k_2, \mathbf{r})$.[11] But at the internal resonance of the cavity formed by S, the set $\hat{n} \cdot \nabla \Re e \psi_n(k_2, \mathbf{r})$ or $\Re e \psi_n(k_2, \mathbf{r})$ is

---

[11] k_2 may be chosen to be k_1 in this case.

incomplete on S as mentioned previously (see Exercises 8.17, 8.19). Then, the Q_{nm} matrix thus derived is ill-conditioned for quite a different reason from those discussed in Subsection 8.2.3 (see Exercise 8.20). Hence, this internal resonance can be overcome by using a complete set to expand the surface sources (see Waterman 1969).

§§8.3.2 The Electromagnetic Wave Case

The extended-boundary-condition method is easily generalized to the vector electromagnetics case for solving the integral equations from (8.1.28) (Waterman 1971; Barber and Yeh 1975; also see papers in Kerker 1988). The vector surface integral equations are

$$
\mathbf{E}_{inc}(\mathbf{r}) = \int_S dS' \left[i\omega\mu_1 \overline{\mathbf{G}}_1(\mathbf{r}, \mathbf{r}') \cdot \hat{n}' \times \mathbf{H}_1(\mathbf{r}') \right.
$$

$$
\left. -\nabla' \times \overline{\mathbf{G}}_1(\mathbf{r}, \mathbf{r}') \cdot \hat{n}' \times \mathbf{E}_1(\mathbf{r}') \right], \ \mathbf{r} \in V_2,
$$

$$(8.3.17)$$

$$
0 = \int_S dS' \left[i\omega\mu_2 \overline{\mathbf{G}}_2(\mathbf{r}, \mathbf{r}') \cdot \hat{n}' \times \mathbf{H}_2(\mathbf{r}') \right.
$$

$$
\left. -\nabla' \times \overline{\mathbf{G}}_2(\mathbf{r}, \mathbf{r}') \cdot \hat{n}' \times \mathbf{E}_2(\mathbf{r}') \right], \ \mathbf{r} \in V_1.
$$

$$(8.3.18)$$

In this case, the unbounded homogeneous-medium dyadic Green's function is expanded as [see Exercise 8.21 and Equation (7.3.40)]

$$
\overline{\mathbf{G}}(\mathbf{r}, \mathbf{r}') = ik \sum_n \boldsymbol{\psi}_n(k, \mathbf{r}_>) \Re e \boldsymbol{\psi}_n(k, \mathbf{r}_<), \quad \mathbf{r}_> \neq \mathbf{r}_<,
$$

$$(8.3.19)$$

where $\boldsymbol{\psi}_n(k, \mathbf{r})$'s are vector wave functions for outgoing waves.

By the same token as (4), the incident wave is expanded as the regular vector wave functions

$$
\mathbf{E}_{inc}(\mathbf{r}) = \sum_n a_n \Re e \boldsymbol{\psi}_n(k_1, \mathbf{r}).
$$

$$(8.3.20)$$

Consequently, (17) and (18) become (Exercise 8.22)

$$
a_n = ik_1 \int_S dS' \left[i\omega\mu_1 \boldsymbol{\psi}_n(k_1, \mathbf{r}') \cdot \hat{n}' \times \mathbf{H}_1(\mathbf{r}') \right.
$$

$$
\left. - \nabla' \times \boldsymbol{\psi}_n(k_1, \mathbf{r}') \cdot \hat{n}' \times \mathbf{E}_1(\mathbf{r}') \right],
$$

$$(8.3.21)$$

$$
0 = ik_2 \int_S dS' \left[i\omega\mu_2 \Re e \boldsymbol{\psi}_n(k_2, \mathbf{r}') \cdot \hat{n}' \times \mathbf{H}_2(\mathbf{r}') \right.
$$

$$
\left. - \nabla' \times \Re e \boldsymbol{\psi}_n(k_2, \mathbf{r}') \cdot \hat{n}' \times \mathbf{E}_2(\mathbf{r}') \right].
$$

$$(8.3.22)$$

Next, we expand

$$\hat{n}' \times \mathbf{E}_2(\mathbf{r}') = \sum_m \alpha_m \hat{n}' \times \Re e\psi_m(k_2,\mathbf{r}'), \qquad \mathbf{r}' \in S, \qquad (8.3.23)$$

$$i\omega\mu_2\hat{n}' \times \mathbf{H}_2(\mathbf{r}') = \sum_m \beta_m \hat{n}' \times \nabla' \times \Re e\psi_m(k_2,\mathbf{r}'), \quad \mathbf{r}' \in S, \qquad (8.3.24)$$

because $\hat{n}' \times \Re e\psi_m(k_2,\mathbf{r}')$ and $\hat{n}' \times \nabla' \times \Re e\psi_m(k_2,\mathbf{r}')$ are complete on S (see Exercise 8.23). Then, Equation (22) becomes

$$0 = \sum_m \int_S dS' \, [\beta_m \Re e\psi_n(k_2,\mathbf{r}') \cdot \hat{n}' \times \nabla' \times \Re e\psi_m(k_2,\mathbf{r}')$$

$$- \alpha_m \nabla' \times \Re e\psi_n(k_2,\mathbf{r}') \cdot \hat{n}' \times \Re e\psi_m(k_2,\mathbf{r}')]. \qquad (8.3.25)$$

To prove $\alpha_m = \beta_m$, we take the volume integral of

$$\nabla' \cdot \{[\nabla' \times \Re e\psi_m(k_2,\mathbf{r}')] \times \Re e\psi_n(k_2,\mathbf{r}') + \Re e\psi_m(k_2,\mathbf{r}') \times [\nabla' \times \Re e\psi_n(k_2,\mathbf{r}')]\} \qquad (8.3.26)$$

over a volume bounded by S and S_2 and note that the above is zero and that the surface integral over S_2 vanishes for $n \neq m$. Hence, we conclude that (see Exercise 8.24)

$$\int_S dS' \, \hat{n}' \cdot [\nabla' \times \Re e\psi_m(k_2,\mathbf{r}')] \times \Re e\psi_n(k_2,\mathbf{r}')$$

$$= \int_S dS' \hat{n}' \cdot [\nabla' \times \Re e\psi_n(k_2,\mathbf{r}')] \times \Re e\psi_m(k_2,\mathbf{r}') \qquad (8.3.27)$$

or that $\alpha_m = \beta_m$ in (25) and then, in (23) and (24). Then, after making use of the continuity of $\hat{n} \times \mathbf{E}$ and $\hat{n} \times \mathbf{H}$ on S, (21) becomes

$$a_n = ik_1 \sum_m \alpha_m \int_S dS' \, [(\mu_1/\mu_2)\hat{n}' \cdot \nabla' \times \Re e\psi_m(k_2,\mathbf{r}') \times \psi_n(k_1,\mathbf{r}')$$

$$- \hat{n}' \cdot \Re e\psi_m(k_2,\mathbf{r}') \times \nabla' \times \psi_n(k_1,\mathbf{r}')]$$

$$= i \sum_m \alpha_m Q_{nm}, \qquad (8.3.28)$$

where

$$Q_{nm} = k_1 \int_S dS' \, [(\mu_1/\mu_2)\hat{n}' \cdot \nabla' \times \Re e\psi_m(k_2,\mathbf{r}') \times \psi_n(k_1,\mathbf{r}')$$

$$- \hat{n}' \cdot \Re e\psi_m(k_2,\mathbf{r}') \times \nabla' \times \psi_n(k_1,\mathbf{r}')]. \qquad (8.3.29)$$

From Equation (28), we can solve for α_m's which then yield the surface unknowns. The matrix Equation (28) suffers from ill-conditioning for the same reason given for the scalar wave case when the object is elongated or the surface of the object is convoluted. In addition, for impenetrable objects, (28) suffers from resonance effects as in the scalar wave case.

§8.4 The Transition and Scattering Matrices

Once a scattering problem is solved, the scattered field everywhere can be found. For instance, in the extended-boundary-condition method for the scalar wave, once the α_m's are found in Equation (8.3.15) by truncating the infinite-dimensional matrix equation, the surface fields are found using (8.3.9) and (8.3.10). The scattered field from (8.1.9) then becomes

$$\phi_{sca}(\mathbf{r}) = \sum_n f_n \psi_n(k_1, \mathbf{r}) = -\sum_{nm} ik_1 \alpha_m \psi_n(k_1, \mathbf{r}) \int_S dS' \left[\Re e\psi_n(k_1, \mathbf{r}') \right.$$

$$\left. \hat{n}' \cdot \nabla' \Re e\psi_m(k_2, \mathbf{r}') \frac{p_2}{p_1} - \Re e\psi_m(k_2, \mathbf{r}') \hat{n}' \cdot \nabla' \Re e \cdot \psi_n(k_1, \mathbf{r}') \right].$$

$$(8.4.1)$$

From the above, we readily deduce that

$$f_n = -i \sum_m \alpha_m \Re eQ_{nm}, \qquad (8.4.2)$$

where Q_{nm} is defined in (8.3.16), and $\Re e$ implies the "regular part of." In other words, $\Re eQ_{nm}$ will convert all the Hankel functions in Q_{nm} into Bessel functions. Consequently, from (8.3.15) and (2), we have

$$\mathbf{a} = i\overline{\mathbf{Q}} \cdot \boldsymbol{\alpha}, \qquad (8.4.3a)$$

$$\mathbf{f} = -i \left(\Re e\overline{\mathbf{Q}} \right) \cdot \boldsymbol{\alpha}, \qquad (8.4.3b)$$

where $\mathbf{a}$, $\boldsymbol{\alpha}$, and $\mathbf{f}$ are vectors containing a_n, α_n, and f_n respectively, and $\overline{\mathbf{Q}}$ is a matrix with elements Q_{mn}. On eliminating $\boldsymbol{\alpha}$ in the above, we obtain

$$\mathbf{f} = - \left(\Re e\overline{\mathbf{Q}} \right) \cdot \overline{\mathbf{Q}}^{-1} \cdot \mathbf{a}. \qquad (8.4.4)$$

A transition matrix $\overline{\mathbf{T}}$ can be defined to relate the scattered wave amplitude to the incoming wave amplitude such that (Waterman 1969, 1971)

$$\mathbf{f} = \overline{\mathbf{T}} \cdot \mathbf{a}, \qquad (8.4.5)$$

where

$$\overline{\mathbf{T}} = - \left(\Re e\overline{\mathbf{Q}} \right) \cdot \overline{\mathbf{Q}}^{-1}. \qquad (8.4.6)$$

Hence, the total field then becomes

$$\phi = \sum_n [a_n \Re e\psi_n(k_1, \mathbf{r}) + f_n \psi_n(k_1, \mathbf{r})]$$

$$= \sum_n \left[a_n \Re e\psi_n(k_1, \mathbf{r}) + \left(\sum_m T_{nm} a_m \right) \psi_n(k_1, \mathbf{r}) \right].$$

$$(8.4.7)$$

Notice that the preceding equation is the same as

$$\phi(\mathbf{r}) = \left[\Re e \boldsymbol{\psi}^t(k_1, \mathbf{r}) + \boldsymbol{\psi}^t(k_1, \mathbf{r}) \cdot \overline{\mathbf{T}}\right] \cdot \mathbf{a}, \tag{8.4.8}$$

where $\boldsymbol{\psi}(k_1, \mathbf{r})$ is a column vector containing $\psi_n(k_1, \mathbf{r})$.

If a scattering matrix $\overline{\mathbf{S}}$ is defined which is related to $\overline{\mathbf{T}}$ as

$$\overline{\mathbf{S}} = \overline{\mathbf{I}} + 2\overline{\mathbf{T}}, \tag{8.4.9}$$

then (8) becomes

$$\phi(\mathbf{r}) = \left[\Re e \boldsymbol{\psi}^t(k_1, \mathbf{r}) - \frac{1}{2}\boldsymbol{\psi}^t(k_1, \mathbf{r}) + \frac{1}{2}\boldsymbol{\psi}^t(k_1, \mathbf{r}) \cdot \overline{\mathbf{S}}\right] \cdot \mathbf{a}. \tag{8.4.10}$$

Consider the fact that

$$\Re e \psi_n(k_1, \mathbf{r}) = \frac{1}{2}\psi_n(k_1, \mathbf{r}) + \frac{1}{2}\psi_n(-k_1, \mathbf{r}), \tag{8.4.11}$$

i.e., a standing wave $\Re e \psi_n(k_1, \mathbf{r})$ can be written as a linear superposition of an outgoing wave $\frac{1}{2}\psi_n(k_1, \mathbf{r})$ plus an incoming wave $\frac{1}{2}\psi_n(-k_1, \mathbf{r})$. Then, (10) becomes

$$\phi(\mathbf{r}) = \frac{1}{2}\left[\boldsymbol{\psi}^t(-k_1, \mathbf{r}) + \boldsymbol{\psi}^t(k_1, \mathbf{r}) \cdot \overline{\mathbf{S}}\right] \cdot \mathbf{a}. \tag{8.4.12}$$

Therefore, the scattering matrix $\overline{\mathbf{S}}$ relates the amplitude of the scattered wave to the incoming wave.

Using reciprocity, it can be proven that $\overline{\mathbf{T}}$ is a symmetric matrix (see Exercise 8.25). Thus,

$$\overline{\mathbf{T}}^t = \overline{\mathbf{T}}, \quad \overline{\mathbf{S}}^t = \overline{\mathbf{S}}. \tag{8.4.13}$$

Moreover, energy conservation implies that (see Exercise 8.26)

$$\overline{\mathbf{S}}^\dagger \cdot \overline{\mathbf{S}} = \overline{\mathbf{I}}, \quad \text{or} \quad \overline{\mathbf{S}}^* \cdot \overline{\mathbf{S}} = \overline{\mathbf{I}}, \tag{8.4.14a}$$

and

$$\overline{\mathbf{T}}^\dagger \cdot \overline{\mathbf{T}} = -\Re e \overline{\mathbf{T}}. \tag{8.4.14b}$$

The above are useful checks for the correctness of the $\overline{\mathbf{T}}$ and $\overline{\mathbf{S}}$ matrices when they are computed. Finally, even though the $\overline{\mathbf{T}}$ and $\overline{\mathbf{S}}$ matrices have been derived here using the EBC solution as an illustration, they can in theory be defined once the scattering solution is known, regardless of the method of solution.

§8.5 The Method of Rayleigh's Hypothesis

A method very closely related to the EBC method is the method of Rayleigh's hypothesis (Rayleigh 1894, 1897, 1907). Even though this method does not involve integral equations, it is worthy of discussion because of its close relationship to the EBC method.

Consider the geometry shown in Figure 8.3.1; the field outside the surface S_1 can be expanded in terms of the incident and scattered waves, which are

$$\phi_{inc}(\mathbf{r}) = \sum_n a_n \Re e \psi_n(k_1, \mathbf{r}), \qquad (8.5.1a)$$

$$\phi_{sca}(\mathbf{r}) = \sum_n f_n \psi_n(k_1, \mathbf{r}). \qquad (8.5.1b)$$

Note that we have expanded the incident field in terms of standing waves but the scattered field in terms of outgoing waves. Next, the field inside the scatterer is expanded again in terms of a standing wave of the form

$$\phi_2(\mathbf{r}) = \sum_n \alpha_n \Re e \psi_n(k_2, \mathbf{r}) \qquad (8.5.2)$$

for $\mathbf{r}$ inside S_2. In addition to this, Rayleigh's hypothesis assumes that (1b) and (2) are valid on S as well. But this is not at all clear because, in the region bounded by S_1 and S, it is not obvious if all the waves are outgoing as expressed by (1b). Moreover, in the region bounded by S_2 and S, it is not obvious if all the waves are standing waves.

In spite of this, we assume the validity of Rayleigh's hypothesis and match boundary conditions on S. Consequently, the continuity of the potential implies

$$\sum_n [a_n \Re e \psi_n(k_1, \mathbf{r}) + f_n \psi_n(k_1, \mathbf{r})] = \sum_n \alpha_n \Re e \psi_n(k_2, \mathbf{r}), \quad \mathbf{r} \in S. \qquad (8.5.3)$$

Furthermore, the boundary condition on the normal derivatives given by (8.1.13) yields

$$\sum_n [a_n p_1 \hat{n} \cdot \nabla \Re e \psi_n(k_1, \mathbf{r}) + f_n p_1 \hat{n} \cdot \nabla \psi_n(k_1, \mathbf{r})]$$

$$= \sum_n \alpha_n p_2 \hat{n} \cdot \nabla \Re e \psi_n(k_2, \mathbf{r}), \quad \mathbf{r} \in S. \qquad (8.5.4)$$

To convert the above into matrix equations, we test Equation (3) by $\hat{n} \cdot \nabla \Re e \psi_m(k_2, \mathbf{r})$ and integrate over S to yield

$$\sum_n \left[a_n \int_S dS \, \hat{n} \cdot \nabla \Re e \psi_m(k_2, \mathbf{r}) \Re e \psi_n(k_1, \mathbf{r}) \right.$$

$$\left. + f_n \int_S dS \, \hat{n} \cdot \nabla \Re e \psi_m(k_2, \mathbf{r}) \psi_n(k_1, \mathbf{r}) \right]$$

$$= \sum_n \alpha_n \int_S dS \, \hat{n} \cdot \nabla \Re e \psi_m(k_2, \mathbf{r}) \Re e \psi_n(k_2, \mathbf{r}). \qquad (8.5.5)$$

Similarly, we test Equation (4) by $\Re e\psi_m(k_2,\mathbf{r})$ and integrate over S to yield

$$\sum_n \left[a_n \frac{p_1}{p_2} \int_S dS\, \Re e\psi_m(k_2,\mathbf{r})\, \hat{n}\cdot\nabla\Re e\psi_n(k_1,\mathbf{r}) \right.$$

$$+ f_n \frac{p_1}{p_2} \int_S dS\, \Re e\psi_m(k_2,\mathbf{r})\, \hat{n}\cdot\nabla\psi_n(k_1,\mathbf{r}) \Bigg]$$

$$= \sum_n \alpha_n \int_S dS\, \Re e\psi_m(k_2,\mathbf{r})\, \hat{n}\cdot\nabla\Re e\psi_n(k_2,\mathbf{r}). \quad (8.5.6)$$

But the right-hand sides of (5) and (6) are equal as a result of (8.3.13). Consequently, we have

$$\sum_n a_n \left[\frac{p_1}{p_2} \int_S dS\, \Re e\psi_m(k_2,\mathbf{r})\hat{n}\cdot\nabla\Re e\psi_n(k_1,\mathbf{r}) \right.$$

$$- \int_S dS\, \hat{n}\cdot\nabla\Re e\psi_m(k_2,\mathbf{r})\Re e\psi_n(k_1,\mathbf{r}) \Bigg]$$

$$= -\sum_n f_n \left[\frac{p_1}{p_2} \int_S dS\, \Re e\psi_m(k_2,\mathbf{r})\, \hat{n}\cdot\nabla\psi_n(k_1,\mathbf{r}) \right.$$

$$- \int_S dS\, \hat{n}\cdot\nabla\Re e\psi_m(k_2,\mathbf{r})\psi_n(k_1,\mathbf{r}) \Bigg]. \quad (8.5.7)$$

Note that the above is the same as

$$\sum_n a_n \Re e Q_{nm} = -\sum f_n Q_{nm}, \quad (8.5.8)$$

which is the same as

$$\Re e\overline{\mathbf{Q}}^t \cdot \mathbf{a} = -\overline{\mathbf{Q}}^t \cdot \mathbf{f}. \quad (8.5.9)$$

Consequently,

$$\mathbf{f} = -\left(\overline{\mathbf{Q}}^t\right)^{-1} \cdot \Re e\overline{\mathbf{Q}}^t \cdot \mathbf{a}, \quad (8.5.10)$$

or the $\overline{\mathbf{T}}$ matrix is

$$\overline{\mathbf{T}} = -\left(\overline{\mathbf{Q}}^t\right)^{-1} \cdot \Re e\overline{\mathbf{Q}}^t. \quad (8.5.11)$$

Observe that Equation (11) is exactly the transpose of Equation (8.4.6) derived by the EBC method. But from the reciprocity condition (8.4.13), the actual $\overline{\mathbf{T}}$ is a symmetric matrix. Therefore, the $\overline{\mathbf{T}}$ matrix derived with

Rayleigh's hypothesis has formally the same error as that derived by the extended-boundary-condition method, even when the $\overline{\mathbf{Q}}$ matrices are truncated (see Exercise 8.27).

The equivalence of EBC and Rayleigh's method has led to much confusion and controversy in the past (Burrow 1969; Millar 1969; Lewin 1970). In particular, the equivalence of Rayleigh's method to the seemingly more rigorous extended-boundary-condition method has been used to establish its legitimacy. However, it is easy to find counterexamples to Rayleigh's hypothesis in the high frequency limit. In this limit, a bouncing ray picture of the waves clearly indicates the existence of incoming waves as well in V_1. Hence, Rayleigh's method is sometimes an approximate method, as is the EBC method as noted earlier.

The EBC method of imposing the extinction theorem on S_1 and S_2 does not imply the extinction of the field everywhere in V_1 and V_2 respectively. However, the exact solution of the surface integral equation yields surface sources that extinct the appropriate field everywhere in V_1 and V_2. This explains the equivalence of these two methods and the same degree of errors in both the solutions. Consequently, the ill-conditioning of the matrix in Rayleigh's method, which gives rise to poor results, is also due to the presence of localized waves or evanescent waves for a highly corrugated or elongated object. Despite its shortfall, Rayleigh' method is attractive because of the simplicity of its derivation compared to the EBC method. [It has been proven for certain surfaces that Equation (1b) does not converge on S (see references in van den Berg 1980).[12]]

§8.6 Scattering by Many Scatterers

Once the $\overline{\mathbf{T}}$ matrix for one scatterer is found, it can be used easily to construct the solution of scattering by many scatterers. But when more than one scatterer is present, there exists multiple scattering between the scatterers. Nonetheless, by applying the translational addition theorem for spherical harmonics or cylindrical harmonics, the solution to such a problem is easily found. In this section, we shall consider first the solution of two scatterers. Then, we shall derive a recursive algorithm for the solution of N scatterers. The N scatterer solution has been presented by Peterson and Ström (1973, 1974a), but the solution we present here will be in a different light (Chew 1989; Chew and Wang 1990; Chew et al. 1990; Wang and Chew 1990; also see Kerker 1988).

§§8.6.1 Two-Scatterer Solution

When two scatterers are present as shown in Figure 8.6.1, we can expand the incident field as

$$\phi_{inc}(\mathbf{r}) = \Re e \psi^t(k_0, \mathbf{r}_0) \cdot \mathbf{a}, \qquad (8.6.1)$$

---

[12] The Rayleigh's hypothesis method has been modified by various scientists, a review of which is given by van den Berg (1980).

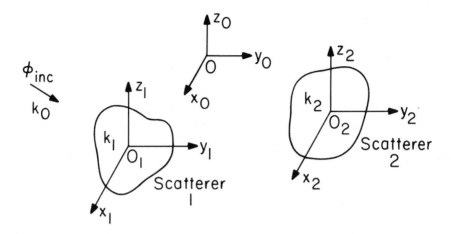

Figure 8.6.1 Two scatterers in the presence of an incident field.

while the scattered field, as

$$\phi_{sca}(\mathbf{r}) = \boldsymbol{\psi}^t(k_0, \mathbf{r}_1) \cdot \mathbf{f}_1 + \boldsymbol{\psi}^t(k_0, \mathbf{r}_2) \cdot \mathbf{f}_2. \qquad (8.6.2)$$

Notice that in the above, the scattered wave from each scatterer is expanded in terms of the outgoing harmonics expressed in the self-coordinates of the scatterers. Fortunately enough, translational formulas exist both for cylindrical and spherical harmonics such that

$$\boldsymbol{\psi}^t(k_0, \mathbf{r}_i) = \Re e \boldsymbol{\psi}^t(k_0, \mathbf{r}_j) \cdot \overline{\boldsymbol{\alpha}}_{ji}, \qquad |\mathbf{r}_j| < d_{ij}, \qquad (8.6.3a)$$

$$\boldsymbol{\psi}^t(k_0, \mathbf{r}_i) = \boldsymbol{\psi}^t(k_0, \mathbf{r}_j) \cdot \overline{\boldsymbol{\beta}}_{ji}, \qquad |\mathbf{r}_j| > d_{ij}, \qquad (8.6.3b)$$

$$\Re e \boldsymbol{\psi}^t(k_0, \mathbf{r}_i) = \Re e \boldsymbol{\psi}^t(k_0, \mathbf{r}_j) \cdot \overline{\boldsymbol{\beta}}_{ji}, \qquad \forall |\mathbf{r}_j|, \qquad (8.6.3c)$$

where d_{ij} is the distance between the O_i and O_j, the origins of the i and j coordinates (Friedman and Russek 1954; Stein 1961; Cruzan 1962; Danos and Maximon 1965; Chew 1989; and Chew et al. 1990; also see Appendix D). These formulas allow expression of the harmonic expansion of the field in one coordinate system in terms of another coordinate system readily. In general, $\overline{\boldsymbol{\beta}}_{ji} = \Re e \overline{\boldsymbol{\alpha}}_{ji}$ where $\Re e$ stands for "the regular part of" (see Exercise 8.28).

Then, using (3a) and (3c), the total field exterior to the scatterers expressed in terms of the coordinates of the first scatterer is

$$\phi(\mathbf{r}) = \Re e \boldsymbol{\psi}^t(k_0, \mathbf{r}_1) \cdot \overline{\boldsymbol{\beta}}_{10} \cdot \mathbf{a} + \boldsymbol{\psi}^t(k_0, \mathbf{r}_1) \cdot \mathbf{f}_1 + \Re e \boldsymbol{\psi}^t(k_0, \mathbf{r}_1) \cdot \overline{\boldsymbol{\alpha}}_{12} \cdot \mathbf{f}_2. \qquad (8.6.4)$$

The first and the third terms in the above can be viewed as the incident field impinging on the scatterer 1, while the second term is the scattered field from scatterer 1. But if the $\overline{\mathbf{T}}$ matrix of the first scatterer when it is isolated is known, we can write a relationship between $\mathbf{a}$, $\mathbf{f}_1$, and $\mathbf{f}_2$ using this $\overline{\mathbf{T}}$ matrix. In other words,

$$\mathbf{f}_1 = \overline{\mathbf{T}}_{1(1)} \cdot \left[\overline{\boldsymbol{\beta}}_{10} \cdot \mathbf{a} + \overline{\boldsymbol{\alpha}}_{12} \cdot \mathbf{f}_2 \right]. \tag{8.6.5}$$

Similarly, for scatterer 2, we have

$$\mathbf{f}_2 = \overline{\mathbf{T}}_{2(1)} \cdot \left[\overline{\boldsymbol{\beta}}_{20} \cdot \mathbf{a} + \overline{\boldsymbol{\alpha}}_{21} \cdot \mathbf{f}_1 \right]. \tag{8.6.6}$$

In the above, $\overline{\mathbf{T}}_{i(1)}$ is the isolated-scatterer $\overline{\mathbf{T}}$ matrix for the i-th scatterer; the parenthesized 1 indicates that it is the one-scatterer $\overline{\mathbf{T}}$ matrix.

Equations (5) and (6) can be solved to yield

$$\mathbf{f}_1 = \left[\overline{\mathbf{I}} - \overline{\mathbf{T}}_{1(1)} \cdot \overline{\boldsymbol{\alpha}}_{12} \cdot \overline{\mathbf{T}}_{2(1)} \cdot \overline{\boldsymbol{\alpha}}_{21} \right]^{-1} \cdot \overline{\mathbf{T}}_{1(1)} \cdot \left[\overline{\boldsymbol{\beta}}_{10} + \overline{\boldsymbol{\alpha}}_{12} \cdot \overline{\mathbf{T}}_{2(1)} \cdot \overline{\boldsymbol{\beta}}_{20} \right] \cdot \mathbf{a}, \tag{8.6.7}$$

$$\mathbf{f}_2 = \left[\overline{\mathbf{I}} - \overline{\mathbf{T}}_{2(1)} \cdot \overline{\boldsymbol{\alpha}}_{21} \cdot \overline{\mathbf{T}}_{1(1)} \cdot \overline{\boldsymbol{\alpha}}_{12} \right]^{-1} \cdot \overline{\mathbf{T}}_{2(1)} \cdot \left[\overline{\boldsymbol{\beta}}_{20} + \overline{\boldsymbol{\alpha}}_{21} \cdot \overline{\mathbf{T}}_{1(1)} \cdot \overline{\boldsymbol{\beta}}_{10} \right] \cdot \mathbf{a}. \tag{8.6.8}$$

Now, from (7) and (8), new $\overline{\mathbf{T}}$ matrices are defined such that

$$\mathbf{f}_1 = \overline{\mathbf{T}}_{1(2)} \cdot \overline{\boldsymbol{\beta}}_{10} \cdot \mathbf{a}, \tag{8.6.9}$$

$$\mathbf{f}_2 = \overline{\mathbf{T}}_{2(2)} \cdot \overline{\boldsymbol{\beta}}_{20} \cdot \mathbf{a}, \tag{8.6.10}$$

where now, $\overline{\mathbf{T}}_{i(2)}$ is a two-scatterer $\overline{\mathbf{T}}$ matrix. It relates the total scattered field due to the i-th scatterer to the incident field amplitude when two scatterers are present. Notice that the equations for $\overline{\mathbf{T}}_{i(2)}$ can be derived by comparing (9) and (10) with (7) and (8). Moreover, the factor $\overline{\boldsymbol{\beta}}_{i0}$ is introduced so that the $\overline{\mathbf{T}}$ matrices are still defined with respect to the self-coordinates of the scatterers.

§§8.6.2 *N-Scatterer Solution—A Recursive Algorithm*

The previous subsection illustrated how the two-scatterer solution can be constructed from the one-scatterer solution. This concept can be further extended to find the scattering solution of $n+1$ scatterers given the solution of the scattering from n scatterers (see Figure 8.6.2). Now, if we define an n-scatterer $\overline{\mathbf{T}}$ matrix $\overline{\mathbf{T}}_{i(n)}$, then the total field external to the n scatterers is of the form

$$\phi(\mathbf{r}) = \Re e\, \boldsymbol{\psi}^t(k_0, \mathbf{r}_0) \cdot \mathbf{a} + \sum_{i=1}^{n} \boldsymbol{\psi}^t(k_0, \mathbf{r}_i) \cdot \overline{\mathbf{T}}_{i(n)} \cdot \overline{\boldsymbol{\beta}}_{i0} \cdot \mathbf{a}. \tag{8.6.11}$$

Similarly, the $(n+1)$-scatterer solution has the form

$$\phi(\mathbf{r}) = \Re e\, \boldsymbol{\psi}^t(k_0, \mathbf{r}_0) \cdot \mathbf{a} + \sum_{i=1}^{n+1} \boldsymbol{\psi}^t(k_0, \mathbf{r}_i) \cdot \overline{\mathbf{T}}_{i(n+1)} \cdot \overline{\boldsymbol{\beta}}_{i0} \cdot \mathbf{a}. \tag{8.6.12}$$

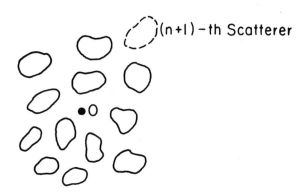

Figure 8.6.2 A recursive algorithm can be derived such that the scattering solution from $n + 1$ scatterers can be obtained from the scattering solution of n scatterers.

The preceding equation can be written more suggestively as

$$\phi(\mathbf{r}) = \Re e \boldsymbol{\psi}^t(k_0, \mathbf{r}_0) \cdot \mathbf{a} + \sum_{i=1}^{n} \boldsymbol{\psi}^t(k_0, \mathbf{r}_i) \cdot \overline{\mathbf{T}}_{i(n+1)} \cdot \overline{\boldsymbol{\beta}}_{i0} \cdot \mathbf{a}$$
$$+ \boldsymbol{\psi}^t(k_0, \mathbf{r}_{n+1}) \cdot \overline{\mathbf{T}}_{n+1(n+1)} \cdot \overline{\boldsymbol{\beta}}_{n+1,0} \cdot \mathbf{a}. \quad (8.6.13)$$

Now, the first and the last terms in Equation (13) can be thought of as incident fields impinging on the $i = 1, \ldots, n$ scatterers. Therefore,

$$\overline{\mathbf{T}}_{i(n+1)} \cdot \overline{\boldsymbol{\beta}}_{i0} \cdot \mathbf{a} = \overline{\mathbf{T}}_{i(n)} \cdot \overline{\boldsymbol{\beta}}_{i0} \cdot \left[\overline{\mathbf{I}} + \overline{\boldsymbol{\alpha}}_{0,n+1} \cdot \overline{\mathbf{T}}_{n+1(n+1)} \cdot \overline{\boldsymbol{\beta}}_{n+1,0} \right] \cdot \mathbf{a}, \quad (8.6.14)$$

where we have used the translation formulas (3) to translate the last terms to the global coordinates $\mathbf{r}_0$. But since $\overline{\mathbf{T}}_{i(n)} \cdot \overline{\boldsymbol{\beta}}_{i0}$ is defined only for a source outside the smallest circle centered at $\mathbf{r}_0 = 0$ and circumscribing the n spheres, the $(n+1)$-th scatterer must be on or outside this circle. In other words, the distances of the scatterers from $\mathbf{r}_0 = 0$ have to be ordered.

Furthermore, the scattered field amplitude from the $(n + 1)$-th scatterer is due to the scattering of the incident field from the other n scatterers via the isolated-scatterer $\overline{\mathbf{T}}$ matrix. Hence, the scattered field amplitude due to the $(n + 1)$-th scatterer is related to the other field amplitudes as

$$\overline{\mathbf{T}}_{n+1(n+1)} \cdot \overline{\boldsymbol{\beta}}_{n+1,0} \cdot \mathbf{a} = \overline{\mathbf{T}}_{n+1(1)} \cdot \left[\overline{\boldsymbol{\beta}}_{n+1,0} + \sum_{i=1}^{n} \overline{\boldsymbol{\alpha}}_{n+1,i} \cdot \overline{\mathbf{T}}_{i(n+1)} \cdot \overline{\boldsymbol{\beta}}_{i0} \right] \cdot \mathbf{a}.$$
$$(8.6.15)$$

Consequently, using (14) in (15), we have

$$
\overline{\mathbf{T}}_{n+1(n+1)} \cdot \overline{\boldsymbol{\beta}}_{n+1,0} = \overline{\mathbf{T}}_{n+1(1)} \cdot \left[\overline{\boldsymbol{\beta}}_{n+1,0} + \sum_{i=1}^{n} \overline{\boldsymbol{\alpha}}_{n+1,i} \cdot \overline{\mathbf{T}}_{i(n)} \cdot \overline{\boldsymbol{\beta}}_{i0} \right.
$$
$$
\left. + \sum_{i=1}^{n} \overline{\boldsymbol{\alpha}}_{n+1,i} \cdot \overline{\mathbf{T}}_{i(n)} \cdot \overline{\boldsymbol{\beta}}_{io} \cdot \overline{\boldsymbol{\alpha}}_{0,n+1} \cdot \overline{\mathbf{T}}_{n+1(n+1)} \cdot \overline{\boldsymbol{\beta}}_{n+1,0} \right]. \quad (8.6.16)
$$

Then, after solving this equation for $\overline{\mathbf{T}}_{n+1(n+1)} \cdot \overline{\boldsymbol{\beta}}_{n+1,0}$, we obtain

$$
\overline{\mathbf{T}}_{n+1(n+1)} \cdot \overline{\boldsymbol{\beta}}_{n+1,0} = \left[\overline{\mathbf{I}} - \overline{\mathbf{T}}_{n+1(1)} \cdot \sum_{i=1}^{n} \overline{\boldsymbol{\alpha}}_{n+1,i} \cdot \overline{\mathbf{T}}_{i(n)} \cdot \overline{\boldsymbol{\beta}}_{io} \cdot \overline{\boldsymbol{\alpha}}_{0,n+1} \right]^{-1}
$$
$$
\cdot \overline{\mathbf{T}}_{n+1(1)} \cdot \left[\overline{\boldsymbol{\beta}}_{n+1,0} + \sum_{i=1}^{n} \overline{\boldsymbol{\alpha}}_{n+1,i} \cdot \overline{\mathbf{T}}_{i(n)} \cdot \overline{\boldsymbol{\beta}}_{i0} \right]. \quad (8.6.17)
$$

But from (14), we have

$$
\overline{\mathbf{T}}_{i(n+1)} \cdot \overline{\boldsymbol{\beta}}_{i0} = \overline{\mathbf{T}}_{i(n)} \cdot \overline{\boldsymbol{\beta}}_{i0} \cdot [\overline{\mathbf{I}} + \overline{\boldsymbol{\alpha}}_{0,n+1} \cdot \overline{\mathbf{T}}_{n+1(n+1)} \cdot \overline{\boldsymbol{\beta}}_{n+1,0}]. \quad (8.6.18)
$$

Therefore, Equations (17) and (18) together constitute the recursive relations enabling one to calculate the $\overline{\mathbf{T}}_{i(n+1)} \cdot \overline{\boldsymbol{\beta}}_{io}$ matrices, $i = 1, \ldots, n+1$, given the $\overline{\mathbf{T}}_{i(n)} \cdot \overline{\boldsymbol{\beta}}_{io}$ matrices, $i = 1, \ldots, n$. Therefore, given the knowledge of the isolated-scatterer $\overline{\mathbf{T}}$ matrices, the N-scatterer solution is constructed recursively, starting from the one-scatterer solution. In this manner, only small matrices determined by the dimensions of the $\overline{\mathbf{T}}$ matrices must be dealt with at each recursion. Consequently, only a small amount of computer memory is required at each recursion, which reduces the number of page-faults in a virtual memory machine.

In the above, if there are N scatterers, and the field around each scatterer is approximated by M harmonics, there are altogether NM unknowns. In this case, $\overline{\mathbf{T}}_{i(n)} \cdot \overline{\boldsymbol{\beta}}_{io} \cdot \mathbf{a}$ in (11) is an M element column vector. But when the scatterers are small, M, the number of unknowns on each scatterer, can be kept small. On the contrary, the number of terms in the translation formulas should be large enough to maintain their accuracy. In other words, $\overline{\mathbf{T}}_{i(n)} \cdot \overline{\boldsymbol{\beta}}_{io}$ need not be square—it should be a $M \times P$ matrix where P is large enough to keep the translation accurate.

In view of this, the dimensions of the matrices in (17) and (18) are indicated as

$$
\overbrace{\overline{\mathbf{T}}_{n+1(n+1)} \cdot \overline{\boldsymbol{\beta}}_{n+1,0}}^{M \times P} = \left[\overline{\mathbf{I}} - \overbrace{\overline{\mathbf{T}}_{n+1(1)}}^{M \times M} \cdot \sum_{i=1}^{n} \overbrace{\overline{\boldsymbol{\alpha}}_{n+1,i}}^{M \times M} \cdot \overbrace{\overline{\mathbf{T}}_{i(n)} \cdot \overline{\boldsymbol{\beta}}_{i0}}^{M \times P} \cdot \overbrace{\overline{\boldsymbol{\alpha}}_{0,n+1}}^{P \times M} \right]^{-1}
$$
$$
\cdot \underbrace{\overline{\mathbf{T}}_{n+1(1)}}_{M \times M} \cdot \left[\underbrace{\overline{\boldsymbol{\beta}}_{n+1,0}}_{M \times P} + \sum_{i=1}^{n} \underbrace{\overline{\boldsymbol{\alpha}}_{n+1,i}}_{M \times M} \cdot \underbrace{\overline{\mathbf{T}}_{i(n)} \cdot \overline{\boldsymbol{\beta}}_{i0}}_{M \times P} \right], \quad (8.6.19)
$$

$$\underbrace{\overline{\mathbf{T}}_{i(n+1)} \cdot \overline{\beta}_{i0}}_{M \times P} = \underbrace{\overline{\mathbf{T}}_{i(n)} \cdot \overline{\beta}_{i0}}_{M \times P} + \big(\underbrace{\overline{\mathbf{T}}_{i(n)} \cdot \overline{\beta}_{i0}}_{M \times P} \cdot \underbrace{\overline{\alpha}_{0,n+1}}_{P \times M}\big) \cdot \underbrace{\overline{\mathbf{T}}_{n+1(n+1)} \cdot \overline{\beta}_{n+1,0}}_{M \times P} \cdot$$

(8.6.20)

In the above, $\overline{\mathbf{T}}_{i(n+1)} \cdot \overline{\beta}_{i0}$ can be regarded as the function to be solved for; it is an $M \times P$ matrix. (The dimensions of the matrices are indicated in the equations above.) Notice that the number of floating point operations required to multiply an $M \times P$ matrix with a $P \times M$ matrix, or an $M \times M$ matrix with an $M \times P$ matrix is equal to M^2P. But since $M \ll P$, the other matrix multiplications and inversions are subdominant. Therefore, at each recursion, the number of floating point operation is $O(nM^2P)$ after counting the dominant matrix multiplications in (19) and (20). Consequently, after applying the recursion relations to N scatterers, the number of cumulative floating point operation is $O(N^2M^2P)$ (Exercise 8.29).

The N-scatterer problem is also expressible as an NM unknown problem by solving NM linear algebraic equations. This would require $O(N^3M^3)$ floating-point operations, however, if these NM linear algebraic equations are solved with Gauss' elimination. On the other hand, if the conjugate gradient method is used here, $O(N^{2+\alpha})$ algorithm (where α depends on the condition number of the matrix) is possible. But still, the conjugate gradient method is an iterative procedure that solves the matrix equation $\overline{\mathbf{A}} \cdot \mathbf{x} = \mathbf{b}$ with a fixed right-hand side. Therefore, it has to be restarted if the right-hand side of the equation changes, and if the incident angle of the wave changes, the equation needs to be solved again. However, the preceding algorithm derived is independent of the incident angle of the incident wave. The reduction in computational effort here can be traced to the fact that the $\overline{\alpha}_{ij}$ or $\overline{\beta}_{ij}$ matrices are the representation of a translation group.

The recursive relations given by (19) and (20) can be further manipulated to a different form by letting $\overline{\alpha}_{n+1,i} = \overline{\alpha}_{n+1,0} \cdot \overline{\beta}_{0i}$. Then, (19) becomes

$$\overline{\mathbf{T}}_{n+1(n+1)} \cdot \overline{\beta}_{n+1,0} = \left[\overline{\mathbf{I}} - \overline{\mathbf{T}}_{n+1(1)} \cdot \overline{\alpha}_{n+1,0} \cdot \left(\sum_{i=1}^{n} \overline{\beta}_{0i} \cdot \overline{\mathbf{T}}_{i(n)} \cdot \overline{\beta}_{i0} \right) \cdot \overline{\alpha}_{0,n+1} \right]^{-1}$$
$$\cdot \overline{\mathbf{T}}_{n+1(1)} \cdot \left[\overline{\beta}_{n+1,0} + \overline{\alpha}_{n+1,0} \cdot \left(\sum_{i=1}^{n} \overline{\beta}_{0i} \cdot \overline{\mathbf{T}}_{i(n)} \cdot \overline{\beta}_{i0} \right) \right]. \quad (8.6.21)$$

Then, an aggregate $\overline{\mathbf{T}}$ matrix for n scatterers can be defined such that

$$\overline{\tau}_{(n)} = \sum_{i=1}^{n} \overline{\beta}_{0i} \cdot \overline{\mathbf{T}}_{i(n)} \cdot \overline{\beta}_{i0}. \quad (8.6.22)$$

And (21) becomes

$$\overline{\mathbf{T}}_{n+1(n+1)} \cdot \overline{\beta}_{n+1,0} = \left[\overline{\mathbf{I}} - \overline{\mathbf{T}}_{n+1(1)} \cdot \overline{\alpha}_{n+1,0} \cdot \overline{\tau}_{(n)} \cdot \overline{\alpha}_{0,n+1} \right]^{-1}$$
$$\cdot \overline{\mathbf{T}}_{n+1(1)} \cdot \left[\overline{\beta}_{n+1,0} + \overline{\alpha}_{n+1,0} \cdot \overline{\tau}_{(n)} \right]. \quad (8.6.23)$$

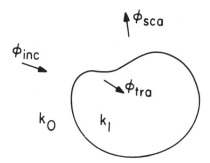

Figure 8.7.1 The one-interface problem for defining the $\overline{\mathbf{T}}$ matrices for the incoming wave case.

Moreover, on multiplying (20) by $\overline{\boldsymbol{\beta}}_{0i}$ and summing over i from 1 to n, we have

$$\overline{\boldsymbol{\tau}}_{(n+1)} = \overline{\boldsymbol{\tau}}_{(n)} + \left[\overline{\boldsymbol{\beta}}_{0,n+1} + \overline{\boldsymbol{\tau}}_{(n)} \cdot \overline{\boldsymbol{\alpha}}_{0,n+1}\right] \cdot \overline{\mathbf{T}}_{n+1(n+1)} \cdot \overline{\boldsymbol{\beta}}_{n+1,0}. \qquad (8.6.24)$$

Now, Equations (23) and (24) constitute the recursion relations expressing $\overline{\boldsymbol{\tau}}_{(n+1)}$ in terms of $\overline{\boldsymbol{\tau}}_{(n)}$. Furthermore, when M multipoles are assumed for each scatterer and P harmonics used for the translation formulas, a count shows that the above is an NMP^2 algorithm. Consequently, if M and P could be kept small, this is a very efficient method of calculating the scattering from many scatterers when N is large. Moreover, an arbitrary shape, inhomogeneous scatterer can be divided into N subscatterers, and its scattering solved by such an algorithm. (The above concepts are easily adapted to the vector electromagnetic scattering problems.)

§8.7 Scattering by Multilayered Scatterers

Thus far, how the $\overline{\mathbf{T}}$ matrix of a single scatterer and many scatterers could be derived recursively has been shown. This concept can be extended to the case of a multilayered scatterer. To do this, it is expedient to elucidate the physics of the scattering at each interface. So, first we shall derive the $\overline{\mathbf{T}}$ matrices for the one interface problem and, ultimately, derive the multi-interface problem from it. Although this solution was originally presented by Peterson and Ström (1974b, 1975) and Wang and Barber (1979), the solution here is considered in a more general sense.

§§8.7.1 One-Interface Problem

For the geometry shown in Figure 8.7.1, the field external to the scatterer is of the form

$$\phi(\mathbf{r}) = \Re e \boldsymbol{\psi}^t(k_0, \mathbf{r}) \cdot \mathbf{a} + \boldsymbol{\psi}^t(k_0, \mathbf{r}) \cdot \overline{\mathbf{R}}_{01} \cdot \mathbf{a}, \qquad (8.7.1)$$

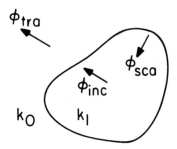

Figure 8.7.2 The one-interface problem for defining the $\overline{\mathbf{T}}$ matrices for the outgoing wave case.

where $\overline{\mathbf{R}}_{01}$ is a matrix equivalent to the $\overline{\mathbf{T}}$ matrix previously defined. In keeping with the spirit of Chapter 3, it is called the reflection matrix here, and $\overline{\mathbf{T}}$ is reserved for a transmission matrix in this section. In addition, the field internal to the scatterer is expressible as

$$\phi_{tra} = \Re e \psi^t(k_1, \mathbf{r}) \cdot \boldsymbol{\alpha}, \tag{8.7.2}$$

where $\boldsymbol{\alpha}$ is a column vector containing α_m's. This internal field is solvable via the EBC method from (8.3.15), which implies that

$$i\overline{\mathbf{Q}} \cdot \boldsymbol{\alpha} = \mathbf{a}, \quad \text{or} \quad \boldsymbol{\alpha} = -i\overline{\mathbf{Q}}^{-1} \cdot \mathbf{a}. \tag{8.7.3}$$

Next, we define a transmission matrix such that

$$\phi_{tra} = \Re e \psi^t(k_1, \mathbf{r}) \cdot \overline{\mathbf{T}}_{01} \cdot \mathbf{a}, \tag{8.7.4}$$

where

$$\overline{\mathbf{T}}_{01} = -i\overline{\mathbf{Q}}^{-1}. \tag{8.7.4a}$$

Now, consider the case where the field is incident at the interface from the inside, as shown in Figure 8.7.2. The incident wave in this case is the outgoing wave. Then, the field internal to the scatterer is of the form

$$\phi(\mathbf{r}) = \phi_{inc}(\mathbf{r}) + \phi_{sca}(\mathbf{r}) = \psi^t(k_1, \mathbf{r}) \cdot \mathbf{a} + \Re e \psi^t(k_1, \mathbf{r}) \cdot \mathbf{f}. \tag{8.7.5}$$

And the field external to the scatterer is

$$\phi_{tra}(\mathbf{r}) = \psi^t(k_0, \mathbf{r}) \cdot \boldsymbol{\alpha}. \tag{8.7.6}$$

The above problem is again solvable by the EBC method (see Exercise 8.30). Therefore, the waves are

$$\phi(\mathbf{r}) = \psi^t(k_1, \mathbf{r}) \cdot \mathbf{a} + \Re e \psi^t(k_1, \mathbf{r}) \cdot \overline{\mathbf{R}}_{10} \cdot \mathbf{a}, \qquad \mathbf{r} \in \text{region} \ 1, \quad (8.7.7a)$$

$$\phi_{tra}(\mathbf{r}) = \psi^t(k_0, \mathbf{r}) \cdot \overline{\mathbf{T}}_{10} \cdot \mathbf{a}, \qquad \mathbf{r} \in \text{region} \ 0. \quad (8.7.7b)$$

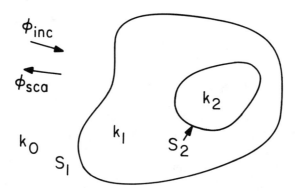

Figure 8.7.3 The two-interface problem.

Consequently, with the canonical problems defined for the one-interface problem, the solution for the many-interface problems is easily derived as shown in the following.

§§8.7.2 Many-Interface Problem

Consider the two-interface problem shown in Figure 8.7.3; the two surfaces S_1 and S_2 may not be concentric. So, it is necessary to express the $\overline{\mathbf{T}}$ matrices of the two interfaces in different coordinate systems. This is easily achieved via the translation formulas described in the previous section.

Consequently, in region 0, the field is

$$\phi_0(\mathbf{r}) = \Re e \psi^t(k_0, \mathbf{r}_1) \cdot \mathbf{a}_0 + \psi^t(k_0, \mathbf{r}_1) \cdot \mathbf{b}_0; \qquad (8.7.8)$$

in region 1, the field is

$$\phi_1(\mathbf{r}) = \Re e \psi^t(k_1, \mathbf{r}_1) \cdot \mathbf{a}_1 + \psi^t(k_1, \mathbf{r}_2) \cdot \mathbf{b}_1; \qquad (8.7.9)$$

and in region 2, the field is

$$\phi_2(\mathbf{r}) = \Re e \psi^t(k_2, \mathbf{r}_2) \cdot \mathbf{a}_2, \qquad (8.7.10)$$

where $\mathbf{r}_1$ is in the coordinates for the first interface S_1, and $\mathbf{r}_2$ is in the coordinates for the second interface S_2.[13] Hence, the scattered fields, which are the second terms in (8) and (9), are expressed in the coordinates of the surfaces that cause the scattering. The transmitted field in (10) is also expressed in the coordinates of the surface that causes the transmission.

---

[13] The coordinates for a surface S should be chosen so that the inscribed and exscribed spheres shown in Figure 8.3.1 are not too far from the surface S in order to avoid ill-conditioned $\overline{\mathbf{T}}$ matrices.

By requiring the outgoing wave in region 0 to be a consequence of the reflection of the incident wave plus the transmission of the outgoing wave in region 1, we have

$$\mathbf{b}_0 = \overline{\mathbf{R}}_{01} \cdot \mathbf{a}_0 + \overline{\mathbf{T}}_{10} \cdot \overline{\boldsymbol{\beta}}_{12} \cdot \mathbf{b}_1. \tag{8.7.11}$$

In the above, the translation formula [see (8.6.3)]

$$\boldsymbol{\psi}^t(k_1, \mathbf{r}_2) \cdot \mathbf{b}_1 = \boldsymbol{\psi}^t(k_1, \mathbf{r}_1) \cdot \overline{\boldsymbol{\beta}}_{12} \cdot \mathbf{b}_1 \tag{8.7.12}$$

is used to translate the outgoing wave in the $\mathbf{r}_2$ coordinates to the $\mathbf{r}_1$ coordinates. The incoming wave in region 1, $\Re e\boldsymbol{\psi}^t(k_1, \mathbf{r}_1) \cdot \mathbf{a}_1$, is a consequence of the transmission of the incoming wave in region 0 plus the reflection of the outgoing wave in region 1. Therefore,

$$\mathbf{a}_1 = \overline{\mathbf{T}}_{01} \cdot \mathbf{a}_0 + \overline{\mathbf{R}}_{10} \cdot \overline{\boldsymbol{\beta}}_{12} \cdot \mathbf{b}_1. \tag{8.7.13}$$

By the same token, we express

$$\Re e\boldsymbol{\psi}^t(k_1, \mathbf{r}_1) \cdot \mathbf{a}_1 = \Re e\boldsymbol{\psi}^t(k_1, \mathbf{r}_2) \cdot \overline{\boldsymbol{\beta}}_{21} \cdot \mathbf{a}_1 \tag{8.7.14}$$

so that (9) becomes

$$\phi_1(\mathbf{r}) = \Re e\boldsymbol{\psi}^t(k_1, \mathbf{r}_2) \cdot \overline{\boldsymbol{\beta}}_{21} \cdot \mathbf{a}_1 + \boldsymbol{\psi}^t(k_1, \mathbf{r}_2) \cdot \mathbf{b}_1. \tag{8.7.15}$$

Then,

$$\mathbf{b}_1 = \overline{\mathbf{R}}_{12} \cdot \overline{\boldsymbol{\beta}}_{21} \cdot \mathbf{a}_1, \tag{8.7.16}$$

where $\overline{\mathbf{R}}_{12}$ is the reflection matrix for waves incident from region 1 onto the interface between regions 1 and 2.

From (11), (13), and (16), it follows that

$$\mathbf{a}_1 = \overline{\mathbf{M}}_{1-} \cdot \overline{\mathbf{T}}_{01} \cdot \mathbf{a}_0, \tag{8.7.17}$$
$$\mathbf{b}_1 = \overline{\mathbf{R}}_{12} \cdot \overline{\boldsymbol{\beta}}_{21} \cdot \overline{\mathbf{M}}_{1-} \cdot \overline{\mathbf{T}}_{01} \cdot \mathbf{a}_0, \tag{8.7.18}$$

and

$$\mathbf{b}_0 = \left[\overline{\mathbf{R}}_{01} + \overline{\mathbf{T}}_{10} \cdot \overline{\boldsymbol{\beta}}_{12} \cdot \overline{\mathbf{R}}_{12} \cdot \overline{\boldsymbol{\beta}}_{21} \cdot \overline{\mathbf{M}}_{1-} \cdot \overline{\mathbf{T}}_{01} \right] \cdot \mathbf{a}_0, \tag{8.7.19}$$

where $\overline{\mathbf{M}}_{1-} = \left[\overline{\mathbf{I}} - \overline{\mathbf{R}}_{10} \cdot \overline{\boldsymbol{\beta}}_{12} \cdot \overline{\mathbf{R}}_{12} \cdot \overline{\boldsymbol{\beta}}_{21} \right]^{-1}$. Similarly, we deduce that the field in region 2 is the transmission of the field in region 1, and hence,

$$\mathbf{a}_2 = \overline{\mathbf{T}}_{12} \cdot \overline{\boldsymbol{\beta}}_{21} \cdot \mathbf{a}_1 = \overline{\mathbf{T}}_{12} \cdot \overline{\boldsymbol{\beta}}_{21} \cdot \overline{\mathbf{M}}_{1-} \cdot \overline{\mathbf{T}}_{01} \cdot \mathbf{a}_0. \tag{8.7.20}$$

With these amplitude coefficients known, the field everywhere is found.

Note that from (19), a generalized reflection matrix can be defined for region 0 such that

$$\widetilde{\overline{\mathbf{R}}}_{01} = \overline{\mathbf{R}}_{01} + \overline{\mathbf{T}}_{10} \cdot \overline{\boldsymbol{\beta}}_{12} \cdot \widetilde{\overline{\mathbf{R}}}_{12} \cdot \overline{\boldsymbol{\beta}}_{21} \cdot \overline{\mathbf{M}}_{1-} \cdot \overline{\mathbf{T}}_{01}. \tag{8.7.21}$$

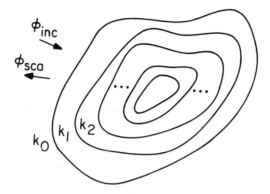

Figure 8.7.4 A multilayered scatterer.

If an inner region is now added to region 2, we need only modify $\overline{\mathbf{R}}_{12}$ to $\widetilde{\overline{\mathbf{R}}}_{12}$ in the above. Consequently, a recursive relation for a many-interface problem as shown in Figure 8.7.4 is expressible as

$$\widetilde{\overline{\mathbf{R}}}_{i,i+1} = \overline{\mathbf{R}}_{i,i+1} + \overline{\mathbf{T}}_{i+1,i} \cdot \overline{\boldsymbol{\beta}}_{i+1,i+2} \cdot \widetilde{\overline{\mathbf{R}}}_{i+1,i+2} \cdot \overline{\boldsymbol{\beta}}_{i+2,i+1} \cdot \overline{\mathbf{M}}_{i+1,-} \cdot \overline{\mathbf{T}}_{i,i+1}, \qquad (8.7.22)$$

where

$$\overline{\mathbf{M}}_{i+1,-} = \left[\overline{\mathbf{I}} - \overline{\mathbf{R}}_{i+1,i} \cdot \overline{\boldsymbol{\beta}}_{i+1,i+2} \cdot \widetilde{\overline{\mathbf{R}}}_{i+1,i+2} \cdot \overline{\boldsymbol{\beta}}_{i+2,i+1} \right]^{-1}, \qquad (8.7.23)$$

and $\widetilde{\overline{\mathbf{R}}}_{i,i+1}$ for the innermost region is zero. Moreover, if the field in region i is written as

$$\phi_i(\mathbf{r}) = \Re e \boldsymbol{\psi}^t(k_i, \mathbf{r}_i) \cdot \mathbf{a}_i + \boldsymbol{\psi}^t(k_i, \mathbf{r}_{i+1}) \cdot \mathbf{b}_i, \qquad (8.7.24)$$

then $\mathbf{a}_i$ could be found recursively as (Exercise 8.31)

$$\mathbf{a}_{i+1} = \overline{\mathbf{M}}_{i+1,-} \cdot \overline{\mathbf{T}}_{i,i+1} \cdot \overline{\boldsymbol{\beta}}_{i+1,i} \cdot \mathbf{a}_i \qquad (8.7.25)$$

with $\mathbf{a}_0$ known. Furthermore, $\mathbf{b}_i$ is related to $\mathbf{a}_i$ as

$$\mathbf{b}_i = \widetilde{\overline{\mathbf{R}}}_{i,i+1} \cdot \overline{\boldsymbol{\beta}}_{i+1,i} \cdot \mathbf{a}_i. \qquad (8.7.26)$$

In this manner, the field everywhere inside the scatterer can be calculated. Note that the translation matrix $\overline{\boldsymbol{\beta}}$ is not necessary if the surfaces are near concentric. However, they are necessary in the example shown in Figure 8.7.3, where it is not possible to expand the $\overline{\mathbf{T}}$ matrices for the two surfaces in one coordinate system. If all the surfaces are concentric circles or spheres, then

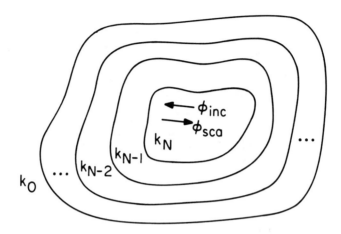

Figure 8.7.5 A multilayered scatterer for the inside-out problem.

the results here reduce to those of Chapter 3. Therefore, we can consider these results a generalization of those in Chapter 3.

On the other hand, if we have an inside-out problem as shown in Figure 8.7.5, the recursive relation for the reflection matrix is

$$\widetilde{\mathbf{R}}_{i+1,i} = \overline{\mathbf{R}}_{i+1,i} + \overline{\mathbf{T}}_{i,i+1} \cdot \overline{\boldsymbol{\beta}}_{i-1,i} \cdot \widetilde{\mathbf{R}}_{i,i-1} \cdot \overline{\boldsymbol{\beta}}_{i,i-1} \cdot \overline{\mathbf{M}}_{i,+} \cdot \overline{\mathbf{T}}_{i+1,i}, \qquad (8.7.27)$$

where

$$\overline{\mathbf{M}}_{i,+} = \left[\overline{\mathbf{I}} - \widetilde{\mathbf{R}}_{i,i-1} \cdot \overline{\boldsymbol{\beta}}_{i,i+1} \cdot \overline{\mathbf{R}}_{i,i+1} \cdot \overline{\boldsymbol{\beta}}_{i,i+1} \right]^{-1}, \qquad (8.7.28)$$

with $\widetilde{\mathbf{R}}_{i+1,i}$ for the outermost region being zero. Moreover, if the field in region i is expressed as in (24), then,

$$\mathbf{b}_i = \overline{\mathbf{M}}_{i,+} \cdot \overline{\mathbf{T}}_{i+1,i} \cdot \overline{\boldsymbol{\beta}}_{i,i+1} \cdot \mathbf{b}_{i+1} \qquad (8.7.29)$$

allowing all $\mathbf{b}_i$'s to be found with $\mathbf{b}_N$ known. In addition, the $\mathbf{a}_i$'s are related to the $\mathbf{b}_i$'s via

$$\mathbf{a}_i = \widetilde{\mathbf{R}}_{i,i-1} \cdot \overline{\boldsymbol{\beta}}_{i,i+1} \cdot \mathbf{b}_i. \qquad (8.7.30)$$

In this manner, all the fields in every region can be found. Note that the $\overline{\mathbf{M}}$ matrices defined above account for multiple reflections in the layered medium (see Exercise 8.31).

Hence, if a source is embedded in one of the layers, the combination of solutions from Figure 8.7.4 and Figure 8.7.5 can be used to calculate the field everywhere. Also, the above algorithm can be easily adapted to vector electromagnetic fields.

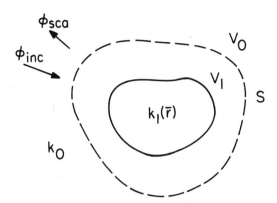

Figure 8.8.1 The scattering from an inhomogeneous scatterer is solvable by the combination of the surface integral equation method and the finite-element method.

§8.8 Surface Integral Equation with Finite-Element Method

The finite-element method (FEM) is versatile for highly inhomogeneous media. Moreover, it generates a sparse matrix for a differential equation economizing on memory requirement. But in using the FEM directly for an infinite domain, a large number of unknowns is usually involved, hence making the memory requirement inordinately large. Fortunately, one way to reduce the size of the problem is to couple the FEM with the surface integral equation method. In this manner, the FEM needs to be applied only over a finite domain drastically reducing the required memory. Such a method is also called the *hybrid method* (Silvester and Hsieh 1971; McDonald and Wexler 1972) or the *unimoment method* (Mei 1974; Chang and Mei 1976; Morgan and Mei 1979).

Consider the scattering by an inhomogeneous scatterer as shown in Figure 8.8.1. In the region exterior to S, we define a Green's function satisfying

$$(\nabla^2 + k_0^2)\, g_0(\mathbf{r}, \mathbf{r}') = -\delta(\mathbf{r} - \mathbf{r}') \tag{8.8.1}$$

and the radiation condition at infinity, and the field satisfying

$$(\nabla^2 + k_0^2)\, \phi_0(\mathbf{r}) = 0. \tag{8.8.2}$$

But interior to S, we define a Green's function satisfying

$$[\nabla \cdot p(\mathbf{r})\nabla + k_1^2(\mathbf{r})]g_1(\mathbf{r}, \mathbf{r}') = -\delta(\mathbf{r} - \mathbf{r}') \tag{8.8.3}$$

and the field satisfying

$$[\nabla \cdot p(\mathbf{r})\nabla + k_1^2(\mathbf{r})]\phi_1(\mathbf{r}) = 0. \tag{8.8.4}$$

Then, it is straightforward to show, as in (8.1.10) and (8.1.11), that

$$\left.\begin{array}{l} \mathbf{r} \in V_0, \ \phi_0(\mathbf{r}) \\ \mathbf{r} \in V_1, \ 0 \end{array}\right\} = \phi_{inc}(\mathbf{r}) - \int_S dS' \left[g_0(\mathbf{r},\mathbf{r}') \, \hat{n}' \cdot \nabla' \phi_0(\mathbf{r}') - \phi_0(\mathbf{r}') \, \hat{n}' \cdot \nabla' g_0(\mathbf{r},\mathbf{r}') \right],$$

$$(8.8.5)$$

and that

$$\left.\begin{array}{l} \mathbf{r} \in V_1, \ \phi_1(\mathbf{r}) \\ \mathbf{r} \in V_0, \ 0 \end{array}\right\} = \int_S dS' \left[g_1(\mathbf{r},\mathbf{r}') \, \hat{n}' \cdot \nabla' \phi_1(\mathbf{r}') - \phi_1(\mathbf{r}') \, \hat{n}' \cdot \nabla' g_1(\mathbf{r},\mathbf{r}') \right].$$

$$(8.8.6)$$

In the above, we have assumed that $p(\mathbf{r}) = 1$ on S. In other words, $p(\mathbf{r}) \neq 1$ only in the scatterer depicted in Figure 8.8.1. Consequently, the above surface integral equations divide the problem into two problems, one internal to S and one external to S. In theory, it could be solved as before. But inside S, the medium is inhomogeneous and $g_1(\mathbf{r},\mathbf{r}')$ is not generally available in closed form.

Therefore, in order to find $g_1(\mathbf{r},\mathbf{r}')$, one presumably can solve (3) with a numerical method like the finite-element or the Galerkin's method. But to derive (6), $g_1(\mathbf{r},\mathbf{r}')$ has to satisfy (3) only for $\mathbf{r}$ and $\mathbf{r}'$ inside S. Hence, Equation (3) needs to be solved only over a finite region. For instance, it could be solved with the imposition of the natural boundary condition $\hat{n} \cdot \nabla' g_1(\mathbf{r},\mathbf{r}') = 0$ on S as discussed in Chapter 5. In this case, we let

$$g_1(\mathbf{r},\mathbf{r}') = \sum_{n=1}^{N} a_n f_n(\mathbf{r}), \quad \mathbf{r},\mathbf{r}' \in V_1, \qquad (8.8.7)$$

where $f_n(\mathbf{r})$ constitutes a basis set that can approximate $g_1(\mathbf{r},\mathbf{r}')$ fairly well. Consequently, the matrix equation corresponding to (3) with the aforementioned natural boundary condition is (see Exercise 8.32)

$$\sum_{n=1}^{N} L_{mn} a_n = b_m, \quad m = 1, \ldots, N, \qquad (8.8.8)$$

where

$$L_{mn} = -\langle \nabla f_m(\mathbf{r}), p(\mathbf{r}) \nabla f_n(\mathbf{r}) \rangle + \langle f_m(\mathbf{r}), k_1^2(\mathbf{r}) f_n(\mathbf{r}) \rangle \qquad (8.8.9a)$$

is symmetric, and

$$b_m = -\langle f_m(\mathbf{r}), \delta(\mathbf{r} - \mathbf{r}') \rangle = -f_m(\mathbf{r}'). \qquad (8.8.9b)$$

As a note, the inner products above are defined as volume integrals in the volume bounded by S, i.e., $\langle f(\mathbf{r}), g(\mathbf{r}) \rangle = \int_{V_1} d\mathbf{r} \, f(\mathbf{r}) \, g(\mathbf{r})$.

Finally, we obtain

$$g_1(\mathbf{r}, \mathbf{r}') = -\mathbf{f}^t(\mathbf{r}) \cdot \overline{\mathbf{L}}^{-1} \cdot \mathbf{f}(\mathbf{r}'), \qquad (8.8.10)$$

where $\overline{\mathbf{L}}$ is a matrix with elements L_{mn} and $\mathbf{f}$ is a column vector containing f_m.

Now, given $g_1(\mathbf{r}, \mathbf{r}')$ and the natural boundary condition, $\hat{n} \cdot \nabla' g_1(\mathbf{r}, \mathbf{r}') = 0$, it satisfies, the upper part of Equation (6) becomes

$$\phi_1(\mathbf{r}) = -\mathbf{f}^t(\mathbf{r}) \cdot \overline{\mathbf{L}}^{-1} \cdot \int_S dS' \, \mathbf{f}(\mathbf{r}') \, \hat{n}' \cdot \nabla' \phi_1(\mathbf{r}'). \qquad (8.8.11)$$

Note that the second term on the right-hand side of (6) vanishes by virtue of $\hat{n} \cdot \nabla' g_1(\mathbf{r}, \mathbf{r}') = 0$ on S. The above and (5) together constitute the integral equations which can be solved for $\phi_1(\mathbf{r})$ and $\hat{n} \cdot \nabla \phi_1(\mathbf{r})$.

To solve (5) and (11), we expand the surface unknowns $\phi_1(\mathbf{r})$ and $\hat{n} \cdot \nabla \phi_1(\mathbf{r})$ as[14]

$$\phi_1(\mathbf{r}) = \sum_{m=1}^{M} c_m \psi_m(\mathbf{r}), \qquad (8.8.12a)$$

$$\hat{n} \cdot \nabla \phi_1(\mathbf{r}) = \sum_{m=1}^{M} d_m \psi_m(\mathbf{r}), \qquad (8.8.12b)$$

where $\psi_m(\mathbf{r})$ constitutes a basis set that can approximate $\phi_1(\mathbf{r})$ and $\hat{n} \cdot \nabla \phi_1(\mathbf{r})$ on S fairly well. The above can be substituted into (11) and tested with $\psi_n(\mathbf{r})$ on S, thereby yielding

$$\sum_{m=1}^{M} \langle \psi_n(\mathbf{r}), \psi_m(\mathbf{r}) \rangle c_m = - \langle \psi_n(\mathbf{r}), \mathbf{f}^t(\mathbf{r}) \rangle \cdot \overline{\mathbf{L}}^{-1} \cdot \sum_{m=1}^{M} d_m \, \langle \mathbf{f}(\mathbf{r}'), \psi_m(\mathbf{r}') \rangle,$$
$$(8.8.13)$$

where the inner product involves a surface integral over S. Furthermore, the above could be written as

$$\langle \boldsymbol{\psi}(\mathbf{r}), \boldsymbol{\psi}^t(\mathbf{r}) \rangle \cdot \mathbf{c} = - \langle \boldsymbol{\psi}(\mathbf{r}), \mathbf{f}^t(\mathbf{r}) \rangle \cdot \overline{\mathbf{L}}^{-1} \cdot \langle \mathbf{f}(\mathbf{r}'), \boldsymbol{\psi}^t(\mathbf{r}') \rangle \cdot \mathbf{d}, \qquad (8.8.14)$$

where $\boldsymbol{\psi}(\mathbf{r})$, $\mathbf{c}$, and $\mathbf{d}$ are column vectors containing $\psi_m(\mathbf{r})$, c_m, and d_m respectively. Alternatively, (14) is equivalent to

$$\overline{\mathbf{F}} \cdot \mathbf{c} = -\overline{\mathbf{A}} \cdot \overline{\mathbf{L}}^{-1} \cdot \overline{\mathbf{A}}^t \cdot \mathbf{d} = -\overline{\mathbf{M}} \cdot \mathbf{d}, \qquad (8.8.15)$$

where

$$\overline{\mathbf{F}} = \langle \boldsymbol{\psi}(\mathbf{r}), \boldsymbol{\psi}^t(\mathbf{r}) \rangle, \qquad (8.8.16a)$$

---

[14] It is not necessary to use the same basis set, $\psi_m(\mathbf{r})$, to expand both surface unknowns as we have done here.

$$\overline{\mathbf{A}} = \langle \boldsymbol{\psi}(\mathbf{r}), \mathbf{f}^t(\mathbf{r}) \rangle, \qquad\qquad (8.8.16b)$$

$$\overline{\mathbf{M}} = \overline{\mathbf{A}} \cdot \overline{\mathbf{L}}^{-1} \cdot \overline{\mathbf{A}}^t, \qquad\qquad (8.8.16c)$$

and $\overline{\mathbf{F}}$ is an $M \times M$ matrix while $\overline{\mathbf{A}}$ is an $M \times N$ matrix where M is usually less than N.

Next, using the continuity of the potential plus the continuity of the normal derivative of the potential, we can express

$$\phi_0(\mathbf{r}) = \sum_{m=1}^{M} c_m \psi_m(\mathbf{r}), \qquad\qquad (8.8.17a)$$

$$\hat{n} \cdot \nabla \phi_0(\mathbf{r}) = \sum_{m=1}^{M} d_m \psi_m(\mathbf{r}). \qquad\qquad (8.8.17b)$$

Then, using the above in the lower half of Equation (5) and testing with $\psi_n(\mathbf{r})$ on S, we have

$$\langle \psi_n(\mathbf{r}), \phi_{inc}(\mathbf{r}) \rangle = \sum_{m=1}^{M} \langle \psi_n(\mathbf{r}), g_0(\mathbf{r}, \mathbf{r}'), \psi_m(\mathbf{r}) \rangle d_m$$

$$- \sum_{m=1}^{M} \langle \psi_n(\mathbf{r}), \hat{n}' \cdot \nabla' g_0(\mathbf{r}, \mathbf{r}'), \psi_m(\mathbf{r}) \rangle c_m, \quad n = 1, \dots, M. \quad (8.8.18)$$

Note that the above is just a matrix equation of the form

$$\phi_{inc} = \overline{\mathbf{g}} \cdot \mathbf{d} - \overline{\mathbf{N}} \cdot \mathbf{c}, \qquad\qquad (8.8.19)$$

where

$$[\phi_{inc}]_n = \langle \psi_n(\mathbf{r}), \phi_{inc}(\mathbf{r}) \rangle, \qquad\qquad (8.8.20a)$$

$$[\overline{\mathbf{g}}]_{nm} = \langle \psi_n(\mathbf{r}), g_0(\mathbf{r}, \mathbf{r}'), \psi_m(\mathbf{r}') \rangle, \qquad\qquad (8.8.20b)$$

$$\left[\overline{\mathbf{N}} \right]_{nm} = \langle \psi_n(\mathbf{r}), \hat{n}' \cdot \nabla' g_0(\mathbf{r}, \mathbf{r}'), \psi_m(\mathbf{r}') \rangle. \qquad\qquad (8.8.20c)$$

Moreover, in the above, we have defined

$$\langle \psi_n(\mathbf{r}), f(\mathbf{r}, \mathbf{r}'), \psi_m(\mathbf{r}') \rangle = \int_S dS \int_S dS' \, \psi_n(\mathbf{r}) f(\mathbf{r}, \mathbf{r}') \psi_m(\mathbf{r}'). \qquad (8.8.20d)$$

Now, (11) and (5) have been reduced to matrix equations (15) and (19) respectively, from which $\mathbf{c}$ and $\mathbf{d}$ can be solved. Therefore, on eliminating $\mathbf{d}$ between (15) and (19), we have

$$\phi_{inc} = - \left[\overline{\mathbf{g}} \cdot \overline{\mathbf{M}}^{-1} \cdot \overline{\mathbf{F}} + \overline{\mathbf{N}} \right] \cdot \mathbf{c}, \qquad\qquad (8.8.21)$$

or

$$\mathbf{c} = -\left[\overline{\mathbf{g}} \cdot \overline{\mathbf{M}}^{-1} \cdot \overline{\mathbf{F}} + \overline{\mathbf{N}} \right]^{-1} \cdot \phi_{inc}, \qquad (8.8.22\text{a})$$

$$\mathbf{d} = -\overline{\mathbf{M}}^{-1} \cdot \overline{\mathbf{F}} \cdot \mathbf{c}. \qquad (8.8.22\text{b})$$

Once $\mathbf{c}$ and $\mathbf{d}$ are found, the surface unknowns can be found through (17) and the field everywhere determined via (5) and (11). This idea is, of course, easily extended to solving vector electromagnetic integral equations, albeit with increased complexity.

Notice that when $g_1(\mathbf{r}, \mathbf{r}')$ was solved for in (3), it was done so with the natural boundary condition $\hat{n} \cdot \nabla g_1(\mathbf{r}, \mathbf{r}') = 0$. This is actually equivalent to a source excitation problem in a cavity with impenetrable walls. Unfortunately, this cavity has resonant frequencies that are purely real in the lossless case. Therefore, at the resonant frequencies of the cavity formed by S, the matrix $\overline{\mathbf{L}}$ becomes singular. When this happens, it is quite difficult to solve for $\overline{\mathbf{M}}$ in (15).

When $\mathbf{c}$ and $\mathbf{d}$ are found in (22), only $\overline{\mathbf{M}}^{-1}$ is required. In this case, the singular value decomposition method may be used to find the inverse of $\overline{\mathbf{L}}$ (see Exercise 8.33). Moreover, the regularization method may be used to find the inverse of $\overline{\mathbf{L}}$ (Exercise 8.34; also see Tikhonov 1963). Alternatively, an impedance boundary condition for the Green's function, instead of the natural boundary condition, will eliminate the resonance problem (see Exercise 8.35), because then the cavity is lossy with a complex resonant frequency.

We have illustrated how the surface-integral-equation method is used to merge a finite-element solution with the solution in the external region. But the surface integral equations yield matrices $\overline{\mathbf{g}}$ and $\overline{\mathbf{N}}$ which are dense, in contrast to the sparse matrix $\overline{\mathbf{L}}$ in (9) generated by the finite-element method. Hence, this precludes the use of a sparse-matrix solver which is usually more efficient than a dense-matrix solver. To remedy this, absorbing boundary conditions defined in Chapter 4 may alternatively be used to merge the finite-element solution with the exterior solution (see Exercise 8.36). This then yields sparse matrices which can be inverted with sparse-matrix solvers.

§8.9 Volume Integral Equations

When a bounded medium is highly inhomogeneous, there are several methods of solving for its scattering solution. One way is to approximate the inhomogeneous medium with N scatterers and seek its scattering solution via the method of Section 8.6. If the inhomogeneous body can be approximated by a multilayered medium, the method expounded in Section 8.7 can be used. Furthermore, the hybrid method of Section 8.8 may be used. An alternative approach is to use volume integral equations where the unknowns in the problem are expressed in terms of volume current flowing in the inhomogeneity. The volume current consists of conduction current as well as displacement current induced by the total electric field. An integral equation can then be formulated from which the total field is solved. We shall first show how such an integral equation can be formulated for the scalar wave equation and later, formulate the integral equation for the electromagnetic wave case. Historically, the volume integral equation method has been developed as early as 1913 by Esmarch (see Born and Wolf 1980, p. 98). This equation is also described by Richmond (1965a, b), Harrington (1968), Poggio and Miller (1973), and Ström (1975).

The volume integral equation offers an alternative physical picture of the mechanism that gives rise to scattering. As such, it provides insight as to how approximate scattering solutions can be obtained, as shall be illustrated in the next section. Furthermore, it can be used to formulate inverse scattering algorithms detailed in the next chapter.

§§8.9.1 Scalar Wave Case

We shall first derive the volume integral equation for the scalar wave case. In this case, the pertinent scalar wave equation is

$$[\nabla^2 + k^2(\mathbf{r})]\phi(\mathbf{r}) = q(\mathbf{r}), \qquad (8.9.1)$$

where $k^2(\mathbf{r}) = \omega^2\mu(\mathbf{r})\epsilon(\mathbf{r})$ represents an inhomogeneous medium over a finite domain V, and $k^2 = k_b^2 = \omega^2\mu_b\epsilon_b$ outside V (see Figure 8.9.1). Next, we define a Green's function satisfying

$$[\nabla^2 + k_b^2]g(\mathbf{r}, \mathbf{r}') = -\delta(\mathbf{r} - \mathbf{r}'). \qquad (8.9.2)$$

Then, Equation (1) can be rewritten as

$$[\nabla^2 + k_b^2]\phi(\mathbf{r}) = q(\mathbf{r}) - [k^2(\mathbf{r}) - k_b^2]\phi(\mathbf{r}). \qquad (8.9.3)$$

Note that the right-hand side of (3) can be considered an equivalent source. Since the Green's function corresponding to the differential operator on the left-hand side of (3) is known, by the principle of linear superposition, we can write

$$\phi(\mathbf{r}) = -\int_{V_s} dV' g(\mathbf{r}, \mathbf{r}')q(\mathbf{r}') + \int_V dV' g(\mathbf{r}, \mathbf{r}')[k^2(\mathbf{r}') - k_b^2]\phi(\mathbf{r}'). \qquad (8.9.4)$$

The first term on the right-hand side is just the field due to the source in the absence of the inhomogeneity, and hence, is the incident field. Therefore, Equation (4) becomes

$$\phi(\mathbf{r}) = \phi_{inc}(\mathbf{r}) + \int_V dV' g(\mathbf{r}, \mathbf{r}')[k^2(\mathbf{r}') - k_b^2]\phi(\mathbf{r}'). \qquad (8.9.5)$$

In the above equation, if the total field $\phi(\mathbf{r}')$ inside the volume V is known, then $\phi(\mathbf{r})$ can be calculated everywhere. But $\phi(\mathbf{r})$ is unknown at this point.

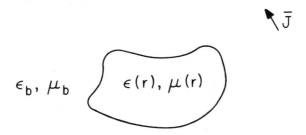

Figure 8.9.1 A current source radiating in the vicinity of a general inhomogeneity.

To solve for $\phi(\mathbf{r})$, an integral equation has to be formulated for $\phi(\mathbf{r})$. To this end, we imposed (5) for $\mathbf{r}$ in V. Then, $\phi(\mathbf{r})$ on the left-hand side and on the right-hand side are the same unknown defined over the same domain. Consequently, Equation (5) becomes the desired integral equation

$$\phi_{inc}(\mathbf{r}) = \phi(\mathbf{r}) - \int_V dV' g(\mathbf{r},\mathbf{r}')[k^2(\mathbf{r}') - k_b^2]\phi(\mathbf{r}'), \quad \mathbf{r} \in V. \tag{8.9.6}$$

In the above, the unknown $\phi(\mathbf{r})$ is defined over a volume V, over which the integration is performed, and hence the name, volume integral equation. Alternatively, the above can be rewritten as

$$\phi_{inc}(\mathbf{r}) = [\mathcal{I} - \mathcal{L}(\mathbf{r},\mathbf{r}')]\phi(\mathbf{r}'), \quad \mathbf{r} \in V, \tag{8.9.7}$$

where $\mathcal{I}$ is an identity operator while $\mathcal{L}$ is the integral operator in (6). It is also a *Fredholm integral equation* of the second kind because the unknown is both inside and outside the integral operator.

§§8.9.2 The Electromagnetic Wave Case

We shall show how the corresponding integral equation can be derived for a finite size, inhomogeneous scatterer for the electromagnetic wave case shown in Figure 8.9.1. First, from Maxwell's equations, it follows that the electric field everywhere satisfies the following equation:

$$\nabla \times \mu^{-1}\nabla \times \mathbf{E}(\mathbf{r}) - \omega^2\epsilon\mathbf{E}(\mathbf{r}) = i\omega\mathbf{J}(\mathbf{r}), \tag{8.9.8}$$

where μ and ϵ are functions of position inside the inhomogeneous region V. Next, subtracting $\nabla \times \mu_b^{-1}\nabla \times \mathbf{E}(\mathbf{r}) - \omega^2\epsilon_b\mathbf{E}(\mathbf{r})$ from both sides of the equation, we have

$$\nabla \times (\mu^{-1} - \mu_b^{-1})\nabla \times \mathbf{E}(\mathbf{r}) - \omega^2(\epsilon - \epsilon_b)\mathbf{E}(\mathbf{r}) = i\omega\mathbf{J}(\mathbf{r}) - \nabla \times \mu_b^{-1}\nabla \times \mathbf{E}(\mathbf{r}) + \omega^2\epsilon_b\mathbf{E}(\mathbf{r}), \tag{8.9.9}$$

To formulate the integral equation, we need the dyadic Green's function to the problem in the absence of the scatterer. The dyadic Green's function satisfies the equation

$$\nabla \times \mu_b^{-1} \nabla \times \overline{\mathbf{G}}(\mathbf{r}, \mathbf{r}') - \omega^2 \epsilon_b \overline{\mathbf{G}}(\mathbf{r}, \mathbf{r}') = \mu_b^{-1} \overline{\mathbf{I}} \delta(\mathbf{r} - \mathbf{r}'). \qquad (8.9.10)$$

Even though μ_b and ϵ_b need not be constant, but when they are constant, the solution to (10) is well known, as discussed in Chapters 1 and 7. Consequently, we can rewrite Equation (9) as

$$\nabla \times \mu_b^{-1} \nabla \times \mathbf{E}(\mathbf{r}) - \omega^2 \epsilon_b \mathbf{E}(\mathbf{r}) = i\omega \mathbf{J}(\mathbf{r}) + \omega^2 (\epsilon - \epsilon_b) \mathbf{E}(\mathbf{r}) - \nabla \times \left(\frac{1}{\mu} - \frac{1}{\mu_b} \right) \nabla \times \mathbf{E}(\mathbf{r}).$$
$$(8.9.11)$$

Physically, the terms on the right-hand side of Equation (11) are effective current sources. Therefore, analogous to (7.4.5a), the solution to (11) is

$$\mathbf{E}(\mathbf{r}) = i\omega \int_V d\mathbf{r}' \, \overline{\mathbf{G}}(\mathbf{r}, \mathbf{r}') \cdot \mu_b \mathbf{J}(\mathbf{r}') + \omega^2 \int_V d\mathbf{r}' \, \overline{\mathbf{G}}(\mathbf{r}, \mathbf{r}') \cdot \mu_b (\epsilon - \epsilon_b) \mathbf{E}(\mathbf{r}')$$

$$- \int_V d\mathbf{r}' \, \overline{\mathbf{G}}(\mathbf{r}, \mathbf{r}') \cdot \mu_b \nabla' \times \left(\frac{1}{\mu} - \frac{1}{\mu_b} \right) \nabla' \times \mathbf{E}(\mathbf{r}'). \qquad (8.9.12)$$

In the above, the first term is just the incident field; hence, (12) becomes

$$\mathbf{E}(\mathbf{r}) = \mathbf{E}_{inc}(\mathbf{r}) + \omega^2 \int_V d\mathbf{r}' \, \overline{\mathbf{G}}(\mathbf{r}, \mathbf{r}') \cdot \mu_b (\epsilon - \epsilon_b) \mathbf{E}(\mathbf{r}')$$

$$- \int_V d\mathbf{r}' \, \overline{\mathbf{G}}(\mathbf{r}, \mathbf{r}') \cdot \mu_b \nabla' \times \left(\frac{1}{\mu} - \frac{1}{\mu_b} \right) \nabla' \times \mathbf{E}(\mathbf{r}'). \qquad (8.9.13)$$

The integrals in (13) are contributions to the field $\mathbf{E}$ from the volume current induced in the scatterer by the total electric field $\mathbf{E}$ and magnetic field $\mathbf{H}$ (note that $\nabla \times \mathbf{E} = i\omega\mu\mathbf{H}$). Hence, the first term is generated by the electric polarization current or displacement current, while the second term is generated by the magnetic polarization charges (see Exercise 8.37). Moreover, when the scatterer is conductive such that $\epsilon = \epsilon' + i\sigma/\omega$, the first integral in (13) is due to the conduction current induced by the field as well. This is obviated by substituting the complex permittivity into (13) and identifying a term proportional to $\sigma\mathbf{E}$ corresponding to conduction currents. On the other hand, if $\mu = \mu_b$, Equation (13) simplifies to

$$\mathbf{E}(\mathbf{r}) = \mathbf{E}_{inc}(\mathbf{r}) + \int_V d\mathbf{r}' \, \overline{\mathbf{G}}(\mathbf{r}, \mathbf{r}') \cdot O(\mathbf{r}') \mathbf{E}(\mathbf{r}'), \qquad (8.9.14)$$

where $O(\mathbf{r}') = \omega^2 (\mu\epsilon - \mu_b \epsilon_b) = k^2(\mathbf{r}') - k_b^2$.

In Equations (13) and (14), the field $\mathbf{E}_{inc}$ is usually known since we know the source $\mathbf{J}$. But the total field $\mathbf{E}(\mathbf{r})$ is unknown, and it is in the integral as well. Therefore, analogous to (6), (14) is a volume integral equation, which can be written as

$$\mathbf{E}(\mathbf{r}) = \mathbf{E}_{inc}(\mathbf{r}) - \overline{\mathcal{L}}(\mathbf{r}, \mathbf{r}') \cdot \mathbf{E}(\mathbf{r}'), \qquad \mathbf{r}' \in V, \quad \mathbf{r} \in V, \qquad (8.9.15)$$

where $\overline{\mathcal{L}}$ is a linear integral operator in (13) or (14). Alternatively, we can write (15) as

$$\mathbf{E}_{inc}(\mathbf{r}) = \left[\overline{\mathcal{I}} - \overline{\mathcal{L}}(\mathbf{r}, \mathbf{r}')\right] \cdot \mathbf{E}(\mathbf{r}'), \qquad \mathbf{r}' \in V, \quad \mathbf{r} \in V, \qquad (8.9.16)$$

where $\overline{\mathcal{I}}$ is an identity operator. Since $\overline{\mathcal{I}} - \overline{\mathcal{L}}$ is a linear operator, we can apply Galerkin's method, or the method of moments to solve (16), as discussed in Chapter 5. Once $\mathbf{E}(\mathbf{r})$ is known inside V, $\mathbf{E}$ can be found everywhere via Equations (13) or (14). Equation (16) is a Fredholm integral equation of the second kind because the unknown is both inside and outside the integral. Equation (14) can also be written in operator form as shown in Subsection 9.3.3 of Chapter 9.

The above derivation is easily generalized to the case where the dyadic Green's function is for layered media discussed in Chapter 7. In this case, the background medium need not be homogeneous.

§§8.9.3 Matrix Representation of the Integral Equation

Given the integral equation in (14), it can be converted into a matrix equation quite easily using the method discussed in Chapter 5, i.e., by projecting the integral operator onto a space spanned by $\mathbf{E}_n(\mathbf{r})$, where $\mathbf{E}_n(\mathbf{r}) = 0$ for $\mathbf{r} \notin V$. To this end, we let

$$\mathbf{E}(\mathbf{r}) = \sum_n a_n \mathbf{E}_n(\mathbf{r}), \qquad \mathbf{r} \in V, \qquad (8.9.17)$$

in (14). Then,

$$\sum_n a_n \mathbf{E}_n(\mathbf{r}) = \mathbf{E}_{inc}(\mathbf{r}) + \sum_n a_n \int_V d\mathbf{r}' \, \overline{\mathbf{G}}(\mathbf{r}, \mathbf{r}') \cdot O(\mathbf{r}') \mathbf{E}_n(\mathbf{r}'). \qquad (8.9.18)$$

The above integral operator acting on $\mathbf{E}_n(\mathbf{r})$ is not a symmetric operator. Nevertheless, it can by symmetrized by multiplying (18) by $O(\mathbf{r})$ (see Exercise 8.38). In this manner, (18) becomes

$$\sum_n a_n O(\mathbf{r}) \mathbf{E}_n(\mathbf{r}) = O(\mathbf{r}) \mathbf{E}_{inc}(\mathbf{r})$$

$$+ \sum_n a_n O(\mathbf{r}) \int_V d\mathbf{r}' \, \overline{\mathbf{G}}(\mathbf{r}, \mathbf{r}') \cdot O(\mathbf{r}') \mathbf{E}_n(\mathbf{r}'). \qquad (8.9.19)$$

Consequently, after dot-multiplying the above by $\mathbf{E}_m(\mathbf{r})$ and integrating as in Galerkin's method, we have

$$\sum_n a_n \langle \mathbf{E}_m, O(\mathbf{r}) \mathbf{E}_n \rangle = \langle \mathbf{E}_m, O(\mathbf{r}) \mathbf{E}_{in} \rangle$$

$$+ \sum_n a_n \left\langle \mathbf{E}_m, O(\mathbf{r}) \int_V d\mathbf{r}' \, \overline{\mathbf{G}}(\mathbf{r}, \mathbf{r}') \cdot O(\mathbf{r}') \mathbf{E}_n(\mathbf{r}') \right\rangle. \qquad (8.9.20)$$

Now, Equation (20) is a matrix equation of the form

$$\sum_n M_{mn} a_n = b_m + \sum_n N_{mn} a_n, \tag{8.9.21}$$

or

$$(\overline{\mathbf{M}} - \overline{\mathbf{N}}) \cdot \mathbf{a} = \mathbf{b}, \tag{8.9.21a}$$

where

$$M_{mn} = \langle \mathbf{E}_m, O(\mathbf{r}) \mathbf{E}_n \rangle, \tag{8.9.22a}$$

$$N_{mn} = \left\langle \mathbf{E}_m, O(\mathbf{r}) \int_V d\mathbf{r}'\, \overline{\mathbf{G}}(\mathbf{r}, \mathbf{r}') \cdot O(\mathbf{r}') \mathbf{E}_n(\mathbf{r}') \right\rangle, \tag{8.9.22b}$$

$$b_m = \langle \mathbf{E}_m, O(\mathbf{r}) \mathbf{E}_{inc} \rangle. \tag{8.9.22c}$$

Moreover, with a finite basis set, we can always invert (21a) to find the unknown $\mathbf{a}$, the column vector that contains the unknowns a_n's.

Because of the singularity of the dyadic Green's function $\overline{\mathbf{G}}(\mathbf{r}, \mathbf{r}')$, (22b) is not well defined (see Chapter 7). This can be remedied, however, by writing the integral as

$$N_{mn} = \int_V d\mathbf{r}\, O(\mathbf{r}) \mathbf{E}_m(\mathbf{r}) \cdot \left(\overline{\mathbf{I}} + \frac{\nabla \nabla}{k_b^2} \right) \cdot \int_V d\mathbf{r}'\, g(\mathbf{r}, \mathbf{r}') O(\mathbf{r}') \mathbf{E}_n(\mathbf{r}'). \tag{8.9.23}$$

Furthermore, using integration by parts on the term that contains $\nabla \nabla$, we have (see Exercise 8.38)

$$N_{mn} = \int_V d\mathbf{r}\, O(\mathbf{r}) \mathbf{E}_m(\mathbf{r}) \cdot \int_V d\mathbf{r}\, g(\mathbf{r}, \mathbf{r}') O(\mathbf{r}') \mathbf{E}_n(\mathbf{r}')$$
$$- \frac{1}{k_b^2} \int_V d\mathbf{r}\, \nabla \cdot [O(\mathbf{r}) \mathbf{E}_m(\mathbf{r})] \int_V d\mathbf{r}' g(\mathbf{r}, \mathbf{r}') \nabla' \cdot [O(\mathbf{r}') \mathbf{E}_n(\mathbf{r}')]. \tag{8.9.24}$$

This is done to circumvent having to integrate the singularity of the dyadic Green's function. Note further that in (21a), $\overline{\mathbf{M}}$ and $\overline{\mathbf{N}}$ matrices are symmetrical. A real-symmetric matrix readily lends itself to being solved by the conjugate gradient method.

Equation (19) is also solvable by testing it with other weighting functions. A popular testing function is the Dirac delta function, as in the method of point matching. In this case, the double integrals in (24) reduce to single integrals, and the effort to compute the matrix element in the matrix equation is greatly reduced.

§8.10 Approximate Solutions of the Scattering Problem

The solution of the volume integral equation usually has to be solved for numerically. This is, in general, computationally intensive because in finding the matrix element N_{mn} in the previous section, we may have to perform a double integration. For many problems, however, especially when the scattering from the inhomogeneity is weak, it suffices to derive approximate solutions to the scattering problem. Therefore, we shall discuss two approximate solutions, the ***Born approximation*** (Born and Wolf 1980, p. 453), which works better at low frequencies, and the ***Rytov approximation*** (Tatarski 1961), which works better at higher frequencies. Moreover, both approximations are weak scatterer approximations.

§§8.10.1 Born Approximation

In the cases when $k^2 - k_b^2$ is small, or where the contrast of the scatterer is weak so that the second term on the right of Equation (8.9.14) is small compared to the first term, we can approximate

$$\mathbf{E}(\mathbf{r}) \simeq \mathbf{E}_{inc}(\mathbf{r}). \tag{8.10.1}$$

Then, the total field in Equation (8.9.14) can be approximately calculated as

$$\mathbf{E}(\mathbf{r}) = \mathbf{E}_{inc}(\mathbf{r}) + \int_V d\mathbf{r}' \, \overline{\mathbf{G}}(\mathbf{r}, \mathbf{r}') \cdot O(\mathbf{r}') \mathbf{E}_{inc}(\mathbf{r}'). \tag{8.10.2}$$

The above is known as the first-order Born approximation. It is also the first-order approximation in the Neumann series expansion of integral equation in (8.9.14), or the Taylor series approximation of $\mathbf{E}(\mathbf{r})$ using $(k^2 - k_b^2)$ as a small parameter (see Exercise 8.39; also see Chapter 9, Subsection 9.3.3).

Since the Born approximation is good only when the second term is much smaller than the first term in (2), we can establish the regime of validity of the Born approximation.[15] First, notice that for the homogeneous background case,

$$\overline{\mathbf{G}}(\mathbf{r}, \mathbf{r}') = \left(\overline{\mathbf{I}} + \frac{\nabla \nabla}{k_b^2} \right) g(\mathbf{r}, \mathbf{r}'). \tag{8.10.3}$$

If the size of the scatterer is of the order L, and $k_b L \ll 1$, then by dimensional analysis (see Exercise 8.40),

$$g(\mathbf{r}, \mathbf{r}') \sim \frac{1}{L}, \quad \nabla \nabla \sim \frac{1}{L^2}. \tag{8.10.4}$$

Then,

$$\overline{\mathbf{G}}(\mathbf{r}, \mathbf{r}') \sim \left(1 + \frac{1}{k_b^2 L^2} \right) \frac{1}{L}, \tag{8.10.5a}$$

$$O(\mathbf{r}) = (k^2 - k_b^2) \sim k_b^2 \Delta \epsilon_r, \tag{8.10.5b}$$

---

[15] The regime of validity of the Born approximation has also been discussed by Keller (1969).

(where $\Delta\epsilon_r = \epsilon/\epsilon_b - 1$) and

$$\int d\mathbf{r}' \sim L^3. \tag{8.10.5c}$$

Therefore, one sees that the second term in (2) is of the order of

$$[(k_b L)^2 + 1]\Delta\epsilon_r E_{inc}. \tag{8.10.6}$$

But since $k_b L \ll 1$, in order for the scattered field to be much smaller than the incident field, the constraint is

$$\Delta\epsilon_r \ll 1, \tag{8.10.7}$$

in the long wavelength limit.

On the other hand, in the short wavelength limit, the controlling factor for the magnitude of $\nabla\nabla$ is not the size of the object, but the field variation inside the object, which is controlled by the wavelength of the field. Then,

$$\nabla\nabla \sim k_b^2. \tag{8.10.8}$$

Furthermore, the phase of the wave as it propagates inside the object becomes important. For example, if $\mathbf{E}_{inc} \sim e^{i\mathbf{k}_b \cdot \mathbf{r}}$, then the total field inside a tenuous inhomogeneity consists of a linear superposition of plane waves of the type $\mathbf{E} \sim e^{i\mathbf{k} \cdot \mathbf{r}}$ (which is motivated by WKB type approximation; see Chapter 2) when $k \to \infty$.[16] Therefore, we can write

$$\mathbf{E} \sim e^{i\mathbf{k}_b \cdot \mathbf{r}} e^{i(\mathbf{k}-\mathbf{k}_b) \cdot \mathbf{r}}$$
$$\sim \mathbf{E}_{inc} e^{i(\mathbf{k}-\mathbf{k}_b) \cdot \mathbf{r}}, \tag{8.10.9}$$

and hence, $\mathbf{E} \simeq \mathbf{E}_{inc}$ only if $(k - k_b)L \ll 1$. Consequently, at high frequencies, the Born approximation is valid only if

$$k_b L \Delta\epsilon_r \ll 1, \quad k_b L \to \infty. \tag{8.10.10}$$

Note further that the above is a much more stringent restriction than (7) (also see Exercise 8.41).

In some applications where there is no charge accumulation (for example, a TM wave impinging on a cylinder), the $\nabla\nabla$ term can be neglected. Then, the problem reduces to a scalar one, and the constraint on the Born approximation from (4) to (6) becomes (Exercise 8.42)

$$k_b^2 L^2 \Delta\epsilon_r \ll 1. \tag{8.10.11}$$

Moreover, in the long wavelength limit, $k_b L \ll 1$, and this constraint could be met even when $\Delta\epsilon_r > 1$. Hence, in this case and the scalar wave case,

---

[16] Strictly speaking, it should be $\mathbf{E} \sim e^{i\int^z k\,dz'}$, but this distinction is not important for this order-of-magnitude argument.

the Born approximation becomes exceedingly good at low frequencies. But at high frequencies, constraint (10) holds true also for scalar waves, since polarization-charge effect is unimportant at high frequencies. Finally, it is important to note that the above constraints are for a three-dimensional space. In a one- or two-dimensional space, they have to be rederived (Exercise 8.42).

The Born approximation is a single-scattering approximation. Moreover, note that in Equation (2), the incident wave enters the scatterer with no distortion, induces the polarization current proportional to $(k^2 - k_b^2)\mathbf{E}_{inc}$, and causes a re-radiation or scattering. Since the incident field is unaffected while it gives rise to a scattered field, the Born approximation violates energy conservation. However, because of the symmetry of the dyadic Green's function, reciprocity is still preserved under the Born approximation.

When a conductive inhomogeneity is in an insulating background, $k^2(\mathbf{r}) \sim i\omega\mu\sigma(\mathbf{r})$ when $\omega \to 0$ and $k_b^2 = \omega^2\mu\epsilon_b$ when $\omega \to 0$. Then, from (3)

$$\overline{\mathbf{G}}(\mathbf{r}, \mathbf{r}') \sim \frac{1}{\omega^2}, \qquad \omega \to 0. \tag{8.10.12}$$

However,

$$O(\mathbf{r}) \sim \omega, \qquad \omega \to 0. \tag{8.10.13}$$

Therefore, the scattered field term in (2) is proportional to $1/\omega$ when $\omega \to 0$. This low-frequency divergence implies that the Born approximation is exceedingly bad at low frequencies when a conductive inhomogeneity is embedded in an insulating background. This happens because when we approximate the induced conduction current (eddy current) in the conductive inhomogeneity by

$$\mathbf{J} = \sigma\mathbf{E} \simeq \sigma\mathbf{E}_{in}, \tag{8.10.14}$$

in the Born approximation, the induced eddy current is terminated abruptly at the insulator/conductor interface. We can see this from the continuity equation, where the charge $\varrho = \nabla \cdot \mathbf{J}/i\omega$ implying that these charges at the interface diverge as $1/\omega$ when $\omega \to 0$, giving rise to this low-frequency divergence if $\nabla \cdot \mathbf{J} \neq 0$. But note that this problem does not arise if both the background and the scatterer are conductive.

§§8.10.2 Rytov Approximation

We have seen in various instances (e.g., Section 2.8, Chapter 2) that the polarization-charge effect is unimportant at high frequencies when the wavelength is much smaller than the size of the inhomogeneity. If this is actually the case, the study of the vector electromagnetic wave equation may be reduced to the study of the scalar wave equation. Therefore, the pertinent equation is then

$$[\nabla^2 + k^2(\mathbf{r})]\phi(\mathbf{r}) = 0. \tag{8.10.15}$$

To derive the Rytov approximation, we first let

$$\phi(\mathbf{r}) = e^{i\psi(\mathbf{r})}. \tag{8.10.16}$$

Then,

$$\nabla\phi(\mathbf{r}) = i\phi(\mathbf{r})\nabla\psi(\mathbf{r}), \tag{8.10.17a}$$

$$\nabla \cdot \nabla\phi(\mathbf{r}) = \{i\nabla^2\psi(\mathbf{r}) - [\nabla\psi(\mathbf{r})]^2\}\phi(\mathbf{r}). \tag{8.10.17b}$$

Using the above in Equation (15), we have

$$i\nabla^2\psi(\mathbf{r}) - (\nabla\psi)^2 + k^2(\mathbf{r}) = 0. \tag{8.10.18}$$

At this point, Equation (18) is still exact but nonlinear. However, we can solve (18) perturbatively by letting

$$\psi(\mathbf{r}) \sim \psi_0(\mathbf{r}) + \psi_1(\mathbf{r}). \tag{8.10.19}$$

Here, $\psi_0(\mathbf{r})$ is assumed to satisfy the equation

$$i\nabla^2\psi_0(\mathbf{r}) - (\nabla\psi_0)^2 + k_b^2(\mathbf{r}) = 0, \tag{8.10.20}$$

i.e., it is the solution in some background medium with wave number k_b. After substituting (19) into (18), consequently, we have

$$i\nabla^2\psi_1(\mathbf{r}) - 2(\nabla\psi_0) \cdot (\nabla\psi_1) - (\nabla\psi_1)^2 + O(\mathbf{r}) = 0, \tag{8.10.21}$$

where $O(\mathbf{r}) = k^2 - k_b^2$. At this point, the above equation is nonlinear, but it could be simplified by using the identity that

$$\nabla^2(\phi_0\psi_1) = \psi_1\nabla^2\phi_0 + 2(\nabla\phi_0) \cdot (\nabla\psi_1) + \phi_0\nabla^2\psi_1, \tag{8.10.22}$$

where $\phi_0 = e^{i\psi_0(\mathbf{r})}$. Since $\nabla^2\phi_0 = -k_b^2\phi_0$ and $\nabla\phi_0 = i(\nabla\psi_0)\,\phi_0$, we have

$$\nabla^2(\phi_0\psi_1) = -k_b^2\psi_1\phi_0 + 2i\phi_0(\nabla\psi_0) \cdot (\nabla\psi_1) + \phi_0\nabla^2\psi_1. \tag{8.10.23}$$

Then, on multiplying (21) by $i\phi_0$ and using (23), we obtain

$$\nabla^2(\phi_0\psi_1) + k_b^2\phi_0\psi_1 = -i\phi_0(\nabla\psi_1)^2 + i\phi_0 O(\mathbf{r}). \tag{8.10.24}$$

Equation (24) is still exact at this point. But if we assume that ψ_1 is small so that $(\nabla\psi_1)^2$ is even smaller, Equation (24) can be approximated as

$$\left(\nabla^2 + k_b^2\right)\phi_0\psi_1 = i\phi_0 O(\mathbf{r}). \tag{8.10.25}$$

The solution to (25) is then

$$\psi_1(\mathbf{r}) = -\frac{i}{\phi_0(\mathbf{r})} \int d\mathbf{r}'\, g(\mathbf{r}, \mathbf{r}')\,\phi_0(\mathbf{r}')O(\mathbf{r}'). \tag{8.10.26}$$

The above approximation is known as the Rytov approximation; the total solution is

$$\phi(\mathbf{r}) \simeq \phi_0(\mathbf{r})e^{i\psi_1(\mathbf{r})}. \tag{8.10.27}$$

The Rytov approximation is valid when the first term on the right of Equation (24) is much smaller than the second term, or

$$(\nabla \psi_1)^2 \ll O(\mathbf{r}). \tag{8.10.28}$$

Moreover, this approximation attempts to correct for the phase of the wave as it propagates through the inhomogeneous media. Hence, it shares some similarity with the WKB approximation.[17]

Applying dimensional analysis, it can be shown from (26) that

$$\psi_1(\mathbf{r}) \sim k_b^2 L^2 \Delta \epsilon_r, \quad \text{when} \quad k_b L \to 0. \tag{8.10.29}$$

Furthermore, assuming that $\nabla \sim 1/L$ when $k_b L \to 0$, then the use of (29) into (28) yields the condition that

$$(k_b L)^2 \Delta \epsilon_r \ll 1, \tag{8.10.30}$$

which is the same as (11). Again, this is dependent on dimensions.

On the other hand, when the frequency tends to infinity, the field inside the inhomogeneity is of the form $e^{i\mathbf{k}\cdot\mathbf{r}}$. Then,

$$\phi(\mathbf{r}) \sim e^{i\mathbf{k}\cdot\mathbf{r}} \sim e^{i\mathbf{k}_b\cdot\mathbf{r}} e^{i(\mathbf{k}-\mathbf{k}_b)\cdot\mathbf{r}} \sim \phi_0 e^{i\psi_1(\mathbf{r})}. \tag{8.10.31}$$

Therefore, $\psi_1(\mathbf{r}) \simeq (\mathbf{k} - \mathbf{k}_b) \cdot \mathbf{r}$, and

$$\psi_1(\mathbf{r}) \sim k_b L \Delta \epsilon_r, \quad k_b L \to \infty. \tag{8.10.32}$$

Moreover, assuming that $\nabla \sim 1/L$ when $k_b L \to \infty$,[18] the use of (32) into (28) yields the constraint

$$\Delta \epsilon_r \ll 1, \tag{8.10.33}$$

which is more relaxed than (10).

In the Rytov approximation, the correction $\psi_1(\mathbf{r})$ occurs as a phase term. But the magnitude of the correction to $\phi_0(\mathbf{r})$ is always unity even when $\psi_1(\mathbf{r})$ is bad. Hence, the approximation breaks down more gracefully compared to the Born approximation. Furthermore, the form given by (27) or (31) is more suitable for the field inside the inhomogeneity. Outside the scatterer, the constraint is again given by (10) for high frequencies (Exercise 8.43).

Note that the Born approximation for the scalar wave equation is of the form

$$\phi_1(\mathbf{r}) = -\int d\mathbf{r}' \, g(\mathbf{r}, \mathbf{r}') O(\mathbf{r}') \phi_0(\mathbf{r}'), \tag{8.10.34}$$

---

[17] The regime of validity of the Rytov approximation has also been discussed by Fried (1967), Brown (1967), Keller (1969), and Crane (1976).

[18] Unlike (8), $\nabla \sim 1/L$ instead of k_b here because ∇ operates on ψ_1, which is the phase variation.

where $\phi_1(\mathbf{r})$ is the scattered field and $\phi_0(\mathbf{r})$ is the incident field. But when $\psi_1(\mathbf{r})$ is very small, we can rewrite (27) as

$$\phi(\mathbf{r}) \cong \phi_0(\mathbf{r}) + i\psi_1(\mathbf{r})\,\phi_0(\mathbf{r}). \qquad (8.10.35)$$

Therefore, $\phi_1(\mathbf{r}) \sim i\psi_1(\mathbf{r})\,\phi_0(\mathbf{r})$ if $\psi_1(\mathbf{r})$ is very small. Then, on multiplying Equation (26) by $i\phi_0(\mathbf{r})$, we recover Equation (34). Hence, the Born and Rytov approximations reduce to the same approximation when the scattered field is very weak.

Both the Born and the Rytov approximations assume that the scattered field is linearly proportional to the inhomogeneity $O(\mathbf{r})$. As such, this linearized approximation makes them particularly suitable for solving the inverse problems when the scatterers are weak scatterers.

Exercises for Chapter 8

8.1 (a) Find another solution of (8.1.3) in V_1 that also satisfies the radiation condition at infinity.

(b) Show that the integral over S_{inf} in (8.1.6) vanishes by virtue of the radiation condition. In other words, if the sources that generate the field are finite in extent, all fields will look like outgoing plane waves when $\mathbf{r} \to \infty$.

8.2 (a) Derive Equation (8.1.11) in the manner of Equation (8.1.10). Does $g_2(\mathbf{r} - \mathbf{r}')$ in (8.1.11) need to satisfy the radiation condition?

(b) Show that (8.1.11) can be simplified by imposing either a homogeneous Dirichlet or Neumann boundary condition for $g_2(\mathbf{r}, \mathbf{r}')$ on S. In this case, explain why $g_2(\mathbf{r}, \mathbf{r}')$ is only defined in V_2, and the lower part of (8.1.11) does not hold anymore.

8.3 (a) For an unbounded homogeneous-medium dyadic Green's function, show that $\nabla \times \overline{\mathbf{G}}(\mathbf{r}, \mathbf{r}') = \nabla \times \overline{\mathbf{I}}\,g(\mathbf{r}, \mathbf{r}') = -\nabla' \times \overline{\mathbf{I}}\,g(\mathbf{r}, \mathbf{r}')$. Hence, show that

$$\left[\nabla \times \overline{\mathbf{G}}(\mathbf{r}, \mathbf{r}')\right]^t = -\left(\nabla' \times \overline{\mathbf{I}}\right)^t g(\mathbf{r}, \mathbf{r}') = \nabla' \times \overline{\mathbf{I}}\,g(\mathbf{r}, \mathbf{r}') = \nabla' \times \overline{\mathbf{G}}(\mathbf{r}', \mathbf{r}).$$

(b) Derive Equation (8.1.27) in a manner similar to deriving (8.1.26).

(c) Show that $\overline{\mathbf{G}}_2(\mathbf{r}, \mathbf{r}')$ need not satisfy the radiation condition in this case.

(d) Show that $\overline{\mathbf{G}}_1(\mathbf{r}, \mathbf{r}')$ in (8.1.28) need not be the homogeneous-medium Green's function.

8.4 (a) Derive the identity in Equation (8.1.37).

(b) Derive Equation (8.1.41) and hence, the integral equations in (8.1.42a) and (8.1.42b).

8.5 Show that the z components of the electromagnetic field, after Fourier transforming according to (8.1.43) to (8.1.45), satisfy (8.1.46).

8.6 Derive Equations (8.1.53) and (8.1.54). Show that $G_1(\rho, \rho')$ and $G_2(\rho, \rho')$ need not be of the form given by (8.1.51). What other forms can you think of?

8.7 Derive the relations (8.1.60) and (8.1.61) and hence, (8.1.62).

8.8 Because of the singularity contained in $\hat{n} \cdot \nabla g(\mathbf{r}, \mathbf{r}')$, Equations (8.1.10) and (8.1.11) are undefined when $\mathbf{r} \in S$. But using the definition of the principal value integral as in (8.2.12), the integrals in (8.1.10) and (8.1.11) can be defined even when $\mathbf{r} \in S$. Show that under such a definition, integral equations similar to (8.1.12a) and (8.1.12b) are

$$\phi_{inc}(\mathbf{r}) = \frac{1}{2}\phi_1(\mathbf{r}) + \fint_S dS' \, \hat{n}' \cdot [g_1(\mathbf{r}, \mathbf{r}')\nabla'\phi_1(\mathbf{r}') - \phi_1(\mathbf{r}')\nabla'g_1(\mathbf{r}, \mathbf{r}')],$$

$$\mathbf{r} \in S,$$

$$0 = \frac{1}{2}\phi_2(\mathbf{r}) - \fint_S dS' \, \hat{n}' \cdot [g_2(\mathbf{r}, \mathbf{r}')\nabla'\phi_2(\mathbf{r}') - \phi_2(\mathbf{r}')\nabla'g_2(\mathbf{r}, \mathbf{r}')],$$

$$\mathbf{r} \in S.$$

8.9 In (8.2.1), show that $\mathcal{L}_{11}$ and $\mathcal{L}_{21}$ are symmetric operators while $\mathcal{L}_{12}$ and $\mathcal{L}_{22}$ are skew-symmetric operators. Hence, show that their matrix representations by Galerkin's method yield symmetric or skew-symmetric matrices.

8.10 (a) Show that Equation (8.2.14) is always uniquely defined, i.e., it is not discontinuous across the surface S.

(b) Derive Equation (8.2.19) similar to the procedure given in (8.2.9) to (8.2.12).

8.11 (a) Show that for the boundary-element method, the tangential magnetic field on the n-th triangular patch can be expanded as

$$\hat{n} \times \mathbf{H}_n = \sum_{i=1}^{3}[h_{i1n}\mathbf{N}_{i1n} + h_{i2n}\mathbf{N}_{i2n}] = \sum_{i=1}^{3}[\mathbf{N}_{i1n}, \mathbf{N}_{i2n}]\begin{bmatrix} h_{i1n} \\ h_{i2n} \end{bmatrix}$$

$$= \sum_{i=1}^{3}\overline{\mathbf{N}}_{in} \cdot \mathbf{h}_{in} = \overline{\mathbf{N}}_n^t \cdot \mathbf{h}_n,$$

where

$$\overline{\mathbf{N}}_n^t = \left[\overline{\mathbf{N}}_{1n}, \overline{\mathbf{N}}_{2n}, \overline{\mathbf{N}}_{3n}\right], \quad \mathbf{h}_n^t = [\mathbf{h}_{1n}^t, \mathbf{h}_{2n}^t, \mathbf{h}_{3n}^t],$$

and

$$\overline{\mathbf{N}}_{in} = [\mathbf{N}_{i1n}, \mathbf{N}_{i2n}], \quad \mathbf{h}_{in}^t = [h_{i1n}, h_{i2n}].$$

$\mathbf{N}_{i1n}$ and $\mathbf{N}_{i2n}$ are as defined in (8.2.23) for the n-th patch. Hence, $\mathbf{h}_n$ is a column vector of length 6.

(b) If a surface S is approximated by a union of N triangular patches, show that the tangential component of the magnetic field can be expanded as

$$\hat{n} \times \mathbf{H} = \sum_{n=1}^{N} \overline{\mathbf{N}}_n^t \cdot \mathbf{h}_n = \overline{\mathbf{N}}^t \cdot \mathbf{h},$$

where $\overline{\mathbf{N}}^t = \left[\overline{\mathbf{N}}_1^t, \cdots, \overline{\mathbf{N}}_N^t \right]$, and $\mathbf{h}^t = [\mathbf{h}_1^t, \cdots, \mathbf{h}_N^t]$. Hence, $\mathbf{h}$ is a column vector of length $6N$.

(c) The elements of $\mathbf{h}$ consist of the normal components of the edge currents at the nodes of the union of triangular patches that approximates S. Since the normal components are continuous from one edge to another, many of the unknowns in $\mathbf{h}$ are redundant. Hence, the actual number of unknowns needed to approximate $\hat{n} \times \mathbf{H}$ is less than $6N$. Convince yourself that the actual number of unknowns is $2M$ where M is the total number of edges on S and that $2M < 6N$.

(d) Given that $\boldsymbol{\eta}$ is a column vector of length $2M$ containing the fundamental unknowns, show that a mapping matrix can be constructed such that

$$\mathbf{h} = \overline{\mathbf{M}}^t \cdot \boldsymbol{\eta},$$

where $\overline{\mathbf{M}}^t$ is a $6N \times 2M$ matrix. Hence, show that

$$\hat{n} \times \mathbf{H} = \overline{\mathbf{N}}^t \cdot \overline{\mathbf{M}}^t \cdot \boldsymbol{\eta}.$$

(e) Given an integral operator $\int_S dS' \, \overline{\mathbf{G}}(\mathbf{r}, \mathbf{r}') \cdot \hat{n} \times \mathbf{H}(\mathbf{r}')$, $\mathbf{r} \in S$, show that its matrix representation using Galerkin's method is given by

$$\overline{\mathbf{M}} \cdot \left\langle \overline{\mathbf{N}}, \int_S dS' \, \overline{\mathbf{G}}(\mathbf{r}, \mathbf{r}') \cdot \overline{\mathbf{N}}^t \right\rangle \cdot \overline{\mathbf{M}}^t,$$

where the inner product is a surface integral on S. What is the dimension of this matrix representation of the integral operator?

8.12 Give an example of $\overline{\mathbf{g}}_n$ in (8.2.27) which is not of a diagonal form, but yet forms a complete set.

8.13 The scattering of a plane wave by a metallic circular cylinder can be solved in closed form:

(a) Using the integral representation of Bessel functions in Chapter 2, Equation (2.2.17), show that a plane wave can be expanded as

$$e^{-ikx} = e^{-ik\rho\cos\phi} = \sum_{n=-\infty}^{\infty} J_n(k\rho)e^{in\phi - in\pi/2}.$$

(b) In two dimensions, the Green's function is [see Equation (3.3.2)]

$$g(\boldsymbol{\rho} - \boldsymbol{\rho}') = \frac{i}{4} H_0^{(1)}(k|\boldsymbol{\rho} - \boldsymbol{\rho}'|) = \frac{i}{4} \sum_{n=-\infty}^{\infty} J_n(k\rho_<) H_n^{(1)}(k\rho_>) e^{in(\phi-\phi')}.$$

Assuming $\phi_{inc}(\mathbf{r})$ is a plane wave as given above, show that (8.2.31) simplifies to

$$\sum_{n=-\infty}^{\infty} J_n(k\rho) e^{in\phi - in\frac{\pi}{2}}$$

$$= \frac{i}{4} \sum_{n=-\infty}^{\infty} J_n(k\rho) e^{in\phi} H_n^{(1)}(ka) a \int_0^{2\pi} d\phi' \, e^{-in\phi'} \hat{n}' \cdot \nabla' \phi_1(\mathbf{r}'), \quad \rho < a$$

for a plane wave incident on a circular metallic cylinder of radius a.

(c) Find the matrix representation of the integral operator above by expanding $\hat{n} \cdot \nabla' \phi_1(\mathbf{r}') = \sum_{m=-M}^{M} a_m e^{im\phi'}$ and testing with $e^{-ip\phi}$ at $\rho = a$. Show that the simplified version of (8.2.31) in the above reduces to

$$J_p(ka) e^{-ip\frac{\pi}{2}} = \frac{i}{4} J_p(ka) H_p^{(1)}(ka) 2\pi a_p.$$

Hence, the matrix representation of the integral operator in (8.2.31) is diagonal in this case.

(d) Show that at the internal resonance of the circular cylinder, both the left-hand side and the right-hand side of the above is zero, rendering a_p undefined.

(e) Show that the resonant sources of (8.2.32) generate no field outside the scatterer. Hence, prove Equation (8.2.33) from the reciprocity theorem (see Chapter 1, Exercise 1.13) for the scalar wave equation. The left-hand side of the equation in (c) vanishes at the internal resonance of the cylinder. Show that (8.2.33) is automatically satisfied for this case.

8.14 A consequence of (8.2.31), by operating on it with $\hat{n} \cdot \nabla$, is that

$$\hat{n} \cdot \nabla \phi_{inc}(\mathbf{r}) = \int_S dS' \, \hat{n} \cdot \nabla g_1(\mathbf{r}, \mathbf{r}') \, \hat{n}' \cdot \nabla' \phi_1(\mathbf{r}'), \quad \mathbf{r} \in S.$$

A combined field integral equation is defined as

$$\phi_{inc}(\mathbf{r}) + \lambda \hat{n} \cdot \nabla \phi_{inc} = \int_S dS' \, g_1(\mathbf{r}, \mathbf{r}') \, \hat{n}' \cdot \nabla' \phi_1(\mathbf{r}')$$

$$+ \lambda \int_S dS' \, \hat{n} \cdot \nabla g_1(\mathbf{r}, \mathbf{r}') \, \hat{n}' \cdot \nabla' \phi_1(\mathbf{r}'), \quad \mathbf{r} \in S.$$

Going through the special case as illustrated in Exercise 8.13, show that the indeterminacy in Exercise 8.13(d) due to internal resonances does not exist for this integral equation if λ is complex.

8.15 For the integral equation in Exercise 8.13(b), we expand

$$\hat{n} \cdot \nabla' \phi_1(\mathbf{r}') = \sum_{m=-M}^{M} a_m e^{im\phi},$$

and test it with $\delta(\rho-a)e^{-ip\phi}$ and $\delta(\rho-a+\Delta)e^{-ip\phi}$ where $p = -M, \cdots, M$ and $\Delta \ll a$. In other words, the integral equation is tested with points on S as well as on points slightly interior to S. Show that the integral equation now reduces to

$$J_p(ka)e^{-ip\frac{\pi}{2}} = \frac{i}{4}J_p(ka)H_p^{(1)}(ka)2\pi a_p,$$
$$p = -M, \cdots, M,$$
$$J_p[k(a-\Delta)]e^{-ip\frac{\pi}{2}} = \frac{i}{4}J_p[k(a-\Delta)]H_p^{(1)}(ka)2\pi a_p,$$
$$p = -M, \cdots, M.$$

The above is a set of overdetermined equations with $4M$ equations but only $2M$ unknowns. Show that the least-square solution of these overdetermined equations eliminates the internal resonance problem encountered in Exercise 8.13, except at the resonance of the annular region bounded by a and $a - \Delta$, which has very high resonant frequencies.

8.16 By using the addition theorem [see Chapter 3, Equation (3.3.2) and Equation (3.7.4)], show that the scalar Green's function for a homogeneous medium can be expanded as in (8.3.3) for both two and three dimensions. What is $\psi_n(k, \mathbf{r})$ for each of these cases?

Hint: It is sometimes more expedient to use $\cos m\phi$ and $\sin m\phi$ rather than $e^{im\phi}$ dependence so that in the lossless case, "regular part of" is the same as "real part of."

8.17 (a) Show that a closed cavity with the boundary condition $\phi_2(\mathbf{r}') = 0$ on S and filled with a material with wavenumber k_2 has a field that satisfies the integral equation

$$0 = \int_S dS' \, g_2(\mathbf{r} - \mathbf{r}') \, \hat{n}' \cdot \nabla' \phi_2(\mathbf{r}'), \quad \mathbf{r} \in V_1,$$

where S and V_1 are the same as that in Figure 8.3.1. Now, using (8.3.3) in the above, show that

$$0 = \int_S dS' \, \Re e \psi_n(k_2, \mathbf{r}') \, \hat{n}' \cdot \nabla' \phi_2(\mathbf{r}'), \quad \text{for all } n.$$

The above integral equation has only a trivial solution for $\hat{n}' \cdot \nabla' \phi_2(\mathbf{r}')$ except at the resonant frequencies of the cavity. This implies that at the nonresonant frequencies of the cavity, the only function that is orthogonal to $\Re e \psi_n(k_2, \mathbf{r}')$ for all n is zero. Consequently, $\Re e \psi_n(k_2, \mathbf{r}')$ is complete on S, except at the resonant frequencies of the cavity.

(b) Similarly, prove that $\hat{n} \cdot \nabla' \Re e \psi_n(k_2, \mathbf{r}')$ is complete on S except at the resonant frequencies of the cavity formed by S filled with a material with wavenumber k_2. What is the boundary condition on the wall of this cavity? Could any of the resonant frequencies in case (a) coincide with the resonant frequencies in this case?

8.18 Derive Equation (8.3.13) from (8.3.12).

8.19 (a) Derive the equivalence of (8.3.15) for an impenetrable scatterer with a homogeneous Dirichlet boundary condition on S. Give the definition for Q_{nm} in this case.

(b) Show that at the internal resonances of the cavity formed by S filled with a material with wavenumber k_2, the matrix Q_{nm} is ill-conditioned.

(c) Show that this problem can be remedied by using a linearly independent set to expand the surface field rather than those suggested by (8.3.9) and (8.3.10).

(d) Explain why the internal resonances do pose a problem for the $\overline{\mathbf{Q}}$ matrix of a penetrable scatterer given by (8.3.16).

8.20 Derive the equivalence of (8.3.7) for a circular metallic cylinder. Show that the internal resonance problem encountered in Exercise 8.13 does not exist here.

8.21 Show that in spherical coordinates, the dyadic Green's function can be expressed as (8.3.19). Identify the vector wave functions (see Chapter 7).

8.22 Derive Equations (8.3.21) and (8.3.22) from Equations (8.3.17) and (8.3.18).

8.23 Show that the set $\hat{n} \times \Re e \psi_m(k_2, \mathbf{r}')$ or $\hat{n} \times \nabla' \times \Re e \psi_m(k_2, \mathbf{r}')$ is complete on the surface S except at the internal resonance of the cavity formed by S and filled with a material with wavenumber k_2.

8.24 Prove the identity (8.3.27).

8.25 From the reciprocity theorem for the scalar wave equation, prove that $\overline{\mathbf{T}}$, and hence $\overline{\mathbf{S}}$, are symmetric matrices.

8.26 (a) From the scalar wave equation $(\nabla^2 + k^2)\phi = 0$, derive the energy conservation theorem that $\int_S dS\, \hat{n} \cdot (\phi \nabla \phi^* - \phi^* \nabla \phi) = 0$ when k is real. Hence, $\mathbf{F} = \phi \nabla \phi^* - \phi^* \nabla \phi$ is an energy flux.

(b) Show that for a lossless medium,

$$\psi_n(-k_1, \mathbf{r}) = \psi_n^*(k_1, \mathbf{r}),$$

where $\psi_n(k_1, \mathbf{r})$ is derived in Exercise 8.16.

(c) Show that $\int_S dS \, \hat{n} \cdot [\boldsymbol{\psi} \nabla \boldsymbol{\psi}^t - (\nabla \boldsymbol{\psi}) \boldsymbol{\psi}^t] = 0$ when S is a closed surface, and $\boldsymbol{\psi}$ is as defined for (8.4.8).

Hint: Derive it first for when S is a circle or a sphere, and deform it to an arbitrary S later. The deformation is allowed because $\nabla \cdot [\boldsymbol{\psi} \nabla \boldsymbol{\psi}^t - (\nabla \boldsymbol{\psi}) \boldsymbol{\psi}^t] = 0$ in a region excluding the origin.

(d) Using (8.4.12), (b), and (c), show that

$$\int_S dS \, \hat{n} \cdot (\phi \nabla \phi^* - \phi^* \nabla \phi) = \frac{1}{4} \mathbf{a}^t \cdot \int_S dS \left[\boldsymbol{\psi}^* \nabla \boldsymbol{\psi}^t - (\nabla \boldsymbol{\psi}^*) \boldsymbol{\psi}^t \right.$$

$$\left. + \overline{\mathbf{S}}^t \cdot (\boldsymbol{\psi} \nabla \boldsymbol{\psi}^\dagger - (\nabla \boldsymbol{\psi}) \boldsymbol{\psi}^\dagger) \cdot \overline{\mathbf{S}}^* \right] \cdot \mathbf{a}^*.$$

(e) Show that

$$\int_S dS \, [\boldsymbol{\psi}^* \nabla \boldsymbol{\psi}^t - (\nabla \boldsymbol{\psi}^*) \boldsymbol{\psi}^t] = ic\overline{\mathbf{I}},$$

where c is a real constant.

Hint: The Wronskian of Bessel functions may be useful here.

(f) Therefore, show that in order for the integral in (d) to vanish to conserve energy, $\overline{\mathbf{S}} \cdot \overline{\mathbf{S}}^\dagger = \overline{\mathbf{S}} \cdot \overline{\mathbf{S}}^* = \overline{\mathbf{I}}$.

8.27 (a) Using Rayleigh's hypothesis method, derive the $\overline{\mathbf{T}}$ matrix for an impenetrable scatterer with a homogeneous Dirichlet boundary condition. Show that this is the transpose of the $\overline{\mathbf{T}}$ matrix derived by the extended-boundary-condition method.

(b) Explain why the error in Rayleigh's hypothesis method and the EBC method should be of the same order if the same number of terms are used in both methods.

8.28 (a) Derive the $\overline{\boldsymbol{\alpha}}_{ji}$ and $\overline{\boldsymbol{\beta}}_{ji}$ matrices in cylindrical coordinates. Show that $\overline{\boldsymbol{\beta}}_{ji} = \Re e \overline{\boldsymbol{\alpha}}_{ji}$, where "$\Re e$" stands for "the regular part of."

(b) Derive the $\overline{\boldsymbol{\alpha}}_{ji}$ and $\overline{\boldsymbol{\beta}}_{ji}$ matrices in spherical coordinates. Show that $\overline{\boldsymbol{\beta}}_{ji} = \Re e \overline{\boldsymbol{\alpha}}_{ji}$ (also see Appendix D).

8.29 Count the number of matrix multiplications required in (8.6.17) and (8.6.18) at each iteration, and that required after N iterations for N scatterers. Show that the number of matrix multiplication is proportional to N^2 when $N \to \infty$.

8.30 Using the EBC method, derive the $\overline{\mathbf{R}}_{10}$ and $\overline{\mathbf{T}}_{10}$ matrices defined in Equations (8.7.7a) and (8.7.7b) for the geometry shown in Figure 8.7.2.

8.31 (a) By expanding (8.7.23) and (8.7.28) into a geometric series, give a physical explanation for each term of the series.

(b) Derive the relationships (8.7.25) and (8.7.26).

8.32 Derive the matrix representation of Equation (8.8.3); hence, derive (8.8.8) and (8.8.9).

8.33 (a) Explain why $\overline{\mathbf{L}}$, and hence $\overline{\mathbf{M}}^{-1}$ in (8.8.15), are singular at the resonant frequencies of the cavity formed by S. Show that for a time harmonic solution, this poses a problem only for a lossless medium filling the cavity.

(b) Show that for a finite vector $\mathbf{d}$, $\mathbf{c}$ is infinite at the resonant frequencies in (8.8.15). At the resonant frequencies of the cavity, however, $\hat{n} \cdot \nabla \phi_1(\mathbf{r}) = 0$, and hence, $\mathbf{d} = 0$ from (8.8.12b). Therefore, $\mathbf{c}$ must be finite while $\mathbf{d} = 0$ at resonances.

(c) At the resonances of the cavity, $\overline{\mathbf{L}}$, which is symmetric, has zero eigenvalues. Using the singular value decomposition method, show that $\overline{\mathbf{L}} = \overline{\mathbf{S}}^t \cdot \overline{\boldsymbol{\lambda}} \cdot \overline{\mathbf{S}}$ where $\overline{\boldsymbol{\lambda}}$ is a diagonal matrix containing the eigenvalues of $\overline{\mathbf{L}}$. Some of these eigenvalues are zero at the resonances of the cavity.

(d) Because of the zero eigenvalues of $\overline{\mathbf{L}}$, its inverse $\overline{\mathbf{L}}^{-1}$ has infinite eigenvalues. Show that $\overline{\mathbf{M}}$ also has infinite eigenvalues at the resonances of the cavity. Hence, $\overline{\mathbf{M}}$ is not computable when $\overline{\mathbf{L}}$ is singular.

(e) Show that by setting the zero eigenvalues of $\overline{\mathbf{L}}$ to a small, nonzero number, $\overline{\mathbf{M}}$ is computable. Instead of having infinite eigenvalues, $\overline{\mathbf{M}}$ has large eigenvalues. In this manner, $\overline{\mathbf{M}}^{-1}$ in (8.8.22) can be found to a degree of accuracy permitted by machine precision.

8.34 A less computationally intensive method of finding $\overline{\mathbf{L}}^{-1}$ to within machine precision is to use the regularization method.

(a) Explain why the internal resonance poses a problem only for lossless media in S. Show that the eigenvalues of $\overline{\mathbf{L}}$ are always real in this case.

(b) If a new $\widetilde{\overline{\mathbf{L}}}$ is defined such that $\widetilde{\overline{\mathbf{L}}} = \overline{\mathbf{L}} + i\delta\overline{\mathbf{I}}$, where $i\delta$ is a pure imaginary number, show that $\widetilde{\overline{\mathbf{L}}}$'s eigenvalues will never be zero. $i\delta$ can be chosen just large enough so that $\widetilde{\overline{\mathbf{L}}}^{-1}$ can be found without overflowing the computer floating-point capability. In this manner, $\overline{\mathbf{M}}$ can be computed, as can $\overline{\mathbf{M}}^{-1}$.

Alternatively, the resonance problem can be alleviated by assuming a small loss in V_1.

8.35 (a) Assuming that an impedance boundary condition such that $\hat{n} \cdot \nabla g_1 = Z g_1$ on S is imposed, formulate the scattering problem by an inhomogeneous body using the finite-element method together with this impedance boundary condition.

(b) Explain why the internal resonance problem would not exist in this case.

8.36 The field in V_1 in Equation (8.8.6) can be decomposed as $\phi_1 = \phi_{inc} + \phi_{sca}$, where ϕ_{sca} is the scattered field due to the inhomogeneity.

(a) Show that if $g_1(\mathbf{r}, \mathbf{r}')$ in (8.8.6) satisfies (8.8.3) plus the radiation condition at infinity, then

$$\int_S dS' \left[g_1(\mathbf{r}, \mathbf{r}') \, \hat{n}' \cdot \nabla' \phi_{sca}(\mathbf{r}') - \phi_{sca}(\mathbf{r}') \, \hat{n}' \cdot \nabla' g_1(\mathbf{r}, \mathbf{r}') \right] = 0, \quad \mathbf{r} \in V_1$$

in Figure 8.8.1. Hence, Equation (8.8.6) can be written as

$$\phi_1(\mathbf{r}) = \int_S dS' \left[g_1(\mathbf{r}, \mathbf{r}') \, \hat{n}' \cdot \nabla' \phi_{inc}(\mathbf{r}') - \phi_{inc}(\mathbf{r}') \, \hat{n}' \cdot \nabla' g_1(\mathbf{r}, \mathbf{r}') \right], \quad \mathbf{r} \in V_1.$$

(b) The above shows that $\phi_1(\mathbf{r})$ inside V_1 is known once the requisite $g_1(\mathbf{r}, \mathbf{r}')$ satisfying the radiation condition is known. Moreover, once $\phi_1(\mathbf{r})$ is known, then ϕ_{sca} is known inside V_1 and on S since $\phi_1(\mathbf{r}) = \phi_{inc}(\mathbf{r}) + \phi_{sca}(\mathbf{r})$. But in (8.8.5), $\phi_0 = \phi_{inc} + \phi_{sca}$ in V_0. Hence, show that if the homogeneous-medium Green's function $g_0(\mathbf{r}, \mathbf{r}')$ satisfies the radiation condition at infinity, then

$$\int_S dS' \left[g_0(\mathbf{r}, \mathbf{r}') \, \hat{n}' \cdot \nabla' \phi_{inc}(\mathbf{r}') - \phi_{inc}(\mathbf{r}') \, \hat{n}' \cdot \nabla' g_0(\mathbf{r}, \mathbf{r}') \right] = 0, \quad \mathbf{r} \in V_0.$$

Hence, from (8.8.5), deduce that

$$\phi_{sca} = -\int_S dS' \left[g_0(\mathbf{r}, \mathbf{r}') \, \hat{n}' \cdot \nabla' \phi_{sca}(\mathbf{r}') - \phi_{sca}(\mathbf{r}') \, \hat{n}' \cdot \nabla' g_0(\mathbf{r}, \mathbf{r}') \right], \quad \mathbf{r} \in V_0.$$

In conclusion, if $g_1(\mathbf{r}, \mathbf{r}')$ is known inside V_1 and on S, then the field everywhere could be found. But the requisite $g_1(\mathbf{r}, \mathbf{r}')$ satisfying the radiation condition can be found approximately using an absorbing boundary condition on S. An absorbing boundary condition may be of the form $\hat{n} \cdot \nabla g_1 = Z g_1$ on S, so that all outgoing waves are absorbed on S emulating the radiation condition. Such a manner of formulating an FEM problem with this boundary condition is described in Exercise 8.35. Moreover, absorbing boundary conditions are described in Chapter 4.

8.37 Explain why the second term in (8.9.13) is related to magnetic polarization charges.

8.38 Derive Equation (8.9.24) from Equation (8.9.23) and show that N_{mn} is a symmetric matrix. Explain why the matrix representation of the integral equation is symmetrized by a multiplication by $O(\mathbf{r})$ as in (8.9.19).

8.39 The Neumann series expansion of an integral equation is the higher-dimensional analogue of the Taylor's series expansion of a scalar function. By using the $k^2 - k_b^2$ as a small parameter, show that the error in (8.10.2) is of higher order.

8.40 (a) Assume an integral of the form

$$I = \int_V d\mathbf{r}'\, g(\mathbf{r}, \mathbf{r}') q(\mathbf{r}'),$$

where $g(\mathbf{r}, \mathbf{r}') = e^{ik_b|\mathbf{r}-\mathbf{r}'|}/4\pi|\mathbf{r} - \mathbf{r}'|$. By letting $\boldsymbol{\eta} = \mathbf{r}/L$ and $\boldsymbol{\eta}' = \mathbf{r}'/L$, where L is the typical size of the volume V (i.e., $V \simeq L^3$), show that

$$I = L^2 \int_{V/L^3} d\boldsymbol{\eta}'\, \frac{e^{ik_b L|\boldsymbol{\eta}-\boldsymbol{\eta}'|}}{4\pi|\boldsymbol{\eta} - \boldsymbol{\eta}'|} q(\boldsymbol{\eta}'L).$$

The integral now is mainly dimensionless except for the dimension of q. If $k_b L \ll 1$ or $L \ll \lambda_b$, then show that

$$I \simeq L^2 \int_{V/L^3} d\boldsymbol{\eta}'\, \frac{q(\boldsymbol{\eta}'L)}{4\pi|\boldsymbol{\eta} - \boldsymbol{\eta}'|} \sim O(L^2)\tilde{q}(\boldsymbol{\eta}'L).$$

When should $\tilde{q}$ be of the same order as q? If so, then I is $O(L^2 q)$.

(b) Show that the above can be obtained more quickly by dimensional analysis, i.e., by assuming that $g(\mathbf{r}, \mathbf{r}') \sim 1/L$, $\int d\mathbf{r}' \sim L^3$.

8.41 (a) For a plane wave at normal incidence on a dielectric slab of thickness L, find the exact solution of the reflected wave.

(b) Derive the approximation of the reflected wave when $\frac{\epsilon}{\epsilon_b} - 1 \to 0$, where ϵ is the permittivity of the dielectric slab and ϵ_b is the permittivity of the background.

(c) Derive the reflected wave using the Born approximation, and show that this result reduces to that in (b) only if (8.10.10) is satisfied.

8.42 (a) For a scalar wave equation, show that (8.10.11) is the constraint for the validity of the Born approximation at low frequencies.

(b) Show that the corresponding constraint for two dimensions is

$$k_b^2 L^2 \ln(kL)\Delta\epsilon_r \ll 1,$$

and that for one dimension is $k_b L \Delta \epsilon_r \ll 1$.

8.43 (a) For a homogeneous dielectric slab of thickness L with a wave normally incident on it, derive the exact solution for the reflected wave as well as the wave inside the slab using the method of Chapter 2.

   (b) Derive the field inside the slab using the Rytov approximation. Show that this result reduces to that of (a) inside the slab when (8.10.33) is satisfied but not outside the slab.

References for Chapter 8

Angkaew, T., M. Matsuhara, and N. Kumagai. 1987. "Finite-element analysis of waveguide modes: A novel approach that eliminates spurious modes." *IEEE Trans. Microwave Theory Tech.* MTT-35: 117–23.

Baker, C. T. H. 1977. *The Numerical Treatment of Integral Equations.* Oxford: Clarendon Press.

Barber, P., and C. Yeh. 1975. "Scattering of electromagnetic waves by arbitrarily shaped dielectric bodies." *Applied Optics* 14(12): 2864–72.

Bates, R. H. T., and D. J. N. Wall. 1977. "Null field approach to scalar diffraction, I. General methods, II. Approximate methods." *Phil. Trans. R. Soc. Lond.* A287: 45–95.

Bolomey, J. C., and W. Tabbara. 1973. "Numerical aspects on coupling between complementary boundary value problems." *IEEE Trans. Antennas Propagat.* AP-21: 356–63.

Bolomey, J. C., and A. Wirgin. 1974. "Numerical comparison of the Green's function and the Waterman and Rayleigh theories of scattering from a cylinder." *Proc. IEE* 121: 794–804.

Born, M., and E. Wolf. 1980. *Principles of Optics.* 6th ed. New York: Pergamon Press. First edition 1959.

Brebbia, C. A. 1978. *The Boundary Element Method for Engineers.* London: Pentech Press.

Brown, W. P., Jr. 1967. "Validity of the Rytov approximation." *J. Opt. Soc. Am.* 57(12): 1539–43.

Burrow, M. L. 1969. "Equivalence of the Rayleigh solution and the extended-boundary-condition solution for scattering problems," *Elect. Lett.* 5(12): 277-78.

Burton, A. J., and G. F. Miller. 1971. "The application of integral equation methods to the numerical solution of some exterior boundary value problems." *Proc. Roy. Soc., Lond.* A323: 201–10.

Chang, S. K., and K. K. Mei. 1976. "Application of the unimoment method to electromagnetic scattering of dielectric cylinders." *IEEE Trans. Antennas Propagat.* AP-24: 35–42.

Chew, W. C. 1989. "An N^2 algorithm for the multiple scattering solution of N scatterers." *Microwave and Optical Technology Letters.* 2(11): 380-83.

Chew, W. C., and Y. M. Wang. 1990. "A fast algorithm for solutions of a scattering problem using a recursive aggregate τ matrix method." *Microwave and Optical Technology Letters.* 3(5): 164-9.

Chew, W. C., J. Friedrich, and R. Geiger. 1990. "A multiple scattering solution for the effective permittivity of a sphere mixture." *IEEE Trans. Geosci. Remote Sensing.* GE-28(2): 207–14.

Chuang, S. L., and J. A. Kong. 1982. "Wave scattering from a periodic dielectric surface for a general angle of incidence." *Radio Science* 17: 545–57.

Crane, R. K. 1976. "Spectra of ionospheric scintillation." *J. Geophys. Res.* 81(3): 2041–50.

Cruzan, O. R. 1962. "Translational addition theorems for spherical vector wave functions." *Quarterly Appl. Math.* 20(1): 33–40.

Danos, M., and L. C. Maximon. 1965. "Multipole matrix elements of the translation operator." *J. Math. Phys.* 6: 766–78.

Elliott, D. 1980. "Integral equations—ninety years on." In *The Application and Numerical Solution of Integral Equations*, ed. R. S. Andersen, F. R. de Hoog, and M. A. Lukas. Netherlands: Sijthoff & Noordhoff.

Fredholm, I. 1903. "Sur une classe d'equations fonctionalles." *Acta Math* 27: 365–90.

Fried, D. L. 1967. "Test of the Rytov approximation." *J. Opt. Soc. Am.* 57(2): 268–69.

Friedman, B., and J. Russek. 1954. "Addition theorems for spherical waves." *Quarterly Appl. Math.* 12(1): 13–23.

Green, G. 1818. "An essay on the application of mathematical analysis to the theories of electricity and magnetism." In *Mathematics Papers of Late George Green*, ed. N. M. Ferrers. 1970. New York: Chelsea.

Harrington, R. F. 1968. *Field Computation by Moment Methods*. New York: Macmillan; Florida: Krieger Publishing, 1983.

Jones, D. S. 1974. "Numerical solution for antenna problems." *Proc. IEE.* 121: 573–82.

Keller, J. B. 1969. "Accuracy and validity of the Born and Rytov approximations," *J. Opt. Soc. Am.* 59: 1003–04.

Kerker, M. ed. 1988. "Selected Papers on Light Scattering." In *Soc. Photo-Opt. Inst. Eng. Milestone Series* 951.

Lewin, L. 1970. "On the restricted validity of point-matching techniques." *IEEE Trans. Microwave Theory Tech.* MTT-18(12): 1041–47.

Mautz, J. R., and R. F. Harrington. 1978. "H-field, E-field, and combined-field solutions for conducting bodies of revolution." *A. E. U.* 32: 157–63.

Mautz, J. R., and R. F. Harrington. 1979. "A combined-source formulation for radiation and scattering from a perfectly conducting body." *IEEE Trans. Antennas Propagat.* AP-27: 445–54.

McDonald, B. H., and A. Wexler. 1972. "Finite-element solution of unbounded field problems." *IEEE Trans. Microwave Theory Tech.* MTT-20: 841–47.

Mei, K. K. 1974. "Unimoment method of solving antenna and scattering problems." *IEEE Trans. Antennas Propagat.* AP-22: 760–66.

Millar, R. F. 1969. "Rayleigh hypothesis in scattering problems." *Elect. Lett.* 5(17): 416-17.

Mittra, R., and C. A. Klein. 1975. "Stability and convergence of moment method solutions." In *Numerical and Asymptotic Techniques in Electromagnetics*, ed. R. Mittra. New York: Springer-Verlag.

Mitzner, K. M. 1968. "Numerical solution of the exterior scattering problem at the eigenfrequencies of the interior problem." URSI Meeting Digest, p. 75, Boston.

Morgan, M. A., and K. K. Mei. 1979. "Finite-element computation of scattering by inhomogeneous penetrable bodies of revolution." *IEEE Trans. Antennas Propagat.* AP-27: 202–14.

Morita, N. 1979. "Resonant solutions involved in the integral equation approach to scattering from conducting and dielectric cylinders." *IEEE Trans. Ant. Propag.* AP-27: 869–71.

Peterson, B., and S. Ström. 1973. "T-matrix for electromagnetic scattering from an arbitrary number of scatterers and representation of E(3)." *Phys. Rev.* D8: 3661–78.

Peterson, B., and S. Ström. 1974a. "Matrix formulation of acoustic scattering from an arbitrary number of scatterers." *J. Acoust. Soc. Am.* 50: 771–80.

Peterson, B., and S. Ström. 1974b. "T-matrix formulation of electromagnetic scattering from multilayered scatterers." *Phys. Rev. D.* 10(8): 2670–84.

Peterson, B., and S. Ström. 1975. "Matrix formulation of acoustic scattering from multilayered scatterers." *J. Acoust. Soc. Am.* 57: 2–13.

Peterson, A. F. "The interior resonance problem associated with surface integral equations of electromagnetics: numerical consequences and a survey of remedies." *Electromagnetics.* To appear in 1990.

Poggio, A. J., and E. K. Miller. 1973. "Integral equation solution of three-dimensional scattering problems." *Computer Techniques for Electromagnetics*, ed. R. Mittra. New York: Pergamon Press; New York: Hemisphere Publishing, 1987.

Rayleigh, J. W. S. 1894. *Theory of Sound*, 2nd ed., vol. II, p. 89 and pp. 297–311. Macmillan; New York: Dover, 1945.

Rayleigh, J. W. S. 1897. "On the incidence of aerial and electromagnetic waves upon small obstacles in the form of ellipsoids or elliptical cylinders, and on the passage of electric waves through a circular aperture in a conducting screen." *Phil. Mag.* 44: 28–52.

Rayleigh, J. W. S. 1907. "On the dynamical theory of gratings." *Proc. Roy. Soc.* (London) A79: 399–416.

Richmond, J. H. 1965a. "Digital computer solutions of the rigorous equations for scattering problems." *Proc. IEEE* 53: 796–804.

Richmond, J. H. 1965b. "Scattering from a dielectric cylinder of arbitrary cross section shape." *IEEE Trans. Antennas Propagat.* AP-13: 334–41.

Schenck, H. A. 1968. *J. Acoust. Soc. Amer.* 44: 41–58.

Silvester, P., and M. S. Hsieh. 1971. "Finite-element solution of two-dimensional exterior field problems." *Proc. IEE* 118: 1743–47.

Stein, S. 1961. "Addition theorems for spherical wave functions." *Quarterly Appl. Math.* 19(1): 15–24.

Ström, S. 1975. "On the integral equation for electromagnetic scattering." *American J. Phys.* 43(12): 1060–69.

Tatarski, V. I. 1961. *Wave Propagation in a Turbulent Medium.* New York: McGraw-Hill.

Tikhonov, A. N. 1963. "Regularization of incorrectly posed problems." *Soviet Math. Dokl.* 4: 1624–27.

van den Berg, P. M. 1980. "Review of some computational techniques in scattering and diffraction." International URSI Symposium on Electromagnetic Waves. August 26-29, Munich.

Wang, D. S., and P. W. Barber. 1979. "Scattering by inhomogeneous nonspherical objects." *Applied Optics* 18(8): 1190–97.

Wang, Y. M., and W. C. Chew. 1989. "Response of the induction logging tool in an arbitrarily shaped borehole." *J. Electromagnetic Waves Applications* 3(7): 621–33.

Wang, Y. M., and W. C. Chew. 1990. "An efficient algorithm for solution of a scattering problem," *Microwave and Optical Technology Letters.* 3(3): 102-6.

Waterman, P. C. 1969. "New formulation of acoustic scattering." *J. Acoust. Soc. Am.* 45: 1417–29.

Waterman, P. C. 1971. "Symmetry, unitarity, and geometry in electromagnetic scattering." *Phys. Rev. D.* 3(4): 825–29.

Werner, P. 1963. *J. Math. Anal. Appl.* 7: 348–95.

Further Readings for Chapter 8

Bagby, J. S., D. P. Nyquist, and B. C. Drachman. 1985. "Integral formulation for analysis of integrated dielectric waveguides." *IEEE Trans. Microwave Theory Tech.* MTT-33: 906–15.

Bahar, E. 1980. "Computations of the transmission and reflection scattering coefficients in an irregular spheroidal model of the earth-ionosphere waveguide." *Radio Sci.* 15(5): 987–1000.

Bahar, E., and M. A. Fitzwater. 1983. "Scattering and depolarization of electromagnetic waves in irregular stratified spheroidal structures of finite conductivity—full wave analysis." *Can. J. Phys.* 61(1): 128–39.

Bossavit, A., and J.-C. Vérité. 1982. "A mixed FEM-BIEM method to solve 3-D eddy-current problems." *IEEE Trans. Magnetics* MAG-18: 431–35.

Bouwkamp, C. J., and H. B. G. Casimir. 1954. "On multipole expansions in the theory of electromagnetic radiation." *Physica* 20: 539–54.

Bowman, J. J., T. B. A. Senior, and P. L. E. Uslenghi. 1969. *Electromagnetic and Acoustic Scattering by Simple Shapes.* Amsterdam: North-Holland.

Bruning, J. H., and Y. T. Lo. 1971. "Multiple scattering of EM waves by spheres, part I and II." *IEEE Trans. Antennas Propagat.* AP-19: 378–400.

Chew, W. C., and R. L. Kleinberg. 1988. "Theory of microinduction measurements." *IEEE Trans. Geosci. Remote Sensing* GE-26: 707–19.

Chu, L. J., and J. A. Stratton. 1941. "Forced oscillations of a prolate spheroid." *J. Appl. Phys.* 12: 241–48.

Chuang, S. L., and J. A. Kong. 1983. "Wave scattering and guidance by dielectric waveguides with periodic surface." *J. Opt. Soc. Am.* 73: 669–79.

Cullen, A. L., O. Özkan, and L. A. Jackson. 1971. "Point-matching technique for rectangular cross-section dielectric rod." *Electron. Lett.* 7: 497–99.

de Ruiter, H. M. 1980. "Integral-equation approach to the computation of modes in an optical waveguide." *J. Opt. Soc. Am.* 70: 1519–24.

DeSanto, J. A. 1974. "Green's function for electromagnetic scattering from a random rough surface." *J. Math. Phys.* 15: 283–88.

Dey, A., and H. F. Morrison. 1979. "Resistivity modelling for arbitrarily shaped two-dimensional structures." *Geophys. Prosp.* 27: 106–36.

Dey, A., and H. F. Morrison. 1979. "Resistivity modelling for arbitrarily shaped three-dimensional structures." *Geophysics* 44: 753–80.

Dey, A., W. H. Meyer, H. F. Morrison, and W. M. Dolan. 1975. "Electric field response of two-dimensional inhomogeneities to unipolar and bipolar electrode configurations." *Geophysics* 40: 630–40.

Eyges, L., P. Gianino, and P. Wintersteiner. 1979. "Modes of dielectric waveguides of arbitrary cross-sectional shape." *J. Opt. Soc. Am.* 69: 1226–35.

Glisson, A., and D. R. Wilton. 1980. "Simple and efficient numerical methods for problems of electromagnetic radiation and scattering from surfaces." *IEEE Trans. Antennas Propagat.* AP-28(5): 593-603.

Goell, J. E. 1969. "A circular-harmonic computer analysis of rectangular dielectric waveguides." *Bell Syst. Tech. J.* 48: 2133–60.

Hochstadt, H. 1973. *Integral Equations, Pure & Applied Mathematics.* New York: Wiley-Interscience.

Hohmann, G. W. 1975. "Three-dimensional induced polarization and electromagnetic modeling." *Geophysics* 40: 309–24.

Holt, A. R., N. K. Uzunoglu, and B. G. Evans. 1978. "An integral equation solution to scattering of electromagnetic radiation by dielectric spheroids and ellipsoids." *IEEE Trans. Antennas Propagat.* AP-26: 706–12.

Ikuno, H., and K. Yasuura. 1978. "Numerical calculation of the scattered field from a periodic deformed cylinder using the smoothing process on the mode-matching method." *Radio Science.* 13(6): 937–46.

Okuno, Y., and K. Yasuura. 1982. "Numerical algorithm based on the mode-matching method with a singular-smoothing procedure for analyzing edge-type scattering problem." *IEEE Trans. Antennas Propagat.* AP-30(4): 580–87.

Ishimaru, A. 1978. *Wave Propagation and Scattering in Random Media*, vols. 1 and 2. New York: Academic Press.

Jin, J. M., and V. V. Liepa. 1988. "Application of hybrid finite element method to electromagnetic scattering from coated cylinders." *IEEE Trans. Antennas Propag.* 36(1): 50–54.

Jin, J. M., V. V. Liepa, and C. T. Tai. 1988. "A volume-surface integral equation for electromagnetic scattering by inhomogeneous cylinders." *J. Electromagnetic Waves and Appl.* 2: 573–88.

Jordan, A. K., and R. H. Lang. 1979. "Electromagnetic scattering patterns from sinusoidal surface." *Radio Sci.* 14: 1077–88.

Lee, K. H., D. F. Pridmore, and H. F. Morrison. 1981. " A hybrid three-dimensional electromagnetic modeling scheme." *Geophysics* 46: 796–805.

Liu, Q. H., and W. C. Chew. 1990. "Surface integral equations technique for the slantingly stratified half-space." *IEEE Trans. Antennas Propag.* Scheduled for May.

Livesay, D. E., and K. M. Chen. 1974. "Electromagnetic fields induced inside arbitrarily shaped biological bodies." *IEEE Trans. Microwave Theory Tech.* MTT-22: 1273–80.

Marin, S. P. 1982. "Computing scattering amplitudes for arbitrary cylinders under incident plane waves." *IEEE Trans. Antennas Propagat.* AP-30: 1045–49.

Miller, E. K., and A. J. Poggio. 1978. "Moment-method techniques in electromagnetics from an applications viewpoint." *Electromagnetic Scattering*, ed. P. L. E. Uslenghi, ch. 9. New York: Academic.

Morgan, M. A., and B. E. Welch. 1986. "The field feedback formulation for electromagnetic scattering computations." *IEEE Trans. Antennas Propagat.* AP-34: 1377–82.

Morita, N. 1978. "Surface integral representations for electromagnetic scattering from dielectric cylinders." *IEEE Trans. Antennas Propagat.* AP-26: 261–66.

Morita, N. 1982. "A method extending the boundary condition for analyzing guided modes of dielectric waveguides of arbitrary cross-sectional shape." *IEEE Trans. Microwave Theory Tech.* MTT-30: 6–12.

Peterson, A. F. 1989. "A comparison of integral, differential, and hybrid methods for TE-wave scattering from inhomogeneous dielectric cylinders." *J. Electromagnetic Waves Applications* 3: 87–106.

Peterson, A. F., and S. P. Castillo. 1989. "Differential equation methods for electromagnetic scattering from inhomogeneous cylinders." *IEEE Trans. Antennas Propagat.* AP-37(5): 601–07.

Peterson, A. F., and P. W. Klock. 1988. "An improved MFIE formulation for TE wave scattering from lossy, inhomogeneous dielectric cylinders." *IEEE Trans. Antennas Propagat.* AP-36: 45–49.

Raiche, A. P. 1974. "An integral equation approach to three-dimensional modeling." *Geophysics, J. Roy. Astr. Soc.* 36: 363–76.

Rao, S. M., D. R. Wilton, and A. W. Glisson. 1982. "Electromagnetic scattering by surfaces of arbitrary shape." *IEEE Trans. Antennas Propagat.* AP-30: 409–18.

Richmond, J. H. 1966. "TE-wave scattering from a dielectric cylinder of arbitrary cross section shape." *IEEE Trans. Antennas Propagat.* AP-14: 460–64.

Richmond, J. H. 1985. "Scattering by thin dielectric strips." *IEEE Trans. Antennas Propagat.* AP-33: 64–68.

Schaubert, D. H., D. R. Wilton, and A. W. Glisson. 1984. "A tetrahedral modeling method for electromagnetic scattering by arbitrarily shaped inhomogeneous dielectric bodies." *IEEE Trans. Antennas Propagat.* AP-32: 77–85.

Shen, C., K. J. Glover, M. I. Sancer, and A. D. Varvatsis. 1989. "The discrete Fourier transform method of solving differential-integral equations in scattering theory." AP37(8): 1032-41.

Shen, L. C., and J. A. Kong. 1984. "Scattering of electromagnetic waves from a randomly perturbed quasiperiodic surface." *J. Appl. Phys.* 56: 10–21.

Shen, L. C., and G. J. Zhang. 1985. "Electromagnetic field due to a magnetic dipole in a medium containing both planar and cylindrical boundaries." *IEEE Trans. Geosci. Remote Sensing* GE-23: 827–33.

Solbach, K., and I. Wolff. 1978. "The electromagnetic fields and phase constants of dielectric image lines." *IEEE Trans. Microwave Theory Tech.* MTT-26: 266–74.

Stodt, J. A., G. W. Hohmann, and S. C. Ting. 1981. "The telluric magnetotelluric method in two- and three-dimensional environments." *Geophysics* 46: 1137–47.

Su, C. C. 1986. "A surface integral equations method for homogeneous optical fibers and coupled image lines of arbitrary cross-sections." *IEEE Trans. Microwave Theory Tech.* MTT-34: 1140–46.

Tabarovsky, L. A. 1975. "Application of integral equations in problems of geoelectric." *Nauka SOAN SSSR*.

Tanaka, K., and M. Kojima. 1988. "Volume integral equations for analysis of dielectric branching waveguides." *IEEE Trans. Microwave Theory Tech.* MTT-36: 1239–45.

Toyoda, I., M. Matsuhara, and N. Kumagai. 1988. "Extended integral equation formulation for scattering problems from a cylindrical scatterer." *IEEE Trans. Antennas Propagat.* AP-36: 1580–86.

Tsai, C. T., H. Massoudi, C. H. Durney, and M. F. Iskander. 1986. "A procedure for calculating fields inside arbitrarily shaped, inhomogeneous dielectric bodies using linear basis functions with the moment method." *IEEE Trans. Microwave Theory Tech.* MTT-34: 1131–39.

van den Berg, P. M. 1981a. "Transition matrix in acoustic scattering by a strip." *J. Acoust. Soc. Am.* 70: 615–19.

van den Berg, P. M. 1981b. "Reflection by a grating: Rayleigh method," *J. Opt. Soc. Am.* 71(10): 1224–29.

van den Berg, P. M., and J. T. Fokkema. 1979. "The Rayleigh hypothesis in the theory of reflection by a grating." *J. Opt. Soc. Am.* 69(1): 27–31.

Waterman, P. C. 1965. "Matrix formulation of electromagnetic scattering." *Proc. IEEE* 53: 805–11.

Weaver, R. L., and Y. H. Pao. 1979. "Application of the transition matrix to a ribbon-shaped scatterer." *J. Acoust. Soc. Am.* 66: 1199–1206.

Whitman, G. M., and F. Schwering. 1977. "Scattering by periodic metal surfaces with sinusoidal height profiles — a theoretical approach." *IEEE Trans. Antennas Propagat.* AP-25: 869–76.

Wilkinson, J. H. 1965. *The Algebraic Eigenvalue Problem.* Oxford: Clarendon.

Wilton, D. R., S. M. Rao, A. W. Glisson, D. H. Schaubert, O. M. Al-Bundak, and C. M. Butler. 1984. "Potential integrals for uniform and linear source distributions on polygonal and polyhedral domains." *IEEE Trans. Antennas Propagat.* AP-32: 276–81.

Xu, X. B., and C. M. Butler. 1987. "Scattering of *TM* excitation by coupled and partially buried cylinders at the interface between two media." *IEEE Trans. Antennas Propagat.* AP-35(5): 529–38.

Yeh, C., S. B. Dong, and W. Oliver. 1975. "Arbitrarily shaped inhomogeneous optical fiber or integrated optical waveguides." *J. Appl. Phys.* 46: 2125–29.

Zare, R. N. 1988. *Angular Momentum—Understanding Spatial Aspects in Chemistry and Physics.* New York: John Wiley & Sons.

CHAPTER 9

INVERSE SCATTERING PROBLEMS

In inverse scattering, one attempts to infer the properties of the scatterer from the scattered field measured outside the scatterer. The ability to infer information on an object without direct contact also expands man's sensory horizon. No doubt, it is a much sought-after capability.

The science of reconstructing an object from some measurement data is known as tomography. Advancement in inverse scattering theory can clearly enhance the science of tomography. Consequently, inverse scattering has wide applications in medical tomography, nondestructive evaluation, target identification, geophysics, seismic exploration, optics, remote sensing, and atmospheric sciences. Unfortunately, the inverse scattering problem is also inherently nonunique. Therefore, a solution often has to be chosen from many possible different solutions in solving the inverse scattering problems. As an added complication, the scattered field is nonlinearly related to the scattering object. This nonlinear relationship exacerbates the difficulty of finding a closed-form solution to the inverse scattering problems.

A general way then to solve the nonlinear problem is via an iterative approach. Most inversion algorithms that find practical applications only solve the problem approximately, for example, by using an approximate linear relationship between the measurement data and the object. In one dimension, inversion algorithms using the Gel'fand-Levitan method and the Marchenko method can unravel the nonlinear relationship between the scatterer and the measurement data, but usually they have limited applications. Moreover, the simplicity of the Gel'fand-Levitan and Marchenko methods in one dimension is not preserved in higher dimensions. Consequently, in higher dimensions, a numerical method is usually needed to solve the nonlinear inverse problem. In this chapter, we shall first discuss the linear inverse scattering problems, and later, the nonlinear ones. First, the nonlinear inverse scattering methods in one dimension, e.g., the Gel'fand-Levitan-Marchenko method and the method of characteristics are presented. Later, nonlinear inverse scattering algorithms proven to work in higher dimensions will be presented.

§9.1 Linear Inverse Problems

In an inverse scattering problem, only field data outside the scatterer are available, for instance, through a measurement scheme shown in Figure 9.1.1. Then, the total field is related to the object via the volume integral equation

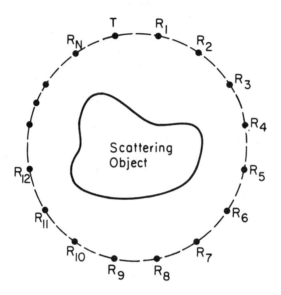

Figure 9.1.1 An example of an inverse scattering experiment.

(see Chapter 8, Section 8.9)

$$\mathbf{E}(\mathbf{r}) = \mathbf{E}_{inc}(\mathbf{r}) + \int_V d\mathbf{r}' \, \overline{\mathbf{G}}(\mathbf{r}, \mathbf{r}') \cdot O(\mathbf{r}')\mathbf{E}(\mathbf{r}'). \qquad (9.1.1)$$

The integral above corresponds to the scattered field and is the only term that contains information on the scatterer, which is described by $O(\mathbf{r}) = k^2(\mathbf{r}) - k_b^2$. It is apparent that the scattered field is a nonlinear functional of $O(\mathbf{r})$, because $\mathbf{E}(\mathbf{r})$ itself is also a functional of $O(\mathbf{r})$. Futhermore, this nonlinearity can be illustrated easily by performing a Neumann series expansion of (1). Then, the presence of the nonlinear terms is apparent in such a Neumann series expansion (see Exercise 9.1; also, see Subsection 9.3.3).

The nonlinear dependence of the scattered field on $O(\mathbf{r})$ is due to the mutual interactions between the induced polarization currents. Moreover, this is also a multiple scattering effect. To show this more clearly, we discuss the following experiment: If the scattered fields of two isolated scatterers are known, then obviously, the total scattered field when the two scatterers are adjacent to each other is not a linear superposition of the isolated scattered fields (see Exercise 9.2). In addition, we need to add a multiply scattered field (see Figure 9.1.2). In essence, the multiple scattering effect precludes the use of the principle of linear superposition to find the total scattered field from many objects. In other words, one cannot solve for the scattered fields

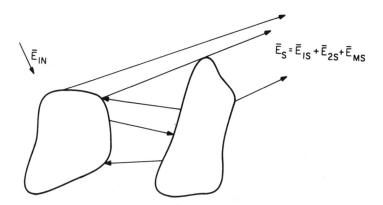

Figure 9.1.2 Multiple scattering between scatterers gives rise to non-linearity in the inverse problem which preclude the use of linear superpositions.

from the two objects independently first, and then, linearly superpose the two solutions later.

The nonlinear relationship of the scattered field to the object preempts the chance for a simple, exact solution of the inverse problem. But if the scattered field is approximated as a linear functional of the object, it vastly simplifies the inverse problem. Moreover, there are several conditions under which the problem could be "linearized." These conditions, for example, are those of the Born and the Rytov approximations. For instance, in the Born approximation, the scattered field amplitude is a linear functional of the object, whereas in the Rytov approximation, the phase perturbation is a linear functional of the object (see Chapter 8). Furthermore, another way of obtaining a linearized relationship between data and the object is to use high-frequency waves. High-frequency waves travel in a ray-like manner through a material. Consequently, with the ray-optics approximation on the wave, the phase delay and the amplitude attenuation of a field are linearly related to the object properties. But such a linear relationship between the phase and object properties is destroyed when multipath effects are important. Unfortunately, this is the case when the frequency of the wave is lower.

In a transient measurement, the high-frequency components usually arrive earlier than the low-frequency components. Then, the pulse-echo methods where the reflected pulse is properly time gated can be used to obtain the early signal arrivals. These early arrivals traverse the object in the shortest time, and hence, have less multipath and multiple scattering effects. Consequently, they are more linearly related to the object than the late arrivals.

In X-ray tomography, the attenuation of an X-ray in a ray path is a linear functional of the object property that it traverses. Furthermore, the ray paths are essentially straight lines due to the high energy of the photons. Moreover, the high momenta of the photons cause little multipath effects. In this case, the back-projection algorithm can be used to reconstruct the object with great success in X-ray tomography. Hence, we shall review the fundamentals of back-projection tomography. A further review of tomographic methods has been given by Mueller, Kaveh, and Wade (1979), and Kak (1979).[1]

§§9.1.1 Back-Projection Tomography

The back-projection algorithm is particularly useful when the measured phase or attenuation is a linear functional of the object, for instance, as in X-ray. The WKB approximation (see Chapter 2) says that at high frequencies, a wave propagates mainly in the forward direction with very little reflections— it is essentially a theory of small reflections. Moreover, the phase shift or attenuation of the wave as it propagates through an inhomogeneous medium at high frequencies is, to a very good approximation, governed by the factor

$$e^{i\omega \int_a^b s(z')\,dz'} \tag{9.1.2}$$

The slowness $s(z)$ is complex if the medium is lossy. In X-ray tomography, only the attenuation is measured, as the wavelength is too short for phase measurement. On the other hand, in ultrasound tomography, one measures the delay of a pulse through a body, which is related to the phase shift and the phase velocity of the object medium (see Exercise 9.3). This delay, using the physical picture of the WKB approximation, is given by

$$\tau = \int_a^b s(z')\,dz', \tag{9.1.3}$$

where $s(z)$ is the slowness of the wave. Moreover, in the above physical picture, the path of the ray through the medium is assumed to be a straight line. This is only true at high frequencies like in X-ray, or if the contrast of the inhomogeneity is small. At ultrasonic frequencies, this physical picture is less accurate; unlike X-ray tomography, ultrasonic tomography with the back-projection algorithm is fraught with more distortions.

If the object is described by its slowness or its attenuation profile $s(x,y)$, a single experiment then yields

$$P(y') = \int_{-\infty}^{\infty} s(x',y')\,dx', \tag{9.1.4}$$

---

[1] Also see Lee and Wade (1986).

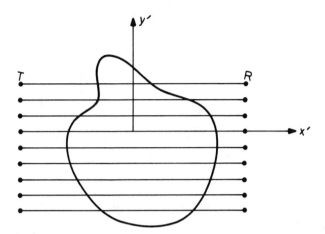

Figure 9.1.3 A projection tomography measurement scheme.

where (x', y') are the coordinates of the laboratory experiment, as shown in Figure 9.1.3. Hence, $P(y')$ is the projection of the function $s(x', y')$. It can be regarded as the shadow cast by $s(x', y')$ on an X-ray film when $s(x', y')$ is illuminated by rays pointing in the x' direction. Moreover, it is related to the Fourier transform of $s(x, y)$. To see this, we write

$$s(x', y') = \frac{1}{(2\pi)^2} \int\!\!\!\int_{-\infty}^{\infty} dk'_x dk'_y \, e^{ik'_x x' + ik'_y y'} S(k'_x, k'_y). \qquad (9.1.5)$$

Then,

$$P(y') = \frac{1}{2\pi} \int_{-\infty}^{\infty} dk'_y \, e^{ik'_y y'} S(0, k'_y). \qquad (9.1.6)$$

Hence, from a single projection, a slice of the Fourier transform of $s(x, y)$, i.e., $S(k'_x, k'_y)$ at $k'_x = 0$, is derivable by inverse Fourier transforming (6). This is also known as the projection-slice theorem (Herman 1980). Consequently,

$$S(0, k'_y) = \int_{-\infty}^{\infty} dy' \, e^{-ik'_y y'} P(y'). \qquad (9.1.7)$$

To get a different slice of the Fourier transform, we need only perform the experiment at a different angle. Therefore, by performing the experiment with angles ranging from 0° to 180°, $S(k_x, k_y)$ will be filled out in the whole

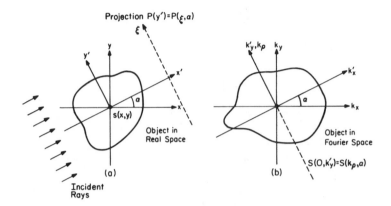

Figure 9.1.4 The object in real space and in Fourier space. A projection, $P(y') = \int s(x', y') dx'$ is related to $S(0, k_y')$, a slice in Fourier space.

Fourier space. Knowing $S(k_x, k_y)$, one can then reconstruct the object $s(x, y)$ (see Figure 9.1.4).

As mentioned previously, the straight-line approximation for an X-ray is an extremely good one, because X-ray photons have high frequencies and momenta and are hardly diffracted by inhomogeneities. But for ultrasound, some diffractions causing non-straight-line projection may ensue, giving rise to distortions in the reconstructed object. Hence, we shall discuss how the diffraction effect is accounted for in a later section on diffraction tomography. Next, we shall discuss Radon transforms, which are the mathematical backbones of back-projection tomography.

§§9.1.2 Radon Transforms

Radon transform was first described by Radon (1917), and was only used for tomographic reconstruction by Bracewell (1956) some forty years later. Radon transforms are very useful in back-projection tomography, because back-projection tomographic reconstruction is expressible as a radon transform. Therefore, we shall first study the two-dimensional Radon transforms and then the three-dimensional Radon transforms (Deans 1983).

(a) Two Dimensions

As shown previously, a slice of the Fourier transform of $s(x', y')$ is obtainable from the projection $P(y')$ via

$$S(0, k_y') = \int\limits_{-\infty}^{\infty} dy'\, e^{-ik_y'y'} P(y'). \qquad (9.1.8)$$

Then, if all the slices of the Fourier transforms are known at different angles, the two-dimensional Fourier transform $S(k_x, k_y)$ is retrievable to reconstruct $s(x, y)$ via

$$s(x, y) = \frac{1}{(2\pi)^2} \iint\limits_{-\infty}^{\infty} dk_x dk_y \, e^{ik_x x + ik_y y} S(k_x, k_y). \qquad (9.1.9)$$

But since the slices have symmetry in cylindrical coordinates, the above is more expediently written in cylindrical coordinates to give (Exercise 9.4)

$$s(\rho, \phi) = \frac{1}{(2\pi)^2} \int_0^\pi d\alpha \int_{-\infty}^{\infty} dk_\rho \, |k_\rho| e^{ik_\rho \rho \cos(\alpha - \phi)} S(k_\rho, \alpha). \qquad (9.1.10)$$

Then, from (8), $S(k_\rho, \alpha)$, for a fixed α, is just a slice of the Fourier transform retrievable from the projection via

$$S(k_\rho, \alpha) = \int_{-\infty}^{\infty} d\xi \, e^{-ik_\rho \xi} P(\xi, \alpha), \qquad (9.1.11)$$

where $P(\xi, \alpha) = P(y')$, the projection obtained for a given α.

Now, after substituting (11) into (10), performing the k_ρ integration first, and using integration by parts, we arrive at

$$s(\rho, \phi) = \frac{1}{2\pi^2} \int_0^\pi d\alpha \, P.V. \int_{-\infty}^{\infty} \frac{\left[\frac{\partial}{\partial \xi} P(\xi, \alpha)\right]}{\rho \cos(\alpha - \phi) - \xi} d\xi. \qquad (9.1.12)$$

To obtain the above, we have made use of the fact that (see Exercise 9.4)

$$P.V. \int_{-\infty}^{\infty} dy \, \frac{e^{-ixy}}{y} = -\pi i \, \text{sgn}(x). \qquad (9.1.13)$$

Moreover, from Fourier transforms, it then follows that

$$P.V. \frac{1}{y} = -\frac{i}{2} \int_{-\infty}^{\infty} dx \, \text{sgn}(x) e^{ixy}, \qquad (9.1.14)$$

and

$$\frac{\partial}{\partial y} P.V. \frac{1}{y} = \frac{1}{2} \int_{-\infty}^{\infty} dx \, |x| e^{ixy}, \qquad (9.1.15)$$

which can be used to obtain (12).

The projection $P(\xi, \alpha)$ can be written as a two-dimensional integral:

$$P(\xi, \alpha) = \iint\limits_{-\infty}^{\infty} dx\, dy\, \delta(\hat{\xi} \cdot \boldsymbol{\rho} - \xi) s(x, y), \qquad (9.1.16)$$

where $\boldsymbol{\rho} = \hat{x} x + \hat{y} y$ and $\hat{\xi}$ is a unit vector in the ξ direction (see Exercise 9.4). Furthermore, in cylindrical coordinates, the above becomes

$$P(\xi, \alpha) = \int\limits_{0}^{2\pi} d\phi \int\limits_{0}^{\infty} \rho\, d\rho\, \delta[\rho \sin(\phi - \alpha) - \xi] s(\rho, \phi). \qquad (9.1.17)$$

Consequently, Equations (17) and (12) together form the Radon-transform pair in two dimensions. They are essential in the two-dimensional back-projection algorithm. Note that Equation (12) implies that the object $s(\rho, \phi)$ can be reconstructed from the measurement $P(\xi, \alpha)$. Moreover, Equation (17) is equivalent to the experiment to obtain the projections of the object $s(\rho, \phi)$.

(b) Three Dimensions

In three dimensions, it is easy to prove in a similar manner that a two-dimensional projection of $s(x', y', z')$ is related to its three-dimensional transform as

$$P(z') = \iint\limits_{-\infty}^{\infty} dx'\, dy'\, s(x', y', z') = \frac{1}{2\pi} \int\limits_{-\infty}^{\infty} dk_z'\, e^{ik_z' z'} S(0, 0, k_z'). \qquad (9.1.18)$$

Hence,

$$S(0, 0, k_z') = \int\limits_{-\infty}^{\infty} dz'\, e^{-ik_z' z'} P(z'). \qquad (9.1.19)$$

Moreover, in spherical coordinates, we can write $s(r, \theta, \phi)$ as

$$s(r, \theta, \phi) = \frac{1}{(2\pi)^3} \int\limits_{0}^{\pi} d\beta \int\limits_{0}^{\pi} d\alpha\, \sin\alpha \int\limits_{-\infty}^{\infty} dk\, k^2 S(k, \beta, \alpha) e^{i\mathbf{k} \cdot \mathbf{r}}. \qquad (9.1.20)$$

In addition, analogous to the two-dimensional Radon transforms, (19) can be written as

$$S(k, \beta, \alpha) = \int\limits_{-\infty}^{\infty} d\xi\, e^{-ik\xi} P(\xi, \beta, \alpha). \qquad (9.1.21)$$

On substitution into (20), we now have

$$s(r, \theta, \phi) = \frac{1}{(2\pi)^3} \int\limits_0^\pi d\beta \int\limits_0^\pi d\alpha \sin\alpha \int\limits_{-\infty}^\infty d\xi \, P(\xi, \beta, \alpha) \int\limits_{-\infty}^\infty dk \, k^2 e^{ik(\hat{k}\cdot\mathbf{r} - \xi)}.$$

$$(9.1.22)$$

Note that in the above, $\hat{k}$ is a unit vector pointing in the direction of $\mathbf{k}$, and hence, is a function of α and β. Furthermore, the innermost integral is integrable as derivatives of a Dirac delta function. Consequently, the $d\xi$ integral can be performed in a closed form. Finally, we have (Exercise 9.5)

$$s(r, \theta, \phi) = \frac{-1}{(2\pi)^2} \int\limits_0^\pi d\beta \int\limits_0^\pi d\alpha \sin\alpha \frac{\partial^2}{\partial^2(\hat{k}\cdot\mathbf{r})} P\left(\hat{k}\cdot\mathbf{r}, \beta, \alpha\right). \qquad (9.1.23)$$

Since

$$P(\xi, \beta, \alpha) = \int d\mathbf{r}' \, \delta\left[\hat{\xi}\cdot\mathbf{r}' - \xi\right] s(\mathbf{r}'), \qquad (9.1.24)$$

(23) and (24) consequently form a Radon-transform pair in three dimensions.

As a final note, a general formula can be derived for inverse Radon transforms in even dimensions while another formula can be derived for inverse Radon transforms in odd dimensions (Exercise 9.5; also see Deans 1983).

§§9.1.3 Diffraction Tomography

In projection tomography, one assumes that waves propagate as straight-line rays with no diffractions. This is no longer true at longer wavelengths, where the diffraction phenomenon is important. Hence, reconstruction using straight-line rays yields distorted images. But this diffraction effect can be incorporated in the context of Born and Rytov approximations (see Chapter 8, Section 8.10) and a reconstruction algorithm developed accordingly. This is also known as diffraction tomography which was first developed by Devaney (1982, 1983).

Consider a transmitter-receiver configuration as shown in Figure 9.1.5. The scattered field using the first-order Born approximation is then[2]

$$\phi_{sca}(\boldsymbol{\rho}_R) = \int d\boldsymbol{\rho}' \, g(\boldsymbol{\rho}_R, \boldsymbol{\rho}') O(\boldsymbol{\rho}') \phi_{inc}(\boldsymbol{\rho}'), \qquad (9.1.25)$$

where two-dimensional scattering is assumed and $O(\boldsymbol{\rho}') = k^2 - k_0^2$ is the object to be reconstructed. Moreover, in two dimensions, the Green's function is [Equation (2.2.4)]

$$g(\boldsymbol{\rho}_R, \boldsymbol{\rho}') = \frac{i}{4} H_0^{(1)}(k_0|\boldsymbol{\rho}_R - \boldsymbol{\rho}'|). \qquad (9.1.26)$$

---

[2] This is obtained by applying the Born approximation [Subsection 8.10.1] to the second term in (8.9.5). For the Rytov approximation, (25) is proportional to the phase perturbation [see (8.10.26)].

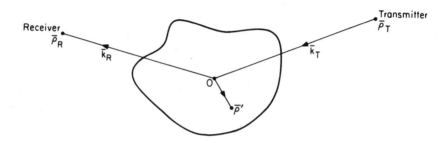

Figure 9.1.5 A transmitter-receiver measurement in diffraction tomography.

But if the receiver is in the far field of the scatterer, then approximately,

$$g(\boldsymbol{\rho}_R, \boldsymbol{\rho}') \simeq \frac{i}{4}\sqrt{\frac{2}{i\pi k_0 \rho_R}}\, e^{ik_0\rho_R - ik_0\hat{\rho}_R \cdot \boldsymbol{\rho}'}. \tag{9.1.27}$$

Also, an incident field generated by a uniform line source is

$$\phi_{inc}(\boldsymbol{\rho}') = \frac{i}{4}H_0^{(1)}(k_0|\boldsymbol{\rho}' - \boldsymbol{\rho}_T|). \tag{9.1.28}$$

And if the transmitter is also in the far field of the object, we have

$$\phi_{inc}(\boldsymbol{\rho}') \simeq \frac{i}{4}\sqrt{\frac{2}{i\pi k_0 \rho_T}}\, e^{ik_0\rho_T - ik_0\hat{\rho}_T \cdot \boldsymbol{\rho}'}. \tag{9.1.29}$$

Finally, after defining $\mathbf{k}_R = k_0\hat{\rho}_R$ and $\mathbf{k}_T = -k_0\hat{\rho}_T$, and substituting the above into (25), we have

$$\phi_{sca}(\boldsymbol{\rho}_R) \simeq \frac{i}{8\pi k_0\sqrt{\rho_T \rho_R}} e^{ik_0(\rho_T + \rho_R)} \int d\boldsymbol{\rho}' e^{-i(\mathbf{k}_R - \mathbf{k}_T)\cdot\boldsymbol{\rho}'} O(\boldsymbol{\rho}'). \tag{9.1.30}$$

Note that now, the integral is a Fourier-transform integral. Consequently,

$$\phi_{sca}(\boldsymbol{\rho}_R) \simeq \frac{i}{8\pi k_0\sqrt{\rho_T \rho_R}} e^{ik_0(\rho_T + \rho_R)} \tilde{O}(\mathbf{k}_R - \mathbf{k}_T), \tag{9.1.31}$$

where $\tilde{O}(\mathbf{k})$ is the Fourier transform of $O(\boldsymbol{\rho})$. Therefore, the scattered field under the Born approximation is related to the Fourier transform of the object.

Observe that the lengths of the vectors $\mathbf{k}_T$ and $\mathbf{k}_R$ are equal to k_0. Hence, $\mathbf{k}_R - \mathbf{k}_T$, or the argument of $\tilde{O}$, can only span a finite space in the Fourier space. For instance, if $\mathbf{k}_T$ is fixed and the receiver is moved around so that $\mathbf{k}_R$

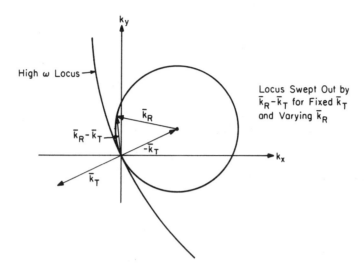

Figure 9.1.6 Locus swept out by $\mathbf{k}_R - \mathbf{k}_T$ for a fixed $\mathbf{k}_T$ and varying $\mathbf{k}_R$.

changes directions, then the locus swept out by $\mathbf{k}_R - \mathbf{k}_T$ is as shown in Figure 9.1.6. Furthermore, if the transmitter is moved around so that $\mathbf{k}_T$ changes direction as well as $\mathbf{k}_R$, then the combination of varying the directions of $\mathbf{k}_T$ and $\mathbf{k}_R$ sweeps out a larger circle of radius $2k_0$ with area $4\pi k_0^2$. Consequently, $\tilde{O}(\mathbf{k})$ is known over the area of this circle. The circle will be larger for larger k_0. Therefore, $O(\boldsymbol{\rho})$ can be reconstructed from $\tilde{O}(\mathbf{k})$ by performing an inverse Fourier transform. But since $\tilde{O}(\mathbf{k})$ is known only for $|\mathbf{k}| \leq k_0$, only a low-pass band-limited version of $O(\boldsymbol{\rho})$ is retrieved in this reconstruction.

From the above discussions, it is clear that another way of filling out the object in the Fourier space is to interrogate it with a broadband signal (see Exercise 9.6).

It is interesting to note that when the frequency becomes very high, the locus swept out by $\mathbf{k}_R - \mathbf{k}_T$ (for $\mathbf{k}_R$ and $\mathbf{k}_T$ to be approximately in the same direction) passes through the origin almost like a straight line. This is reminiscent of the projection-slice theorem, indicating that projection tomography is a special case of diffraction tomography. Hence, in the high-frequency limit, we need only perform forward scattering measurements with the receivers sweeping through small angles, and a straight-line slice in the Fourier space is recovered. This experiment can be performed as shown in Figure 9.1.7. Here, the transmitter is far from the object, so the receivers need only sweep out a small angle in the forward scattering direction. Note that the rays will almost be parallel, as in the case of projection tomography.

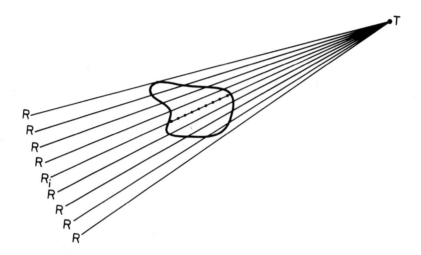

Figure 9.1.7 The limit when diffraction tomography becomes projection tomography.

Under the Born approximation in the high-frequency limit, all the scattering centers, indicated by dots on the object in Figure 9.1.7, will scatter in phase (coherently) at the receiver R_i. But all the other scattering centers will scatter incoherently to the receiver R_i. Hence, R_i measures the cumulative amplitudes of all the scattered waves from all the scattering centers aligned between T and R_i, or R_i measures a projection of the object.

Under the Rytov approximation, all the scattering centers are aligned to contribute cumulatively to the phase perturbation of the wave. But at lower frequencies, this physical picture does not hold anymore, as diffraction effect is more important.

§§9.1.4 Finite-Source Effect

In the previous subsection, we have derived the equations for diffraction tomography using a far-field approximation. The far-field approximation is a limiting one, since for it to be valid,

$$\rho_T \gg \rho', \qquad \rho_R \gg \rho', \tag{9.1.32}$$

or the transmitter and the receiver have to be far apart compared to the size of the scatterer. This is not practical when the scatterers are large or the background medium is lossy.

In fact, the source and the receiver effects can be treated when they are finite in size. As before, we shall consider a point-source transmitter, but without the far-field approximations. Therefore, the incident field or the source field is given by Equation (28). Then, using the plane-wave representation for the Hankel function [Chapter 2, Equation (2.2.11)], the incident field becomes

$$\phi_{inc}(\boldsymbol{\rho}') = \frac{i}{4\pi} \int\limits_{-\infty}^{\infty} dk_x \frac{1}{k_y} e^{ik_x(x'-x_T)+ik_y|y'-y_T|}. \qquad (9.1.33)$$

Similarly, the plane-wave representation for the Green's function is

$$g(\boldsymbol{\rho}_R, \boldsymbol{\rho}') = \frac{i}{4\pi} \int\limits_{-\infty}^{\infty} dk_x' \frac{1}{k_y'} e^{ik_x'(x_R-x')+ik_y'|y_R-y'|}. \qquad (9.1.34)$$

Consequently, we can substitute (33) and (34) into (25) to obtain

$$\phi_{sca}(\boldsymbol{\rho}_R, \boldsymbol{\rho}_T) = \frac{-1}{16\pi^2} \int\limits_{-\infty}^{\infty} \frac{dk_x}{k_y} \int\limits_{-\infty}^{\infty} \frac{dk_x'}{k_y'} e^{ik_x'x_R-ik_xx_T}$$

$$\int dx'dy' \, e^{i(k_x-k_x')x'+ik_y|y'-y_T|+ik_y'|y_R-y'|}O(x',y'). \qquad (9.1.35)$$

Note that the scattered field is a function of both the transmitter and the receiver positions. But in forward scattering experiments, $y' > y_T$ and $y_R > y'$ so that the modulus signs in (35) can be removed to arrive at

$$\phi_{sca}(\boldsymbol{\rho}_R, \boldsymbol{\rho}_T) = \frac{-1}{16\pi^2} \int\limits_{-\infty}^{\infty} \frac{dk_x}{k_y} \int\limits_{-\infty}^{\infty} \frac{dk_x'}{k_y'} e^{ik_x'x_R+ik_y'y_R-ik_xx_T-ik_yy_T}$$

$$\cdot \int dx'dy' \, e^{i(k_x-k_x')x'+i(k_y-k_y')y'}O(x',y'). \qquad (9.1.36)$$

Now, if $\phi_{sca}(\boldsymbol{\rho}_R, \boldsymbol{\rho}_T)$ is measured along a line in the x direction, with the transmitter also aligned in the x direction as shown in Figure 9.1.8, we can transform $\phi_{sca}(\boldsymbol{\rho}_R, \boldsymbol{\rho}_T)$ in the x_R and x_T variables to obtain

$$\phi_{sca}(k_x', y_R, -k_x, y_T) = \frac{-1}{4} \frac{e^{ik_y'y_R-ik_yy_T}}{k_yk_y'} \tilde{O}(-k_x + k_x', -k_y + k_y'). \qquad (9.1.37)$$

In this case, the Fourier transform of the measured field is related to the Fourier transform of the object $\tilde{O}$.

From Equation (37), note that one cannot make much use of the evanescent spectrum corresponding to the case when k_y' and k_y are purely imaginary. This happens when $k_x' > k_0$ and $k_x > k_0$, as seen from the dispersion relationships $k_x'^2 + k_y'^2 = k_0^2$ and $k_x^2 + k_y^2 = k_0$. Therefore, it is reasonable to assume

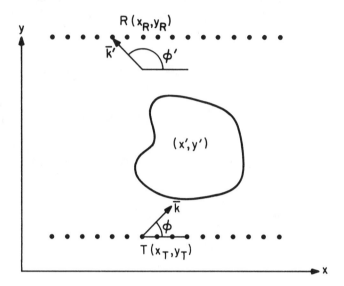

Figure 9.1.8 The finite-source effect in diffraction tomography.

that the directions of **k** and **k**$'$ only sweep from $0°$ to $180°$. Since $\tilde{O}$ is a function of **k**$' - $**k**, the locus swept out by **k**$' - $**k** with varying **k**$'$ for a fixed **k** is as shown in Figure 9.1.9. But if the directions of **k** and **k**$'$ vary from $0°$ to $180°$, the area swept out by **k**$' - $**k** consists of two disks as shown in Figure 9.1.9 (see Exercise 9.7). Therefore, a forward scattering experiment alone is not enough to reconstruct the object well since the data in the spectral domain is not complete. Nevertheless, a band-limited reconstruction is possible.

On the other hand, if a backscattering experiment is performed instead, only the signs of k'_y in Equation (37) need to be changed. In this case, the area swept out by **k**$' - $**k** includes a semicircle as well in the lower half $k_x k_y$ plane (Exercise 9.7). To fill out a full circle, the experimental setting is rotated so that the two disks sweep out a circle of radius $2k_0$. Alternatively, the transmitters and receivers can be switched to sweep out a full circle on the $k_x k_y$ plane. In this manner, more Fourier data can be collected in the Fourier space.

§§9.1.5 Nonuniqueness of the Solution

The inverse scattering problem is inherently nonunique (see Bleistein and Cohen 1977; Devaney 1978; Devaney and Sherman 1982). A hint of this nonuniqueness is seen in diffraction tomography, where the reconstructed object is a low-pass filtered version of the object. Therefore, in the diffraction tomography algorithm, if two different objects have different high-frequency

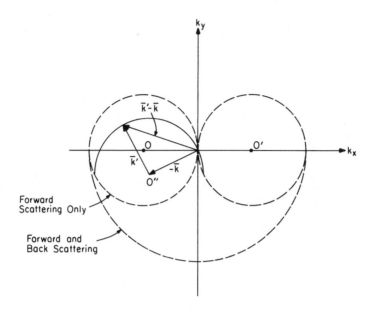

Figure 9.1.9 Locus swept out in the **k**-space in a forward and backward scattering experiment only.

components outside the range of the low-pass filter, but the same spectral components within the range of the low-pass filter, the reconstructed objects would be indistinguishable. As the frequency increases, however, the domain of this low-pass filter becomes larger, and the problem of nonuniqueness becomes less severe. But in general, the tomographic algorithms described so far cannot reconstruct features of the object that are much less than a wavelength.

An inverse scattering problem is nonunique even outside the linear regime. Consider the scattering by two objects which differ only by a small notch on the surface. When the interrogating field has wavelengths larger than the notch dimension, the difference in the scattered field is only in some localized region around the notch. (The reason is that the notch gives rise to evanescent waves which are highly localized.) Since the two objects yield essentially the same scattered field at some distance away from the objects, the reconstruction of the object is clearly nonunique.

Because the high-frequency spectrum of the object is buried in the evanescent spectrum of the scattered field, the object will be better reconstructed if the evanescent spectrum is utilized as well. This is actually done in some measurement scheme like ultrasonic microscopy.

§9.2 One-Dimensional Inverse Problems

As we have discussed previously, an inverse scattering problem is inherently nonlinear. Moreover, the nonlinearity is a result of multiple scattering within an object. The nonlinear inverse scattering problems do not have closed-form solutions in general. But in one dimension, the problem has been thoroughly studied by a large number of scientists. Many of these algorithms are layer-stripping algorithms, in which the reconstruction is done in layers. In addition, they also rely on the propagation of singularities, which is only possible for hyperbolic type partial differential equations. Unfortunately, many of the one-dimensional techniques are not easily generalized to higher dimensions. Nonetheless, they are mathematically elegant. Hence, we shall study several of these one-dimensional profile reconstruction techniques in this section (also see Habashy and Mittra 1987).

§§9.2.1 The Method of Characteristics

The method of characteristics is a time domain method for profile reconstruction (Bube and Burridge 1983; Sezginer 1985). It relies on the causality of a physical signal to successively strip off the layers of a profile for reconstruction. Hence, the method of characteristics can be viewed as a layer-stripping algorithm. Furthermore, it also exploits the propagation of singularities in a wave equation, as we shall see.

The *characteristics* of a partial differential equation (PDE) are lines along which the singularity or the discontinuity of a solution propagates (Sommerfeld 1949; Garabedian 1964). Consider a general PDE

$$A\phi_{zz} + 2B\phi_{zt} + C\phi_{tt} = \Phi(\phi, \phi_t, \phi_z, z, t), \qquad (9.2.1)$$

where A, B, and C may be functions of z and t.[3] (In the above, a subscript variable denotes a partial derivative with respect to that variable.) If the above equation has a characteristic curve along which the discontinuity of the solution to (1) propagates, then in the vicinity of (z_0, t_0), the curve is approximately

$$z = v(z_0, t_0)t + a(z_0, t_0), \qquad (9.2.2)$$

i.e., a straight line passing through (z_0, t_0) as shown in Figure 9.2.1. If $\phi(z, t)$ has discontinuity, the solution to (1) in the vicinity of (z_0, t_0) is of the form

$$\phi(z, t) \simeq u(z - vt - a), \quad z \simeq vt + a, \qquad (9.2.3)$$

where $u(z)$ is a unit-step function. Then, using (3) in (1), and keeping only the most singular terms, we have

$$\phi_{zz} \sim u''(z - vt - a), \quad \phi_{zt} \sim -v\, u''(z - vt - a), \quad \phi_{tt} \sim v^2 u''(z - vt - a). \qquad (9.2.4)$$

---

[3] When A, B, and C are functions of ϕ, ϕ_t, and ϕ_z as well, the PDE is termed quasilinear.

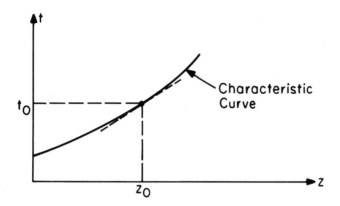

Figure 9.2.1 A sample characteristic curve of a partial differential equation.

These most singular terms must cancel each other in (1) in order for the equation to be satisfied since the right-hand side of (1) is regular. Consequently (Exercise 9.8),

$$A - 2Bv + Cv^2 = 0. \tag{9.2.5}$$

From the above,

$$v = \frac{B \pm \sqrt{B^2 - AC}}{C}. \tag{9.2.6}$$

Note that condition (5) is imposed only in the vicinity of (z_0, t_0). Hence, A, B, and C are functions of (z_0, t_0) only.

In the above, v will be real-valued if $B^2 \geq AC$, and complex-valued if $B^2 < AC$. Hence, the characteristic curve only exists if $B^2 \geq AC$. Accordingly, the PDE is classified depending on the ratio of B^2 to AC:

(i) A PDE is **hyperbolic** if $B^2 > AC$. There are two possible values of v for every (z_0, t_0). Namely, there are two characteristic curves that pass through every (z_0, t_0). The wave equation is an example of a hyperbolic equation (Exercise 9.8).

(ii) A PDE is **parabolic** if $B^2 = AC$. Then, v only has one value for every (z_0, t_0). Only one characteristic curve passes through (z_0, t_0). The diffusion equation and the Schrödinger equation are examples of parabolic equations.

(iii) A PDE is **elliptic** if $B^2 < AC$. Then, v is complex and the discontinuity described by (3) does not exist. In this case, a characteristic curve along which a singularity would propagate is absent. Laplace's equation (with x replacing t) is an example of an elliptic equation.

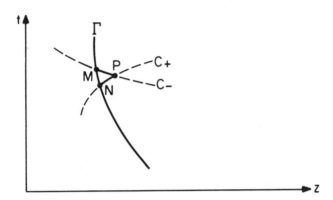

Figure 9.2.2 The use of the method of characteristics to find the field at P given Cauchy data at M and N on Γ.

In the hyperbolic type PDE's, the singularities propagate along the characteristics. But for the elliptic type PDE's, there are no characteristics at all. The parabolic type PDE's are in between these two cases. They have characteristic curves, but it can be shown that solutions of parabolic type PDE's are always smooth and analytic, as is the case for the solutions of diffusion equations. Hence, the magnitudes of the discontinuities in a parabolic equation are always zero.

If a characteristic curve exists for a PDE, it can be shown easily that (Exercise 9.8)

$$Av^{-1}\frac{d\phi_z}{dt} + Cv\frac{d\phi_t}{dz} = A\phi_{zz} + 2B\phi_{zt} + C\phi_{tt} \qquad (9.2.7)$$

along the characteristic curve where $v = dz/dt$. Hence, using (7) in (1), it becomes

$$Ad\phi_z + Cvd\phi_t = \Phi dz \qquad (9.2.8)$$

along a characteristic curve. Consequently, if ϕ and its normal derivative $\partial\phi/\partial n$ are specified on a curve Γ (also known as Cauchy data) where Γ is not a characteristic curve, then (8) can be used to find ϕ, ϕ_z, and ϕ_t everywhere via a finite-difference scheme. To see this, note that when ϕ is known on Γ, so is $\partial\phi/\partial s$, where s is along the direction of the Γ curve. From $\partial\phi/\partial n$ and $\partial\phi/\partial s$, ϕ_z and ϕ_t are derivable on Γ. Hence, Φ is also known on Γ, and (8) can then be used to solve for ϕ, ϕ_z, and ϕ_t in the neighborhood of Γ.

For a hyperbolic type PDE, there are two possible values of $v = v_\pm$ from (6). To find ϕ_z and ϕ_t at P in Figure 9.2.2, which is at a small distance from Γ, we first identify two characteristics passing through P. Then, using a finite-difference scheme, the two unknowns, ϕ_z and ϕ_t, are derivable at P from the two equations:

$$A \, d\phi_z + C v_+ \, d\phi_t = \Phi dz, \tag{9.2.9a}$$

$$A \, d\phi_z + C v_- \, d\phi_t = \Phi dz, \tag{9.2.9b}$$

which follow from (8). Furthermore, since

$$d\phi = \phi_z \, dz + \phi_t \, dt \tag{9.2.10}$$

and that ϕ is known on Γ, it is straightforward to integrate (10) along C_+ or C_-, the two characteristic curves shown in Figure 9.2.2, to obtain ϕ at P. In this manner, the value of ϕ, ϕ_z, and ϕ_t are found in the neighborhood of Γ. Moreover, this process can be continued until ϕ, ϕ_z, and ϕ_t are known everywhere. As a result, the problem is then solved like an initial value problem; it is also known as the *Cauchy initial value problem* (Exercise 9.9; also see Ames 1977).

The above illustrates the use of the method of characteristics to find the forward solution of a problem. More importantly, it can easily be adapted to solve the profile inversion problem. To illustrate this, consider the equation governing a TE plane wave at oblique incidence to a conductive, one-dimensional profile, which is[4]

$$\mu(\mu^{-1}\phi_z)_z + \left[\frac{\omega^2}{c^2} - k_x^2 + i\omega\mu\sigma \right] \phi = 0 \tag{9.2.11}$$

in the frequency domain, where $c = 1/\sqrt{\mu\epsilon}$. By letting $k_x = \frac{\omega}{c}\sin\theta$, the above becomes

$$\phi_{zz} - (\ln\mu)_z \phi_z + \left[\frac{\omega^2}{c^2}\cos^2\theta + i\omega\mu\sigma \right] \phi = 0, \tag{9.2.12}$$

where we have written $\mu(\mu^{-1}\phi_z)_z$ as $\phi_{zz} - (\ln\mu)_z\phi_z$. To apply the method of characteristics, however, we need the time-domain equivalence of the above equation, which is (see Exercise 9.10)

$$\phi_{zz} - (\ln\mu)_z \phi_z - \frac{1}{c_e^2}\phi_{tt} - \mu\sigma\phi_t = 0, \tag{9.2.13}$$

where $c_e = c/\cos\theta$ is the effective phase velocity for an obliquely incident plane wave.

Equation (13) is a hyperbolic type PDE for TE waves.[5] From (6), the characteristics of the above equation are given by (Exercise 9.8b)

$$\frac{dz}{dt} = \pm c_e. \tag{9.2.14}$$

---

[4] See Chapter 2, Equation (2.1.8a), where ϵ has been replaced by a complex $\tilde{\epsilon} = \epsilon + i\frac{\sigma}{\omega}$.

[5] For a lossless TM wave case, the equation is similar to (13), but when the medium has conductive loss, the equation is a lot more complicated than (13).

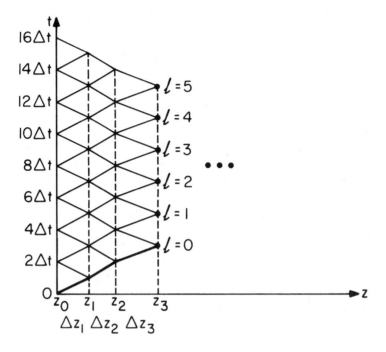

Figure 9.2.3 The finite-difference grid for the method of characteristics.

It is along these characteristics that the singularities of (13) would propagate.
In particular, along the characteristics $dz/dt = c_e$, the discontinuity $u(z -
c_e t)$ propagates in $+z$ direction forward in time, but along $dz/dt = -c_e$,
the discontinuity $u(z + c_e t)$ propagates in $+z$ direction backward in time.
Alternatively, one might say that the discontinuity propagates forward in
time, but in the $-z$ direction.

Along the characteristics, in accordance with (8), (13) becomes

$$d\phi_z - \frac{1}{c_e} d\phi_t - \Phi(z,t)\, dz = 0, \quad \text{on } C_+, \qquad (9.2.15a)$$

$$d\phi_z + \frac{1}{c_e} d\phi_t - \Phi(z,t)\, dz = 0, \quad \text{on } C_-, \qquad (9.2.15b)$$

where C_+ are characteristics with positive slopes, while C_- are characteristics
with negative slopes, and $\Phi(z,t) = (\ln \mu)_z \phi_z + \mu \sigma \phi_t$. Consequently, Equations
(15a) and (15b) can be used to find the field everywhere given the Cauchy
data on Γ. For example, on the $z = 0$ line, if the Cauchy data (ϕ and its
normal derivative ϕ_z) are available, then (15) can be used to calculate ϕ
everywhere if c_e, μ, and σ are known. Consequently, Equations (15a) and
(15b) can be discretized along the characteristics by letting

$$\phi_m^l = \phi[z_m, (m + 2l)\Delta t], \quad c_{em} = c_e(z_m), \qquad (9.2.16a)$$

$$\Delta z_m = z_{m+1} - z_m = c_{em}\,\Delta t. \tag{9.2.16b}$$

Here, l denotes the l-th positive-slope characteristic starting from $t = 0$, and ϕ_m^l is a point at z_m and on the l-th characteristic (see Figure 9.2.3). Furthermore, by discretizing (15a) and (15b) along the characteristics using a forward-differencing scheme, we have

$$\phi_{z,m+1}^l - \phi_{z,m}^l - \frac{1}{c_{em}}\left(\phi_{t,m+1}^l - \phi_{t,m}^l\right) - \Phi_m^l \Delta z_m = 0, \quad \text{on } C_+,$$
$$\tag{9.2.17a}$$

$$\phi_{z,m+1}^l - \phi_{z,m}^{l+1} + \frac{1}{c_{em}}\left(\phi_{t,m+1}^l - \phi_{t,m}^{l+1}\right) - \Phi_m^{l+1} \Delta z_m = 0, \quad \text{on } C_-.$$
$$\tag{9.2.17b}$$

Equations (17a) and (17b) can then be used to propagate ϕ_t and ϕ_z at $z = 0$ to the region where $z > 0$.

If μ and σ are known, but c_e is unknown, the method of characteristics can be used to yield c_e by invoking causality. Since ϕ, ϕ_z, and ϕ_t are zero for $t < 0$, then ϕ, ϕ_z, and ϕ_t are also zero for the region below the $l = 0$ characteristic. In other words, the wavefront travels along the $l = 0$ characteristic. Therefore,

$$\phi_{z,m}^0 = \phi_{t,m}^0 = 0, \quad \forall\, m. \tag{9.2.18}$$

The subsequent use of (18) in (17a) implies that the left-hand side of (17a) equals zero. Moreover, using (18) in (17b) yields the condition that

$$-\phi_{z,m}^1 - \frac{1}{c_{em}}\phi_{t,m}^1 - \Phi_m^1 c_{em}\,\Delta t = 0, \quad \forall\, m, \tag{9.2.19}$$

where we have replaced Δz_m with $c_{em}\Delta t$. The above yields a quadratic equation for c_{em} in terms of $\phi_{z,m}^1$, $\phi_{t,m}^1$, and Φ_m^1, where

$$\Phi_m^1 = (\ln\mu)_{z,m}\phi_{z,m}^1 + \mu_m\sigma_m\phi_{t,m}^1.$$

Hence, (19) becomes

$$-\phi_{z,m}^1 - \frac{1}{c_{em}}\phi_{t,m}^1 - [(\ln\mu)_{z,m}\phi_{z,m}^1 + \mu_m\sigma_m\phi_{t,m}^1]c_{em}\,\Delta t = 0, \quad \forall\, m.$$
$$\tag{9.2.20}$$

The above equation is first used to solve for c_{em} with $m = 0$ since $\phi_{z,0}^l$ and $\phi_{t,0}^l$ are known. Then, the value of c_{eo} found can be used to propagate the Cauchy data from $m = 0$ to $m = 1$, and the cycle is repeated again. Moreover, if $\mu =$ constant and $\sigma = 0$, the above could be simplified to yield $c_{em} = -\phi_{t,m}^1/\phi_{z,m}^1$.

On the other hand, if c_{em} and σ_m are known, a difference equation can be set up to yield μ. Moreover, if c_{em}, μ_m, and $(\ln\mu)_{z,m}$ are known, then σ_m could be found. But still, the above equation does not lend itself easily to a simultaneous reconstruction of c_{em} and σ_m. To perform a simultaneous reconstruction, first note that $c_e = c/\cos\theta$. Then, by choosing plane waves with different θ's, more than one equation like (20) can be derived for every m. This then allows the simultaneous reconstruction of c and σ. But it is usually harder to reconstruct σ than c because at early time (i.e., along the $l = 0$ characteristic), the last term on the left-hand side of (13), $\mu\sigma\phi_t$, is subdominant. In other words, the physics of the wavefront propagation (or the characteristic curve) is controlled by the ϕ_{zz} and ϕ_{tt} terms in (13). The method of characteristics exploits information buried in this wavefront which is little affected by the $\mu\sigma\phi_t$ term. As a last note, the above can easily be generalized to wave equations in other coordinate systems (see Exercise 9.11).

Many other algorithms using similar layer-stripping ideas have been explored in the literature. For instance, when the second-order PDE is converted to two first-order systems, and the propagation of singularity and causality are imposed along the characteristics, the method is known as the Schur or the Cholesky algorithm (see Levy 1985). An inhomogeneous profile, in addition, can also be approximated by a finely layered profile with equal travel time and the subsurface reflections modeled as a time series. Then, by invoking causality, the properties of the fine layers can be unraveled using the Goupillaud algorithm (see Exercise 9.12; also see Goupillaud 1961; Aki and Richards 1980).

§§9.2.2 Transformation to a Schrödinger-like Equation

Much of the work on inverse scattering theory is based upon the Schrödinger equation of quantum scattering. Therefore, it is expedient to relate one-dimensional wave equations to a Schrödinger-like equation on which a wealth of literature on inverse scattering has been written. For instance, with the Schrödinger-like equation, the linear Gel'fand-Levitan and Marchenko integral equations are derivable for inverse scattering.

A Schrödinger-like equation in one dimension, after normalization, has the basic form[6]

$$\frac{\partial^2}{\partial t^2}\phi(\zeta, t) - \frac{\partial^2}{\partial\zeta^2}\phi(\zeta, t) + V(\zeta)\phi(\zeta, t) = 0, \qquad (9.2.21)$$

where $V(\zeta)$ is a potential. Moreover, in the frequency domain, this reduces to

$$\frac{d^2}{d\zeta^2}\phi(\zeta) + [\omega^2 - V(\zeta)]\phi(\zeta) = 0. \qquad (9.2.22)$$

It turns out that most one-dimensional wave equations can be transformed to the form (22) via the Liouville transformation (see Ware and Aki 1969; Berryman and Greene 1980).

To see this, consider a one-dimensional wave equation of the form

$$\epsilon\frac{d}{dz}\epsilon^{-1}\frac{d}{dz}\phi(z) + \omega^2\mu\epsilon\phi(z) = 0. \qquad (9.2.23)$$

---

[6] A Schrödinger equation has a first derivative in time.

The preceding equation governs the propagation of TM electromagnetic plane waves through a layered medium. [For TE waves, the equation is of a similar form (see Chapter 2).] Now, by letting

$$dz = v\,d\zeta, \qquad \zeta = \int^{z} v^{-1}(z')\,dz', \tag{9.2.24}$$

where $v = 1/\sqrt{\mu\epsilon}$, Equation (23) becomes

$$\left[\epsilon v^{-1}\frac{d}{d\zeta}\epsilon^{-1}v^{-1}\frac{d}{d\zeta} + \frac{\omega^2}{v^2}\right]\phi(z) = 0. \tag{9.2.25}$$

This then reduces to

$$\left[\frac{d}{d\zeta}\eta\frac{d}{d\zeta} + \omega^2\eta\right]\phi(z) = 0, \tag{9.2.26}$$

where $\eta = \sqrt{\frac{\mu}{\epsilon}}$. Furthermore, by letting $\psi = \eta^{\frac{1}{2}}\phi$, we have

$$\eta\frac{d\phi}{d\zeta} = \left(\eta\frac{d}{d\zeta}\eta^{-\frac{1}{2}}\right)\psi + \eta^{\frac{1}{2}}\frac{d\psi}{d\zeta}, \tag{9.2.27a}$$

$$\begin{aligned}\frac{d}{d\zeta}\eta\frac{d\phi}{d\zeta} &= \left(\frac{d}{d\zeta}\eta\frac{d}{d\zeta}\eta^{-\frac{1}{2}}\right)\psi + \left(\eta\frac{d}{d\zeta}\eta^{-\frac{1}{2}}\right)\frac{d\psi}{d\zeta}\\ &\quad + \left(\frac{d\eta^{\frac{1}{2}}}{d\zeta}\right)\frac{d\psi}{d\zeta} + \eta^{\frac{1}{2}}\frac{d^2\psi}{d\zeta^2}.\end{aligned} \tag{9.2.27b}$$

It can be shown that the second term and the third term in the above cancel each other. Therefore, (26) becomes

$$\eta^{\frac{1}{2}}\frac{d^2\psi}{d\zeta^2} + \left(\frac{d}{d\zeta}\eta\frac{d}{d\zeta}\eta^{-\frac{1}{2}}\right)\psi + \omega^2\eta^{\frac{1}{2}}\psi = 0. \tag{9.2.28}$$

But since

$$\frac{d}{d\zeta}\eta\frac{d}{d\zeta}\eta^{-\frac{1}{2}} = -\frac{d^2}{d\zeta^2}\left(\eta^{\frac{1}{2}}\right), \tag{9.2.29}$$

Equation (28) becomes

$$\frac{d^2\psi(\zeta)}{d\zeta^2} + [\omega^2 - V(\zeta)]\psi(\zeta) = 0, \tag{9.2.30}$$

where

$$V(\zeta) = \frac{1}{\eta^{\frac{1}{2}}}\frac{d^2}{d\zeta^2}\eta^{\frac{1}{2}}. \tag{9.2.31}$$

For a dispersionless medium, $V(\zeta)$ is independent of frequency and (30) becomes (21) in the time domain (also see Exercise 9.13).

If an inversion algorithm is available to reconstruct $V(\zeta)$ in (30), then from (31), $\eta^{\frac{1}{2}}$ can be retrieved by solving [7]

$$\frac{d^2}{d\zeta^2}\eta^{\frac{1}{2}}(\zeta) - V(\zeta)\eta^{\frac{1}{2}}(\zeta) = 0. \qquad (9.2.32)$$

Once $\eta(\zeta) = \sqrt{\frac{\mu}{\epsilon}}$ is found, we can find $v(\zeta) = 1/\sqrt{\mu\epsilon}$ provided that either μ or ϵ is known. Subsequently, a relationship between z and ζ can be found such that

$$z = \int_0^\zeta v(\zeta')\, d\zeta'. \qquad (9.2.33)$$

In the above, we assume $\zeta = 0$ maps to $z = 0$. Once z is known as a function of ζ, $\eta(z)$ and $v(z)$ can be found quite easily.

Now, a number of techniques can be used to solve the inverse problem corresponding to Equation (21), including the method of characteristics discussed previously. Next, we shall discuss how linear integral equations for the inverse problem can be derived for the one-dimensional Schrödinger-like equation corresponding to (21).

§§9.2.3 The Gel'fand-Levitan Integral Equation

Now, we shall present the derivation of the Gel'fand-Levitan integral equation in the time domain using a method developed by Burridge (1980). Consider the following equation:

$$\frac{\partial^2\phi(\zeta,t)}{\partial t^2} - \frac{\partial^2\phi(\zeta,t)}{\partial\zeta^2} + V(\zeta)\phi(\zeta,t) = 0, \qquad (9.2.34)$$

with the boundary condition

$$-\frac{\partial}{\partial\zeta}\phi(0,t) + h\phi(0,t) = \delta'(t). \qquad (9.2.35)$$

The corresponding boundary condition in the frequency domain is then

$$-\frac{\partial}{\partial\zeta}\phi(0,\omega) + h\phi(0,\omega) = -i\omega. \qquad (9.2.36)$$

In light of Figure 9.2.4, the boundary condition (35) or (36) is equivalent to having a sheet source at $\zeta = 0$. Moreover, this sheet source is backed by an impedance boundary with the impedance boundary condition (see Exercise 9.14)[8]

$$-\frac{\partial}{\partial\zeta}\phi(0,t) + h\phi(0,t) = 0. \qquad (9.2.37)$$

---

[7] Equation (94) later offers an alternative method of finding $\eta(\zeta)$ in the Gel'fand-Levitan-Marchenko method.

[8] Note that the impedance boundary condition, together with the radiation condition when $\zeta \to \infty$, uniquely determines the solution (see Chapter 1).

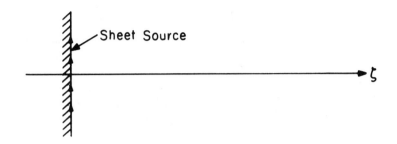

Figure 9.2.4 The one-dimensional inverse problem that can be solved by the Gel'fand-Levitan method.

In general, the solution to (34) is

$$\phi(\zeta, t) = \delta(t - \zeta) - K_1(\zeta, t). \tag{9.2.38}$$

In other words, the impulsive source at $\zeta = 0$ generates a delta function pulse that propagates through the medium. Note that if $V(\zeta) = 0$ in Equation (34) and $h = 0$ in Equation (35), then

$$\phi(\zeta, t) = \delta(t - \zeta) \tag{9.2.39}$$

satisfies (34) with the boundary condition (35). Hence, the mere presence of $V(\zeta)$ and h results in the extra term $K_1(\zeta, t)$ in (38). Namely, $K_1(\zeta, t)$ is the scattered field due to the inhomogeneous potential $V(\zeta)$ and the "impedance" h in the presence of the incident field (or primary field) given by (39).

By substituting (38) into (34) and (35), we can easily show that

$$\frac{\partial^2}{\partial t^2} K_1(\zeta, t) - \frac{\partial^2}{\partial \zeta^2} K_1(\zeta, t) + V(\zeta) K_1(\zeta, t) - V(\zeta) \delta(t - \zeta) = 0, \tag{9.2.40}$$

and

$$-\frac{\partial}{\partial \zeta} K_1(\zeta, t) \Big|_{\zeta=0} + h K_1(0, t) - \delta(t) h = 0. \tag{9.2.41}$$

Furthermore, in the vicinity of $t = \zeta$, the most singular terms in (40) should cancel each other. Then, by equating the most singular terms in (40), we have

$$\frac{\partial^2}{\partial t^2} K_1(\zeta, t) - \frac{\partial^2}{\partial \zeta^2} K_1(\zeta, t) = V(\zeta) \delta(t - \zeta) \tag{9.2.42}$$

when $t \simeq \zeta$.

The above is reminiscent of the Born approximation[9]: the scattered field $K_1(\zeta, t)$ is due to the induced source caused by $\delta(t - z)$ impinging upon

---

[9] Unlike the Born approximation for wave equations discussed in Chapter 8, the approximation in (42) is increasingly good when $t \to \zeta$ irrespective of the size of $V(\zeta)$.

$V(\zeta)$. Furthermore, Equation (42) also implies that $K_1(\zeta, t)$ is a discontinuous function when $t \simeq \zeta$. Therefore, let us assume that

$$K_1(\zeta, t) = f(\zeta, t) u(t - \zeta), \quad t \simeq \zeta, \qquad (9.2.43)$$

where $f(\zeta, t)$ is a smooth function while $u(t-\zeta)$ is the Heaviside step function. Then, using (43) in (42), we find that

$$\frac{\partial^2}{\partial t^2} K_1(\zeta, t) = \frac{\partial^2}{\partial t^2}[f(\zeta, t)] u(t - \zeta) + 2\frac{\partial}{\partial t}[f(\zeta, t)]\, \delta(t - \zeta)$$
$$+ f(\zeta, t)\, \delta'(t - \zeta), \quad t \simeq \zeta, \qquad (9.2.44)$$

and

$$\frac{\partial^2}{\partial \zeta^2} K_1(\zeta, t) = \frac{\partial^2}{\partial \zeta^2}[f(\zeta, t)] u(t - \zeta) - 2\frac{\partial}{\partial \zeta}[f(\zeta, t)]\, \delta(t - \zeta)$$
$$+ f(\zeta, t)\, \delta'(t - \zeta), \quad t \simeq \zeta. \qquad (9.2.45)$$

In addition, after using (44) and (45) in (42), and then matching the most singular terms, we have

$$2\left[\frac{\partial}{\partial t} + \frac{\partial}{\partial \zeta}\right] f(\zeta, t)\bigg|_{t=\zeta} = 2\frac{d}{d\zeta}[f(\zeta, \zeta)] = V(\zeta). \qquad (9.2.46)$$

But since $f(\zeta, \zeta) = K_1(\zeta, \zeta)$, we arrive at the relationship

$$V(\zeta) = 2\frac{d}{d\zeta} K_1(\zeta, \zeta). \qquad (9.2.47)$$

Finally, after substituting (43) into (41), and matching the most singular terms, we have

$$h = K_1(0, 0). \qquad (9.2.48)$$

From the above analysis, it is evident the presence of the inhomogeneous potential $V(\zeta)$ gives rise to $K_1(\zeta, t)$ in (38). It could be thought of as the "wake" of the impulse function $\delta(t - \zeta)$ as it propagates through the inhomogeneous potential $V(\zeta)$. It is in this "wake," $K_1(\zeta, t)$, that the information on $V(\zeta)$ is embedded as expressed by Equation (47). The Gel'fand-Levitan integral equation allows one to solve for a function similar to $K_1(\zeta, t)$ as a solution from which we can find $V(\zeta)$ via (47).

To derive the Gel'fand-Levitan integral equation, first, we define a noncausal Green's function, $G(\zeta, t)$, satisfying

$$\frac{\partial^2}{\partial t^2} G(\zeta, t) - \frac{\partial^2}{\partial \zeta^2} G(\zeta, t) + V(\zeta) G(\zeta, t) = 0, \qquad (9.2.49)$$

with the boundary conditions

$$-\frac{\partial}{\partial \zeta} G(0, t) + h G(0, t) = 0, \qquad (9.2.50a)$$

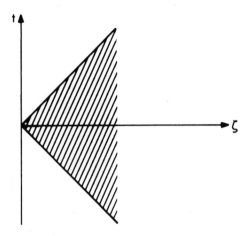

Figure 9.2.5 The support of the function $G(\zeta, t)$ in the (ζ, t) space.

$$G(0, t) = 2\delta(t). \tag{9.2.50b}$$

The preceding equations comprise the Cauchy data from which $G(\zeta, t)$ can be deduced for all ζ and t. But since $G(\zeta, -t)$ is also a solution to (49) satisfying the boundary conditions (50), we have $G(\zeta, t) = G(\zeta, -t)$ from the uniqueness of the solution to (49) with the specific boundary conditions (50). In other words, $G(\zeta, t)$ is an even function of t. Hence, $G(\zeta, t)$ cannot be causal. In fact, it can be shown that $G(\zeta, t) \neq 0$ only for $|t| < \zeta$, as shown in Figure 9.2.5 (see Exercise 9.15).[10] Consequently, $G(\zeta, t)$ has the general form

$$G(\zeta, t) = \delta(\zeta - t) + \delta(\zeta + t) + K(\zeta, t), \tag{9.2.51}$$

where $K(\zeta, t) \neq 0$, $|t| < \zeta$. Moreover, via the same analysis as before, by matching the most singular terms in the equation, we deduce that (Exercise 9.16)

$$V(\zeta) = 2\frac{d}{d\zeta}K(\zeta, \pm\zeta), \tag{9.2.52a}$$

and

$$h = K(0, 0). \tag{9.2.52b}$$

---

[10] The support of $G(\zeta, t)$ is also clear from identifying the characteristics of the hyperbolic equation (49). The Cauchy data at a point $(t, \zeta) = (0, 0)$ will only affect a triangular region shown in Figure 9.2.5.

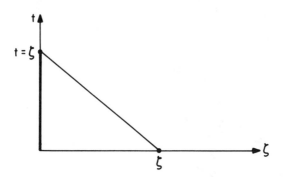

Figure 9.2.6 The value of $\phi(\zeta, t = 0)$ is affected by the values of $\phi(\zeta = 0, t = \zeta)$ and $\phi(\zeta = 0, t)$ for $0 < t < \zeta$.

Now, if $\phi(\zeta, t)$ satisfies (34) and (37), then by the principle of linear superposition,

$$\phi(\zeta, t) = \frac{1}{2} \int\limits_{-\infty}^{\infty} d\tau \, G(\zeta, t - \tau) \phi(0, \tau). \qquad (9.2.53)$$

This is clearly a solution of (34) satisfying the boundary condition (37). Furthermore, for $t = 0$, the above becomes

$$\phi(\zeta, 0) = \frac{1}{2} \int\limits_{-\infty}^{\infty} d\tau \, G(\zeta, -\tau) \phi(0, \tau). \qquad (9.2.54)$$

Then, using (51) in (54), we have

$$\phi(\zeta, 0) = \frac{1}{2} \left[\phi(0, \zeta) + \phi(0, -\zeta) + \int\limits_{-\zeta}^{\zeta} d\tau \, K(\zeta, -\tau) \phi(0, \tau) \right]. \qquad (9.2.55)$$

Moreover, if $\phi(\zeta, t)$ is even in t, (55) simplifies to

$$\phi(\zeta, 0) = \phi(0, \zeta) + \int\limits_{0}^{\zeta} d\tau \, K(\zeta, \tau) \phi(0, \tau). \qquad (9.2.56)$$

Equation (56) says that the value of $\phi(\zeta, 0)$ is related to the value of $\phi(0, \zeta)$ and the values of $\phi(0, t)$ for $0 < t < \zeta$ as shown in Figure 9.2.6.

In an actual experiment, however, the field is causal such that

$$\phi_1(\zeta, t) = \delta(t - \zeta) - K_1(\zeta, t), \qquad (9.2.57)$$

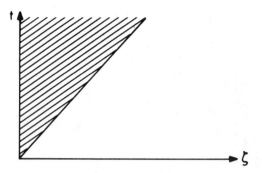

Figure 9.2.7 The support of $\phi_1(\zeta, t)$ which is a causal function.

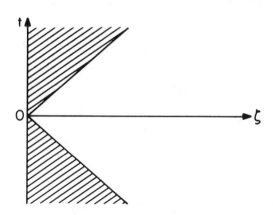

Figure 9.2.8 The support of $F(\zeta, t)$ which is an even function of t.

and this is shown in Figure 9.2.7. Since $\phi(\zeta, t)$ in (56) has to be even in t, an even function of t can be constructed by defining

$$F(\zeta, t) = \phi_1(\zeta, t) + \phi_1(\zeta, -t). \qquad (9.2.58)$$

The support of $F(\zeta, t)$ is shown in Figure 9.2.8. Furthermore, another even function of t can be constructed as

$$F(\zeta, t, T) = \frac{1}{2}\left[F(\zeta, t - T) + F(\zeta, t + T)\right], \qquad (9.2.59)$$

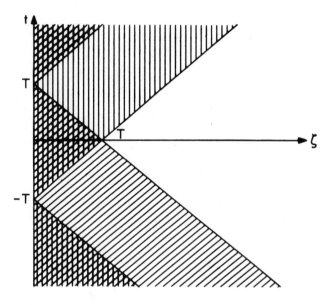

Figure 9.2.9. The support of $F(\zeta, t, T)$ which is a solution to (34).

whose support is shown in Figure 9.2.9. It is easy to show that

$$F(0, t, T) = \delta(t - T) + \delta(t + T) + f(t, T), \qquad (9.2.60)$$

where

$$f(t, T) = -\frac{1}{2}\{K_1(0, |t - T|) + K_1(0, |t + T|)\}. \qquad (9.2.61)$$

Note that $F(\zeta, t, T)$ is still a solution of (34) satisfying boundary condition (37). More importantly, it is an even function of t. Therefore, we can use $F(\zeta, t, T)$ as $\phi(\zeta, t)$ in Equation (56), which is intended for even functions of t. For $\zeta > T$, $\phi(\zeta, 0) = F(\zeta, 0, T) = 0$ (see Figure 9.2.9). Consequently, we have

$$0 = F(0, \zeta, T) + \int\limits_{0}^{\zeta} d\tau \, K(\zeta, \tau) F(0, \tau, T), \qquad \zeta > T. \qquad (9.2.62)$$

Since

$$F(0, \zeta, T) = f(\zeta, T), \qquad\qquad \zeta > T, \qquad (9.2.63a)$$
$$F(0, t, T) = \delta(t - T) + f(t, T), \quad 0 < t < \zeta, \qquad (9.2.63b)$$

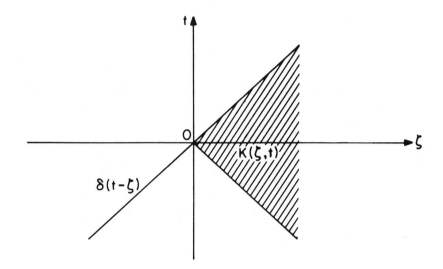

Figure 9.2.10 The support of $G(\zeta, t)$ in Equation (65) for the Marchenko method.

we have from (62)

$$0 = f(\zeta, T) + \int_0^\zeta d\tau \, K(\zeta, \tau) f(\tau, T) + K(\zeta, T), \qquad (9.2.64)$$

where $0 < T < \zeta$. Equation (64) is the linear Gel'fand-Levitan equation. In this equation, $f(\zeta, T)$ given by (61) can be constructed from the measurement data (57) via (58) to (60). Consequently, we can solve the above for the unknown $K(\zeta, t)$ using well-known methods to solve linear integral equations (see Exercise 9.17; also see Berryman and Greene 1980). Once $K(\zeta, t)$ is known, we can find $V(\zeta)$ from (52a).

§§9.2.4 The Marchenko Integral Equation

The Marchenko integral equation is in many ways similar to the Gel'fand-Levitan integral equation. However, in the Marchenko integral equation, instead of an impenetrable impedance boundary condition at $\zeta = 0$, we have $V(\zeta) = 0$, $\zeta < 0$. Hence, a wave is sent from the left to probe the potential $V(\zeta)$. Then, the potential $V(\zeta)$, which is like a half-space, is reconstructed from the reflected wave for $\zeta < 0$.

First, let us define a noncausal Green's function

$$G(\zeta, t) = \delta(t - \zeta) + K(\zeta, t), \qquad (9.2.65)$$

whose support is shown in Figure 9.2.10. $G(\zeta, t)$ is a solution of (34) with a right-going wave for $\zeta < 0$. Notice that if $G(\zeta, t)$ is a solution of (34), so is $G(\zeta, -t)$. However, $G(\zeta, -t)$ now represents a solution with a left-going wave for $\zeta < 0$. Consequently, if a solution to

$$\frac{\partial^2}{\partial t^2}\phi(\zeta, t) - \frac{\partial^2}{\partial \zeta^2}\phi(\zeta, t) + V(\zeta)\phi(\zeta, t) = 0 \qquad (9.2.66)$$

consists of left-going waves only for $\zeta < 0$, it can be written as a linear superposition of $G(\zeta, -t)$. In particular, since $G(0, -t) = \delta(t)$, by the principle of linear superposition, we can write

$$\phi(\zeta, t) = \int\limits_{-\infty}^{\infty} d\tau\, G(\zeta, -\tau)\phi(0, t-\tau). \qquad (9.2.67)$$

On substituting $G(\zeta, -\tau)$ from (65) into (67), we have

$$\phi(\zeta, t) = \phi(0, t+\zeta) + \int\limits_{-\zeta}^{\zeta} K(\zeta, -\tau)\phi(0, t-\tau)\, d\tau$$

$$= \phi(0, t+\zeta) + \int\limits_{-\zeta}^{\zeta} K(\zeta, \tau)\phi(0, t+\tau)\, d\tau. \qquad (9.2.68)$$

In the above, the last equality follows from a simple change of variable from τ to $-\tau$.

A physical experiment has to be causal. An example of a causal solution to (66) is given by

$$\phi_1(\zeta, t) = \delta(t-\zeta) + K_1(\zeta, t), \qquad (9.2.69)$$

where the support of $K_1(\zeta, t)$ is shown in Figure 9.2.11. But if $\phi_2(\zeta, t)$ is defined such that

$$\phi_2(\zeta, t) = \phi_1(\zeta, t) - G(\zeta, t) = K_1(\zeta, t) - K(\zeta, t), \qquad (9.2.70)$$

then $\phi_2(\zeta, t)$ consists purely of left-going waves for $\zeta < 0$. As a result, $\phi_2(\zeta, t)$ can be used in (68) to yield

$$K_1(\zeta, t) - K(\zeta, t) = K_1(0, t+\zeta) + \int\limits_{-\zeta}^{\zeta} K(\zeta, \tau)K_1(0, t+\tau)\, d\tau. \qquad (9.2.71)$$

Moreover, if the above equation is imposed for $\zeta > t$, then $K_1(\zeta, t) = 0$ on the left, and

$$-K(\zeta, t) = K_1(0, t+\zeta) + \int\limits_{-t}^{\zeta} K(\zeta, \tau)K_1(0, t+\tau)\, d\tau, \qquad \zeta > t. \qquad (9.2.72)$$

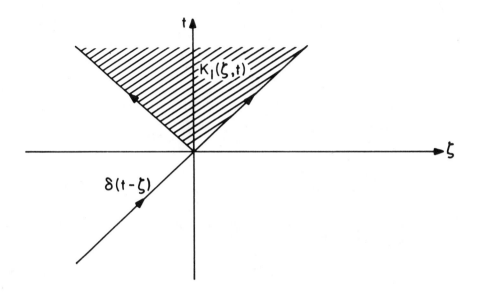

Figure 9.2.11 The support of $\phi_1(\zeta,t)$ or $K_1(\zeta,t)$ in Equation (69).

Equation (72) constitutes the Marchenko integral equation. Here, $K_1(0,t)$, defined by (69), is known since it can be provided by an actual causal experiment. Equation (72) then is a linear integral equation from which we can find $K(\zeta,t)$ (Exercise 9.17). Once $K(\zeta,t)$ is found, $V(\zeta)$ is obtainable via

$$V(\zeta) = 2\frac{d}{d\zeta}K(\zeta,\zeta). \qquad (9.2.73)$$

§§9.2.5 The Gel'fand-Levitan-Marchenko Integral Equation

An integral equation that encompasses both the Gel'fand-Levitan equation and the Marchenko equation as special cases is also derivable (Balanis 1972; Habashy and Mittra 1987). It is called the Gel'fand-Levitan-Marchenko (GLM) integral equation. First, let us consider the solution to the Schrödinger-like equation (34) subject to the boundary conditions

$$\phi(0,t) = \delta(t) + R(t), \qquad (9.2.74a)$$

$$\frac{\partial}{\partial\zeta}\phi(0,t) = -\delta'(t) + r'(t). \qquad (9.2.74b)$$

The above then constitutes a Cauchy initial value problem.

Next, consider two independent solutions of (34), $G_1(\zeta,t)$ and $G_2(\zeta,t)$, with the boundary conditions that

$$G_1(0,t) = \delta(t), \qquad (9.2.75a)$$

$$\frac{\partial}{\partial\zeta}G_1(0,t) = -\delta'(t), \qquad (9.2.75b)$$

and

$$G_2(0, t) = \delta(t), \qquad (9.2.76a)$$

$$\frac{\partial}{\partial \zeta} G_2(0, t) = \delta'(t). \qquad (9.2.76b)$$

Hence, $G_1(\zeta, t)$ and $G_2(\zeta, t)$ are noncausal Green's functions with support over $0 \leq |t| \leq \zeta$ as shown in Figure 9.2.5.[11] Alternatively,

$$G_1(\zeta, t) = \delta(t - \zeta) + K_1(\zeta, t), \qquad (9.2.77a)$$

$$G_2(\zeta, t) = \delta(t + \zeta) + K_2(\zeta, t). \qquad (9.2.77b)$$

$K_1(\zeta, t)$ and $K_2(\zeta, t)$ are present due to nonzero $V(\zeta)$. In other words, they are "wakes" generated by $\delta(t \pm \zeta)$ propagating through the inhomogeneous medium. But since $G_2(\zeta, -t)$ is also a solution to (34) with the boundary conditions given by (75), we must have

$$G_1(\zeta, t) = G_2(\zeta, -t) \qquad (9.2.78)$$

because of the uniqueness of the Cauchy initial value problem.

Consequently, using the principle of linear superposition, a general solution to (34) is expressible as

$$\phi(\zeta, t) = \int_{-\infty}^{\infty} d\tau \, A(t - \tau) G_1(\zeta, \tau) + \int_{-\infty}^{\infty} d\tau \, B(t - \tau) G_2(\zeta, \tau). \qquad (9.2.79)$$

Furthermore, with the boundary conditions provided by Equations (75) and (76) in (79) at $\zeta = 0$, we have

$$\phi(0, t) = A(t) + B(t), \qquad (9.2.80a)$$

$$\frac{\partial}{\partial \zeta} \phi(0, t) = -\frac{\partial}{\partial t} A(t) + \frac{\partial}{\partial t} B(t). \qquad (9.2.80b)$$

Then, using the boundary conditions (74), we have

$$A(t) + B(t) = \delta(t) + R(t), \qquad (9.2.81a)$$

$$-\frac{\partial}{\partial t} A(t) + \frac{\partial}{\partial t} B(t) = -\delta'(t) + r'(t). \qquad (9.2.81b)$$

Equation (81b) can be integrated from $-\infty$ to t, and assuming that $A(-\infty) = B(-\infty) = r(-\infty) = 0$ (a valid assumption), we have

$$-A(t) + B(t) = -\delta(t) + r(t). \qquad (9.2.82)$$

---

[11] The support of $G_1(\zeta, t)$ and $G_2(\zeta, t)$ can be appreciated by considering the Cauchy data given by (75) and (76).

Consequently, from (81a) and (82), we have

$$A(t) = \delta(t) + \frac{1}{2}[R(t) - r(t)], \tag{9.2.83a}$$

$$B(t) = \frac{1}{2}[R(t) + r(t)]. \tag{9.2.83b}$$

Moreover, with the definition that

$$S_1(t) = \frac{1}{2}[R(t) - r(t)], \tag{9.2.84a}$$

$$S_2(t) = \frac{1}{2}[R(t) + r(t)], \tag{9.2.84b}$$

Equation (79) becomes

$$\phi(\zeta, t) = G_1(\zeta, t) + \int_{-\infty}^{\infty} d\tau\, S_1(t - \tau)G_1(\zeta, \tau) + \int_{-\infty}^{\infty} d\tau\, S_2(t - \tau)G_2(\zeta, \tau). \tag{9.2.85}$$

The above is the mathematical expression for propagating the Cauchy data at $\zeta = 0$ to a region where $\zeta > 0$. But for $\phi(\zeta, t)$ to be causal, $\phi(\zeta, t) = 0$ when $\zeta > t$. Therefore, using this causality condition plus the definition of $G_1(\zeta, t)$ and $G_2(\zeta, t)$ given by (77), we have

$$0 = K_1(\zeta, t) + S_1(t - \zeta) + S_2(t + \zeta) + \int_{-\zeta}^{\zeta} d\tau\, [S_1(t - \tau) + S_2(t + \tau)]K_1(\zeta, \tau), \quad \zeta > t. \tag{9.2.86}$$

Equation (78) has been used to combine the two integrals in (85) into one integral in (86). Equation (86) is the Gel'fand-Levitan-Marchenko equation. In (86), $S_1(t)$ and $S_2(t)$ are known since they are obtainable from $\phi(0, t)$ and $\frac{\partial}{\partial \zeta}\phi(0, t)$, the measurement data. Hence, the unknown $K_1(\zeta, t)$ can be solved for from (86). Then, $V(\zeta)$ is retrievable using (47).

To retrieve the Gel'fand-Levitan equation from (86), we let $r(t) = 0$. Then $S_1(t) = S_2(t) = \frac{1}{2}R(t)$. As a result, Equation (86) becomes

$$0 = K_1(\zeta, t) + \frac{1}{2}[R(t - \zeta) + R(t + \zeta)]$$

$$+ \frac{1}{2}\int_{-\zeta}^{\zeta} d\tau\, [R(t - \tau) + R(t + \tau)]K_1(\zeta, \tau), \quad \zeta > t. \tag{9.2.87}$$

The second integral can be folded to yield

$$0 = K_1(\zeta, t) + \frac{1}{2}[R(t - \zeta) + R(t + \zeta)]$$

$$+ \frac{1}{2}\int_{0}^{\zeta} d\tau\, [R(t - \tau) + R(t + \tau)][K_1(\zeta, \tau) + K_1(\zeta, -\tau)], \quad \zeta > t. \tag{9.2.88}$$

Moreover, by letting t become $-t$, Equation (88) becomes

$$0 = K_1(\zeta, -t) + \frac{1}{2}[R(-t-\zeta) + R(-t+\zeta)]$$

$$+ \frac{1}{2}\int_0^\zeta d\tau\,[R(-t-\tau) + R(-t+\tau)][K_1(\zeta,\tau) + K_1(\zeta,-\tau)], \quad \zeta > -t. \tag{9.2.89}$$

In (74), $R(t)$ is nonzero only if $V(\zeta) \neq 0$. Hence, $R(t)$ represents the reflected field from the inhomogeneous potential $V(\zeta)$. Because of causality, $R(t) = 0$, $t < 0$. Using this fact, and by adding (88) and (89), we finally have

$$0 = K(\zeta, t) + \frac{1}{2}[R(\zeta + t) + R(\zeta - t)]$$

$$+ \frac{1}{2}\int_0^\zeta d\tau\,[R(|t-\tau|) + R(t+\tau)]K(\zeta,\tau), \quad \zeta > t, \tag{9.2.90}$$

where $K(\zeta, t) = K_1(\zeta, t) + K_1(\zeta, -t)$ is a symmetric function of t. Equation (90) is the same as Equation (64), the Gel'fand-Levitan equation.

To retrieve the Marchenko integral equation, we let $r(t) = R(t)$. In this case, $S_1(t) = 0$, $S_2(t) = R(t)$. Then, (86) becomes

$$0 = K_1(\zeta, t) + R(\zeta + t) + \int_{-\zeta}^\zeta d\tau\,R(t+\tau)K_1(\zeta,\tau), \quad \zeta > t. \tag{9.2.91}$$

Note that the above is the same as (72).

The Gel'fand-Levitan equation, with $r(t) = 0$, is equivalent to probing a one-dimensional profile with an impressed electric current sheet source on a perfect magnetic wall. On this wall, the reflected electric field $R(t) \neq 0$, while the reflected magnetic field (analogous to $\frac{\partial \phi}{\partial \zeta}$) is zero.[12] On the other hand, the Marchenko integral equation, with $R(t) = r(t)$, is equivalent to probing a one-dimensional profile for $\zeta > 0$, where $V(\zeta) = 0$ for $\zeta < 0$. It is equivalent to sending a plane wave from the left to probe $V(\zeta)$ for $\zeta > 0$ (see Exercise 9.18).

Once the $K(\zeta, t)$ functions are found in the Gel'fand-Levitan-Marchenko method, a more direct method of obtaining η compared to that given by (32) is available, since (32) is the same as (30) with $\omega = 0$. We have shown before that the general solution to (30) in the time domain is given by (79). Now, transforming (79) to the frequency domain, we deduce that the general

---

[12] In this case, the Schrödinger-like equation is a transformed TE wave equation which is the dual of Equation (23).

solution to (32) is then

$$\phi(\zeta, \omega) = A(\omega) \int_{-\zeta}^{\zeta} e^{i\omega\tau} G_1(\zeta, \tau) \, d\tau + B(\omega) \int_{-\zeta}^{\zeta} e^{i\omega\tau} G_2(\zeta, \tau) \, d\tau. \qquad (9.2.92)$$

Setting $\omega = 0$, the general solution of (32) is

$$\eta^{\frac{1}{2}}(\zeta) = (A + B) \int_{-\zeta}^{\zeta} G_1(\zeta, \tau) \, d\tau, \qquad (9.2.93)$$

where (78) is being used to combine the two integrals in (92). Finally, after making use of (77a) for $G_1(\zeta, \tau)$, we have

$$\eta^{\frac{1}{2}}(\zeta) = \eta^{\frac{1}{2}}(0) \left[1 + \int_{-\zeta}^{\zeta} K_1(\zeta, \tau) \, d\tau \right]. \qquad (9.2.94)$$

The above gives $\eta(\zeta)$ directly in terms of $K_1(\zeta, t)$.

The method of characteristics, the Gel'fand-Levitan-Marchenko method, and layer-stripping methods are essentially high-frequency methods that rely on causality and the propagation of singularities. Since singularities manifest themselves as high-frequency components in the frequency domain, we can study (22), the frequency-domain version of (21), more closely. By looking at Equation (22), one can see that when $\omega \to \infty$, the inhomogeneous potential $V(\zeta)$ is subdominant. Under this condition, the Born approximation becomes increasingly good when $\omega \to \infty$.[13] Hence, a very simple, linear relationship exists between the first-order scattered field and the potential. This relationship is expressed by (47). Note that $K_1(\zeta, \zeta)$ is at the wavefront of the "wake," and hence, contains the most high-frequency information.

§9.3 Higher-Dimensional Inverse Problems

The methods discussed in the previous section have limited applications as they are appropriate only for one-dimensional inverse problems. Furthermore, they cannot be used for low-frequency inversion. For a lossy, conductive medium, low frequency also makes the equation more diffusive, and hence, parabolic-like. Then, methods which rely on the propagation of singularities would not work. Even though the Gel'fand-Levitan-Marchenko integral equation has been generalized to higher dimensions for Schrödinger equations, no numerical simulation has been obtained (Newton 1980a, 1980b, 1981). Moreover, in higher dimensions, it has not been possible to transform a wave equation into a Schrödinger-like equation as in the case of one dimension.

---

[13] Note that this is peculiar of the Schrödinger-like equation, but not the wave equation (see Chapter 8).

In order to formulate a theory for a general, higher-dimensional, nonlinear inverse problem, we have to use a more general integral equation. But then, the integral equation is nonlinear in the object function being sought. Hence, iterative methods must be used to solve the nonlinear equation. Consequently, we shall discuss the use of the distorted Born iterative and the Born iterative methods to solve such a nonlinear integral equation in this section. Such methods have been used to solve the one-dimensional inverse scattering problems (see reference list for this chapter). In addition, they have also been extended to higher dimensions (Johnson and Tracy 1983a, 1983b; Wang and Chew 1989; Chew and Wang 1990).

§§9.3.1 Distorted Born Iterative Method

Previously, we have learnt that an electromagnetic scattering problem can be formulated as the solution of the integral equation

$$\mathbf{E}(\mathbf{r}) = \mathbf{E}_{inc}(\mathbf{r}) + \int_V d\mathbf{r}' \, \overline{\mathbf{G}}(\mathbf{r}, \mathbf{r}', \epsilon_b) \cdot [k^2(\mathbf{r}') - k_b^2] \mathbf{E}(\mathbf{r}'), \qquad (9.3.1)$$

where ϵ_b is the permittivity of the background, $k_b^2 = \omega^2 \mu \epsilon_b$, and $k^2(\mathbf{r}) = \omega^2 \mu \epsilon(\mathbf{r})$ is a function of position, and it is the function to be sought. In the above, $\overline{\mathbf{G}}(\mathbf{r}, \mathbf{r}', \epsilon_b)$ is the dyadic Green's function, which is the solution of the equation

$$\nabla \times \nabla \times \overline{\mathbf{G}}(\mathbf{r}, \mathbf{r}', \epsilon_b) - k_b^2 \, \overline{\mathbf{G}}(\mathbf{r}, \mathbf{r}', \epsilon_b) = \overline{\mathbf{I}} \delta(\mathbf{r} - \mathbf{r}'). \qquad (9.3.2)$$

Furthermore, ϵ_b need not be a constant of position, and k_b^2 is the estimate of the actual function $k^2(\mathbf{r})$. In (1), $\mathbf{E}_{inc}(\mathbf{r})$ is the field present when $k^2(\mathbf{r}) = k_b^2(\mathbf{r})$, i.e., it is the total field in the presence of the inhomogeneity $k_b^2(\mathbf{r})$. In addition, the measurement data are available only outside the scatterer in the inverse scattering problem. Therefore, we have at our disposal only

$$\mathbf{E}_{sca}(\mathbf{r}) = \mathbf{E}(\mathbf{r}) - \mathbf{E}_{inc}(\mathbf{r}), \qquad \mathbf{r} \in S, \qquad (9.3.3)$$

where S is some surface outside V (see Figure 9.3.1). It is only the scattered field $\mathbf{E}_{sca}$ that bears information on the scatterer. Hence, we can write (1) as

$$\mathbf{E}_{sca}(\mathbf{r}) = \int_V d\mathbf{r}' \, \overline{\mathbf{G}}(\mathbf{r}, \mathbf{r}', \epsilon_b) \cdot [k^2(\mathbf{r}') - k_b^2] \mathbf{E}(\mathbf{r}'), \qquad \mathbf{r} \in S. \qquad (9.3.4)$$

At this point, the preceding integral equation cannot be used to solve for $k^2(\mathbf{r})$ since $\mathbf{E}(\mathbf{r})$ is unknown. Furthermore, $\mathbf{E}(\mathbf{r})$ is a function of $k^2(\mathbf{r})$ so that the integral is a nonlinear functional of $k^2(\mathbf{r})$. To linearize the problem, we let $\mathbf{E}(\mathbf{r}) = \mathbf{E}_{inc}(\mathbf{r})$; hence, (4) becomes

$$\mathbf{E}_{sca}(\mathbf{r}) = \int_V d\mathbf{r}' \, \overline{\mathbf{G}}(\mathbf{r}, \mathbf{r}', \epsilon_b) \cdot [k^2(\mathbf{r}') - k_b^2] \mathbf{E}_{inc}(\mathbf{r}'), \qquad \mathbf{r} \in S. \qquad (9.3.5)$$

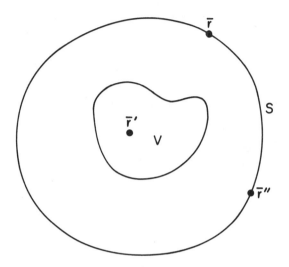

Figure 9.3.1 An inverse scattering experiment where the measurement data are obtained at **r** on S and the transmitter is at $\mathbf{r}''$.

Now, the preceding equation is an integral equation linear in $k^2(\mathbf{r}) - k_b^2$. Moreover, the error in the above equation can be easily shown to be of the order $(k^2 - k_b^2)^2$. In addition, for the special case where $\mathbf{E}_{inc}(\mathbf{r}')$ is generated by a point source **a** located at $\mathbf{r}'' \in S$, we can write

$$\mathbf{E}_{inc}(\mathbf{r}') = \overline{\mathbf{G}}(\mathbf{r}', \mathbf{r}'', \epsilon_b) \cdot \mathbf{a}. \tag{9.3.6}$$

Then, (5) becomes

$$\mathbf{E}_{sca}(\mathbf{r}, \mathbf{r}'') = \int_V d\mathbf{r}' \, \overline{\mathbf{G}}(\mathbf{r}, \mathbf{r}', \epsilon_b) \cdot \overline{\mathbf{G}}(\mathbf{r}', \mathbf{r}'', \epsilon_b) \cdot \mathbf{a} \left[k^2(\mathbf{r}') - k_b^2\right], \quad \mathbf{r} \in S, \quad \mathbf{r}'' \in S,$$
$$\tag{9.3.7}$$

or

$$\mathbf{E}_{sca}(\mathbf{r}, \mathbf{r}'') = \int_V d\mathbf{r}' \, \mathbf{M}(\mathbf{r}, \mathbf{r}'', \mathbf{r}', \epsilon_b)[k^2(\mathbf{r}') - k_b^2], \quad \mathbf{r} \in S, \quad \mathbf{r}'' \in S.$$
$$\tag{9.3.7a}$$

Since $k^2(\mathbf{r})$ is a three-dimensional function with support on V, a single measurement of $\mathbf{E}_{sca}(\mathbf{r}, \mathbf{r}'')$ for **r** on the surface S for a fixed $\mathbf{r}''$ is not sufficient to generate enough data to solve for $k^2(\mathbf{r})$. In other words, we do not expect to retrieve information accurately on a three-dimensional function from a two-dimensional function. Therefore, data for a range of **r** and $\mathbf{r}''$ are needed in order to reconstruct $k^2(\mathbf{r})$ accurately (see Exercise 9.19). The problem may still be ill-posed, however, because the operator $\mathbf{M}(\mathbf{r}, \mathbf{r}'', \mathbf{r}', \epsilon_b)$, which maps the object function, $k^2(\mathbf{r}') - k_b^2$, to the data space is ill-posed. This is especially true for the high spectral components of $k^2(\mathbf{r}) - k_b^2$. The reason is that these high spectral components generate evanescent waves, which are rapidly decaying from the object. Hence, there are spectral components of $k^2(\mathbf{r}) - k_b^2$ that are mapped onto exponentially small values on S, the measurement surface (also see Subsection 9.1.5). Because of this, the integral equation (7a) cannot be solved with the usual techniques. For instance, if (7a) is converted into a matrix equation, the values of $k^2(\mathbf{r}) - k_b^2$ obtained will be unbounded due to the ill-posed nature of the equation (see Baker 1977).

One way to overcome this problem is to use the method of regularization to solve the equation (Tikhonov and Arsenin 1977). In this method, the solution of (7a) is sought via an optimization procedure. Hence, a functional of $k^2(\mathbf{r}) - k_b^2$ is first defined as

$$I = \delta_t \int_V d\mathbf{r}' \, |k^2(\mathbf{r}') - k_b^2|^2$$

$$+ \sum_{i,j} \left| \mathbf{E}_{sca}(\mathbf{r}_i, \mathbf{r}_j) - \int_V d\mathbf{r}' \, \mathbf{M}(\mathbf{r}_i, \mathbf{r}_j, \mathbf{r}', \epsilon_b)[k^2(\mathbf{r}') - k_b^2] \right|^2. \quad (9.3.8)$$

In the above, $\mathbf{r}_i$ and $\mathbf{r}_j$ are the locations of the receivers and transmitters respectively. Note that the first term in (8) corresponds to the norm of $k^2(\mathbf{r}) - k_b^2$. But the second term in (8) is the error between the measurement data and that predicted by the linearized model. An optimal solution for $k^2(\mathbf{r}) - k_b^2$ is sought in order to minimize I. Here, δ_t is a tuning parameter chosen to weigh the relative importance of the first term and the second term. In this manner, the optimal solution will minimize the error between the measurement data and the norm of $k^2(\mathbf{r}) - k_b^2$. Because of this, the values of $k^2(\mathbf{r}) - k_b^2$ obtained will be bounded.

As an illustration, we expand

$$k^2(\mathbf{r}) - k_b^2 = \sum_n a_n b_n(\mathbf{r}), \quad (9.3.9)$$

where $b_n(\mathbf{r})$ is a basis set that approximates $k^2(\mathbf{r}) - k_b^2$ fairly well. Then using (9) in (8), we have

$$I = \delta_t \sum_{n,m} a_n a_m^* \int_V d\mathbf{r}' \, b_n(\mathbf{r}') b_m^*(\mathbf{r}')$$

$$+ \sum_{i,j} \left| \mathbf{E}_{sca}(\mathbf{r}_i, \mathbf{r}_j) - \sum_n a_n \int_V d\mathbf{r}' \, \mathbf{M}(\mathbf{r}_i, \mathbf{r}_j, \mathbf{r}', \epsilon_b) b_n(\mathbf{r}') \right|^2. \quad (9.3.10)$$

Note that the above is of the form

$$I = \delta_t \sum_{n,m} a_n a_m^* B_{mn} + \sum_k \left| \mathbf{E}_k - \sum_n a_n \mathbf{L}_{kn} \right|^2, \quad (9.3.11)$$

where

$$B_{mn} = \int_V d\mathbf{r}' \, b_n(\mathbf{r}') b_m^*(\mathbf{r}'), \qquad (9.3.12a)$$

$$\mathbf{E}_k = \mathbf{E}_{sca}(\mathbf{r}_i, \mathbf{r}_j), \qquad (9.3.12b)$$

$$\mathbf{L}_{kn} = \int d\mathbf{r}' \, \mathbf{M}(\mathbf{r}_i, \mathbf{r}_j, \mathbf{r}', \epsilon_b) b_n(\mathbf{r}'). \qquad (9.3.12c)$$

Furthermore, in the above, we have used the k index to replace the i and j indices. A subsequent minimization of (11) with respect to a_n yields (see Exercise 9.20)

$$0 = \delta_t \sum_n a_n B_{mn} - \sum_k \mathbf{E}_k \cdot \mathbf{L}_{km}^* + \sum_k \sum_n a_n \mathbf{L}_{kn} \cdot \mathbf{L}_{km}^*. \qquad (9.3.13)$$

This is a matrix equation of the form

$$0 = \delta_t \overline{\mathbf{B}} \cdot \mathbf{a} - \mathbf{C} + \overline{\mathbf{P}} \cdot \mathbf{a}, \qquad (9.3.14)$$

where

$$\left[\overline{\mathbf{B}}\right]_{mn} = B_{mn}, \qquad [\mathbf{C}]_m = \sum_k \mathbf{L}_{km}^* \cdot \mathbf{E}_k, \qquad (9.3.15a)$$

$$\left[\overline{\mathbf{P}}\right]_{mn} = \sum_k \mathbf{L}_{kn} \cdot \mathbf{L}_{km}^*, \qquad [\mathbf{a}]_n = a_n. \qquad (9.3.15b)$$

Consequently, Equation (14) could be solved to yield

$$\mathbf{a} = \left[\overline{\mathbf{P}} + \delta_t \overline{\mathbf{B}}\right]^{-1} \cdot \mathbf{C}. \qquad (9.3.16)$$

Once $\mathbf{a}$ is found, we can find $k^2(\mathbf{r}) - k_b^2$ from (9).

The above method only allows us to find $k^2(\mathbf{r})$ approximately; however, we can use this new $k^2(\mathbf{r})$ as the estimate k_b^2. Then, a new $\overline{\mathbf{G}}(\mathbf{r}, \mathbf{r}', \epsilon_b)$ that corresponds to this new k_b^2 in (2) has to be found. This new Green's function can be found, for instance, with the methods described in Chapter 8. It is the arbitrary point-source response in an inhomogeneous medium. For example, this arbitrary point-source response is solvable with the volume integral equation approach. Consequently, with the new $\overline{\mathbf{G}}(\mathbf{r}, \mathbf{r}', \epsilon_b)$, the whole process is repeated until the $k^2(\mathbf{r})$ found produces the same data as the measurement data. This iterative procedure is like Newton's method in solving a nonlinear integral equation, except that the solution is regularized at every iteration.

The above method is realized quite easily in one dimension because the Green's function for a one-dimensional inhomogeneity is easily found (Chapter 2). In higher dimensions, however, the Green's function has to be found via numerical methods: the finite-element or the integral equation method (Chapters 5 and 8). Hence, the implementation of this method can become numerically intensive for higher-dimensional problems.

In addition to regularizing the amplitude of $k^2(\mathbf{r}) - k_b^2$ as in (8), the derivatives of $k^2(\mathbf{r}) - k_b^2$ can also be regularized (see Exercise 9.21). Since the derivatives enhance the high spectral components of $k^2(\mathbf{r}) - k_b^2$, regularizing the derivatives attenuates the object's high spectral components. Moreover, since it is the high spectral components of $k^2(\mathbf{r}) - k_b^2$ that give rise to ill-conditioning, this is sometimes preferable to regularizing the magnitude.

Another method that may be used to solve (7a) is via the method of singular-value decomposition (Golub and Kahan 1965; Wilkinson 1965). If (7a) is converted into a matrix equation without the regularization term in the first term of (8), then the matrix equation we obtain is similar to (14) but with $\delta_t = 0$, that is,

$$\overline{\mathbf{P}} \cdot \mathbf{a} = \overline{\mathbf{C}}. \tag{9.3.17}$$

However, $\overline{\mathbf{P}}$ is ill-conditioned, and it is not possible to find its inverse. Moreover, because $\overline{\mathbf{P}}$ is ill-conditioned, it has eigenvalues that are very small or zero. But since $\overline{\mathbf{P}}$ is Hermitian, we can decompose $\overline{\mathbf{P}}$ as

$$\overline{\mathbf{P}} = \overline{\mathbf{S}} \cdot \overline{\boldsymbol{\lambda}} \cdot \overline{\mathbf{S}}^\dagger, \tag{9.3.18}$$

where $\overline{\mathbf{S}}$ is a unitary matrix, and $\overline{\boldsymbol{\lambda}}$ is a diagonal matrix containing the eigenvalues of $\overline{\mathbf{P}}$, which are real in this case. The inverse of $\overline{\mathbf{P}}$ is then

$$\overline{\mathbf{P}}^{-1} = \overline{\mathbf{S}} \cdot \overline{\boldsymbol{\lambda}}^{-1} \cdot \overline{\mathbf{S}}^\dagger. \tag{9.3.19}$$

However, because of the zero or near-zero eigenvalues of $\overline{\mathbf{P}}$, $\overline{\boldsymbol{\lambda}}^{-1}$ is unbounded. But if we remove the small eigenvalues of $\overline{\mathbf{P}}$ and define

$$\overline{\mathbf{P}}^{-1} = \overline{\mathbf{S}} \cdot \widetilde{\overline{\boldsymbol{\lambda}}}^{-1} \cdot \overline{\mathbf{S}}^\dagger, \tag{9.3.20}$$

where $\widetilde{\overline{\boldsymbol{\lambda}}}$ contains only the nonsmall eigenvalues of $\overline{\mathbf{P}}$, then $\widetilde{\overline{\boldsymbol{\lambda}}}^{-1}$ is bounded. In this case, the solution to (17) can be defined as

$$\mathbf{a} = \overline{\mathbf{S}} \cdot \widetilde{\overline{\boldsymbol{\lambda}}}^{-1} \cdot \overline{\mathbf{S}}^\dagger \cdot \overline{\mathbf{C}}. \tag{9.3.21}$$

Such a solution of $\mathbf{a}$ is also known as the least-norm solution, because the $\mathbf{a}$ thus obtained has the minimum norm, i.e., $|\mathbf{a}|$ is minimum (see Exercise 9.22).

In the regularization method, as shown in Equation (16), the inverse of $\overline{\mathbf{P}} + \delta_t \overline{\mathbf{B}}$ is sought instead of $\overline{\mathbf{P}}$. If $\overline{\mathbf{B}}$ is a well-conditioned matrix, $\overline{\mathbf{P}} + \delta_t \overline{\mathbf{B}}$ is also well-conditioned. In other words, we have essentially padded the zero eigenvalues of $\overline{\mathbf{P}}$ with nonzero values by adding a term $\delta_t \overline{\mathbf{B}}$ to $\overline{\mathbf{P}}$.

In two- and three-dimensional problems, finding the inverse of $\overline{\mathbf{P}} + \delta_t \overline{\mathbf{B}}$ in the regularization method or finding the singular value decomposition of $\overline{\mathbf{P}}$ usually involves a number of floating-point operations of the order of N^3.

But because of the large number of unknowns involved in two- and three-dimensional problems, they become prohibitively intensive in the use of computational resources. Alternatively, the conjugate gradient method can be used to invert $\overline{\mathbf{P}} + \delta_t \overline{\mathbf{B}}$.

§§9.3.2 Born Iterative Method

In the distorted Born iterative method, the left-hand side of Equation (7) becomes smaller as k_b^2 approaches $k^2(\mathbf{r})$ as the iteration step increases. So, if the measurement data $\hat{\mathbf{E}}(\mathbf{r})$ are contaminated with noise so that

$$\hat{\mathbf{E}}(\mathbf{r}) = \mathbf{E}(\mathbf{r}) + n(\mathbf{r}), \qquad (9.3.22)$$

where $n(\mathbf{r})$ is the noise, then

$$\hat{\mathbf{E}}_{sca}(\mathbf{r}) = \mathbf{E}(\mathbf{r}) - \mathbf{E}_{inc}(\mathbf{r}) + n(\mathbf{r}). \qquad (9.3.23)$$

Hence, as the iteration step improves on $\mathbf{E}_{inc}(\mathbf{r})$ so that it is closer to $\mathbf{E}(\mathbf{r})$, $\hat{\mathbf{E}}_{sca}(\mathbf{r})$ is swamped by noise $n(\mathbf{r})$. Then, the iteration procedure may diverge if no precautionary step to counter this is taken.

A more robust method of solving the inverse scattering problem is to use the Born iterative method (Wang and Chew 1989). In this method, we write Equation (1) as

$$\mathbf{E}(\mathbf{r}) = \mathbf{E}_{inc}(\mathbf{r}) + \int_V d\mathbf{r}' \, \overline{\mathbf{G}}_0(\mathbf{r}, \mathbf{r}') \cdot [k^2(\mathbf{r}') - k_0^2] \mathbf{E}(\mathbf{r}'), \qquad (9.3.24)$$

where k_0^2 is a constant wavenumber for the homogeneous background. Here, $\overline{\mathbf{G}}_0(\mathbf{r}, \mathbf{r}')$ is a homogeneous Green's function with wavenumber k_0. Hence, $\mathbf{E}_{inc}$ is the field in the absence of the scatterer, but $\mathbf{E}(\mathbf{r})$ is the total field in the presence of the scatterer. Analogous to (4), Equation (24) is written as

$$\mathbf{E}_{sca}(\mathbf{r}) = \int_V d\mathbf{r}' \, \overline{\mathbf{G}}_0(\mathbf{r}, \mathbf{r}') \cdot [k^2(\mathbf{r}') - k_0^2] \mathbf{E}(\mathbf{r}'), \quad \mathbf{r} \in S \qquad (9.3.25)$$

and

$$\mathbf{E}_{sca}(\mathbf{r}) = \mathbf{E}(\mathbf{r}) - \mathbf{E}_{inc}(\mathbf{r}), \quad \mathbf{r} \in S, \qquad (9.3.25a)$$

which is available from the measured field on S.

In the Born iterative method, an estimate of $k^2(\mathbf{r})$ is first made. Next, the total $\hat{\mathbf{E}}(\mathbf{r})$ in the object corresponding to this estimate is found by solving the forward scattering problem. Then, we may write

$$\hat{\mathbf{E}}(\mathbf{r}') = \overline{\mathbf{G}}(\mathbf{r}', \mathbf{r}'', \hat{\epsilon}) \cdot \mathbf{a}, \qquad (9.3.26)$$

which is due to a point source at $\mathbf{r}''$, and $\hat{\epsilon}$ is the guessed profile. In this manner, (25) becomes

$$\mathbf{E}_{sca}(\mathbf{r}, \mathbf{r}') = \int_V d\mathbf{r}' \, \overline{\mathbf{G}}_0(\mathbf{r}, \mathbf{r}') \cdot \overline{\mathbf{G}}(\mathbf{r}', \mathbf{r}'', \hat{\epsilon}) \cdot \mathbf{a} \, [k^2(\mathbf{r}) - k_0^2],$$

$$\mathbf{r} \in S, \mathbf{r}'' \in S. \quad (9.3.27)$$

In a similar manner, a new estimate of $k^2(\mathbf{r})$ can be obtained from this equation via the methods described in (8) to (21). This new estimate can again be used to update $\mathbf{E}(\mathbf{r})$. The iterative procedure is repeated until $k^2(\mathbf{r})$ and $\mathbf{E}_{sca}$ do not change. Note that in this method, $\mathbf{E}_{sca}$ remains a finite number and is less susceptible to contamination by noise.

The difference between the distorted Born iterative method and the Born iterative method is that in the distorted Born iterative method, the Green's function that propagates the field from a point in the object, $\mathbf{r}'$, to the receiver point, $\mathbf{r}$, is updated at each step. This is not so in the Born iterative method. It can be shown that the error estimate for the distorted Born approximation is of the second order, while that for the Born approximation is of the first order. Hence, the distorted Born iterative method converges faster than the Born iterative method (see Exercise 9.23; also, see the next subsection). The Born iterative method, however, is more robust.

§§9.3.3 Operator Forms of the Scattering Equations

Even though no new problems would be solved by expressing the scattering equation in an operator form, new insight into the problem is often obtained by doing so. Consider first the scattering equation for the Born iterative method given by (24). Its operator form would be (see Chapter 5; also, see Tsang and Kong 1980)

$$\mathcal{E}\rangle = \mathcal{E}_{inc}\rangle + \overline{\mathcal{G}}_0 \cdot \overline{\mathcal{O}} \cdot \mathcal{E}\rangle, \qquad (9.3.28)$$

where $\mathcal{E}\rangle$ is a state vector representing the $\mathbf{E}$ field, $\overline{\mathcal{G}}_0$ is a free-space Green's function operator, and $\overline{\mathcal{O}}$ is an operator related to the object function $k^2(\mathbf{r}) - k_0^2$. The operator form expresses the scattering equation in a form independent of the linear vector spaces. But the coordinate-space representation of the above equation can be retrieved by evaluating its inner product with a coordinate vector $\mathbf{r}\rangle$. Then, (28) becomes

$$\langle \mathbf{r}, \mathcal{E}\rangle = \langle \mathbf{r}, \mathcal{E}_{inc}\rangle + \langle \mathbf{r}, \overline{\mathcal{G}}_0 \cdot \overline{\mathcal{O}} \cdot \mathcal{E}\rangle. \qquad (9.3.29)$$

Furthermore, by noting that an identity operator is (see Chapter 5) $\overline{\mathcal{I}} = \int d\mathbf{r}\, \mathbf{r}\rangle\langle\mathbf{r}$, we could insert it into the above to yield

$$\langle \mathbf{r}, \mathcal{E}\rangle = \langle \mathbf{r}, \mathcal{E}_{inc}\rangle + \int d\mathbf{r}'d\mathbf{r}''\, \langle \mathbf{r}, \overline{\mathcal{G}}_0, \mathbf{r}'\rangle\langle \mathbf{r}', \overline{\mathcal{O}}, \mathbf{r}''\rangle\langle \mathbf{r}'', \mathcal{E}\rangle. \qquad (9.3.30)$$

Next, on identifying

$$\langle \mathbf{r}, \mathcal{E}\rangle = \mathbf{E}(\mathbf{r}), \qquad \langle \mathbf{r}, \mathcal{E}_{inc}\rangle = \mathbf{E}_{inc}(\mathbf{r}), \qquad (9.3.31a)$$
$$\langle \mathbf{r}, \overline{\mathcal{G}}_0, \mathbf{r}'\rangle = \mathbf{G}_0(\mathbf{r}, \mathbf{r}'), \qquad \langle \mathbf{r}', \overline{\mathcal{O}}, \mathbf{r}''\rangle = \delta(\mathbf{r}' - \mathbf{r}'')[k^2(\mathbf{r}') - k_0^2], \qquad (9.3.31b)$$

the above is the same as (24). Hence, the coordinate-space representation of the operator $\overline{\mathcal{O}}$ is diagonal as is evident from (31b). Moreover, $\overline{\mathcal{O}}$ is related to the object function to be reconstructed in the inverse scattering problem. Other than expressing it in coordinate space, (28) can be expressed in any space, e.g., the spectral space, by a process similar to (29) to (31) (see Exercise 9.24).

Now, we can find the solution of (28) in the operator sense. Consequently, we have, by solving (28) for $\mathcal{E}\rangle$,

$$\mathcal{E}\rangle = \left(\overline{I} - \overline{G}_0 \cdot \overline{\mathcal{O}}\right)^{-1} \cdot \mathcal{E}_{inc}\rangle. \tag{9.3.32}$$

Using (32) back in (28), we have

$$\mathcal{E}\rangle = \mathcal{E}_{inc}\rangle + \overline{G}_0 \cdot \overline{\mathcal{O}} \cdot \left(\overline{I} - \overline{G}_0 \cdot \overline{\mathcal{O}}\right)^{-1} \cdot \mathcal{E}_{inc}\rangle. \tag{9.3.33}$$

Moreover, notice now that the scattered field, which is the second term on the right of (33), is nonlinearly dependent on $\overline{\mathcal{O}}$, the operator for the object function. This is also the underlying reason for the nonlinearity of the inverse scattering problem. In addition, the factor $\left(\overline{I} - \overline{G}_0 \cdot \overline{\mathcal{O}}\right)^{-1}$ is due to multiple scattering. This becomes obvious if it is expanded in a geometrical series. By doing so, (33) becomes

$$\mathcal{E}_{sca}\rangle = \left[\overline{G}_0 \cdot \overline{\mathcal{O}} + \overline{G}_0 \cdot \overline{\mathcal{O}} \cdot \overline{G}_0 \cdot \overline{\mathcal{O}} + \overline{G}_0 \cdot \overline{\mathcal{O}} \cdot \overline{G}_0 \cdot \overline{\mathcal{O}} \cdot \overline{G}_0 \cdot \overline{\mathcal{O}} + \cdots\right] \cdot \mathcal{E}_{inc}\rangle. \tag{9.3.34}$$

It is clear that the n-th term in the above series corresponds to the incident field being scattered n times in the object. Moreover, the term $\overline{\mathcal{O}} \cdot \mathcal{E}\rangle$ can be considered the current induced by the field in the object.

Alternatively, (33) could be written as

$$\mathcal{E}_{sca}\rangle = \overline{T} \cdot \mathcal{E}_{inc}\rangle, \tag{9.3.35}$$

where

$$\overline{T} = \overline{G}_0 \cdot \overline{\mathcal{O}} \cdot \left(\overline{I} - \overline{G}_0 \cdot \overline{\mathcal{O}}\right)^{-1} = \left(\overline{I} - \overline{G}_0 \cdot \overline{\mathcal{O}}\right)^{-1} \cdot \overline{G}_0 \cdot \overline{\mathcal{O}}, \tag{9.3.36}$$

is a transition operator. The latter equality in (36) is obvious from (34).

In the inverse scattering problem, $\mathcal{E}_{sca}\rangle$ is only known at some distance remote from the object. Even though the scattered field is a nonlinear functional of $\overline{\mathcal{O}}$, the object function can nevertheless be solved for using an optimization or estimation procedure described previously. For instance, in the Born iterative method, given the measured scattered field $\hat{\mathcal{E}}_{sca}\rangle$, the object function is sought such that the norm

$$\|\hat{\mathcal{E}}_{sca}\rangle - \mathcal{E}_{sca}\rangle\| \tag{9.3.37}$$

is minimum, or

$$\left\|\hat{\mathcal{E}}_{sca}\rangle - \overline{G}_0 \cdot \overline{\mathcal{O}} \cdot \left(\overline{I} - \overline{G}_0 \cdot \overline{\mathcal{O}}\right)^{-1} \cdot \mathcal{E}_{inc}\rangle\right\|, \tag{9.3.38a}$$

which is equivalent to

$$\left\| \hat{\mathcal{E}}_{sca} \rangle - \overline{\mathcal{G}}_0 \cdot \overline{\mathcal{O}} \cdot \mathcal{E} \rangle \right\|, \tag{9.3.38b}$$

being minimum. Since the above is a nonlinear functional of $\overline{\mathcal{O}}$, it results in a nonlinear optimization problem. To reduce it to a linear problem, the total field in the object, which is $\mathcal{E}\rangle = \left(\overline{\mathcal{I}} - \overline{\mathcal{G}}_0 \cdot \overline{\mathcal{O}} \right)^{-1} \cdot \mathcal{E}_{inc}\rangle$, is approximated iteratively. First, $\mathcal{E}\rangle$ is assumed to be approximately $\mathcal{E}_{inc}\rangle$. Then, the object function $\overline{\mathcal{O}}$ is solved for as a linear optimization problem in (38). The new object function thus found is then used to estimate $\mathcal{E}\rangle = \left(\overline{\mathcal{I}} - \overline{\mathcal{G}}_0 \cdot \overline{\mathcal{O}} \right)^{-1} \cdot \mathcal{E}_{inc}\rangle$, the total field inside the object. This process is repeated until (38) is minimized. Moreover, the regularization procedure is used at every estimation to minimize the norm of $\overline{\mathcal{O}}$ as illustrated in the previous subsection.

Note that the minimization of (38) can be done in any linear vector space. Moreover, the process needed to find the operator $\left(\overline{\mathcal{I}} - \overline{\mathcal{G}}_0 \cdot \overline{\mathcal{O}} \right)^{-1}$ is precisely that needed for the direct scattering problem. For instance, the matrix representation, $\langle f_m, \overline{\mathcal{G}}_0 \cdot \overline{\mathcal{O}}, f_n \rangle$, of the operator $\overline{\mathcal{G}}_0 \cdot \overline{\mathcal{O}}$ can be found first by using some basis set, and subsequently, the inverse of the matrix representation $\langle f_m, \left(\overline{\mathcal{I}} - \overline{\mathcal{G}}_0 \cdot \overline{\mathcal{O}} \right), f_n \rangle$ can be found. This, in turn, can be used to approximate the matrix representation of $\langle f_m, \left(\overline{\mathcal{I}} - \overline{\mathcal{G}}_0 \cdot \overline{\mathcal{O}} \right)^{-1}, f_n \rangle$. But note that this is just equivalent to using the method of moments to solve the direct scattering problem corresponding to the integral equation in (24).

In the distorted Born iterative method, the incident field is incident in the presence of an inhomogeneous background medium, and hence, the corresponding Green's function is that of an inhomogeneous medium, i.e., a point source response in the inhomogeneous background medium. The corresponding scattering equation from (1) is

$$\mathcal{E}\rangle = \mathcal{E}_{b,inc}\rangle + \overline{\mathcal{G}}_b \cdot \delta\overline{\mathcal{O}} \cdot \mathcal{E}\rangle, \tag{9.3.39}$$

where

$$\mathcal{E}_{b,inc}\rangle = \mathcal{E}_{inc}\rangle + \overline{\mathcal{G}}_0 \cdot \overline{\mathcal{O}}_b \cdot \mathcal{E}_{b,inc}\rangle, \tag{9.3.40a}$$

$$\overline{\mathcal{G}}_b = \overline{\mathcal{G}}_0 + \overline{\mathcal{G}}_0 \cdot \overline{\mathcal{O}}_b \cdot \overline{\mathcal{G}}_b, \tag{9.3.40b}$$

$$\overline{\mathcal{O}} = \overline{\mathcal{O}}_b + \delta\overline{\mathcal{O}}. \tag{9.3.40c}$$

In the above, the coordinate-space representation of $\overline{\mathcal{O}}_b$ is a diagonal matrix with diagonal elements $k_b^2(\mathbf{r}) - k_0^2$. In addition, (39) reduces to (28) upon substituting the solutions of (40) into (39) (see Exercise 9.25). Note that $\mathcal{E}\rangle = \mathcal{E}_{b,inc}\rangle$ if $\delta\overline{\mathcal{O}} = 0$. Moreover, $\mathcal{E}_{b,inc}\rangle$ is obtained by solving (40a), implying that it is the field present when the inhomogeneous background is present. Furthermore, $\overline{\mathcal{G}}_b$ is obtained by solving (40b) indicating that it is the point-source response in an inhomogeneous background k_b^2.

In the inverse scattering problem, $\delta\overline{\mathcal{O}}$ is the unknown to be sought. First, Equation (39) could be solved for $\mathcal{E}\rangle$ as in (32). Then, (39) becomes

$$\mathcal{E}\rangle = \mathcal{E}_{b,inc}\rangle + \overline{\mathcal{G}}_b \cdot \delta\overline{\mathcal{O}} \cdot \left(\overline{\mathcal{I}} - \overline{\mathcal{G}}_b \cdot \delta\overline{\mathcal{O}} \right)^{-1} \cdot \mathcal{E}_{b,inc}\rangle. \tag{9.3.41}$$

Moreover, the scattered field, defined as the difference between $\mathcal{E}\rangle$ and $\mathcal{E}_{b,inc}\rangle$, is

$$\mathcal{E}_{b,sca}\rangle = \overline{\mathcal{G}}_b \cdot \delta\overline{\mathcal{O}} \cdot \left(\overline{\mathcal{I}} - \overline{\mathcal{G}}_b \cdot \delta\overline{\mathcal{O}}\right)^{-1} \cdot \mathcal{E}_{b,inc}\rangle. \qquad (9.3.42)$$

Given the measurement data $\hat{\mathcal{E}}_{sca}\rangle$, then, since $\overline{\mathcal{O}}_b$ is known, $\hat{\mathcal{E}}_{b,sca}\rangle$ can be constructed from the data. Consequently, an optimization problem can be set up to minimize

$$\left\| \hat{\mathcal{E}}_{b,sca}\rangle - \overline{\mathcal{G}}_b \cdot \delta\overline{\mathcal{O}} \cdot \left(\overline{\mathcal{I}} - \overline{\mathcal{G}}_b \cdot \delta\overline{\mathcal{O}}\right)^{-1} \cdot \mathcal{E}_{b,inc}\rangle \right\|. \qquad (9.3.43)$$

The above can be solved for $\delta\overline{\mathcal{O}}$ by assuming first that

$$\mathcal{E}\rangle = \left(\overline{\mathcal{I}} - \overline{\mathcal{G}}_b \cdot \delta\overline{\mathcal{O}}\right)^{-1} \cdot \mathcal{E}_{b,inc}\rangle \approx \mathcal{E}_{b,inc}\rangle. \qquad (9.3.43a)$$

Next, $\delta\overline{\mathcal{O}}$ is ascertained by a linear optimization procedure. Given $\delta\overline{\mathcal{O}}$, a new $\overline{\mathcal{O}}$ given by (40c) can be used as the new $\overline{\mathcal{O}}_b$, and hence, yielding new $\overline{\mathcal{G}}_b$ and $\mathcal{E}_{b,inc}\rangle$. Consequently, the process can be repeated until (43) is minimized.

Note that $\delta\overline{\mathcal{O}}$ becomes smaller when $\overline{\mathcal{O}}$ and $\overline{\mathcal{O}}_b$ becomes closer, implying that the approximation (43a) becomes increasingly good. Moreover, the approximation in (43a), when used in (43), is to effect a second-order error in (43). This is unlike the Born iterative method. Hence, the distorted Born iterative method has a second-order convergence, whereas the Born iterative method has only a first-order convergence. In the Born iterative method, a first-order error in the estimate of $\mathcal{E}\rangle$ via the use of (32), and its subsequent use in (38b) effects a first-order error in (38b).

Exercises for Chapter 9

9.1 By assuming $O(\mathbf{r})$ to be a small parameter, perform a Neumann series expansion of (9.1.1) and show that the scattered field is a nonlinear functional of $O(\mathbf{r})$. Give a physical interpretation to each term of the Neumann series expansion (also, see Subsection 9.3.3).

9.2 Using the methods in Chapter 8, derive the scattering from one scatterer and two scatterers using the $\overline{\mathbf{T}}$ matrix formulation. Show that the two-scatterer solution is not a linear superposition of a one-scatterer solution.

9.3 Show that the phase shift of a high-frequency signal, given by Equation (9.1.2), is related to the time delay of the signal through an inhomogeneous medium.

9.4 (a) Derive Equation (9.1.10) from Equation (9.1.9) by changing (9.1.9) into cylindrical coordinates.

   (b) Prove the identity (9.1.13) and, hence, (9.1.15). Then, derive (9.1.12).

   (c) Show that the project $P(\xi, \alpha)$ can be written as (9.1.16) and (9.1.17).

9.5 (a) Derive Equations (9.1.23) and (9.1.24) for three-dimensional Radon transforms.

(b) Derive a formula for inverse Radon transforms in even dimensions, which is a generalization of (9.1.12).

(c) Do the same for odd dimensions, which is a generalization of (9.1.23).

9.6 Assume that you have a transducer, which can be used as a transmitter and a receiver. The transducer can generate a broadband pulse with frequency components in the range $0 < \omega < \Omega$, and can receive in the same frequency range. Describe how you can use the transducer to generate the Fourier data of an object in the manner of diffraction tomography. What is the resolution of the reconstructed image of the object?

9.7 (a) Convince yourself that the Fourier data obtainable in the forward scattering experiment in Figure 9.1.9 consist of two disks as shown.

(b) If a backscattering experiment is included as well, show that the additional Fourier data available consist of a semicircle in the lower half plane.

9.8 (a) Derive Equation (9.2.5).

(b) Derive the equation for the characteristic curves of the wave equation, which is the hyperbolic equation $\left(\frac{\partial^2}{\partial z^2} - \frac{1}{c^2}\frac{\partial^2}{\partial t^2}\right)\phi(z,t) = 0$.

(c) Show that Equation (9.2.7) is true along the characteristic curves of a PDE, and that (9.2.8) follows from (9.2.1).

9.9 (a) Find the equivalence of (9.2.9a) and (9.2.9b) for the wave equation

$$\left(\frac{\partial^2}{\partial z^2} - \frac{1}{c^2}\frac{\partial^2}{\partial t^2}\right)\phi(z,t) = 0.$$

(b) If $\phi(z,t)$ and $\phi_t(z,t)$ are known for all z at $t = 0$, describe how you would use the method of characteristics to solve the Cauchy initial value problem.

9.10 (a) A field $\psi(z,x,t)$ satisfies the equation

$$\mu\frac{\partial}{\partial z}\mu^{-1}\frac{\partial}{\partial z}\psi + \frac{\partial^2}{\partial x^2}\psi - \frac{1}{c^2}\frac{\partial^2}{\partial t^2}\psi - \mu\sigma\frac{\partial}{\partial t}\psi = 0.$$

By writing

$$\psi(z,x,t) = \frac{1}{(2\pi)^2}\int_{-\infty}^{\infty} dk_x\, e^{ik_x x}\int_{-\infty}^{\infty} d\omega\, e^{-i\omega t}\phi(z,k_x,\omega),$$

show that $\phi(z,k_x,\omega)$ satisfies (9.2.11).

(b) If we let $k_x = \frac{\omega}{c}\cos\theta$, and define

$$\phi(z,\theta,t) = \frac{1}{2\pi}\int_{-\infty}^{\infty} d\omega\, e^{-i\omega t}\phi\left(z,\frac{\omega}{c}\cos\theta,\omega\right),$$

show that $\phi(z, \theta, t)$ satisfies (9.2.13). Find a relationship between $\phi(z, \theta, t)$ and $\psi(z, x, t)$.

9.11 In cylindrical coordinates, a wave satisfies the following equation

$$\frac{1}{\rho} \frac{\partial}{\partial \rho} \rho \frac{\partial}{\partial \rho} \phi(\rho, t) - \frac{1}{c^2(\rho)} \frac{\partial^2}{\partial t^2} \phi(\rho, t) = 0.$$

Using the method of characteristics, find a layer-stripping algorithm for $c^2(\rho)$ given the Cauchy data at $\rho = 0$ for all t.

9.12 Assume that an inhomogeneous profile consists of dispersionless fine layers with equal travel time. A plane wave consisting of a delta function in the time domain is normally incident on this inhomogeneous profile.

(a) Show that the reflected wave consists of a time series of impulse functions at equal intervals with different amplitudes.

(b) Derive an algorithm from which you can unravel the wave impedance of each layer using causality.

9.13 Show that if the medium is a conductive lossy medium, i.e., $\epsilon = \epsilon' + \frac{i\sigma}{\omega}$, then the corresponding equation of (9.2.30) in the time domain is not (9.2.21). What is the corresponding equation in the time domain in this case?

9.14 (a) Explain why the boundary condition defined by (9.2.36) is equivalent to a sheet source backed by an impedance boundary.

Hint: A transmission line analogy may be useful here.

(b) Explain why the radiation condition that a wave is outgoing when $\zeta \to \infty$ is equivalent to the causality condition in the time domain.

9.15 By using the method of characteristics, show that the Cauchy data provided by (9.2.50a) and (9.2.50b) affect the triangular region shown in Figure 9.2.5. Show that $G(\zeta, t)$ is of the form given by (9.2.51).

9.16 By going through an analysis similar to (9.2.40) to (9.2.46), prove the equality (9.2.52a) and (9.2.52b).

9.17 (a) By discretizing Equation (9.2.64), describe how you would solve the resultant equation for $K(\zeta, t)$.

(b) Similarly, describe how you would solve Equation (9.2.72) for $K(\zeta, t)$.

9.18 (a) Show that the Gel'fand-Levitan equation given by (9.2.90) is equivalent to probing a one-dimensional profile with an impressed sheet source on a perfect magnetic wall.

(b) Similarly, show that the Marchenko integral equation (9.2.91) is equivalent to probing a one-dimensional profile for $\zeta > 0$ where $V(\zeta) = 0$, $\zeta < 0$.

9.19 Nonradiating sources may give rise to nonuniqueness in the inverse scattering problem.

(a) Define an arbitrary field $\psi(\mathbf{r})$, which has finite support, i.e., $\psi(\mathbf{r}) = 0$, $\mathbf{r} \notin V$, $\psi(\mathbf{r}) \neq 0$, $\mathbf{r} \in V$. Derive a source $S(\mathbf{r})$ that will generate this field $\psi(\mathbf{r})$. What is the support of $S(\mathbf{r})$?

(b) Because of the existence of such nonradiating sources, show that

$$\int d\mathbf{r}' \, g(\mathbf{r}, \mathbf{r}') S(\mathbf{r}') = 0, \quad \mathbf{r} \notin V,$$

where $g(\mathbf{r}, \mathbf{r}')$ satisfies $(\nabla^2 + k_b^2) g(\mathbf{r}, \mathbf{r}') = \delta(\mathbf{r} - \mathbf{r}')$. Hence, show that the integral operator in the above has a nullspace.

(c) Show that the inverse scattering equation for a scalar wave problem is

$$\phi_{sca}(\mathbf{r}) = \int\limits_V d\mathbf{r}' \, g(\mathbf{r}, \mathbf{r}') [k^2(\mathbf{r}') - k_b^2] \phi_{inc}(\mathbf{r}').$$

(d) Show that the integral operator $\int d\mathbf{r}' \, g(\mathbf{r}, \mathbf{r}') \phi_{inc}(\mathbf{r}')$, which operates on $k^2(\mathbf{r}') - k_b^2$, has a nullspace. Give an example of a state vector which is in the nullspace of this integral operator.

(e) Show that this nullspace can be reduced or squeezed by using more measurement data from different transmitter locations. In this case, the inverse scattering equation is

$$\phi_{sca}(\mathbf{r}, \mathbf{r}'') = \int\limits_V d\mathbf{r}' \, g(\mathbf{r}, \mathbf{r}') [k^2(\mathbf{r}') - k_b^2] \phi_{inc}(\mathbf{r}', \mathbf{r}''),$$

where $\mathbf{r}''$ denotes the locations of the transmitter.

9.20 (a) Derive Equation (9.3.13) from (9.3.11).

(b) If $b_n(\mathbf{r})$ is chosen to be pulse functions in (9.3.9), show that $\mathbf{B}$ in (9.3.14) is a diagonal matrix.

9.21 Another way of regularizing (9.3.8) is to regularize the derivatives. Then, (9.3.8) becomes

$$I = \delta_t \int\limits_V d\mathbf{r}' \left| \nabla [k^2(\mathbf{r}') - k_b^2] \right|^2$$

$$+ \sum_{i,j} \left| \mathbf{E}_{sca}(\mathbf{r}_i, \mathbf{r}_j) - \int\limits_V d\mathbf{r}' \, \mathbf{M}(\mathbf{r}_i, \mathbf{r}_j, \mathbf{r}', \epsilon_b) [k^2(\mathbf{r}') - k_b^2] \right|^2.$$

The derivative ∇ can be approximated by a finite-difference approximation if $\mathbf{r}'$ is discretized into a grid. The integral can be replaced by some integration rules. Derive a corresponding matrix equation for the optimal solution of the above functional.

9.22 The matrix $\overline{\mathbf{P}}$ in (9.3.17) has zero eigenvalues due to the inherent non-uniqueness of the inverse problem using the distorted Born iterative method.

(a) Show that **a** obtained by (9.3.21), where the zero eigenvalues of $\overline{\lambda}$ have been removed, is the least-norm solution.

(b) Show that the regularization method is like padding the zero eigenvalues of $\overline{\mathbf{P}}$ with a small nonzero value.

9.23 The Newton-Raphson method of finding the root of the equation $f(x) = 0$ is to start with a guess x_0. Then, $f(x)$ is Taylor series expanded around x_0 to yield $f(x) \simeq f(x_0) + (x - x_0)f'(x_0) \simeq 0$. Hence, the root to $f(x) = 0$ is approximately given by $x \simeq x_0 - \frac{f(x_0)}{f'(x_0)}$.

(a) Show that when x_0 is very close to the exact answer x_e, the error in the Newton-Raphson method is of second order.

(b) Show that the distorted Born iterative method is analogous to (in one dimension) solving the equation

$$m = m_0(\epsilon_b) + g(\epsilon_b)\,(\epsilon - \epsilon_b)\,\phi(\epsilon),$$

where m, $m_0(\epsilon_b)$, and $g(\epsilon_b)$ are known, and ϕ as a function of ϵ is known, and the unknown to be sought is ϵ. Show that the error in the distorted Born iterative method is of higher order and, hence, is analogous to the Newton-Raphson method in higher dimensions.

(c) Another way of solving for the root of the equation $x = h(x)$ is via an iterative method. The n-th estimated root can be expressed in terms of the $(n-1)$-th estimated root as $x_n = h(x_{n-1})$. Prove that a necessary condition for the convergence of such a method is that $|h'(x_e)| < 1$.

(d) Show that the Born iterative method is the multidimensional equivalent of the method in part (c). Given the N-dimensional equation $\mathbf{x} = \mathbf{h}(\mathbf{x})$, if the method of part (c) is used to find its root, what is the necessary criterion for convergence?

9.24 (a) Find the representation of (9.3.28) in the spectral space. What do you notice is peculiar about $\overline{\mathcal{G}}_0$ and $\overline{\mathcal{O}}$ in the spectral space?

(b) Next, find the representation of (9.3.28) in a space spanned by the eigenfunction of $\overline{\mathcal{G}}_0$, i.e., $\mathcal{F}_n\rangle$ such that $\overline{\mathcal{G}}_0, \mathcal{F}_n\rangle = \lambda \mathcal{F}_n\rangle$. Would the matrix representation of $\overline{\mathcal{G}}_0$ be diagonal? What about the matrix representation of $\overline{\mathcal{O}}$?

9.25 For the distorted Born iterative method:

(a) Show that $\mathcal{E}_{b,inc}\rangle = \left(\overline{\mathcal{I}} - \overline{\mathcal{G}}_0 \cdot \overline{\mathcal{O}}_b\right)^{-1} \cdot \mathcal{E}_{inc}\rangle$ and $\overline{\mathcal{G}}_b = \left(\overline{\mathcal{I}} - \overline{\mathcal{G}}_0 \cdot \overline{\mathcal{O}}_b\right)^{-1} \cdot \overline{\mathcal{G}}_0$ from (9.3.40).

(b) Show that (9.3.39) reduces to (9.3.28) after substituting the solution of (9.3.40) into it.

(c) Show that by letting $\overline{\mathcal{O}} = \overline{\mathcal{O}}_b + \delta\overline{\mathcal{O}}$ in (9.3.28), it reduces to (9.3.39). Hence, Equation (9.3.39) is like Taylor series expanding (9.3.28) about the value $\overline{\mathcal{O}}_b$.

(d) Show that in the optimization of both (9.3.38) and (9.3.43), the operator $\left(\overline{\mathcal{I}} - \overline{\mathcal{G}}_0 \cdot \overline{\mathcal{O}}\right)^{-1}$ is required, i.e., both require the solution of the direct scattering problem. Hence, if the matrix representation of $\left(\overline{\mathcal{I}} - \overline{\mathcal{G}}_0 \cdot \overline{\mathcal{O}}\right)$ is found with N basis functions as in the method of moments so that its inverse can be found, convince yourself that the computational effort in finding both (9.3.38) and (9.3.43) are about the same.

References for Chapter 9

Aki, K., and P. G. Richards. 1980. *Quantitative Seismology.* New York: Freeman.

Ames, W. F. 1977. *Numerical Methods for Partial Differential Equations.* New York: Academic Press.

Baker, C. T. H. 1977. *The Numerical Treatment of Integral Equations.* Oxford: Clarendon.

Balanis, G. N. 1972. "The plasma inverse problem." *J. Math. Phys.* 13: 1001–05.

Berryman, J. G., and R. R. Greene. 1980. "Discrete inverse methods for elastic waves in layered media." *Geophysics* 45: 213–33.

Bleistein N., and J. K. Cohen. 1977. "Nonuniqueness in the inverse source problem in acoustics and electromagnetics." *J. Math. Phys.* 18: 194–201.

Bracewell, R. N. 1956. "Strip integration in radioastronomy." *Aust. J. Phys.* 9: 198–217.

Bube, K. P., and R. Burridge. 1983. "The one-dimensional problem of reflection seismology." *SIAM Review* 25(4): 497–559.

Burridge, R. 1980. "The Gel'fand-Levitan, the Marchenko, and the Gopinath-Sondhi integral equation of inverse scattering theory, regarded in the context of inverse impulse-response problems." *Wave Motion* 2: 305–23.

Chew, W. C., and Y. M. Wang. 1990. "Reconstruction of two-dimensional permittivity distribution using the distorted Born iterative method." *IEEE Trans. Med. Imag.* 9(2): 218-25.

Deans, S. R. 1983. *The Radon Transform and Some of Its Applications.* New York: John Wiley & Sons.

Devaney, A. J. 1978. "Nonuniqueness in the inverse scattering problem." *J. Math. Phys.* 19(7): 1526–31.

Devaney, A. J. 1982. "A filtered backpropagation algorithm for diffraction tomography." *Ultrasonic Imaging* 4: 336–60.

Devaney, A. J. 1983. "A computer simulation study of diffraction tomography." *IEEE Trans. Biomed. Eng.* 30: 377–86.

Devaney, A. J., and G. C. Sherman. 1982. "Nonuniqueness in inverse source and scattering problems." *IEEE Trans. Antennas Propagat.* 8: 1034–42.

Garabedian, P. R. 1964. *Partial Differential Equations.* New York: John Wiley & Sons.

Golub, G. H., and W. Kahan. 1965. "Calculating the singular values and pseudo inverse of a matrix." *SIAM J. Numer. Anal.* 2: 205.

Goupillaud, P. L. 1961. "An approach to inverse filtering of near-surface layer effects from seismic records." *Geophysics* 26: 754–60.

Habashy, T. M., and R. Mittra. 1987. "On some inverse methods in electromagnetics." *J. of Electromagnetic Waves and Applications* 1: 25–28.

Herman, G. T. 1980. *Image Reconstruction from Projections.* New York: Academic Press.

Johnson, S. J., and M. L. Tracy. 1983a. "Inverse scattering solutions by a sinc basis, multiple source, moment method – part I: theory." *Ultrasonic Imaging* 5: 361–75.

Johnson, S. J., and M. L. Tracy. 1983b. "Inverse scattering solutions by a sinc basis, multiple source, moment method – part II: numerical evaluations." *Ultrasonic Imaging* 5: 376–92.

Kak, A. C. 1979. "Computerized tomography with X-ray, emission, and ultrasound sources." *Proc. IEEE* 67(9): 1245–72.

Lee, H., and G. Wade. 1986. *Modern Acoustical Imaging.* New York: IEEE Press.

Levy, B. C. 1985. "Layer by layer reconstruction methods for the earth resistivity from direct current measurements." *IEEE Trans. Geosci. Remote Sensing* GE-23: 841–50.

Mueller, R. K., M. Kaveh, and G. Wade. 1979. "Reconstructive tomography and applications to ultrasonics." *Proc. IEEE* 67: 567–87.

Newton, R. G. 1980a. "Inverse scattering. I. One dimension." *J. Math. Phys.* 21(3): 493–505.

Newton, R. G. 1980b. "Inverse scattering. II. Three dimensions." *J. Math. Phys.* 21(7): 1698–1715.

Newton, R. G. 1981. "Inverse scattering. III. Three dimensions, continued." *J. Math. Phys.* 22(10): 2191–2200.

Radon, J. 1917. "Uber die bestimmung von funktionen durch ihre intergralwerte langs gewisser mannigfaltigkaiten." *Berichte Saechsische Akademie der Wissenschaften* 69: 262–77.

Sezginer, A. 1985. *Forward and Inverse Problems in Transient Electromagnetic Fields.* Ph.D. thesis, M.I.T.

Sommerfeld, A. 1949. *Partial Differential Equations in Physics.* New York: Academic Press.

Tikhonov, A. N., and V. Y. Arsenin. 1977. *Solution of Ill-posed Problems.* Washington DC: V. H. Winston and Sons.

Tsang, L., and J. A. Kong. 1980. "Multiple scattering of electromagnetic waves by random distribution of discrete scatterers with coherent potential and quantum mechanical formulism." *J. Appl. Phys.* 15: 3465-85.

Wang, Y. M., and W. C. Chew. 1989. "An iterative solution of two-dimensional electromagnetic inverse scattering problem." *Int. Jour.*

Imaging Sys. Tech. 1: 100–08.

Ware, J. A., and K. Aki. 1969. "Continuous and discrete inverse-scattering problems in a stratified elastic medium." *J. Acoust. Soc. Am.* 45(4): 911.

Wilkinson, J. H. 1965. *The Algebraic Eigenvalue Problem.* Oxford: Clarendon.

Further Readings for Chapter 9

Angell, T., and R. Kleinman. 1987. "A new algorithm for inverse scattering problems." *Meth. Verf. Math. Physics* 33: 41–57.

Azimi, M., and A. C. Kak. 1983. "Distortion in diffraction tomography caused by multiple scattering." *IEEE Trans. Med. Imaging* 2: 176–95.

Blok, H., and A. G. Tijhuis. 1983. "One-dimensional time domain inverse scattering applicable to electromagnetic imaging." *Proceed. NATO Ad. Res. Workshop*, BAD, September, Windheim.

Bojarski, N. N. 1980. "One-dimensional direct and inverse scattering in causal space." *Wave Motion* 2: 115–24.

Bolomey, J.-Ch., Ch. Durix, and D. Lesselier. 1979. "Determination of conductivity profiles by time-domain reflectometry." *IEEE Transactions Antennas Propagat.* 27: 244–48.

Bruckstein, A. M., and T. Kailath. 1987. "Inverse scattering for discrete transmission-line models." *SIAM Review* 29: 359–89.

Chadan, K., and P. C. Sabatier. 1977. *Inverse Problems in Quantum-Scattering Theory.* New York: Springer-Verlag.

Chew, W. C., and S. L. Chuang. 1984. "Profile inversion of a planar medium with a line source or a point source." *Proceedings of IGARSS'84 Symposium*, August, Strasbourg, France.

Coen, S. 1981. "Inverse scattering of a layered and dispersionless dielectric half-space, Part I: reflection data from plane waves at normal incidence." *IEEE Trans. Antennas Propagat.* AP-29: 298–306.

Coen, S., and W. Yu. 1981. "The inverse problem of the direct current conductivity profile of layered earth." *Geophysics* 46: 1702–13.

Coen, S., K. K. Mei, and D. J. Angelakos. 1981. "Inverse scattering technique applied to remote sensing of layered media." *IEEE Trans. Antennas Propagat.* AP-29(2): 298–306.

Colton, D., and P. Monk. 1988. "The inverse scattering problem for time harmonic acoustic waves in an inhomogeneous medium." *Q. J. Mech. Appl. Math.* 41: 97–125.

Crase, E., A. Pica, M. Noble, J. McDonald, and A. Tarantola. 1990. "Robust nonlinear waveform inversion: Application to real data." *Geophysics.* 55(5): 527-38.

Deift, P., and E. Trubowitz. 1979. "Inverse scattering on the line." *Comm. Pure., Appl. Math.* 32: 121–251.

Devaney, A. J. 1983. "A computer simulation study of diffraction tomography." *IEEE Trans. Biomed. Eng.* 30: 377–86.

Devaney, A. J., and E. Wolf. 1973. "Radiating and non-radiating classical current distributions and the fields they generate." *Phys. Rev. D* 8: 1044–47.

Farhat, N. H., C. L. Werner, and T. H. Chu. 1984. "Prospect for three-dimensional projective and tomographic imaging radar network." *Radio Sci.* 19: 1347–55.

Fawcett, J. 1984. "On the stability of inverse scattering problems." *Wave Motion* 6: 489–99.

Frank, M., and C. A. Balanis. 1989. "Method for improving the stability of electromagnetic geophysical inversions." *IEEE Trans. Geosci. Remote Sensing* GE-27: 339–43.

Ge, D. B. 1987a. "Reconstruction of conductivity profiles from pulse response." *IEEE Trans. Antennas Propagat.* AP-35: 1185–87.

Ge, D. B. 1987b. "An iterative technique in one-dimensional profile inversion." *Inverse Problems* 3: 399–406.

Ghodgonkar, D. K., O. P. Gandhi, and M. J. Hagmann. 1983. "Estimation of complex permittivities of three-dimensional inhomogeneous biological bodies." *IEEE Trans. Microwave Theory Tech.* MTT-31: 442–46.

Gilbert, F. 1971. "Ranking and winnowing gross earth data for inversion and resolution." *Geophys. J. R. Astron. Soc.* 23: 125–28.

Habashy, T. M., W. C. Chew, and E. Y. Chow. 1986. "Simultaneous reconstruction of permittivity and conductivity profiles in a radially inhomogeneous slab." *Radio Sci.* 21: 635–45.

Jaggard, D. L., and P. V. Frangos. 1987. "The electromagnetic inverse scattering problem for layered dispersionless dielectrics." *IEEE Trans. Antennas Propagat.* AP-35: 934–46.

Jaggard, D. L., and Y. Kim. 1985. "Accurate one-dimensional inverse scattering using a nonlinear renormalization technique." *J. Opt. Soc. Am.* A2: 1922–30.

Jaggard, D. L., and K. E. Olson. 1985. "Numerical reconstruction for dispersionless refractive profiles." *J. Opt. Soc. Am.* A2: 1931–36.

Jaulent, M. 1976. "Inverse scattering problems for absorbing media." *J. Math. Phys.* 17: 1351–60.

Kay, I., and H. E. Moses. 1982. *Inverse Scattering Papers, 1955–1963.* Math. Sci. Press.

Keller, J. B. 1969. "Accuracy and validity of the Born and Rytov approximations." *J. Opt. Soc. Am.* 59: 1003–04.

Kristensson, G., and R. J. Krueger. 1986a. "Direct and inverse scattering in the time domain for a dissipative wave equation. Part I: scattering operators." *J. Math. Phys.* 27: 1667–82.

Kristensson, G., and R. J. Krueger. 1986b. "Direct and inverse scattering in the time domain for a dissipative wave equation. Part II: simultaneous reconstruction of dissipation and phase velocity profiles." *J. Math. Phys.* 27: 1683–93.

Kristensson, G., and R. J. Krueger. 1987. "Direct and inverse scattering in the time domain for a dissipative wave equation. Part III: scattering operators in the presence of a phase velocity mismatch." *J. Math. Phys.* 28: 360–70.

Ladouceur, H. D., and A. K. Jordan. 1985. "Renormalization of an inverse-scattering theory for inhomogeneous dielectrics." *J. Opt. Soc. Am.* A2: 1916–21.

Langenberg, K. J. 1986. *Applied Inverse Problems.* Kassel, Germany: Fachgebiet Theoretische Elektrotechnik der Gesamthochschule Kassel.

Lee, C. Q. 1982. "Wave propagation and profile inversion in lossy inhomogeneous media." *Proc. IEEE* 70: 219–28.

Lesselier, D. 1978. "Determination of index profiles by time domain reflectometry." *J. Optics* 9: 349–58.

Lesselier, D. 1982. "Optimization techniques and inverse problems: reconstruction of conductivity profiles in the time domain." *IEEE Trans. Antennas Propagat.* AP-30: 59–65.

Lewis, R. M. 1969. "Physical optics inverse diffraction." *IEEE Trans. Antennas Propagat.* AP-17: 308–14.

Mendel, J. M., and F. Habibi-Ashrafi. 1980. "A survey of approaches to solving inverse problems for lossless layered media systems." *IEEE Trans. Geosci. Remote Sensing* GE-18: 320–30.

Mensa, D., G. Heiddbreder, and G. Wade. 1983. "Coherent doppler tomography for microwave imaging." *Proc. IEEE* 71: 245–61.

Mora, P. 1987. "Nonlinear two-dimensional elastic inversion of multioffset seismic data." *Geophysics,* 52(9): 1211-28.

Moses, H. E. 1959. "Solution of Maxwell's equations in terms of a spinor notation: The direct and inverse problem." *Phys. Rev.* 113(6): 1670–79.

Mostafavi, M., and R. Mittra. 1972. "Remote probing of inhomogeneous media using parameter optimization techniques." *Radio Sci.* 7: 1105–11.

Ney, M. M., A. M. Smith, and S. S. Stuchly. 1984. "A solution of electromagnetic imaging using pseudoinverse transformation." *IEEE Trans. Medical Imaging* MI-3: 155–162.

Porter, R. P. 1970. "Diffraction-limited scalar image formation with holograms of arbitrary shape." *J. Opt. Soc. Am.* 60: 1051–59.

Porter, R. P., and A. J. Devaney. 1982. "Holography and the inverse source problem." *J. Opt. Soc. Am.* 72: 327–30.

Reed, M., and B. Simon. 1972. *Functional Analysis.* New York: Academic.

Robinson, E. A., and S. Treitel. 1980. *Geophysical Signal Analysis.* Englewood Cliffs, NJ: Prentice-Hall.

Roger, A., D. Maystre, and M. Cadilhac. 1978. "On a problem of inverse scattering in optics: The dielectric inhomogeneous medium." *J. Opt.* 9(2): 83–90.

Rose, J. H., M. Cheney, and B. DeFacio. 1984. "The connection between time- and frequency-domain three-dimensional inverse scattering methods." *J. Math. Phys.* 25: 2995–3000.

Rose, J. H., M. Cheney, and B. DeFacio. 1985. "Three-dimensional inverse scattering: Plasma and variable velocity wave equations." *J. Math. Phys.* 26: 2803–13.

Ross, G., M. A. Fiddy, and M. Nieto-Vesperinas. 1979. "The inverse scattering problem in structural determinations." *Inverse Scattering Problems*, ed. H. P. Baltes. New York: Springer-Verlag.

Sabatier, P. C. 1983. "Theoretical considerations for inverse scattering." *Radio Sci.* 18: 1–18.

Sarkar, T. K., D. D. Weiner, and V. K. Jain. 1981. "Some mathematical considerations in dealing with the inverse problem." *IEEE Trans. Antennas Propagat.* AP-29: 373–79.

Schaubert, D. H., and R. Mittra. 1977. "A spectral domain method for remotely probing stratified media." *IEEE Trans. Antennas Propagat.* AP-25: 261–65.

Slaney, M., A. C. Kak, and L. E. Larsen. 1984. "Limitations of imaging with first-order diffraction tomography." *IEEE Trans. Microwave Theory Tech.* MTT-32: 860–74.

Tabbara, W. 1979. "Reconstruction of permittivity profiles from a spectral analysis of the reflection coefficient." *IEEE Trans. Antennas Propagat.* AP-27(2): 241–48.

Tabbara, W., B. Duchêne, Ch. Pichot, D. Lesselier, L. Chommeloux, and N. Joachimowicz. 1988. "Diffraction tomography: contribution to the analysis of applications in microwaves and ultrasonics." *Inverse Problem* 4: 305–31.

Tarantola, A. 1984. "The seismic reflection inverse problem." in *Inverse Problems of Acoustic and Elastic Waves,* eds. F. Sentosa, Y. H. Pao, W. Symes, and C. Holland. Soc. Industr. Appl. Math.

Tijhuis, A. G. 1981. "Iterative determination of permittivity and conductivity profiles of a dielectric slab in the time domain." *IEEE Trans. Antennas Propagat.* AP-29: 239–45.

Tijhuis, A. G. 1987. *Electromagnetic Inverse Profiling: Theory and Numeri-*

cal Implementation. Utrecht, The Netherlands: VNU Science Press.

Tijhuis, A. G., and C. Van der Worm. 1984. "Iterative appproach to the frequency-domain solution of the inverse-scattering problem for an inhomogeneous lossless dielectric slab." *IEEE Trans. Antennas Propagat.* AP-32(7): 711–16.

Tijhuis, A. 1989. "Born-type reconstruction of material parameters of an inhomogeneous lossy dielectric slab from reflected-field data." *Wave Motion* 11: 151–73.

Tricomi, F. G. 1985. *Integral Equations.* New York: Dover Publications.

Twomey, S. 1977. *Introduction to the Mathematics of Inversion in Remote Sensing and Indirect Measurements.* New York: Elsevier Scientific.

Weston, V. H. 1972. "On the inverse problem for a hyperbolic dispersive partial differential equation." *J. Math. Phys.* 13: 1952–56.

Weston, V. H. 1974. "On inverse scattering." *J. Math. Phys.* 15: 209–13.

Weston, V. H., and R. J. Krueger. 1972. "On the inverse scattering problem for a hyperbolic dispersive partial differential equation. II." *J. Math. Phys.* 14: 406–08.

Wiggins, R. A. 1972. "The general linear inverse problem: Implication of surface waves and free oscillations for earth structure." *Rev. Geophys.* 10: 251–85.

Wolf, E. 1969. "Three-dimensional structure determination of semitransparent objects from holographic data." *Opt. Commun.* 1: 153–56.

Wolf, E., and R. P. Porter. 1986. "On the physical contents of some integral equations for inverse scattering from inhomogeneous objects." *Radio Sci.* 21: 627.

Yagle, A. E., and B. C. Levy. 1986. "Layer-stripping solutions of multidimensional inverse scattering problems." *J. Math. Phys.* 27: 1701–10.

APPENDIX A

Some Useful Mathematical Formulas

A.1 Useful Vector Identities

$$\mathbf{a} \cdot (\mathbf{b} \times \mathbf{c}) = \mathbf{b} \cdot (\mathbf{c} \times \mathbf{a}) = \mathbf{c} \cdot (\mathbf{a} \times \mathbf{b}), \tag{A.1}$$

$$\mathbf{a} \times (\mathbf{b} \times \mathbf{c}) = \mathbf{b}(\mathbf{a} \cdot \mathbf{c}) - \mathbf{c}(\mathbf{a} \cdot \mathbf{b}), \tag{A.2}$$

$$\nabla \times \nabla \psi = 0, \tag{A.3}$$

$$\nabla \cdot \nabla \times \mathbf{A} = 0, \tag{a.4}$$

$$\nabla \cdot (\psi \mathbf{A}) = \mathbf{A} \cdot \nabla \psi + \psi \nabla \cdot \mathbf{A}, \tag{A.5}$$

$$\nabla \times (\psi \mathbf{A}) = \nabla \psi \times \mathbf{A} + \psi \nabla \times \mathbf{A}, \tag{A.6}$$

$$\nabla \cdot (\mathbf{A} \times \mathbf{B}) = \mathbf{B} \cdot \nabla \times \mathbf{A} - \mathbf{A} \cdot \nabla \times \mathbf{B}, \tag{A.7}$$

$$\nabla(\mathbf{A} \cdot \mathbf{B}) = (\mathbf{A} \cdot \nabla)\mathbf{B} + (\mathbf{B} \cdot \nabla)\mathbf{A} + \mathbf{A} \times \nabla \times \mathbf{B} + \mathbf{B} \times \nabla \times \mathbf{A}, \tag{A.8}$$

$$\nabla \times (\mathbf{A} \times \mathbf{B}) = (\mathbf{B} \cdot \nabla)\mathbf{A} - (\mathbf{A} \cdot \nabla)\mathbf{B} + \mathbf{A}\nabla \cdot \mathbf{B} - \mathbf{B}\nabla \cdot \mathbf{A}, \tag{A.9}$$

$$\nabla \times \nabla \times \mathbf{A} = \nabla\nabla \cdot \mathbf{A} - \nabla^2 \mathbf{A}. \tag{A.10}$$

In Cartesian coordinates, $\nabla^2 \mathbf{A}$ can be decomposed as

$$\nabla^2 \mathbf{A} = \hat{x}\nabla^2 A_x + \hat{y}\nabla^2 A_y + \hat{z}\nabla^2 A_z, \tag{A.11}$$

because ∇^2 commutes with $\hat{x}$, $\hat{y}$, and $\hat{z}$, i.e., $\nabla^2 \hat{x} = \hat{x}\nabla^2$ and so on. This is not true in other curvilinear coordinates; hence, this decomposition is not allowed.

A.2 Gradient, Divergence, Curl, and Laplacian in Rectangular, Cylindrical, Spherical, and General Orthogonal Curvilinear Coordinate Systems

(a) Rectangular System; x, y, z:

$$\nabla\psi = \frac{\partial \psi}{\partial x}\hat{x} + \frac{\partial \psi}{\partial y}\hat{y} + \frac{\partial \psi}{\partial z}\hat{z}, \tag{A.12}$$

$$\nabla \cdot \mathbf{A} = \frac{\partial A_x}{\partial x} + \frac{\partial A_y}{\partial y} + \frac{\partial A_z}{\partial z}, \tag{A.13}$$

$$\nabla \times \mathbf{A} = \left(\frac{\partial A_z}{\partial y} - \frac{\partial A_y}{\partial z}\right)\hat{x} + \left(\frac{\partial A_x}{\partial z} - \frac{\partial A_z}{\partial x}\right)\hat{y} + \left(\frac{\partial A_y}{\partial x} - \frac{\partial A_x}{\partial y}\right)\hat{z}, \tag{A.14}$$

$$\nabla^2\psi = \frac{\partial^2 \psi}{\partial x^2} + \frac{\partial^2 \psi}{\partial y^2} + \frac{\partial^2 \psi}{\partial z^2}. \tag{A.15}$$

(b) Cylindrical System; ρ, ϕ, z:

$$\nabla\psi = \frac{\partial\psi}{\partial\rho}\hat{\rho} + \frac{1}{\rho}\frac{\partial\psi}{\partial\phi}\hat{\phi} + \frac{\partial\psi}{\partial z}\hat{z}, \qquad (A.16)$$

$$\nabla\cdot\mathbf{A} = \frac{1}{\rho}\frac{\partial}{\partial\rho}(\rho A_\rho) + \frac{1}{\rho}\frac{\partial A_\phi}{\partial\phi} + \frac{\partial A_z}{\partial z}, \qquad (A.17)$$

$$\nabla\times\mathbf{A} = \left(\frac{1}{\rho}\frac{\partial A_z}{\partial\phi} - \frac{\partial A_\phi}{\partial z}\right)\hat{\rho} + \left(\frac{\partial A_\rho}{\partial z} - \frac{\partial A_z}{\partial\rho}\right)\hat{\phi} + \frac{1}{\rho}\left(\frac{\partial}{\partial\rho}(\rho A_\phi) - \frac{\partial A_\rho}{\partial\phi}\right)\hat{z},$$
$$(A.18)$$

$$\nabla^2\psi = \frac{1}{\rho}\frac{\partial}{\partial\rho}\left(\rho\frac{\partial\psi}{\partial\rho}\right) + \frac{1}{\rho^2}\frac{\partial^2\psi}{\partial\phi^2} + \frac{\partial^2\psi}{\partial z^2}. \qquad (A.19)$$

(c) Spherical System; r, θ, ϕ:

$$\nabla\psi = \frac{\partial\psi}{\partial r}\hat{r} + \frac{1}{r}\frac{\partial\psi}{\partial\theta}\hat{\theta} + \frac{1}{r\sin\theta}\frac{\partial\psi}{\partial\phi}\hat{\phi}, \qquad (A.20)$$

$$\nabla\cdot\mathbf{A} = \frac{1}{r^2}\frac{\partial}{\partial r}(r^2 A_r) + \frac{1}{r\sin\theta}\frac{\partial}{\partial\theta}(\sin\theta A_\theta) + \frac{1}{r\sin\theta}\frac{\partial A_\phi}{\partial\phi}, \qquad (A.21)$$

$$\nabla\times\mathbf{A} = \frac{1}{r\sin\theta}\left[\frac{\partial}{\partial\theta}(\sin\theta A_\phi) - \frac{\partial A_\theta}{\partial\phi}\right]\hat{r} + \frac{1}{r}\left[\frac{1}{\sin\theta}\frac{\partial A_r}{\partial\phi} - \frac{\partial}{\partial r}(r A_\phi)\right]\hat{\theta}$$
$$+ \frac{1}{r}\left[\frac{\partial}{\partial r}(r A_\theta) - \frac{\partial A_r}{\partial\theta}\right]\hat{\phi}, \quad (A.22)$$

$$\nabla^2\psi = \frac{1}{r^2}\frac{\partial}{\partial r}\left(r^2\frac{\partial\psi}{\partial r}\right) + \frac{1}{r^2\sin\theta}\frac{\partial}{\partial\theta}\left(\sin\theta\frac{\partial\psi}{\partial\theta}\right) + \frac{1}{r^2\sin^2\theta}\frac{\partial^2\psi}{\partial\phi^2}. \qquad (A.23)$$

(d) General Orthogonal Curvilinear Coordinate System; x_1, x_2, x_3:

The metric coefficients (h_1, h_2, h_3) in a general orthogonal curvilinear coordinate system are defined by

$$ds_i = h_i\,dx_i; \quad i = 1 \text{ or } 2, \text{ or } 3, \qquad (A.24)$$

where ds_i denotes a differential length in the direction of dx_i. Moreover, the variable, x_i may not have the dimension of length. One way of finding the metric coefficients is to express the rectangular variables in terms of the variables of that system:

$$x = y(x_1, x_2, x_3),$$

$$y = y(x_1, x_2, x_3),$$

$$z = z(x_1, x_2, x_3).$$

Then

$$ds_i = \left[\left(\frac{\partial x}{\partial x_i} \right)^2 + \left(\frac{\partial y}{\partial x_i} \right)^2 + \left(\frac{\partial z}{\partial x_i} \right)^2 \right]^{1/2} dx_i, \quad i = 1, 2, 3. \qquad (A.25)$$

Hence,

$$h_i = \left[\left(\frac{\partial x}{\partial x_i} \right)^2 + \left(\frac{\partial y}{\partial x_i} \right)^2 + \left(\frac{\partial z}{\partial x_i} \right)^2 \right]^{1/2}. \qquad (A.26)$$

For instance, in an elliptical coordinate system,

$$x = c \cosh u \cos v, \qquad (A.27)$$

$$y = c \sinh u \sin v. \qquad (A.28)$$

If (x_1, x_2, x_3) represent (u, v, z), then by applying (26), we have

$$h_1 = h_2 = c(\sinh^2 u \cos^2 v + \cosh^2 u \sin^2 v)^{1/2} = c(\cosh^2 u - \cos^2 v)^{1/2},$$
$$(A.29)$$
$$h_3 = 1. \qquad (A.30)$$

In general, for any orthogonal curvilinear coordinate system,

$$\nabla \psi = \sum_{i=1}^{3} \frac{1}{h_i} \frac{\partial \psi}{\partial x_i} \hat{x}_i, \qquad (A.31)$$

$$\nabla \cdot \mathbf{A} = \frac{1}{\Delta} \sum_{i=1}^{3} \frac{\partial}{\partial x_i} \left(\frac{\Delta A_i}{h_i} \right), \quad \Delta = h_1 h_2 h_3, \qquad (A.32)$$

$$\nabla \times \mathbf{A} = \frac{1}{\Delta} \left| \begin{pmatrix} h_1 \hat{x}_1 & h_2 \hat{x}_2 & h_3 \hat{x}_3 \\ \frac{\partial}{\partial x_1} & \frac{\partial}{\partial x_2} & \frac{\partial}{\partial x_3} \\ h_1 A_1 & h_2 A_2 & h_3 A_3, \end{pmatrix} \right|, \qquad (A.33)$$

$$\nabla^2 \psi = \frac{1}{\Delta} \sum_{i=1}^{3} \frac{\partial}{\partial x_i} \left(\frac{\Delta}{h_i^2} \frac{\partial \psi}{\partial x_i} \right). \qquad (A.34)$$

A.3 Useful Integral Identities

In the following formulas, V is a volume bounded by a closed surface S. The unit vector $\hat{n}$ is normal to S and points outward.

(a) Gradient Identity:

$$\oint_V \nabla \phi \, dV = \oint_S \phi \hat{n} \, dS. \qquad (A.35)$$

(b) Gauss' Divergence Theorem:

$$\oint_V \nabla \cdot \mathbf{A}\, dV = \oint_S \mathbf{A} \cdot \hat{n}\, dS. \tag{A.36}$$

(c) Vector Stokes' Theorem:

$$\oint_V \nabla \times \mathbf{A}\, dV = \oint_S \hat{n} \times \mathbf{A}\, dS. \tag{A.37}$$

(d) First Form of Green's Theorem:

$$\oint_V [\phi_1 \nabla^2 \phi_2 + \nabla \phi_1 \cdot \nabla \phi_2]\, dV = \oint_S \hat{n} \cdot \phi_1 \nabla \phi_2\, dS. \tag{A.38}$$

(e) Second Form of Green's Theorem:

$$\oint_V [\phi_1 \nabla^2 \phi_2 - \phi_2 \nabla^2 \phi_1]\, dV = \oint_S \hat{n} \cdot (\phi_1 \nabla \phi_2 - \phi_2 \nabla \phi_1)\, dS. \tag{A.39}$$

(f) Vector Green's Theorem:

$$\oint_V [\mathbf{P} \cdot \nabla \times \nabla \times \mathbf{Q} - \mathbf{Q} \cdot \nabla \times \nabla \times \mathbf{P}]\, dV$$

$$= \oint_S [\mathbf{Q} \times \nabla \times \mathbf{P} - \mathbf{P} \times \nabla \times \mathbf{Q}] \cdot \hat{n}\, dS. \tag{A.40}$$

The above may all be proved from Gauss' divergence theorem.

(g) Stokes' Theorem:

If S is an unclosed surface bounded by a contour C, then

$$\int_S (\nabla \times \mathbf{A}) \cdot \hat{n}\, dS = \oint_C \mathbf{A} \cdot d\mathbf{l}, \tag{A.41}$$

$$\int_S \hat{n} \times \nabla \phi\, dS = \oint_C \phi\, d\mathbf{l}. \tag{A.42}$$

(h) Gauss' Theorem in Two Dimensions:

$$\int_S (\nabla \cdot \mathbf{A})\, dS = \oint_C \mathbf{A} \cdot \hat{n}\, dl, \tag{A.43}$$

The above identities for tensors and dyads can also be readily established (see Appendix B).

A.4 Integral Transforms

(a) Fourier:

$$f(x) = \frac{1}{2\pi} \int_{-\infty}^{\infty} dy\, e^{ixy} \tilde{f}(y), \tag{A.44}$$

$$\tilde{f}(y) = \int_{-\infty}^{\infty} dx\, e^{-ixy} f(x), \tag{A.45}$$

$$\delta(x - x') = \frac{1}{2\pi} \int_{-\infty}^{\infty} dy\, e^{i(x-x')y}. \tag{A.46}$$

(b) Cylindrical Hankel:

$$f(\rho) = \int_0^{\infty} d\lambda\, \lambda J_n(\lambda\rho) \tilde{f}(\lambda), \tag{A.47}$$

$$\tilde{f}(\lambda) = \int_0^{\infty} d\rho\, \rho J_n(\lambda\rho) f(\rho), \tag{A.48}$$

$$\frac{\delta(\rho - \rho')}{\rho} = \int_0^{\infty} d\lambda\, \lambda J_n(\lambda\rho) J_n(\lambda\rho'), \tag{A.49}$$

where $J_n(x)$ is a cylindrical Bessel function of n-th order.

(c) Spherical Hankel:

$$f(r) = \left(\frac{2}{\pi}\right)^{1/2} \int_0^{\infty} d\lambda\, \lambda^2 j_n(\lambda r) \tilde{f}(\lambda), \tag{A.50}$$

$$\tilde{f}(\lambda) = \left(\frac{2}{\pi}\right)^{1/2} \int_0^{\infty} dr\, r^2 j_n(\lambda r) f(r), \tag{A.51}$$

$$\frac{\delta(r - r')}{r^2} = \frac{2}{\pi} \int\limits_0^\infty d\lambda\, \lambda^2 j_n(\lambda r) j_n(\lambda r'), \tag{A.52}$$

where $j_n(x)$ is a spherical Bessel function of n-th order.

(d) Hilbert:

$$g(t) = \frac{1}{\pi} P.V. \int\limits_{-\infty}^\infty d\tau\, \frac{f(\tau)}{\tau - t}, \tag{A.53}$$

$$f(\tau) = -\frac{1}{\pi} P.V. \int\limits_{-\infty}^\infty dt\, \frac{g(t)}{t - \tau}, \tag{A.54}$$

$$\delta(t - t') = -\frac{1}{\pi^2} P.V. \int\limits_{-\infty}^\infty d\tau\, \frac{1}{\tau - t} P.V. \frac{1}{t' - \tau}, \tag{A.55}$$

where $P.V.\frac{1}{x}$ is regarded as a generalized function (see Appendix C).

(e) Radon:

A Radon transform in an n-dimensional space is defined as

$$\tilde{f}(p, \hat{\xi}) = \int\limits_{-\infty}^\infty f(\mathbf{x})\, \delta\left(p - \hat{\xi} \cdot \mathbf{x}\right) d\mathbf{x}, \tag{A.56}$$

where $\mathbf{x}$ is a position vector in an n-dimensional space, $\hat{\xi}$ is a unit vector, and $d\mathbf{x}$ implies a volume integral in an n-dimensional space. The inverse Radon transform is different in even and odd dimensions. In even dimensions, it is

$$f(\mathbf{x}) = \frac{1}{(2\pi i)^n} \int\limits_{|\xi|=1} d\xi \int\limits_{-\infty}^\infty dp\, \frac{1}{p - \hat{\xi} \cdot \mathbf{x}} \left(\frac{\partial}{\partial p}\right)^{n-1} \tilde{f}(p, \hat{\xi}), \tag{A.57}$$

where the $d\xi$ integral implies integrating over all angles of $\hat{\xi}$, i.e., $d\xi$ is an elemental area on the surface of a unit sphere. In odd dimensions,

$$f(\mathbf{x}) = \frac{1}{2} \frac{1}{(2\pi i)^{n-1}} \int\limits_{|\xi|=1} d\xi \left(\frac{\partial}{\partial p}\right)^{n-1} \tilde{f}(p, \hat{\xi})\Big|_{p=\xi \cdot \mathbf{x}}. \tag{A.58}$$

APPENDIX B

Review of Tensors

A tensor represents a physical quantity that is invariant under a coordinate rotation. For instance, a scalar is a zeroth rank (or order) tensor, while a vector is a first rank tensor. Moreover, when the coordinate system is changed from (x_1, x_2, x_3) to (x'_1, x'_2, x'_3),[1] all the components of a vector **a** will be differently defined in the new coordinate system, but the direction of the vector **a** remains unchanged in space. A vector can be written as a linear superposition of three unit vectors, i.e.,

$$\mathbf{a} = a_1\hat{x}_1 + a_2\hat{x}_2 + a_3\hat{x}_3 = a'_1\hat{x}'_1 + a'_2\hat{x}'_2 + a'_3\hat{x}'_3, \tag{B.1}$$

where $\hat{x}_i$ denotes a unit vector in the x_i direction. Alternatively, the above could also be written using indicial notation as

$$\mathbf{a} = a_i\hat{x}_i = a'_i\hat{x}'_i, \tag{B.2}$$

where summation over repeated indices is implied, i.e., $a_i\hat{x}_i = \sum_{i=1}^{3} a_i\hat{x}_i$. It is clear that $a_i = \mathbf{a} \cdot \hat{x}_i$. Furthermore, a_i is also known as the matrix representation of a vector. The relationship between a_i and a'_i is

$$a'_i = T_{ij}a_j, \tag{B.3}$$

where $T_{ij} = \hat{x}'_i \cdot \hat{x}_j$ is the rotation matrix.

Since the length of a vector remains unchanged under a coordinate rotation,

$$a'_i a'_i = T_{ij}a_j T_{ik}a_k = a_j a_j. \tag{B.4}$$

In other words,

$$T_{ij}T_{ik} = \delta_{jk}. \tag{B.5}$$

The above implies that $\overline{\mathbf{T}}^t \cdot \overline{\mathbf{T}} = \overline{\mathbf{I}}$ or $\overline{\mathbf{T}}^{-1} = \overline{\mathbf{T}}^t$. Hence, a rotation matrix is unitary.

A second rank (or order) tensor linearly relates a vector to a vector. The most general linear relationship between two vectors is

$$\mathbf{b} = \overline{\tau} \cdot \mathbf{a}. \tag{B.6}$$

An example of this is the permittivity tensor that relates $\mathbf{D} = \overline{\epsilon} \cdot \mathbf{E}$. Since $\overline{\epsilon}$ represents a physical quantity, it is invariant under a coordinate rotation.

A simple example of a second rank tensor is the *dyad* **ab**, which is formed from two vectors **a** and **b**. It has the property that when it operates on a vector **c**, it yields

$$\mathbf{ab} \cdot \mathbf{c} = \mathbf{a}(\mathbf{b} \cdot \mathbf{c}). \tag{B.7}$$

---

[1] (x_1, x_2, x_3) represents (x, y, z) here so that indicial notation could be used.

Hence, a dyad takes the projection of a vector **c** onto **b**, and multiplies it to **a**. It is a linear mapping of a vector to another vector. Furthermore, when **a** and **b** are expressed in the Cartesian coordinates, the dyad **ab** can be thought of as a matrix, namely,

$$\mathbf{ab} = \begin{bmatrix} a_1 b_1 & a_1 b_2 & a_1 b_3 \\ a_2 b_1 & a_2 b_2 & a_2 b_3 \\ a_3 b_1 & a_3 b_2 & a_3 b_3 \end{bmatrix}. \tag{B.8}$$

In indicial notation, $\mathbf{ab} = a_i b_j$. Note that $\text{tr}(\mathbf{ab}) = \mathbf{a} \cdot \mathbf{b}$, where $\text{tr}(\mathbf{A})$ is the trace of the matrix $\overline{\mathbf{A}}$ which is equal to the sum of its diagonal elements. Notice that a dyad is also the outer product between two vectors.

Not all second rank tensors are dyads as the dyad **ab** has a nullspace of rank two, since given a vector **b** in a three-dimensional space, there exists a two-dimensional plane orthogonal to **b**. All vectors on this plane are in the nullspace of **ab**. All second rank tensors, however, can be written as a linear superposition of nine dyads. For example, given a tensor $\overline{\tau}$, it could be written as

$$\overline{\tau} = \tau_{ij} \hat{x}_i \hat{x}_j, \tag{B.9}$$

where there are nine possible values of τ_{ij}. It is clear that $\tau_{ij} = \hat{x}_i \cdot \overline{\tau} \cdot \hat{x}_j$, which is also known as the matrix representation of the tensor $\overline{\tau}$ under the basis $\{\hat{x}_1, \hat{x}_2, \hat{x}_3\}$.[2] Hence, as a matrix, the tensor $\overline{\tau}$ is

$$\overline{\tau} = \begin{bmatrix} \tau_{11} & \tau_{12} & \tau_{13} \\ \tau_{21} & \tau_{22} & \tau_{23} \\ \tau_{37} & \tau_{32} & \tau_{33} \end{bmatrix}. \tag{B.10}$$

In a rotated coordinate system (x_1', x_2', x_3'), we have

$$\overline{\tau} = \tau_{ij}' \hat{x}_i' \hat{x}_j'. \tag{B.11}$$

Equating (9) and (11) yields

$$\tau_{ij}' \hat{x}_i' \hat{x}_j' = \tau_{kl} \hat{x}_k \hat{x}_l. \tag{B.12}$$

Dot-multiplying the above by $\hat{x}_i'$ and $\hat{x}_j'$ gives

$$\tau_{ij}' = \tau_{kl} (\hat{x}_i' \cdot \hat{x}_k)(\hat{x}_l \cdot \hat{x}_j') = T_{ik} \tau_{kl} T_{jl}, \tag{B.13}$$

where T_{ik} is the same as that in (3). Therefore, a second rank tensor is transformed with two $\overline{\mathbf{T}}$ matrices under a coordinate rotation, namely,

$$\overline{\tau}' = \overline{\mathbf{T}} \cdot \overline{\tau} \cdot \overline{\mathbf{T}}^t, \tag{B.14}$$

---

[2] When tensors are expressed in Cartesian coordinates, they are also called Cartesian tensors.

where $\overline{\tau}'$ is the matrix representation of $\overline{\tau}$ in the (x_1', x_2', x_3') coordinates.

Given a matrix representing a tensor, it can be easily shown that its eigenvalues, its trace, and its determinants are scalars invariant under a coordinate rotation. Hence, given two tensors, $\overline{\tau}$ and $\overline{S}$, then

$$\text{tr}(\overline{\tau} \cdot \overline{S}) = \tau_{ij} S_{ji} = \overline{\tau} : \overline{S}. \tag{B.15}$$

The double dot product here indicates a summation over both indices of the tensors.

The matrix representing a second rank tensor can be symmetric or antisymmetric. A symmetric tensor can be diagonalized by a proper choice of a coordinate system, since the eigenvectors of a symmetric matrix are orthogonal. Hence, for a symmetric tensor, there exist three orthogonal bases $\hat{e}_1$, $\hat{e}_2$, and $\hat{e}_3$ such that

$$\overline{\tau} \cdot \hat{e}_1 = \lambda_1 \hat{e}_1, \tag{B.16a}$$

$$\overline{\tau} \cdot \hat{e}_2 = \lambda_2 \hat{e}_2, \tag{B.16b}$$

$$\overline{\tau} \cdot \hat{e}_3 = \lambda_3 \hat{e}_3. \tag{B.16c}$$

The matrix representation of the tensor under these bases is then

$$\hat{e}_i \cdot \overline{\tau} \cdot \hat{e}_j = \lambda_i \delta_{ij}, \tag{B.17}$$

which is a diagonal matrix.

An example of an antisymmetric tensor is given by the tensor $\overline{a}$ such that

$$\overline{a} \cdot \mathbf{b} = \mathbf{a} \times \mathbf{b}. \tag{B.18}$$

Here, the matrix $\overline{a}$ is

$$\overline{a} = \begin{bmatrix} 0 & -a_3 & a_2 \\ a_3 & 0 & -a_1 \\ -a_2 & a_1 & 0 \end{bmatrix}. \tag{B.19}$$

All second rank tensors can be expressed as a linear superposition of symmetric and antisymmetric tensors.

A third rank tensor linearly relates a vector to a second rank tensor, i.e.,

$$\overline{\tau} = \overline{\overline{\beta}} \cdot \mathbf{a}, \tag{B.20}$$

where $\overline{\overline{\beta}}$ is a third rank tensor. An example of a third rank tensor is the **_triad_** **abc**. All third rank tensors can be written as a superposition of the twenty-seven triads $\hat{x}_i \hat{x}_j \hat{x}_k$, i.e.,

$$\overline{\overline{\beta}} = \beta_{ijk} \hat{x}_i \hat{x}_j \hat{x}_k. \tag{B.21}$$

Then, it is clear that $\beta_{ijk} = \hat{x}_j \cdot \left(\hat{x}_i \cdot \overline{\overline{\beta}} \right) \cdot \hat{x}_k$.

A third rank tensor transforms as

$$\beta'_{ijk} = T_{il} T_{jm} T_{kn} \beta_{lmn}. \tag{B.22}$$

Hence, the rank (or order) of a tensor is indicated by the number of indices it has, and under a coordinate rotation, it transforms as

$$\beta'_{i'_1,i'_2,i'_3,\cdots,i'_n} = T_{i'_1,i_1} T_{i'_2,i_2} T_{i'_3,i_3} \cdots T_{i'_n,i_n} \beta_{i_1,i_2,i_3,\cdots,i_n}. \tag{B.23}$$

A commonly encountered third rank tensor is the **alternating tensor** or the **Levi-Civita tensor**, defined as

$$\epsilon_{ijk} = \begin{cases} 1, & \text{if } i,j,k \text{ is an even permutation of 1,2,3,} \\ -1, & \text{if } i,j,k \text{ is an odd permutation of 1,2,3,} \\ 0, & \text{if any two of } i,j,k \text{ are equal.} \end{cases} \tag{B.24}$$

With this tensor, then

$$(\mathbf{a} \times \mathbf{b})_i = \epsilon_{ijk} a_j b_k \tag{B.25}$$

or

$$\mathbf{a} \times \mathbf{b} = \overline{\overline{\epsilon}} : \mathbf{ab}. \tag{B.26}$$

Moreover, the alternating tensor can also be used to define the cross product between a second rank tensor and a vector. For example,

$$[\mathbf{ab} \times \mathbf{c}]_{ij} = a_i (\mathbf{b} \times \mathbf{c})_j = a_i \epsilon_{jkl} b_k c_l \tag{B.27}$$

or

$$[\overline{\tau} \times \mathbf{c}]_{ij} = \tau_{ik} \epsilon_{jkl} c_l = -\tau_{ik} \epsilon_{kjl} c_l. \tag{B.28}$$

Consequently,

$$\overline{\tau} \times \mathbf{c} = -\overline{\tau} \cdot \overline{\overline{\epsilon}} \cdot \mathbf{c}. \tag{B.29}$$

Similarly, it follows that

$$\mathbf{c} \times \overline{\tau} = -\mathbf{c} \cdot \overline{\overline{\epsilon}} \cdot \overline{\tau}. \tag{B.30}$$

From (25), ϵ_{ijk} is also the number

$$\epsilon_{ijk} = \hat{x}_i \cdot (\hat{x}_j \times \hat{x}_k). \tag{B.31}$$

Consequently,

$$\begin{aligned}
\epsilon_{ijk} \epsilon_{ilm} &= (\hat{x}_i \cdot \hat{x}_j \times \hat{x}_k)(\hat{x}_i \cdot \hat{x}_l \times \hat{x}_m) \\
&= (\hat{x}_j \times \hat{x}_k) \cdot \hat{x}_i \hat{x}_i \cdot (\hat{x}_l \times \hat{x}_m) \\
&= (\hat{x}_j \times \hat{x}_k) \cdot (\hat{x}_l \times \hat{x}_m) \\
&= (\hat{x}_j \times \hat{x}_k) \times \hat{x}_l \cdot \hat{x}_m \\
&= [\hat{x}_k(\hat{x}_j \cdot \hat{x}_l) - \hat{x}_j(\hat{x}_k \cdot \hat{x}_l)] \cdot \hat{x}_m \\
&= \delta_{km} \delta_{jl} - \delta_{jm} \delta_{kl}. \tag{A.32}
\end{aligned}$$

The second rank identity tensor $\bar{\mathbf{I}}$ and the third rank tensor $\bar{\bar{\epsilon}}$ are known as isotropic tensors since their forms are invariant under a coordinate rotation. A fourth rank tensor which is isotropic can be shown to be of the form

$$c_{ikmp} = \mu_1 \, \delta_{ik} \, \delta_{mp} + \mu_2 \, \delta_{im} \, \delta_{kp} + \mu_3 \, \delta_{ip} \, \delta_{km}. \tag{B.33}$$

Moreover, a fourth rank tensor linearly relates two second rank tensors. For example, the stress and strain tensors in an elastic solid are related via a fourth rank tensor. For an isotropic medium, the fourth rank tensor is of the form given by (33).

Differentiation of a vector field generates a higher rank tensor. For example, $\nabla \mathbf{v}$ is a second rank tensor, or

$$\nabla \mathbf{v} = \frac{\partial v_j}{\partial x_i} \hat{x}_i \hat{x}_j. \tag{B.34}$$

It follows that $\mathrm{tr}(\nabla \mathbf{v}) = \nabla \cdot \mathbf{v}$.

Since

$$\nabla \times \mathbf{v} = \epsilon_{ijk} \frac{\partial v_k}{\partial x_j} \hat{x}_i, \tag{B.35}$$

it follows that

$$\nabla \times (\nabla \times \mathbf{v}) = \epsilon_{ijk}\epsilon_{klm} \frac{\partial^2 v_m}{\partial x_j \partial x_l} \hat{x}_i = \epsilon_{kij}\epsilon_{klm} \frac{\partial^2 v_m}{\partial x_j \partial x_l} \hat{x}_i$$

$$= (\delta_{il}\,\delta_{jm} - \delta_{im}\,\delta_{jl}) \frac{\partial^2 v_m}{\partial x_j \, \partial x_l} \hat{x}_i$$

$$= \left[\frac{\partial^2 v_j}{\partial x_j \partial x_i} - \frac{\partial^2 v_i}{\partial x_j^2} \right] \hat{x}_i = \nabla(\nabla \cdot \mathbf{v}) - \nabla^2 \mathbf{v}. \tag{B.36}$$

Hence, indicial notation, together with the alternating tensor, can be used to prove vector identities readily.

Integral theorems are also simplified in tensor notation. For example, Gauss' divergence theorem is

$$\oint_S \mathbf{A} \cdot \hat{n} \, dS = \oint_V \nabla \cdot \mathbf{A} \, dV. \tag{B.37}$$

By letting $\mathbf{A} = \tau_{ijk\cdots}\hat{x}_p$, where $\tau_{ijk\cdots}$ is a tensor of arbitrary rank, the above becomes

$$\oint_S \tau_{ijk\cdots} n_p \, dS = \oint_V \frac{\partial \tau_{ijk\cdots}}{\partial x_p} \, dV. \tag{B.38}$$

Now, letting $\tau_{ijk\cdots} = v_i$, a vector, we have

$$\oint_S v_i n_p \, dS = \oint_V \frac{\partial v_i}{\partial x_p} \, dV, \tag{B.39}$$

or

$$\oint_{S} dS \,\hat{n}\mathbf{v} = \oint_{V} dV \,\nabla\mathbf{v}. \qquad (B.40)$$

Taking the trace of the above yields Gauss' divergence theorem again.

Furthermore, by letting $\tau_{ijk\cdots} = \phi$, a scalar, we have

$$\oint_{S} \phi n_{p} \,dS = \oint_{V} \frac{\partial\phi}{\partial x_{p}} dV \qquad (B.41)$$

or

$$\oint_{S} \phi\hat{n} \,dS = \oint_{V} \nabla\phi \,dV. \qquad (B.42)$$

The above is a gradient identity [see Equation (A.35)]. By letting $\tau_{ijk\cdots} = \tau_{ij}$, a second rank tensor, we have

$$\oint_{S} \tau_{ij} n_{p} \,dS = \oint_{V} \frac{\partial\tau_{ij}}{\partial x_{p}} \,dV. \qquad (B.43)$$

Then, if we let $p = i$, the above becomes

$$\oint_{S} dS \,\hat{n} \cdot \overline{\boldsymbol{\tau}} = \oint_{V} dV \,\nabla \cdot \overline{\boldsymbol{\tau}}, \qquad (B.44)$$

which is the divergence theorem for tensors. Similar to (37), from Stokes' theorem that

$$\oint_{C} \mathbf{A} \cdot d\mathbf{l} = \int_{S} \nabla \times \mathbf{A} \cdot d\mathbf{S}, \qquad (B.45)$$

we conclude that

$$\oint_{C} \tau_{ijk\cdots} \,dx_{i} = \int_{S} \epsilon_{ijk} \frac{\partial\tau_{klm\cdots}}{\partial x_{j}} \,dS_{i}. \qquad (B.46)$$

Furthermore, by letting $\tau_{ijk\cdots} = \phi\,\delta_{ik}$, we have [see Equation (A.42)]

$$\oint_{C} \phi \,d\mathbf{l} = \int_{S} d\mathbf{S} \times \nabla\phi. \qquad (B.47)$$

And, by letting $\tau_{ijk\cdots} = \epsilon_{ijk} v_{j}$, we have

$$\oint_{C} d\mathbf{l} \times \mathbf{v} = \int_{S} \nabla\mathbf{v} \cdot d\mathbf{S} - \int_{S} \nabla \cdot \mathbf{v} dS. \qquad (B.48)$$

In this manner, integral theorems for vector and tensor fields are easily established.

APPENDIX C

Generalized Functions

Generalized functions are expanded definitions of functions to include functions which cannot be described in a classical sense. Examples of generalized functions are the Dirac delta function and its derivatives. Despite their nonclassical characters, engineers and physicists have appreciated these functions as a limiting case of a sequence of functions. For example, after defining

$$f_n(x) = e^{-nx^2} \left(\frac{n}{\pi}\right)^{1/2}, \tag{C.1}$$

a Dirac delta function is then

$$\delta(x) = \lim_{n\to\infty} f_n(x) = \lim_{n\to\infty} e^{-nx^2} \left(\frac{n}{\pi}\right)^2. \tag{C.2}$$

Hence, $\delta(x)$ is a function that is zero everywhere, and infinite at $x = 0$. Furthermore, $\int\limits_{-\infty}^{\infty} \delta(x)\,dx = 1$, and

$$\int\limits_{-\infty}^{\infty} \delta(x)f(x)\,dx = f(0). \tag{C.3}$$

A Dirac delta function is a simple way to describe, e.g., a point charge which has infinite charge density, but a finite net charge. This function, however, is not acceptable in a classical sense as it is defined only at one point.

If the use of a delta function defies understanding in a physical application, one can usually gain a better insight by studying a sequence of functions (also known as a delta family or a delta sequence). Many aspects of generalized functions, however, are difficult to understand intuitively. For example, it is well known that the Fourier representation of

$$\delta(x) = \frac{1}{2\pi} \int\limits_{-\infty}^{\infty} dy\, e^{ixy}, \tag{C.4}$$

and

$$\delta'(x) = \frac{i}{2\pi} \int\limits_{-\infty}^{\infty} dy\, y\, e^{ixy}. \tag{C.5}$$

The right-hand sides of these equations are nonconvergent integrals; hence, it is not clear what the equalities in the above mean. But the equivalences in (4) and (5) as generalized functions can be established readily using distribution theory.

In Chapter 5, we have shown that a function $f(x)$ can be thought of as a vector in an infinite dimensional space. One way to define such a vector is to define the value of $f(x)$ at each x. Another way is to seek out a complete set of independent functions $\phi_n(x)$ that spans a space. Then $f(x)$ can be equivalently defined by the numbers $\langle f(x), \phi_n(x) \rangle$ for all n.[1] For example, if $f(x)$ is a function in the L_2 space, then $\phi_n(x)$ is a set of functions that spans the L_2 space, which is a Hilbert space. The above is a generalized way of defining a function $f(x)$. But it is not general enough for the following reason: Since $\phi_n(x)$ is from L_2 space, it can be a discontinuous function. Hence, $f(x)$ cannot be a Dirac delta function $\delta(x)$, for if $\phi_n(x)$ is discontinuous at $x = 0$, then $\langle \delta(x), \phi_n(x) \rangle$ is undefined.

Consequently, in order to expand the class of functions that can be defined in this manner, it is expedient to choose a subset of functions from the L_2 space. If these functions are well behaved and smooth, then the class of functions $f(x)$ that can be defined in this manner can be expanded beyond the class of functions in L_2. A classic example of a smooth function within the interval $a \leq x \leq a$ is

$$\phi_0(x) = \begin{cases} e^{-a^2/(a^2 - x^2)}, & |x| \leq a, \\ 0, & |x| > a. \end{cases} \tag{C.6}$$

This function is smooth, and vanishes at $x = \pm a$. Furthermore, its derivative of any order is smooth and vanishes at $x = \pm a$. Moreover, after defining the n-th derivative of such a function to be $\phi_n(x)$, the $\phi_n(x)$'s form a linearly independent set. They span the space of a class of functions which are infinitely smooth and is a subspace of the L_2 space. Hence, a generalized way of defining a function f is by the numbers $\langle f, \phi_n \rangle$ for all n.

The action of $f(x)$ on $\phi_n(x)$, i.e., $\langle f, \phi_n \rangle$ for all n, is said to generate a distribution, and a function is defined in a distributional sense in this manner. The set of functions $\phi_n(x)$ is also known as testing functions for defining a generalized function. Subsequently, we shall denote $\phi_n(x)$ by $\phi(x)$, which represents all possible infinitely smooth functions. Since the set of testing functions is extremely smooth, a large class of functions and their derivatives can be defined in this manner. For a generalized function, $f(x)$, f, and $\langle f, \phi \rangle$ are often used interchangeably.

With this definition, the derivative of a generalized function $f'(x)$ is defined by the distribution

$$\langle f', \phi \rangle = \int_{-a}^{a} f'(x)\, \phi(x)\, dx, \tag{C.7}$$

---

[1] $f(x)$ and $\phi(x)$ are assumed real here. Generalization to complex functions is straightforward.

where ϕ is all possible functions from the class of smooth functions. Using integration by parts, the above is just $-\int_{-a}^{a} f(x)\phi'(x)\,dx$. In other words,

$$\langle f', \phi \rangle = -\langle f, \phi' \rangle. \tag{C.8}$$

Similarly, the m-th derivative of a generalized function is defined by the distribution $\langle f^{(m)}, \phi \rangle$, which, by repeated integration by parts, is just

$$\langle f^{(m)}, \phi \rangle = (-1)^m \langle f, \phi^{(m)} \rangle. \tag{C.9}$$

Since $\phi(x)$ is an extremely smooth function, the inner product $\langle f, \phi^{(m)} \rangle$ exists for all m. Hence, the m-th derivative of a generalized function could be defined.

If a generalized function $u(x)$ is defined such that

$$u(x) = \begin{cases} 1, & x \le 0, \\ 0, & x < 0, \end{cases} \tag{C.10}$$

it is easy to show that the generalized function $\delta(x) = u'(x)$ with the distribution

$$\langle \delta, \phi \rangle = \phi(0). \tag{C.11}$$

Furthermore, the derivative of a delta function, $\delta'(x)$, is a generalized function with the distribution

$$\langle \delta', \phi \rangle = -\phi'(0). \tag{C.12}$$

In this manner, derivatives of all orders can be defined for $u(x)$ and hence, for $\delta(x)$.

The interval in (6) can be readily extended to $(-\infty, \infty)$. Ordinarily, the m-th derivative of a delta function, $\delta^{(m)}(x)$, has meaning only via its action on a function $\phi(x)$, yielding $(-)^m \phi^{(m)}(0)$. Such an action $\langle \delta^{(m)}, \phi \rangle$ is undefined if $\phi^{(m)}(0)$ is undefined.

It can be easily shown by a change of variable that

$$\left\langle f\left(\frac{x}{a}\right), \phi(x) \right\rangle = a\langle f(x), \phi(ax) \rangle. \tag{C.13}$$

Hence,

$$\left\langle \delta\left(\frac{x}{a}\right), \phi(x) \right\rangle = a\langle \delta(x), \phi(ax) \rangle = a\phi(0) = \langle a\delta(x), \phi(x) \rangle. \tag{C.14}$$

Consequently, $\delta\left(\frac{x}{a}\right) = a\,\delta(x)$ in a distributional sense.

The following are some important properties of generalized functions:

(i) The generalized functions $f(x)$ and $h(x)$ are equal in $|x| \le a$ if they generate the same distribution with $\phi(x)$, i.e.,

$$\langle f, \phi \rangle = \langle h, \phi \rangle. \tag{C.15}$$

(ii) A generalized function $f_m(x)$ converges to $f(x)$, i.e., $\lim\limits_{m \to \infty} f_m(x) = f(x)$ if

$$\lim_{m \to \infty} \langle f_m, \phi \rangle = \langle f, \phi \rangle. \tag{C.16}$$

(iii) For generalized functions, if $\lim\limits_{m \to \infty} f_m(x) = f(x)$, then $\lim\limits_{m \to \infty} f'_m(x) = f'(x)$. The above can be proven easily, because $\langle f', \phi \rangle = -\langle f, \phi' \rangle$, and $\langle f'_m, \phi \rangle = -\langle f_m, \phi' \rangle$. From property (ii), $\lim\limits_{m \to \infty} f_m(x) = f(x)$ implies that $\langle f, \phi' \rangle = \lim\limits_{m \to \infty} \langle f_m, \phi' \rangle$. Hence, $\langle f', \phi \rangle = \lim\limits_{m \to \infty} \langle f'_m, \phi \rangle$.

(iv) Consequently, it is easy to show that for generalized functions, if

$$\lim_{m \to \infty} f_m(x) = f(x),$$

then

$$\lim_{m \to \infty} f_m^{(l)}(x) = f^{(l)}(x). \tag{C.17}$$

Property (iv) is one of the most important properties of generalized functions. It can be used to establish divergent Fourier series as generalized functions. For example, if a function $f(x)$ is defined by a Fourier series expansion in the interval $|x| \le a$ by

$$f(x) = \sum_{k=-\infty}^{\infty} a_k e^{i\frac{k\pi}{a}x}, \tag{C.18}$$

then $f_m(x)$ can be defined as

$$f_m(x) = \sum_{k=-m}^{m} a_k e^{i\frac{k\pi}{a}x}. \tag{C.19}$$

Now, if $f(x)$ and $f_m(x)$ are regarded as generalized functions, it is clear that $\lim\limits_{m \to \infty} f_m(x) = f(x)$. Taking the l-th derivative of (19) yields

$$f_m^{(l)}(x) = \sum_{k=-m}^{m} a_k \left(\frac{ik\pi}{a}\right)^l e^{i\frac{k\pi}{a}x}. \tag{C.20}$$

From property (iv) which implies that $\lim\limits_{m \to \infty} f_m^{(l)}(x) = f^{(l)}(x)$, we conclude that

$$f^{(l)}(x) = \sum_{k=-\infty}^{\infty} a_k \left(\frac{ik\pi}{a}\right)^l e^{i\frac{k\pi x}{a}}. \tag{C.21}$$

The right-hand side of (21) could very well be a divergent series even though (18) may be convergent. But when both sides of (21) are generalized functions, the equality in (21) is readily established. Moreover, when the interval

$(-a, a) \rightarrow (-\infty, \infty)$, a Fourier series becomes a Fourier transform. In this case, the right-hand side of (21) could very well be a divergent integral.

With the above concepts, we can establish the meaning of the equalities in (4) and (5). They are only equal if both sides of (4) and (5) are regarded as generalized functions.

There are several ways to find the Fourier transforms of generalized functions. One way is to define a function with a parameter n as in $f_n(x)$ in Equation (1). Then, the Fourier transform of $f_n(x)$ is easily found to be

$$\tilde{f}_n(y) = \int\limits_{-\infty}^{\infty} dx\, e^{-ixy} e^{-nx^2} \left(\frac{n}{\pi}\right)^{1/2} = e^{-y^2/4n}. \tag{C.22}$$

Since $\lim\limits_{n \to \infty} \tilde{f}_n(y) = 1$, the Fourier transform of $\delta(x) = 1$.

Another example is the Fourier transform of a $\mathrm{sgn}(x)$ function. To find its Fourier transform, we define a function

$$f_n(x) = \mathrm{sgn}(x) e^{-|x|/n} \tag{C.23}$$

whose Fourier transform can be found.

$$\tilde{f}_n(y) = \int\limits_{0}^{\infty} dx\, e^{-iyx} e^{-x/n} - \int\limits_{-\infty}^{0} dx\, e^{-iyx + x/n}$$

$$= \frac{-1}{(-iy - 1/n)} + \frac{-1}{(-iy + 1/n)} = \frac{-2iy}{y^2 + \frac{1}{n^2}}. \tag{C.24}$$

The function $-2iy/(y^2 + 1/n^2)$ can only be regarded as a generalized function when $n \to \infty$. By choosing an appropriate test function $\tilde{\phi}(y)$, we have

$$\lim_{n \to \infty} \langle \tilde{f}_n, \tilde{\phi} \rangle = \lim_{n \to \infty} \int\limits_{-\infty}^{\infty} dy \left(\frac{-2iy}{y^2 + \frac{1}{n^2}} \right) \tilde{\phi}(y)$$

$$= \lim_{\epsilon \to 0} \lim_{n \to \infty} \left[\int\limits_{-\infty}^{-\epsilon} dy \frac{-2iy\tilde{\phi}(y)}{y^2 + \frac{1}{n^2}} \right.$$

$$\left. + \int\limits_{\epsilon}^{\infty} dy \frac{-2iy\tilde{\phi}(y)}{y^2 + \frac{1}{n^2}} + \int\limits_{-\epsilon}^{\epsilon} \frac{-2iy\tilde{\phi}(k_x)}{y^2 + \frac{1}{n^2}} \right]. \tag{C.25}$$

In the above, ϵ and $1/n$ can always be made small enough so that the last term on the right-hand side of (25) can be evaluated as

$$\lim_{\epsilon \to 0} \lim_{n \to \infty} \tilde{\phi}(0) \int\limits_{-\epsilon}^{\epsilon} dy \frac{-2iy}{y^2 + \frac{1}{n^2}} = 0. \tag{C.26}$$

Consequently,

$$\lim_{n \to \infty} \langle \tilde{f}_n, \tilde{\phi} \rangle = \lim_{\epsilon \to 0} \left[\int_{-\infty}^{-\epsilon} dy \frac{-2i}{y} \tilde{\phi}(y) + \int_{\epsilon}^{\infty} dy \frac{-2i}{y} \tilde{\phi}(y) \right]$$

$$= P.V. \int_{-\infty}^{\infty} dy \frac{-2i}{y} \tilde{\phi}(y), \tag{C.27}$$

where $P.V.$ stands for the Cauchy principal value integral, which is the definition of the integral in (27). A generalized function $P.V.\frac{1}{y}$ can be defined with the property that its action on a test function implies a Cauchy principal value integral. Then, in a distributional sense,

$$\lim_{n \to \infty} \tilde{f}_n(y) = -2iP.V.\frac{1}{y}. \tag{C.28}$$

Therefore, as generalized functions, we can equate

$$-2iP.V.\frac{1}{y} = \int_{-\infty}^{\infty} dx \, \text{sgn}(x) \, e^{-ixy}. \tag{C.29}$$

Note that both $P.V.\frac{1}{y}$ and $\text{sgn}(x)$ are not ordinary functions.

The Fourier transform relationship of generalized functions can also be established by Parseval's theorem, which requires that

$$\langle f, \phi \rangle = \frac{1}{2\pi} \langle \tilde{f}^*, \tilde{\phi} \rangle. \tag{C.30}$$

If the test function $\phi(x)$ is an extremely smooth function that decays exponentially when $x \to \infty$ (for example, it may be a Hermite-Gaussian polynomial), then $\tilde{\phi}(y)$ is also an extremely smooth function that decays exponentially fast when $y \to \infty$. In this case, $\tilde{\phi}(y)$ is also a legitimate test function for generating a distribution in the Fourier space.

Using Parseval's theorem, we have

$$\langle \text{sgn}(x), \phi(x) \rangle = \frac{i}{\pi} \left\langle P.V.\frac{1}{y}, \tilde{\phi}(y) \right\rangle. \tag{C.31}$$

We can rewrite the right-hand side of the above as

$$\frac{i}{\pi} \left\langle P.V.\frac{1}{y}, \int_{-\infty}^{\infty} dx \, e^{-ixy} \phi(x) \right\rangle = \frac{i}{\pi} P.V. \int_{-\infty}^{\infty} dy \frac{1}{y} \int_{-\infty}^{\infty} dx \, e^{-ixy} \phi(x)$$

$$= \int_{-\infty}^{\infty} dx \, \phi(x) \frac{i}{\pi} P.V. \int_{-\infty}^{\infty} dy \frac{1}{y} e^{-ixy}. \tag{C.32}$$

The above then implies the Fourier transform relationship that

$$\text{sgn}(x) = -\frac{i}{\pi} P.V. \int\limits_{-\infty}^{\infty} dy \frac{1}{y} e^{ixy}. \tag{C.33}$$

Given a train of delta functions

$$f(x) = \sum_{k=-\infty}^{\infty} \delta(x - 2k\pi), \tag{C.34}$$

which is a periodic function of x with period 2π, a Fourier series expansion yields

$$\sum_{k=-\infty}^{\infty} \delta(x - 2k\pi) = \frac{1}{2\pi} \sum_{n=-\infty}^{\infty} e^{inx}. \tag{C.35}$$

Note that both sides of the equation are periodic functions of x. The left side is a generalized function, while the right side is a nonconvergent series. Hence, they are equal only in a distributional sense. Note that by taking the Fourier transform of (35), one obtains

$$\tilde{f}(y) = \sum_{k=-\infty}^{\infty} e^{-2ik\pi y} = \sum_{n=-\infty}^{\infty} \delta(y - n). \tag{C.36}$$

Then, by letting $y = y'/2\pi$, and using the fact that $\delta(y'/2\pi) = 2\pi\,\delta(y')$, (36) can be made similar to (35). The above shows that the Fourier transform of a delta function train is a delta function train.

Using Parseval's theorem given by (30), and $f(x)$ and $\tilde{f}(y)$ given by (34) and (36), we have

$$\sum_{k=-\infty}^{\infty} \phi(2k\pi) = \frac{1}{2\pi} \sum_{n=-\infty}^{\infty} \tilde{\phi}(n). \tag{C.37}$$

The above is known as the Poisson's summation formula. If a series $\phi(2k\pi)$ is slowly convergent, i.e., $\phi(x)$ diminishes slowly as $x \to \infty$, then $\tilde{\phi}(y) \to 0$ rapidly as $y \to \infty$, and the series on the right-hand side of (37) is rapidly convergent. Equation (37) is useful for accelerating the convergence of a series, if the terms in the series can be identified with $\phi(x)$ whose Fourier transform $\tilde{\phi}(y)$ is known.

Given a differential equation

$$\mathcal{L}\,u = f, \tag{C.38}$$

where $\mathcal{L}$ is a differential operator, an adjoint operator $\mathcal{L}^a$ is defined as

$$\langle \mathcal{L}\,u, \phi \rangle = \langle u, \mathcal{L}^a \phi \rangle. \tag{C.39}$$

Now, if f is a generalized function, (38) has no meaning in a classical sense. However, a solution u of (38) can be sought such that

$$\langle u, \mathcal{L}^a \phi \rangle = \langle f, \phi \rangle, \tag{C.40}$$

where ϕ is from the space of infinitely smooth functions. Then u that satisfies (39) satisfies (38) in a distributional sense. Note that in a classical sense, u that satisfies (38) has to be smooth and differentiable, since $\mathcal{L}$ is a differential operator. But u in (39) need not be differentiable. A u that satisfies (39) also satisfies (38) in a **weak sense**, or u is a **weak solution** of (38).

APPENDIX D

Addition Theorems

The use of addition theorems arises in a number of solutions of the scattering problem. They express wave functions in one coordinate system in terms of the wave functions of another coordinate system which is linearly translated from the first one.

In cylindrical coordinates, the addition theorem is [see (3.3.2), and Exercise 3.9]

$$H_0^{(1)}(k_\rho|\boldsymbol{\rho} - \boldsymbol{\rho}'|) = \sum_{n=-\infty}^{\infty} J_n(k_\rho\rho_<)H_n^{(1)}(k_\rho\rho_>)e^{in(\phi-\phi')}, \qquad (D.1)$$

where $\rho_<$ is the smaller of ρ and ρ' and $\rho_>$ is the larger of ρ and ρ' (see Figure D.1). By using the raising operator given by Equation (2.2.16) of Chapter 2, it can be readily established that

$$H_m^{(1)}(k_\rho|\boldsymbol{\rho} - \boldsymbol{\rho}'|)e^{im\phi''} = \begin{cases} \displaystyle\sum_{n=-\infty}^{\infty} J_{n-m}(k_\rho\rho')H_n^{(1)}(k_\rho\rho)e^{in\phi-i(n-m)\phi'}, & \rho > \rho', \\ \displaystyle\sum_{n=-\infty}^{\infty} H_{n-m}^{(1)}(k_\rho\rho')J_n(k_\rho\rho)e^{in\phi-i(n-m)\phi'}, & \rho < \rho'. \end{cases}$$

$$(D.2)$$

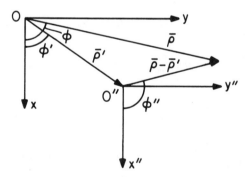

Figure D.1 Translation in the cylindrical coordinate system.

Moreover, by taking the regular part of (2) on both sides of the equation,

$$J_m(k_\rho|\boldsymbol{\rho} - \boldsymbol{\rho}'|)e^{im\phi''} = \sum_{n=-\infty}^{\infty} J_{n-m}(k_\rho\rho')J_n(k_\rho\rho)e^{in\phi-i(n-m)\phi'}, \qquad (D.3)$$

which establishes the addition theorem for Bessel functions.

In spherical coordinates, the addition theorems could be established by the use of the plane-wave expansion formula (Stratton 1941)

$$e^{i\mathbf{k}\cdot\mathbf{r}} = \sum_{lm} 4\pi i^l Y_{lm}^*(\theta_k, \phi_k) Y_{lm}(\theta, \phi) j_l(kr), \qquad (D.4)$$

where l is summed from 0 to $+\infty$ and m is summed from $-l$ to $+l$, and $\mathbf{k}$ is pointing in the (θ_k, ϕ_k) direction while $\mathbf{r}$ is pointing in the (θ, ϕ) direction in spherical coordinates. By letting

$$\mathbf{r} = \mathbf{r}' + \mathbf{r}'', \qquad (D.5)$$

then

$$e^{i\mathbf{k}\cdot\mathbf{r}} = e^{i\mathbf{k}\cdot\mathbf{r}'} e^{i\mathbf{k}\cdot\mathbf{r}''}. \qquad (D.6)$$

Next, on using (4) in (6), one obtains

$$\sum_{lm} 4\pi i^l Y_{lm}^*(\theta_k, \phi_k) Y_{lm}(\theta, \phi) j_l(kr)$$

$$= \sum_{l'm'} 4\pi i^{l'} Y_{l'm'}^*(\theta_k, \phi_k) Y_{l'm'}(\theta', \phi') j_{l'}(kr')$$

$$\cdot \sum_{l''m''} 4\pi i^{l''} Y_{l''m''}^*(\theta_k, \phi_k) Y_{l''m''}(\theta'', \phi'') j_{l''}(kr''). \qquad (D.7)$$

After multiplying both sides by $Y_{lm}(\theta_k, \phi_k)$ and integrating over (θ_k, ϕ_k), we have

$$i^l Y_{lm}(\theta, \phi) j_l(kr)$$

$$= \sum_{l'm'} \sum_{l''m''} 4\pi i^{(l'+l'')} Y_{l'm'}(\theta', \phi') j_{l'}(kr') Y_{l''m''}(\theta'', \phi'') j_{l''}(kr'')$$

$$\cdot \int d\Omega_k Y_{lm}(\theta_k, \phi_k) Y_{l'm'}^*(\theta_k, \phi_k) Y_{l''m''}^*(\theta_k, \phi_k), \qquad (D.8)$$

where $\int d\Omega_k = \int_0^{2\pi} \int_0^{\pi} \sin\theta_k d\theta_k d\phi_k$. The integral can be expressed in terms of Gaunt coefficients as

$$\int d\Omega_k Y_{lm}(\theta_k, \phi_k) Y_{l'm'}^*(\theta_k, \phi_k) Y_{l''m''}^*(\theta_k, \phi_k)$$

$$= (-1)^m [(2l+1)(2l'+1)(2l''+1)/4\pi]^{1/2} \begin{pmatrix} l & l' & l'' \\ 0 & 0 & 0 \end{pmatrix} \begin{pmatrix} l & l' & l'' \\ -m & m' & m'' \end{pmatrix}, \qquad (D.9)$$

where $\begin{pmatrix} j_1 & j_2 & j_3 \\ m_1 & m_2 & m_3 \end{pmatrix}$ is the Wigner 3-j symbol. It is related to the Clebsch-Gordon Coefficients as (Merzbacher 1970; Abramowitz and Stegun 1965)

$$\begin{pmatrix} j_1 & j_2 & j_3 \\ m_1 & m_2 & m_3 \end{pmatrix} = \frac{(-1)^{j_1-j_2-m_3}}{\sqrt{2j_3+1}} (j_1 m_1 j_2 m_2 | j_1 j_2 j_3, -m_3). \qquad (D.10)$$

It is nonzero only if $m_3 = -m_1 - m_2$ and if $j_1 + j_2 \geq j_3 \geq |j_1 - j_2|$. Furthermore, $\begin{pmatrix} l & l' & l'' \\ 0 & 0 & 0 \end{pmatrix}$ is nonzero only if $l + l' + l''$ is an even integer. Consequently,

$$
\begin{aligned}
Y_{lm}&(\theta, \phi) j_l(kr) \\
&= \sum_{l'm'} Y_{l'm'}(\theta', \phi') j_{l'}(kr') \sum_{l''} 4\pi i^{(l'+l''-l)} Y_{l'',m-m'}(\theta'', \phi'') j_{l''}(kr'') \\
&\quad \cdot (-1)^m [(2l+1)(2l'+1)(2l''+1)/4\pi]^{1/2} \\
&\qquad \cdot \begin{pmatrix} l & l' & l'' \\ 0 & 0 & 0 \end{pmatrix} \begin{pmatrix} l & l' & l'' \\ -m & m' & m-m' \end{pmatrix}, \quad (D.11)
\end{aligned}
$$

where $Y_{l''m''}(\theta'', \phi'') = 0$ if $m'' > l''$. It can be shown that[1]

$$
\int_0^\infty dk \, k^2 \frac{j_l(kr_0) j_l(kr)}{k^2 - k_0^2} = \frac{\pi i}{2} k_0 j_l(k_0 r_0) h_l^{(1)}(k_0 r), \quad r > r_0', \qquad (D.12a)
$$

$$
\begin{aligned}
\int_0^\infty dk \, k^2 &\frac{j_l(kr_0) j_{l'}(kr') j_{l''}(kr'')}{k^2 - k_0^2} = \frac{\pi i}{2} k_0 j_l(k_0 r_0) \\
&\cdot \begin{cases} j_{l'}(k_0 r') h_{l''}^{(1)}(k_0 r''), & r'' > r' + r_0, \\ j_{l''}(k_0 r'') h_{l'}^{(1)}(k_0 r'), & r' > r'' + r_0, \end{cases} \qquad (D.12b)
\end{aligned}
$$

where k_0 has a small positive imaginary part (r_0' can be chosen arbitrarily small so that it becomes insignificant in the above inequalities). Using the above, we can multiply (11) by $\frac{k^2 j_l(kr_0)}{k^2 - k_0^2}$ and integrate over k from 0 to ∞ to yield

$$
\begin{aligned}
Y_{lm}&(\theta, \phi) h_l^{(1)}(k_0 r) \\
&= \sum_{l'm'} Y_{l'm'}(\theta', \phi') h_{l'}^{(1)}(k_0 r') \sum_{l''} 4\pi i^{(l'+l''-l)} Y_{l'',m-m'}(\theta'', \phi'') j_{l''}(k_0 r'') \\
&\cdot (-)^m [(2l+1)(2l'+1)(2l''+1)/4\pi]^{1/2} \begin{pmatrix} l & l' & l'' \\ 0 & 0 & 0 \end{pmatrix} \begin{pmatrix} l & l' & l'' \\ -m & m' & m-m' \end{pmatrix}, \\
&\hspace{8cm} r' > r'', \qquad (D.13a)
\end{aligned}
$$

$$
\begin{aligned}
Y_{lm}&(\theta, \phi) h_l^{(1)}(k_0 r) \\
&= \sum_{l'm'} Y_{l'm'}(\theta', \phi') j_{l'}(k_0 r') \sum_{l''} 4\pi i^{(l'+l''-l)} Y_{l'',m-m'}(\theta'', \phi'') h_{l''}^{(1)}(k_0 r'') \\
&\cdot (-)^m [(2l+1)(2l'+1)(2l''+1)/4\pi]^{1/2} \begin{pmatrix} l & l' & l'' \\ 0 & 0 & 0 \end{pmatrix} \begin{pmatrix} l & l' & l'' \\ -m & m' & m-m' \end{pmatrix}, \\
&\hspace{8cm} r' < r''. \qquad (D.13b)
\end{aligned}
$$

---

[1] These identities can be derived in a manner similar to (7.3.35) of Chapter 7, in addition to using the fact that $l + l' + l''$ must be even.

It can further be shown that the preceding formulas hold true if $h_n^{(2)}(x)$ takes the place of $h_n^{(1)}(x)$. To derive the expressions for $h_n^{(2)}(x)$, k_0 is assumed to have a small negative imaginary part.

In summary, for $z_l(x)$ equals either $h_l^{(1)}(x)$, $h_l^{(2)}(x)$, $y_l(x)$, or $j_l(x)$, we can write

$$Y_{lm}(\theta, \phi)z_l(k_0r) = \begin{cases} \displaystyle\sum_{l'm'} Y_{l'm'}(\theta', \phi')z_{l'}(k_0r')\beta_{l'm',lm}, & r' > r'', \\[2ex] \displaystyle\sum_{l'm'} Y_{l'm'}(\theta', \phi')j_{l'}(k_0r')\alpha_{l'm',lm}, & r' < r'', \end{cases} \qquad \text{(D.14)}$$

where

$$\beta_{l'm',lm} = \sum_{l''} 4\pi i^{(l'+l''-l)} Y_{l'',m-m'}(\theta'', \phi'')j_{l''}(k_0r'')$$

$$\cdot (-1)^m[(2l + 1)(2l' + 1)(2l'' + 1)/4\pi]^{1/2}$$

$$\cdot \begin{pmatrix} l & l' & l'' \\ 0 & 0 & 0 \end{pmatrix} \begin{pmatrix} l & l' & l'' \\ -m & m' & m - m' \end{pmatrix}, \qquad \text{(D.15a)}$$

$$\alpha_{l'm',lm} = \sum_{l''} 4\pi i^{(l'+l''-l)} Y_{l'',m-m'}(\theta'', \phi'')z_{l''}(k_0r'')$$

$$\cdot (-1)^m[(2l + 1)(2l' + 1)(2l'' + 1)/4\pi]^{1/2}$$

$$\cdot \begin{pmatrix} l & l' & l'' \\ 0 & 0 & 0 \end{pmatrix} \begin{pmatrix} l & l' & l'' \\ -m & m' & m - m' \end{pmatrix}. \qquad \text{(D.15b)}$$

Note that when $z_l(x) = j_l(x)$, $\beta_{l'm',lm} = \alpha_{l'm',lm}$. Since $l + l' + l''$ has to be even, $l + l' \geq l'' \geq |l - l'|$, l'' increments by a step of 2 in the above.

In electromagnetic scattering problems, addition theorems for the vector wave functions are useful for solving for the scattering solution from multiple scatterers. If

$$\psi_{lm}(\mathbf{r}) = Y_{lm}(\theta, \phi)z_l(kr), \qquad \text{(D.16)}$$

where $z_l(x)$ can either be $j_l(x)$, $y_l(x)$, $h_l^{(1)}(x)$, or $h_l^{(2)}(x)$, a vector wave function can be defined as

$$\mathbf{M}_{lm}(\mathbf{r}) = \nabla \times [\mathbf{r}\psi_{lm}(\mathbf{r})] = [\nabla\psi_{lm}(\mathbf{r})] \times \mathbf{r}. \qquad \text{(D.17)}$$

The latter equality follows since $\nabla \times \mathbf{r} = 0$. Under a coordinate translation, we can write

$$\psi_{lm}(\mathbf{r}) = \sum_{l'm'} \psi_{l'm'}(\mathbf{r}')\alpha_{l'm',lm}, \qquad \text{(D.18))}$$

where $\mathbf{r} = \mathbf{r}' + \mathbf{r}''$. The gradient operator ∇ is invariant under coordinate translation. Hence,

$$\mathbf{M}_{lm}(\mathbf{r}) = \sum_{l'm'} [\nabla\psi_{l'm'}(\mathbf{r}') \times (\mathbf{r}' + \mathbf{r}'')]\alpha_{l'm',lm}$$

$$= \sum_{l'm'} \mathbf{M}_{l'm'}(\mathbf{r}')\alpha_{l'm',lm} + \sum_{l'm'} \nabla\psi_{l'm'}(\mathbf{r}') \times \mathbf{r}''\alpha_{l'm',lm}. \qquad \text{(D.19)}$$

The above function is divergence free. Therefore, the second term in (D.19) can be expressed as a linear combination of $\mathbf{M}$ and $\mathbf{N}$ functions. Consequently,

$$\mathbf{M}_{lm}(\mathbf{r}) = \sum_{l'm'}[\mathbf{M}_{l'm'}(\mathbf{r}')A_{l'm',lm} + \mathbf{N}_{l'm'}(\mathbf{r}')B_{l'm',lm}]. \qquad (D.20)$$

Since $\mathbf{N} = \frac{1}{k}\nabla \times \mathbf{M}$, and $\mathbf{M} = \frac{1}{k}\nabla \times \mathbf{N}$, we also have

$$\mathbf{N}_{lm}(\mathbf{r}) = \sum_{l'm'}[\mathbf{N}_{l'm'}(\mathbf{r}')A_{l'm',lm} + \mathbf{M}_{l'm'}(\mathbf{r}')B_{l'm',lm}]. \qquad (D.21)$$

The coefficients $A_{l'm',lm}$ and $B_{l'm',lm}$ have been worked out by Stein (1961) and Cruzan (1962). The results are simplified by Bruning and Lo (1969). The sign error in Cruzan's work has also been pointed out by Tsang and Kong (1982), and Tsang et al. (1985). They are given in the notation used here as

$$A_{l'm',lm} = i^{l'-l}\frac{4\pi}{2l'(l'+1)}\sum_{l''}i^{l''}[l(l+1) + l'(l'+1) - l''(l''+1)]$$
$$\cdot A(m,l,-m',l',l'')z_{l''}(kr'')Y_{l'',m-m'}(\theta'',\phi''), \qquad (D.22)$$

$$B_{l'm',lm} = i^{l'-l}\frac{4\pi}{2l'(l'+1)}\sum_{l''}i^{l''}$$
$$\cdot B(m,l,-m',l',l'',l''-1)z_{l''}(kr'')Y_{l'',m-m'}(\theta'',\phi''), \qquad (D.23)$$

where

$$B(m,l,-m',l',l'',l''-1)$$
$$= \left(\frac{2l''+1}{2l''-1}\right)^{1/2}\left\{-[(l'-m')(l'+m'+1)(l''+m-m')(l''+m-m'-1)]^{1/2}\right.$$
$$\cdot A(m,l,-m'-1,l',l''-1)$$
$$+ [(l'+m')(l'-m'+1)(l''-m+m')(l''-m+m'+1)]^{1/2}$$
$$\cdot A(m,l,-m'+1,l',l''-1)$$
$$\left.+2m'[(l''-m+m')(l''+m-m')]^{1/2}A(m,l,-m',l',l''-1)\right\}, \qquad (D.23a)$$

and

$$A(m,l,-m',l',l'') = (-1)^m[(2l+1)(2l'+1)(2l''+1)/4\pi]^{1/2}$$
$$\cdot\begin{pmatrix}l & l' & l'' \\ 0 & 0 & 0\end{pmatrix}\begin{pmatrix}l & l' & l'' \\ -m & m' & m-m'\end{pmatrix}. \qquad (D.23b)$$

Another form given by Stein (1961) which is more convenient for computation is

$$B_{l'm',lm} = \frac{ikr''\cos\theta''}{2l'(l'+1)}2m'\alpha_{l'm',lm}$$
$$+ \frac{ikr''\sin\theta''}{2l'(l'+1)}\left[e^{i\phi''}[(l'-m')(l'+m'+1)]^{1/2}\alpha_{l',m'+1,lm}\right.$$
$$\left.+e^{-i\phi''}[(l'+m')(l'-m'+1)]^{1/2}\alpha_{l',m'-1,lm}\right]. \qquad (D.24)$$

In the above, $z_{l'}(kr')$ in $\mathbf{M}_{l'm'}(\mathbf{r}')$ and $\mathbf{N}_{l'm'}(\mathbf{r}')$ and $z_{l''}(kr'')$ in (22) and (23) switch roles as in (14) and (15). That is, if $r' < r''$, $z_{l'}(kr') = j_{l'}(kr')$ in $\mathbf{M}_{l'm'}(\mathbf{r}')$ and $\mathbf{N}_{l'm'}(\mathbf{r}')$, and $z_{l''}(kr'')$ in (22) and (23) is the same type of spherical Bessel function as $z_l(kr)$ in (16). But if $r' > r''$, then, $z_{l''}(kr'') = j_{l''}(kr'')$ in (22) and (23), and $z_{l'}(kr')$ in $\mathbf{M}_{l'm'}(\mathbf{r}')$ and $\mathbf{N}_{l'm'}(\mathbf{r}')$ is the same type of spherical Bessel function as $z_l(kr)$ in (16).

Recurrence relations for determining $A(m, l, -m', l', l'')$ are also given by Cruzan (1962), and Bruning and Lo (1969).

As a final note, to be consistent with the Clebsch-Gordon coefficients defined by convention, the spherical harmonics should be defined as

$$Y_{lm}(\theta, \phi) = \sqrt{\frac{(l-m)!}{(l+m)!}\frac{2l+1}{4\pi}} P_l^m(\cos\theta)e^{im\phi}, \quad m \geq 0, \qquad (\text{D.25})$$

and $Y_{l,-m}(\theta, \phi) = (-1)^m Y_{lm}^*(\theta, \phi)$ (Messiah 1958, p. 495).

References for the Appendices

Abramowitz, M., and I. A. Stegun. 1965. *Handbook of Mathematical Functions*. New York: Dover Publications.

Bruning, J. H., and Y. T. Lo. May 1969. "Multiple scattering by spheres." Technical Report, Dept. of Elect. Eng., Univ. of Illinois, Urbana.

Bruning, J. H., and Y. T. Lo. 1971. "Multiple scattering of EM waves by spheres, part I and II." *IEEE Trans. Antennas Propagat.* AP-19: 378–400.

Cruzan, O. R. 1962. "Translational addition theorems for spherical vector wave functions." *Quarterly Appl. Math.* 20(1): 33–40.

Messiah, A., 1958. *Quantum Mechanics*. Amsterdam: North Holland.

Stein, S. 1961. "Addition theorems for spherical wave functions." *Quarterly Appl. Math.* 19(1): 15–24.

Stratton, J. A. 1941. *Electromagnetic Theory*. New York: McGraw-Hill.

Tsang, L., and J. A. Kong. 1982. "Effective propagation constant for coherent electromagnetic waves in media embedded with dielectric scatterers." *J. Appl. Phys.* 53(11): 7162–73.

Tsang, L., J. A. Kong, and R. T. Shin. 1985. *Theory of Microwave Remote Sensing*. New York: John Wiley & Sons.

INDEX